Geophysical Monograph Series

Geophysical Monograph Series

175 **A Continental Plate Boundary: Tectonics at South Island, New Zealand** *David Okaya, Tim Stem, and Fred Davey (Eds.)*
176 **Exploring Venus as a Terrestrial Planet** *Larry W. Esposito, Ellen R. Stofan, and Thomas E. Cravens (Eds.)*
177 **Ocean Modeling in an Eddying Regime** *Matthew Hecht and Hiroyasu Hasumi (Eds.)*
178 **Magma to Microbe: Modeling Hydrothermal Processes at Oceanic Spreading Centers** *Robert P. Lowell, Jeffrey S. Seewald, Anna Metaxas, and Michael R. Perfit (Eds.)*
179 **Active Tectonics and Seismic Potential of Alaska** *Jeffrey T. Freymueller, Peter J. Haeussler, Robert L. Wesson, and Göran Ekström (Eds.)*
180 **Arctic Sea Ice Decline: Observations, Projections, Mechanisms, and Implications** *Eric T. DeWeaver, Cecilia M. Bitz, and L.-Bruno Tremblay (Eds.)*
181 **Midlatitude Ionospheric Dynamics and Disturbances** *Paul M. Kintner, Jr., Anthea J. Coster, Tim Fuller-Rowell, Anthony J. Mannucci, Michael Mendillo, and Roderick Heelis (Eds.)*
182 **The Stromboli Volcano: An Integrated Study of the 2002–2003 Eruption** *Sonia Calvari, Salvatore Inguaggiato, Giuseppe Puglisi, Maurizio Ripepe, and Mauro Rosi (Eds.)*
183 **Carbon Sequestration and Its Role in the Global Carbon Cycle** *Brian J. McPherson and Eric T. Sundquist (Eds.)*
184 **Carbon Cycling in Northern Peatlands** *Andrew J. Baird, Lisa R. Belyea, Xavier Comas, A. S. Reeve, and Lee D. Slater (Eds.)*
185 **Indian Ocean Biogeochemical Processes and Ecological Variability** *Jerry D. Wiggert, Raleigh R. Hood, S. Wajih A. Naqvi, Kenneth H. Brink, and Sharon L. Smith (Eds.)*
186 **Amazonia and Global Change** *Michael Keller, Mercedes Bustamante, John Gash, and Pedro Silva Dias (Eds.)*
187 **Surface Ocean–Lower Atmosphere Processes** *Corinne Le Quèrè and Eric S. Saltzman (Eds.)*
188 **Diversity of Hydrothermal Systems on Slow Spreading Ocean Ridges** *Peter A. Rona, Colin W. Devey, Jérôme Dyment, and Bramley J. Murton (Eds.)*
189 **Climate Dynamics: Why Does Climate Vary?** *De-Zheng Sun and Frank Bryan (Eds.)*
190 **The Stratosphere: Dynamics, Transport, and Chemistry** *L. M. Polvani, A. H. Sobel, and D. W. Waugh (Eds.)*
191 **Rainfall: State of the Science** *Firat Y. Testik and Mekonnen Gebremichael (Eds.)*
192 **Antarctic Subglacial Aquatic Environments** *Martin J. Siegert, Mahlon C. Kennicut II, and Robert A. Bindschadler*
193 **Abrupt Climate Change: Mechanisms, Patterns, and Impacts** *Harunur Rashid, Leonid Polyak, and Ellen Mosley-Thompson (Eds.)*
194 **Stream Restoration in Dynamic Fluvial Systems: Scientific Approaches, Analyses, and Tools** *Andrew Simon, Sean J. Bennett, and Janine M. Castro (Eds.)*
195 **Monitoring and Modeling the Deepwater Horizon Oil Spill: A Record-Breaking Enterprise** *Yonggang Liu, Amy MacFadyen, Zhen-Gang Ji, and Robert H. Weisberg (Eds.)*
196 **Extreme Events and Natural Hazards: The Complexity Perspective** *A. Surjalal Sharma, Armin Bunde, Vijay P. Dimri, and Daniel N. Baker (Eds.)*
197 **Auroral Phenomenology and Magnetospheric Processes: Earth and Other Planets** *Andreas Keiling, Eric Donovan, Fran Bagenal, and Tomas Karlsson (Eds.)*
198 **Climates, Landscapes, and Civilizations** *Liviu Giosan, Dorian Q. Fuller, Kathleen Nicoll, Rowan K. Flad, and Peter D. Clift (Eds.)*
199 **Dynamics of the Earth's Radiation Belts and Inner Magnetosphere** *Danny Summers, Ian R. Mann, Daniel N. Baker, and Michael Schulz (Eds.)*
200 **Lagrangian Modeling of the Atmosphere** *John Lin (Ed.)*
201 **Modeling the Ionosphere-Thermosphere** *Jospeh D. Huba, Robert W. Schunk, and George V Khazanov (Eds.)*
202 **The Mediterranean Sea: Temporal Variability and Spatial Patterns** *Gian Luca Eusebi Borzelli, Miroslav Gacic, Piero Lionello, and Paola Malanotte-Rizzoli (Eds.)*
203 **Future Earth - Advancing Civic Understanding of the Anthropocene** *Diana Dalbotten, Gillian Roehrig, and Patrick Hamilton (Eds.)*
204 **The Galápagos: A Natural Laboratory for the Earth Sciences** *Karen S. Harpp, Eric Mittelstaedt, Noémi d'Ozouville, and David W. Graham (Eds.)*
205 **Modeling Atmospheric and Oceanic Flows: Insightsfrom Laboratory Experiments and Numerical Simulations** *Thomas von Larcher and Paul D. Williams (Eds.)*
206 **Remote Sensing of the Terrestrial Water Cycle** *Venkat Lakshmi (Eds.)*
207 **Magnetotails in the Solar System** *Andreas Keiling, Caitríona Jackman, and Peter Delamere (Eds.)*
208 **Hawaiian Volcanoes: From Source to Surface** *Rebecca Carey, Valerie Cayol, Michael Poland, and Dominique Weis (Eds.)*
209 **Sea Ice: Physics, Mechanics, and Remote Sensing** *Mohammed Shokr and Nirmal Sinha (Eds.)*
210 **Fluid Dynamics in Complex Fractured-Porous Systems** *Boris Faybishenko, Sally M. Benson, and John E. Gale (Eds.)*
211 **Subduction Dynamics: From Mantle Flow to Mega Disasters** *Gabriele Morra, David A. Yuen, Scott King, Sang Mook Lee, and Seth Stein (Eds.)*
212 **The Early Earth: Accretion and Differentiation** *James Badro and Michael Walter (Eds.)*
213 **Global Vegetation Dynamics: Concepts and Applications in the MC1 Model** *Dominique Bachelet and David Turner (Eds.)*
214 **Extreme Events: Observations, Modeling and Economics** *Mario Chavez, Michael Ghil, and Jaime Urrutia-Fucugauchi (Eds.)*
215 **Auroral Dynamics and Space Weather** *Yongliang Zhang and Larry Paxton (Eds.)*
216 **Low-Frequency Waves in Space Plasmas** *Andreas Keiling, Dong-Hun Lee, and Valery Nakariakov (Eds.)*
217 **Deep Earth: Physics and Chemistry of the Lower Mantle and Core** *Hidenori Terasaki and Rebecca A. Fischer (Eds.)*
218 **Integrated Imaging of the Earth: Theory and Applications** *Max Moorkamp, Peter G. Lelievre, Niklas Linde, and Amir Khan (Eds.)*
219 **Plate Boundaries and Natural Hazards** *Joao Duarte and Wouter Schellart (Eds.)*
220 **Ionospheric Space Weather: Longitude and Hemispheric Dependences and Lower Atmosphere Forcing** *Timothy Fuller-Rowell, Endawoke Yizengaw, Patricia H. Doherty, and Sunanda Basu (Eds.)*
221 **Terrestrial Water Cycle and Climate Change: Natural and Human-Induced Impacts** *Qiuhong Tang and Taikan Oki (Eds.)*
222 **Magnetosphere-Ionosphere Coupling in the Solar System** *Charles R. Chappell, Robert W. Schunk, Peter M. Banks, James L. Burch, and Richard M. Thorne (Eds.)*
223 **Natural Hazard Uncertainty Assessment: Modeling and Decision Support** *Karin Riley, Peter Webley, and Matthew Thompson (Eds.)*
224 **Hydrodynamics of Time-Periodic Groundwater Flow: Diffusion Waves in Porous Media** *Joe S. Depner and Todd C. Rasmussen (Eds.)*
225 **Active Global Seismology** *Ibrahim Cemen and Yucel Yilmaz (Eds.)*

Geophysical Monograph 226

Climate Extremes

Patterns and Mechanisms

S.-Y. Simon Wang
Jin-Ho Yoon
Christopher C. Funk
Robert R. Gillies
Editors

This Work is a copublication of the American Geophysical Union and John Wiley and Sons, Inc.

AGU
American Geophysical Union

WILEY

This Work is a co-publication between the American Geophysical Union and John Wiley & Sons, Inc.

This edition first published 2017 by John Wiley & Sons, Inc., 111 River Street, Hoboken, NJ 07030, USA and the American Geophysical Union, 2000 Florida Avenue, N.W., Washington, D.C. 20009

Published under the aegis of the AGU Publications Committee
Brooks Hanson, Director of Publications
Robert van der Hilst, Chair, Publications Committee
For details about the American Geophysical Union visit us at www.agu.org.

Wiley Global Headquarters
111 River Street, Hoboken, NJ 07030, USA

For details of our global editorial offices, customer services, and more information about Wiley products visit us at www.wiley.com.

Library of Congress Cataloging-in-Publication data is available

ISBN: 9781119067849

Cover image: Earth Globe Image Credit: NASA JPL
Cover design by Wiley

Set in 10/12pt Times New Roman by SPi Global, Pondicherry, India

10 9 8 7 6 5 4 3 2 1

CONTENTS

CONTRIBUTORS

Mathew Barlow
University of Massachusetts Lowell
Environmental Earth and Atmospheric Sciences
Lowell, Massachusetts, USA

Emily J. Becker
Climate Prediction Center/NOAA and
INNOVIM LLC
College Park, Maryland, USA

Gerald D. Bell
Climate Prediction Center
NOAA/NWS/NCEP
College Park, Maryland, USA

Eric S. Blake
National Hurricane Center
National Oceanic and Atmospheric Administration
Miami, Florida, USA

Grant Branstator
National Center for Atmospheric Research
Boulder, Colorado, USA

Forest Cannon
University of California Santa Barbara
Geography
Santa Barbara, California, USA

Jen-Ping Chen
Department of Atmospheric Sciences
National Taiwan University
Taipei, Taiwan

Mengyan Chen
State Key Laboratory of Tropical Oceanography
South China Sea Institute of Oceanology
Chinese Academy of Sciences
Guangzhou, Guangdong, China

Mingyue Chen
NOAA/NWS/NCEP Climate Prediction Center
College Park, Maryland, USA

Adam J. Clark
NOAA/OAR National Severe Storms Laboratory
Norman, Oklahoma, USA

Dim Coumou
Institute for Environmental Studies (IVM)
Department of Water and Climate Risk
VU University Amsterdam, Amsterdam,
The Netherlands; *and*
Potsdam Institute for Climate Impact Research
Earth System Analysis
Potsdam, Germany

Markus G. Donat
ARC Centre of Excellence for Climate
System Science
Climate Change Research Centre
University of New South Wales
Sydney, NSW, Australia

Aaron Donohoe
Karlsruhe Institute of Technology
Institute of Meteorology and Climate Research
Department Troposphere Research
Karlsruhe,Germany; *and*
Lamont-Doherty Earth Observatory
Columbia University
New York, NY, USA

Chris Funk
US Geological Survey and
University of California Santa Barbara
Geography
Santa Barbara, California, USA

Robert R. Gillies
Utah Climate Center, Utah State University
Logan, Utah, USA; *and* Department of Plants,
Soils, and Climate, Utah State University
Logan, Utah, USA

Lawrence Gloeckler
Department of Atmospheric and Environmental Sciences
University at Albany
Albany, New York, USA

Ed Hawkins
National Centre for Atmospheric Science
Department of Meteorology
University of Reading
Reading, UK

Andrew Hoell
NOAA Earth System Research Laboratory
Physical Sciences Division
Boulder, Colorado, USA

Huang-Hsiung Hsu
Research Center for Environmental Changes
Academia Sinica, Taipei, Taiwan

Charles Jones
Department of Geography and Earth
Research Institute
University of California Santa Barbara
Santa Barbara, California, USA

David J. Karoly
ARC Centre of Excellence for Climate System Science
School of Earth Science
University of Melbourne
Melbourne, Australia

Andrew D. King
ARC Centre of Excellence for Climate
System Science
School of Earth Science
University of Melbourne
Melbourne, Australia

Philip J. Klotzbach
Department of Atmospheric Science
Colorado State University
Fort Collins, Colorado, USA

Kai Kornhuber
Potsdam Institute for Climate Impact Research
Earth System Analysis
University of Potsdam
Faculty of Science
Potsdam, Germany

Randal D. Koster
NASA Goddard Space Flight Center
Global Modeling and Assimilation Office
Greenbelt, Maryland, USA

Arun Kumar
NOAA/NWS/NCEP Climate Prediction Center
College Park, Maryland, USA

Tzu-Hsien Kuo
Department of Atmospheric Sciences
National Taiwan University
Taipei, Taiwan

Chia-Wei Lan
Department of Atmospheric Sciences
National Taiwan University
Taipei, Taiwan

Jascha Lehmann
Potsdam Institute for Climate Impact Research
Earth System Analysis
University of Potsdam
Faculty of Science
Potsdam, Germany

Sophie C. Lewis
ARC Centre of Excellence for Climate
System Science
Fenner School of Environment and Society
The Australian National University
Acton, Australia

Min-Hui Lo
Department of Atmospheric Sciences
National Taiwan University
Taipei, Taiwan

Bradfield Lyon
School of Earth and Climate Sciences and Climate
Change Institute
University of Maine
Orono, Maine, USA

Kyle MacRitchie
NASA Goddard Space Flight Center, GES-DISC
Greenbelt, Maryland, USA

Houk Paek
Department of Earth System Science
University of California
Irvine, California, USA

Kathy V. Pegion
Department of Atmospheric, Oceanic, and
Earth Sciences and Center for Ocean-Land-Atmosphere
Studies
George Mason University
Fairfax, Virginia, USA

Sarah E. Perkins
ARC Centre of Excellence for Climate System Science
Climate Change Research Centre
University of New South Wales
Sydney, NSW, Australia

Vladimir Petoukhov
Institute for Environmental Studies (IVM)
Department of Water and Climate Risk
VU University Amsterdam
Amsterdam,
The Netherlands; *and*
Potsdam Institute for Climate Impact Research
Earth System Analysis
Potsdam, Germany

Paul E. Roundy
Department of Atmospheric and Environmental Sciences
University at Albany
Albany, New York, USA

Naoko Sakaeda
NOAA Earth System Research Laboratory
Physical Sciences Division
Boulder, Colorado, USA

Mark A. Saunders
Department of Space and Climate Physics
University College London
Holmbury St. Mary
Dorking, Surrey, UK

Siegfried D. Schubert
NASA Goddard Space Flight Center
Global Modeling and Assimilation Office
Greenbelt, Maryland, USA

Christopher Selman
Department of Atmospheric, Oceanic, and Earth Sciences and Center for Ocean-Land-Atmosphere Studies
George Mason University
Fairfax, Virginia, USA

Jana Sillmann
Center for International Climate and Environmental Research, Oslo
Cicero Senter for Klimaforskning
Oslo, Norway

Haiyan Teng
National Center for Atmospheric Research
Boulder, Colorado, USA

Nicolas Vigaud
International Research Institute for Climate and Society
The Earth Institute, Columbia University
Palisades, New York, USA

Timo Vihma
Finnish Meteorological Institute
Helsinki, Finland

Aiko Voigt
Lamont Doherty Earth Observatory
Columbia University
New York, NY, USA

Hailan Wang
NASA Goddard Space Flight Center
Global Modeling and Assimilation Office
Greenbelt, Maryland, USA

Shih-Yu (Simon) Wang
Utah Climate Center, Utah State University
Logan, Utah, USA; *and*
Department of Plants, Soils, and Climate
Utah State University
Logan, Utah, USA

Xin Wang
State Key Laboratory of Tropical Oceanography
South China Sea Institute of Oceanology
Chinese Academy of Sciences
Guangzhou, Guangdong, China

Scott J. Weaver
Environmental Defense Fund
Washington, DC, USA

Hao-Wei Wey
Department of Atmospheric Sciences
National Taiwan University
Taipei, Taiwan

Song Yang
School of Atmospheric Sciences
Sun Yat-sen University
GuangZhou, Guangdong, China;
Guangdong Province Key Laboratory for Climate Change and Natural Disaster Studies

Jinho Yoon
School of Earth Sciences and Environmental Engineering
Gwangju Institute of Science and Technology
Repulic of Korea

Jin-Yi Yu
Department of Earth System Science
University of California
Irvine, California, USA

PREFACE

Increasing extreme weather and climate events have resulted in tremendous loss of life and property worldwide. As of the writing of this volume, 2016 was the warmest year for globally averaged temperatures since records began. Additionally, the list of record-breaking extreme heat and heavy precipitations events is steadily growing. Driven by the global increase in extreme event impacts, and increasing awareness about their causes, the past decade has seen a significant surge of academic interest in analyses of weather and climate extremes. The rapidly emerging science of extreme event attribution to that of climate change is uncovering the physical bases as to how weather and climate fluctuations are intensifying in a warming climate; this new science will have potential to improve upon the prediction of extreme weather and climate events.

Examples of emerging theories include *Arctic amplification,* which targets the effect of accelerated Arctic warming on interfering the jet stream, *tropical expansion,* which describes the poleward shift of the thermally driven circulation from the equator, the increasing reference to the *Clausius-Clapeyron relationship* quantifying the moisture-holding capacity in the warming air, and changing teleconnection processes originating from the tropical oceans. However, these mechanisms can contradict each other or be overgeneralized. Controversial theories (such as recent debates regarding Arctic amplification) can confuse the general public and other stakeholders who are not familiar with meteorology and climate dynamics. Likewise, overgeneralized concepts such as global warming being a sole cause for the entirety of observed climate extremes, do little to advance the prediction of extreme events. It is the editors' belief that finding the varied pathways through which extreme events happen provides the modus operandi toward improved comprehension and prediction and will facilitate mitigation of climate extremes.

This book is designed to highlight the different pathways whereby climate mechanisms engender extreme events as well as harness that information to improve extreme event prediction. Since the beginning of the 21st century, a great deal of research effort has been undertaken toward uncovering the physical mechanisms related to the changing pattern of climate extremes. The goal of this book is to be inclusive of the diverse, even contradictory, theories by (1) integrating the different views and approaches in research of climate extremes and (2) transitioning the systematically acquired scientific knowledge to benefit the greater scientific community and society at large. Aimed to be a resource for researchers and graduate students, this book includes four parts of the climate extremes research: forcings, processes, regionality, and prediction.

Forcings – The need for effective and scientifically defensible ways for the attribution analysis of climate extreme events was highlighted by the National Academies of Sciences' 2016 (NAS) report. The first important step toward extreme events attribution is establishing a sound understanding of natural variability that represents the different origins of climate forcing, which can modify a region's climate pattern and extreme weather. Examples included in this book are the changing pattern of the El Niño–Southern Oscillation (ENSO), Arctic-midlatitude interactions, role of anthropogenic aerosols, shifting dynamics of internal atmospheric waves, and land-surface processes.

Processes – The NAS report also highlights the development of appropriate ways to tease apart the interplay between meteorological and human-made factors in the face of extreme events. Doing so requires intimate knowledge of meteorological processes linking internal and external climate forcings to climate extremes. Presented in this book is a collection of six chapters on the various meteorological processes leading to certain climate extremes, including the timing in which human-made effects on climate extremes emerge, recent changes in temperature extremes over land, forced shifts in tropical precipitation, influences from various modes of tropical intraseasonal variability, warming of the western Pacific Ocean, and the influence of atmospheric waves embedded in jet streams. The interplay between these processes and forcings is complex, but it is possible to identify pathways in which climate extremes are modulated.

Regionality – The impact of climate extremes on humanity and ecosystems occurs at the regional level. It is therefore important to understand the extent to which individual event types are connected to their respective locations and the processes through which they evolve. Examples of regional studies included are drought occurrences in Austria, East Africa, southwest Asia, North America (specifically California), and tornadoes in the United States. Each of these regional extreme event types has changed in a way that links to different oceanic, atmospheric, and surface-hydrological processes as well as teleconnection sources. Of essence is the fact that external forcing(s) of anthropogenic origin affects these different regional extreme events through different pathways.

Prediction – Recent extreme events, such as intense drought, heat waves, historic local storms, extensive flooding, and cold spells have become increasingly severe and costly. While a few of those events were anticipated days before they happened, most were not; this suggests a knowledge gap in predicting and mitigating for future occurrences of extreme events. In another 2016 NAS report that focuses on modeling, the importance in developing the subseasonal to seasonal (S2S) prediction is highlighted and backed by a decade of research progress in identifying prolonged episodes of extreme drought and severe weather outbreaks. Chapters collected in this book include forecast advancement toward ENSO precursors, hurricanes, subseasonal variations, temperature and precipitation extremes, and US tornadoes. These chapters illuminate the progress toward developing actionable information to foresee the likelihood of a climate extreme, and so get ready for the event and its consequences.

S.-Y. Simon Wang
Jin-Ho Yoon
Christopher C. Funk
Robert R. Gillies

ACKNOWLEDGMENTS

All chapters in this book were peer reviewed following AGU journal standards. Therefore, our sincere thanks go to the many anonymous reviewers and chapter coauthors who spent extra time in helping maintain the integrity of this book. Mary Grace Hammond of Wiley did a wonderful job in handling the many tedious processes and challenging communications in editing this book; she will be missed. Rituparna Bose, the publishing editor, planted a seed in 2013 that grew up to be a comprehensive book project. Thanks also go to Scott Weaver who helped proofread the preface. Special thanks to Johan Lorck, a climate-concerned citizen from Paris, France, for his encouragement and support even at difficult times in Paris.

Part I
Forcings of Climate Extremes

1

The Changing El Niño–Southern Oscillation and Associated Climate Extremes

Jin-Yi Yu[1], Xin Wang[2], Song Yang[3,4], Houk Paek[1], and Mengyan Chen[2]

ABSTRACT

The El Niño–Southern Oscillation (ENSO) is one of the most powerful climate phenomena that produce profound global impacts. Extensive research since the 1970s has resulted in a theoretical framework capable of explaining the observed properties and impacts of the ENSO and predictive models. However, during the most recent two decades there have been significant changes observed in the properties of ENSO that suggest revisions are required in the existing theoretical framework developed primarily for the canonical ENSO. The observed changes include a shift in the location of maximum sea surface temperature variability, an increased importance in the underlying dynamics of coupled ocean-atmosphere process in the subtropical Pacific, and different remote atmospheric teleconnection patterns that give rise to distinct climate extremes. The causes of these recent changes in ENSO are still a matter of debate but have been attributed to both global warming and natural climate variability involving interactions between the Pacific and Atlantic oceans. The possible future changes of ENSO properties have also been suggested using climate model projections.

1.1. INTRODUCTION

El Niño–Southern Oscillation (ENSO) is a prominent climate phenomenon in the tropical Pacific that can disrupt global atmospheric and oceanic circulation patterns and exert profound impacts on global climate and socioeconomic activities. Since *Bjerknes* [1969] first recognized that ocean-atmosphere coupling was a fundamental characteristic of ENSO, a tremendous amount of effort has been expended by the research community to describe and understand the complex nature of this phenomenon and its underlying generation mechanisms. By the 1990s, the effort had led to the developments of successful theoretical frameworks that could explain the major features observed during ENSO events [*Neelin et al.*, 1998] and useful forecast systems had been formulated to predict ENSO evolution with significant lead times of up to three seasons [*Latif et al.*, 1998]. The typical ENSO impacts in various parts of the globe had also been extensively examined and documented. Through these efforts, it was also recognized that ENSO properties were not the same among events and could change from one decade to another. The diversity of ENSO characteristics attracted renewed interest at the beginning of the 21st century (see *Capotondi et al.* [2015], for a summary of these efforts), when it became obvious that the central location of the

[1] *Department of Earth System Science, University of California, Irvine, California, USA*

[2] *State Key Laboratory of Tropical Oceanography, South China Sea Institute of Oceanology, Chinese Academy of Sciences, Guangzhou, Guangdong, China*

[3] *School of Atmospheric Sciences, Sun Yat-sen University, GuangZhou, Guangdong, China*

[4] *Guangdong Province Key Laboratory for Climate Change and Natural Disaster Studies*

Climate Extremes: Patterns and Mechanisms, Geophysical Monograph 226, First Edition.
Edited by S.-Y. Simon Wang, Jin-Ho Yoon, Christopher C. Funk, and Robert R. Gillies.

sea surface temperature (SST) anomalies associated with ENSO appeared to be moving from the tropical eastern Pacific near the South American coast to locations near the International Dateline in the tropical central Pacific. Most of the El Niño events that have occurred so far in the 21st century developed primarily in the central Pacific [*Lee and McPhaden*, 2010; *Yu et al.*, 2012b; *Yu et al.*, 2015a; *Yeh et al.*, 2015]. The fact that El Niño can sometimes occur in the eastern Pacific, sometimes in the central Pacific, and sometimes simultaneously in both portions of the Pacific has suggested that there may exist more than one type of ENSO, whose generation mechanisms and associated climate extremes may be different. The changes in ENSO observed during the recent decades have motivated the research community to revisit the conventional views of ENSO properties, dynamics, and global impacts.

1.2. CHANGES IN ENSO PROPERTIES

1.2.1. Flavors of ENSO

Slow or interdecadal changes have been observed in ENSO properties, including its intensity, period, and propagation direction [*Gu and Philander*, 1995; *Wang and Wang,* 1996; *Torrence and Webster*, 1999; *An and Wang*, 2000; *Fedorov and Philander*, 2000; *Wang and An*, 2001; *Timmermann*, 2003; *An and Jin*, 2004; and many others]. The amplitude of ENSO-associated SST anomalies, for example, was found to be stronger at the beginning and the end of the twentieth century, but weaker in between [*Gu and Philander*, 1995; *Wang and Wang*, 1996]. The propagation direction of ENSO SST anomalies has also alternated between eastward, westward, and standing during the past few decades [*Wang and An*, 2001; *An and Jin*, 2004]. Its recurrence frequency changed from about 2 to 3 yr before 1975 to a longer frequency of 4 to 6 yr afterward [*An and Wang*, 2000]. In recent decades, one of the most noticeable changes in ENSO properties has been the displacement of the central location of ENSO SST anomalies [*Larkin and Harrison*, 2005; *Ashok et al.*, 2007; *Kao and Yu,* 2009; *Kug et al.*, 2009]. ENSO is characterized by interannual SST variations in the equatorial eastern and central Pacific. In the canonical ENSO events portrayed by *Rasmusson and Carpenter* [1982], SST anomalies first occur near the South American coast and then spread westward along the equator. However, ENSO events characterized by SST anomalies primarily in the equatorial central Pacific, and which spread eastward [*An and Wang*, 2000] also occur. *Trenberth and Stepaniak* [2001] recognized that the different SST evolutions of ENSO events cannot be fully described with a single index such as the Niño-3 (5°S–5°N, 90°–150°W) SST index. They defined a trans-Niño index, which measures the SST gradient along the equator by taking the difference between normalized Niño-1 + 2 (10°S–0°, 80°–90°W) and Niño-4 (5°S–5°N, 160°E–150°W) SST indices to help discriminate the different SST evolutions. Their study implies that different types of ENSO may be better identified by contrasting SST anomalies between the eastern and central Pacific. A similar conclusion was reached by *Yu and Kao* [2007] in their analysis of the persistence barrier of Niño indices in the central-to-eastern Pacific (i.e., Niño-1 + 2, Niño-3, Niño-3.4 (5°S–5°N, 120°–170°W), and Niño-4 SST indices). They found different decadal changes between the indices in the equatorial central and eastern Pacific. They also found that the decadal changes in ocean heat content variations along the equatorial Pacific coincide with the decadal changes in the SST persistence barriers in the eastern Pacific, but not with those in the central Pacific. This finding led them to suggest that there are two types of ENSO: An Eastern-Pacific (EP) type located primarily in the tropical eastern Pacific and whose generation mechanism involves equatorial thermocline variations, and a Central-Pacific (CP) type located in the central tropical Pacific and whose generation is less sensitive to thermocline variations. A detailed comparison of these two types of ENSO was presented in *Kao and Yu* [2009], where they identified the spatial structure, temporal evolution, and underlying dynamics of these two types of events. While this "two types of ENSO" point of view is adopted in this chapter, it should be noted that debate remains (for example, see the discussion in *Capotondi et al.* [2015]) as to whether these two types are really dynamically distinct.

Examples of the EP and CP types of ENSO are shown in Figure 1.1, which displays the SST anomaly patterns during the peak phase of the 1977–1978 and 1997–1998 El Niño events. During the 1997–1998 El Niño (Fig. 1.1a), SST anomalies were mostly located in the eastern part of the tropical Pacific, extending from the South American coast around 80°W to 160°W, covering the Niño-1 + 2 and Niño-3 regions. During the 1977–1978 El Niño (Fig. 1.1b), SST anomalies were mostly concentrated in the equatorial central Pacific from 160°E to 120°W, covering the Niño-3.4 and Niño-4 regions. Some early studies had already noticed the existence of a group of ENSO events that developed in the central Pacific around the International Dateline [e.g., *Weare et al.*, 1976; *Fu et al.*, 1986; *Hoerling and Kumar,* 2002; *Larkin and Harrison,* 2005a]. *Larkin and Harrison* [2005a] suggested that this group of ENSO events can produce different impacts on the US climate from conventional ENSO events and referred to them as the Dateline El Niño. *Ashok et al.* [2007] also focused on this group of ENSO events and termed them El Niño Modoki. Furthermore, *Wang and Wang* [2013] classified El Niño Modoki into two subtypes: El Niño Modoki I and El Niño Modoki II because

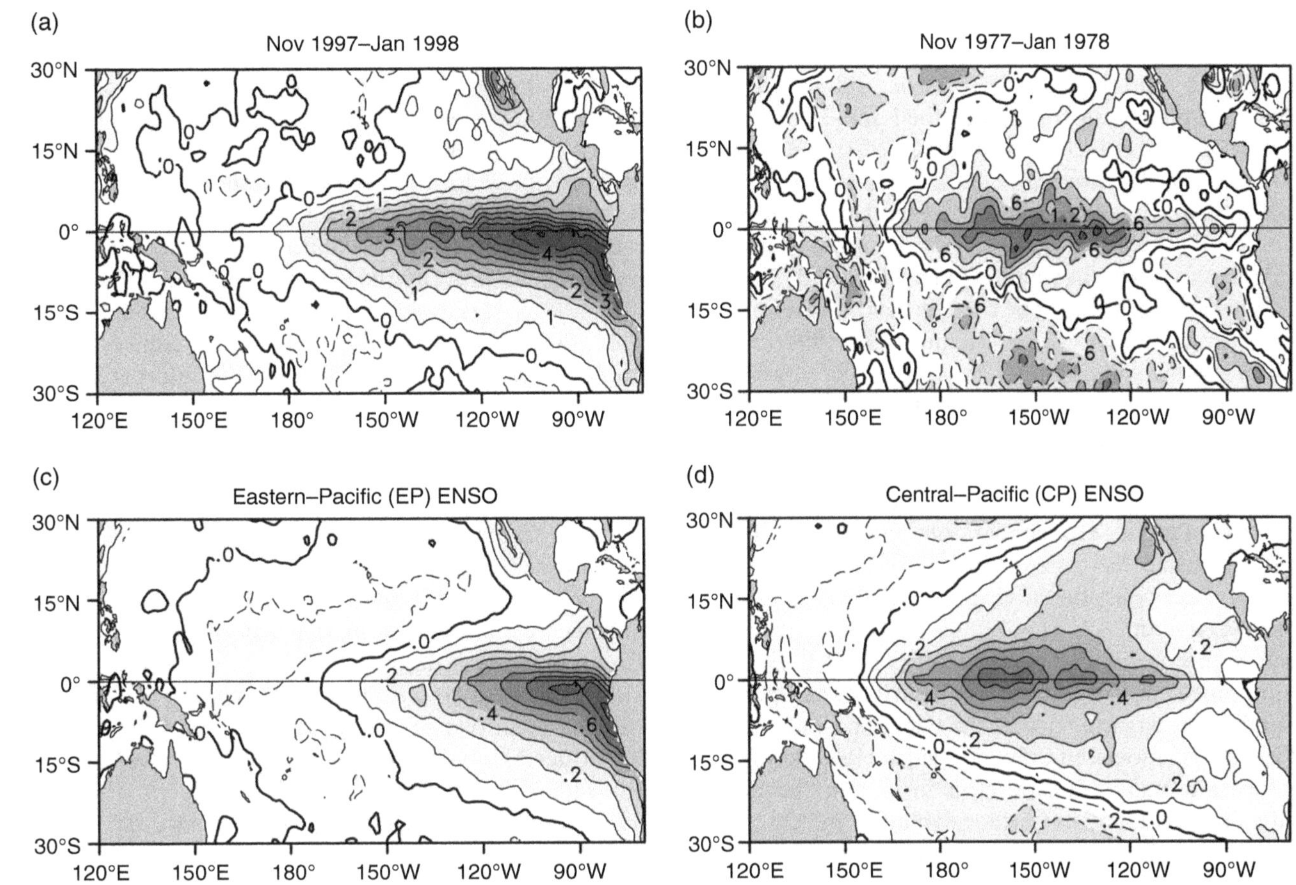

Figure 1.1 SST anomalies during the peak phase (November–January) of (a) the 1997–1998 El Niño event and (b) the 1977–1978 El Niño event. The leading EOF (empirical orthogonal function) patterns obtained from a regression-EOF analysis [*Kao and Yu*, 2009] are shown in (c) for the Eastern-Pacific type of ENSO and in (d) for the Central-Pacific type of ENSO. The contour intervals are 0.5 °C for (a), 0.3 °C for (b), and 0.1 °C for (c) and (d). The Hadley Centre Sea Ice and Sea Surface Temperature dataset (HadISST) [*Rayner et al.*, 2003] for the period 1948–2010 is used in these calculations.

they show significantly different impacts on rainfall in southern China and typhoon landfall activity. These two subtypes of El Niño Modoki have distinct origins and patterns of SST anomalies during their developing phase. *Kug et al.* [2009] also named this ENSO-type warm pool El Niño and the other type cold tongue El Niño.

1.2.2. Distinct Characteristics of the Eastern-Pacific (EP) and Central-Pacific (CP) Types of ENSO

Significant differences were found between the structure and evolution of the EP and CP types of ENSO [*Kao and Yu*, 2009; *Kug et al.*, 2009]. The typical SST anomaly patterns associated with these two types of ENSO are shown in Figure 1.1c and d, which are obtained by applying a regression-EOF (empirical orthogonal function) analysis [*Kao and Yu,*2009; *Yu and Kim*, 2010a] to tropical Pacific SST anomalies. In this method, SST anomalies regressed on the Niño-1 + 2 SST index (representing the EP ENSO influence) were first removed from the original monthly SST anomalies, and then an EOF analysis was applied to the residual SST anomalies to obtain the spatial pattern of the CP ENSO. Similarly, SST anomalies regressed on the Niño-4 index (representing the CP ENSO influence) were first removed from the original SST anomalies, and then an EOF analysis was applied to identify the leading structure of the EP ENSO. The principal components (PCs) associated with these patterns are referred to, respectively, as the CP ENSO index and the EP ENSO index. The EOF patterns show that the EP ENSO-related SST anomalies are mostly located in the eastern equatorial Pacific and connected to the coast of South America (Fig. 1.1c), while the CP ENSO has most of its SST anomalies confined in the central Pacific with extension into the northeastern subtropical Pacific (Fig. 1.1d). Although other studies also proposed different identification methodologies for separating the two types of ENSO, including weighted averaged SST anomalies in the tropical western, central, and eastern Pacific [*Ashok et al.*, 2007]; relative magnitudes of the Niño-3 and Niño-4 indices [*Kug et al.*, 2009; *Yeh et al.*, 2009]; a simple transformation of the Niño-3 and Niño-4 indices [*Ren and Jin*, 2011] or two PCs of tropical Pacific SST anomalies [*Takahashi et al.*, 2011]; and mean upper ocean temperature anomalies in CP and EP regions [*Yu et al.*, 2011], their corresponding SST anomaly patterns are similar to those shown in Figure 1.1c and d. The SST anomaly features revealed by the regression-EOF analysis were observed in the 1977–1978 and 1997–1998 El Niño events as shown in Figure 1.1a and b. The meridional extension into the subtropics of the CP ENSO SST anomalies is an important feature, because it resembles the Pacific Meridional Mode (PMM) [*Chiang and Vimont*, 2004] that is the leading coupled variability mode of the subtropical Pacific. This connection indicates that the CP ENSO dynamic has a stronger tie to atmospheric and oceanic process outside the tropical Pacific than does the EP ENSO dynamic.

Below the ocean surface (Fig. 1.2), the EP ENSO is accompanied by significant subsurface temperature anomalies propagating across the Pacific basin, while the CP ENSO is associated only with subsurface ocean temperature anomalies that develop in situ in the central Pacific. The different subsurface evolution indicates that, in contrast to the EP ENSO, the underlying dynamics of the CP ENSO seem to be less dependent on thermocline variations [*Kao and Yu*, 2009; *Kug et al.*, 2009], which is consistent with the suggestion of *Yu and Kao* [2007]. *McPhaden* [2008] already noticed that the variations in the equatorial Pacific Ocean heat content (i.e., the so-called warm water volume; *Meinen and McPhaden* [2000]) tended to lead the ENSO SST anomalies by about two to three seasons in the 1980s and 1990s but by only one season after 2000. *McPhaden* [2012] attributed this decrease to the shift of ENSO from the EP type to the CP type in recent decades. The decreased lead time implies a reduced impact of thermocline feedback to SST anomalies for the CP ENSO and may explain the reduced predictability observed in recent decades [*Barnston et al.*, 2012] as this type of ENSO occurred more frequently. The weakened relationship between ENSO SST anomalies and the underlying thermocline variation indicates that the CP ENSO dynamic differs from the traditional ENSO dynamic, such as the delayed oscillator theory [*Suarez and Schopf*, 1988; *Battisti and Hirst*, 1989] or the recharged oscillator theory [e.g., *Wyrtki* 1975; *Zebiak* 1989; *Jin* 1997] that emphasize the subsurface ocean dynamics in governing ENSO SST evolution.

In the atmosphere, wind stress and precipitation anomaly patterns associated with the two types of ENSO are also different. While the EP El Niño is associated with significant westerly anomalies over a large part of the tropical Pacific, the westerly anomalies associated with the CP El Niño have a smaller spatial scale and are centered in the equatorial central-to-western Pacific [*Kao and Yu*, 2009; *Kug et al.*, 2009]. This more westward location of the westerly anomalies in the CP El Niño is consistent with the location of its SST anomalies. Significant easterly anomalies also appear over the tropical eastern Pacific during the CP El Niño. Theses anomalies were suggested to prevent the development of positive SST anomalies in this region during CP El Niño events [*Ashok et al.*, 2007; *Kug et al.*, 2009]. These differences in wind stress anomalies are important to the interpretation of the underlying dynamics of these two types of ENSO. Moreover, positive anomalies of precipitation associated with the EP El Niño typically extend from the equatorial eastern-to-central Pacific where the largest SST anomalies are located. For the CP El Niño, the precipitation anomalies are characterized by a dipole pattern, with positive anomalies located mainly in the western Pacific and negative anomalies in the eastern Pacific [*Kao and Yu*, 2009; *Kug et al.*, 2009]. The different precipitation patterns associated with these two types of ENSO imply that the locations of associated convective heating and the induced global teleconnections are different as well.

In addition to the central locations of their SST anomalies, the two types of ENSO also differ in their evolution, amplitude, and recurrence frequency. The propagation of SST anomalies is weaker and less apparent in the CP type of ENSO than in the EP type of ENSO [*Kao and Yu*, 2009]. Also, there is no identifiable phase-reversal signal in the composite analysis of the CP ENSO SST anomalies, implying that this type of ENSO does not behave like a cycle or oscillation as is often the case for the EP ENSO. Discussions of whether ENSO should be viewed as an event or a cycle [e.g., *Kessler*, 2002] might be aided by recognizing that there are two types of ENSO: one behaves more as an event and the other more as a cycle. *Sun and Yu* [2009] pointed out that the amplitude of the CP El Niño tends to be smaller than that of the EP El Niño, while the CP La Niña is often stronger than the EP La Niña. The reason for this difference in behavior between the El Niño and La Niña phases of these two types of ENSO has yet to be explained. *Sun and Yu* [2009], however, proposed an ENSO-mean state interaction mechanism to suggest that the amplitude asymmetries between the El Niño and La Niña phases of these two types of ENSO enable ENSO to modify the tropical Pacific mean state, which can give rise to a 10–15 yr modulation cycle in ENSO amplitude. Strong El Niño events can occur every 10–15 yr due to this cycle, which is consistent with the time interval between the last three strongest El Niño events: the 1972–1973, 1982–1983, and 1997–1998 El Niño events. For the recurrence frequency, it is well known that ENSO events tend to occur every 2

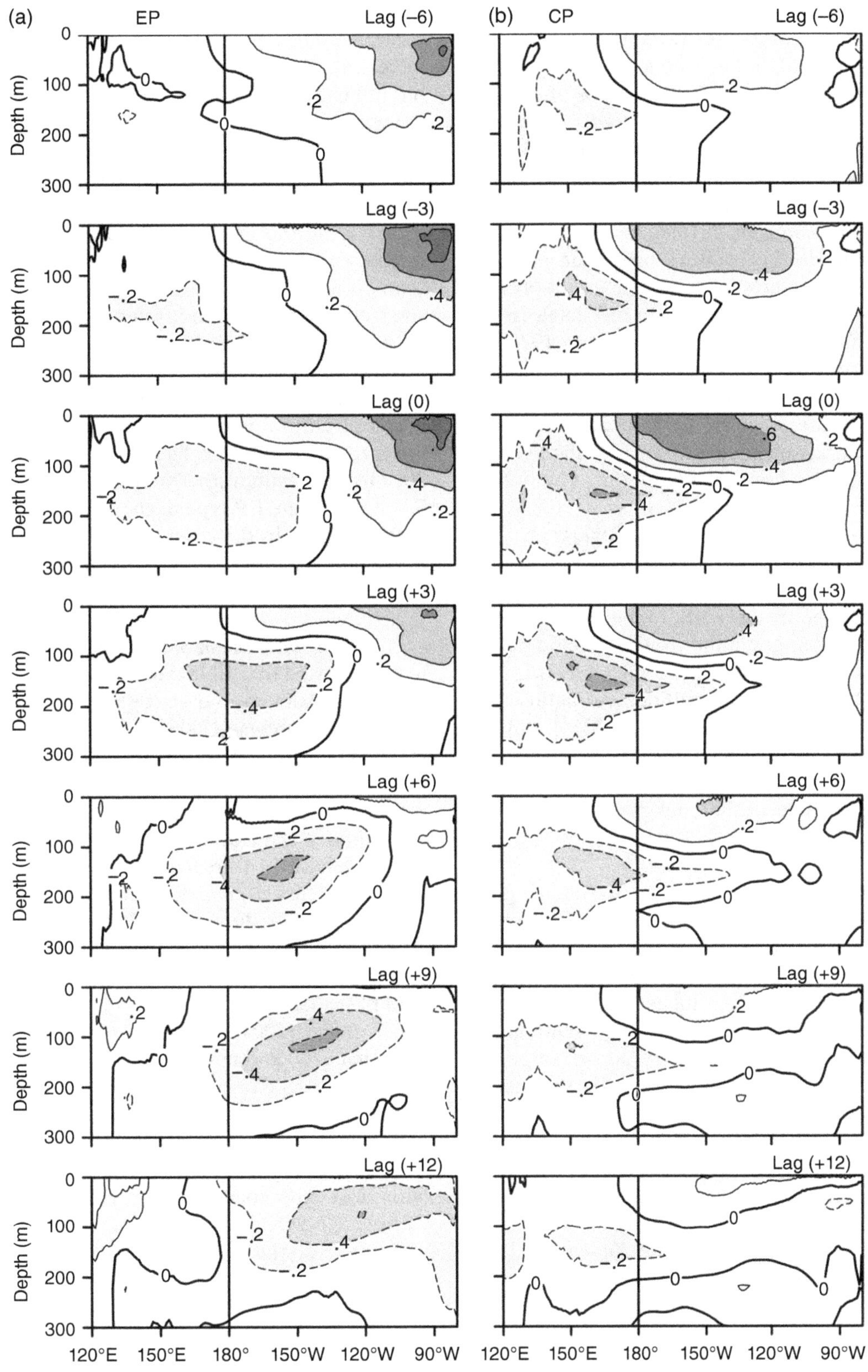

Figure 1.2 Evolution of equatorial (5°S–5°N) subsurface ocean temperature anomalies associated with the (a) EP ENSO and (b) CP ENSO. Values shown are temperature anomalies regressed on the EP and CP ENSO indices from a lag of −6 mo to a lag of +12 mo. The contour interval is 0.2 °C. Ocean temperature assimilation from the German Estimating the Circulation and Climate of the Ocean (GECCO) [*Kohl et al.*, 2006] for the period 1970–2001 is used.

to 7 yr. Spectral analyses have suggested that there are two dominant bands of ENSO variability: a quasibiennial (~2 yr) band and a low-frequency (3–7 yr) band [*Rasmusson and Carpenter,* 1982; *Rasmusson et al.*, 1990; *Barnett*, 1991; *Gu and Philander,* 1995; *Jiang et al.*, 1995; *Wang and Wang,* 1996]. The EP ENSO is dominated more by the low-frequency band, while the CP ENSO is dominated by the quasibiennial band [*Kao and Yu,* 2009; *Yu et al.*, 2010; *Yu and Kim*, 2010a]. As indicated above, El Niño events in the 21st century have been mostly of the CP type, and it is interesting to note that they occurred about every 2 yr in 2002, 2004, 2006, and 2009. In contrast, the El Niño events observed during the 1970s and 1980s tended to occur every 4 yr. If model projections are correct about the more frequent occurrence of the CP ENSO in the future warmer world [*Yeh et al.*, 2009; *Kim and Yu,* 2012], we should expect El Niño events in the coming decades to become more frequent but weaker.

1.2.3. Possible Causes for the Change in ENSO Properties

The recent change in the ENSO type may be due to background state changes in the tropical Pacific that are either caused by anthropogenic warming [*Yeh et al.*, 2009; *Kim and Yu*, 2012] or decadal/multidecadal natural variability [*McPhaden et al.*, 2011; *Yeh et al.,* 2011]. Random fluctuations have also been suggested as possibly giving rise to the change in ENSO type on decadal timescales even without a change in the background state of the tropical Pacific [*Newman et al.*, 2011].

1.2.3.1. Anthropogenic Warming. *Yeh et al.* [2009] and *Kim and Yu* [2012] used multiple model datasets of phases 3 and 5 of the Coupled Model Intercomparison Project (CMIP3 and CMIP5) [*Meehl et al.*, 2007; *Taylor et al.*, 2012] to examine the influence of anthropogenic warming on the ENSO type. *Yeh et al.* [2009] analyzed CMIP3 model projections to show that anthropogenic warming can flatten the equatorial Pacific thermocline (due to a weakened Walker circulation) resulting in a shift of the thermocline feedback mechanism from eastern to central Pacific and increasing the occurrence of ENSO in the central Pacific. They suggested that the same mechanism may be at work in recent decades to increase the occurrence frequency of the CP ENSO. *Kim and Yu* [2012] compared the amplitude ratio of these two types of ENSO in CMIP5 models. They found that the intensity of the CP ENSO increases steadily from the preindustrial to the historical simulations and the future projections, but the intensity of the EP ENSO decreases. The CP-to-EP ENSO intensity ratio, as a result, is projected to increase.

This global warming view has been challenged on two fronts. First, *Lee and McPhaden* [2010] found that not only has the occurrence of the CP El Niño increased in recent decades but its intensity has also significantly increased. However, the intensity of CP La Niña events has not changed during the same period. As a result, they argued, the warming trend observed in the central Pacific is a result of the increasing frequency and intensity of the CP El Niño, rather than the other way around. Second, *McPhaden et al.* [2011] found the Pacific background state during the most recent decade (2000–2010) has a steeper equatorial thermocline and stronger trade winds than during earlier decades (1980–1999), which are opposite from the background states changes expected by *Yeh et al.* [2009] should anthropogenic warming be the cause for the change in the ENSO type. This inconsistency implies that the change in ENSO type may be a manifestation of natural variability. *McPhaden et al.* [2011] also interpreted the Pacific background state during 2000–2010 to be a result, rather than a cause, of the shift of the ENSO from the EP type to the CP type. Their argument is that change in the central location of the ENSO SST anomalies can project onto the changes in the background states.

1.2.3.2. ENSO–Mean State Interaction. Earlier studies have suggested that ENSO can force changes in the tropical Pacific background state [e.g., *Rodgers et al.*, 2004; *Schopf and Burgman*, 2006; *Sun and Yu,* 2009; *Choi et al.*, 2009]. Due to nonlinearities in ENSO dynamics, it is known that asymmetries exist in the intensity, frequency, or spatial structures of the warm (El Niño) and cold (La Niña) phases of the cycle. For example, strong El Niño events tend to reach a larger intensity than strong La Niña events [*Hannachi et al.*, 2003; *An and Jin,* 2004; *Duan et al.*, 2008; *Frauen and Dommenget,* 2010; *Su et al.*, 2010]. Also, strong El Niño events tend to be located more in the eastern Pacific while strong La Niña events tend to be located more in the central Pacific [*Monahan,* 2001; *Hsieh,* 2004; *Rodgers et al.,* 2004; *Schopf and Burgman,* 2006; *Sun and Yu,* 2009]. Because of these asymmetries, the positive and negative anomalies associated with the El Niño and La Niña phases may produce a nonzero residual effect on the time-mean state of the tropical Pacific. *Sun and Zhang* [2006] also conducted numerical experiments using a coupled model with and without ENSO to show that ENSO works as a basin-scale mixer on the time-mean thermal stratification in the upper equatorial Pacific. The mean state change associated with decadal ENSO variability is one of the leading modes of tropical Pacific decadal variability in many coupled atmosphere-ocean models [*Rodgers et al.*, 2004; *Yeh and Kirtman*, 2004; *Yu and Kim,* 2011b].

Sun and Yu [2009] suggested an ENSO–mean state interaction mechanism to explain how the ENSO type may shift between the EP and CP types on decadal timescales.

A key element in their hypothesized mechanism is that strong El Niño events tend to be of the EP type, whereas strong La Niña events tend to be of the CP type. In contrast, weak El Niño events tend to be of the CP type, whereas weak La Niña events tend to be of the EP type. They verified this reversed spatial asymmetry characteristic using observed ENSO events during the 1880–2006 period in two reconstructed SST datasets. During strong ENSO periods or decades, strong El Niño events are located in the eastern tropical Pacific and strong La Niña events in the central tropical Pacific. This spatial asymmetry between El Niño and La Niña can gradually reduce the background east-west SST gradient to weaken the ocean-atmosphere coupling and to shift the background state toward one favoring a weak ENSO regime. Conversely, during the weak ENSO periods or decades, weak El Niño events are located in the eastern tropical Pacific and weak La Niña events in the central tropical Pacific, which can gradually increase the east-west SST gradient, strengthen the ocean-atmosphere coupling, and shift the background state toward one favoring a strong ENSO regime. The reversed spatial asymmetries between El Niño and La Niña enable ENSO to force the tropical Pacific mean state in opposite ways between the strong and weak ENSO regimes, acting as a restoring force to push the mean state back and forth between states that sustain strong and weak ENSO intensities. ENSO can also alternate between the EP and CP types on decadal timescales. *Choi et al.* [2012] have noted such an ENSO–mean state interaction mechanism and its associations with the ENSO types and Pacific background states in coupled model simulations.

1.2.3.3. The Pacific Decadal Oscillation. Decadal changes in ENSO properties can also be associated with forcing external to the tropical Pacific, such as that related to slow SST oscillations in the North Pacific, Atlantic, or Indian oceans [e.g., *Gu and Philander*, 1997; *Behera and Yamagata*, 2003; *Verdon and Franks*, 2006; *Luo et al.*, 2005; *Dong et al.*, 2006]. One of the most well-known forcing of this kind is related to the Pacific Decadal Oscillation (PDO) [*Zhang et al.*, 1997; *Mantua et al.*, 1997], which is a leading mode of decadal SST variability in the North Pacific. Earlier studies considered the PDO a physical mode of climate variability that results from atmosphere-ocean coupling within the North Pacific [e.g., *Latif and Barnet*, 1994; *Alexander and Deser*, 1995]. More recent views consider it not a single physical mode but a combination of several processes operating on various timescales (see a review by *Newman et al.* [2015]). It has been suggested that the PDO is capable of modulating the ENSO properties by influencing the tropical Pacific background states through ocean subduction, thermocline ventilation, or atmospheric teleconnections [e.g., *McCreary and Lu*, 1994; *Gu and Philander*, 1995; *Wang and Weisberg*, 1998; *Zhang et al.*, 1998; *Barnett et al.*, 1999; *Pierce et al.*, 2000; *Chang et al.*, 2001; *McPhaden and Zhang*, 2002]. During the period of reliable instrumental records, the PDO shifted its phase twice: from a negative to a positive phase around 1976–1977 and back to a negative phase around 1999–2000. Along with the 1976–1977 phase shift, the background SSTs increased in the eastern and central tropical Pacific. This background state change was accompanied by noticeable changes in ENSO frequency and intensity [e.g., *Trenberth and Hoar*, 1996; *Rajagopolan et al.*, 1997]. Specifically, ENSO amplitude increased after the shift [*An and Wang*, 2000], and the ENSO frequency increased from being dominated by a quasi-2 yr band to a quasi-4 yr band [*Wang and Wang*, 1996]. In addition, ENSO SST anomalies were found to propagate westward before the shift but eastward after the shift [*Wang and An*, 2001], and there were more frequent El Niño events and fewer La Niña events after the shift [*Trenberth and Hoar*, 1996]. However, there was no noticeable change in the ENSO type across the 1976–1977 climate shift.

The 1999–2000 phase shift of the PDO is closer to the time that the occurrence frequency of the CP ENSO began to increase, which was suggested to be sometime around the 1980s [*Ashok et al.*, 2007], the 1990s [*Yu et al.*, 2012a], or the beginning of the 21st century [*Lee and McPhaden*, 2010; *McPhaden et al.*, 2011]. Associated with this PDO phase shift, a late-1990s climate shift has also been suggested to occur in the Pacific Rim [e.g., *Minobe*, 2000; *Peterson and Schwing*, 2003]. Dramatic changes around 1999 were reported in the North Pacific sea level pressure (SLP), SST and sea level patterns [*Minobe*, 2000; *Cummins et al.*, 2005], and in the northeastern Pacific ecosystems [*Peterson and Schwing*, 2003]. In the western North Pacific, tropical cyclone activity abruptly increased after 2000 [*Tu et al.*, 2009; *He et al.*, 2015]. Over the neighboring continents, precipitation patterns changed after the late 1990s over North and South America [*Huang et al.*, 2005] and over China [*Zhu et al.*, 2011; *Xu et al.*, 2015]. Surface temperatures in the Tibetan Plateau suddenly increased during the late 1990s [*Xu et al.*, 2009]. A La Niña-like background pattern appeared in the tropical Pacific [e.g., *McPhaden et al.*, 2011; *Chung and Li*, 2013; *Hu et al.*, 2013; *Xiang et al.*, 2013; *Kumar and Hu*, 2014], characterized by a steeper equatorial thermocline slope [*McPhaden et al.*, 2011] and strengthened Walker circulation [*Chung and Li*, 2013; *Dong and Lu*, 2013]. The breakdown of the subsurface ocean variability–ENSO relationship [*McPhaden*, 2008, 2012; *Horri et al.*, 2012] and a coherent weakening of tropical ocean-atmosphere variability [*Hu et al.*, 2013] after 2000 are likely associated with the decreases in ENSO forecast skill since 1999–2000 [*Barnston et al.*, 2012; *Xue et al.*,

2013]. Beyond the Pacific, changes in the SST and sea ice variability over the Bering and Chukchi seas [*Minobe,* 2002; *Yeo et al.,* 2014] and in the southern ocean SST variability modes [*Yeo and Kim,* 2015] have also been found after the late 1990s.

1.2.3.4. The Atlantic Multidecadal Oscillation. In addition to the PDO-related late-1990s climate shift, there is evidence that the Pacific also experienced a basin-wide climate shift in the early-1990s that can also be responsible for the change in the type of ENSO. As shown in Figure 1.1, the EP ENSO has an SST anomaly pattern that spans the tropical eastern and central Pacific. During this type of ENSO, SST variations in the central Pacific (i.e., the Niño-4 region) have an in-phase association with the SST variations in the eastern Pacific (i.e., the Niño-3 region). During the CP ENSO, the association between SST anomalies in the central Pacific and the eastern Pacific is weaker. Instead, the SST variations in the tropical central Pacific are more related to SST variations in the northeastern subtropical Pacific that are induced by an interannual atmospheric variability mode, the so-called North Pacific Oscillation (NPO) [*Walker and Bliss,* 1932; *Rogers,* 1981; *Linkin and Nigam,* 2008]. The NPO is the second leading mode of the North Pacific SLP variability and is characterized by out-of-phase SLP fluctuations between the Aleutian low and the Pacific subtropical high. Based on these features, *Yu et al.* [2012a] used three oceanic and atmospheric indices and found that the SST variations in the tropical central Pacific (i.e., the Niño-4 index) are strongly related to the SST variations in the tropical eastern Pacific (i.e., the Niño-3 index) before the early 1990s but more closely related to the SLP variations associated with NPO (i.e., the NPO index) afterward. They concluded that the change in ENSO from the EP type to the CP type occurred in the early 1990s. This climate shift resulted in a change in ENSO location from the eastern Pacific to the central Pacific. *Yu et al.* [2015a] further revealed that the early-1990s climate shift was related to the Atlantic Multidecadal Oscillation (AMO) [*Schlesinger and Ramankutty,* 1994; *Kerr,* 2000], which changed from a negative phase to a positive phase in the early 1990s as well. The AMO is characterized by slow changes in SST in the North Atlantic and is believed to have a period of 60–80 yr. Although there is still ongoing debate on the causes of the AMO, it is agreed that the slow variability mode influences climate not only within the Atlantic basin but also across the entire Northern Hemisphere, including the Pacific basin [e.g., *Dong et al.,* 2006; *Zhang and Delworth,* 2007]. The physical processes that link the NPO and AMO to the CP ENSO are described in detail in section 1.3.

Evidence of the early-1990s climate shift in the Pacific has also been found in other climate phenomena of the low-latitude Pacific. In the western Pacific, upper ocean currents and sea level structure also changed around this time, with the north equatorial current and the north equatorial countercurrent migrating southward [*Qiu and Chen,* 2012] and sea level rise accelerating to a rate not seen in the preceding several decades [*Merrifield,* 2011]. These oceanic state changes were accompanied by changes in trade wind patterns [*Qiu and Chen,* 2012]. *Sui et al.* [2007] found that the interannual variability of the western Pacific subtropical high was dominated by frequencies of 3–5 yr before the early 1990s but 2–3 yr afterward. The relationships between the Asian monsoon and the other monsoon systems in the Northern Hemisphere (e.g., the North American and North African monsoons) changed in such a way that the interannual rainfall variability produced by these the monsoon systems decreased dramatically after 1993 [*Lee et al.,* 2014]. The relationships between the East Asian winter monsoon and ENSO also undergo low-frequency oscillation [*Zhou et al.,* 2007]. Moreover, the typical drought pattern in China was replaced by a new pattern after the early 1990s [*Qian et al.,* 2014]. These recent studies together indicate that the Pacific climate shift in the early 1990s affected the ENSO location, the monsoons, the subtropical high, and the subtropical ocean gyre.

1.2.3.5. Random Forcing. Other mechanisms have also been put forth to explain the increased occurrence of the CP ENSO during the past few decades. *Newman et al.* [2011] used a linear inverse modeling technique to suggest that natural random variations in the atmosphere can project onto two particular initial SST anomaly patterns, each of which can eventually develop into either the EP or CP type of ENSO through thermocline and zonal advection feedbacks, respectively. They argued that such natural random variations can result in a slow alternation in the type of ENSO without the need for changes in the background state.

1.3. CHANGES IN ENSO DYNAMICS

1.3.1. The Different Generation Mechanisms for the Two Types of ENSO

The fundamental dynamics of ENSO have been traditionally believed to reside mostly within the tropical Pacific [*Neelin et al.,* 1998; *Philander* 1999; *Wang et al.,* 2015; and the references therein]. The prevailing ENSO theories emphasize the delayed response of subsurface ocean to atmospheric forcing for the oscillation of ENSO. In this view, the delayed oceanic response results from the propagation and reflection of oceanic waves along the equator in the delayed oscillator theory [*Suarez and Schopf* 1988; *Battisti and Hirst* 1989], the adjustment of the nonequilibrium between zonal-mean thermocline depth and wind forcing in the recharged oscillator theory [e.g., *Wyrtki* 1975; *Zebiak* 1989; *Jin* 1997], or the lagged response of western Pacific wind stress to ENSO SST anomalies in the western Pacific oscillator theory [*Weisberg and Wang* 1997; *Wang et al.*, 1999]. Variations in ENSO properties (such as its low-frequency and quasibiennial periodicity) can be produced by these delayed atmosphere-ocean coupling processes and their interactions with the annual cycle or by the nonlinear chaotic characteristics of these ENSO dynamics [*Jin et al.*, 1994; *Jin* 1996, 1997; *Tziperman et al.*, 1994, 1995; *Chang et al.*, 1995, 1996; *Mantua and Battisti* 1995; *An and Jin*, 2001]. From this point of view, the various types of ENSO do not necessarily represent distinct ENSO dynamics. However, recent studies of the EP and CP types of ENSO have begun to suggest that these two types of ENSO are likely to have different underlying dynamics [e.g., *Kao and Yu*, 2009; *Yu et al.*, 2011; *Yu and Kim*, 2011a; *S.-T. Kim et al.*, 2012; *Yu et al.*, 2012a; *Lin et al.*, 2014]. In this framework, the EP ENSO is produced by the traditional ENSO dynamics, while the generation of the CP ENSO relies more on forcing from the subtropical Pacific and its interaction with ocean mixed-layer dynamics in the equatorial central Pacific.

As shown in Figure 1.2, the composite CP ENSO is associated with in situ subsurface ocean temperature anomalies that do not show the basinwide propagation characteristics typical of the delayed oscillator theory [*Suarez and Schopf* 1988; *Battisti and Hirst* 1989]. *Yu and Kim* [2010b] further examined the thermocline variations associated with nine major CP El Niño events since 1960. Three of the CP El Niño events (i.e., 1968–1969, 1990–1991, and 1991–1992) they examined undergo a prolonged decaying phase (Fig. 1.3a) that decays slowly and is followed by a warm event in the eastern Pacific. Figure 1.3d shows that the mean thermocline depth (represented by the 20 °C isotherm) for this group of events (i.e., Group 1) is deeper than normal in the equatorial Pacific (i.e., recharged state), which later gives rise to an EP El Niño to prolong the decay of these three CP El Niño events. The other three CP El Niño events (1963–1964, 1977–1978, and 1987–1988) occurred in the presence of shallow-than-normal thermoclines (i.e., discharged state; Group 2 of Fig. 1.3d). This group of CP El Niño events is followed by a La Niña event in the eastern Pacific, resulting in an abrupt termination of the CP El Niño (Fig. 1.3b). The last three events (1994–1995, 2002–2003, and 2004–2005) occurred with near-normal thermocline depths (i.e., neutral state; Group 3 of Fig. 1.3d). This group of events undergoes a symmetric-decaying pattern, whose SST anomalies grow and decay symmetrically with respect to a peak phase (Fig. 1.3c). Therefore, CP El Niño events can occur when the equatorial Pacific thermocline is above, below, or near the normal depth, as also suggested by a recent modeling study [*Fedorov et al.*, 2014]. This finding further confirms that the generation of CP ENSO does not depend on thermocline dynamics. However, these different thermocline structures do affect whether a warming, cooling, or neutral event may occur in the eastern Pacific as the CP event decays. Depending on the thermocline structure, the eastern Pacific warming, or cooling may interact with the SST evolution of the CP event and give rise to the three distinct evolution patterns of CP El Niño. Although the thermocline variations are not crucial to the generation and the dynamics of the CP type of El Niño, the state of the thermocline at the peak phase of a CP ENSO event can potentially be used to predict when and how the CP ENSO event will terminate. It is the interplay between the dynamics of the CP and EP types of ENSO that makes the thermocline information useful for the prediction of both types of SST variations.

1.3.2. The Central-Pacific ENSO: Extratropical Atmospheric Forcing Mechanism

As mentioned, for the CP El Niño the equatorial westerly anomalies appear to the west of the positive SST anomalies in the central Pacific and the equatorial easterly anomalies to the east. *Ashok et al.* [2007] argued that the thermocline variations induced by this wind anomaly pattern are responsible for the generation of the CP ENSO. The equatorial westerly anomalies induce downwelling Kelvin waves propagating eastward and the equatorial easterly anomalies induce downwelling Rossby waves propagating westward and, together, they deepen the thermocline in the central Pacific to produce the CP El Niño. *Kug et al.* [2009] emphasized the fact that the equatorial easterly anomalies can suppress warming in the eastern Pacific during a CP El Niño event by enhancing upwelling and surface evaporation. However, they also argued that the mean depth of thermocline in the central Pacific is relatively deep and the wind-induced thermocline variations may not be efficient in producing the CP SST anomalies. Instead, they suggested that ocean advection is responsible for the development of the central Pacific warming. *Yu et al.* [2010] also concluded that the SST anomalies of the CP ENSO undergo rapid intensification through ocean advection processes. However, they showed that the initial establishment of the CP SST anomalies is related to forcing from the extratropical atmosphere and subsequent atmosphere-ocean coupling in the subtropics.

Yu et al. [2010] performed an analysis of the near-surface ocean temperature budget to identify the physical processes involved in the development of the CP ENSO. They showed that the SST anomalies associated

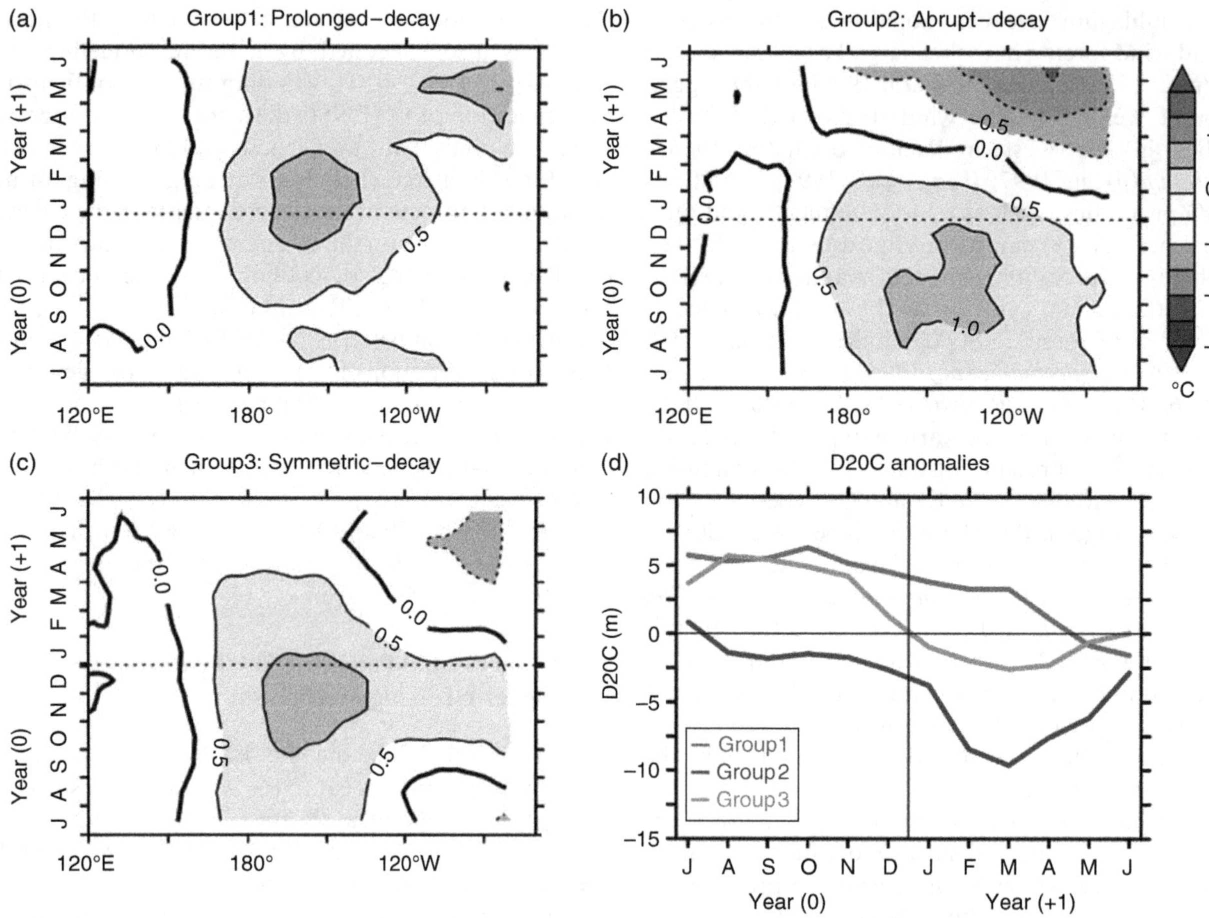

Figure 1.3 Evolution of composite equatorial Pacific (5°S–5°N) SST anomalies for (a) the group of CP El Niño events that has a prolonged decay (i.e., 1968–1969, 1990–1991, and 1991–1992), (b) the group that has an abrupt decay (i.e., 1963–1964, 1977–1978, and 1987–1988), and (c) the group that has a symmetric decay (i.e., 1994–1995, 2002–2003, and 2004–2005). The evolution is shown from July of the El Niño year (0) to June of the following year (+1) and is calculated from the Extended Reconstruction of Historical Sea Surface Temperature dataset (ERSST) [*Smith et al.*, 2008]. Panel (d) shows the evolution of the depth anomalies of the zonal-mean 20 °C isotherm (D20C) at the equator (5°S–5°N) for the three groups of events. The zonal mean is averaged between 120°E and 80°W and is calculated using the Simple Ocean Data Assimilation Reanalysis data (SODA) [*Carton et al.*, 2000]. Adapted from *Yu and Kim* [2010b].

with the CP ENSO undergo rapid intensification in the equatorial central Pacific through local atmosphere-ocean interactions, which involve only mixed layer ocean dynamics and are similar to those described for the slow-SST mode of ENSO [*Neelin* 1991; *Neelin and Jin* 1993]. This mode requires surface wind anomalies to induce anomalous air-sea fluxes and ocean advection processes to intensify the initial SST perturbation. *Yu et al.* [2010] suggested that the initial SST anomalies in the central equatorial Pacific were, however, established by forcing from the extratropical atmosphere. This suggestion is supported by regressions of the Pacific SST anomalies onto the CP ENSO index. As shown in Figure 1.4, the regressed positive SST anomalies appear first in the northeastern subtropical Pacific and later spread toward the central equatorial Pacific to initiate ENSO event in the central Pacific.

The possibility of subtropical precursors in triggering ENSO events has long been recognized. However, recent studies on CP ENSO dynamics have begun to emphasize that these subtropical processes are not just precursors of ENSO but an essential element in the generation of CP ENSO events. As revealed by Figure 1.4, the CP ENSO is preceded by SST anomalies in the northeastern Pacific, which are characterized by a band of SST anomalies extending typically from Baja California toward the equatorial central Pacific. As the subtropical SST anomalies approach the equatorial Pacific, El Niño events develop and begin to grow. Several recent studies [e.g., *Vimont et al.*, 2003; *Anderson*, 2004; *Yu and Kim*, 2011a]

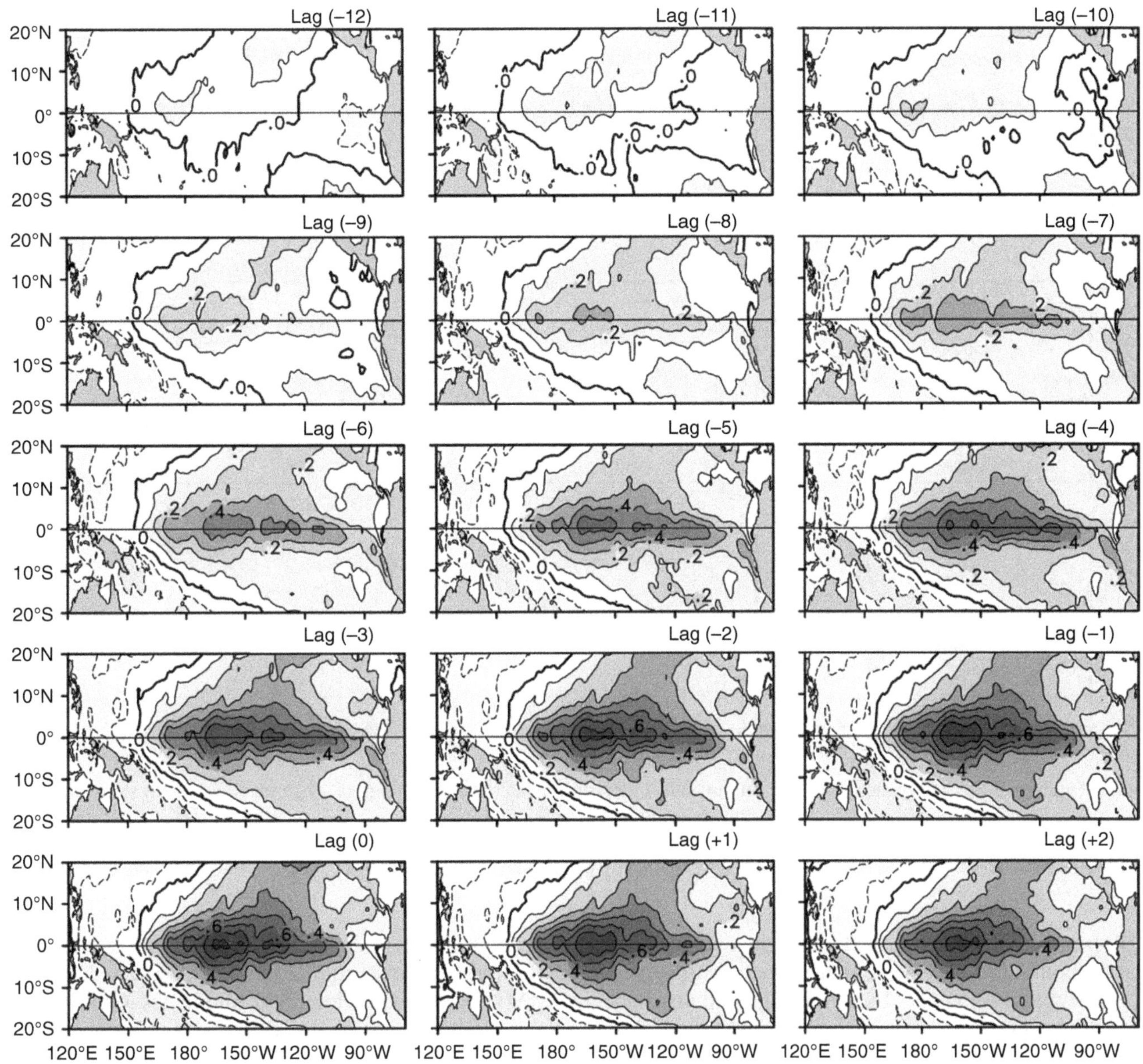

Figure 1.4 Evolution of tropical Pacific SST anomalies of the CP ENSO. Values shown are SST anomalies regressed on the CP ENSO index from a lag of −12 mo to a lag of +2 mo. The contour interval is 0.1 °C. The HadISST dataset for the period 1948–2010 is used.

have suggested that the initial warming off Baja California is forced by atmospheric fluctuations, particularly those associated with NPO. When the Pacific SST anomalies are regressed onto the NPO index (Fig. 1.5), it is obvious that the NPO first induces positive SST anomalies off Baja California via surface heat fluxes. The SST anomalies then extend southwestward toward the tropical Pacific. After the SST anomalies reach the tropical Pacific, an El Niño event occurs and develops there. The evolution shown in Figure 1.5 is very similar to the evolution of SST anomalies shown in Figure 1.4 preceding the development of a CP El Niño event. This similarity offers evidence that there exists a close relationship between the subtropical Pacific precursors and CP ENSO.

Ocean-atmosphere coupling in the subtropical Pacific is crucial for the spread of the NPO-induced SST anomalies southward into the deep tropics. The coupling processes begin as the SST anomalies off Baja California feed back onto and modify near surface winds via convection. The wind anomalies induced by the convection tend to be located to the southwest of initial subtropical SST anomalies [*Xie and Philander,* 1994], where new positive

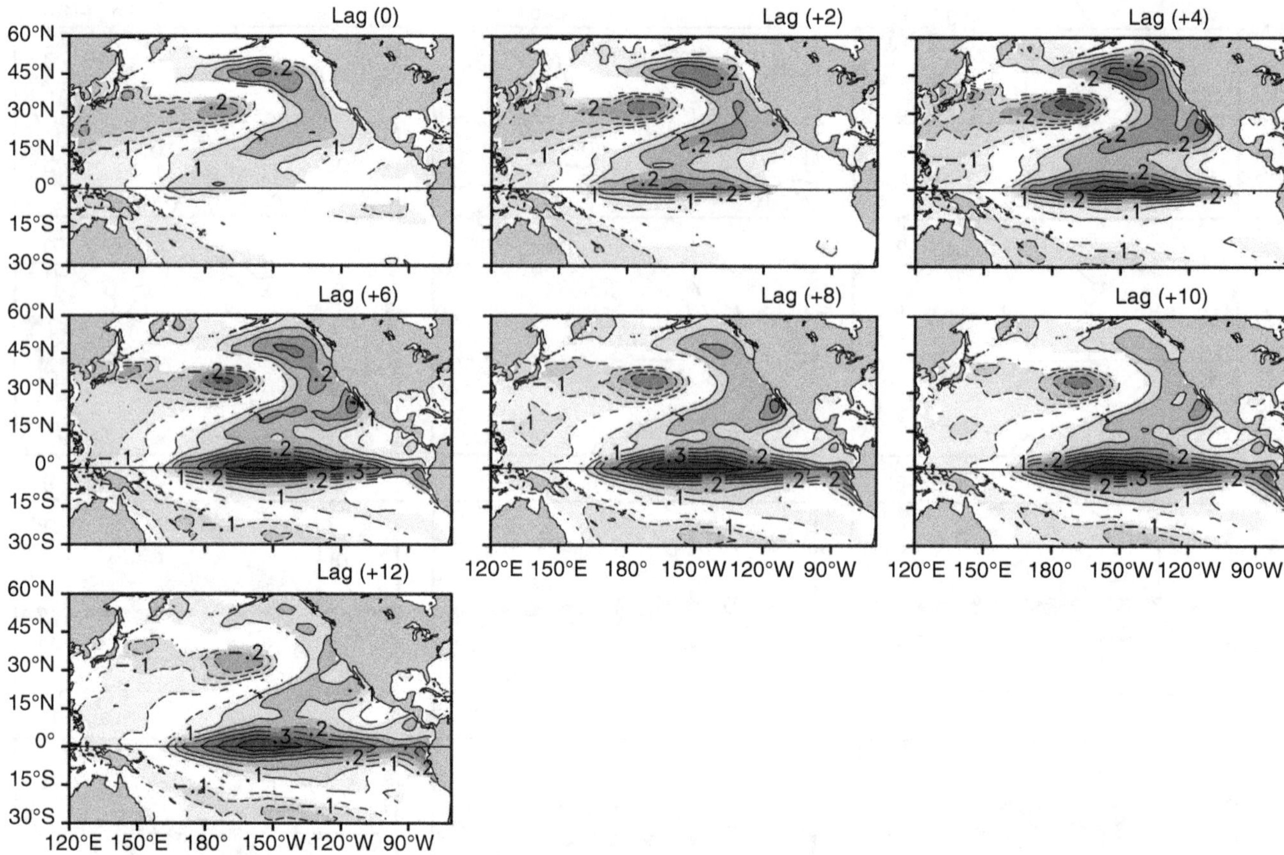

Figure 1.5 SST anomalies regressed on the NPO index from a lag of 0 mo to a lag of +12 mo. The contour interval is 0.05 °C and only the coefficients exceeding the 99% confidence level are shown. Positive lags denote that the NPO leads the SST anomalies. The National Centers for Environmental Prediction/National Center for Atmospheric Research (NCEP/NCAR) reanalysis [*Kalnay et al.,* 1996] and HadISST datasets for the period of 1948–2010 are used.

SST anomalies can be formed through a reduction in evaporation. The atmosphere then continues to respond to the new SST anomalies by producing wind anomalies farther southwestward. Through this wind-evaporation-SST (WES) feedback [*Xie and Philander,* 1994], the SST anomalies initially induced by the NPO off Baja California can extend southwestward into the deep tropics. This series of subtropical Pacific coupling processes is referred to as the seasonal footprinting mechanism [*Vimont et al.,* 2001, 2003, and 2009]. This mechanism also offers a way to explain how the subtropical SST anomalies can be sustained from boreal winter, when extratropical atmospheric variability (e.g., the NPO) is largest, to the following spring or summer to excite CP ENSO events.

As mentioned, the SST anomaly pattern of the subtropical precursor to ENSO (in particular, the CP ENSO) strongly resembles the PMM that is the leading coupled variability mode of the subtropical Pacific. A strong association between the spring PMM index and the following winter's ENSO index was demonstrated in *Chang et al.* [2007], who found that a majority of El Niño events over the past four decades were preceded by SST and surface wind anomalies similar to those associated with the PMM. Recently, the winter SST anomalies in the western North Pacific have been suggested as a subtropical ENSO precursor, which precede ENSO development in the following year [*S.-Y. Wang et al.,* 2012, 2013]. There are several ways to link these ENSO precursor-associated anomalies to the generation of ENSO events in the equatorial Pacific. One explanation emphasizes how the surface wind anomalies associated with the subtropical precursor can directly or indirectly (through the reflection of off-equatorial Rossby wave in the western Pacific) excite downwelling Kelvin waves along the equatorial thermocline that propagate eastward to trigger El Niño events in the eastern Pacific [e.g., *Alexander et al.,* 2010; *S.-Y. Wang et al.,* 2012, 2013]. This view appears more related to the onset of EP ENSO events. However, when the subtropical precursor reaches the equatorial Pacific, it can also force a CP ENSO event by interacting with the local ocean mixed layer. Figure 1.6 illustrates this view of how the underlying dynamics of the two types of ENSO

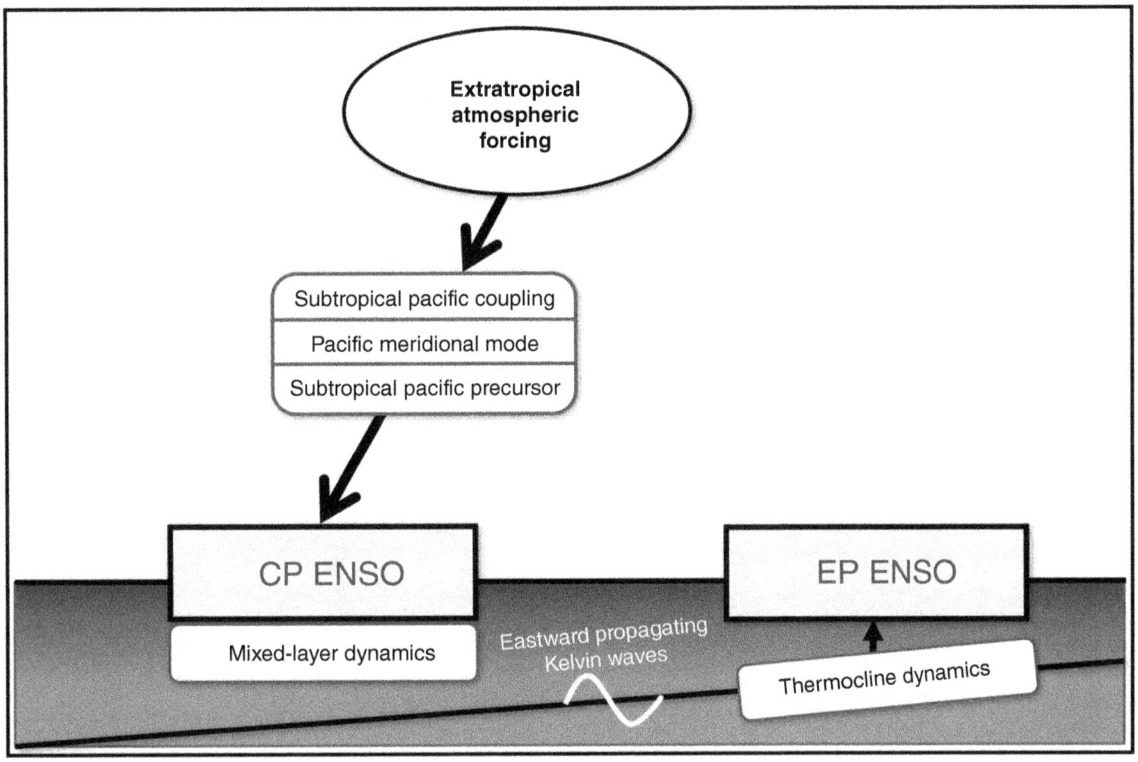

Figure 1.6 A schematic diagram illustrating the possible relationships between the subtropical Pacific precursors and the two types of ENSO (CP and EP). (*See insert for color representation of the figure.*)

can interact with the subtropical Pacific precursor. In this view, the generation of the CP ENSO is more related to mixed layer ocean dynamics. *Dommenget* [2010] and *Clement et al.* [2011] have demonstrated that ENSO-like events can be produced in coupled models where the ocean component consists of a mixed layer only without any thermocline dynamics. The generation of the EP ENSO is considered more related to the thermocline dynamics depicted by the delayed-oscillator [*Suarez and Schopf* 1988; *Battisti and Hirst* 1989] and charge-recharged oscillator theories [e.g., *Wyrtki* 1975; *Zebiak* 1989; *Jin* 1997].

1.3.3. Links Between the Atlantic Multidecadal Oscillation and the Central-Pacific ENSO

The strength of the subtropical Pacific coupling, therefore, plays a key role in determining how efficiently the subtropical Pacific precursors can penetrate into the tropics to excite ENSO events, particularly CP ENSO events. As mentioned, *Yu et al.* [2012a] found that the early 1990s is the time when ENSO changed from the EP type to the CP type. *Yu et al.* [2015a] further confirmed that the subtropical Pacific coupling was stronger after the early 1990s. They analyzed the subtropical Pacific coupling strength by examining the correlation coefficient between the SST and surface wind anomalies associated with the PMM. As shown in Figure 1.7, the coupling strength was at a relatively constant level before and after the 1976–1977 climate shift, which suggests that the phase change in the PDO cannot affect the subtropical Pacific coupling. In contrast, the subtropical Pacific coupling strength increased after the early-1990s climate shift. The stronger subtropical Pacific coupling has made it easier for the subtropical precursors to penetrate into and to influence the deep tropics, and, as a result, the occurrence of CP ENSO events increases. As mentioned, the early-1990s climate shift is related to the phase change of the AMO. *Yu et al.* [2015a] coupled an atmospheric general circulation model (AGCM) with a slab ocean model to demonstrate the relationship between the AMO and the CP ENSO. In the coupled model, they prescribed a positive phase of the AMO SST anomalies in one experiment and a negative phase of the AMO in the other experiment. They showed that the Pacific SST variability produced by the coupled model increased most in the tropical central Pacific when a positive AMO was prescribed in the Atlantic. They also laid out the chain of events to explain how the AMO influenced the CP ENSO. First, a switch of the AMO to its positive phase in the early 1990s led to an intensification of the Pacific subtropical high. The intensified high resulted in stronger-than-average

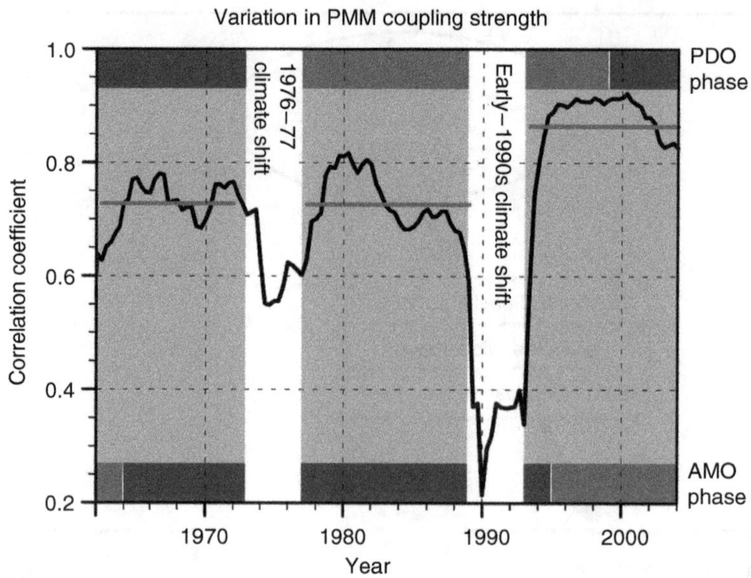

Figure 1.7 Coefficients of the 10 yr running correlation between the PMM-SST and PMM-wind indices in boreal spring (March–May). The ERSST and NCAR/NCEP reanalysis datasets are used. The red line indicates the mean of the correlation coefficients during each period. The shading at the top and bottom show the phases (positive red/negative blue) of the PDO and the AMO, respectively. Modified from *Yu et al.* [2015a].

background trade winds that enhanced the WES feedback mechanism, strengthening the subtropical Pacific coupling between the atmosphere and ocean, making the subtropical Pacific precursors more capable of penetrating into the deep tropics, and ultimately leading to increased occurrence of CP ENSO events. The study of *Yu et al.* [2015a] suggests that the recent shift from more conventional EP to more CP type of ENSO events can thus be at least partly understood as a Pacific Ocean response to the phase change in the AMO.

1.3.4. The Roles of the Hadley and Walker Circulations

The emergence of CP ENSO represents a change in ENSO dynamic from one that emphasizes the tropical Pacific coupling processes to another that emphasizes the subtropical Pacific forcing. This recent change in ENSO dynamic can also be understood via the different roles played by the Walker and Hadley circulations in the ENSO dynamic, as noted by *Yu et al.* [2010]. Figure 1.8 shows the correlation coefficients between SLP anomalies and the EP and CP ENSO indices. The EP ENSO is associated with a southern-oscillation pattern characterized by out-of-phase SLP anomalies between the eastern and western tropical Pacific. The SLP variations over the maritime continent region are linked to SLP over the eastern equatorial Pacific through the Walker circulation. The SLP anomaly pattern associated with CP ENSO does not really resemble the Southern Oscillation. Instead, the SLP variations over the maritime continent are linked to SLP variations in the subtropics of both the North and South Pacific, suggesting that a connection through local Hadley circulation may be more important. Therefore, *Yu et al.* [2010] argued that the tropical Pacific Ocean interacts with the Walker circulation to produce EP ENSO but the local Hadley circulation to produce CP ENSO. The alternation between the EP and CP types of ENSO can be viewed as a result of changes in the relative strengths of the Hadley and Walker circulations. A stronger Walker circulation may lead to a dominance of the EP type, while a stronger Hadley circulation may favor the CP type. This view was further examined in *Yu et al.* [2012a], in which they compared the strength of the time-mean Walker and Hadley circulations before and after the early-1990s climate shift. Consistent with the view of *Yu et al.* [2010], the period of postclimate shift has a stronger mean Hadley circulation compared to the period of preclimate shift. The stronger Hadley circulation could help the NPO-induced SST variations to spread into the tropical central Pacific. In association with the changes in the strength of mean atmospheric circulations, the tropical central Pacific has become more responsive to the forcing and influence from the extratropical atmosphere, which has resulted in more frequent occurrence of CP ENSO events. The relative dominance of the two types of ENSO may be related to the relative strength of the Hadley and Walker circulations. As for the mean strength of the Walker circulation, it is in debate whether it has weakened [e.g., *Vecchi et al.,* 2006; *Vecchi and Soden,* 2007; *Xie et al.,* 2010; *Tokinaga et al.,* 2012; *Yu et al.,* 2012a] or strengthened

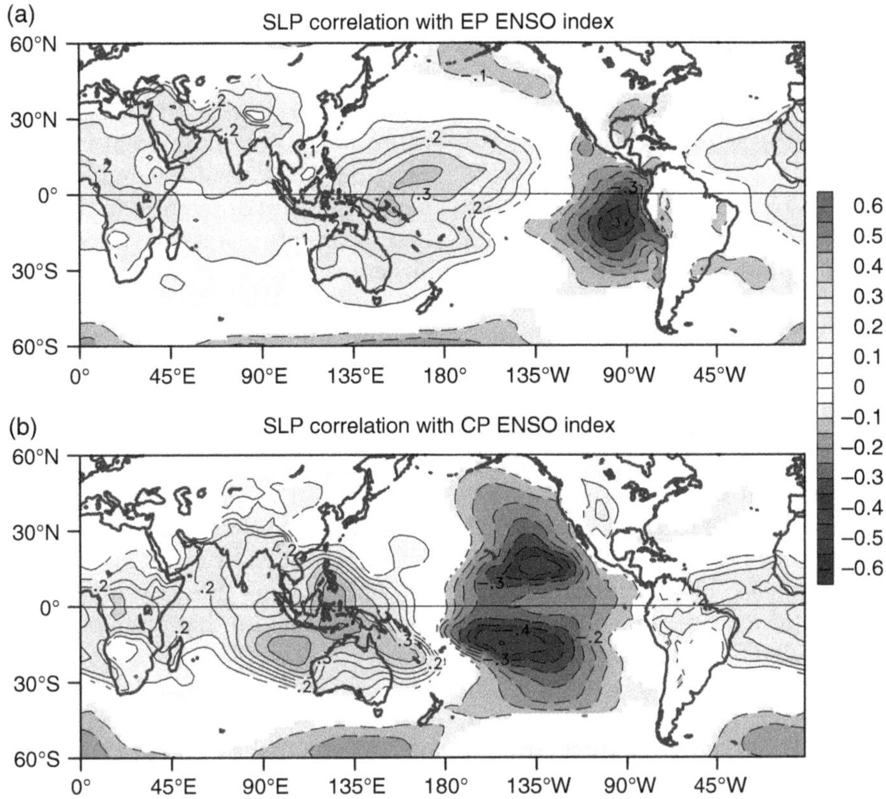

Figure 1.8 Correlation of SLP anomalies with (a) the EP and (b) CP ENSO indices. Only the coefficients exceeding the 95% confidence level are shown. The NCEP/NCAR reanalysis dataset for the period of 1948–2010 is used.

[e.g., *Dong and Lu,* 2013; *Hu et al.*, 2013; *L'Heureux et al.*, 2013] in recent decades, although it is more generally agreed that the circulation has strengthened in the most recent decade [e.g., *England et al.*, 2014].

1.3.5. Other Possible Mechanisms

In addition to the extratropical atmospheric forcing mechanism, other mechanisms have been proposed to explain the generation of the two types of ENSO. For example, *Yu et al.* [2009] used ocean-atmosphere coupled GCM (CGCM) experiments to suggest that the CP ENSO can be excited by forcing associated with the Asian and Australian monsoons. *Chung and Li* [2013] argued that the La Niña-like (stronger zonal SST gradient) decadal background state since 1999 is responsible for the more frequent occurrence of CP El Niño. Their numerical experiments showed that with the increased background zonal SST gradient, the anomalous wind and precipitation response to EP or CP SST forcing shifts to the west, leading to a westward shift in the SST anomaly tendency that resembles the CP ENSO. *Ren and Jin* [2013] found that with the decadal signal removed, the ocean heat content shows recharge-discharge processes throughout the life cycles of both ENSO types. They showed that the different zonal locations of the SST and easterly wind anomalies and the associated thermocline feedback contribute to the growth and phase transitions of the two ENSO types. This view of varying thermocline feedbacks for the two ENSO types is consistent with the study of *Hu et al.* [2012b] who suggested that the subsurface ocean temperature anomalies are weaker in the CP ENSO than in the EP ENSO. *Hu et al.* [2012b] also found differences in the westerly wind bursts (WWBs) between the two types of ENSO. They suggested that the CP (EP) ENSO is followed by weaker and more westward confined (stronger and eastward extended) WWBs along the equatorial Pacific, which are associated with suppressed (enhanced) air-sea interactions over the cold tongue/Intertropical Convergence Zone and weaker (stronger) thermocline feedbacks. *Fedorov et al.* [2014] and *Hu et al.* [2014] also reached a similar conclusion on the differences in the importance of WWB for the two ENSO types. These two modeling studies showed that the EP El Niño develops when WWBs are imposed in an initial recharge ocean state, while the CP El Niño develops in a neutral ocean state in the presence of a WWB or in a recharge state without a WWB. All these WWB studies suggest that properly accounting for the preceding westerly wind signal and air-sea interaction may be necessary for accurate ENSO predictions [*Hu et al.*, 2012b; *Fedorov et al.*, 2014; *Hu et al.*, 2014; *D. Chen et al.*, 2015].

1.4. CHANGES IN ENSO TELECONNECTIONS AND ASSOCIATED CLIMATE EXTREMES

The recent identification of the two distinct types of ENSO offers a new way to consider the impact of ENSO on global climate. There are indications that the atmospheric response to ENSO events can be sensitive to the exact location of their SST anomalies [e.g., *Mo and Higgins* 1998; *Hoerling and Kumar*, 2002; *Barsugli and Sardeshmukh*, 2002; *DeWeaver and Nigam*, 2004; *Larkin and Harrison*, 2005b]. An increasing number of recent studies have looked into the different impacts produced by these two types of ENSO on various parts of global climate. Many of these studies have revealed that the CP ENSO can affect global climate differently from the EP ENSO [e.g., *Larkin and Harrison*, 2005b], indicating that one should not simply use the existing understanding of the conventional EP ENSO to anticipate the climate impacts and teleconnections associated with the CP ENSO. A systematic investigation of the global and regional impacts of the CP ENSO is required. In this section, we use the differing impacts of the two types of ENSO on global climate to illustrate how ENSO's teleconnection and associated climate extremes have been changing in the past few decades.

1.4.1. North America

It is well recognized that North American climate is significantly impacted by ENSO [e.g., *Ropelewski and Halpert*, 1986, 1989; *Kiladis and Diaz*, 1989; *Livezey et al.*, 1997; *Cayan et al.*, 1999; *Larkin and Harrison*, 2005b; and many others]. However, the existence of two types of ENSO was not taken into account in most of the past studies. The El Niño impact on US winter temperatures is traditionally characterized as a north-south dipole pattern, in which above normal temperatures are found over the northern states and below normal temperatures over the southern states [e.g., *Ropelewski and Halpert*, 1986]. *Larkin and Harrison* [2005a] first pointed out that seasonal US weather anomalies associated with the two types of El Niño are quite different. *Mo* [2010] considered a priori the two types of ENSO to examine ENSO's impacts on North American climate. The author contrasted the impacts of ENSO on air temperature and precipitation over the United States during the period 1915–1960 when the EP type dominated and the period 1962–2006 when the CP type had increased in importance, and concluded that the EP El Niño produced a north-south contrast pattern in the air temperature variations, similar to the traditional view. For the CP El Niño, the author found the temperature variations are characterized by more of an east-west contrast pattern with warming over the Pacific Northwest and cooling in the Southeast. *Yu et al.* [2012b] also found that the two types of ENSO produced different impacts on US winter temperature. Their result is similar to that of *Mo* [2010] but with significant differences in the details. *Yu et al.* [2012b] first separately regressed winter (January–March) surface air temperature anomalies to the EP and CP ENSO indices to show that, during EP El Niño events (Fig. 1.9a), positive winter temperature anomalies are concentrated mostly over the northeastern part of the United States and negative anomalies are most obvious over the southwestern states. During CP El Niño events (Fig. 1.9b), the warm anomalies are located in the northwestern United States and the cold anomalies are centered in the southeastern United States. The US temperature anomaly patterns are rotated by about 90° between these two types of El Niño. Adding these two impact patterns together can result in a pattern that resembles the traditional El Niño impact on the US winter temperature (i.e., a warm-north and cold-south pattern). It indicates that the traditional view of El Niño impact is likely a mixture of the impacts of the two types of El Niño. *Yu et al.* [2012b] was able to reproduce these two different impact patterns in forced model experiments performed with the Community Atmosphere Model version 4 (CAM4) [*Neale et al.*, 2013] from the National Center for Atmospheric Research (NCAR) (shown in Fig. 1.9c and d], in which SST anomalies characteristic of the EP El Niño and CP El Niño were separately used to force the model in two different sets of ensemble experiments. The regressed impact patterns can also be identified in the four strongest EP El Niño events (in 1997, 1982, 1972, and 1986) and four of the top five strongest CP El Niño events (in 2009, 1957, 2002, and 2004). Both *Mo* [2010] and *Yu et al.* [2012b] argued that the cause of the different impacts on US winter temperature is the different wave train patterns excited by these two types of El Niño in the extratropical atmosphere. The winter atmosphere produces a Pacific–North American (PNA) teleconnection pattern [*Wallace and Gutzler*, 1981] during the CP El Niño, which consists of a positive anomaly center extending from eastern Alaska to the northwestern United States and a negative anomaly center over the southeastern United States, resulting in a warm-northwest and cold-southeast pattern of temperature anomalies. During the EP El Niño, the winter atmosphere produces a poleward wave train emanating from the tropical eastern Pacific, across the southwestern United States, and into the northeastern United States, leading to the cold-southwest and warm-northeast pattern in US winter temperatures. This wave train pattern resembles the Tropical North Hemisphere (TNH) pattern [*Mo and Livezey*, 1986].

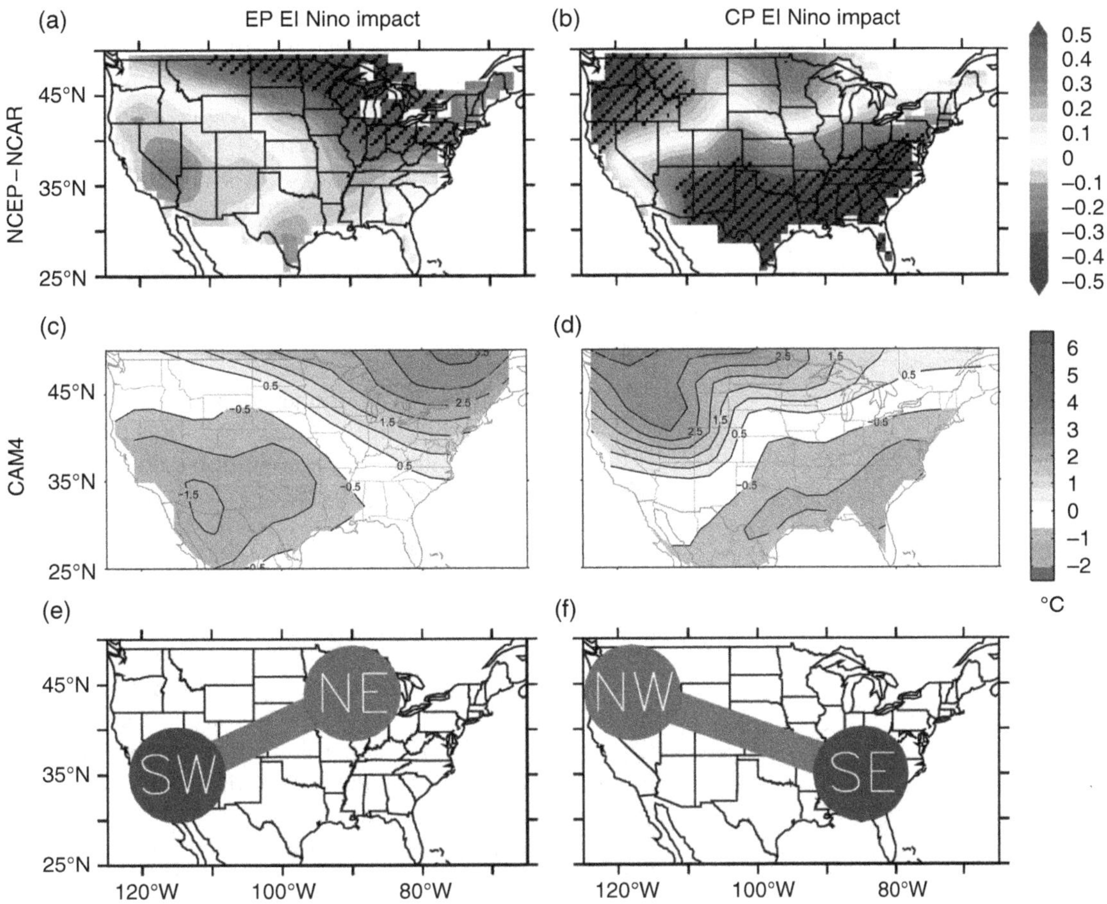

Figure 1.9 EP (left panels) and CP (right panels) El Niño impacts on US winter (January–March) surface air temperatures. Observed temperature anomalies regressed onto the (a) EP and (b) CP ENSO indices. Coefficients exceeding the 90% confidence level are dotted. The NCEP/NCAR reanalysis dataset for the period 1950–2010 is used. Temperature patterns reproduced by the Community Atmosphere Model version 4 (CAM4) in the (c) EP and (d) CP experiments. Only differences exceeding the 90% confidence level are shaded. Schematic diagrams of the (e) EP and (f) CP El Niño impacts on temperatures. Adapted from *Yu et al.* [2012b]. (*See insert for color representation of the figure.*)

Concerning the US winter precipitation, the classical view of El Niño's impact is one of negative precipitation anomalies in the northern United States and positive precipitation anomalies in the southern United States due to a southward displacement of the tropospheric jet stream and associated winter storm activity [e.g., *Ropelewski and Halpert*, 1986, 1989; *Kiladis and Diaz*, 1989; *Dettinger et al.*, 1998; *Mo and Higgins*, 1998; *Cayan et al.*, 1999]. *Mo* [2010] suggested that, during the EP El Niño events, precipitation decreases in the Pacific Northwest but increases in northern California. For the CP El Niño, its influence on precipitation increases in the Southwest but decreases in the Ohio valley compared to the EP El Niño. *Yu and Zou* [2013] regressed US winter precipitation anomalies on the EP and CP ENSO indices to identify the impact patterns (Fig. 1.10a and b). Both types of El Niño produce dry-north and wet-south anomaly patterns, similar to the seesaw pattern that has traditionally been used to describe El Niño impacts on US winter precipitation. However, the dry anomalies produced by the CP El Niño are of larger magnitude and cover larger areas than those produced by the EP El Niño. For example, the dry anomalies cover only the Great Lakes region during the EP El Niño, but extend southwestward through the Ohio-Mississippi valley toward the Gulf Coast during the CP El Niño. In contrast, the wet anomalies tend to have smaller magnitudes during the CP El Niño than during the EP El Niño, a phenomenon that is most obvious over the southeast United States. In Figure 1.10, a and b indicate that the CP El Niño tends to intensify the dry anomalies but weaken the wet anomalies of the impact pattern produced by the EP El Niño. This important difference is clearly revealed in Figure 1.10c, where the precipitation anomalies regressed on the EP El Niño were subtracted from the anomalies regressed in the CP El Niño.

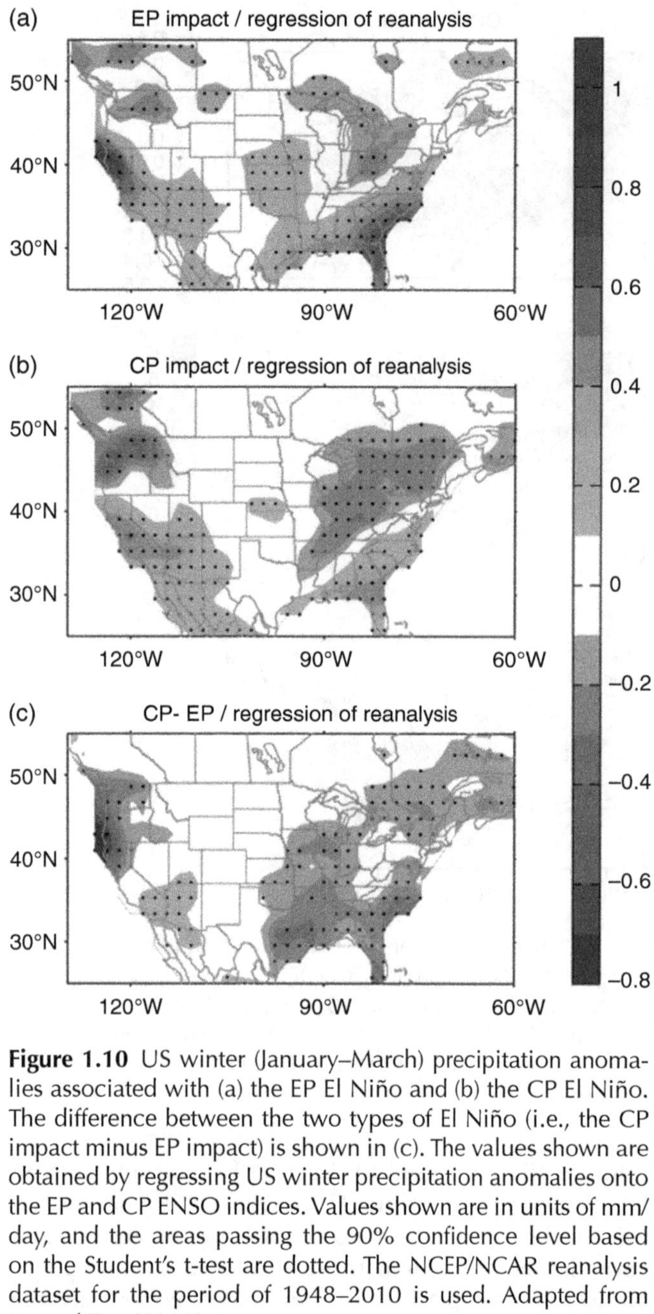

Figure 1.10 US winter (January–March) precipitation anomalies associated with (a) the EP El Niño and (b) the CP El Niño. The difference between the two types of El Niño (i.e., the CP impact minus EP impact) is shown in (c). The values shown are obtained by regressing US winter precipitation anomalies onto the EP and CP ENSO indices. Values shown are in units of mm/day, and the areas passing the 90% confidence level based on the Student's t-test are dotted. The NCEP/NCAR reanalysis dataset for the period of 1948–2010 is used. Adapted from *Yu and Zou* [2013].

The negative values in Figure 1.10c indicate that a shift in El Niño from the EP type to the CP type enhances the dry anomalies over the Pacific Northwest and along the Ohio-Mississippi valley and weakens the wet anomalies over the Southeast. As a result, the emerging CP El Niño produces an overall drying effect during the US winter. The different impacts on winter precipitation can lead to an asymmetric response between the two types of ENSO in the springtime soil water hydrology over the Mississippi River basin as shown by *Liang et al.* [2014], who also indicated that the winter to spring memory was carried through by subsurface hydrological storage processes.

Winter precipitation over the United States is primarily associated with winter storms, whose tracks are controlled by the locations of tropospheric jet streams. *Yu and Zou* [2013] related the enhanced drying effect of the CP ENSO to a more southward displacement of the tropospheric jet streams. Previous studies have suggested that such an equatorward shift of the tropospheric jet streams during El Niño events results from El Niño-induced Rossby wave trains and strengthening of the Hadley circulation [e.g., *Wang and Fu*, 2000; *Seager et al.*, 2003; *Lu et al.*, 2008]. As the jet streams shift southward, winter storms shift south as well, leading to a dry-north and wet-south pattern of precipitation anomalies during the El Niño. *Yu and Zou* [2013] found that the jet streams were displaced more southward during CP El Niño than during EP El Niño. The increased southward displacements of the jet streams cause dry anomalies over the northern United States (including the Northwest and Ohio-Mississippi valley), which expand and intensify more significantly during the CP El Niño than during the EP El Niño. Similarly, the wet anomalies over the southern United States expand over the Southwest and extend into Mexico during the CP El Niño. However, the same southward displacements over the East Coast push the core of the jet stream (and therefore the storm tracks) south of the US continent and into the Gulf of Mexico and the Caribbean, which results in only a small area of wet anomalies left in the southeast United States during the CP El Niño. One major implication of the findings of *Yu and Zou* [2013] is that the droughts occurring in the Ohio-Mississippi valley and Pacific Northwest during El Niño years may be intensified as El Niño becomes more frequently of the CP type and that the Southeast should not expect as much enhancement of winter precipitation during El Niño years as in the past. In addition to cold season impacts, *Liang et al.* [2015] suggested that the two ENSO types have different influences on the warm season (May–June) Great Plains low-level jet (GPLLJ) during their decaying phases. They found that the EP El Niño intensifies the GPLLJ while CP El Niño weakens the GPLLJ, the latter leading to drier conditions in the central United States.

Capotondi and Alexander [2010] have shown that CMIP3 models are capable of reproducing the general features of the observed precipitation pattern over North America despite the coarse resolutions of these global models. The six models they analyzed captured the observed precipitation maxima in the American Northwest and Southeast and the lower values in the central United States. Their analysis also showed that the seasonal cycles of precipitation produced by the CMIP3 models in the Great Plains were realistic, with a rainy season that starts in early summer and slowly winds down toward winter. *Mo* [2010] also showed that the CMIP3 simulations could be useful in identifying the CP and EP ENSO's impacts on US winter climate. A series of studies organized by the National Oceanic and Atmospheric Administration (NOAA) Modeling, Analysis, Predictions, and Projections (MAPP) Program's CMIP5 Task Force have also demonstrated that CMIP5 models are useful in understanding the climate over North America [*Sheffield et al.*, 2013a, 2013b, 2014; *Maloney et al.*, 2014]. These recent studies demonstrated that CMIP simulations were useful for studying the regional climate impacts produced by global-scale climate phenomena. The performance of coupled climate models in simulating the EP and CP types of ENSO was also examined for both the CMIP3 models [*Yu and Kim*, 2010a] and the CMIP5 models [*Kim and Yu*, 2012]. Compared to the CMIP3 models, the CMIP5 models are found to better simulate the observed spatial patterns of the two types of ENSO and to have a significantly smaller intermodel diversity in ENSO intensity. The improved performance of the CMIP5 models is particularly apparent in the simulation of EP ENSO intensity, although it is still more difficult for the models to reproduce the observed EP ENSO intensity than the observed CP ENSO intensity. The performance of CMIP5 models in simulating the different impacts of the two types of ENSO on US winter climate was also examined in *Zou et al.* [2014], where 30 CMIP5 preindustrial simulations were examined to conclude that the CMIP5 models can simulate the observed EP El Niño impacts but not the observed CP El Niño impacts. The responses of observed US winter temperature to the EP El Niño are reasonably simulated by most of the CMIP5 models. The northeast-warm and southwest-cold pattern can be seen in many of the CMIP5 models. However, the observed northwest-warm and southeast-cold response pattern to the CP ENSO can only be found in some of the CMIP5 models. In some of the other CMIP5 models, the US winter temperature anomalies are even opposite from the observed.

Why are CMIP5 models less successful in simulating CP ENSO impact on the US winter climate? Heating associated with deep convection is a key process by which ENSO influences the atmosphere. The intensity and location of deep convection can be represented by the outgoing longwave radiation (OLR). The importance of using OLR anomalies to understand the atmospheric teleconnections from the tropics has been emphasized in many earlier studies [e.g., *Heddinghaus and Krueger*, 1981; *Lau and Chan*, 1983, 1985; *Gruber and Krueger*, 1984; *Ardanuy and Kyle*, 1986; *Chiodi and Harrison*, 2013]. *Zou et al.* [2014] looked into the relationships between SST and OLR anomalies for the two types of El Niño and examined how well the observed relationships were simulated in the CMIP5 models. It was found that during the EP ENSO, the strongest SST anomalies are located in the eastern equatorial Pacific and can easily influence the strength of the Walker circulation, giving rise to basinwide OLR anomalies. The modeled atmospheric responses to the EP El Niño are thus not sensitive to SST anomaly structure and can be well simulated by most of the CMIP5 models. In contrast, the SST anomalies of the CP El Niño are located in the central equatorial Pacific and can induce only local OLR anomalies to the west of the SST anomalies. The modeled atmospheric responses to the CP El Niño are, therefore, different among the models depending on the simulated magnitudes and locations of the CP El Niño SST anomalies. The finding reported by *Zou et al.* [2014] implies that it may become more challenging for climate models to simulate and predict ENSO's impacts on US climate as ENSO changes from the EP type to the CP type.

1.4.2. Western Pacific and Indian Oceans

ENSO events also produce significant impacts on the climate of the western Pacific and the Indian Ocean basin. In the conventional point of view, El Niño events shift tropical convection eastward to the equatorial central-to-eastern Pacific resulting in an anomalous Walker circulation that has an anomalous descending branch over the western Pacific, Maritime Continent, and northern Australia (Fig. 1.8a). This anomalous descending motion then induces two modes of Indian Ocean (IO) variability through surface heat fluxes and oceanic Rossby waves [e.g., *Klein et al.*, 1999; *Venzke et al.*, 2000; *Alexander et al.*, 2002; *Xie et al.*, 2002; *Yu* and Lau, 2004; *Yu et al.*, 2005; *Liu and Alexander*, 2007]. One of the modes is the Indian Ocean Dipole [*Saji et al.*, 1999; *Webster et al.*, 1999] that has one SST anomaly center in the tropical western Indian Ocean and the other center in the southeast Indian Ocean, and the other mode is the Indian Ocean basinwide warming (IOBW). The IOD tends to peak during El Niño's developing phase during autumn (September–November; SON) and IOBW peaks during El Niño's decaying phase during spring (March–May) [e.g., *Schott et al.*, 2009]. These ENSO-induced Indian Ocean variability modes are known to affect Indian summer monsoon, Australia and East Africa, the western North Pacific, and Southern Hemisphere high latitudes [e.g., *Saji et al.*, 1999; *Ashok et al.*, 2001, 2003; *Lau and Nath*, 2004; *Yang et al.*, 2007; *Xie et al.*, 2009]. Associated with the different Walker circulation response to two ENSO types (Fig. 1.8a and b), *Wang and Wang* [2014] suggested that the EP El Niño and some CP El Niño (El Niño Modoki I) events induce the positive IOD, while the other CP El Niño (El Niño Modoki II) events produce the negative

IOD. However, *Yu et al.* [2015b] argued that the EP El Niño events are more associated with the IOD while the CP El Niño events are not. *Paek et al.* [2015] showed that the EP El Niño events tend to be accompanied by basinwide warming in the Indian Ocean but not the CP El Niño events. The different Indian Ocean responses induced by the two types of ENSO bring different ENSO impacts to the neighboring regions. *Kumar et al.* [2006] used a historical rainfall record and AGCM experiments to suggest that the CP El Niño is more effective in enhancing subsidence over the Indian subcontinent during Indian summer monsoon season (June–September) leading to more severe droughts than during EP El Niños. *Yadav et al.* [2013] found that the EP El Niño leads to more precipitation over the north and central India regions during El Niño peak season (December–February). *Ratman et al.* [2014] reported significant drought in austral summer over southern Africa during EP El Niños but not during CP El Niños. Significant decreases in rainfall in Australia have been reported during CP El Niño events compared to the EP El Niño events [*Wang and Hendon*, 2007; *Lim et al.*, 2009; *Taschetto and England*, 2009].

1.4.3. East Asia

ENSO has profound impacts on East Asian climate variability such as the East Asian monsoon [e.g., *Zhang et al.*, 1999; *Chang et al.*, 2000; *Zhou and Chan*, 2007; *Li and Yang*, 2010], East Asian rainfall [e.g., *McBride et al.*, 2003; *Juneng and Tangang*, 2005; *Zhou et al.*, 2009], western Pacific tropical cyclone (TC) activity [e.g., *Chan*, 2000; *Chia and Ropelewski*, 2002; *Camargo and Sobel*, 2005; *Ho et al.*, 2004], and SSTs over the South China Sea [e.g., *He and Guan*, 1997; *Xie et al.*, 2003; *Wang et al.*, 2006; *Zhang et al.*, 2010]. The anomalous western North Pacific subtropical high (WNPSH) or low-level Philippine Sea anticyclone is a potentially important atmospheric component connecting ENSO and East Asian climate as suggested by several authors [e.g., *Zhang et al.*, 1999; *Wang et al.*, 2000; *Lau and Nath*, 2006; *Chou et al.*, 2009]. ENSO induces SST anomalies in the western Pacific to initiate the WNPSH during the ENSO developing seasons and the WNPSH variability is sustained until the following summer (June–August) through local air-sea interactions [*Wang et al.*, 2000], or the IO SST anomalies (i.e., IOBW) to excite a Kelvin wave to the east maintaining the WNPSH variability [*Yang et al.*, 2007; *Xie et al.*, 2009], or the tropical central Pacific SST anomalies that produce the WNPSH as a Rossby wave response to the west of the SST anomalies [*B. Wang et al.*, 2013], or the SST anomalies over the maritime continent that give rise to a local Hadley circulation descending into the WNPSH region [*Sui et al.*, 2007].

Recently, *Zhang et al.* [2011] contrasted the different impacts of the two ENSO types on the WNPSH. They found that during the El Niño developing phase in autumn, an intensified WNPSH for the EP El Niño but a weakened WNPSH for the CP El Niño. *Yuan et al.* [2012] further documented that from the El Niño developing summer to decaying summer, the EP El Niño is accompanied by a strong WNPSH that extends from the Indian Ocean to the east of the Philippines, while the CP El Niño is accompanied by a weak, short-lived WNPSH that is confined to the west of the Philippines. They also argued that the IO SST anomalies are responsible for the persistence of WNPSH into the ENSO following summer for the EP El Niño, but not for the CP El Niño. *Paek et al.* [2015] suggested in their observational and AGCM study, that the 3–5-yr band of summer WNPSH variability is more related to the EP ENSO and its SST anomalies over the Indian Ocean, whereas the 2–3-yr band of WNPSH variability is related to CP ENSO and its SST anomalies in the central Pacific and maritime continent. These different WNPSH responses to the two ENSO types are then shown to modulate East Asian climate. For example, southern China experiences increased rainfall during the EP El Niño but decreased rainfall during the CP El Niño due to the different WNPSH response and associated moisture transport into the region in El Niño developing phase in autumn [*Zhang et al.*, 2011], mature phase in winter [*Weng et al.*, 2009], and decaying phase in spring [*Feng and Li*, 2011]. During El Niño decay in summer, *Feng et al.* [2011] found increased rainfall south to the Yangtze River and decreased rainfall in the northern Yangtze–Huaihe river region for the EP El Niño, and increased rainfall in the Huaihe River–Yellow River region and decreased rainfall in southern China for the CP El Niño. *J.-S. Kim et al.* [2012] showed a significant increase in precipitation over South Korea and southern Japan during the CP El Niño that represents conditions opposite of those found for the EP El Niño. *Weng et al.* [2007] and *Yuan and Yang* [2012] also found that the different WNPSH can exert a different impact on the summer and winter monsoon over East and Southeast Asia, which implies difficulty in predicting East Asian monsoon variability based on the conventional monsoon-ENSO relationships [e.g., *Yang and Jiang*, 2014]. The WNPSH and associated convection also modulate TC activity over the South China Sea and western North Pacific during the El Niño developing phase in summer and autumn. The CP El Niño tends to enhance TC activity in the regions but the EP El Niño suppresses TC activity, which increases the TC landfall in Taiwan, South China, Korea, and Japan particularly in autumn [*Chen and Tam*, 2010; *Chen*, 2011; *Hong et al.*, 2011; *Kim et al.*, 2011; *Wang et al.*, 2014]. In the South China Sea, anomalous net surface heat flux associated with the different atmospheric circulation responses to the two ENSO types leads to a strong warm basin mode during the EP El Niño and a warm semibasin mode during the CP El Niño [*Liu et al.,* 2014].

1.4.4. Europe and the Atlantic Ocean

ENSO can influence the Atlantic/European sector via the tropospheric temperature mechanism [*Chiang and Sobel,* 2002] and the atmospheric bridge mechanism [*Klein et al.,* 1999]. The tropospheric temperature mechanism involves an ENSO-induced tropical tropospheric warming that spreads to the Atlantic as a Kelvin wave response to ENSO forcing, which then increases SST anomalies there [e.g., *Yulaeva and Wallace,* 1994; *Enfield and Mayer,* 1997; *Chiang and Sobel*, 2002]. In the atmospheric bridge mechanism, ENSO weakens the Walker circulation and excites a PNA teleconnection pattern that reduces trade wind and evaporation in the tropical North Atlantic, leading to a warming there [e.g., *Lau and Nath,* 2004; *Klein et al.,* 1999; *Alexander et al.,* 2002]. The canonical El Niño is known to suppress Atlantic TC activity through changes in vertical wind shear over the main TC development region [e.g., *Gray*, 1984; *Goldenberg and Shapiro,* 1996] and changes in tropospheric and surface temperatures [*Tang and Neelin,* 2004]. Recently, *Kim et al.* [2009] found that in contrast to the EP El Niño, the CP El Niño tends to enhance Atlantic TC activity and increase landfall potential along the coast of the Gulf of Mexico and Central America, which is associated with the different impacts on the vertical wind shear in the main TC development region by the two ENSO types. In contrast, *S.-K. Lee et al.* [2010] and *Larson et al.* [2012] suggested that the CP El Niño has an insubstantial impact on Atlantic TC activity, which is associated with a weak warming of the tropical North Atlantic during the CP El Niño and a strong warming during the EP El Niño [*Rodrigues et al.,* 2011; *Amaya and Foltz*, 2014]. *Graf and Zanchettin* [2012] found different impacts on European winter temperatures. They showed that EP El Niño events lead to weak warming while CP El Niño events lead to significant cooling. They further attributed the European cooling to the negative phase of North Atlantic Oscillation (NAO) through the CP ENSO-NAO teleconnection in the troposphere.

1.4.5. Arctic

For the Northern Hemisphere high-latitude and Arctic climate, the ENSO signal can be transmitted there via a stratospheric pathway. Canonical El Niño events tend to be accompanied by a weakened Northern Hemisphere polar vortex. The positive PNA pattern and a deepened Aleutian low in the troposphere during El Niño may enhance planetary wave activities and increase the upward propagation of the waves into the stratosphere [e.g., *Garfinkel and Hartmann,* 2008; *Garfinkel et al.,* 2010; *Nishii et al.,* 2010], where they dissipate and result in a weakened polar vortex and more frequent polar stratospheric warmings [e.g., *Sassi et al.,* 2004; *García-Herrera et al.,* 2006; *Manzini et al.,* 2006]. The warm anomalies then descend to the troposphere in boreal winter through wave–mean flow interactions [e.g., *Holton*, 1976; *McIntyre,* 1982; *Zhou et al.,* 2002; *Manzini et al.,* 2006]. This weakened polar vortex and descending warm anomalies excite the Arctic Oscillation (AO) and the NAO [e.g., *Thompson and Wallace,* 1998; *Baldwin and Dunkerton*, 1999; *Zhou et al.,* 2002; *Kunz et al.,* 2009]. Recently, differing polar teleconnections between the two types of ENSO have been reported. *Hegyi and Deng* [2011] used post-1979 observations to suggest that in contrast to the EP El Niño, the CP El Niño can lead to a stronger polar vortex and a positive phase of the AO. They argued that the positive AO tendency associated with the CP El Niño events in recent decades has contributed to decadal changes in the Arctic precipitation. *Xie et al.* [2012] also reported a weakened midlatitude upward propagation of wave activity during the 1979–2010 CP El Niños, leading to a stronger and colder polar vortex. In contrast, *Graf and Zanchettin* [2012] showed both types of El Niño since 1948 were accompanied by a weakened polar vortex. *Zubiaurre and Calvo* [2012] and *Sung et al.* [2014] found a polar response (that is not statistically significant) to the CP El Niño that is related to a delayed deepening of the Aleutian low and weakened upward wave propagation to stratosphere. *Garfinkel et al.* [2013] compared the different composites used in previous studies [*Hegyi and Deng*, 2011; *Graf and Zanchettin*, 2012; *Xie et al.,* 2012; *Zubiaurre and Calvo*, 2012] to show that the diverse CP ENSO-Arctic teleconnections are very sensitive to the selection of CP El Niño events from the short observational record. *Iza and Calvo* [2015] emphasized the role of stratospheric sudden warming (SSW) events for the downward propagation. They found that with the SSWs removed, the different EP and CP El Niño responses are robust regardless of the CP El Niño definition and the composite size. In a modeling study, *Xie et al.* [2012] showed that whether the CP El Niño induces a strengthened or a weakened vortex depends on the phase of the quasi-biennial oscillation (QBO), while other studies [*Garfinkel et al.,* 2013; *Hurwitz et al.,* 2013; *Hurwitz et al.,* 2014] found a consistent weakening of the polar vortex under CP Niño SST forcing using a single AGCM. In the CMIP5 multimodel simulations, *Hurwitz et al.* [2014] found that the multimodel means show a weakening of the polar vortex during the EP El Niño in boreal winter, and a nonsignificant strengthening of the vortex during the CP El Niño.

1.4.6. Antarctica

In the Southern Hemisphere, ENSO has been suggested to influence mid-to-high-latitude climate [e.g., *Karoly*, 1989; *Mo*, 2000; *Yuan*, 2004], often via the two leading Southern Hemisphere atmospheric variability modes: the southern annular mode (SAM) [*Hartmann and Lo*, 1998; *Thompson and Wallace,* 2000] and the Pacific–South American teleconnection (PSA) [*Mo and Ghil*, 1987] pattern. For example, previous studies found that ENSO, together with the SAM and PSA, influenced South American rainfall [e.g., *Ropelewski and Halpert,* 1987, 1989; *Grimm*, 2003], the Southern Ocean SSTs [e.g., *Ciasto and Thompson,* 2008; *Yeo and Kim*, 2015], Antarctic sea-ice concentrations [e.g., *Liu et al.*, 2004; *Stammerjohn et al.*, 2008; *Yuan and Li,* 2008], and Antarctic surface air temperatures [e.g., *Kwok and Comiso,* 2002; *Ding et al.*, 2011; *Schneider et al.*, 2012]. Recently, several studies have begun to examine the different impacts of the two types of ENSO on the Southern Hemisphere climate. The westward shift of tropical Pacific convection during the CP El Niño can excite Rossby wave train (i.e., PSA) toward higher latitudes, which then results in suppressed storm track activity over Australia and enhanced activity over central Argentina [*Ashok et al.*, 2009], a northward displacement of the South American precipitation pattern [*Hill et al.*, 2009], the melting of Antarctic ice by inducing a stationary anticyclone and enhancing the eddy heat flux into the region [*T. Lee et al.*, 2010], and stronger Antarctic sea-ice variability [*Song et al.*, 2011]. In addition to the PSA pattern, early studies showed that the ENSO signals can be transmitted to high latitudes through a tropospheric pathway mechanism in which an ENSO-induced equatorward shift in the subtropical jet can change the midlatitude eddy activity and its associated eddy–mean flow interactions to modulate the SAM [e.g., *Seager et al.*, 2003; *L'Heureux and Thompson*, 2006; *Lu et al.*, 2008; *Chen et al.*, 2008]. Recent studies have used observations and GCM experiments to also suggest a stratospheric pathway mechanism in which the CP ENSO can affect the upward propagation of planetary waves into the stratosphere and induce polar temperature and vortex anomalies that subsequently descend into the troposphere to excite the SAM [e.g., *Mechoso et al.*, 1985; *Hurwitz et al.*, 2011a,b; *Lin et al.*, 2012; *Zubiaurre and Calvo,* 2012; *Son et al.*, 2013; *Evtushevsky et al.*, 2014].

Fogt and Bromwich [2006] noticed that the relationships among ENSO, PSA, and SAM can change from decade to decade, which is attributed to the strengthened ENSO–Southern Hemisphere influences from the 1980s to the 1990s. More recently, *Yu et al.* [2015b] further suggested a shift in the ENSO–Southern Hemisphere teleconnections occurred in the early 1990s. They conducted a 15 yr running correlation analysis among the SAM, PSA, and ENSO indices during austral spring (SON) for the period 1948–2014. For an ENSO index, they used cold tongue index (CTI; SST anomalies averaged over 6°S–6°N and 180°–90°W). They found that the correlation coefficient between the CTI and SAM index (Fig. 1.11e) dramatically increased from insignificantly small to significantly large values after the early 1990s, while the CTI-PSA correlation (Fig. 1.11f) stayed at significant values throughout the analysis period. The SST anomaly pattern correlated with the SAM or PSA index after the early 1990s is shown to be similar to the CP ENSO-SST anomaly pattern [*Kao and Yu,*, 2009], whereas the SST anomaly pattern correlated with the PSA index before the early 1990s is more similar to the pattern for the EP ENSO. Their composite analyses also showed both types of ENSO (Fig. 11a and b) can excite the PSA (Fig. 1.11d) while only CP type can excite the SAM (Fig. 1.11c) through tropospheric and stratospheric pathway mechanisms. The EP ENSO is found to induce the Indian Ocean Dipole but not the CP ENSO (Fig. 1.11a and b). *Yu et al.* [2015b] showed that the different influences of the two types of ENSO on the Indian Ocean SST anomalies enable these two ENSO types to produce difference influences on the SAM through eddy–mean flow interactions. They further suggested that the different ENSO–Southern Hemisphere teleconnections and more frequent occurrence of the CP ENSO in recent decades have led to a more in-phase SAM-PSA relationship after the early 1990s, which consequently impacted the post-1990s Antarctic climate in some aspects: an increase in the geopotential height anomalies over the Amundsen-Bellingshausen seas, a stronger Antarctic sea-ice dipole structure [e.g., *Yuan and Martinson*, 2001] between the Ross Sea and Weddell Sea, and a shift in the phase relationships of surface air temperature anomalies among the Antarctic Peninsula, East and West Antarctica. The findings of *Yu et al.* [2015b] imply that the linkages between ENSO and Antarctic climate depend on the dominant ENSO type and have become tighter during the recent two decades and will likely remain so in the coming decades if the dominance of the CP ENSO persists.

1.5. ENSO IN THE FUTURE

The changes in ENSO in the future warmer world are one major concern of the climate research community due to the profound impacts produced by ENSO. Climate models are generally used in ENSO projections; however, the behavior of ENSO differs strongly from model to model and is thus highly uncertain. Despite a number of studies, there is still no consensus on how ENSO SST variability will change in response to global warming [i.e., *Meehl et al.*, 1993; *Knutson et al.*, 1997; *Collins*, 2000b;

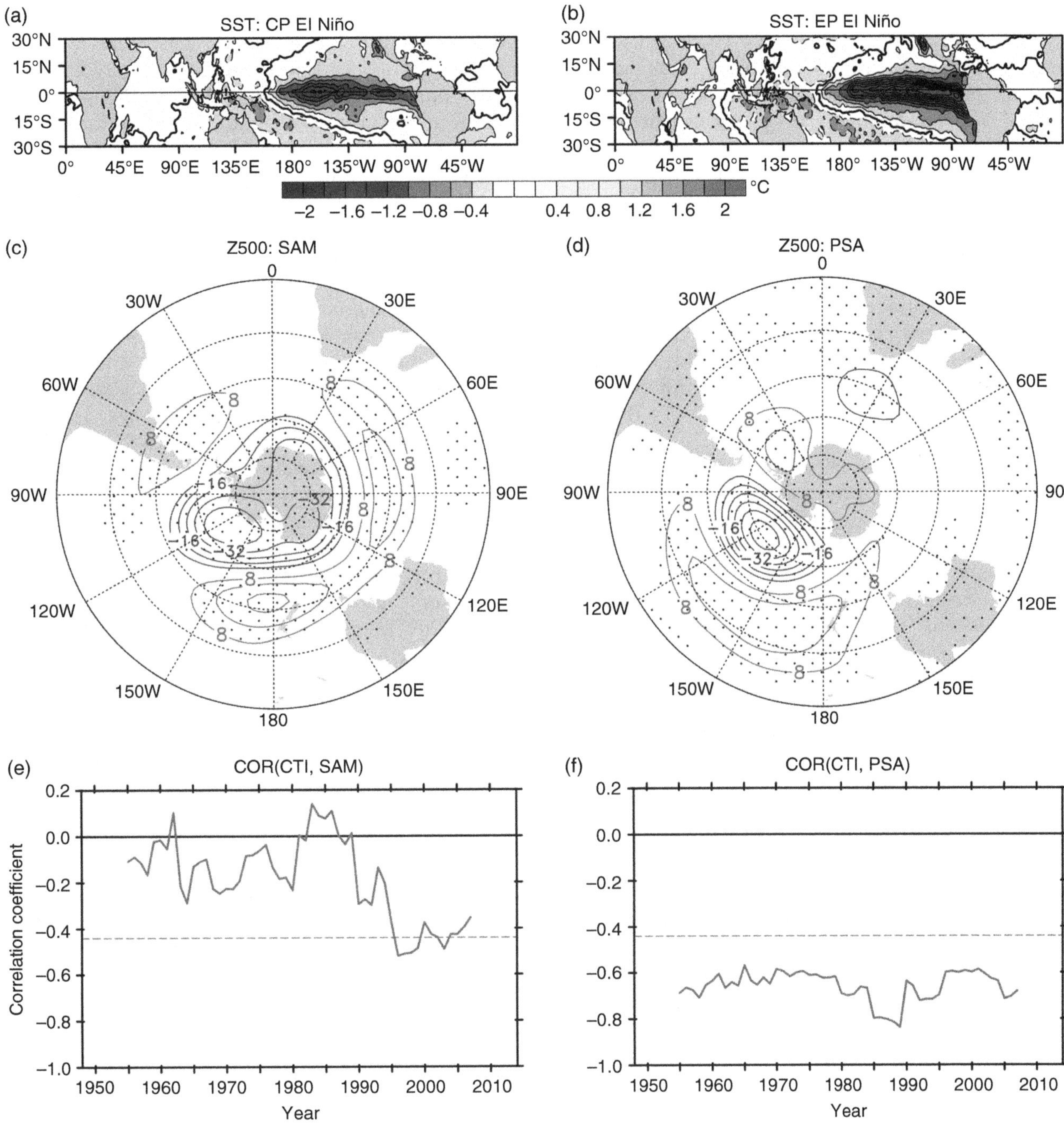

Figure 1.11 Upper panels show the composite SST anomalies for the (a) CP El Niño and (b) EP El Niño events during austral spring (SON). Middle panels show the 500 hPa geopotential height (Z500) anomaly structures of the (c) SAM and (d) PSA. Dots indicate areas where the values exceed the 95% confidence interval using a two-tailed Student's t-test. Contour interval is 8 m. Lower panels show the 15 yr running correlation coefficients (e) between the cold tongue index (CTI) and SAM index, and (f) between the CTI and PSA index. The green lines indicate the 90% confidence interval using a two-tailed Student's t-test. Modified from *Yu et al.* [2015b].

Zelle et al., 2005; *Guilyardi*, 2006; *Merryfield*, 2006; *An et al.*, 2008; *Yeh et al.*, 2009; *Newman et al.*, 2011; *Kim and Yu*, 2012; *Hu et al.*, 2012a; *Stevenson*, 2012; *Kim et al.*, 2014; *Cai et al.*, 2014; *L. Chen et al.*, 2015]. Model projections have yet to be conclusive on whether ENSO activity will be enhanced, weakened, or remain the same. The uncertainties come at least from three sources. The first source is related to the diverse abilities of climate models to represent the physical processes responsible for determining the characteristics of ENSO. The second source comes from the different model projections of how the mean state may change under warming scenarios, which can modulate the strengths of the atmosphere-ocean feedbacks that characterize ENSO activity [e.g., *Collins*, 2000a, b]. The third source of uncertainty is related to the fact that intrinsic ENSO modulation can occur on interdecadal or intercentennial timescales even without external forcing [e.g., *Wittenberg*, 2009], which makes it difficult to determine if the ENSO changes produced in model projections are really global-warming related.

Early studies analyzing ENSO behavior in response to increasing greenhouse gas (GHG) were mostly based on a single CGCM [*Meehl et al.*, 1993; *Tett*, 1995; *Knutson et al.*, 1997; *Timmermann et al.*, 1999; *Collins*, 2000a, b; *Zelle et al.*, 2005]. For example, *Meehl et al.* [1993] used a CGCM with a coarse horizontal resolution to indicate little change in ENSO amplitude or period as the global atmosphere warms. A similar conclusion was obtained by *Tett* [1995] using a different climate model. *Knutson et al.* [1997] suggested that the responses of ENSO to global warming in models can depend on the model resolution. In coarse-resolution models, the impact of increased GHG on ENSO is unlikely to be clearly distinguishable from the climate system "noise" in the near future. *Timmermann et al.* [1999] used a model with a horizontal resolution high enough to adequately reproduce the narrow upwelling and low-frequency waves on the equator to suggest that ENSO events will become more frequent, stronger, and skew toward the La Niña phase as warming progresses. By comparing the ENSO behaviors under an identical 4 x CO2 scenario using two versions of the Hadley Centre coupled model, *Collins* [2000a] found significant changes in ENSO amplitude, frequency, and phase locking to the seasonal cycle in one version but not in the other version. Such discriminations are attributed to the differences in the mean pattern of climate change brought about by subtle changes in the physical parameterization between the two versions of the model. *Collins* [2000a], therefore, suggested that the uncertainties in the projection of future ENSO behavior also come from the model projections of the future mean climate states. The tropical Pacific mean state was projected to become more El Niño-like (i.e., with greater warming in the east than in the west) [e.g., *Meehl and Washington*, 1996; *Timmermann et al.*, 1999], more La Niña-like [e.g., *Noda et al.*, 1999], or uniformly warm [e.g., *Meehl et al.*, 2000] by various models, each of which may give rise to different changes in ENSO. In addition to the mean background SST state and its zonal gradient, the vertical contrast between the mean surface and subsurface temperatures (i.e., the upper ocean stratification) in the tropical Pacific is another mean state variable of crucial importance in determining future ENSO changes [*An et al.*, 2008].

To reduce the uncertainty related to the diverse model dynamics and physics, multimodel ensembles have been increasingly used for ENSO projections in recent years. This is particularly so after the CMIP3 and CMIP5 multimodel outputs were made available to the research community and were analyzed for the Intergovernmental Panel on Climate Change (IPCC) Assessment Reports. Among the studies that analyzed the CMIP3 models, *van Oldenborgh et al.* [2005] suggested that there are no statistically significant changes in the amplitude of ENSO variability in the projections. *Guilyardi* [2006] found that models capable of realistically simulating SST-wind feedback tend to project a significant increase in ENSO amplitude under elevated CO_2. In contrast to the change in ENSO amplitude, there were no clear indications of an El Niño frequency change. *Merryfield* [2006] suggested that changes in ENSO amplitude under CO_2 doubling are associated with the meridional width of the zonal wind stress response to ENSO in models. The models were found to have a 5% decrease in ENSO periods, which he argued resulted from an increase in equatorial wave speed through an associated speedup of the delayed-oscillator feedback. *Yeh and Kirtman* [2007] suggested that whether the model ENSO is in the linear or nonlinear regime determines the changes in ENSO statistics among various climate change projections.

Although the CMIP5 models represent an improvement over the CMIP3 models in many aspects (i.e., resolution, physics, etc.), the response of ENSO to global warming as seen in the CMIP5 output has not resolved this issue. The response of ENSO amplitude to global warming in CMIP5 shows great diversity from model to model, which is similar to that found in the CMIP3 models [*L. Chen et al.*, 2015]. Many studies found no consistent changes across the CMIP5 projections in the location, magnitude, frequency, or temporal evolution of ENSO events [*Stevenson*, 2012; *Taschetto et al.*, 2014]. *An and Choi* [2015] suggested that the feedback between ENSO and tropical climate change induced by global warming is very weak, which may be the reason why the global warming trend cannot significantly modify ENSO amplitude. Other CMIP5 studies have shown that greenhouse warming can alter both the amplitude of ENSO [*Kim et al.*, 2014] and the frequency of extreme ENSO events [*Cai et al.*, 2014, 2015]. *Kim et al.* [2014] analyzed the nine CMIP5 CGCMs that best reproduce the observed Bjerknes coupled stability index to project that, under global warming, ENSO amplitude is likely to increase after 2010, remain essentially unchanged until approximately 2040, and then decrease again afterward. *Cai et al.* [2014] suggested that global warming will increase the occurrence of extreme El Niño events, due to a faster warming in the eastern equatorial Pacific that enables deep convection to occur more easily in the region and provide a coupled feedback to grow the ENSO amplitude. *Cai et al.* [2015] found an increased occurrence of extreme La Niña events under global warming due to an increased warming of the maritime continent compared to the equatorial central Pacific that enables convection to move to the maritime continent more easily to trigger the Bjerknes feedback favoring the extreme events.

After the recent recognition of different types of ENSO, how the ENSO type responds to future warming has also been explored in literature. Modeling studies have suggested that anthropogenic greenhouse gas forcing can be one of the causes for the increased occurrence of the CP El Niño [*Yeh et al.*, 2009; *Collins et al.*, 2010; *Na et al.*, 2011; *Kim and Yu*, 2012]. *Yeh et al.* [2009] used the CMIP3 models to suggest that the occurrence frequency of the CP ENSO will increase in the future, while *Kim and Yu,* [2012] used CMIP5 models to suggest the amplitude of CP ENSO will increase too. *Na et al.* [2011] projected that the occurrence frequency of the CP El Niño during 2007–2056 will be 2.5 times that in the 1980–2006 period.

In addition to the response of ENSO properties to global warming, changes in ENSO teleconnection under global warming scenarios have attracted substantial attention [i.e., *Meehl and Teng*, 2007; *Kug et al.*, 2010; *Zou et al.*, 2014]. Based on a six-member multimodel ensemble, *Meehl and Teng* [2007] identify some possible changes in the future El Niño teleconnections. Among them, the anomalous low in the North Pacific is suggested to weaken and shift eastward and northward in future events. This location shift may lead to a decreased anomalous warming over northern North America, increased cooling over southern North America, and enhanced precipitation in the Pacific Northwest compared to present-day El Niño event teleconnections. *Kug et al.* [2010] also found that the atmospheric teleconnection patterns over the Northern Hemisphere may shift eastward due to an eastward migration of the convection centers in the tropical Pacific associated with the future ENSO events. *Zou et al.* [2014] also found the ENSO-forced Pacific–North American teleconnection pattern to become stronger and more eastward displaced in the warming projections. These are just a few examples of possible ways that the changing ENSO may impact a future warmer world.

1.6. SUMMARY

The changing properties of the ENSO observed during the past two decades and the recent identification of two distinct types of ENSO have prompted the climate research community to revisit traditional views of ENSO dynamics and impacts and how they may change as global climate changes. A better understanding of ENSO diversity is important for the further development of models used for the prediction of ENSO and its associated climate extremes. One source of uncertainty in climate predictions and projections is associated with whether or not modern climate models can realistically simulate both types of ENSO and the alternation between them, and capture their different impacts. If ENSO dynamics are conclusively shown to be changing, the ENSO prediction models and strategies have to be revised to include this changing dynamic. For instance, successful prediction and modeling of the ENSO in the recent and coming decades may depend more on a better understanding and improved skill in the modeling of the subtropical Pacific precursors and their underlying generation mechanisms. Prediction systems based on this framework would be different from the prediction and modeling systems the climate research community has relied on since the 1980s and 1990s for the conventional ENSO that emphasize subsurface ocean dynamics in the equatorial Pacific. *Larson and Kirtman* [2014] have reported some skill in using the PMM to forecast ENSO events with the North American multimodel ensemble (NMME) experiments. *Yang and Jiang* [2014] showed that the National Centers for Environmental Prediction (NCEP) Climate Forecast System had substantial skill in predicting both types of ENSO and the relationship between the EP ENSO and the Asian monsoon. However, the skill in predicting the relationship between CP ENSO and the monsoon is low. Nevertheless, in order to utilize the subtropical precursors, particularly the PMM, to forecast ENSO events, coupled atmosphere-ocean models have to be able to realistically simulate the precursor events. *Lin et al.* [2014] examined 23 CMIP5 models to conclude that the PMM structure can be reasonably simulated in most of the coupled models. However, the so-called seasonal footprinting mechanism that sustains an equatorward extension of the PMM is not well simulated in a majority of the CMIP5 models. Therefore, it is necessary to improve the subtropical Pacific coupling in these models in order for them to apply them successfully in forecasts of ENSO occurrence. Currently, paleoclimate proxy records have mostly been interpreted from the point of view that there is a single type of ENSO (EP) and one set of conventional global impacts (such as precipitation) to derive the history of ENSO. The view that the two types of ENSO can produce distinct global teleconnection offers an additional way to interpret the paleoclimate proxies. A deeper understanding of how the changing El Niño affects global precipitation may also offer different ways to interpret paleoclimate proxies, which may lead to a new understanding of the history of Earth's climate and new implications for future climate changes.

ACKNOWLEDGMENTS

This work was supported by the National Science Foundation's Climate and Large Scale Dynamics Program (Grants AGS-1233542 and AGS-1505145), the National Key Research Program of China (Grant 2014CB953900), the National Oceanic and Atmospheric Administration's Modeling, Analysis, Predictions, and Projections Program (Grant NA11OAR4310102), and the National Natural Science Foundation of China (Grants 41422601, 41376025, 41690123 and 41690120).

REFERENCES

Alexander, M. A., and C. Deser (1995), A mechanism for the recurrence of wintertime midlatitude SST anomalies, *J. Phys. Oceanogr.*, *25*, 122–137.

Alexander, M. A., D. J. Vimont, P. Chang, and J. D. Scott (2010), The impact of extratropical atmospheric variability on ENSO: Testing the seasonal footprinting mechanism using coupled model experiments, *J. Climate*, *23*, 2885–2901; doi:10.1175/2010JCLI3205.1.

Alexander, M. A., I. Blade, M. Newman, J. R. Lanzante, N.-C. Lau, and J. D. Scott (2002), The atmospheric bridge: The influence of ENSO teleconnections on air-sea interaction over the global oceans, *J. Climate*, *15*, 2205–2231.

Amaya, D.J., and G.R. Foltz (2014), Impacts of canonical and Modoki El Niño on tropical Atlantic SST, *J. Geophys. Res. Oceans*, *119* (2), 777–789.

An, S.-I., and B. Wang (2000), Interdecadal change of the structure of ENSO mode and its impact on the ENSO frequency, *J. Climate*, *13*, 2044–2055.

An S.-I., and F.-F. Jin (2001), Collective role of zonal advective and thermocline feedbacks in ENSO mode, *J. Climate*, *14*, 3421–3432; doi: http://dx.doi.org/10.1175/1520-0442[2001]014<3421:CROTAZ>2.0.CO;2.

An, S.-I., and F.-F. Jin (2004), Nonlinearity and Asymmetry of ENSO, *J. Climate*, *17*, 2399–2412; doi: http://dx.doi.org/10.1175/1520-0442[2004]017<2399:NAAOE>2.0.CO;2.

An, S-I., and J. Choi (2015), Why the twenty-first century tropical Pacific trend pattern cannot significantly influence ENSO amplitude? *Climate Dyn.*, *44* (1–2), 133–146.

An, S.-I., J.-S. Kug, Y.-G. Ham, and I.-S. Kang (2008), Successive modulation of ENSO to the future greenhouse warming, *J. Climate*, *21*, 3–21.

Anderson, B. T. (2004), Investigation of a large-scale mode of ocean atmosphere variability and its relation to tropical Pacific sea surface temperature anomalies, *J. Climate*, *17*, 1089–4098; doi: 10.1175/1520-0442[2004]017<4089:IOALMO>2.0.CO;2.

Ardanuy, P. E., and H. L. Kyle (1986), Observed perturbations of the Earth's radiation budget: A response to the El Chichón stratospheric aerosol layer?, *J. Climate Appl. Meteor.*, *25*, 505–516; doi:10.1175/1520-0450[1986]025<0505:OPOTER>2.0.CO;2.

Ashok, K., C. Tam, and W. Lee (2009), ENSO Modoki impact on the Southern Hemisphere storm track activity during extended austral winter, *Geophys. Res. Lett.*, *36*, L12705; doi:10.1029/2009GL038847.

Ashok, K., Z. Gaun, and T. Yamagata (2001), Impact of the Indian ocean dipole on the relationship between the Indian monsoon rainfall and ENSO, *Geophys. Res. Lett.*, *28*, 4459–4502.

Ashok, K., Z. Gaun, and T. Yamagata (2003), Impact of the Indian Ocean dipole on the Australian winter rainfall. *Geophys. Res. Lett.*, *30*, 1821; doi:10.1029/2003GL017926.

Ashok, K., S. K. Behera, S. A. Rao, H. Weng, and T. Yamagata (2007), El Niño Modoki and its possible teleconnection, *J. Geophys. Res.*, *112*, C11007; doi:10.1029/2006JC003798.

Baldwin, M. P., and T. J. Dunkerton (1999), Propagation of the Arctic Oscillation from the stratosphere to the troposphere, *J. Geophys. Res.*, *104*, 30937–30946; doi:10.1029/1999JD900445.

Barnett, T. P. (1991), The interaction of multiple time scales in the tropical climate system, *J. Climate*, *4*, 269–285; doi: http://dx.doi.org/10.1175/1520-0442[1991]004<0269:TIOMTS>2.0.CO;2.

Barnett, T. P., D. W. Pierce, M. Latif, D. Dommenget, and R. Saravanan (1999), Interdecadal interactions between the tropics and midlatitudes in the Pacific basin, *Geophys. Res. Lett.*, *26*, 615–618.

Barnston, A. G., M. K. Tippett, M. L. L'Heureux, S. Li, and D. G. DeWitt (2012), Skill of real-time seasonal ENSO model predictions during 2002–11: Is our capability increasing?, *Bull. Amer. Meteor. Soc.*, *93*, 631–651.

Barsugli, J. J., and P. D. Sardeshmukh (2002), Global atmospheric sensitivity to tropical SST anomalies throughout the Indo-Pacific basin, *J. Climate*, *15*, 3427–3442.

Battisti, D. S., and A. C. Hirst (1989), Interannual variability in the tropical atmosphere-ocean model: Influence of the basic state, ocean geometry and nonlinearity, *J. Atmos. Sci.*, *45*, 1687–1712; doi:10.1175/1520-0469%281989%29046<1687%3AIVIATA>2.0.CO%3B2.

Behera, S. K., and T. Yamagata (2003), Influence of the Indian Ocean dipole on the Southern Oscillation, *J. Meteor. Soc. Japan*, *81*, 169–177.

Bjerknes, J. (1969), Atmospheric teleconnections from the equatorial Pacific, *Monthly Weather Rev.*, *97*, 163–172; doi: http://dx.doi.org/10.1175/1520-0493[1969]097<0163:ATFTEP>2.3.CO;2.

Cai, W., G. Wang, A. Santoso, M. J. McPhaden, L. Wu, F.-F. Jin, A. Timmermann, M. Collins, G. Vecchi, M. Lengaigne, M. H. England, D. Dommenget, K. Takahashi, and E. Guilyardi (2015), Increased frequency of extreme La Niña events under greenhouse warming, *Nature Climate Change*, *5*, 132–137; doi:10.1038/nclimate2492.

Cai, W., S. Borlace, M. Lengaigne, P. van Rensch, M. Collins, G. Vecchi, A. Timmermann, A. Santoso, M. J. McPhaden, L. Wu, M. H. England, G. Wang, E. Guilyardi, and Fei-Fei Jin (2014), Increasing frequency of extreme El Niño events due to greenhouse warming. *Nature Climate Change*, *4*, 111–116; doi: 10.1038/nclimate2100.

Camargo, S. J., and A. H. Sobel (2005), Western North Pacific tropical cyclone intensity and ENSO, *J. Climate*, *18*, 2996–3006.

Capotondi, A., and M. A. Alexander (2010), Relationship between precipitation in the Great Plains of the United States and global SSTs: Insights from the IPCC-AR4 models, *J. Climate*, *23*, 2941–2958.

Capotondi, A., A. T. Wittenberg, M. Newman, E. Di Lorenzo, J.-Y. Yu, P. Braconnot, J. Cole, B. Dewitte, B. Giese, E. Guilyardi, F.-F. Jin, K. Karnauskas, B. Kirtman, T. Lee, N. Schneider, Y. Xue, and S.-W. Yeh (2015), Understanding ENSO diversity, *Bull. Amer. Meteor. Soc.*; doi:10.1175/BAMS-D-13-00117.1.

Carton, J. A., G. A. Chepurin, X. Cao, and B. S. Giese (2000), A simple ocean data assimilation analysis of the global upper ocean 1950–1995, Part I: Methodology, *J. Phys. Oceanogr.*, *30*, 294–309; doi:10.1175/1520-0485[2000]030<0294:ASODAA>2.0.CO;2.

Cayan, D. R., K. T. Redmond, and L. G. Riddle, 1999), ENSO and hydrologic extremes in the western United States, *J. Climate*, *12*, 2881–2893.

Chan, J. C. L. (2000), Tropical cyclone activity over the western North Pacific associated with El Niño and La Niña events, *J. Climate*, *13*, 2960–2972.

Chang, C. P., Y. Zhang, and T. Li (2000), Interannual and interdecadal variations of the East Asian summer monsoon and tropical Pacific SSTs, Part I: Roles of the subtropical ridge, *J. Climate*, *13*, 4310–4325.

Chang, P., B. S. Giese, L. Ji, H. F. Seidel, and F. Wang (2001), Decadal change in the south tropical Pacific in a global assimilation analysis, *Geophys. Res. Lett.*, *28*, 3461–3464.

Chang, P., B. Wang, T. Li, and J. Lin (1995), Interactions between the seasonal cycle and El Niño–Southern Oscillation in an intermediate coupled ocean–atmosphere model, *J. Atmos. Sci.*, *52*, 2353–2372.

Chang, P., L. Ji, H. Li, and M. Flügel (1996), Chaotic dynamics versus stochastic processes in El Niño–Southern Oscillation in coupled ocean-atmosphere models, *Physica D: Nonlinear Phenomena*, *98* (2), 301–320.

Chang, P., L. Zhang, R. Saravanan, D. J. Vimont, J. C. H. Chiang, L. Ji, H. Seidel, and M. K. Tippett (2007), Pacific meridional mode and El Niño–Southern Oscillation, *Geophys. Res. Lett.*, *34*, L16608; doi:10.1029/2007GL030302.

Chen, D., T. Lian, C. Fu, M. A. Cane, Y. Tang, R. Murtugudde, X. Song, Q. Wu, and L. Zhou (2015), Strong influence of westerly wind bursts on El Niño diversity, *Nat. Geosci.*, *8* (5), 339–345.

Chen, G. (2011), How does shifting pacific ocean warming modulate on tropical cyclone frequency over the South China Sea?, *J. Climate*, *24*, 4695–4700.

Chen, G., and C.-Y. Tam (2010), Different impacts of two kinds of Pacific Ocean warming on tropical cyclone frequency over the western North Pacific, *Geophys. Res. Lett.*, *37*, L01803; doi:10.1029/2009GL041708.

Chen, G., J. Lu, and D. M. W. Frierson (2008), Phase speed spectra and the latitude of surface westerlies: Interannual variability and global warming trend, *J. Climate*, *21*, 5942–5959.

Chen, L., T. Li, and Y. Yu (2015), Causes of strengthening and weakening of ENSO amplitude under global warming in four CMIP5 models, *J. Climate*, *28*, 3250–3274; doi:10.1175/JCLI-D-14-00439.1.

Chia, H. H., and C. F. Ropelewski (2002), The interannual variability in the genesis location of tropical cyclones in the northwest Pacific, *J. Climate*, *15*, 2934–2944.

Chiang, J. C. H., and D. J. Vimont (2004), Analogous Pacific and Atlantic meridional modes of tropical atmosphere-ocean variability, *J. Climate*, *17*, 4143–4158; doi:10.1175/JCLI4953.1.

Chiang, J. C., and Sobel, A. H. (2002), Tropical tropospheric temperature variations caused by ENSO and their influence on the remote tropical climate, *J. Climate*, *15*, 2616–2631.

Chiodi, A. M., and D. E. Harrison (2013), El Niño impacts on seasonal U.S. Atmospheric circulation, temperature, and precipitation anomalies: The OLR-event perspective, *J. Climate*, *26*, 822–837; doi: http://dx.doi.org/10.1175/JCLI-D-12-00097.1.

Choi, J., S. -I. An, and S. W. Yeh (2012), Decadal amplitude modulation of two types of ENSO and its relationship with the mean state, *Climate Dyn.*, *38* (11–12), 2631–2644.

Choi, J., S. -I. An, B. Dewitte, and W. W. Hsieh (2009), Interactive feedback between the tropical Pacific Decadal Oscillation and ENSO in a coupled general circulation model, *J Climate*, *22*, 6597–6611.

Chou, C., L.-F. Huang, J.-Y. Tu, L. Tseng, and Y.-C. Hsueh (2009), El Niño impacts on precipitation in the western North Pacific–East Asian sector, *J. Climate*, *22*, 2039–2057.

Chung, P.-H., and T. Li (2013), Interdecadal relationship between the mean state and El Niño types, *J. Climate*, *26*, 361–379; doi:10.1175/JCLI-D-12-00106.1.

Ciasto, L. M., and D. W. J. Thompson (2008), Observations of large scale ocean-atmosphere interaction in the Southern Hemisphere, *J. Climate*, *21*, 1244–1259; doi:10.1175/2007JCLI1809.1.

Clement, A., P. DiNezio, and C. Deser (2011), Rethinking the ocean's role in the Southern Oscillation, *J. Climate*, *24*, 4056–4072; doi:10.1175/2011JCLI3973.1.

Collins, M. (2000a), Understanding uncertainties in the response of ENSO to greenhouse warming, *Geophys. Res. Lett.*, *27*, 3509–3513.

Collins, M. (2000b), The El Niño Southern Oscillation in the second Hadley centre coupled model and its response to greenhouse warming, *J. Climate*, *13*, 1299–1312.

Collins, M., and coauthors (2010), The impact of global warming on the tropical Pacific Ocean and El Niño, *Nat. Geosci.*, *3*, 391–397; doi:10.1038/NGEO868.

Cummins, P. F., G. S. E. Lagerloef, and G. Mitchum (2005), A regional index of northeast Pacific variability based on satellite altimeter data, *Geophys. Res. Lett.*, *32*, L17607; doi:10.1029/2005GL023642.

Dettinger, M. D., D. R. Cayan, H. F. Diaz, and D. M. Meko (1998), North-south precipitation patterns in western North America on interannual-to-decadal timescales, *J. Climate*, *11*, 3095–3111.

DeWeaver, E., and S. Nigam (2004), On the forcing of ENSO teleconnections by anomalous heating and cooling, *J. Climate*, *17*, 3225–3235.

Ding, Q., E. J. Steig, D. S. Battisti, and M. Kuettel (2011), Winter warming in West Antarctica caused by central tropical Pacific warming, *Nat. Geosci.*, *4*, 398–403; doi:10.1038/ngeo1129.

Dommenget, D. (2010), The slab ocean El Niño, *Geophys. Res. Lett.*, *37*, L20701; doi:10.1029/2010/GL044888.

Dong, B., R. T. Sutton, and A. A. Scaife (2006), Multidecadal modulation of El Niño–Southern Oscillation (ENSO) variance by Atlantic Ocean sea surface temperatures, *Geophys. Res. Lett.*, *33*, L08705; doi:10.1029/2006GL025766.

Dong, B. W., and R. Y. Lu (2013), Interdecadal enhancement of the Walker circulation over the tropical Pacific in the late 1990s, *Adv. Atmos. Sci.*, *30* (2), 247–262; doi: 10.1007/s00376-012-2069-9.

Duan, W., H. Xu, and M. Mu (2008), Decisive role of nonlinear temperature advection in El Niño and La Niña amplitude asymmetry, *J. Geophys. Res.*, *113*, C01014; doi:10.1029/2006JC003974.

Enfield, D. B., and D. A. Mayer (1997), Tropical Atlantic sea surface temperature variability and its relation to El Niño–Southern Oscillation, *J. Geophys. Res.*, *102*, 929–945; doi:10.1029/96JC03296.

England, M. H., S. McGregor, P. Spence, G. A. Meehl, A. Timmermann, W. Cai, A. S. Gupta, M. J. McPhaden, A. Purich, and A. Santoso (2014), Recent intensification of wind-driven circulation in the Pacific and the ongoing warming hiatus, *Nature Climate Change*, *4*, 222–227; doi:10.1038/nclimate2106.

Evtushevsky, O. M., V. O. Kravchenko, L. L. Hood, and G. P. Milinevsky (2014), Teleconnection between the central tropical Pacific and the Antarctic stratosphere: Spatial patterns and time lags, *Climate Dyn.*, 1–15.

Fedorov, A. V., and S. G. Philander (2000), Is El Niño changing?, *Science*, *288*, 1997–2002.

Fedorov, A. V., S. Hu, M. Lengaigne, and E. Guilyardi (2014), The impact of westerly wind bursts and ocean initial state on the development, and diversity of El Niño events, *Climate Dyn.*, *44* (5–6), 1381–1401.

Feng, J., and J. Li (2011), Influence of El Niño Modoki on spring rainfall over South China, *J. Geophys. Res.*, *116*, D13102; doi:10.1029/2010JD015160.

Feng, J., W. Chen, C.-Y. Tam, and W. Zhou (2011), Different impacts of El Niño and El Niño Modoki on China rainfall in the decaying phases, *Int. J. Climatol.*, *31*, 2091–2101; doi:10.1002/joc.2217.

Fogt, R. L., and D. H. Bromwich (2006), Decadal variability of the ENSO teleconnection to the high-latitude South Pacific governed by coupling with the southern annular mode, *J. Climate*, *19*, 979–997; doi:10.1175/JCLI3671.1.

Frauen, C., and D. Dommenget (2010), El Niño and La Niña amplitude asymmetry caused by atmospheric feedbacks, *Geophys. Res. Lett.*, *37*, L18801; doi:10.1029/2010GL044444.

Fu, C., H. F. Diaz, J. O. Fletcher (1986), Characteristics of the response of sea surface temperature in the central Pacific associated with warm episodes of the Southern Oscillation. *Monthly Weather Rev.*, *114*, 1716–1738.

García-Herrera, R., N. Calvo, R. R. Garcia, and M. A. Giorgetta (2006), Propagation of ENSO temperature signals into the middle atmosphere: A comparison of two general circulation models and ERA-40 reanalysis data, *J. Geophys. Res.*, *111*, D06101; doi:10.1029/2005JD006061.

Garfinkel, C. I., and D. L. Hartmann (2008), Different ENSO teleconnections and their effects on the stratospheric polar vortex, *J. Geophys. Res.*, *113*, D18114; doi:10.1029/2008JD009920.

Garfinkel, C. I., D. L. Hartmann, and F. Sassi (2010), Tropospheric precursors of anomalous Northern Hemisphere stratospheric polar vortices, *J. Climate*, *23*, 3282–3299; doi:10.1175/2010JCLI3010.1.

Garfinkel, C. I., M. M. Hurwitz, D. W. Waugh, and A. H. Butler (2013), Are the teleconnections of central Pacific and eastern Pacific El Niño distinct in boreal wintertime?, *Climate Dyn.*, *41*, 1835–1852; doi:10.1007/s00382-012-1570-2.

Goldenberg, S. B., and L. J. Shapiro (1996), Physical mechanisms for the association of El Niño and West African rainfall with Atlantic major hurricane activity, *J. Climate*, *9*, 1169–1187; doi:10.1175/1520-0442[1996] 009<1169:PMFTAO>2.0.CO;2.

Graf, H. F., and D. Zanchettin (2012), Central Pacific El Niño, the "subtropical bridge," and Eurasian climate, *J. Geophys. Res.*, *117*, D01102; doi:10.1029/2011JD016493.

Gray, W. M. (1984), Atlantic seasonal hurricane frequency. Part I: El Niño and 30-mb quasi-biennial oscillation influences. *Monthly Weather Rev.*, *112*, 1649–1668.

Grimm, A. M. (2003), The El Niño impact on the summer monsoon in Brazil: Regional processes versus remote influences, *J. Climate*, *16*, 263–280.

Gruber, A., and A. F. Krueger (1984), The status of the NOAA outgoing longwave radiation data set, *Bull. Amer. Meteor. Soc.*, *65*, 958–962; doi:10.1175/1520-0477[1984]065<0958:TSOTNO>2.0.CO;2.

Gu, D., and S. G. H. Philander (1995), Secular changes of annual and interannual variability in the tropics during the past century, *J. Climate*, *8*, 864–876; doi: http://dx.doi.org/10.1175/1520-0442[1995]008<0864:SCOAAI>2.0.CO;2.

Gu, D., and S. G. H. Philander (1997), Interdecadal climate fluctuations that depend on exchanges between the tropics and extratropics, *Science*, *275*, 805–807.

Guilyardi, E. (2006), El Niño-mean state-seasonal cycle interactions in a multi-model ensemble, *Climate Dyn.*, *26*, 329–348.

Hannachi, A., D. Stephenson, and K. Sperber (2003), Probability-based methods for quantifying nonlinearity in the ENSO, *Climate Dyn.*, *20*, 241–256.

Hartmann, D. L., and F. Lo (1998), Wave-driven zonal flow vacillation in the Southern Hemisphere, *J. Atmos. Sci.*, *55*(8), 1303–1315.

He, H., J. Yang, D. Gong, R. Mao, Y. Wang, and M. Gao (2015), Decadal changes in tropical cyclone activity over the western North Pacific in the late 1990s, *Climate Dyn.*, 1–13.

He, Y., and C. Guan (1997), Interannual and interdecadal variability in heat content of the upper ocean of South China Sea, *Adv. Atmos. Sci.*, *16*, 23–29.

Heddinghaus, T. R., and A. F. Krueger (1981), Annual and interannual variations in outgoing longwave radiation over the tropics, *Monthly Weather Rev.*, *109*, 1208–1218; doi: http://dx.doi.org/10.1175/1520-0493[1981]109<1208:AAIVIO>2.0.CO;2.

Hegyi, B. M., and Y. Deng (2011), A dynamical fingerprint of tropical Pacific sea surface temperatures on the decadal-scale variability of cool-season Arctic precipitation, *J. Geophys. Res.*, *116*, D20121; doi:10.1029/2011JD016001.

Hill, K. J., A. S. Taschetto, and M. H. England (2009), South American rainfall impacts associated with inter-El Niño variations, *Geophys. Res. Lett.*, *36*, L19702; doi:10.1029/2009GL040164.

Ho, C. H., J. J. Baik, J. H. Kim, D. Y. Gong, and C. H. Sui (2004), Interdecadal changes in summertime typhoon tracks, *J. Climate*, *17*, 1767–1776.

Hoerling, M. P., and A. Kumar (2002), Atmospheric response patterns associated with tropical forcing, *J. Climate*, *15*, 2184–2203.

Holton, J. R. (1976), A semi-spectral numerical model for wave-mean flow interactions in the stratosphere: applications to sudden stratospheric warmings, *J. Atmos. Sci.*, *33*, 1639–1649.

Hong, C.-C., Y.-H. Li, T. Li, and M.-Y. Lee (2011), Impacts of central Pacific and eastern Pacific El Nino on tropical cyclone tracks over the western North Pacific, *Geophys. Res. Lett.*, *38*, L16712; doi:10.1029/2011GL048821.

Horii, T., I. Ueki, and K. Hanawa (2012), Breakdown of ENSO predictors in the 2000s: Decadal changes of recharge/discharge-SST phase relation and atmospheric intraseasonal forcing, *Geophys. Res. Lett.*, *39*, L10707; doi:10.1029/2012GL051740.

Hsieh, W. W. (2004), Nonlinear multivariate and time series analysis by neural network methods, *Rev. Geophys.*, *42*, RG1003; doi:10.1029/2002RG000112.

Hu, S., A. V. Fedorov, M. Lengaigne, and E. Guilyardi (2014), The impact of westerly wind bursts on the diversity and predictability of El Niño events: an ocean energetics perspective. *Geophys. Res. Lett.*,*41*,4654–4663;doi:10.1002/2013GL058954.

Hu, Z.-Z., A. Kumar, B. Jha, and B. Huang (2012a), An analysis of forced and internal variability in a warmer climate in CCSM3, *J. Climate*, *25* (7), 2356–2373; doi: 10.1175/JCLI-D-11-00323.1.

Hu, Z.-Z., A. Kumar, B. Jha, W. Wang, B. Huang, and B. Huang (2012b), An analysis of warm pool and cold tongue El Niños: Air-sea coupling processes, global influences, and recent trends, *Climate Dyn.*, *38* (9–10), 2017–2035; doi: 10.1007/s00382-011-1224-9.

Hu, Z.-Z., A. Kumar, H.-L. Ren, H. Wang, M. L'Heureux, and F.-F. Jin (2013), Weakened interannual variability in the tropical Pacific Ocean since 2000, *J. Climate*, *26* (8), 2601–2613; doi: 10.1175/JCLI-D-12-00265.1.

Huang, H. P., R. Seager, and Y. Kushnir (2005), The 1976/77 transition in precipitation over the Americas and the influence of tropical sea surface temperature, *Climate Dyn.*, *24* (7–8), 721–740.

Hurwitz, M. M., C. I. Garfinkel, P. A. Newman, and L. D. Oman (2013), Sensitivity of the atmospheric response to warm pool El Niño events to modeled SSTs and future climate forcings, *J. Geophys. Res. Atmos.*, *118* (13), 371–13,382; doi:10.1002/2013JD021051.

Hurwitz, M. M., I. S. Song, L. D. Oman, P. A. Newman, A. M. Molod, S. M. Frith, and J. E. Nielsen (2011b), Response of the Antarctic stratosphere to warm pool El Niño events in the GEOS CCM, *Atmos. Chem. Phys.*, *11*, 9659–9669; doi:10.5194/acp-11-9659-2011.

Hurwitz, M. M., N. Calvo, C. I. Garfinkel, A. H. Butler, S. Ineson, C. Cagnazzo, E. Manzini, and C. Peña-Ortiz (2014), Extra-tropical atmospheric response to ENSO in the CMIP5 models, *Climate Dyn.*, *43* (12), 3367–3376; doi:10.1007/s00382-014-2110-z.

Hurwitz, M. M., P. A. Newman, L. D. Oman, and A. M. Molod (2011a), Response of the Antarctic stratosphere to two types of El Niño events, *J. Atmos. Sci.*, *68*, 812–822.

Iza, M., and N. Calvo (2015), Role of stratospheric sudden warmings on the response to central Pacific El Niño, *Geophys. Res. Lett.*, *42*, 2482–2489; doi:10.1002/2014GL062935.

Jiang, N., J. D. Neelin, and M. Ghil (1995), Quasi-quadrennial and quasi-biennial variability in the equatorial Pacific, *Climate Dyn.*, *12*, 101–112.

Jin, F.-F. (1996), Tropical ocean-atmosphere interaction, the pacific cold tongue, and the El Niño Southern Oscillation, *Science*, *274*, 76–78; doi: 10.1126/science.274.5284.76.

Jin, F.-F. (1997), An equatorial recharge paradigm for ENSO, I. Conceptual model, *J. Atmos. Sci.*, *54*, 811–829.

Jin, F.-F., J. D. Neelin and M. Ghil (1994), El Niño on the devil's staircase: Annual and subharmonic steps to chaos, *Science*, *264*, 70–72; doi: 10.1126/science.264.5155.70.

Juneng, L., and F. T. Tangang (2005), Evolution of ENSO-related rainfall anomalies in Southeast Asia region and its relationship with atmosphere ocean variations in Indo-Pacific sector, *Climate Dyn.*, *25*(4), 337–350; doi:10.1007/s00382-005-0031-6.

Kalnay, E., et al. (1996), The NCEP/NCAR 40-year reanalysis project, *Bull. Amer. Meteor. Soc.*, *77*, 437–471.

Kao, H. Y., and J. -Y. Yu (2009), Contrasting eastern-Pacific and central-Pacific types of ENSO, *J. Climate*, *22*, 615–632; doi:10.1175/2008JCLI2309.1.

Karoly, D. J. (1989), Southern Hemisphere circulation features associated with El Niño–Southern Oscillation events, *J. Climate*, *2*, 1239–1252; doi:10.1175/1520-0442[1989]002, 1239:SHCFAW.2.0.CO;2.

Kerr, R. A. (2000), A North Atlantic climate pacemaker for the centuries, *Science*, *288*, 1984–1986; doi:10.1126/science.288.5473.1984.

Kessler, W. S. (2002), Is ENSO a cycle or a series of events?, *Geophys. Res. Lett.*,*29*[23], 2125; doi:10.1029/2002GL015924.

Kiladis, G. N., and H. F. Diaz (1989), Global climatic anomalies associated with extremes in the Southern Oscillation, *J. Climate*, *2*, 1069–1090.

Kim, H.-M., P. J. Webster, and J. A. Curry (2009), Impact of shifting patterns of Pacific Ocean warming on north Atlantic tropical cyclones, *Science*, *325*, 77–80.

Kim, H. -M., P. J. Webster, and J. A. Curry (2011), Modulation of North Pacific tropical cyclone activity by three phases of ENSO, *J. Climate*, *24*, 1839–1849.

Kim, J.-S., W. Zhou, X. Wang, and S. Jain (2012), El Niño Modoki and the summer precipitation variability over South Korea: A diagnostic study, *J. Meteor. Soc. Japan*, *90*, 673–684; doi: 10.2151/jmsj.2012-507.

Kim, S. -T., and J.-Y. Yu (2012), The two types of ENSO in CMIP5 models, *Geophys. Res. Lett.*, *39*, L11704; doi:10.1029/2012GL052006.

Kim, S. -T., J.-Y. Yu, A. Kumar, and H. Wang (2012), Examination of the two types of ENSO in the NCEP CFS model and its extratropical associations, *Monthly Weather Rev.*, *140*, 1908–1923; doi:10.1175/MWR-D-11-00300.1.

Kim, S.-T., W. Cai, F.-F. Jin, A. Santoso, L. Wu, E. Guilyardi, and S.-I. An (2014), Response of El Niño sea surface temperature variability to greenhouse warming, *Nature Climate Change*, *4*, 786–790; doi: 10.1038/NCLIMATE2326.

Kinter, J. L., K. Miyakoda, and S. Yang (2002), Recent change in the connection from the Asian monsoon to ENSO, *J. Climate*, *15*, 1203–1215.

Klein, S. A., B. J. Soden, and N.-C. Lau (1999), Remote sea surface temperature variations during ENSO: Evidence for a tropical atmospheric bridge, *J. Climate*, *12*, 917–932.

Knutson, T. R., S. Manabe, and D. Gu (1997), Simulated ENSO in a global coupled ocean-atmosphere model: Multidecadal amplitude modulation and CO2 sensitivity, *J. Climate*, *10*, 131–161.

Kohl, A., D. Dommenget, K. Ueyoshi, and D. Stammer (2006), The global ECCO 1952 to 2001 ocean synthesis, *ECCO Tech. Rep.*, *40*.

Kug, J.-S., F.-F. Jin, and S.-I. An (2009), Two types of El Niño events: Cold tongue El Niño and warm pool El Niño, *J. Climate*, *22*, 1499–1515; doi:10.1175/2008JCLI2624.1.

Kug, J.-S., J. Choi, S.-I. An, F.-F. Jin, and A. T. Wittenberg (2010), Warm pool and cold tongue El Niño events as simulated by the GFDL CM2.1 coupled GCM, *J. Climate*, *23*, 1226–1239; doi:10.1175/2009JCLI3293.1.

Kumar, A., and Z.-Z. Hu (2014), Interannual and interdecadal variability of ocean temperature along the equatorial Pacific in conjunction with ENSO, *Climate Dyn.*, *42* (5–6), 1243–1258; doi: 10.1007/s00382-013-1721-0.

Kumar, K. K., B. Rajagopatan, M. Hoerling, G. Bates, and M. Cane (2006), Unraveling the mystery of Indian monsoon failure during El Niño, *Science*, *314*, 115–119.

Kunz, T., K. Fraedrich, and F. Lunkeit (2009), Impact of synoptic-scale wave breaking on the NAO and its connection with the stratosphere in ERA-40, *J. Climate;* doi:10.1175/2009JCLI2750.1.

Kwok, R., and J. C. Comiso (2002), Spatial patterns of variability in Antarctic surface temperature: Connections to the South Hemisphere annular mode and the Southern Oscillation, *Geophys. Res. Lett.*, *29*, 1705; doi:10.1029/2002GL015415.

Larkin, N. K., and D. E. Harrison (2005a), On the definition of El Niño and associated seasonal average U.S. weather anomalies, *Geophys. Res. Lett.*, *32*, L13705; doi:10.1029/2005GL022738.

Larkin, N. K., and D. E. Harrison (2005b), Global seasonal temperature and precipitation anomalies during El Niño autumn and winter, *Geophys. Res. Lett.*, *32*, L16705; doi:10.1029/2005GL022860.

Larson, S., S.-K. Lee, C. Wang, E.-S. Chung, and D. Enfield (2012), Impacts of non-canonical El Nino patterns on Atlantic hurricane activity, *Geophys. Res. Lett.*, *39*, L14706; doi:10.1029/2012GL052595.

Larson, S. M., and B. P. Kirtman (2014), The Pacific meridional mode as an ENSO precursor and predictor in the North American multimodel ensemble, *J. Climate*, *27*, 7018–7032; doi: http://dx.doi.org/10.1175/JCLI-D-14-00055.1.

Latif, M., and T. P. Barnett (1994), Causes of decadal climate variability over the North Pacific and North America, *Science*, *266*, 634–637.

Latif, M., D. Anderson, T. Barnett, M. Cane, R. Kleeman, A. Leetmaa, J. O'Brien, A. Rosati, and E. Schneider (1998), A review of the predictability and prediction of ENSO, *J. Geophys. Res.*, *103* (C7), 14375–14393; doi:10.1029/97JC03413.

Lau, K.-M., and P. H. Chan (1983), Short-term climate variability and atmospheric teleconnections from satellite-observed outgoing longwave radiation. Part I: Simultaneous relationships, *J. Atmos. Sci.*, *40*, 2735–2750; doi:10.1175/1520-0469[1983]040<2735:STCVAA>2.0.CO;2.

Lau, K.-M., and P. H. Chan (1985), Aspects of the 40–50 day oscillation during the northern winter as inferred from outgoing longwaveradiation, *Monthly. Weather Rev.*, *113*, 1889–1909; doi:10.1175/1520-0493[1985]113<1889:AOTDO D>2.0.CO;2.

Lau, N. C., and M. J. Nath (2004), Coupled GCM simulation of atmosphere-ocean variability associated with zonally asymmetric SST changes in the tropical Indian Ocean, *J. Climate*, *17*, 245–265.

Lee, E.-J., K.-J. Ha, and J.-G. Jhun (2014), Interdecadal changes in interannual variability of the global monsoon precipitation and interrelationships among its subcomponents, *Climate Dyn.*, *42*, 2585–2601.

Lee, S.-K., C. Wang, and D. B. Enfield (2010), On the impact of central Pacific warming events on Atlantic tropical storm activity, *Geophys. Res. Lett.*, *37*, L17702; doi:10.1029/2010GL044459.

Lee, T., and coauthors (2010), Record warming in the South Pacific and western Antarctica associated with the strong central Pacific El Niño in 2009–10, *Geophys. Res. Lett.*, *37*, L19704; doi:10.1029/2010GL044865.

Lee, T., and M. J. McPhaden (2010), Increasing intensity of El Niño in the central-equatorial Pacific, *Geophys. Res. Lett.*, *37*, L14603; doi:10.1029/2010GL044007.

L'Heureux, M. L., and D. W. J. Thompson (2006), Observed relationships between the El Niño–Southern Oscillation and the extratropical zonal-mean circulation, *J. Climate*, *19*, 276–287; doi:10.1175/JCLI3617.1.

L'Heureux, M. L., S. Lee, and B. Lyon (2013), Recent multidecadal strengthening of the Walker circulation across the tropical Pacific, *Nature Climate Change*, *3*, 571–576.

Li, Y., and S. Yang (2010), A dynamical index for the east Asian winter monsoon, *J. Climate*, *23*, 4255–4262.

Liang, Y.-C., J.-Y. Yu, M.-H. Lo, and C. Wang (2015), The changing influence of El Niño on the Great Plains low-level jet, *Atmosph. Sci. Lett.*; doi: 10.1002/asl.590.

Liang, Y.-C., M.-H. Lo, and J.-Y. Yu (2014), Asymmetric responses of land hydroclimatology to two types of El Niño in the Mississippi River basin, *Geophys. Res. Lett.*, *41*, 582–588; doi:10.1002/2013GL058828.

Lim, E.-P., and H. H. Hendon, D. Hudson, G. Wang, and O. Alves (2009), Dynamical forecast of inter–El Niño variations of tropical SST and Australian spring rainfall, *Monthly Weather Rev.*, *137*, 3796–3810; doi:10.1175/2009MWR2904.1.

Lin, C.-Y., J.-Y. Yu, and H. H. Hsu (2014), CMIP5 model simulations of the Pacific meridional mode and its connection to the two types of ENSO, *Int. J. Climatol.*; doi:10.1002/joc.4130.

Lin, P., Q. Fu, D. L. Hartmann (2012), Impact of tropical SST on stratospheric planetary waves in the Southern Hemisphere, *J. Climate*, *25*, 5030–5046; doi:10.1175/JCLI-D-11-00378.1.

Linkin, M. E., and S. Nigam (2008), The north Pacific Oscillation-west Pacific teleconnection pattern: Mature-phase structure and winter impacts, *J. Climate*, *21*, 1979–1997; doi:10.1175/2007JCLI2048.1.

Liu, J., J. A. Curry, and D. G. Martinson (2004), Interpretation of recent Antarctic sea ice variability, *Geophys. Res. Lett.*, *31*, L02205; doi:10.1029/2003GL018732.

Liu, Q.-Y., D. Wang, X. Wang, Y. Shu, Q. Xie, J. Chen (2014), Thermal variations in the South China Sea associated with the eastern and central Pacific El Niño events and their mechanisms, *J. Geophys. Res. Oceans*, *119*, 8955–8972; doi:10.1002/2014JC010429.

Liu, Z., and M. Alexander (2007), Atmospheric bridge, oceanic tunnel, and global climatic teleconnections, *Rev. Geophys.*, *45*, RG2005; doi:10.1029/2005RG00017.

Livezey, R. E., M. Masutani, A. Leetmaa, A., H. Rui, M. Ji, and A. Kumar (1997), Teleconnective response of the Pacific–North American region atmosphere to large central equatorial Pacific SST anomalies, *J. Climate*, *10*, 1787–1820.

Lu, J., G. Chen, and D. M. W. Frierson (2008), Response of the zonal mean atmospheric circulation to El Niño versus global warming, *J. Climate*, *21*, 5835–5851.

Luo, J. J., S. Masson, S. Behera, S. Shingu, and T. Yamagata (2005), Seasonal climate predictability in a coupled OAGCM using a different approach for ensemble forecasts, *J. Climate*, *18*, 4474–4497.

Maloney, E. D., and coauthors (2014), North American climate in CMIP5 experiments: Part III: Assessment of twenty-first-century projections, *J. Climate*, *27*, 2230–2270; doi: http://dx.doi.org/10.1175/JCLI-D-13-00273.1.

Mantua, N. J., and D. S. Battisti (1995), Aperiodic variability in the Zebiak-cane coupled ocean-atmosphere model: Air-sea interactions in the western equatorial Pacific, *J. Climate*, *8*, 2897–2927; doi: http://dx.doi.org/10.1175/1520-0442[1995]008<2897:AVITZC>2.0.CO;2.

Mantua, N. J., S. R. Hare, Y. Zhang, J. M. Wallace, and R. C. Francis (1997), A Pacific interdecadal climate oscillation with impacts on salmon production, *Bull. Amer. Meteor. Soc.*, *78*, 1069–1079.

Manzini, E., M. A. Giorgetta, M. Esch, L. Kornblueh, and E. Roeckner (2006), The influence of sea surface temperatures on the northern winter stratosphere: Ensemble simulations with the MAECHAM5 model, *J. Climate*, *19*, 3863–3881; doi:10.1175/JCLI3826.1.

McBride, J., M. Haylock, and N. Nicholls (2003), Relationships between the maritime continent heat source and the El Niño–Southern Oscillation phenomenon, *J. Climate*, *16* (17), 2905–2914; doi:10.1175/1520–0442 [2003]016<2905:RBTMCH>2.0.CO;2.

McCreary, J. P., and P. Lu (1994), Interaction between the subtropical and equatorial ocean circulations: The subtropical cell, *J. Phys. Oceanogr.*, *24*, 466–497.

McIntyre, M. E. (1982), How well do we understand the dynamics of stratospheric warming?, *J. Meteor. Soc. Japan*, *60*, 37–64.

McPhaden, M. J. (2008), Evolution of the 2006–2007 El Niño: The role of intraseasonal to interannual time scale dynamics, *Adv. Geosci.*, *14*, 219–230; doi:10.5194/adgeo-14-219-2008.

McPhaden, M. J. (2012), A 21st century shift in the relationship between ENSO SST and warm water volume anomalies, *Geophys. Res. Lett.*, *39*, L09706; doi:10.1029/2012GL051826.

McPhaden, M. J., and D. Zhang (2002), Slowdown of the meridional overturning circulation in the upper Pacific Ocean, *Nature*, *415*, 603–608.

McPhaden, M. J., T. Lee, and D. McClurg (2011), El Niño and its relationship to changing background conditions in the tropical Pacific Ocean, *Geophys. Res. Lett.*, *38*, L15709; doi:10.1029/2011GL048275.

Mechoso, C. R., D. L. Hartmann, and J. D. Farrara (1985), Climatology and interannual variability of wave, mean-flow interaction in the Southern Hemisphere, *J. Atmos. Sci.*, *42*, 2189–2206.

Meehl, G. A., and H. Teng (2007), Multi-model changes in El Niño teleconnections over North America in a future warmer climate, *Climate Dyn.*, *29*, 779–790; doi:10.1007/s00382-007-0268-3.

Meehl, G. A., and W. M. Washington (1996), El Niño-like climate change in a model with increased atmospheric CO2 concentrations, *Nature*, *382* (6586), 56–60.

Meehl, G. A., et al. (2007), The WCRP CMIP3 multimodel dataset: A new era in climate change research, *Bull. Amer. Meteorol. Soc.*, *88*, 1383–1394; doi:10.1175/BAMS-88-9-1383.

Meehl, G. A., G. W. Branstator, and W. M. Washington (1993), Tropical Pacific interannual variability and CO2 climate change, *J. Climate*, *6*, 42–63.

Meehl, G. A., W. D. Collins, B. A. Boville, J. T. Kiehl, T. M. L.Wigley, and J. M. Arblaster (2000), Response of the NCAR climate system model to increased CO2 and the role of physical processes, *J. Climate*, *13* (11), 1879–1898.

Meinen, C. S., and M. J. McPhaden (2000), Observations of warm water volume changes in the equatorial Pacific and their relationship to El Niño and La Niña, *J. Climate*, *13*, 3551–3559; doi:10.1175/1520-0442[2000] 013<3551:OOWWVC>2.0.CO;2.

Merrifield, M. A. (2011), A shift in western tropical Pacific sea level trends during the 1990s, *J. Climate*, *24*, 4126–4138; doi: http://dx.doi.org/10.1175/2011JCLI3932.1.

Merryfield, W. J. (2006), Changes to ENSO under CO2 doubling in a multimodel ensemble, *J. Climate*, *19*, 4009–4027.

Minobe, S. (2000), Spatio-temporal structure of the pentadecadal variability over the North Pacific, *Prog. Oceanogr.*, *47*, 381–408; doi:10.1016/S0079-6611[00]00042-2.

Minobe, S. (2002), Interannual to interdecadal changes in the Bering Sea and concurrent 1998/99 changes over the North Pacific, *Prog. Oceanogr.*, *55*, 45–64; doi:10.1016/S0079-6611[02]00069-1.

Mo, K. C. (2000), Relationships between interdecadal variability in the Southern Hemisphere and sea surface temperature anomalies, *J. Climate*, *13*, 3599–3610.

Mo, K. C. (2010), Interdecadal modulation of the impact of ENSO on precipitation and temperature over the United States, *J. Climate*, *23*, 3639–3656; doi:10.1175/2010JCLI3553.1.

Mo, K. C., and M. Ghil (1987), Statistics and dynamics of persistent anomalies, *J. Atmos. Sci.*, *44*, 877–902; doi:10.1175/1520-0469[1987]044,0877:SADOPA.2.0.CO;2.

Mo, K. C., and R. E. Livezey (1986), Tropical-extratropical geopotential height teleconnections during the Northern Hemisphere winter, *Monthly Weather Rev.*, *114*, 2488–2515.

Mo, K. C., and R. W. Higgins (1998), Tropical convection and precipitation regimes in the western United States, *J. Climate*, *11*, 2404–2423.

Monahan, A. H. (2001), Nonlinear principal component analysis: Tropical Indo-Pacific sea surface temperature and sea level pressure, *J. Climate*, *14*, 219–233.

Na, H., B.-G. Jang, W.-M. Choi, and K.-Y. Kim (2011), Statistical simulations of the future 50-year statistics of cold-tongue

El Niño and warm-pool El Niño, *Asia-Pacific J. Atmos. Sci.*, *47*, 223–233; doi:10.1007/s13143-011-0011-1.

Neale, R. B., J. Richter, S. Park, P. H. Lauritzen, S. J. Vavrus, P. J. Rasch, and M. Zhang (2013), The mean climate of the community atmosphere model (CAM4) in forced SST and fully coupled experiments, *J. Climate*, *26*, 5150–5168; doi:10.1175/JCLI-D-12-00236.1.

Neelin, J. D. (1991), The slow sea surface temperature mode and the fast-wave limit: Analytic theory for tropical interannual oscillations and experiments in a hybrid coupled model, *J. Atmos. Sci.*, *48*, 584–606.

Neelin, J. D., and F.-F. Jin (1993), Modes of interannual tropical ocean-atmosphere interaction-a unified view, II, Analytical results in the weak coupling limit, *J. Atmos. Sci.*, *50*, 3504–3522.

Neelin, J. D., D. S. Battisti, A. C. Hirst, F.-F. Jin, Y. Wakata, T. Yamagata, and S. E. Zebiak (1998), ENSO theory, *J. Geophys. Res.*, *103* (C7), 14261–14290; doi:10.1029/97JC03424.

Newman, M., et al. (2015), The Pacific decadal oscillation, revisited, *Bull. Amer. Meteor. Soc.* (submitted).

Newman, M., S.-I. Shin, and M. A. Alexander (2011), Natural variation in ENSO flavors. *Geophys. Res. Lett.*, L14705; doi:10.1029/2011GL047658.

Nishii, K., H. Nakamura, and Y. J. Orsolini (2010), Cooling of the wintertime Arctic stratosphere induced by the western Pacific teleconnection pattern, *Geophys. Res. Lett.*, *371*, L13805; doi:10.1029/2010GL043551.

Noda, A., K. Yoshimatsu, S. Yukimoto, K. Yamahuchi, and S. Yamaki (1999), Relationship between natural variability and CO2-induced warming pattern: MRI AOGCM experiment, in *10th Symposium on Global Change Studies*, American Meteorological Society Publication.

Paek, H., J.-Y., Yu, J.-W. Hwu, M.-M. Lu, and T. Gao (2015), A source of AGCM bias in simulating the western Pacific subtropical high: Different sensitivities to the two types of ENSO, *Monthly Weather Rev.*, *143*, 2348–2362.

Peterson, W. T., and F. B. Schwing (2003), A new climate regime in northeast pacific ecosystems, *Geophys. Res. Lett.*, *30*, 1896; doi:10.1029/2003GL017528.

Philander, S. G. (1999), A review of tropical ocean–atmosphere interactions, *Tellus A*, *51* (1), 71–90.

Pierce, D. W., T. P. Barnett, and M. Latif (2000), Connections between the Pacific Ocean tropics and midlatitudes on decadal timescales, *J. Climate*, *13*, 1173–1194.

Qian, C., J. -Y. Yu, and G. Chen (2014), Decadal summer drought frequency in China: The increasing influence of the Atlantic multi-decadal oscillation, *Environ. Res. Lett.*, *9* (12), 124004; doi:10.1088/1748-9326/9/12/124004.

Qiu, B., and S. Chen (2012), Multidecadal sea level and gyre circulation variability in the northwestern tropical Pacific Ocean, *J. Phys. Oceanogr.*, *42*, 193–206; doi: http://dx.doi.org/10.1175/JPO-D-11-061.1.

Rajagopolan, B., U. Lall, and M. Cane (1997), Anomalous ENSO occurrences: An alternate view, *J. Climate*, *10*, 2351–2357.

Rasmusson, E. M., and T. H. Carpenter (1982), Variations in tropical sea surface temperature and surface wind fields associated with the Southern Oscillation/El Niño, *Monthly Weather Rev.*, *110*, 354–384; doi: http://dx.doi.org/10.1175/1520-0493[1982]110<0354:VITSST>2.0.CO;2.

Rasmusson, E. M., X. Wang, and C. F. Ropelewski (1990), The biennial component of ENSO variability, *J. Mar. Syst.*, *1*, 71–90; doi:10.1016/0924-7963[90]90153-2.

Ratnam, J. V., S. K. Behera, Y. Masumoto, and T. Yamagata (2014), Remote effects of El Niño on Modoki events on the austral summer precipitation of southern Africa, *J. Climate*, *27*, 3802–3815; doi:10.1175/JCLI-D-13-00431.1.

Rayner, N. A., et al. (2003), Global analyses of sea surface temperature, sea ice, and night marine air temperature since the late nineteenth century, *J. Geophys. Res.*, *108* (D14), 4407.

Ren, H.-L., and F.-F. Jin (2011), Niño indices for two types of ENSO, *Geophys. Res. Lett.*, *38*, L04704; doi: 10.1029/2010GL046031.

Ren, H.-L., and F.-F. Jin (2013), Recharge oscillator mechanisms in two types of ENSO, *J. Climate*, *26*, 6506–6523. doi:10.1175/JCLI-D-12-00601.1.

Rodgers, K. B., P. Friederichs, and M. Latif (2004), Tropical Pacific decadal variability and its relation to decadal modulations of ENSO, *J. Climate*, *17*, 3761–3774.

Rodrigues, R. R., R. J. Haarsma, E. J. D. Campos, T. Ambrizzi (2011), The impacts of inter-El Niño variability on the tropical Atlantic and northeast Brazil climate, *J. Climate*, *24*, 3402–3422.

Rogers, J. C. (1981), The North Pacific Oscillation, *Int. J. Climatol.*, *1*, 39–57; doi:10.1002/joc.3370010106.

Ropelewski, C. F., and M. S. Halpert (1986), North American precipitation and temperature patterns associated with the El Niño/Southern Oscillation, *Monthly Weather Rev.*, *114*, 2352–2362.

Ropelewski, C. F., and M. S. Halpert (1987), Global and regional scale precipitation patterns associated with the El Nino Southern Oscillation, *Monthly Weather Rev.*, *115*, 1606–1626.

Ropelewski, C. F., and M. S. Halpert (1989), Precipitation patterns associated with the high index phase of the Southern Oscillation, *J. Climate*, *2*, 268–284.

Saji, N. H., B. N. Goswami, P. N. Vinayachandran, and T. Yamagata (1999), A dipole mode in the tropical Indian Ocean, *Nature*, *401*, 360–363.

Sassi, F., D. Kinnison, B. A. Boville, R. R. Garcia, and R. Roble (2004), Effect of El Niño–Southern Oscillation on the dynamical, thermal, and chemical structure of the middle atmosphere, *J. Geophys. Res.*, *109*, D17108; doi:10.1029/2003JD004434.

Schlesinger, M. E., and N. Ramankutty (1994), An oscillation in the global climate system of period 65–70 years, *Nature*, *367*, 723–726; doi:10.1038/367723a0.

Schneider, D. P., Y. Okumura, and C. Deser (2012), Observed Antarctic interannual climate variability and tropical linkages, *J. Climate*, *25*, 4048–4066; doi:10.1175/JCLI-D-11-00273.1.

Schopf, P. S., and R. J. Burgman (2006), A simple mechanism for ENSO residuals and asymmetry, *J. Climate*, *19*, 3167–3179.

Schott. F. A., S. P. Xie, and J. McCreary (2009), Indian Ocean circulation and climate variability, *Rev. Geophys.*, *47*, RG1002; doi:10.1029/2007RG000245.

Seager, R., N. Harnik, Y. Kushnir, W. Robinson, and J. Miller (2003), Mechanisms of hemispherically symmetric climate variability, *J. Climate*, *16*, 2960–2978.

Sheffield, J., and coauthors (2013a), North American climate in CMIP5 experiments, Part I: Evaluation of historical simulations of continental and regional climatology, *J. Climate*, *26*, 9209–9245; doi: http://dx.doi.org/10.1175/JCLI-D-12-00592.1.

Sheffield, J., and coauthors (2013b), North American climate in CMIP5 experiments, Part II: Evaluation of 20th century intraseasonal to decadal variability, *J. Climate*, *23*, 9247–9290.

Sheffield, J., and coauthors (2014), Regional climate processes and projections for North America: CMIP3/CMIP5 differences, attribution and outstanding issues, *NOAA Tech. Rep., OAR CPO-2*.

Smith, T. M., R. W. Reynolds, T. C. Peterson, and J. Lawrimore (2008), Improvements to NOAA's historical merged land-ocean surface temperature analysis (1880–2006), *J. Climate*, *21*, 2283–2296.

Son, S. -W., A. Purich, H. H. Hendon, B. M. Kim, and L. M. Polvani (2013), Improved seasonal forecast using ozone hole variability? *Geophys. Res. Lett.*, *40*, 6231–6235; doi:10.1002/2013GL057731.

Song, H.-J., E. Choi, G.-H. Lim, Y. H. Kim, J.-S. Kug, and S.-W. Yeh (2011), The central Pacific as the export region of the El Niño–Southern Oscillation sea surface temperature anomaly to Antarctic sea ice, *J. Geophys. Res.*, *116*, D21113; doi:10.1029/2011JD015645.

Stammerjohn, S. E., D. G. Martinson, R. C. Smith, X. Yuan, and D. Rind (2008), Trends in Antarctic annual sea ice retreat and advance and their relation to El Niño–Southern Oscillation and southern annular mode variability, *J. Geophys. Res.*, *113*, C03S90; doi:10.1029/2007JC004269.

Stevenson, S., B. Fox-Kemper, M. Jochum, R. Neale, C. Deser, and G. Meehl (2012), Will there be a significant change to El Niño in the twenty-first century?, *J. Climate*, *25*, 2129–2145; doi:10.1175/JCLI-D-11-00252.1.

Su, J., R. Zhang, T. Li, X. Rong, J.-S. Kug, and C.-C. Hong (2010), Causes of the El Niño and La Niña amplitude asymmetry in the equatorial eastern Pacific, *J. Climate*, *23*, 605–617.

Suarez, M. J., and P. S. Schopf (1988), A delayed action oscillator for ENSO, *J. Atmos. Sci.*, *45*, 3283–3287; doi:10.1175/1520-0469%281988%29045<3283%3AADAOFE>2.0.CO%3B2.

Sui, C. H., P. H. Chung, and T. Li (2007), Interannual and interdecadal variability of the summertime western North Pacific subtropical high, *Geophys. Res. Lett.*, *34*, L11701; doi:10.1029/2006GL029204.

Sun, D.-Z., and T. Zhang (2006), A regulatory effect of ENSO on the timemean thermal stratification of the equatorial upper ocean, *Geophys. Res. Lett.*, *33*, L07710; doi:10.1029/2005GL025296.

Sun, F., and J. -Y. Yu (2009), A 10–15-yr modulation cycle of ENSO intensity, *J. Climate*, *22*, 1718–1735; doi: http://dx.doi.org/10.1175/2008JCLI2285.1.

Sung, M.-K., B.-M. Kim, and S.-I. An (2014), Altered atmospheric responses to eastern Pacific and central Pacific El Niños over the North Atlantic region due to stratospheric interference, *Climate Dyn.*, *42*, 159–170; doi:10.1007/s00382-012-1661-0.

Takahashi, K., A. Montecinos, K. Goubanova, and B. Dewitte (2011), ENSO regimes: Reinterpreting the canonical and Modoki El Niño, *Geophys. Res. Lett.*, *38*, L10704; doi:10.1029/2011GL047364.

Tang, B. H., and J. D. Neelin (2004), ENSO influence on Atlantic hurricanes via tropospheric warming, *Geophys. Res. Lett.*, *31*, L24204; doi:10.1029/2004GL021072.

Taschetto, A. S., and M. H. England (2009), El Niño Modoki impacts on Australian rainfall, *J. Climate*, *22*, 3167–3174.

Taschetto, A. S., A. Sen Gupta, N. C. Jourdain, A. Santoso, C. C. Ummenhofer, and M. H. England (2014), Cold tongue and warm pool ENSO events in CMIP5: Mean state and future projections, *J. Climate*, *27*, 2861–2885; doi:10.1175/JCLI-D-13-00437.1.

Taylor, K. E., R. J. Stouffer, and G. A. Meehl (2012), An overview of CMIP5 and the experiment design, *Bull. Amer. Meteor. Soc.*, *93*, 485–498.

Tett, S. (1995), Simulation of El Niño–Southern Oscillation-like variability in a global AOGCM and its response to CO2 increase, *J. Climate*, *8* (6), 1473–1502.

Thompson, D. W. J., and J. M. Wallace (1998), The Arctic Oscillation signature in the wintertime geopotential height and temperature fields, *Geophys. Res. Lett.*, *25*, 1297–1300; doi:10.1029/98GL00950.

Thompson, D. W. J., and J. M. Wallace (2000), Annular modes in the extratropical circulation, Part I: Month-to-month variability, *J. Climate*, *13*, 1000–1016.

Timmermann, A. (2003), Decadal ENSO amplitude modulations: A nonlinear mechanism, *Global Planet. Change*, *37*, 135–156.

Timmermann, A., and F.-F. Jin (2002), A nonlinear mechanism for decadal El Niño amplitude changes, *Geophys. Res. Lett.*, *29*, 1003; doi:10.1029/2001GL013369.

Timmermann, A., J. Oberhuber, A. Bacher, M. Esch, M. Latif, and E. Roeckner (1999), Increased El Niño frequency in a climate model forced by future greenhouse warming, *Nature*, *398*, 694–697.

Tokinaga, H., S.-P. Xie, C. Deser, Y. Kosaka, and Y. M. Okumura (2012), Slowdown of the Walker circulation driven by tropical Indo-Pacific warming, *Nature*, *491*, 439–443; doi:10.1038/nature11576.

Torrence, C., and P. J. Webster (1999), Interdecadal changes in the ENSO-monsoon system, *J. Climate*, *12*, 2679–2690.

Trenberth, K. E., and D. P. Stepaniak (2001), Indices of El Niño evolution, *J. Climate*, *14*, 1697–1701. doi: http://dx.doi.org/10.1175/1520-0442[2001]014<1697:LIOENO>2.0.CO;2.

Trenberth, K. E., and T. J. Hoar (1996), The 1990–1995 El Niño–Southern Oscillation event: Longest on record, *Geophys. Res. Lett.*, *23*, 57–60.

Tu, J., C. Chou, and P. Chu (2009), The abrupt shift of typhoon activity in the vicinity of Taiwan and its association with western North Pacific–East Asian climate change, *J. Climate*, *22*, 3617–3628.

Tziperman, E., L. Stone, M. Cane, and H. Jarosh (1994), El Niño chaos: Overlapping of resonances between the seasonal cycle and the Pacific Ocean-atmosphere oscillator, *Science*, *264*, 72.

Tziperman, E., M. Cane, and S. E. Zebiak (1995), Irregularity and locking to the seasonal cycle in an ENSO prediction model as explained by the quasi-periodicity route to chaos, *J. Atmos. Sci.*, *52*, 293–306.

van Oldenborgh, G. J., S. Y. Philip, and M. Collins (2005), El Niño in a changing climate: A multi-model study, *Ocean Sci.*, *1*, 81–95; doi:10.5194/os-1-81-2005.

Vecchi, G. A., and B. J. Soden (2007), Global warming and the weakening of the tropical circulation, *J. Climate*, *20*, 4316–40.

Vecchi, G. A., et al. (2006), Weakening of tropical Pacific atmospheric circulation due to anthropogenic forcing, *Nature*, *441*, 73–76.

Venzke, S., M. Latif, and A. Villwock (2000), The coupled GCM ECHO-2, Part II: Indian Ocean response to ENSO, *J. Climate*, *13*, 1317–1383.

Verdon, D. C., and S. W. Franks (2006), Long-term behaviour of ENSO: Interactions with the PDO over the past 400 years inferred from paleoclimate records, *Geophys. Res. Lett.*, *33*, L06712; doi:10.1029/2005GL025052.

Vimont, D. J., D. S. Battisti, and A. C. Hirst (2001), Footprinting: A seasonal connection between the tropics and mid-latitudes, *Geophys. Res. Lett.*, *28*, 3923–3926; doi:10.1029/2001GL013435.

Vimont, D. J., J. M. Wallace, and D. S. Battisti (2003), The seasonal footprinting mechanism in the Pacific: Implications for ENSO, *J. Climate*, *16*, 2668–2675; doi:10.1175/1520-0442%282003%29016<2668%3ATSFMIT>2.0.CO%3B2.

Vimont, D. J., M. Alexander, and A. Fontaine (2009), Midlatitude excitation of tropical variability in the Pacific: The role of thermodynamic coupling and seasonality, *J. Climate*, *22* (3), 518–534; doi:10.1175/2008JCLI2220.1.

Walker, G. T., and E. W. Bliss (1932), World weather V Mem, *R. Meteor. Soc.*, *4*, 53–84.

Wallace, J. M., and D. S. Gutzler (1981), Teleconnections in the potential height field during the Northern Hemisphere winter, *Monthly Weather Rev.*, *109*, 784–812.

Wang, B., and S. -I. An (2001), Why the properties of El Niño change in the late 1970s?, *Geophys. Res. Lett.*, *28*, 3709–3712.

Wang, B., and Y. Wang (1996), Temporal structure of the Southern Oscillation as revealed by a waveform and a wavelet transform, *J. Climate*, *9*, 1586–1598. doi: http://dx.doi.org/10.1175/1520-0442[1996]009<1586:TSOTSO>2.0.CO;2.

Wang, B., B. Xiang, and J.-Y. Lee (2013), Subtropical high predictability establishes a promising way for monsoon and tropical storm predictions, *PNAS*, *110*, 2718–2722.

Wang, B., R. Wu, and X. Fu (2000), Pacific–East Asian teleconnection: How does ENSO affect East Asian climate?, *J. Climate*, *13*, 1517–1536.

Wang, C., and R. H. Weisberg (1998), Climate variability of the coupled tropical-extratropical ocean-atmosphere system, *Geophys. Res. Lett.*, *25*, 3979–3982.

Wang, C., and X. Wang (2013), Classifying El Niño Modoki I and II by different impacts on rainfall in southern China and typhoon tracks, *J. Climate*, *26*, 1322–1338; doi: 10.1175/JCLI-D-12-00107.1.

Wang, C., C. Deser, J.-Y. Yu, P. DiNezio, and A. Clement (2015), El Nino and Southern Oscillation (ENSO): A review, A chapter for Springer book: *Coral Reefs of the Eastern Pacific*, in press.

Wang, C., R. H. Weisberg, and J. I. Virmani (1999), Western Pacific interannual variability associated with the El Niño–Southern Oscillation, *J. Geophys. Res.*, *104*, 5131–5149. DOI: 10.1029/1998JC900090.

Wang, D., Q. Liu, R. X. Huang, Y. Du, and T. Qu (2006), Interannual variability of the South China Sea throughflow inferred from wind data and an ocean data assimilation product, *Geophys. Res. Lett.*, *33*, L14605; doi:10.1029/2006GL026316.

Wang, G., and H. H. Hendon (2007), Sensitivity of Australian rainfall to inter–El Niño variations, *J. Climate*, *20*, 4211–4226.

Wang, H., and R. Fu (2000), Influences of ENSO SST anomalies and winter storm tracks on the interannual variability of upper-troposphere water vapor over the Northern Hemisphere extratropics, *J. Climate*, *13*, 59–73.

Wang, S.-Y., M. L'Heureux, and H. -H. Chia (2012), ENSO prediction one year in advance using western North Pacific sea surface temperatures, *Geophys. Res. Lett.*, *39*, L05702; doi:10.1029/2012GL050909.

Wang, S.-Y., M. L'Heureux, and J. H. Yoon (2013), Are greenhouse gases changing ENSO precursors in the western North Pacific?, *J. Climate*, *26* (17), 6309–6322.

Wang, X., and C. Wang (2014), Different impacts of various El Niño events on the Indian Ocean dipole, *Climate Dyn.*, *42*, 991–1005; doi: 10.1007/s00382-013-1711-2.

Wang, X., W. Zhou, C. Li, and D. Wang (2014), Comparison of the impact of two types of El Niño on tropical cyclone genesis over the South China Sea, *Int. J. Climatol.*, *34*, 2651–2660; doi: 10.1002/joc.3865.

Weare, B. C., A. R. Navato, and R. E. Newell (1976), Empirical orthogonal analysis of Pacific sea surface temperatures, *J. Phys. Ocean*, *6*, 671–678.

Webster, P. J., A. M. Moore, J. P. Loschnigg, and R. R. Leben (1999), Coupled ocean-atmosphere dynamics in the Indian Ocean during 1997–98, *Nature*, *401*, 356–360.

Weisberg, R. H., and C. Wang (1997), A western Pacific oscillator paradigm for the El Niño–Southern Oscillation, *Geophy. Res. Lett.*, *24*, 779–782; doi: 10.1029/97GL00689.

Weng, H., K. Ashok, S. K. Behera, S. A. Rao, and T. Yamagada (2007), Impacts of recent El Nino Modoki on droughts/floods in the Pacific rim during boreal summer, *Climate Dyn.*, *29*, 113–129; doi:10.1007/s00382-007-0234-0.

Weng, H., S. K. Behera, and T. Yamagada (2009), Anomalous winter climate conditions in the Pacific rim during recent El Nino Modoki and El Nino events, *Climate Dyn.*, *32*, 663–674; doi:10.1007/s00382-008-0394-6.

Wittenberg, A. T. (2009), Are historical records sufficient to constrain ENSO simulations?, *Geophys. Res. Lett.*, *36*, L12702; doi:10.1029/2009GL038710.

Wyrtki, K. (1975), El Niño: The dynamic response of the equatorial Pacific Ocean to atmospheric forcing, *J. Phys. Oceanogr.*, *5*, 572–584; doi:10.1175/1520-0485%281975%29005<0572%3AENTDRO>2.0.CO%3B2.

Xiang, B., B. Wang, and T. Li (2013), A new paradigm for the predominance of standing Central Pacific warming after the late 1990s, *Climate Dyn.*, *41* (2), 327–340; doi: 10.1007/s00382-012-1427-8.

Xie, F., J. Li, W. Tian, J. Feng, and Y. Huo (2012), Signals of El Niño Modoki in the tropical tropopause layer and stratosphere, *Atmos. Chem. Phys.*, *12*, 5259–5273; doi:10.5194/acp-12-5259-2012.

Xie, S.-P., and S. G. H. Philander (1994), A coupled ocean-atmosphere model of relevance to the ITCZ in the eastern Pacific, *Tellus*, *46A*, 340–350; doi:10.1034/j.1600-0870.1994.t01-1-00001.x.

Xie, S.-P., C. Deser, G. A. Vecchi, J. Ma, H. Teng, and A. T. Wittenberg (2010), Global warming pattern formation: Sea surface temperature and rainfall, *J. Climate*, *23*, 966–986; doi:10.1175/2009JCLI3329.1.

Xie, S.-P., H. Annamalai, F. Schott, and J. P. McCreary Jr. (2002), Origin and predictability of South Indian Ocean climate variability, *J. Climate*, *15* (8), 864–874.

Xie, S. -P., K. Hu, J. Hafner, H. Tokinaga, Y. Du, G. Huang, and T. Sampe (2009), Indian Ocean capacitor effect on Indo–Western Pacific climate during the summer following El Niño, *J. Climate*, *22* (3).

Xie, S.-P., Q. Xie, D. Wang, and W. T. Liu (2003), Summer upwelling in the South China Sea and its role in regional climate variations, *J. Geophys. Res.*, *108* (C8), 3261; doi:10.1029/2003JC001867.

Xu, B., J. Cao, J. Hansen, T. Yao, D. Joswia, N. Wang, G. Wu, M. Wang, H. Zhao, W. Yang, X. Liu, and J. He (2009), Black soot and the survival of Tibetan glaciers, *Proc. Nat. Acad. Sci.*, *106*, 22114.

Xu, Z., K. Fan, and H. Wang (2015), Decadal variation of summer precipitation over China and associated atmospheric circulation after the late 1990s, *J. Climate*, *28* (10), 4086–4106.

Xue, Y., M. Chen, A. Kumar, Z.-Z. Hu, and W. Wang (2013), Prediction skill and bias of tropical Pacific sea surface temperatures in the NCEP climate forecast system version 2, *J. Climate*, *26*, 5358–5378; doi:10.1175/JCLI-D-12-00600.1.

Yadav, R. K., D. A. Ramu, and A. P. Dimri (2013), On the relationship between ENSO patterns and winter precipitation over north and central India, *Global Planet. Change*, *107*, 50–58; doi:10.1016/j.gloplacha.2013.04.006.

Yang, J., Q. Liu, S.-P. Xie, Z. Liu, and L. Wu (2007), Impact of the Indian Ocean SST basin mode on the Asian summer monsoon, *Geophys. Res. Lett.*, *34*, L02708; doi:10.1029/2006GL028571.

Yang, S., and X. Jiang (2014), Prediction of eastern and central Pacific ENSO events and their impacts on East Asian climate by the NCEP climate forecast system, *J. Climate*, *27*, 4451–4472.

Yeh, S.-W., and B. P. Kirtman (2004), Tropical Pacific decadal variability and ENSO amplitude modulation in a CGCM, *J. Geophys. Res.*, *109*, C11009; doi:10.1029/2004JC002442.

Yeh, S. -W., and B. P. Kirtman (2007), ENSO amplitude changes due to climate change projections in different coupled models, *J. Climate*, *20*, 203–217.

Yeh, S.-W., B. P. Kirtman, J.-S. Kug, W. Park, and M. Latif (2011), Natural variability of the central Pacific El Niño event on multi-centennial timescales, *Geophys. Res. Lett.*, *38*, L02704; doi:10.1029/2010GL045886.

Yeh, S. W., J. S. Kug, B. Dewitte, M. H. Kwon, B. P. Kirtman, and F.-F. Jin (2009), El Niño in a changing climate, *Nature*, *461*, 511–514; doi:10.1038/nature08316.

Yeh, S.-W., X. Wang, C. Wang, and B. Dewitte (2015), On the relationship between the North Pacific climate variability and the central Pacific El Niño, *J. Climate*, *28*, 663–677; doi: http://dx.doi.org/10.1175/JCLI-D-14-00137.1.

Yeo, S. -R., and K. -Y. Kim (2015), Decadal changes in the Southern Hemisphere sea surface temperature in association with El Niño–Southern Oscillation and southern annular mode, *Climate Dyn.*, 1–16, http://dx.doi.org/10.1007/s00382-015-2535-z.

Yeo, S. -R., K. Y. Kim, S. W. Yeh, B. M. Kim, T. Shim, and J. G. Jhun (2014), Recent climate variation in the Bering and Chukchi seas and its linkages to large-scale circulation in the Pacific, *Climate Dyn.*, *42* (9–10), 2423–2437.

Yu, J.-Y., and H.-Y. Kao (2007), Decadal changes of ENSO persistence barrier in SST and ocean heat content indices: 1958–2001, *J. Geophys. Res.*, *112*, D13106; doi:10.1029/2006JD007654.

Yu, J.-Y., and K. M. Lau (2004), Contrasting Indian Ocean SST variability with and without ENSO influence: A coupled atmosphere-ocean GCM study, *Meteor. Atmos. Phys.*, *90*, 179–191; doi:10.1007/s00703-004-0094-7.

Yu, J.-Y., and S. T. Kim (2010a), Identification of central-Pacific and eastern-Pacific types of ENSO in CMIP3 models, *Geophys. Res. Lett.*, *37*, L15705; doi:10.1029/2010GL044082.

Yu, J.-Y., and S. T. Kim (2010b), Three evolution patterns of central-Pacific El Niño, *Geophys. Res. Lett.*, *37*, L08706; doi:10.1029/2010GL042810.

Yu, J.-Y., and S. T. Kim (2011a), Relationships between extratropical sea level pressure variations and the central-Pacific and eastern-Pacific types of ENSO, *J. Climate*, *24*, 708–720; doi:10.1175/2010JCLI3688.1.

Yu, J.-Y., and S.-T. Kim (2011b), Reversed spatial asymmetries between El Niño and La Niña and their linkage to decadal ENSO modulation in CMIP3 models, *J. Climate*, *24*, 5423–5434.

Yu, J.-Y., and Y. Zou (2013), The enhanced drying effect of Central-Pacific El Niño on US winter, *Environ. Res. Lett.*, *8*; doi:10.1088/1748-9326/8/1/014019.

Yu, J.-Y., H.-Y. Kao, and T. Lee (2010), Subtropics-related interannual sea surface temperature variability in the equatorial central Pacific, *J. Climate*, *23*, 2869–2884; doi: http://dx.doi.org/10.1175/2010JCLI3171.1.

Yu, J.-Y., H.-Y. Kao, T. Lee, and S. T. Kim (2011), Subsurface ocean temperature indices for central-Pacific and eastern-Pacific types of El Niño and La Niña events, *Theor. App. Climatol.*; doi: 10.1007/s00704-010-0307-6.

Yu, J.-Y., F. Sun, and H.-Y. Kao (2009), Contributions of Indian Ocean and monsoon biases to the excessive biennial ENSO in CCSM3, *J. Climate*, *22*, 1850–1858.

Yu, J.-Y., H. Paek, E. S. Saltzman, and T. Lee (2015b), The early-1990s change in ENSO-PSA-SAM relationships and its impact on Southern Hemisphere climate, *J. Climate*; doi: 10.1175/JCLI-D-15-0335.1, in press.

Yu, J.-Y., M.-M. Lu, and S. T. Kim (2012a), A change in the relationship between tropical central Pacific SST variability and the extratropical atmosphere around 1990, *Environ. Res. Lett.*, *7*, 034025; doi:10.1088/1748-9326/7/3/034025.

Yu, J.-Y., Y. Zou, S. T. Kim, and T. Lee (2012b), The changing impact of El Niño on US winter temperatures, *Geophys. Res. Lett.*, *39*, L15702; doi:10.1029/2012GL052483.

Yu, J.-Y., P. -K. Kao, H. Paek, H. -H. Hsu, C. -W. Hung, M. -M. Lu, and S. -I. An (2015a), Linking emergence of the Central-Pacific El Niño to the Atlantic multi-decadal oscillation, *J. Climate*, *28*, 651–662; doi:10.1175/JCLI-D-14-00347.1.

Yu, W., B. Xiang, L. Liu, and N. Liu (2005), Understanding the origins of interannual thermocline variations in the tropical Indian Ocean, *Geophys. Res. Lett.*, *32*, L24706; doi:10.1029/2005GL024327.

Yuan, X. (2004), ENSO-related impacts on Antarctic sea ice: A synthesis of phenomenon and mechanisms, *Antarctic Sci.*, *16* (04), 415–425.

Yuan, X., and C. Li (2008), Climate modes in southern high latitudes and their impacts on Antarctic sea ice, *J. Geophys. Res.*, *113*, C06S91; doi:10.1029/2006JC004067.

Yuan, X., and D. G. Martinson (2001), The Antarctic dipole and its predictability, *Geophys. Res. Lett.*, *28*, 3609–3612.

Yuan, Y., and S. Yang (2012), Impacts of different types of El Niño on East Asian climate: Focus on ENSO cycles, *J. Climate*, *25*, 7702–7722.

Yuan, Y., S. Yang, and Z. Zhang (2012), Different evolutions of the Philippine Sea anticyclone between eastern and central Pacific El Niño: Possible effect of Indian Ocean SST, *J. Climate*, *25*, 7867–7883.

Yulaeva, E., and J. M. Wallace (1994), The signature of ENSO in global temperature and precipitation fields derived from the microwave sounding unit, *J. Climate*, *7*, 1719–1736; doi:10.1175/1520-0442[1994] 007<1719:TSOEIG>2.0.CO;2.

Zebiak, S. E. (1989), Ocean heat content variability and El Niño cycles, *J. Phys. Oceanogr.*, *19*, 475–486; doi:10.1175/1520-0485%281989%29019<0475%3AOHCVAE>2.0.CO%3B2.

Zelle, H., G. J. van Oldenborgh, G. Burgers, and H. Dijkstra (2005), El Niño and greenhouse warming: Results from ensemble simulations with the NCAR CCSM, *J. Climate*, *18*, 4669–4683.

Zhang, L., L. Wu, X. Lin, and D. Wu (2010), Modes and mechanisms of sea surface temperature low-frequency variations over the coastal China Seas, *J. Geophys. Res.*, *115*, C08031; doi:10.1029/2009JC006025.

Zhang, R., and T. L. Delworth (2007), Impact of the Atlantic multidecadal oscillation on North Pacific climate variability, *Geophys. Res. Lett.*, *34*, L23708; doi:10.1029/2007GL031601.

Zhang, R., A. Sumi, and M. Kimoto (1999), A diagnostic study of the impact of El Nino on the precipitation in China, *Adv. Atmos. Sci.*, *16*, 229–241.

Zhang, W., F. Jin, J. Li, and H. Ren (2011), Contrasting impacts of two-type El Niño over the western north Pacific during boreal autumn, *J. Meteor. Soc. Japan.*, *89*, 563–569.

Zhang, X. B., J. Sheng, and A. Shabbar (1998), Modes of interannual and interdecadal variability of Pacific SST, *J. Climate*, *11*, 2556–2569.

Zhang, Y., J. M. Wallace, and D. S. Battisti (1997), ENSO-like interdecadal variability: 1900–93, *J. Climate*, *10*, 1004–1020.

Zhou, L.-T., C.-Y. Tam,W. Zhou, and J. C. L. Chan (2009), Influence of South China Sea SST and the ENSO on winter rainfall over South China, *Adv. Atmos. Sci.*, *27*, 832–844.

Zhou, S., A. J. Miller, J. Wang, J. K. Angell (2002), Downward-propagating temperature anomalies in the preconditioned polar stratosphere, *J. Climate*, *15*, 781–792.

Zhou, W., and J. C. L. Chan (2007), ENSO and the South China Sea summer monsoon onset, *Int. J. Climatol.*, *27*, 157–167.

Zhou W., X. Wang, T. J. Zhou, C. Y. Li, and J. C. L. Chan (2007), Interdecadal variability of the relationship between the East Asian winter monsoon and ENSO, *Meteor. Atmos. Phys.*, *98*, 283–293; doi: 10.1007/s00703-007-0263-6.

Zhou, Z. Q., S. P. Xie, X. T. Zheng, Q. Y. Liu, and H. Wang (2014), Global warming–induced changes in El Niño teleconnections over the North Pacific and North America, *J. Climate*, *27*, 9050–9064.

Zhu, Y., W. Wang, W. Zhou, and J. Ma (2011), Recent changes in the summer precipitation patterns in East China and the background circulation, *Climate Dyn.*, *36*, 1463–1473.

Zou, Y., J.-Y. Yu, T. Lee, M.-M. Lu, and S. T. Kim (2014), CMIP5 model simulations of the impacts of the two types of El Niño on US winter temperature, *J. Geophys. Res. Atmos.*, *119*, 3076–3092; doi:10.1002/2013JD021064.

Zubiaurre, I., and N. Calvo (2012), The El Niño–Southern Oscillation [ENSO] Modoki signal in the stratosphere, *J. Geophys. Res.*, *117*, D04104; doi:10.1029/2011JD016690.

2

Weather Extremes Linked to Interaction of the Arctic and Midlatitudes

Timo Vihma

ABSTRACT

A review is presented on Arctic forcing to midlatitude weather extremes and midlatitude forcing to Arctic weather extremes. Several studies have addressed the following mechanisms: reduction of the meridional temperature gradient, southward shift and weakening and increased meandering of the polar jet stream, changes in storm tracks and blocking, planetary wave trains generated by anomalous surface heating, and wave trapping due to resonance with surface forcing. These mechanisms undoubtedly affect weather extremes, but uncertainty remains on the contributions of Arctic and lower latitudes to the mechanisms. Among robust findings is that Arctic amplification contributes to the southward shift and, at least in summer, weakening of the jet stream. A major part of the Arctic tropospheric warming is, however, driven by lower latitudes. Wave trains originating from the Arctic affect summer precipitation in East Asia and northwestern Europe and winter cold spells in East Asia. North American cold winters have occurred due to amplified waves and high-latitude blockings, partly originating from warm anomalies in the Arctic. Attribution of the extremes is hampered by short datasets and small signal-to-noise ratios. Better understanding is needed on circulation response to climate change and reduction of model uncertainty.

2.1. INTRODUCTION

From the point of view of societies around the world, the most serious effects of climate change are supposed to be felt through changes in extreme weather. To plan adaptation, there is a strong need to continuously improve estimates for the future changes in the occurrence, duration, and intensity of extreme weather events. High extremes in air temperature and precipitation have already become increasingly common [*Donat et al.*, 2013]. Summer events have included heat waves and droughts in the United States [*Peterson et al.*, 2013] and central-southern Europe and Russia [*Coumou and Rahmsdorf*, 2012], as well as rains and floods in northern Europe [*Screen*, 2013; *Huntingford et al.*, 2014] and East Asia [*Seo et al.*, 2012]. However, despite the global warming, the occurrence of cold winter months has increased since 1999 over land areas from 20°to 50°N [*Cohen et al.*, 2014], reflecting a pattern of a warm Arctic surrounded by cold continents in winter [*Overland et al.*, 2011]. Cold, snow-rich winters have been common particularly in eastern North America and central-eastern Eurasia [*Overland et al.*, 2015; *Kim et al.*, 2014]. Throughout midlatitudes, recent weather extremes have often been associated with the persistence of particular circulation patterns [*Franzke*, 2013; *Hoskins and Woollings*, 2015].

There is a lot of scientific and public interest toward causes of extreme weather events. The attributions presented by scientists are, however, often restricted to explanation of the large-scale circulation pattern related to an extreme event without a deeper causal analysis on the external forcing factors of the event. A wider public

Finnish Meteorological Institute, Helsinki, Finland

Climate Extremes: Patterns and Mechanisms, Geophysical Monograph 226, First Edition.
Edited by S.-Y. Simon Wang, Jin-Ho Yoon, Christopher C. Funk, and Robert R. Gillies.

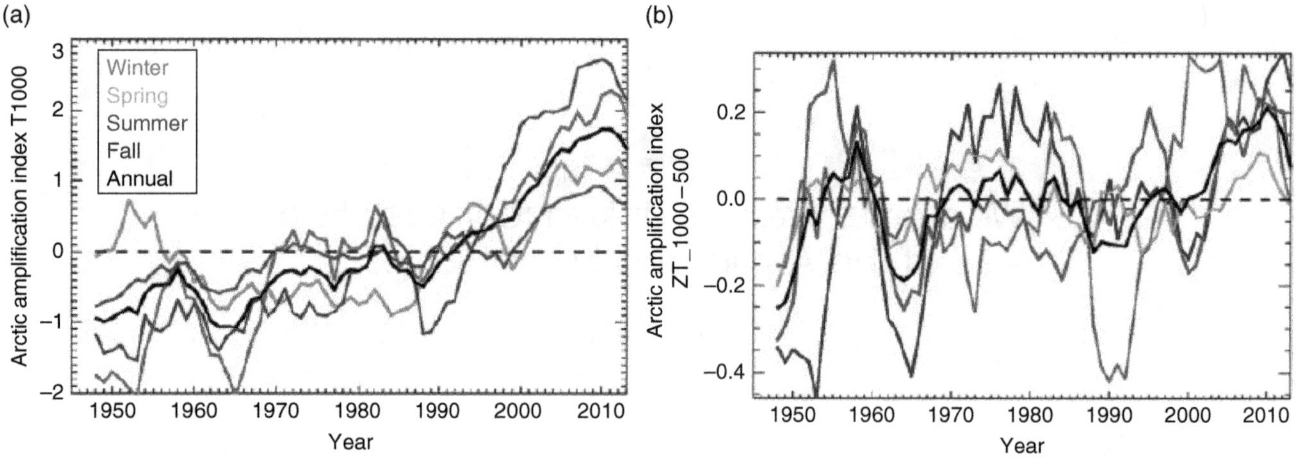

Figure 2.1 Arctic amplification seasonal time series (relative to 1948–2013 mean, °C) between the Arctic (70–90°N) and midlatitudes (30–60°N) based on two metrics: (a) differences in 1000 hPa temperature anomalies (relative to 1948–2013 mean, °C) between the Arctic (70–90°N) and midlatitudes (30–60°N), and (b) differences in 1000–500 hPa thickness anomalies (percent of mean for each zone) between the two zones. Data were obtained from the NCEP/NCAR Reanalysis at http://www.esrl.noaa.gov/psd/. Reproduced from *Francis and Vavrus* [2015], with permission via the Creative Commons Attribution 3.0 license http://creativecommons.org/licenses/by/3.0/legalcode. (*See insert for color representation of the figure.*)

seems particularly interested in the contribution of anthropogenic climate warming on extreme weather events. In the case of events driven by internal dynamics of the atmosphere instead of direct thermal forcing, attribution to climate change is, however, usually very difficult and liable to errors. This is due to the small signal-to-noise ratio: modeled changes in large-scale circulation due to anthropogenic climate warming are generally small, nonrobust, and not well understood, whereas the natural variability is large [*Trenberth et al.*, 2015].

In any case, recent rapid changes in the Arctic have often been considered as reasons for anomalous weather patterns in northern midlatitudes [see *Vihma*, 2014; *Cohen et al.*, 2014; *Overland et al.*, 2015; and *Gao et al.*, 2015, for recent reviews]. The climate warming in the Arctic has been 2–4 times faster than the global mean [*Screen et al.*, 2012]; the phenomenon is called the "Arctic amplification" (Fig. 2.1). The air temperature increase has been both a reason for and a consequence of the changes in the Arctic cryosphere. Since the early 1980s, the Arctic sea-ice extent has decreased by roughly 50% in summer and autumn [*Cavalieri and Parkinson*, 2012], and the winter sea-ice thickness in the central Arctic has also decreased by approximately 50% [*Kwok and Rothrock*, 2009]. In May and June, the rate of decrease in the Northern Hemisphere terrestrial snow cover has been even much faster (Fig. 2.2). The reductions in sea-ice extent and thickness have resulted in increased fluxes of sensible and latent heat from the ocean to the atmosphere. This has increased air temperatures and decreased static stability in the lower troposphere, making the marine Arctic more liable to cyclogenesis [*Jaiser et al.*, 2013]. The decrease of terrestrial snow cover in spring and early summer has enhanced the warming of the land surface and, hence, the lower troposphere. According to several studies, the Arctic warming with decline of sea ice and snow has contributed to changes in the polar front jet stream (also called the "midlatitude jet stream," hereafter simply "jet stream"), near-surface zonal and meridional winds, cyclone activity and tracks, and planetary waves (Rossby waves). Changes in the Arctic have been neither spatially nor seasonally uniform. In summer, the sea-ice decline has been largest in the Pacific sector of the Arctic Ocean, but in winter in the Barents and Kara seas and the marginal seas at lower latitudes [*Cavalieri and Parkinson*, 2012]. Since 1990, the winter near-surface air temperature increase has been largest in the Atlantic-Eurasian sector of the Arctic Ocean and in the Greenland–Baffin Island region [*Cohen et al.*, 2014], while the summer warming has been much weaker. Due to these differences in the Arctic changes and topographic differences within midlatitudes, we cannot expect similar effects on different midlatitude regions.

In this chapter, I review the proposed impacts of the Arctic on extreme weather events in midlatitudes (section 2.2) and the impacts of midlatitudes and tropics on extremes in the Arctic (section 2.3). The focus is on observed weather extremes, but the projected future extremes are also discussed. The interest toward Arctic-midlatitude linkages is strongly growing, driven above all by the potential for improved extended-range forecasts at midlatitudes [*Walsh*, 2014; *Hall et al.*, 2015] and the

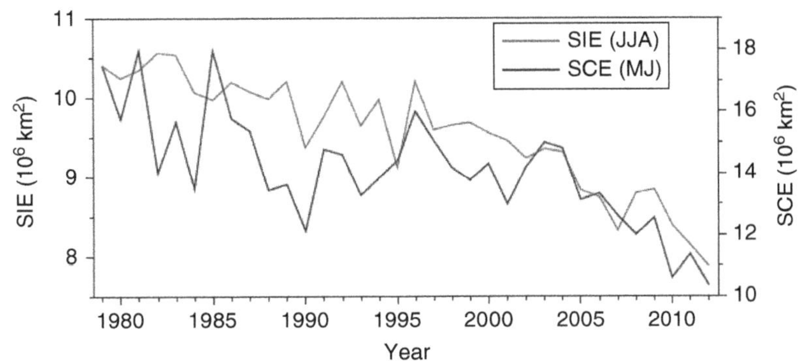

Figure 2.2 Summer (JJA) Arctic SIE and early summer (May and June) Northern Hemisphere SCE. Data span 1979–2012. Note the different vertical axes for SIE and SCE. From *Tang et al.* [2014]; reprinted by permission from Macmillan Publishers Ltd.

devastating effects of numerous midlatitude weather extremes experienced in recent years [*Coumou and Rahmstorf*, 2012; *Coumou et al.*, 2015]. In addition to the scientific community, the wider public is also paying attention to the potential linkages of Arctic changes and recent extreme weather events [*Hamilton and Lemcke-Stampone*, 2014]. The topic is challenging, as strong changes in the Arctic are recent and the data record is accordingly short for robust statistical analyses. Further, a variety of methods have been applied by different groups carrying out data analyses and model experiments. Contrasting conclusions have been made based on different data, approaches, metrics, and interpretations [*Overland et al.*, 2015]. Further, the formulation of research questions has been variable and not always entirely clear [*Barnes and Screen*, 2015].

2.2. ARCTIC EFFECTS ON MIDLATITUDE EXTREMES

Attribution of individual extreme weather events to changes in the Arctic is very difficult. However, the increasing frequency of occurrence and the remarkable persistence of many of the extreme events are in many papers partly attributed to the warming and melting of the Arctic. Below I review the mechanisms most commonly suggested; these are schematically introduced in Figure 2.3. The mechanisms addressed interact via complex and, in many cases, poorly known processes. Different studies have addressed different mechanisms, but the points of view are often complementary. An example by *Hoskins and Woollings* [2015] is that a strong jet stream shifted to the north can also be interpreted as an amplified stationary Rossby wave pattern, or a developing blocking, or the situation can be characterized in terms of a positive phase of the North Atlantic Oscillation/Arctic Oscillation (NAO/AO).

2.2.1. Overall Warming

Probably the most evident effect of Arctic warming on midlatitude weather is that cold-air outbreaks from the Arctic are no more as cold as they used to be on average [*Serreze et al.*, 2011; *Screen*, 2014]. Midlatitude temperatures during southerly winds have increased much less, as the subtropics have warmed much less than the Arctic. As a result, cold days (occurring during northerly winds) have warmed more than warm days (during southerly winds), which has decreased the daily temperature variability [*Screen*, 2014]. The cooling effect of wintertime cold-air outbreaks has decreased also due to the fact that land areas have warmed faster than the seas. Accordingly, even if wind anomalies are unchanged, wintertime air temperature anomalies originating from thermal advection are becoming weaker [*de Vries et al.*, 2012; *Masato et al.*, 2014; *Schneider et al.*, 2015]. For example, *Cattiaux et al.* [2010] estimated that without climate warming the cold winter 2009–2010 in Europe, dominated by cold-air outbreaks related to extremely negative NAO, would have been even more severe. Despite the climate warming, however, in East Asia winter cold-air outbreaks have in this century become more frequent, stronger, and longer lasting than in the 1990s [*Kim et al.*, 2014], which is further discussed in section 2.2.3.

Considering the future, on the basis of large ensembles of model simulations, *Screen et al.* [2015a] concluded that the projected anthropogenic climate warming gradually decreases the likelihood of North American cold extremes that occur in timescales up to 9 days. The occurrence and intensity of extreme weather events will be affected by numerous factors. *Screen et al.* [2015b] isolated the expected contribution of future sea-ice decline by carrying out atmospheric general circulation model experiments forced by projected Arctic sea-ice loss. The result suggested

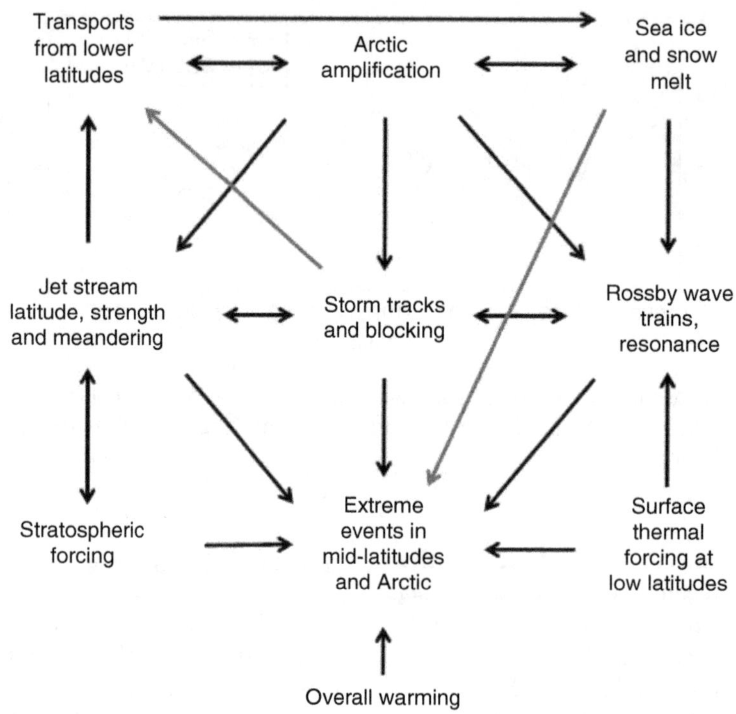

Figure 2.3 Suggested causal linkages between processes associated with Arctic-midlatitude linkages and extreme weather events.

strong regional differences, with a decrease in the occurrence and duration of cold extremes in central and eastern North America but an increase in central Asia and an increase of hot and wet extremes in high latitudes. Note that these are not projections for the future climate but for the isolated effects of sea-ice decline.

Considering extreme weather events in smaller spatial scales, the Arctic warming is expected to cause changes in the occurrence of polar lows. With the sea-ice pack retreating farther north, polar lows will become more common in the Barents and Chukchi seas and less common in the North Atlantic as well as Norwegian and Labrador seas [*Kolstad and Bracegirdle*, 2008; *Zahn and von Storch*, 2010].

2.2.2. Meridional Temperature Gradient and Jet Stream Latitude and Speed

A major effect of the Arctic amplification is that it decreases the north-south temperature gradient, hence reducing tropospheric zonal winds (including the jet stream) via the thermal wind effect [*Newson*, 1973; *Warshaw and Rapp*, 1973; *Overland and Wang*, 2010; *Francis and Vavrus*, 2012; 2015; *Cvijanovic and Caldeira*, 2015]. The reduction of midlatitude zonal winds is robust particularly in summer (Fig. 2.4), which may be due to the fact that, although Arctic temperatures have increased most in winter, the largest changes in the north-south temperature gradient within midlatitudes have occurred in summer [*Coumou et al.*, 2015]. *Hoskins and Woollings* [2015] note, however, that the thermal wind balance is generally shaped by complex eddy-mean flow feedbacks, and a weaker jet stream is not necessarily a straightforward result of Arctic amplification.

Several studies based on observations, reanalyses, and various kinds of atmospheric and coupled models have suggested that the Arctic amplification and sea-ice decline increase the probability for circulation patterns resembling the negative phase of AO and NAO in winter [see *Vihma*, 2014; and *Hoskins and Woollings*, 2015, for summaries]. In the lower troposphere, climate warming is faster in the Arctic than at lower latitudes, which tends to induce an equatorward shift of the jet stream and storm tracks in midlatitudes, typically associated with a negative phase of NAO/AO [*Cohen et al.*, 2014; *Hall et al.*, 2015]. According to *Hoskins and Woollings* [2015], the equatorward shift of the jet stream is in general the dominant robust midlatitude response to the Arctic amplification. In winter, a negative NAO is typically associated with low temperatures in large parts of Eurasia. For example, during the extremely cold midlatitude winter of 2009–2010, the NAO index was the lowest recorded since winter 1824 [*Cattiaux et al.,* 2010]. In summer, according to *Screen* [2013], wet (dry) summers in northern Europe tend to occur when the jet stream is displaced to the south (north).

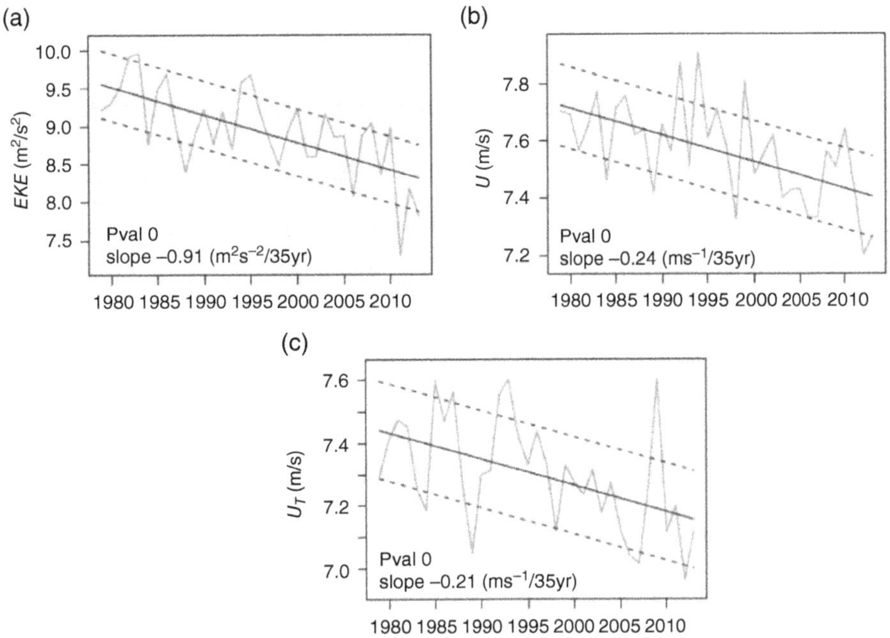

Figure 2.4 Weakening summer circulation in the midlatitudes. Absolute changes in (a) eddy kinetic energy (EKE), (b) zonal wind (U), and (c) thermal wind (U_T) over 1979–2013 in summer (June, July, and August). Variables are calculated at 500 hPa and averaged over 35°N to 70°N and all longitudes, with gray lines plotting observations, solid black lines the linear trend, and dashed black lines the ±1 residual standard error range. Slope and p-values for the trend estimates are given in the panels. From *Coumou, Lehmann, and Beckmann* [2015]; reprinted with permission from AAAS.

Conditions in the Arctic are naturally only one of the factors affecting the jet stream. The others include stratospheric circulation, North Atlantic sea surface temperatures, the El Niño–Southern Oscillation, the Quasi-Biennial Oscillation, solar variability, and volcanic eruptions [*Hall et al.*, 2015]. Further, the potential of the Arctic to change the jet stream latitude is limited by the contrasting warming trends in the lower and upper troposphere. In the lower troposphere, the Arctic is warming faster than lower latitudes, whereas in the upper troposphere, the tropics are warming faster than higher latitudes. The former (latter) is expected to induce an equatorward (poleward) shift of the jet streams and storm tracks in midlatitudes [*Held*, 1993; *Hall et al.*, 2015].

2.2.3. Jet Stream Meandering, Blocking, and Wave Trains

The reduction of zonal winds potentially results in a wavier jet stream (i.e., having a larger meridional amplitude) and slower eastward motion of weather patterns [*Francis and Vavrus*, 2012, 2015; *Tang et al.*, 2013, 2014; *Francis and Skific*, 2015]. These relationships are, however, not necessarily robust but depend on the exact methodology and metrics applied [*Cohen et al.*, 2014; *Barnes*, 2013; *Barnes and Screen*, 2015]. The maximum meridional amplitude of jet stream patterns has not necessarily increased (as it is geographically restricted), but there is a growing evidence of an increase in the frequency of occurrence of amplified jet stream patterns [*Francis and Skific*, 2015; *Francis and Vavrus*, 2015; *Overland et al.*, 2015]. A wavier jet stream with slowly moving weather patterns favors the occurrence of extreme weather events [*Screen and Simmonds*, 2014; *Coumou et al.*, 2015], above all when considering extremes on monthly instead of daily timescales.

A jet stream with meandering high-amplitude patterns is usually associated with blocking events [*Woollings et al.*, 2010; *Hanna et al.*, 2014a], but the effects of Arctic amplification and weaker zonal winds on blocking activity are not clear [*Overland et al.*, 2015]. According to *Barnes et al.* [2014], recent trends in blocking activity have not been statistically significant. Further, on the basis of climate model experiments [*Barnes et al.*, 2012; *Masato et al.*, 2013; *Dunn-Sigouin and Son*, 2013], a consensus is emerging of a general decrease in blocking occurrence under climate change [*Hoskins and Woollings*, 2015]. This suggests that changes in the Arctic are not the primary factor controlling the occurrence of midlatitude blockings. However, the general decrease does not mean that blocking occurrence will decrease everywhere. According to *Masato et al.* [2013] and *Dunn-Sigouin and*

Son [2013], the European blocking is shifting eastward, which causes a local increase in blocking over western Russia, in particular in summer. Further, the model experiments by *Mori et al.* [2014] and *Screen et al.* [2015a] suggest strengthening of the Siberian High during the 21st century.

Extreme weather patterns are favored particularly in cases of resonance of Rossby waves. *Petoukhov et al.* [2013] suggested that trapping of Rossby waves (with zonal wave numbers 6, 7, or 8) due to quasi-stationary thermal and orographic forcing may have contributed to heat waves in Europe (summer 2003), Russia (2010), and the United States (2011), as well as to floods in Pakistan (2010). *Petoukhov et al.* [2013] argued that the Arctic amplification, reduction of midlatitude westerlies, and steepening of the northern flank of the subtropical jet stream may favor the trapping of Rossby waves. Also, *Tang et al.* [2014], *Screen and Simmonds* [2014], and *Coumou et al.* [2014, 2015] have demonstrated the role of quasi-stationary or slowly moving waves for summer heat extremes. The occurrence of high-amplitude wave-resonance events has increased during the period of the Arctic amplification since the late 1990s [*Coumou et al.*, 2014, 2015]. *Coumou et al.* [2015] further point out that the eddy kinetic energy of fast-moving waves has steadily decreased, consistent with more persistent summer weather. Accordingly, weather variability has decreased on short timescales but increased on monthly timescales. Blocking and occurrence of trapped quasi-stationary Rossby waves is associated with a low cyclone activity. According to *Kyselý* [2008] and *Coumou et al.* [2015], low cyclone activity over Europe in recent years has led to more persistent weather, contributing to prolonged heat waves. For example, during the Russian heat wave in 2010, the blocking lasted a long time [*Dole et al.*, 2011] with extremely low eddy kinetic energy around Moscow in July [*Coumou et al.*, 2015].

Of special interest for Arctic-midlatitude linkages are high-latitude blockings typically located near Greenland or northeastern Siberia [*Davini*, 2013; *Hanna et al.*, 2014b]. Rather than entirely blocking the jet stream, they divert it southward [*Woollings et al.*, 2010]. Greenland blocking typically occurs simultaneously with an upstream trough located near Labrador and a ridge farther west over North America [*Overland et al.*, 2015]. In such conditions, which have been common since 2000 (particularly in Decembers 2009, 2010, and 2012), the jet stream is strongly meandering with high meridional amplitude, favoring long-lasting northwesterly cold-air outbreaks causing extreme winter weather in eastern North America [*Francis and Vavrus*, 2015]. *Overland et al.* [2015] concluded that the ridge-trough pattern that has prevailed over North America throughout many of the early winters since 2007 is due to the combined effects of internal dynamic processes, local surface temperature anomalies, low-level temperature advection, and remote influences. On the basis of data analyses supported by idealized climate model experiments and extensive analyses of CMIP5 multimodel simulations, *Kug et al.* [2015] suggested that warming over the East Siberian and Chukchi seas was accompanied by local development of an anomalous anticyclone and a downstream development of a midlatitude trough, resulting in cold-air advection to North America.

According to *Kug et al.* [2015], an analogous mechanism has linked warm anomalies over the Barents-Kara seas to cold winters over East Asia. *Kim et al.* [2014], *Sato et al.* [2014], and *Mori et al.* [2014] attributed the strong cold-air outbreaks to a dominating effect of a negative phase of the AO/NAO associated with episodic southeastward-propagating wave trains from the Barents-Kara seas to East Asia. This has resulted in persistent cold spells in East Asia during several winters since 2007, which have simultaneously been exceptionally warm in the Arctic. According to *Hori M. E., et al.* [2011], a typical evolution is such that a wave train develops over Eurasia 5 days before a cold-air outbreak in Japan. *Kim et al.* [2014] paid particular attention to the effect of the Barents/Kara seas on upward propagation of Rossby waves, which weaken the stratospheric polar vortex, favoring a negative phase of AO. Also, *Jaiser et al.* [2013] and *Peings and Magnusdottir* [2014] stressed the role of Rossby wave penetration to the stratosphere and subsequent generation of teleconnections between the Arctic and midlatitudes.

Considering summer, precipitation extremes in China have been associated with wave trains generating interseasonal response times to high-latitude anomalies [*Zhao et al.*, 2004; *Wu et al.*, 2009, 2013; *Guo et al.*, 2014; *Uotila et al.*, 2014]. Surface forcing from Atlantic and Pacific sea surface temperatures (SSTs), Eurasian soil water content, Arctic sea-ice anomalies, and Eurasian snow cover anomalies have triggered Rossby wave trains, affecting summer precipitation in China. The results of different studies, based on different methodologies, have, however, been partly contradicting with respect to the sign and location of the precipitation anomalies.

Also in Europe, extreme weather events have been attributed to wave trains. *Screen* [2013] proposed that modified storm tracks in the North Pacific and a stationary wave train from the Pacific to Atlantic were the mechanisms that transmitted the effects of winter/spring sea-ice loss in the Sea of Okhotsk to May-June weather over northwestern Europe, where six consecutive wet summers were observed from 2007 to 2012. In addition to the Sea of Okhotsk, sea-ice loss in the Hudson Bay and the Labrador Sea contributed to the generation of the wave train. Also, it is suggested that a wave train pattern may link winter sea-ice concentration in the Baffin Bay and Labrador Sea to summer atmospheric circulation over the mid- to high latitudes of Eurasia. Further, *Lee et al.* [2011] concluded that the Arctic amplification probably contributed to the exceptionally wet, stormy, and mild winter of 2013–2014 in the United Kingdom. See section 2.3 for the attribution of Arctic amplification by *Lee et al.* [2011].

As a summary of section 2.2, there is increasing evidence of the effects of Arctic changes on meridional temperature gradient, jet stream properties, storm tracks, and blockings as well as Rossby wave trains and resonance. Challenges remain, however, in understanding the relative importance of the Arctic and lower-latitude forcing on these circulation features.

2.3. MIDLATITUDE EFFECTS ON ARCTIC EXTREMES

Considering recent extreme air temperatures in the Arctic, the most extensive study is that by *Matthes et al.* [2015], who focused on warm and cold spells in winter and summer during the period 1979–2013. They detected, among other things, a widespread decrease of cold spells in both winter and summer of up to 4 days per decade, but mostly small and statistically insignificant changes in warm spells. Considering individual extreme weather events, among the most well-known ones are the record strong summer cyclones over the Arctic Ocean in August 2012 [*Simmonds and Rudeva*, 2012] and near Alaska in November 2011 [*Overland et al.*, 2012]. Air temperature extremes include the exceptionally warm April 2007 near the North Pole [*Vihma et al.*, 2008], unprecedented monthly air temperature anomalies of up to +4°C over the Arctic Ocean in October–December 2008 [*Overland et al.*, 2009] and over northern Canada during the first half of 2010 [*Overland et al.*, 2010], and extreme monthly temperature anomalies of +10°C in Alaska in early 2014 [*Overland et al.*, 2014]. Precipitation extremes in the marine Arctic are poorly known, and in the terrestrial Arctic, changes in their occurrence and intensity have been spatially variable [*Vihma et al.*, 2016]. Climate models project that precipitation extremes, including rain events in winter, will increase during this century [*McAfee et al.*, 2013].

Much of the near-surface warming trend in the Arctic is attributable to sea-ice decline [*Screen et al.*, 2012]. Hence, it has inevitably contributed to the occurrence of warm extremes in air temperature. Considering midlatitude effects, *Overland et al.* [2014] attributed the warm extremes in early 2014 to a high-amplitude jet stream that brought warm air from the Pacific to Alaska. In general, however, not much attention has been paid to attribution of individual extreme events in the Arctic. Instead, there has been increasing activity in exploring lower-latitude contribution to the trends observed in the Arctic atmosphere.

The Arctic amplification is partly due to local processes in the Arctic, related to the surface albedo, clouds, water vapor, aerosols, the shape of the temperature profile, and the small heat capacity of the shallow atmospheric boundary layer [*Pithan and Mauritsen*, 2014; *Esau and Zilitinkevich*, 2010; *Vihma et al.*, 2014a]. In addition, heat transport in the ocean [*Polyakov et al.*, 2010; *Döscher et al.*, 2014] as well as heat and moisture transports in the atmosphere [*Lu and Cai*, 2008; *Graversen et al.*, 2008; *Serreze et al.*, 2009; *Kapsch et al.*, 2013] strongly contribute to the Arctic amplification. For example, in the climate model experiments by *Lu and Cai* [2008], the sensible heat transport was vital in generating the Arctic amplification. Further, on the basis of atmospheric general circulation model experiments, *Screen et al.* [2012] concluded that the Arctic near-surface warming has been mostly driven by changes in local sea-ice cover and SSTs. However, the majority of observed warming aloft is explained by SST changes at lower latitudes and related changes in the temperature of air masses originating from the south [*Screen et al.*, 2012]. Hence, transports from midlatitudes directly contribute to the occurrence of warm temperatures and precipitation extremes in the Arctic.

Also, more complex teleconnections from lower latitudes to the Arctic have been recently identified. The Barents and Kara seas have been among the Arctic regions with most rapid climate warming during recent decades. According to *Sato et al.* [2014], a major factor for this warming originates from the Gulf Stream region, where a sea surface temperature front has shifted poleward, contributing to the sea-ice decline over the Barents Sea. Further, *Sato et al.* [2014] concluded that the Rossby wave trains that have caused cold spells in Eastern Asia (section 2.2.3) are triggered over the Gulf Stream region when the SST front is shifted northward, and the role of the Barents Sea is to amplify the waves.

Considerng the tropics, the analyses of *Lee et al.* [2011] suggest that convection over the tropical Indian and western Pacific oceans plays a major role in the Arctic amplification via exciting hemispheric-scale circulation patterns that result in enhanced stationary-eddy heat flux and adiabatic warming over the ice-covered parts of the Arctic Ocean. The teleconnection between enhanced tropical convection and Arctic warming is, according to *Lee et al.* [2011], characterized by an increase in the frequency of 250 hPa streamfunction patterns resembling the positive phase of the Pacific–North American (PNA) pattern, including a wave train over the North Pacific and northwestern Canada. The effect of tropical oceans on Arctic warming was emphasized also by *Ding et al.* [2014].

On the basis of observations and ERA-Interim reanalysis, they concluded that the recent common occurrence of negative NAO phase (associated with high temperatures in Greenland and northeast Canada) is a response to anomalous Rossby wave train activity originating from a specific SST pattern in the tropical Pacific.

Perlwitz et al. [2015] addressed changes in the Arctic and midlatitudes in October–December between 2003 and 2012 and 1979 and 1988. According to their model experiments (applying a coarse vertical resolution), the sea-ice loss is the most important factor only for near-surface warming in the Arctic, but it can only explain some 20% of the warming of the 1000–500 hPa layer. For the latter, the most important contributors are decadal variations and SST changes outside the Arctic. Accordingly, these results do not support a strong causal linkage from Arctic sea-ice decline to weaker zonal winds. Instead, according to *Perlwitz et al.* [2015], the decadal change in midlatitude zonal winds at the 500 hPa level is mostly due to internal atmospheric variability and decadal ocean variability. However, this does not invalidate the suggested [*Francis and Vavrus*, 2012, 2015; *Overland et al.*, 2015] causal chain of events from a reduced temperature gradient to weaker westerly winds, potentially wavier jet stream, slower eastward motion of weather patterns, and more likely extreme weather events. Even if the Arctic amplification at higher altitudes is mostly due to increased heat transport from lower latitudes, it results in a weaker meridional temperature gradient, which feeds back to lower latitudes via the thermal wind effect.

2.4. DISCUSSION AND CONCLUSIONS

Changes in the meridional temperature gradient, jet stream properties, storm tracks, and blocking as well as wave trains and resonance are important for weather extremes (Fig. 2.3). These circulation features affect weather both in the Arctic and at lower latitudes and are affected by forcing from these regions, but the relative importance of forcing from the Arctic and lower latitudes is still poorly known. The Arctic Amplification interacts with sea ice and snow melt and increased transports from lower latitudes (and with several small-scale processes not shown in Fig. 2.3) and affects the jet stream properties, storm tracks, blocking, and Rossby waves, which further have a direct effect on the occurrence, strength, and duration of extreme weather events (Fig. 2.3). Other factors that have both direct and indirect effects on extreme events include global warming, stratospheric forcing, sea ice and snow melt in the Arctic, and surface thermal forcing at lower latitudes (Fig. 2.3), the latter including changes in SST and terrestrial snowpack as well as soil temperature and moisture.

Despite the challenges in data analyses and model experiments, a consensus is emerging on the following issues. First, a warmer Arctic, in particular the Barents and Kara seas, favors the occurrence of cold spells in East Asia [*Kim et al.*, 2014; *Sato et al.*, 2014; *Mori et al.*, 2014; *Kug et al.*, 2015; *Overland et al.*, 2015]. Second, the recent cold, snow-rich winters in North America are partly driven by the warmer Arctic [*Tang et al.*, 2013; *Francis and Vavrus*, 2015; *Francis and Skific*, 2015; *Kug et al.*, 2015], but during this century, the probability of cold winters is expected to gradually decrease in North America [*Screen et al.*, 2015a] and elsewhere in northern midlatitudes [*Petoukhov and Semenov*, 2010; *Deser et al.*, 2010; *Yang and Christensen*, 2012; *Peings and Magnusdottir*, 2014]. Third, changes in the high Arctic have probably less evident influence on northern Europe [*Overland et al.*, 2015], but summer temperature and precipitation in northern Europe seem to be affected by SST and sea-ice changes at latitudes around 60°N (Labrador Sea, Sea of Okhotsk, and Hudson Bay). It is relevant that surface temperature anomalies upstream of prevailing westerlies have a stronger effect on North European weather than anomalies upstream of more rarely occurring winds, such as northerly winds from the central Arctic.

The open questions and gaps in knowledge include, among others, the following. A major problem is that large-scale circulation responses to surface thermodynamic forcing are, in general, less direct and robust than thermodynamic responses [*Hoskins and Woollings*, 2015]. Regional and seasonal differences in the future evolution of the jet stream and blocking are a major unknown issue from the point of view of extreme weather events. This is associated with the uncertainty in the causal linkage from weaker zonal winds to a wavier jet stream with more frequent blockings. Considering individual extreme events, the jet stream has often been regarded as a cause and a midlatitude temperature extreme as a consequence. However, not only climatologically but also during an individual event, the strength and orientation of the jet stream depend on the temperature field. There is also a need to better understand the wave-guiding ability of the jet stream [*Manola et al.*, 2013]. Further, quantitative understanding of the role of cyclogenesis, cyclolysis, cyclone tracks, and their interaction with larger-scale conditions is rather limited, compared to the importance of cyclones on extreme weather. Moreover, surprisingly little attention has been put on the potential effects of the Pacific Decadal Oscillation (PDO) on changes in the Arctic. The importance of PDO to global climate is well known, and it includes signals from the Pacific to Europe [*Vihma et al.*, 2014b], potentially across the central Arctic [*Trenberth and Fasullo*, 2013].

An advance in the field of Arctic-midlatitude interactions requires a better understanding of the mechanisms of circulation response and reduction of model uncertainty. Considering understanding of the mechanisms by means of data analyses, the small signal-to-noise ratios in short time series make it difficult to distinguish between the contributions of Arctic thermodynamic forcing and inherent interannual and decadal variability of the atmosphere [*Hopsch et al.*, 2012; *Screen et al.*, 2013], in particular in cold seasons [*Coumou et al.*, 2015]. This calls for development of new metrics and methodologies suitable to analyze and attribute episodic events [*Overland et al.*, 2015; *Francis and Skific*, 2015]. Among promising computational methodologies is genetic programming [*Stanislawska et al.*, 2012].

In any case, model experiments are needed to support the interpretation of weak signals emerging from data analyses. To reduce the uncertainty of model experiments, there is an obvious need for more experiments with larger ensembles and a greater diversity of models. From the point of view of effects of surface thermal forcing, it is crucial that models have a high vertical resolution. The number of vertical levels of the order of 30 is not sufficient for successful simulations of boundary layer processes in the Arctic [*Byrkjedal et al.*, 2008], but most studies on Arctic-midlatitude linkages have applied an even smaller number; in the 34 models reviewed by *Gao et al.* [2015], the mean number of levels was only 24. In particular, in the case of reduced sea-ice cover, the increased surface heat fluxes have to work against the stable background stratification in the Arctic inversion layer, and a successful simulation of the vertical distribution of heat requires a high resolution [*Lüpkes et al.*, 2012]. The vertical distribution of heat is important, among other things, for the thermal wind effect on the zonal wind and for triggering of Rossby waves [*Hoskins and Karoly*, 1981]. Strong conclusions on the existence or nonexistence of effects of sea-ice decline have, however, been made on the basis of experiments applying a coarse vertical resolution.

With improved understanding of the complex, interactive physical processes, improved models, and more extensive multimodel experiments with large ensembles, it will be possible to improve seasonal weather forecasts for northern midlatitudes and the Arctic. Wintertime cold extremes still continue to be a major issue and, under warming climate, forecasting of summer heat waves will require increasingly more attention.

ACKNOWLEDGMENTS

The work was supported by the Academy of Finland (contract 283101). I thank Klaus Dethloff, Jennifer Francis, Edward Hanna, Richard Hall, Seong-Joong Kim, James Overland, James Screen, Ted Shepherd, and John Walsh for fruitful discussions.

REFERENCES

Barnes, E. A. (2013), Revisiting the evidence linking Arctic amplification to extreme weather in midlatitudes, *Geophys. Res. Lett., 40*; doi: 10.1002.grl.50880.

Barnes, E. A., and J. Screen (2015), The impact of Arctic warming on the midlatitude jet stream: Can it? Has it? Will it?, *WIREs Climate Change, 6*; doi: 10.1002/wcc.337.

Barnes, E. A., E. Dunn-Sigouin, G. Masato, and T. Woollings (2014), Exploring recent trends in Northern Hemisphere blocking, *Geophys. Res. Lett., 41*; doi: 10.1002/2013GL058745.

Barnes, E. A., J. Slingo, and T. Woollings (2012), A methodology for the comparison of blocking climatologies across indices, models and climate scenarios, *Climate Dyn., 38*, 2467–2481.

Byrkjedal, Ø., I. N. Esau, and N. G. Kvamstø (2008), Sensitivity of simulated wintertime Arctic atmosphere to vertical resolution in the ARPEGE/IFS model, *Climate Dyn., 30*, 687–701; doi: 10.1007/s00382-007-0316-z.

Cattiaux, J., R. Vautard, C. Cassou, P. Yiou, V. Masson-Delmotte, and F. Codron (2010), Winter 2010 in Europe: A cold extreme in a warming climate, *Geophys. Res. Lett., 37*, L20704.

Cavalieri, D. J., and C. L. Parkinson (2012), Arctic sea ice variability and trends, 1979–2010, *Cryosphere, 6*, 881–889.

Cohen, J., J. A. Screen, J. C. Furtado, M. Barlow, D. Whittleston, D. Coumou, J. Francis, K. Dethloff, D. Entekhabi, J. Overland, and J. Jones (2014), Recent Arctic amplification and extreme midlatitude weather, *Nat. Geosci., 7*, 627–637.

Coumou, D., and S. Rahmstorf (2012), A decade of weather extremes, *Nature Climate Change*; doi: 10.1038/NCLIMATE1452.

Coumou, D., J. Lehmann, and J. Beckmann (2015), The weakening summer circulation in the Northern Hemisphere midlatitudes, *Science, 348*, 324–327, doi: 10.1126/science.1261768.

Coumou, D., V. Petoukhov, S. Rahmstorf, S. Petri, and H. J. Schellnguber (2014), Quasi-resonant circulation regimes and hemispheric synchronization of extreme weather in boreal summer, *Proc. Nat. Acad. Sci.*, 12331–12336.

Cvijanovic, I., and K. Caldeira (2015), Atmospheric impacts of sea ice decline in CO2 induced global warming, *Climate Dyn., 44*, 1173–1186; doi: 10.1007/s00382-015-2489-1.

Davini, P. (2013), Atmospheric blocking and winter midlatitude climate variability, Università Ca' Foscari Venezia, http://hdl.handle.net/10579/2241.

Deser, C., R. Tomas, M. Alexander, and D. Lawrence (2010), The seasonal atmospheric response to projected Arctic sea ice loss in the late twenty-first century, *J. Climate, 23*, 333–351.

De Vries, H., R. J. Haarsma, and W. Hazeleger (2012), Western European cold spells in current and future climate, *Geophys. Res. Lett., 39*, L04706.

Ding, Q., J. M. Wallace, D. S. Battisti, E. J. Steig, A. J. E. Gallant, H.-J. Kim, and L. Geng (2014), Tropical forcing of the recent rapid Arctic warming in northeastern Canada and Greenland, *Nature, 509*, 209–212; doi: 10.1038/nature13260.

Dole, R., M. Hoerling, J. Perlwitz, J. Eischeid, P. Pegion, T. Zhang, X.-W. Quan, T. Xu, and D. Murray (2011), Was there a basis for anticipating the 2010 Russian heat wave?, *Geophys. Res. Lett., 38*, L06702; doi: 10.1029/2010GL046582.

Donat, M. G., L. V. Alexander, H. Yang, I. Durre, R. Vose, R. J. H. Dunn, K. M. Willett, E. Aguilar, M. Brunet, J. Caesar, B. Hewitson, C. Jack, A. M. G. Klein Tank, A. C. Kruger, J. Marengo, T. C. Peterson, M. Renom, C. Oria Rojas, M. Rusticucci, J. Salinger, A. S. Elrayah, S. S. Sekele, A. K. Srivastava, B. Trewin, C. Villarroel, L. A. Vincent, P. Zhai, X. Zhang, and S. Kitching (2013), Updated analyses of temperature and precipitation extreme indices since the beginning of the twentieth century: The HadEX2 dataset, *J. Geophys. Res. Atmos.*, *118*, 2098–2118; doi: 10.1002/jgrd.50150.

Döscher, R., T. Vihma, and E. Maksimovich (2014), Recent advances in understanding the Arctic climate system state and change from a sea ice perspective: A review, *Atmos. Chem. Phys.*, *14*, 13571–13600; doi: 10.5194/acp-14-13571-2014.

Dunn-Sigouin, E., and S.-W. Son (2013), Northern Hemisphere blocking frequency and duration in the CMIP5 models, *J. Geophys. Res. Atmos.*, *118*, 1179–1188.

Esau, I., and S. Zilitinkevich (2010), On the role of the planetary boundary layer depth in climate system, *Adv. Sci. Res.*, *4*, 63–69.

Francis, J. A., and N. Skific (2015), Evidence linking rapid Arctic warming to mid-latitude weather patterns, *Phil. Trans. R. Soc. A*, *373*; doi:10.1098/rsta.2014.0170.

Francis, J. A., and S. J. Vavrus (2012), Evidence linking Arctic amplification to extreme weather in midlatitudes, *Geophys. Res. Lett.*, *39*, L06801.

Francis, J. A. and S. J. Vavrus (2015), Evidence for a wavier jet stream in response to rapid Arctic warming, *Environ. Res. Lett.*, *10*; doi:10.1088/1748-9326/10/1/014005.

Franzke, C. L. (2013), Persistent regimes and extreme events of the North Atlantic atmospheric circulation, *Philosophical Transactions of the Royal Society of London A: Mathematical, Physical and Engineering Sciences*, *371*, 20110471.

Gao, Y., J. Sun, F, Li, S. He, S. Sandven, Q. Yan, Z. Zhang, K. Lohmann, N. Keenlyside,, T. Furevik,, and L. Suo (2015), Arctic sea ice and Eurasian climate: A review, *Adv. Atmos. Sci.*, *32*, 92–114.

Graversen, R. G., T. Mauritsen, M. Tjernström, E. Källen, and G. Svensson (2008), Vertical structure of recent Arctic warming, *Nature*, *451*, 53–56.

Guo, D., Y. Gao, I. Bethke, D. Gong, O. M. Johannessen, and H. Wang (2014), Mechanism on how the spring Arctic sea ice impacts the East Asian summer monsoon, *Theor. Appl. Climatol.*, *115*, 107–119; doi: 10.1007/s00704-013-0872-6.

Hall, R., R. Erdélyi, E. Hanna, J. M. Jones, and A. A. Scaife (2015), Drivers of North Atlantic Polar Front jet stream variability, *Int. J. Climatol.*, *35*, 1697–1720; doi: 10.1002/joc.4121.

Hamilton, L. C., and M. Lemcke-Stampone (2014), Arctic warming and your weather: public belief in the connection, *Int. J. Climatol.*, *34*, 1723–1728; doi: 10.1002/joc.3796.

Hanna, E., T. E. Cropper, P. D. Jones, A. A. Scaife, and R. Allan (2014b), Recent seasonal asymmetric in the NAO (a marked summer decline and increased winter variability) and associated changes in the AO and Greenland Blocking Index, *Int. J. Climatol.*, *34*; doi:10.1002/joc.4157.

Hanna, E., X. Fettweis, S. H. Mernild, J. Cappelen, M. H. Ribergaard, C. A. Shuman, K. Steffen, L. Wood, and T. L. Mote (2014a), Atmospheric and oceanic climate forcing of the exceptional Greenland ice sheet surface melt in summer 2012, *Int. J. Climatol.*, *34*, 1022–1037; doi:10.1002/joc.3743.

Held, I. (1993), Large scale dynamics and global warming., *Bull. Amer. Meteor. Soc.*, *74*, 228–241.

Hopsch, S., J. Cohen, and K. Dethloff (2012), Analysis of a link between fall Arctic sea ice concentration and atmospheric patterns in the following winter, *Tellus A*, *64*, 18624.

Hori, M.E., et al. "Recurrence of intraseasonal cold air outbreak during the 2009/2010 winter in Japan and its ties to the atmospheric condition over the Barents-Kara Sea." *SOLA 7* (2011): 25–28.

Hoskins, B., and T. Woollings (2015), Persistent extratropical regimes and climate extremes, *Curr. Climate Change Rep.*; doi: 10.1007/s40641-015-0020-8.

Hoskins, B. J., and D. J. Karoly (1981), The steady linear response of a spherical atmosphere to thermal and orographic forcing, *J. Atmos. Sci.*, *38*, 1179–1196.

Huntingford, C., T. Marsh, A. A. Scaife, E. J. Kendon, J. Hannaford, A. L. Kay, M. Lockwood, C. Prudhomme, N. S. Reynard, S. Parry, J. A. Lowe, J. A. Screen, H. C. Ward, M. Roberts, P.A. Stott, V.A. Bell, M. Bailey, A. Jenkins, T. Legg, F. E. L. Otto, N. Massey, N. Schaller, J. Slingo, and M. R. Allen (2014), Potential influences in the United Kingdom's floods of winter 2013/14, *Nature Climate Change*, *4*, 769–777.

Jaiser, R., K. Dethloff, and D. Handorf (2013), Stratospheric response to Arctic sea ice retreat and associated planetary wave propagation changes, *Tellus*, *65A*, 19375; doi: 10.3402/tellusa.v65i0.19375.

Kapsch, M. L., R. G. Graversen, and M. Tjernström (2013), Springtime atmospheric energy transport and the control of Arctic summer sea-ice extent, *Nature Climate Change*, *3*, 744–748.

Kim, B.-M., S.-W. Son, S.-K. Min, J.-H. Jeong, S.-J. Kim, X. Zhang, T. Shim, and J.-H. Yoon (2014), Weakening of the stratospheric polar vortex by Arctic sea-ice loss, *Nature Communications*, *5*, 4646; doi:10.1038/ncomms5646.

Kolstad, E. W., and T. J. Bracegirdle (2008), Marine cold-air outbreaks in the future: an assessment of IPCC AR4 model results for the Northern Hemisphere, *Climate Dyn.*, *30*, 871–885.

Kug, J.-S., J.-H. Jeong, Y.-S. Jang, B.-M. Kim, C. K. Folland, S-K. Min, and S.-W. Son (2015), Two distinct influences of Arctic warming on cold winters over North America and East Asia, *Nat. Geosci.*, *8*, 759–762; doi:10.1038/ngeo2517.

Kwok, R., and D. A. Rothrock (2009), Decline in Arctic sea ice thickness from submarine and ICESat records: 1958–2008, *Geophys, Res, Lett.*, *36*, L15501.

Kyselý, J. (2008), Influence of the persistence of circulation patterns on warm and cold temperature anomalies in Europe: analysis over the 20th century, *Global Planet. Change*, *62*, 147–163.

Lee, S., T. Gong, N. Johnson, S. B. Feldstein, and D. Pollard (2011), On the possible link between tropical convection and the Northern Hemisphere Arctic surface air temperature change between 1958 and 2001, *J. Climate*, *24*, 4350–4367.

Lu, J., and M. Cai (2008), A new framework for isolating individual feedback processes in coupled general circulation climate models, Part I: Formulation, *Climate Dyn.*, *32*, 873–885.

Lüpkes, C., T. Vihma, G. Birnbaum, S. Dierer, T. Garbrecht, V.M. Gryanik, M. Gryschka, J. Hartmann, G. Heinemann, L. Kaleschke, S. Raasch, H. Savijärvi, K.H. Schlünzen, and U. Wacker (2012), Mesoscale modeling of the Arctic atmospheric boundary layer and its interaction with sea ice, in *Arctic Climate Change: The ACSYS Decade and Beyond*, edited by P. Lemke and H.-W. Jacobi, Atmospheric and Oceanographic Sciences Library, vol. *43*; doi: 10.1007/978-94-007-2027-5.

Manola, I., F. Selten, H. Vries, and W. B. Hazeleger (2013), Waveguidability of idealized jets, *J. Geophys. Res. Atmos.*, *118*, 10–432.

Masato, G., B. J. Hoskins, and T. Woollings (2013), Winter and summer Northern Hemisphere blocking in CMIP5 models, *J. Climate*, *26*, 7044–7059.

Masato, G., T. Woollings, and B. J. Hoskins (2014), Structure and impact of atmospheric blocking over the Euro-Atlantic region in present day and future simulations, *Geophys. Res. Lett.*, *41*, 1051–1058.

Matthes, H., A. Rinke, and K. Dethloff (2015), Recent changes in Arctic temperature extremes: warm and cold spells during winter and summer, *Environ. Res. Lett.*, *10*, 114020; doi:10.1088/1748-9326/10/11/114020.

McAfee, S. A., J. E. Walsh, and T. S. Rupp (2013), Statistically downscaled projections of snow/rain partitioning for Alaska, *Hydrolog. Processes*, *28*, 3930–3946; doi: 1844 10.1002/hyp.9934.

Mori, M., M. Watanabe, H. Shiogama, J. Inoue, and M. Kimoto (2014), Robust Arctic sea-ice influence on the frequent Eurasian cold winters in past decades, *Nat. Geosci.*, *7*, 869–873; doi:10.1038/NGEO2277.

Newson, R. L. (1973), Response of a general circulation model of the atmosphere to removal of the Arctic ice-cap, *Nature*, *241*, 39–40.

Overland, J., E. Hanna, I. Hanssen-Bauer, S.-J. Kim, J. Walsh, M. Wang, U. S. Bhatt (2014), Air temperature (in Arctic Report Card 2014), http://www.arctic.noaa.gov/reportcard

Overland, J., J. Francis, R. Hall, E. Hanna, S.-J. Kim, and T. Vihma (2015), The Melting Arctic and Midlatitude Weather Patterns: Are they connected? *J. Climate*, *28*, 7917–7932; doi: 10.1175/JCLI-D-14-00822.1.

Overland, J., J. Key, B.-M. Kim, S.-J. Kim, Y. Liu, J. Walsh, M. Wang, and U. Bhatt (2012), Air temperature, atmospheric circulation and clouds (in Arctic Report Card 2012), http://www.arctic.noaa.gov/reportcard

Overland, J., M. Wang, and J. Walsh (2009), Atmosphere (in Arctic Report Card 2009), http://www.arctic.noaa.gov/reportcard.

Overland, J., M. Wang, and J. Walsh (2010), Atmosphere (in Arctic Report Card 2010), http://www.arctic.noaa.gov/reportcard.

Overland, J. E., and M. Wang (2010), Large-scale atmospheric circulation changes are associated with the recent loss of Arctic sea ice, *Tellus A.*, *62*, 1–9.

Overland, J. E., K. R. Wood, and M. Wang (2011), Warm Arctic–cold continents: Impacts of the newly open Arctic Sea, *Polar Res.*, *30*, 15787; doi: 10.3402/polar.v30i0.15787.

Peings, Y., and G. Magnusdottir (2014), Response of the wintertime Northern Hemisphere atmospheric circulation to current and projected Arctic sea ice decline: a numerical study with CAM5, *J. Climate;* doi:10.1175/JCLI-D-13-00272.1.

Perlwitz, J., M. Hoerling, and R. Dole (2015), Arctic tropospheric warming: Causes and linkages to lower latitudes, *J. Climate*, *28*, 2154–2167; doi: 10.1175/JCLI-D-14-00095.1.

Peterson, T. C., M. P. Hoerling, P. A. Stott and S. Herring, Eds. (2013), Explaining extreme events of 2012 from a climate perspective, *Bull. Amer. Meteor. Soc.*, *94*, S1–S74.

Petoukhov, V., and V. A. Semenov (2010), A link between reduced Barents-Kara sea ice and cold winter extremes over northern continents, *J. Geophys. Res.*, *115*, D21111; doi:10.1029/2009JD013568.

Petoukhov, V., S. Rahmstorf, S. Petri, and H. J. Schellnhuber (2013), Quasiresonant amplification of planetary waves and recent Northern Hemisphere weather extremes, *Proc. Natl. Acad. Sci.*, *110*, 5336–5341.

Pithan, F., and T. Mauritsen (2014), Arctic amplification dominated by temperature feedbacks in contemporary climate models, *Nat. Geosci.*, *7*, 181–184; doi:10.1038/ngeo2071.

Polyakov, I. V., L. A. Timokhov, V. A. Alexeev, S. Bacon, I. A. Dmitrenko, L. Fortier, and J. Toole (2010), Arctic ocean warming contributes to reduced polar ice cap, *J. Phys. Oceanogr.*, *40*, 2743–2756.

Sato, K., J. Inoue, and M. Watanabe (2014), Influence of the Gulf Stream on the Barents Sea ice retreat and Eurasian coldness during early winter, *Environ. Res. Lett.*, *9*, 084009.

Schneider, T., T. Bischoff, and H. Płotka (2015), Physics of changes in synoptic midlatitude temperature variability, *J. Climate*, *28*, 2312–2331.

Screen, J. A. (2013), Influence of Arctic sea ice on European summer precipitation, *Environ. Res. Lett.*, *8*, 044015.

Screen, J. A. (2014), Arctic amplification decreases temperature variance in northern mid- to high-latitudes, *Nature Climate Change*, *4*, 577–582.

Screen, J. A., and I. Simmonds (2014), Amplified midlatitude planetary waves favour particular regional weather extremes, *Nature Climate Change*, *4*, 704–709.

Screen, J. A., C. Deser, and I. Simmonds (2012), Local and remote controls on observed Arctic warming, *Geophys. Res. Lett.*, *39*, L10709; doi:10.1029/2012GL051598.

Screen, J. A., C. Deser, and L. Sun (2015a), Reduced risk of North American cold extremes due to continued Arctic sea ice loss, *Bull. Amer. Meteorol. Soc.;* doi:10.1175/BAMS-D-14-00185.1.

Screen, J. A., C. Deser, and L. Sun (2015b), Projected changes in regional climate extremes arising from Arctic sea ice loss, *Environ. Res. Lett.*, *10*, 084006.

Screen, J. A., I. Simmonds, C. Deser, and R. Tomas (2013), The atmospheric response to three decades of observed Arctic sea ice loss, *J. Climate*, *26*, 1230–1248.

Seo, K.-H., J.-H. Son, S.-E. Lee, T. Tomita, and H.-S. Park (2012), Mechanisms of an extraordinary East Asian summer monsoon event in July 2011, *Geophys. Res. Lett.*, *39*, L05704; doi:10.1029/2011GL050378.

Serreze, M. C., A. P. Barrett, and J. J. Cassano (2011), Circulation and surface controls on the lower tropospheric temperature field of the Arctic, *J. Geophys. Res.*, *116*, D07104.

Serreze, M. C., A. P. Barrett, J. C. Stroeve, D. N. Kindig, and M M. Holland (2009), The emergence of surface-based Arctic amplification, *Cryosphere*, *3*, 11–19.

Simmonds, I., and I. Rudeva (2012), The great Arctic cyclone of August 2012, *Geophys. Res. Lett.*, *39*, L23709; doi:10.1029/2012GL054259.

Stanislawska, K., K. Krawiec, and Z. W. Kundzewicz (2012), Modelling global temperature changes with genetic programming, *Computer & Mathematics with Applications*, *64*, 3717–3728.

Tang, Q., X. Zhang, and J. A. Francis (2014), Extreme summer weather in Northern midlatitudes linked to a vanishing cryosphere, *Nature Climate Change*, *4*, 45–50; doi:10.1038/nclimate2065.

Tang, Q., X. Zhang, X. Yang, and J. A. Francis (2013), Cold winter extremes in northern continents linked to Arctic sea ice loss, *Environ. Res. Lett.*, *8*, 014036.

Trenberth, K. E., and J. T. Fasullo (2013), An apparent hiatus in global warming?, *Earth's Future*; doi:10.1002/2013EF000165.

Trenberth, K. E., J. T. Fasullo, and T. G. Shepherd (2015), Attribution of climate extreme events, *Nature Climate Change*, *5*, 725–730; doi:10.1038/nclimate2657.

Uotila, P., A. Karpechko, and T. Vihma (2014), Links between Arctic sea ice and extreme summer precipitation in China: An alternative view, *Adv. Polar Sci.*, *25*, 222–233; doi:10.13679/j.advps.2014.4. 00222.

Vihma, T. (2014), Effects of Arctic sea ice decline on weather and climate: a review, *Surv. Geophys.*, *35*, 1175–1214, DOI 10.1007/s10712-014-9284-0.

Vihma, T., B. Cheng, P. Uotila, L. Wei, and T. Qin (2014b), Linkages between Arctic sea ice cover, large-scale atmospheric circulation, and weather and ice conditions in the Gulf of Bothnia, Baltic Sea, *Adv. Polar Sci.*, *25*, 289–299; doi:10.13679/j.advps.2014.4.00289.

Vihma, T., J. Jaagus, E. Jakobson, and T. Palo (2008), Meteorological conditions in the Arctic Ocean in spring and summer 2007 as recorded on the drifting ice station Tara, *Geophys. Res. Lett.*, *35*, L18706; doi: 10.1029/2008GL034681.

Vihma, T., J. Screen, M. Tjernström, B. Newton, X. Zhang, V. Popova, C. Deser, M. Holland, and T. Prowse (2016), The atmospheric role in the Arctic water cycle: A review on processes, past and future changes, and their impacts, *J. Geophys. Res. Biogeosci.*, *121*; doi:10.1002/2015JG003132.

Vihma, T., R. Pirazzini, I. Fer, I. A. Renfrew, J. Sedlar, M. Tjernström, C. Lüpkes, T. Nygård, D. Notz, J. Weiss, D. Marsan, B. Cheng, G. Birnbaum, S. Gerland, D. Chechin, and J. C. Gascard (2014a), Advances in understanding and parameterization of small-scale physical processes in the marine Arctic climate system: A review, *Atmos. Chem. Phys.*, *14*, 9403–9450; doi:10.5194/acp-14-9403-2014.

Walsh, J. E. (2014), Intensified warming of the Arctic: Causes and impacts on middle latitudes, *Glob. Planet Change*, *117*, 52–63; doi:10.1016/j.gloplacha.2014.03.003.

Warshaw, M., and R. P. Rapp (1973), An experiment on the sensitivity of a global circulation model, *J. Appl. Meteor.*, *12*, 43–49.

Woollings, T., A. Hannachi, and B. Hoskins (2010), Variability of the North Atlantic eddy-driven jet stream, *Quart. J. Roy. Meteor. Soc.*, *136*, 856–868; doi:10.1002/qj.625.

Wu, B., R. Zhang, B. Wang, and R. D'Arrigo (2009), On the association between spring Arctic sea ice concentration and Chinese summer rainfall, *Geophys. Res. Lett.*, *36*, L09501.

Yang, S., and J. H. Christensen (2012), Arctic sea ice reduction and European cold winters in CMIP5 climate change experiments, *Geophys. Res. Lett.*, *39*, L20707.

Zahn, M. and H. von Storch (2010), Decreased frequency of North Atlantic polar lows associated with future climate warming, *Nature*, *467*, 309–312.

Zhao, P., X. Zhang, X. Zhou, M. Ikeda, and Y. Yin (2004), The sea ice extent anomaly in the North Pacific and its impact on the East Asian summer monsoon rainfall, *J. Climate*, *17*, 3434–3447.

3

Impact of Aerosols on Regional Changes in Climate Extremes

Jana Sillmann[1] and Jinho Yoon[2]

ABSTRACT

This chapter provides a short introduction of aerosol effects on radiation and clouds and their impacts on climate by alternating surface temperatures and affecting the hydrological cycle through changes in the atmospheric and surface energy budgets. Some examples of aerosol impacts on regional climate aspects, such as monsoon, tropical storms and the Artic are provided and mitigation scenarios for emission reductions of aerosols are discussed. The chapter will further outline current knowledge and modelling efforts regarding aerosol impacts on temperature and precipitation extremes and discuss future research needs.

3.1. INTRODUCTION

Aerosols compose an important component of the climate system through their complex and multiple interactions with clouds and radiation. In general, the term *aerosol* refers to airborne liquid or solid particles of various size, shape and composition, which originate from natural or anthropogenic sources. In contrast to greenhouse gases (GHGs), such as carbon dioxide (CO_2), aerosols are not homogeneous distributed in the troposphere. Aerosols have a very short lifetime (one day to a couple of weeks), which strongly determines their temporal and spatial effect on weather and climate [*Brasseur et al.*, 2003]. Aerosols can affect the climate directly and indirectly through diverse processes and contribute the largest uncertainty to estimates and interpretations of changes in the Earth's energy budget [*Boucher et al.*, 2013]. In general, aerosols act on various scales in the climate system, from very small-scale processes in terms of cloud microphysics to large-scale processes in terms of affecting surface albedo or surrounding air masses. For instance, aerosols play a critical role in the formation and development of liquid and ice clouds, and affect precipitation by changing cloud microphysics. Aerosols deposited on snow and glaciers can alter the surface albedo and accelerate the melting of snow or glacier. Depending on the type, aerosols can have a significant cooling or warming effect and influence regional and global climate dynamics both with respect to global warming and the hydrological cycle [e.g., *Ramanathan et al.*, 2001; *Kaufmann et al.*, 2002; *Ramanathan and Feng*, 2009; *Fiore et al.*, 2012].

Anthropogenic aerosol emissions interact with natural continental and maritime sources of aerosols, such as sea salt, dust, and pollen [e.g., *de Gouw et al.*, 2005; *Stier et al.*, 2006; *Chernoff and Bertram*, 2010]. The distribution and composition of atmospheric aerosols change as a consequence of increasing urbanization, societal changes in land and water use as well as other societal developments [*Streets et al.*, 2004; *Moss et al.*,

[1] *Center for International Climate and Environmental Research, Oslo; Cicero Senter for Klimaforskning, Oslo, Norway*

[2] *School of Earth Sciences and Environmental Engineering, Gwangju Institute of Science and Technology, Repulic of Korea*

Climate Extremes: Patterns and Mechanisms, Geophysical Monograph 226, First Edition.
Edited by S.-Y. Simon Wang, Jin-Ho Yoon, Christopher C. Funk, and Robert R. Gillies.

2010]. Evidence is growing that anthropogenic aerosol emissions can have both local and global impacts [e.g., *Meehl et al.* 2008; *Wang et al.*, 2009; *Bollasina et al.*, 2011]. Increasing anthropogenic aerosol emissions from Asia, for instance, may have regionally counteracted warming induced by increasing greenhouse gases [e.g., *Kauffman et al.*, 2011]. Since these anthropogenic aerosols inadvertently emitted as part of fossil fuel burning are also associated with adverse health effects, they become more and more subject to policy regulations. However, deliberate emission of aerosols into the atmosphere as geoengineering strategy to counteract global warming is currently also under discussion [e.g., *Rasch et al.*, 2008; *Sillmann et al.*, 2015; *Kristjánsson et al.*, 2015].

This chapter provides a short introduction on direct and indirect aerosol effects and their impact on regional climate by alternating surface temperatures and affecting the hydrological cycle through changes in the atmospheric and surface energy budgets. Examples of regional impacts of aerosols on climate and human health will be discussed. The chapter further outlines current knowledge and modeling efforts regarding the impact of aerosols on climate extremes, in particular temperature and precipitation extremes, and discusses future research needs.

3.2. DIRECT AND INDIRECT EFFECTS OF AEROSOLS ON CLOUDS AND RADIATION

Aerosol forcing in the climate system can be divided into two main components, the direct and indirect effects (Fig. 3.1), also referred to as aerosol-radiation interaction and aerosol-cloud interaction [*Boucher et al.*, 2013]. Direct effects of aerosols are related to the absorption and scattering of shortwave and longwave radiation. The scattering of solar radiation by aerosols results in the cooling of the Earth's surface, the absorption leads to a warming of the surrounding atmosphere. The scattering and absorption of solar radiation further causes dimming at the surface [e.g., *Wild et al.*, 2005], which reduces the amount of solar energy available for evaporation, and consequently leads to a slow-down of the hydrological cycle [e.g., *Ramanathan et al.*, 2001; *Hartmann et al.*, 2013].

Direct effects of aerosols on the climate system are much better understood and quantified than the indirect effects [*Boucher et al.*, 2013]. Indirect effects comprise the impact of aerosols on cloud microphysics, such as changes in radiative properties of clouds or their frequency and lifetime. In particular, aerosols can act as nuclei on which water condenses to form cloud droplets that will eventually precipitate. A high concentration of aerosols in the atmosphere leads to clouds that contain many more and, hence, smaller water droplets than clouds with the same water content formed in cleaner air. Clouds with more droplets are more reflective (i.e., have a higher albedo), which induces a negative radiative forcing, and is described as the first indirect, the cloud-albedo, or the Twomey effect [*Twomey*, 1977]. Radiative forcing is generally defined at the top of the atmosphere [*Forster et al.*, 2007].

The second indirect effect, or more commonly used the "cloud lifetime effect," refers to the effect of aerosols on the height, cover, and water content of a cloud. Finally, the absorption of solar radiation by some aerosols modifies the air temperature, its relative humidity. and the vertical stability of the surrounding air column. This affects the formation and lifetime of clouds, and is referred to as

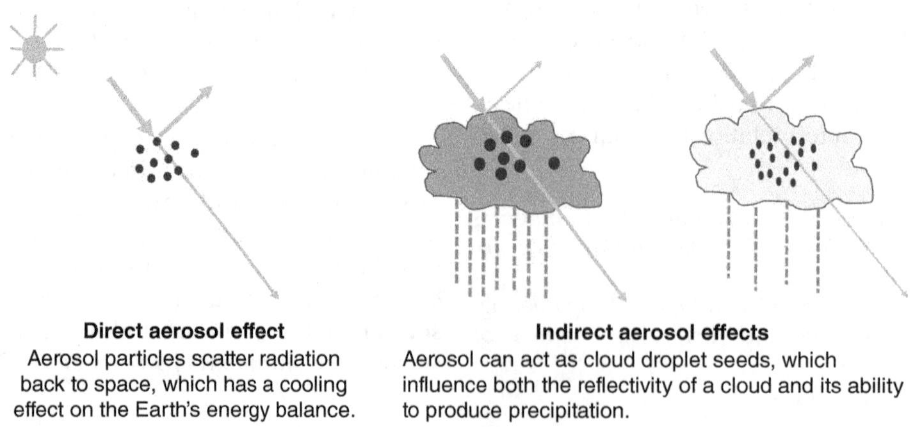

Figure 3.1 Direct and indirect effects of aerosols on radiation and clouds. Simplified schematic illustration adapted and redrawn from http://caice.ucsd.edu/; see *Boucher et al.* [2013] for more detailed illustration and discussion.

the semidirect effect of aerosols [e.g., *Hansen et al.*, 1997; *Ackerman et al.*, 2000; *Koch and DelGenio*, 2010].

Observational studies have been crucial in the progress made on understanding aerosol effects. State-of-the-art global climate models (GCMs) remain an essential tool for describing aerosol effects on the climate, although the direct and indirect effects of aerosol forcing still constitute a large uncertainty in climate model simulations of the radiative forcing [e.g., *Boucher et al.*, 2013]. In current climate system modeling studies, the three main types of aerosols mostly considered are sulphate (SO_4), black carbon (BC). and organic carbon (OC). These aerosols arise from different natural and anthropogenic sources and can affect radiation and clouds in different ways [see, for instance, *Brasseur et al.*, 2003]. The main effect of SO_4 aerosols is the scattering of incoming solar radiation back to space through direct and indirect aerosol effects for a cooling effect [*Myhre et al.*, 2013]. This induces a negative radiative forcing, but with highly heterogeneous distribution because of regional variations in the concentration of SO_4 aerosols in the atmosphere [*Boucher et al.*, 2013].

By contrast, the main effect of BC is its strong absorption of solar radiation, which tends to warm the surrounding air mass and, hence, induces a positive radiative forcing through the direct aerosol effect. An additional small positive forcing is caused by the deposit of black carbon on snow, which reduces the albedo of the snow. Furthermore, absorbing aerosols, such as BC, can modify cloud distribution by suppressing convective activity and thus influence hydrology [*Rosenfeld et al.*, 2008]. Relevant mechanisms related to aerosols, which may affect precipitation, are discussed in *Khain* [2009], who proposes a classification scheme of aerosol effects on precipitation.

In general, precipitation changes induced by a forcing agent can be understood from energy budget analysis of the atmosphere [*Allen and Ingram*, 2002; *O'Gorman et al.*, 2012; *Pendergrass and Hartmann*, 2014]. In this respect, the precipitation change can be divided into a fast precipitation and a slow precipitation change [*Bala et al.*, 2010]. The slow precipitation change is a response to surface temperature change from the forcing agent and the fast precipitation change is found to depend on the atmospheric absorption [*Andrews et al.*, 2010; *Kvalevåg et al.*, 2013]. BC has a strong atmospheric absorption and reduces the precipitation on a fast timescale, whereas SO_4 aerosols impact the atmospheric absorption less than most other forcing agents, such as CO_2 [*Andrews et al.*, 2010; *Kvalevåg et al.*, 2013; *Ming et al.*, 2010; *Pendergrass and Hartmann*, 2012].

Model simulations have further shown that the altitude at which the absorbing aerosols interact with the clouds plays an important role in their effect on precipitation [*Ban-Weis et al.*, 2012; *Samset and Myhre*, 2015]. Primary OC aerosols are mainly emitted together with BC and have approximately the same lifetime, but OC scatters sunlight to a much larger extent than BC. OC has, therefore, a cooling effect, which can partly cancel out the warming effect of BC depending on the emission source. OC can also form as secondary organic aerosol [*Kanakidou et al.*,, 2005].

3.3. AEROSOL IMPACT ON REGIONAL CLIMATE CHANGE

3.3.1. Monsoon Climate and Tropical Storms

The impact of the direct and indirect effects of anthropogenic aerosols will be felt most notably in regions of high emission, such as Southeast Asia, where regional anthropogenic aerosol burdens can lead to changes in the monsoon circulation [e.g., *Bollasina et al.*, 2011; *Ganguly et al.*, 2012] and a weakening of the hydrological cycle resulting in a reduction of the availability of fresh water [e.g., *Ramanathan et al.*, 2001; *Hartmann et al.*, 2013]. There are a couple of plausible hypotheses proposed to explain the potential role of aerosols affecting monsoon climate, especially hydrological cycle. Some of the hypotheses are complementary or contradictory to each other. For example, the Elevated Heat Pump (EHP) hypothesis by *Lau and Kim* [2006] was debated by *Nigam and Bollasina* [2010] and *Lau and Kim* [2011]. It is difficult to draw a clear conclusion due to uncertainty in both satellite observations and climate models. Another important aspect is to understand the impact of natural aerosols such as dust and sea salt on the monsoon climate. For example, *Vinoj et al.* [2014] demonstrated that dust aerosols over the Arabian Peninsula and the Indian Ocean can lead to stronger monsoon rain based on both observations and climate models. Recent studies also propose a potential impact of aerosols on tropical storms, both on the intensity and size of the storm [*Evan et al.*, 2011; *Wang et al.*, 2012; *Evan et al.*, 2012; *Rosenfeld et al.*, 2012; *Hazra et al.*, 2013; *Herbener et al.*, 2014; *Wang et al.*, 2014].

3.3.2. Arctic

Black carbon has also received significant attention for its potentially strong impact on the Arctic, where it leads to increased temperatures and accelerated ice and snow melt [*AMAP*, 2011; *Flanner*, 2013; *Sand et al.*, 2015]. A recent modeling study by *Chuwah et al.* [2016] has found that BC and SO_4 can have opposite effects on surface temperature in the Arctic. In particular, annual mean surface temperatures in the Arctic can increase by up to 1°C if SO_4 is decreased as in *Chuwah et al.* [2016], who used scenarios in which SO_4 concentrations, for instance, in eastern China are reduced by up to about 5 μg m^{-3}. In contrast, reductions in BC can result in surface cooling of the Arctic, which can locally be as high as 0.5°C in the annual mean and 1°C in winter.

The impact of aerosol emissions is not limited to the region where they are emitted. Studies show that long-range transport from Europe, Russia, and North America presently dominate the Arctic BC concentration at low altitudes, while emissions from Asia reach the Arctic at high altitudes [*Stohl et al.*, 2006; *Shindell et al.*, 2008; *Lund et al.*, 2014]. Even without reaching the Arctic region, aerosols can impose a significant impact on Arctic climate through large-scale teleconnections. For instance, local heating of BC emissions, with associated changes in atmospheric stability, can affect poleward transport patterns of moisture and energy through latent heat. Several studies [*Shindell et al.*, 2012a; *Sand et al.*, 2013] point to a significant sensitivity of Arctic temperature response to BC radiative forcing exerted at lower latitudes.

3.4. MITIGATION SCENARIOS FOR AEROSOL EMISSIONS

In addition to their climate impact, anthropogenic aerosols cause also a variety of adverse health impacts [e.g., *WHO* 2003; *Silva et al.*, 2013] that have motivated policy regulations in many regions with high aerosol emissions in the past, which consequently will lead or have already led to decreasing aerosol burdens in the affected regions. Politically, a reduction of local or regional air pollution (i.e., aerosols) may be particularly promising because of large cobenefits in terms of improved air quality and because the benefits of mitigating aerosol emissions are seen more quickly than those for CO_2 mitigation [*Shindell et al.*, 2012b; *Sand et al.*, 2015]. A realistic assessment of future climate change will thus rely on our capability to simulate trends in both GHG concentrations and aerosol emissions and understand the combined impact of GHGs and aerosols on climate [*Kloster et al.*, 2010].

Therefore, scenarios have been developed to study the combined mitigation effect. For instance, the European Union ENSEMBLES project has developed an aggressive mitigation scenario known as E1 [*Lowe et al.*, 2009; *Johns et al.*, 2011], which has a lower sulfur burden based on mitigation policies compared to the A1B scenario of the Special Report on Emission Scenarios (SRES) [*Nakicenovic et al.*, 2000]. Other scenario studies [*Kloster et al.*, 2010; *Kulmala et al.*, 2011] based on an aerosol emission inventory provided by IIASA (International Institute for Applied System Analysis) [*Cofala et al.*, 2007] consider two possible future developments: current legislation (CLE) and most feasible reduction (MFR). These scenarios include anthropogenic SO_4, BC, and OC emissions for present day (2000) and a future scenario for 2030 based on the B2 SRES scenario. *Chuwah et al.* [2016] explore HIGH and LOW aerosol-forcing scenarios using aerosol concentrations based on a set of scenarios similar to the most recent representative concentration pathways emission scenario, RCP6.0 [*Masui et al.*, 2011], developed using the IMAGE integrated assessment model [*Bouwman et al.*, 2006].

3.5. AEROSOL EFFECT ON TEMPERATURE AND PRECIPITATION EXTREMES

Climate extremes, such as very warm days and nights, extreme precipitation amounts, and dry spells, will intensify under increased greenhouse gas concentrations as projected by large ensembles of GCMs [*Collins et al.*, 2013; *Sillmann et al.*, 2013a]. Extreme precipitation, for instance, will increase about three times as much as mean precipitation under different GHG emission scenarios [*Sillmann et al.*, 2013a]. Considerable uncertainty exists in the estimated multimodel-average increase in annual extreme precipitation, which ranges from 4 to 10% K^{-1} according to *Kharin et al.* [2013]. The scaling of extreme precipitation with temperature may be much more complex than is implied by the Clausius-Clapeyron relationship, with considerable regional patterns due to various dynamic and thermodynamic mechanisms [e.g., *Westra et al.*, 2012; *Prein et al.*, 2015]. In particular, the effect of inhomogeneous aerosol burdens in the atmosphere can lead to regionally diversified patterns of changes in mean and extreme precipitation [e.g., *Pall et al.*, 2007; *Chen et al.*, 2011].

A decrease in aerosol emissions worldwide as prescribed by the MFR scenario (see section 3.3) compared to present-time climate will result in changing precipitation patterns and increasing temperatures everywhere in the world including Europe (see Figs. 2c and 3 in *Kloster et al.*, 2010]. *Sillmann et al.* [2013b], using the same scenario, show that global aerosol reduction greatly enforces the general warming effect due to greenhouse gases and results in significant increases of temperature and precipitation extremes in Europe. This is represented in the significant increase of temperature extremes, as represented by the climate extremes indices hottest summer day temperature (TXx), tropical nights (TR), and coldest winter night (TNn) shown in Figure 3.2. In the MFR scenario, a stronger increase in these indices is found compared to the other scenarios CLEmod and GHG. The warming pattern across Europe, however, depends strongly on the season and index under consideration. TXx and TR increase predominately in the Mediterranean region, whereas the TNn increase in central and northern Europe with highest rates in northeastern Europe. In CLEmod and GHG, TNn and TR show similar patterns

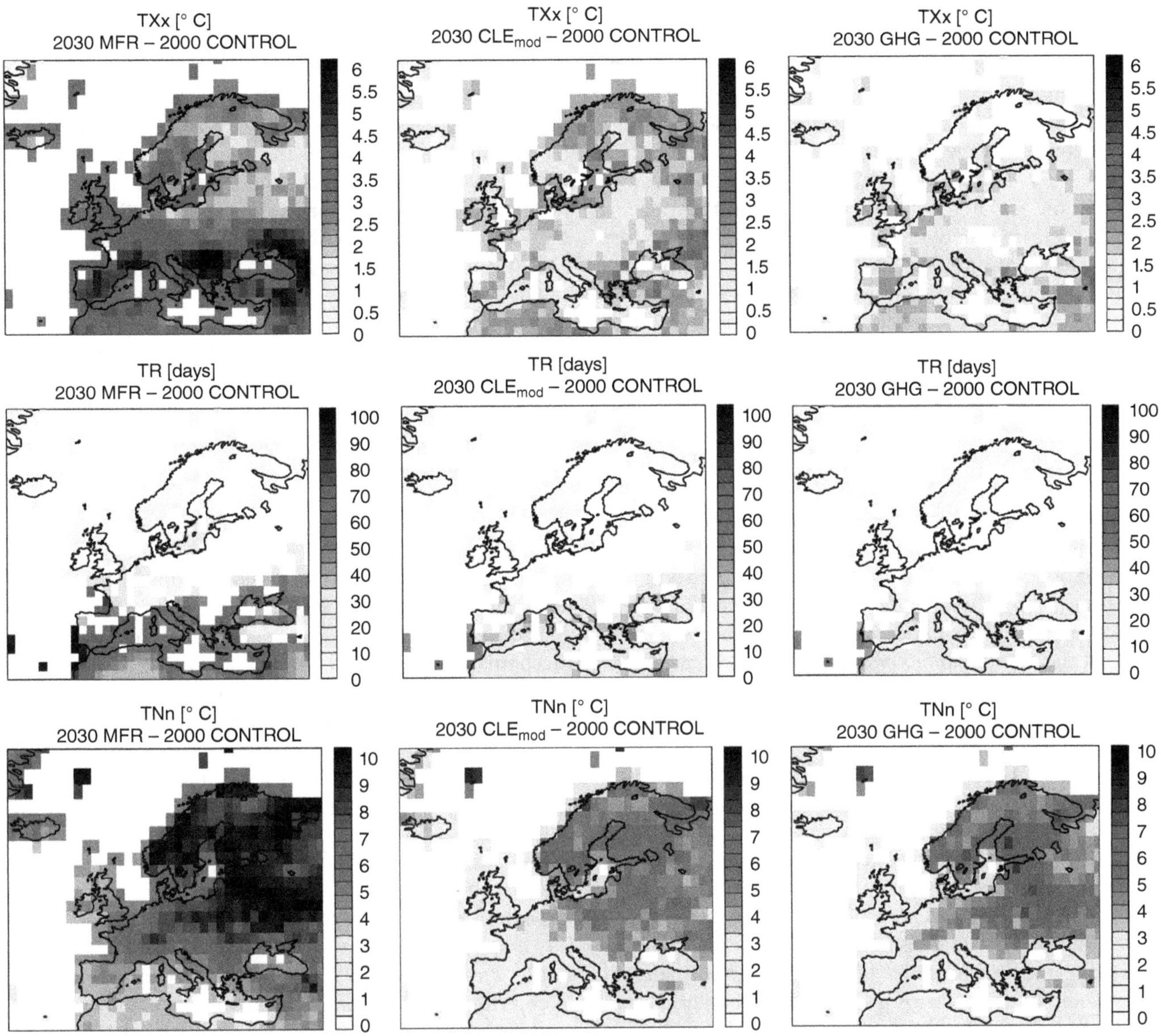

Figure 3.2 Changes in extreme temperature indices, hottest day of the year (TXx), tropical nights (TR), and coldest night of the year (TNn) as simulated in three future scenarios [*Kloster et al.*, 2010]: (left panels) maximum feasible reduction (MFR); (middle panels) MFR in Europe only, and current legislation emission in the rest of the world (CLEmod); and (right panels) GHG only emission scenario B2-SRES (GHG), compared to present climate conditions (CONTROL). Ocean grid cells and changes not statistically significant at the 5% significance level are left white. From *Sillmann et al.* [2013b].

of changes. This is not the case for TXx, which significantly increases particularly in northern parts of Europe in CLEmod, but not in GHG. This indicates that regional aerosol reductions will have most effect on hottest daytime temperatures, when the diming effect of scattering aerosols is removed.

Caesar and Lowe [2012], using E1 and A1B SRES scenario (see section 3.3), also find that temperature related extremes are markedly reduced under a mitigation scenario compared with the nonmitigation scenarios and identify regional differences in terms of benefits from mitigation efforts. For instance, for TXx, the largest avoided changes occur in the northern midlatitudes, particularly over Russia, but also Canada, the eastern United States, and Europe where avoided changes are in the order of 3°C or above (scenarios based on the SRES A1B). In general, geographical patterns of change for each extremes index are relatively stable across the different emissions scenario projections and the magnitude of change of temperature extremes exhibits a strong relationship with global mean temperature change [*Tebaldi et al.*, 2006; *Collins et al.*, 2013].

Sillmann et al. [2013b] further discuss that aerosols exert a considerable effect on precipitation patterns and that precipitation changes due to greenhouse gases can also be amplified by reductions in aerosol emissions. *Hodnebrog et al.* [2016] carefully studied the effect of local BC and OC aerosol emissions from anthropogenic biomass burning activities on southern African dry season precipitation. This study showed that improving air quality by reducing biomass burning emissions of BC and OC would contribute to mitigating reduced rainfall rates in southern Africa. A study by *Stjern et al.* [2011] looking at aerosol impacts on total and light precipitation in Europe further revealed that a potential aerosol signal can be found in changes of light precipitation, particularly for wind directions associated with the largest reductions in SO_4 levels.

However, changes in precipitation extremes appear with high regional variability in the sign of precipitation change. Some GCMs (e.g., HadGEM2) appear to have a greater response to aerosol forcing than others due to differences in the representation of the indirect aerosol effect or BC in the models [*Johns et al.*, 2011; *Caesar and Lowe*, 2012]. *Levy et al.* [2013] found that both temperature and precipitation show much stronger responses when aerosol indirect effects are included.

Estimates of the hydrological sensitivity, which refers to the relationship between precipitation change per unit global mean temperature change, have been included in various studies [see, for instance, *Collins et al.*, 2013], but little work has been done in comparisons under different emissions scenarios and aerosols. *Shiogama et al.* [2010] show a robust emissions scenario dependency, with lower sensitivities occurring under higher greenhouse gas and aerosol emissions scenarios. *Pendergrass et al.* [2015] find that in most models, the rate of extreme precipitation increase per degree warming does not depend on the emissions scenario, in contrast to mean precipitation. They conclude that the increase of extreme precipitation depends only on the magnitude of warming in these models and not on the composition of the change in forcing.

Figure 3.3 illustrates in particular changes in the precipitation-based climate extreme indices maximum consecutive 5-day rainfall (RX5d), very wet days (R95p), and consecutive dry days (CDD) for Europe for the MFR, CLEmod, and GHG scenario (see section 3.3). The areas with significant changes are substantially reduced compared to the temperature extremes shown in Figure 3.2. The strongest changes can be seen in MFR with increases in the wet extremes across central and northern parts for RX5d and increases particularly in northern parts of Europe for R95p. Increases in the dry extremes (CDD) are concentrated around the Mediterranean region. In the CLEmod and GHG scenario, a coherent pattern of increase can be seen particularly for RX5d, mainly in northern Europe, whereas only scattered grid boxes indicate significant changes in R95p and CDD across Europe.

While *Caesar and Lowe* [2012] and *Sillmann et al.* [2013b] look specifically at the effect of aerosol emission reduction on temperature and precipitation extremes, these studies do not separate the different effects of SO_4, OC, and BC aerosols. A recent study by *Chuwah et al.* [2016] did not consider effects on extremes, but shows that a decrease in SO_4 concentrations alone results in surface warming, while reduction in BC concentrations might decrease or increase the local surface temperature depending on the location. They conclude that different, but plausible, air pollution control policies can have substantial local climate effects and induce remote responses through dynamic teleconnections.

3.6. FUTURE RESEARCH NEEDS

Although there is an increasing number of studies investigating the impact of scattering and absorbing aerosols on regional and global climate change, the exact mechanisms through which aerosols affect regional and global climate dynamics remain largely uncertain and need further investigation. In general, GHG warming causes a spin-up of the hydrological cycle, whereas aerosol direct and indirect cooling can cause a slowdown, and the potential influence of aerosols on global and regional climate could be far more significant than currently estimated. Particularly climate forcers with short lifetimes, such as black carbon and sulphate aerosols, play an important role for changes in mean and extreme temperature and precipitation on regional scales [e.g., *Koren et al.*, 2012; *Altaratz et al.*, 2010], but the mechanisms involved in determining the impact of aerosols particularly on precipitation are currently not well understood [*Tao et al.*, 2012]. Furthermore, the nonlinearity of aerosol-precipitation interactions combined with large natural internal climate variability complicates the detection of very small signals induced by aerosol forcing [*Stjern et al.*, 2011].

Currently available studies are based on simulations of a single climate model [e.g., *Caesar and Lowe,* 2012; *Sillmann et al.*, 2013; *Chuwah et al.*, 2016]. Coordinated multimodel simulations of aerosol scenarios are urgently needed to carefully assess model-related uncertainties, which are also important for the quantification of the aerosol effect on climate extremes. A larger number of ensemble simulations from individual GCMs would also allow for an investigation of the contribution of natural internal climate variability versus the effect of external forcing mechanisms and the difference in their extreme precipitation and temperature response. A multimodel intercomparison project (PDRMIP, Precipitation Driver

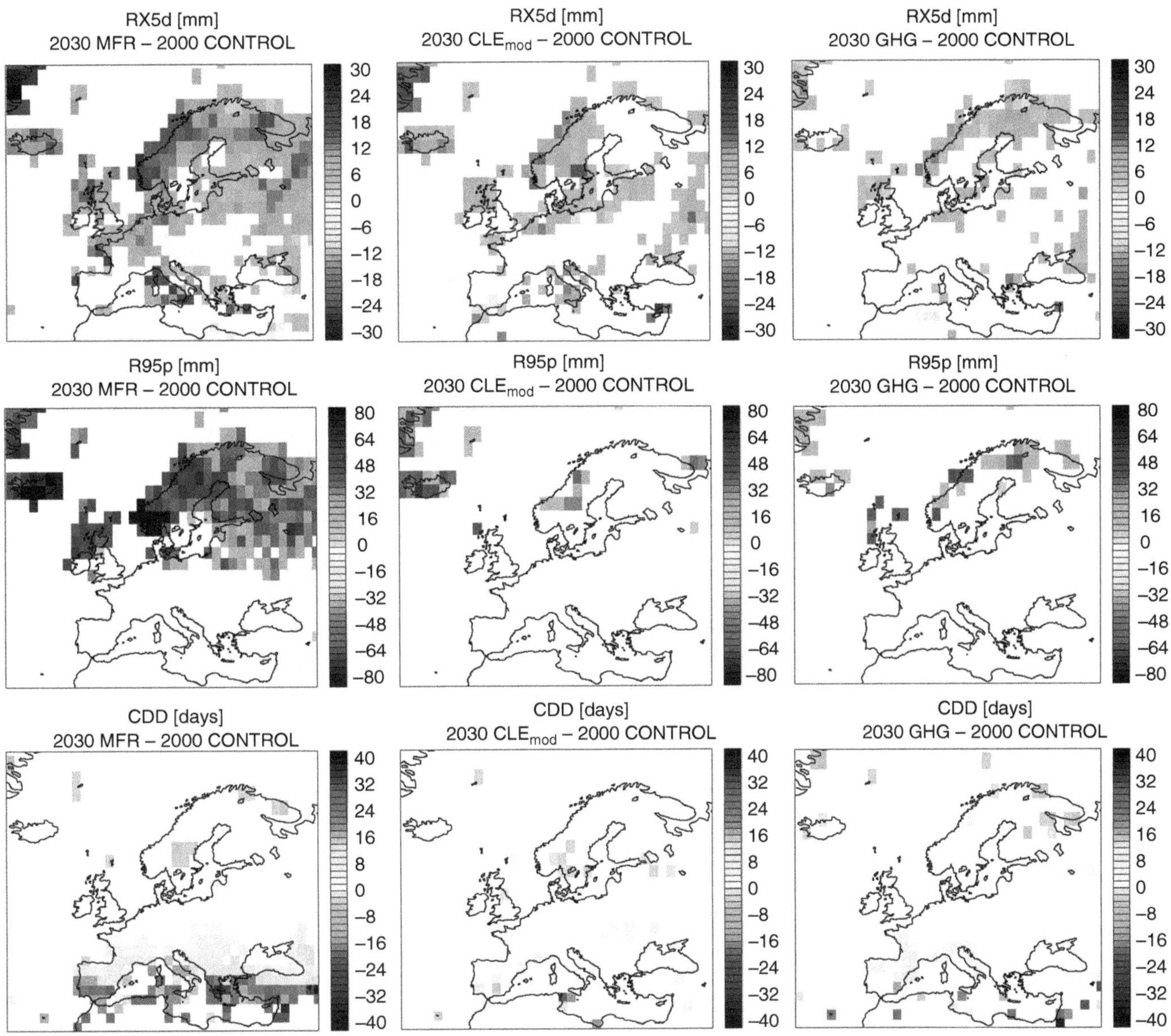

Figure 3.3 As Fig. 3.2, but for changes in extreme precipitation indices, maximum consecutive 5 day rainfall (RX5d), very wet days (R95p), and consecutive dry days (CDD). From *Sillmann et al.* [2013b].

Response Model Intercomparison Project) launched in 2013 is meant to tackle these challenges (*http://www.cicero.uio.no/en/PDRMIP*). PDRMIP includes several dedicated analyses of predefined aerosol scenarios simulated with at least 10 different GCMs to gain a better understanding of the importance of internal and external drivers in determining changes in the energy budget and mean and extreme precipitation [*Myhre et al.*, 2017].

While progress in these model simulations and their interpretation can support our understanding of the potential effects of direct and indirect aerosol forcing on clouds and precipitation in a coupled global system, it should be kept in mind that there are also many uncertainties related to the distribution and concentrations of aerosols in observations [*Boucher et al.*, 2013]. In addition, there are many simplifications in the models that limit their ability to realistically simulate the indirect effects of aerosols [*Feingold et al.*, 2009]. Thus, there remains a substantial need and diverse opportunities for future research in understanding and modeling of aerosol impacts on regional changes in climate extremes.

ACKNOWLEDGMENTS

J. Sillmann is supported by the Research Council of Norway funded projects NAPEX (no. 229778) and CiXPAG (no. 244551). We thank Luca Pozolli for reproducing Figures 3.2 and 3.3.

REFERENCES

Ackerman, A. S., O. B. Toon, D. E. Stevens, A. J. Heymsfield, V. Ramanathan, and E. J. Welton (2000), Reduction of tropical cloudiness by soot, *Science*, *288*, 1042–1047; doi:10.1126/science.288.5468.1042.

Allen, M, R., and W, J, Ingram (2002), Constraints on future changes in climate and the hydrologic cycle, *Nature*, *419*, 224–232.

Altaratz, O., et al. (2010), Lightning response to smoke from Amazonian fires, *Geophys. Res. Lett.*, *37*, 10.1029/2010gl042679.

AMAP (2011), *The Impact of Black Carbon on Arctic Climate*, by P. K. Quinn, A. Stohl, A. Arneth, T. Berntsen, J. F. Burkhart, J. Christensen, M. Flanner, K. Kupiainen, H. Lihavainen, M. Shepherd, V. Shevchenko, H. Skov, and V. Vestreng, Actic Monitoring and Assessment Programme (AMAP), Oslo.

Andrews, T., P. Forster, O. Boucher, N. Bellouin, and A. Jones (2010), Precipitation, radiative forcing and global temperature change, *Geophys. Res. Lett.*, *37*, L14701.

Bala, G., K. Caldeira, and R. Nemani (2010), Fast versus slow response in climate change: implications for the global hydrological cycle, *Climate Dyn.*, *35*, 423–434.

Ban-Weiss, G. A., L. Cao, G. Bala, and K. Caldeira (2012), Dependence of climate forcing and response on the altitude of black carbon aerosols, *Climate Dyn.*, *38*, 897–911; doi:10.1007/s00382-011-1052-y.

Bollasina, M., Y. Ming, and V. Ramaswamy (2011), Anthropogenic aerosols and the weakening of the South Asian summer monsoon, *Science*, *334*, 502–505.

Boucher, O., D. Randall, P. Artaxo, C. Bretherton, G. Feingold, P. Forster, V.-M. Kerminen, Y. Kondo, H. Liao, U. Lohmann, P. Rasch, S.K. Satheesh, S. Sherwood, B. Stevens and X.Y. Zhang (2013), Clouds and aerosols, in *Climate Change 2013: The Physical Science Basis, Contribution of Working Group I to the Fifth Assessment Report of the Intergovernmental Panel on Climate Change*, edited by T. F. Stocker, D. Qin, G.-K. Plattner, M. Tignor, S. K. Allen, J. Boschung, A. Nauels, Y. Xia, V. Bex and P. M. Midgley, Cambridge University Press, Cambridge, UK, and New York, USA.

Bouwman, A. F., T. Kram, and K.Klein Goldewijk (2006), *Integrated Modeling of Global Environmental Change, An Overview of IMAGE 2.4*, Netherlands Environmental Assessment Agency (MNP), Bilthoven.

Brasseur, G. P., R. G. Prinn, and A. A. P. Pszenny, eds. (2003), *Atmospheric Chemistry in a Changing World: An Integration and Synthesis of a Decade of Tropospheric Chemistry Research*, Global Change, The IGBP Series, Springer.

Caesar, J., and J.-A. Lowe (2012), Comparing the impacts of mitigation versus non-intervention scenarios on future temperature and precipitation extremes in the HadGEM2 climate model, *J. Geophys. Res.*, *117*, D15109; doi:10.1029/2012JD017762.

Chen, G., et al. (2011), Testing the Clausius-Clapeyron constraint on the aerosol-induced changes in mean and extreme precipitation, *Geophys. Res. Lett.*, *38*, 10.1029/2010gl046435.

Chernoff, D. I., and A. K. Bertram (2010), Effects of sulfate coatings on the ice nucleation properties of a biological ice nucleus and several types of minerals, *J. Geophys. Res. Atmos.*, *115*, D20205; doi:10.1029/2010JD014254.

Chuwah, C., T. van Noije, D.P. van Vuuren, P. Le Sager, and W. Hazeleger (2016), Global and regional climate impacts of future aerosol mitigation in an RCP6.0-like scenario in EC-Earth, *Climatic Change*, *134*, 1–14; doi:10.1007/s10584-015-1525-9.

Cofala, J., M. Amann, Z. Klimont, K. Kupiainen, and L. Hglund-Isaksson (2007), Scenarios of global anthropogenic emissions of air pollutants and methane until 2030, *Atmos. Environ.*, *41*(38), 8486–8499.

Collins, M., R. Knutti, J. Arblaster, J.-L. Dufresne, T. Fichefet, P. Friedlingstein, X. Gao, W.J. Gutowski, T. Johns, G. Krinner, M. Shongwe, C. Tebaldi, A. J. Weaver, and M. Wehner (2013), Long-term climate change: Projections, commitments and irreversibility, in *Climate Change 2013: The Physical Science Basis, Contribution of Working Group I to the Fifth Assessment Report of the Intergovernmental Panel on Climate Change*, edited by T. F. Stocker, D. Qin, G.-K. Plattner, M. Tignor, S. K. Allen, J. Boschung, A. Nauels, Y. Xia, V. Bex, and P.M. Midgley, Cambridge University Press, Cambridge, UK, and New York, USA.

de Gouw, J. A., A. M. Middlebrook, C. Warneke, P. D. Goldan, W. C. Kuster, J. M. Roberts, F. C. Fehsenfeld, D. R. Worsnop, M. R. Canagaratna, A. A. P. Pszenny, W. C. Keene, M. Marchewka, S. B. Bertman, and T. S. Bates (2005), Budget of organic carbon in a polluted atmosphere: Results from the New England air quality study in 2002, *J. Geophys. Res. Atmos.*, *110*, D16305; doi:10.1029/2004JD005623.

Evan, A. T., J. P. Kossin, C. E. Chung, and V. Ramanathan (2011), Strengthening of Arabian Sea tropical cyclones and the South Asian atmospheric brown cloud, *Nature*, *479*, 94–97; doi:10.1038/nature10552.

Evan, A. T., J. P. Kossin, C. E. Chung, and V. Ramanathan (2012), Reply to "How have tropical storms over the Arabian Sea intensified?", *Nature*, *489*, E2–E3; doi:10.1038/nature11471.

Feingold, G., et al. (2009), Effects of pollution aerosol and biomass burning on clouds and precipitation: Numerical modeling studies, in *Aerosol Pollution Impact on Precipitation: A Scientific Review*, edited by Z. Levin and R.W. Cotton, Springer.

Fiore, A. M., et al. (2012), Global air quality and climate, *Chem. Soc. Rev.*, *41*, 6663–6683.

Flanner, M. G. (2013), Arctic climate sensitivity to local black carbon, *J. Geophys. Res. Atmos.*, *118*(4), 1840–1851; doi: 10.1002/jgrd.50176.

Forster, P., et al. (2007), Changes in atmospheric constituents and in radiative forcing, *Climate Change 2007: The Physical Science Basis. Contribution of Working Group I to the Fourth Assessment Report of the Intergovernmental Panel on Climate Change*, Cambridge University Press, Cambridge, UK, and New York, USA.

Ganguly, D., et al. (2012), Climate response of the South Asian monsoon system to anthropogenic aerosols, *J. Geophys. Res.*, *117*(D13209); doi: 10.1029/2012JD017508.

Hansen, J., M. Sato, and R. Ruedy (1997), Radiative forcing and climate response, *J. Geophys. Res.*, *102*, 6831–6864; doi:10.1029/96JD03436.

Hartmann D. L., A. M. G. Klein Tank, M. Rusticucci, et al. (2013), Observations: Atmosphere and surface climate change 2013, in *The Physical Science Basis*, edited by T. F. Stocker, et al., 159–254, Cambridge Univ. Press, Cambridge.

Hazra, A., P. Mukhopadhyay, S. Taraphdar, J.-P. Chen, and W. R. Cotton (2013), Impact of aerosols on tropical cyclones: An investigation using convection-permitting model simulation, *J. Geophys. Res. Atmos.*, *118*, 7157–7168; doi:10.1002/jgrd.50546.

Herbener, S. R., S. C. van den Heever, G. G. Carrió, S. M. Saleeby, and W. R. Cotton (2014), Aerosol indirect effects on idealized tropical cyclone dynamics, *J. Atmos. Sci.*, *71*, 2040–2055; doi:10.1175/JAS-D-13-0202.1.

Hodnebrog, Ø., G. Myhre, P.M. Forster, J. Sillmann, and B. H. Samset (2016), Local biomass burning is the dominant cause of observed drying in southern Africa, *Nature Communication,* in press.

Johns, T. C., et al. (2011), Climate change under aggressive mitigation: The ENSEMBLES multi-model experiment, *Climate Dyn.*, *37*, 1975–2003; doi:10.1007/s00382-011-1005-5.

Kanakidou, M., J. H. Seinfeld, S. N. Pandis, I. Barnes, F. J. Dentener, M. C. Facchini, R. Van Dingenen, B. Ervens, A. Nenes, C. J. Nielsen, E. Swietlicki, J. P. Putaud, Y. Balkanski, S. Fuzzi, J. Horth, G. K. Moortgat, R. Winterhalter, C. E. L. Myhre, K. Tsigaridis, E. Vignati, E. G. Stephanou, and J. Wilson (2005), Organic aerosol and global climate modeling: A review, *Atmos. Chem. Phys.*, *5*, 1053–1123; doi:10.5194/acp-5-1053-2005.

Kauffman, R. K., H. Kauppi, M. L. Mann, and J. H. Stock (2011), Reconciling anthropogenic climate change with observed temperature 1998–2008, *Proceedings of the National Academy of Sciences*, *108,* 11790–11793; doi:10.1073/pnas.1102467108.

Kaufman, Y. J., D. Tanré, and O. Boucher (2002), A satellite view of aerosols in the climate system, *Nature*, *419*, 215–223; doi:10.1038/nature01091.

Khain, A. P. (2009), Notes on state-of-the-art investigations of aerosol effects on precipitation: A critical review, *Environ. Res. Lett.*, *4*, 015004; doi:10.1088/1748-9326/4/1/015004.

Kharin, V. V., et al. (2013), Changes in temperature and precipitation extremes in the CMIP5 ensemble, *Climate Change*, *1–13*, 10.1007/s10584-013-0705-8.

Kloster, S., F. Dentener, J. Feichter, F. Raes, U. Lohmann, E. Roeckner, and I. Fischer-Bruns (2010), A GCM study of future climate response to aerosol pollution reductions, *Climate Dyn.*, *34*, 1177–1194.

Koch, D., and A. D. Del Genio (2010), Black carbon absorption effects on cloud cover: Review and synthesis, *Atmos. Chem. Phys.*, *10*, 7685–7696; doi:10.5194/acp-10-7685-2010.

Koren, I., et al. (2012), Aerosol-induced intensification of rain from the tropics to the mid-latitudes, *Nat. Geosci.*, *5* (2), 118–122, 10.1038/ngeo1364.

Kristjánsson, J. E., H. Muri, and H. Schmidt (2015), The hydrological cycle response to cirrus cloud thinning, *Geophys. Res. Lett.*, *42*, 10,807–10,815; doi:10.1002/2015GL066795.

Kulmala, M., et al. (2011), General overview: European Integrated project on Aerosol Cloud Climate and Air Quality Interactions (EUCAARI), Integrating aerosol research from nano to global scales, *Atmos. Chem. Phys.*, *11*, 13,061–130,143.

Kvalevåg, M. M., B. H. Samset, and G. Myhre (2013), Hydrological sensitivity to greenhouse gases and aerosols in a global climate model, *Geophys. Res. Lett.*, *40*, 1432–1438.

Lau, K. M., and K. M. Kim (2006), Observational relationships between aerosol and Asian monsoon rainfall, and circulation, *Geophys. Res. Lett.*, *33*, L21810; doi:10.1029/2006GL027546.

Lau, K. M., and K. M. Kim (2011), Comment on " 'Elevated heat pump' hypothesis for the aerosol-monsoon hydroclimate link: 'Grounded' in observations?" by S. Nigam and M. Bollasina, *J. Geophys. Res.*, *116*, D07203; doi:10.1029/2010JD014800.

Levy, H., II, L. W. Horowitz, M. D. Schwarzkopf, Y. Ming, J.-C. Golaz, V. Naik, and V. Ramaswamy (2013), The roles of aerosol direct and indirect effects in past and future climate change, *J. Geophys. Res. Atmos.*, *118*, 4521–4532; doi:10.1002/jgrd.50192.

Lowe, J. A., C. D. Hewitt, D. P. vanVuuren, T. C. Johns, E. Stehfest, J.-F. Royer, and P. J. van derLinden (2009), New study for climate modeling, analyses, and scenarios, *EOS Trans. AGU*, *90*(21), 181; doi:10.1029/2009EO210001.

Lund, M. T., et al. (2014), Global and regional climate impacts of black carbon and co-emitted species from the on-road diesel sector, *Atmos. Environ.*, *98*, 50–58; doi:10.1016/j.atmosenv.2014.08.033.

Masui, T., et al. (2011), An emission pathway for stabilization at 6 Wm^{-2} radiative forcing, *Climatic Change*, *109*, 59; doi:10.1007/s10584-011-0150-5.

Meehl, G. A., J. M. Arblaster, and W. D. Collins (2008), Effects of black carbon aerosols on the Indian monsoon, *J Climate*, *21*, 2869–2882; doi: 10.1175/2008JCLI2362.1.

Ming, Y., V. Ramaswamy, and G. Persad (2010), Two opposing effects of absorbing aerosols on global-mean precipitation, *Geophys. Res. Lett.*, *37*, L13701.

Moss, R. H., J. A. Edmonds, K. A. Hibbard, M. R. Manning, S. K. Rose, D. P. van Vuuren, T. R. Carter, S. Emori, M. Kainuma, T. Kram, G. A. Meehl, J. F. B. Mitchell, N. Nakicenovic, K. Riahi, S. J. Smith, R. J. Stouffer, A. M. Thomson, J. P. Weyant, and T. J. Wilbanks (2010), The next generation of scenarios for climate change research and assessment, *Nature*, *463*, 747–756; doi:10.1038/nature08823.

Myhre, G., C. E. L. Myhre,, B. H. Samset, and T. Storelvmo (2013), Aerosols and their relation to global climate and climate sensitivity, *Nat. Ed. Knowl.*, *4*(5), 7.

Myhre, G., P. M. Forster, B. H. Samset, Ø. Hodnebrog, J. Sillmann, T. Andrews, O. Boucher, G. Faluvegi, D. Fläschner, T. Iversen, M. Kasoar, V. Kharin, A. Kirkevåg, J.-F. Lamarque, D. Olivié, T. Richardson, D. Shindell, K. P. Shine, Camilla W. Stjern, T. Takemura, A. Voulgarakis, and F. Zwiers (2017), PDRMIP: A Precipitation Driver and Response Model Intercomparison Project, *Protocol and preliminary results, BAMS*, in press, doi:10.1175/BAMS-D-16-0019.1.

Nakicenovic, N., et al. (2000), *Special Report on Emission Scenarios*, Cambridge University Press, Cambridge, UK.

Nigam, S., and M. Bollasina (2010), "Elevated heat pump" hypothesis for the aerosol-monsoon hydroclimate link: "Grounded" in observations?, *J. Geophys. Res.*, *115*, D16201; doi:10.1029/2009JD013800.

O'Gorman, P. A., R. P. Allan, M. P. Byrne, and M. Previdi (2012), Energetic constraints on precipitation under climate change, *Surv. Geophys.*, *33*, 585–608.

Pall, P., et al. (2007), Testing the Clausius-Clapeyron constraint on changes in extreme precipitation under CO2 warming, *Climate Dyn.*, *28*(4), 351–363, 10.1007/s00382-006-0180-2.

Pendergrass, A. G., and D. L. Hartmann (2012), Global-mean precipitation and black carbon in AR4 simulations, *Geophys. Res. Lett.*, *39*, L01703.

Pendergrass, A. G., and D. L. Hartmann (2014), The atmospheric energy constraint on global-mean precipitation change, *J. Climate*, *27*, 757–768.

Pendergrass, A. G., F. Lehner, B. M. Sanderson, and Y. Xu (2015), Does extreme precipitation intensity depend on the emissions scenario?, *Geophys. Res. Lett.*, *42*, 8767–8774; doi:10.1002/2015GL065854

Prein, A. F., et al. (2015), A review on regional convection-permitting climate modeling: Demonstrations, prospects, and challenges, *Rev. Geophys.*, *53*, 323–361; doi:10.1002/2014 RG000475.

Quinn, P. K., T. S. Bates, E. Baum, N. Doubleday, A. M. Fiore, M. Flanner, A. Fridlind, T. J. Garrett, D. Koch, S. Menon, D. Shindell, A. Stohl, and, S. G Warren (2008), Short-lived pollutants in the Arctic: Their climate impact and possible mitigation strategies, *Atmos. Chem. Phys.*, *8*, 1723–1735; doi:10.5194/acp-8-1723-2008.

Ramanathan, V., and Y. Feng (2009), Air pollution, greenhouse gases and climate change: Global and regional perspectives, *Atmos. Environ.*, *43*, 37–50.

Ramanathan, V., P. J. Crutzen, J. T. Kiehl, and D. Rosenfeld (2001), Atmosphere-aerosols, climate, and the hydrological cycle, *Science*, *294*, 2119–2124; doi:10.1126/science.1064034.

Rasch, P. J., S. Tilmes, R. Turco, A. Robock, L. Oman, C.-C. Chen, G. Stenchikov, and R. R. Garcia (2008), A review of stratospheric sulfate aerosols for geoengineering, *Phil. Trans. R. Soc. Lond.*, *366*, 4007–4037; doi:10.1098/rsta2008.0131.

Rosenfeld, D., et al. (2008), Flood or drought: How do aerosols affect precipitation?, *Science*, *321*, 1309–1313; doi:10.1126/science.1160606.

Rosenfeld, D., W. L. Woodley, A. Khain, W. R. Cotton, G. Carrió, I. Ginis, and J. H. Golden (2012), Aerosol effects on microstructure and intensity of tropical cyclones, *Bull. Amer. Meteor. Soc.*, *93*, 987–1001; doi: http://dx.doi.org/10.1175/BAMS-D-11-00147.1.

Samset, B. H., and G. Myhre (2015), Climate response to externally mixed black carbon as a function of altitude, *J. Geophys. Res. Atmos.*, *120*, 2913–2927; doi:10.1002/2014jd022849.

Sand, M., et al. (2015), Response of Arctic temperature to changes in emissions of short-lived climate forcers, *Nature Climate Change*; doi: 10.1038/nclimate2880.

Sand, M., T. K. Berntsen, J. E. Kay, J. F. Lamarque, Ø. Seland, and A. Kirkevåg (2013), The Arctic response to remote and local forcing of black carbon, *Atmos. Chem. Phys.*, *13*, 211–224; doi: 10.5194/acp-13-211-2013.

Shindell, D., et al. (2008), Climate forcing and air quality change due to regional emissions reductions by economic sector, *Atmos. Chem. Phys.*, *8*, 7101–7113; doi: 10.5194/acp-8-7101–2008.

Shindell, D. T., et al. (2012a), Precipitation response to regional radiative forcing, *Atmos. Chem. Phys.*, *12*, 6969–6982; doi: 10.5194/acp-12-6969-2012.

Shindell, D. T., et al. (2012b), Simultaneously mitigating near-term climate change and improving human health and food security, *Science*, *335*, 183–189; doi:10.1126/science.1210026.

Shiogama, H., et al. (2010), Emission scenario dependencies in climate change assessments of the hydrological cycle, *Climate Change*, *99*, 321–329; doi:10.1007/s10584-009-9765-1.

Sillmann, J., et al. (2013b), Aerosol effect on climate extremes in Europe under different future scenarios, *Geophys. Res. Lett.*, *40*, 2290–2295; doi: 10.1002/grl.50459.

Sillmann, J., V. V. Kharin, F. W. Zwiers, X. Zhang, and D. Bronaugh (2013a), Climate extremes indices in the CMIP5 multi-model ensemble, Part 2: Future climate projections, *J. Geophys. Res. Atmos.*, *118*, 2473–2493; doi: 10.1002/jgrd.50188.

Sillmann, J., T. Lenton, A. Levermann, K. Ott, M. Hulme, F. Benduhn, and J. Horton (2015),Climate emergencies do not justify engineering the climate, *Nature Climate Change*, *5*, 290–292; doi:10.1038/nclimate2539.

Silva, R.A., et al. (2013), Global premature mortality due to anthropogenic outdoor air pollution and the contribution of past climate change, *Environ. Res. Lett.*, *8*, 034005; doi:10.1088/1748–9326/8/3/034005.

Stier, P., J. H. Seinfeld, S. Kinne, J. Feichter, and O. Boucher (2006), Impact of nonabsorbing anthropogenic aerosols on clear-sky atmospheric absorption, *J. Geophys. Res. Atmos.*, *111*, D18201; doi:10.1029/2006JD007147.

Stjern, C. W., A. Stohl, and J. E. Kristjánsson (2011), Have aerosols affected trends in visibility and precipitation in Europe?, *J. Geophys. Res.*, *116*, D02212; doi:10.1029/2010JD014603.

Stohl, A. (2006), Characteristics of atmospheric transport into the Arctic troposphere, *J. Geophys. Res. Atmos.*, *111*(D11); doi: 10.1029/2005JD006888.

Streets, D. G., T. C. Bond, T. Lee, and C. Jang (2004), On the future of carbonaceous aerosol emissions, *J. Geophys. Res. Atmos.*, *109*, D24212; doi:10.1029/2004JD004902.

Tao, W. K., et al. (2012), Impact of aerosols on convective clouds and precipitation, *Rev. Geophys.*, *50*, 10.1029/2011rg000369.

Tebaldi, C., K. Hayhoe, J. M. Arblaster, and G. A. Meehl (2006), Going to the extremes: An intercomparison of model-simulated historical and future changes in extreme events, *Climate Change*, *79*, 185–211; doi:10.1007/s10584-006-9051-4.

Twomey, S. (1977), The influence of pollution on the shortwave albedo of clouds, *J. Atmos. Sci.*, *34*, 1149–52; doi:10.1175/1520-0469(1977)034<1149:TIOPOT>2.0.CO;2.

Vinoj, V., P. J. Rasch, H. L. Wang, J. H. Yoon, P. L. Ma, K. Landu, and B. Singh (2014), Short-term modulation of Indian summer monsoon rainfall by west Asian dust, *Nat. Geosci.*, *7*(4), 308–313. 10.1038/ngeo2107.

Wang, B., S. Xu, and L. Wu (2012), Intensified Arabian Sea tropical storm, *Nature*, *489*, E1–E2; doi:10.1038/nature/11470.

Wang, C., D. Kim, A. M. L. Ekman, M. C. Barth, and P. J. Rasch (2009), Impact of Anthropogenic aerosols on Indian summer monsoon, *Geophys. Res. Lett.*, *36*, L21704; doi:10.1029/2009GL040114.

Wang, Y., K.-H. Lee, Y. Lin, M. Levy, and R. Zhang (2014), Distinct effects of anthropogenic aerosols on tropical cyclones, *Nature Climate Change*, *4*, 368–373; doi:10.1038/nclimate2144.

Westra, S., et al. (2012), Global increasing trends in annual maximum daily precipitation, *J. Climate*, 10.1175/JCLI-D-12-00502.1.

WHO (2003), Health Aspects of Air Pollution with Particulate Matter, Ozone and Nitrogen Dioxide, Report on a WHO Working Group, Bonn, Germany, Report Nr. EUR/03/5042688.

Wild, M., H. Gilgen, A. Roesch, et al. (2005), From dimming to brightening: Decadal changes in solar radiation at Earth's surface, *Science*, *308*, 847–850.

4

Weakened Flow, Persistent Circulation, and Prolonged Weather Extremes in Boreal Summer

Dim Coumou[1], Kai Kornhuber[2], Jascha Lehmann[2], and Vladimir Petoukhov[1]

ABSTRACT

Changes in atmospheric circulation can strongly alter the frequency and/or magnitude of high-impact extreme weather events. Here we address the link between circulation changes and the occurrence of long-lasting heat waves in the Northern Hemisphere summer. We show that boreal summer circulation has seen pronounced changes in circulation over the last decades, possibly related to rapid warming of the Arctic. Generally, the midlatitude zonal mean flow has weakened and also the kinetic energy associated with transient synoptic eddies has reduced. At the same time, for some wave numbers, we see an increased occurrence-frequency of high-amplitude quasi-stationary waves. We argue that this increase in frequency is associated with a recent cluster of resonance events that can create such high-amplitude waves. The reduction in amplitude of fast-moving transient waves and the more-frequent occurrence of high-amplitude quasi-stationary waves both favor more persistent weather conditions. We present statistical evidence of links between such persistent upper-level flow and the occurrence of heat extremes at the surface.

4.1. INTRODUCTION

4.1.1. Detected Changes in Midlatitude Summer Circulation

The Northern Hemisphere midlatitudes have seen significant changes in the large-scale summer circulation over the last decades [*Overland et al.*, 2012; *Francis and Vavrus*, 2015; *Coumou et al.*, 2015]. The zonal mean zonal wind (or "jet") has weakened by about 5% (see Fig. 4.1), a signal that is seen in different reanalysis products and at different pressure levels [*Coumou et al.*, 2015]. The driver behind this weakening might be the much faster warming in the Arctic as compared to the rest of the hemisphere, a phenomenon known as Arctic amplification (Fig. 4.2; see also Chapter 2). This reduces the pole-to-equator temperature gradient and thereby the thermally driven jet (top panel Fig. 4.1). However, a strict direction of causality has not been shown. In principle it could be the other way around, too, whereby a change in midlatitude circulation alters the poleward heat transport, which would give rise to more rapid warming in the Arctic [*Cohen et al.*, 2014; *Walsh*, 2014]. For example, an alternative hypothesis has been proposed where the warmer Arctic is a consequence of midlatitude circulation changes, which would increase the exchanges of cold air from the Arctic with warmer air from lower latitudes, triggered by teleconnections originating from the tropical Pacific [*Trenberth et al.*, 2014; *Palmer*, 2014]. Irrespective of the underlying drivers, changes in midlatitude circulation are expected to affect day-to-day weather variability and possibly extreme weather events.

[1] *Institute for Environmental Studies (IVM), Department of Water and Climate Risk, VU University Amsterdam, Amsterdam, The Netherlands; Potsdam Institute for Climate Impact Research, Earth System Analysis, Potsdam, Germany*

[2] *Potsdam Institute for Climate Impact Research, Earth System Analysis, University of Potsdam, Faculty of Science, Potsdam, Germany*

Climate Extremes: Patterns and Mechanisms, Geophysical Monograph 226, First Edition.
Edited by S.-Y. Simon Wang, Jin-Ho Yoon, Christopher C. Funk, and Robert R. Gillies.

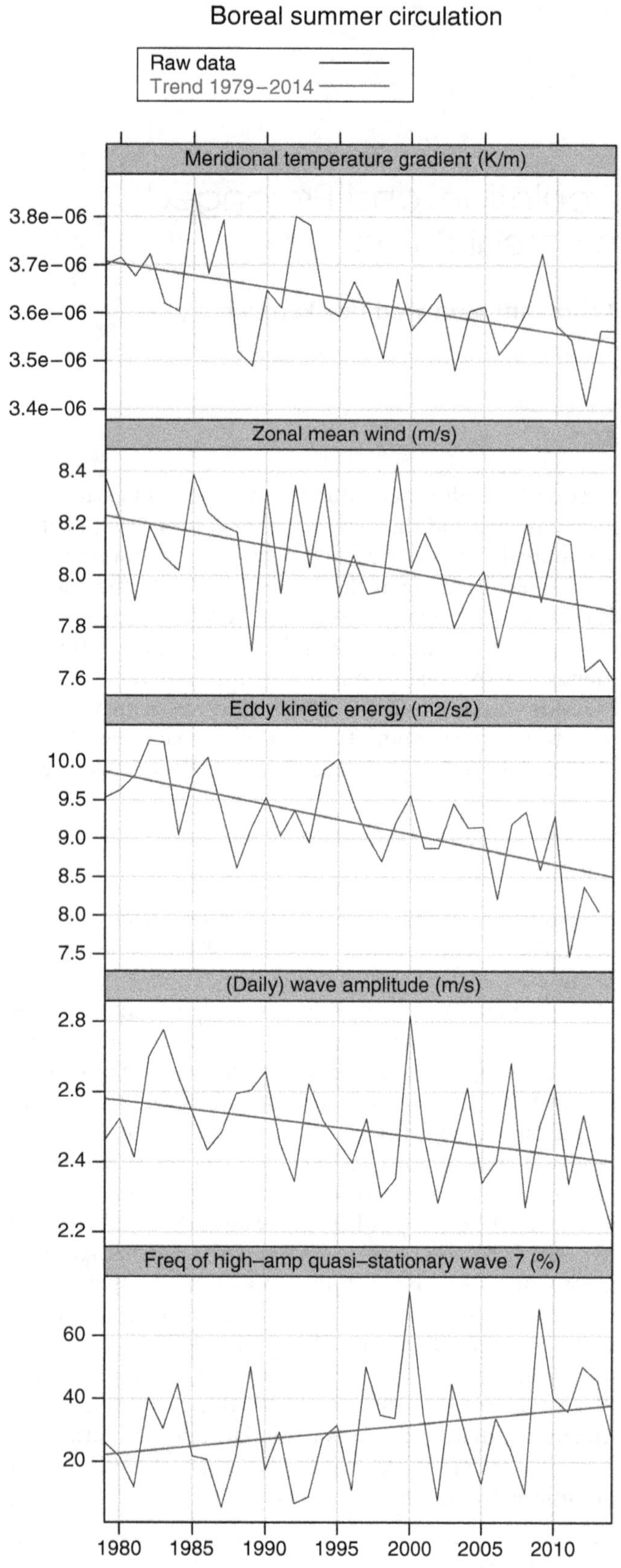

In conjunction with the summer jet, the kinetic energy associated with transient synoptic-scale weather systems has seen a significant weakening as well [*Coumou et al.*, 2015]. This kinetic energy can for example be extracted by bandpass filtering daily wind field data, because synoptic activity has typical timescales of 2 to 5 days, bringing weather variability on subweekly timescales. Typical wave lengths of synoptic transients are of the order of 2000–3500 km, or wave numbers from 6 to 10, with relatively fast phase speeds (i.e., eastward propagation) of the order of 6–12 m/s. These waves are *free* waves in the sense that they do not require any forcing but are a direct product of the atmospheric instability in the midlatitudes. Thus, in the absence of any temperature or orographic forcing, as for example on a so-called aquaplanet, these waves would still emerge [*Schneider et al.*, 2014]. Based on linear Rossby wave theory, one would expect the phase speed of these waves to become smaller under a weakened background flow. Starting from the linearized nondivergent barotropic vorticity equation [*Pedlosky*, 1979] without any thermal or orographic forcing (i.e., an equation describing adiabatic free atmospheric waves), an equation for the phase speed c can be derived:

$$c = U - \frac{\beta}{\left(k^2 + l^2\right)} \tag{4.1}$$

where U denotes the zonal mean zonal wind, β the Rossby parameter, and k and l, respectively, the zonal and meridional wave numbers. Thus, these *free* waves essentially flow on top of the background flow (U) and a weakening of the latter results in slower phase speeds. However, so far, no robust changes in phase speed have been detected in the observations. In contrast, significant trends have

Figure 4.1 Weakening summer circulation in the Northern Hemisphere midlatitudes over 1979–2014. Absolute changes in meridional temperature gradient, zonal mean wind, eddy kinetic energy (EKE), daily wave amplitudes for waves 6–10, and frequency of high-amplitude quasi-stationary wave 7. Variables are based on ERA-interim reanalysis, calculated at 500 mb, averaged over 35°N–65°N and all longitudes, for June-July-August. Plotted are actual observations and the linear trend, which are all significant at 95% confidence except for the linear trend in the bottom panel, which is significant at 90% confidence only. Similar trends are found at other pressure levels and in other reanalyses [*Coumou et al.*, 2014, 2015]. *See electronic version for color representation.*

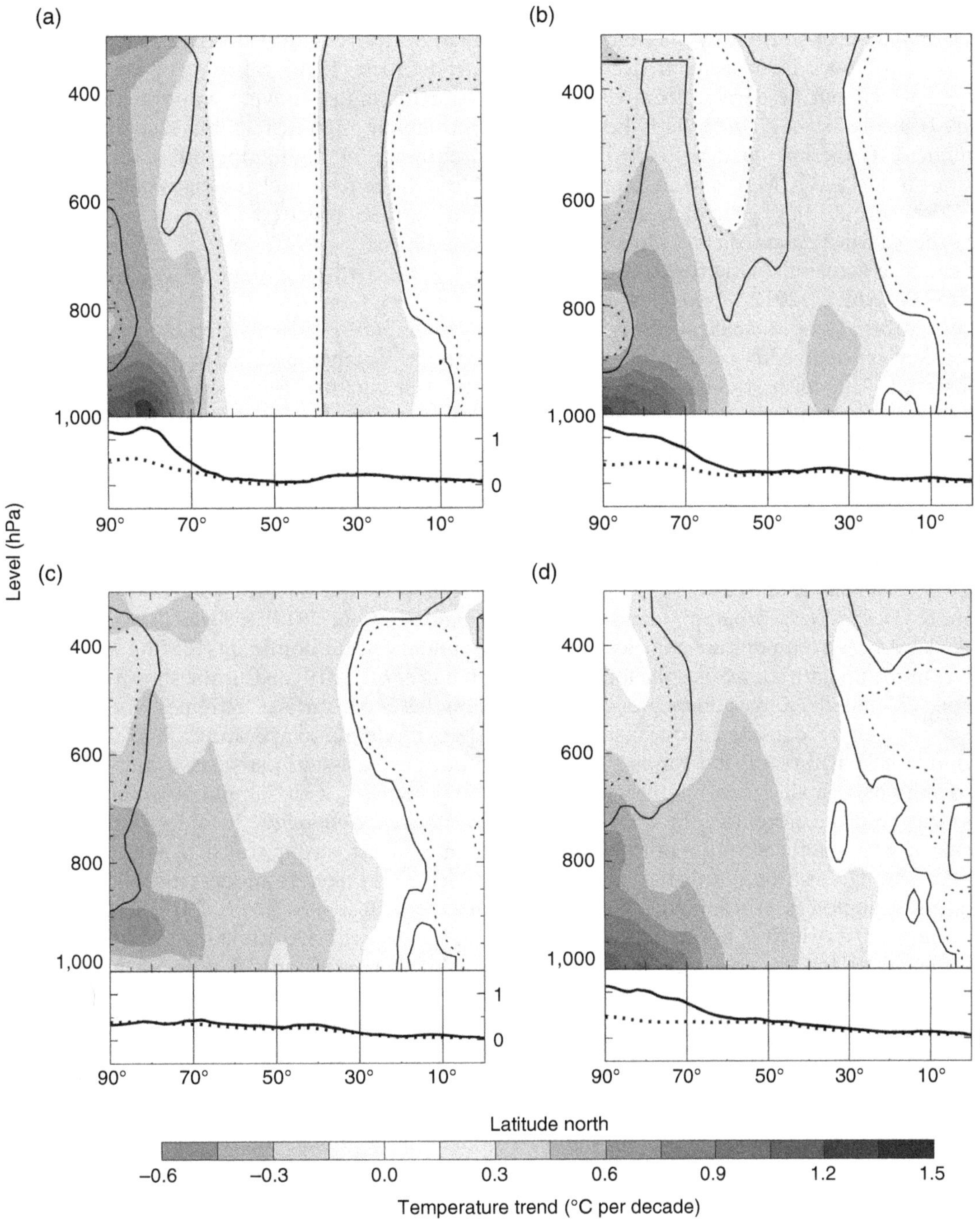

Figure 4.2 Arctic amplification of temperature trends over 1979–2014 for each season. Trends in zonal mean temperature for (a) winter (December–February), (b) spring (March–May), (c) summer (June–August), and (d) autumn (September–November) based on the ERA-interim reanalysis. Black contours indicate regions with significance at the 99% (solid lines) and 95% (dotted lines) confidence levels. The line graphs show trends averaged over the lower atmosphere (950–1000 hPa; solid lines) and over the entire atmospheric column (300–1000 hPa; dotted lines). From *Cohen et al.* [2014].

been detected in the amplitude of these waves, defined as the magnitude of the north-south flow (meridional wind in units of m/s). Amplitudes are expected to decline when the jet weakens, which can be derived from theoretical considerations [*Coumou et al.*, 2011]. This behavior is also seen in numerical experiments of both idealized general circulation models (GCMs) and comprehensive GCMs [*Schneider et al.*, 2014]. In general, transient eddies are forced by the jet via vertical shear (and thus baroclinicity) but can also accelerate it via the eddy-driven jet [*Woollings and Blackburn*, 2012; *Cohen et al.*, 2014; *Lehmann et al.*, 2014]. Thus, a weakening of the jet is expected to weaken synoptic eddy activity, in agreement with the observed long-term trends. As shown in Figure 4.1, statistically significant reductions are seen in both the mean amplitude of wave numbers 6–10 as well as in the total kinetic energy associated with synoptic eddies (eddy kinetic energy, or EKE). This summertime weakening of both the jet and EKE is also a very robust signal in future projections of CMIP5 climate models [*O'Gorman*, 2010; *Lehmann et al.*, 2014].

In addition to free waves, orography and thermal anomalies (like land-ocean temperature contrasts) generate *forced* waves in the midlatitudes. Since the forcing patterns change on substantially longer timescales of several weeks, these waves can be considered quasi-stationary. Due to the large-scale nature of the forcing patterns, forced waves tend to have smaller wave numbers (<6). The forced-component (and therefore also the quasi-stationary component) of wave numbers 6–10 is normally small. Note that free waves can also be quasi-stationary when the second term in equation (4.1) is similar in magnitude as the zonal mean jet (U), which is normally the case for wave 4. Thus, in general, the low wave number regime of the flow is quite different from the high wave number regime, with the latter dominated by fast moving waves. Only when averaging over longer time spans, like 15 days to monthly means, one effectively removes this transient wave component and the quasi-stationary component remains. For waves 6–10, this quasi-stationary component has generally a low magnitude and also little significant trends can be detected. Nevertheless, the trend in the highest amplitudes of the quasi-stationary component of wave 7 is upward. The lowest panel of Figure 4.1 plots how often the seventieth percentile wave-amplitude in 15-day running mean wind field data is exceeded, showing a moderate upward trend. This is also consistent with the reported increase in frequency of high-amplitude, quasi-stationary wave 7 [*Coumou et al.*, 2014]. Though this signal is seen in different reanalyses and at different pressure levels [*Coumou et al.*, 2014], it should be noted that statistically it is only marginally significant [*Wang et al.*, 2013]. Nevertheless, it is consistent with the recent cluster of resonance events involving wave 7 as discussed later.

In summary, we can conclude that boreal summer circulation has changed in three important ways over the last decades: (1) A weakening of the zonal mean jet, (2) a reduction in fast-moving synoptic wave activity and (3) a (marginally significant) increase in frequency of the occurrence of high-amplitude quasi-stationary wave 7. These three detectable changes all point at more persistent circulation patterns.

4.1.2. Midlatitude Rossby Waves and Extreme Weather

In general, persistent circulation patterns can lead to extreme weather and in particular to prolonged heat waves in summer. Slow wave propagation prolongs the prevailing weather conditions at the surface and can therefore lead to extremes on timescales of weeks: One day with temperatures over 30°C in western Europe is not unusual, but 10 or 20 days in a row will be. It has been demonstrated that high-amplitude quasi-stationary waves in the atmosphere are statistically linked to extreme weather at the surface [*Screen and Simmonds,* 2014; *Coumou et al.*, 2014]. Especially regions at the western boundary of the continents like the western United States [*Wang et al.*, 2010] or Europe show the strongest association between surface extremes and upper-level waves. Here, moderate temperatures tend to be associated with reduced quasi-stationary wave amplitudes and extreme temperatures with amplified quasi-stationary waves [*Screen and Simmonds*, 2014].

In contrast, strong transient wave activity, as captured by EKE, is linked to moderate surface temperatures and vice versa [*Coumou et al.*, 2015; *Lehmann and Coumou*, 2015]. Over most continental regions affected by storm tracks, there is a significant negative correlation between monthly EKE and surface temperature (Fig. 4.3). Thus, the hottest summers are associated with extremely low EKE, while mild summers are associated with more pronounced EKE. Again the western boundaries of the continents, notably Europe, are especially sensitive but also the eastern United States and western Asia. This makes sense as these regions are most directly influenced by the storm tracks. If EKE is low, this implies that weather variability on 2–5 day timescales is low, which is consistent with blocking anticyclones associated with quasi-stationary waves. In summer, transient waves transport moist and cool air from the oceans to the continents bringing relief during periods of oppressive heat [*Kyselý and Huth*, 2005; *Kyselý,* 2008]. Thus, low transient wave activity implies that cool maritime air masses become less frequent creating favorable conditions for blocking anticyclonic flow over continents and the buildup of heat and drought conditions. Amplifying local feedback mechanisms, such as soil-moisture feedbacks, can further drive temperatures upward, inhibit cloud formation, and

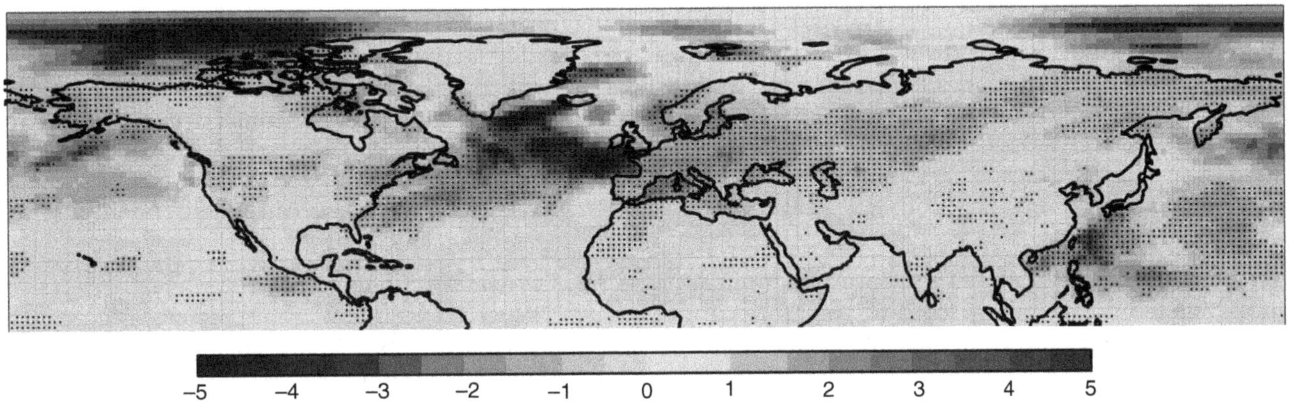

Figure 4.3 Regression slope between EKE and near-surface temperature in summer months (June–August) in units of m^2s^{-2}/K. Both variables were calculated from ERA-interim data and linearly detrended at each grid point. Significant negative correlation (stippling indicating 95% confidence) exists over most storm-track–affected regions. From *Coumou et al.* [2015].

thereby enhance the anticyclonic flow over the continents [*Schär et al.*, 2004; *Alexander*, 2010; *Mueller and Seneviratne*, 2012]. Thus, an absence of substantial transient wave activity likely prolongs the duration of blocking weather systems, as for example during the Russian heat wave of 2010. The record-breaking July temperatures over Moscow in that year were associated with extremely low EKE [*Coumou et al.*, 2015].

In summary, the observed trends in boreal summer circulation (Fig. 4.1) favor more persistent weather in summer and hence heat extremes on timescales of several weeks.

4.2. RESONANT CIRCULATION REGIMES

The observed reduction in transient wave activity seems to be a direct response of the weakening jet. Climate models project both summer jet and EKE to weaken under future high-emission scenarios and the projected relative changes in these quantities are consistent with those observed [*Coumou et al.*, 2015]. However, the observed increase in frequency of high-amplitude quasi-stationary wave 7 cannot be directly linked to the weakening jet.

To explain this increased frequency, a highly nonlinear, dynamical mechanism is invoked: resonance. This mechanism, which was first proposed by *Petoukhov et al.* [2013], can amplify the quasi-stationary component of waves 6, 7, or 8. As stated above, these wave numbers normally have a fast eastward propagation and their quasi-stationary component tends to be weak. Forcing of these waves is normally weak in the midlatitudes and their energy is effectively dispersed toward the poles and equator. However, under specific conditions, their wave energy can be trapped in midlatitudinal waveguides, with only weak dispersion at the lateral boundaries. If this happens and if the forcing for the trapped wave is large enough, then resonance can strongly magnify the wave amplitude. There are thus two important criteria that need to be fulfilled for resonance to occur: (1) waveguide formation, which traps the energy of wave 6, 7, or 8 in the midlatitudes, and (2) reasonably large forcing of the trapped free wave.

4.2.1. Waveguide

The jet stream can guide waves in the zonal direction and thereby trap them inside the midlatitudes. Whether a particular wave is trapped depends on the shape of the zonal mean zonal wind (U) and the wave number itself (k). Based on the linearized stationary barotropic vorticity equation at the equivalent barotropic level (in the mid/upper troposphere at about 300–500 hPa), an expression for the square of the meridional wavenumber of the trapped wave (l^2) can be derived [*Petoukhov et al.*, 2013]:

$$l^2(U,k,\varphi)=\frac{2\Omega cos^3\varphi}{aU}-\frac{cos^2\varphi}{a^2U}\frac{d^2U}{d\varphi^2}+\frac{sin\varphi cos\varphi}{a^2U}\frac{dU}{d\varphi}+\frac{1}{a^2}-\left(\frac{k}{a}\right)^2 \quad (4.2)$$

Here a is the Earth's radius, Ω its angular velocity, and φ latitude. Note that l may be a real or purely imaginary number and thus l^2 can change sign. For a waveguide to form for wave number k, two turning points are needed at two sufficiently distant latitudinal locations. At these turning points, the energy of wave k gets reflected back toward the midlatitudes. For this to occur, l^2 has to be positive inside the waveguide and change sign at the turning points themselves while U has to be positive (i.e., eastward) inside and directly outside the waveguide. The latter is normally the case outside the tropics and thus it is especially the change in sign of l^2 that is important. Finally, the turning points have to be located at a sufficiently large distance (several degrees latitude) so that the waveguide exceeds the characteristic scale of the relevant Airy function, which governs the latitudinal boundary conditions of the waveguide [*Petoukhov et al.*, 2013].

4.2.2. Wave Forcing

Trapping of a quasi-stationary free wave in a midlatitude waveguide not necessary leads to amplification of that wave: Forcing is needed as well. When a waveguide exists for a synoptic wave with wavenumber k the following approximate equation for the amplitude of the forced quasi-stationary component is valid [*Petoukhov et al.*, 2013]:

$$A_m = \frac{A_{m,F}}{\sqrt{\left[\left(\frac{k}{a}\right)^2 - \left(\frac{m}{a}\right)^2\right]^2 + \left(\frac{L}{a^2} + \frac{R_0^2}{L}\right)^2 \left(\frac{m}{a}\right)^2}} \quad (4.3)$$

Here R_0 is the Rossby number for eddies and L is the characteristic Rossby radius, which are both taken as constants. $A_{m,F}$ is the amplitude of the forcing at the equivalent barotropic level (in units of m/s) for wavenumber m, and is a function of temperature and orography. Equation (4.3) shows that when k is close to m (i.e., when the forcing pattern is close to that of the trapped free wave), the denominator becomes small and thus A_m large. This is the essence of resonant amplification. Note that the forcing $A_{m,F}$ can thus be moderate (although not zero) for resonance and hence high wave amplitudes to occur.

4.2.3. Resonance Occurrence

Based upon the above described criteria, resonance events in the Northern Hemisphere have been identified to occur primarily in summer and also spring when the background flow is relatively weak. In these seasons, so-called double jet flow configurations, characterized by two peaks in the zonal mean zonal wind, are much more likely to occur. Such double jet states favor waveguide formation due to the unusual shape of the zonal mean zonal wind across latitudes, which primarily determines l^2. From the thermal wind equation it follows that in order to generate double jet states, sharp poleward temperature gradients are needed at mid (~45°N) and subpolar (~70°N) latitudes with weaker gradients in between. Weak thermal gradients around 50°N–65°N are unlikely to occur in winter and autumn when the equator to pole temperature difference is much larger. This is likely the underlying reason that double jet states and hence resonance events are limited to summer and spring, when the overall equator to pole temperature drop is much more moderate. Likewise, the cluster of resonance events in recent summers (see section 4.3.2, Cluster of Resonance Events) might thus be related to the significant reduction in the meridional temperature gradient around 50°N over the last decades (Fig. 4.1).

4.3. REAL EVENTS

4.3.1. Case Study: Summer 2010

The summer of 2010 saw anomalous large-scale flow patterns over the Eurasian continent, which has been identified as a key factor behind both the Russian heat wave and Pakistan flooding that summer [*Schubert et al.*, 2011; *Schneidereit et al.*, 2012; *Lau and Kim*, 2012; *Tachibana et al.*, 2010; *Petoukhov et al.*, 2013; *Schubert et al.*, 2014]. These extreme events illustrated how destructive extreme weather can be to societies: The death toll in Moscow has been estimated at 11,000 and drought-caused grain-harvest losses of 30%, leading to a Russian export ban on wheat [*Coumou and Rahmstorf*, 2012]. At the same time Pakistan was hit by the worst flooding in its history, which affected approximately one-fifth of its total land area and 20 million people [*Hong et al.*, 2011; *Coumou and Rahmstorf*, 2012]. The heat wave over western Russia was due to a blocking high, which resulted in record temperatures exceeding the extreme European heat wave of 2003 in amplitude and spatial extent [*Dole et al.*, 2011; *Rahmstorf and Coumou*, 2011; *Barriopedro et al.*, 2011]. In fact, based upon a global heat wave index, the event was quantified as the strongest heat wave ever [*Russo et al.*, 2014]. Especially striking was the long lifetime of the blocking event: Roughly 30% of summer days in western Russia were considered as blocking days where normally this is only about 10% [*Schneidereit et al.*, 2012]. The anticyclonic flow anomaly was associated with a stationary Rossby wave train extending over the full Eurasian continent and beyond. In fact, several studies have highlighted the hemispheric nature of these flow anomalies [*Tachibana et al.*, 2010; *Schubert et al.*, 2011; *Petoukhov et al.*, 2013]. The trough downstream of the western Russia anticyclone was instrumental in triggering anomalously heavy rainfall over northern Pakistan leading to flooding of the Indus River [*Lau and Kim*, 2012].

Figure 4.4 shows Hovmöller diagrams of a set of zonal mean dynamic variables during the summer of 2010 over the Northern Hemisphere. In early July, a pronounced second peak in the zonal mean zonal wind emerges at about 70°N creating a clear double jet flow configuration, which persists until half of August (Fig. 4.4a). This shift in the zonal mean flow regime is associated with a pronounced warming of regions around 60°N, that is, the approximate latitude of Moscow. In fact, in the single jet regime (June and late August), pronounced positive temperature anomalies are seen over the Arctic Ocean (North

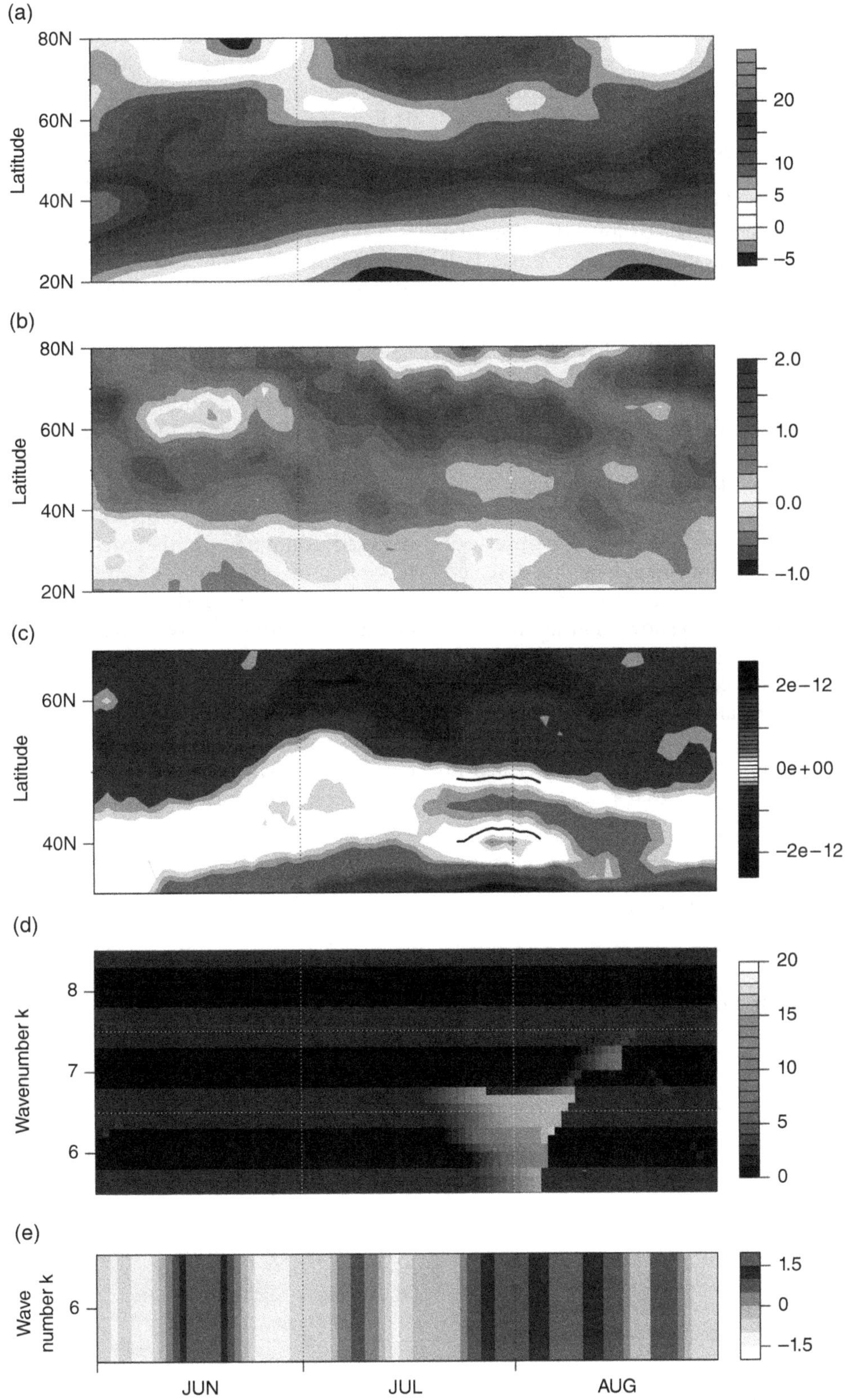

Figure 4.4 Double jet and resonant amplification during summer of 2010. Northern Hemisphere Hovmöller diagrams during summer 2010 for (a) zonal wind, (b) temperature anomaly, and (c) $l^2(k = 6)$. Variables are calculated from the ERA-interim reanalysis using a 15 day running mean low-frequency bandpass filter and averaged over all longitudes. Temperature anomalies from climatology (panel b) were calculated at each grid point before averaging over all longitudes. The black lines in panel (c) indicate the emergence of the two turning points of the waveguide for wave 6. Panel (d) plots the number of consecutive days for which resonance conditions were fulfilled for waves 6–8, and panel (e) plots the observed amplitude of wave 6. (*See insert for color representation of the figure.*)

of ~70°N) and further south (between 40°–50°N), while in the double jet regime they are concentrated around 60°N (Fig. 4.4b). The zonal mean flow changes result in the formation of a waveguide by mid-July, which becomes pronounced in late July and early August (Fig. 4.4c). Two turning points can be distinguished in Figure 4.4c: one at 40°N and one at about 50°N (solid black lines), that is, placed roughly in between the latitudes of the heat wave in Moscow and the flooding in Pakistan. Following the criteria defined above, l^2 is positive between those latitudes and changes sign at the turning points themselves. Further, the zonal mean flow is clearly positive over all latitudes north of about 30°N including the position of the waveguide. In addition, the turning points are located at sufficiently large distance and thus all criteria for a waveguide trapping waves with wave numbers close to 6 in the midlatitudes were fulfilled. Since also the forcing A_F for wave 6 was sufficiently large (see equation [4.3]), the trapped wave underwent resonant amplification (Fig. 4.4d). This amplification of wave 6 is also observed: It starts by mid-July when the waveguide is first detected and maximum amplitudes are reached in early August (red in Fig. 4.4e). This panel also shows that in mid-June similarly high wave amplitudes are observed which were not related to the resonance mechanism, highlighting that of course other mechanisms can also cause high-amplitude waves.

This analysis, and more detailed analyses described in [*Petoukhov et al.*, 2013], illustrate that resonance played an important role in creating high-amplitude quasi-stationary Rossby waves from mid-July to early August 2010, which was the most severe period of the Russian heat wave and which includes the period of flooding in Pakistan.

4.3.2. Cluster of Resonance Events

Based upon the criteria discussed above and analyzing monthly data for July and August, *Petoukhov et al.* [2013] identified 19 resonance months since 1980 (Fig. 4.5). Many of these months were associated with memorable extreme weather events, including severe heat waves and flooding events in Europe and heat waves and droughts in the United States. Coinciding with the onset of rapid Arctic amplification in 2000, there appears to be a cluster of resonance events though its significance cannot be assessed. Nevertheless, in the post-2000 period, the frequency of July or August resonance months has almost doubled compared to the pre-2000 period. Especially, resonance involving waves 7 and 8 seems to have occurred more frequently recently. Spectral analysis of daily data of July and August days, *Coumou et al.* [2014] showed that there indeed have been significant changes with more frequent high-amplitude quasi-stationary waves with these wave numbers. In the post-2000 period, the number of days with a quasi-stationary wave 7 with amplitudes larger than 3 m/s increased by 30%, and, with amplitudes larger than 5 m/s, have more than doubled [*Coumou et al.*,

7–2011	Heat wave in the United States
7/8–2010	Russian heat wave and Pakistan flood
7–2006	European heat wave
8–2004	Winter like temperatures in Northern Europe
8–2003	European summer 2003 heat wave
8–2002	Elbe and Danube floods in Europe
7–2000	Floods in northern Italy and the Tisza basin, heat wave in the southern U.S.
7/8–1997	Great European flood, floods in Pakistan and western U.S.
7–1994	Heat wave in southern Europe
7–1993	Unprecedented flood in the U.S.
7–1989	Widespread drought in U.S.
8–1987	Severe drought in the southeastern U.S.
8–1984	Severe heat and drought in the U.S.
7/8–1983	Severe heat and drought in U.S. mid-west

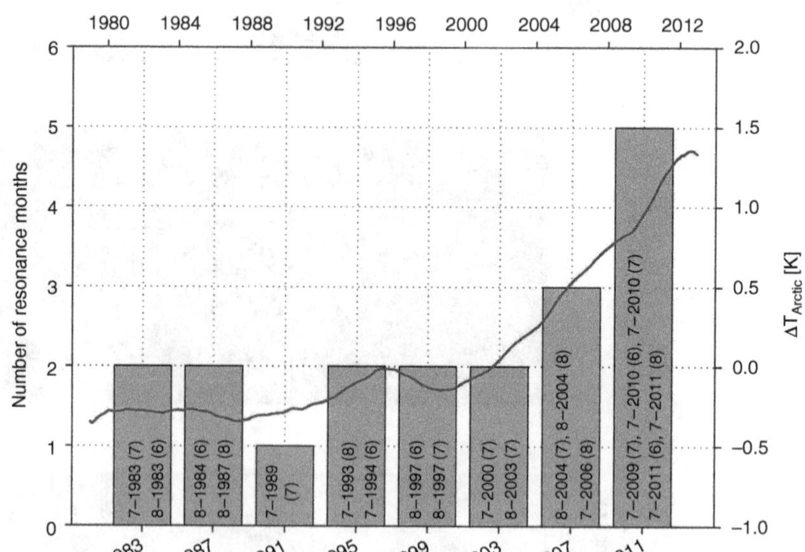

Figure 4.5 Number of identified July and August resonance events [*Petoukhov et al.,* 2013] binned together in 4 yr periods from 1980 to 2011. Text in the grey bars indicates the actual months and wave numbers involved (in brackets) and the table on the left lists the associated extreme weather events. The red line plots the difference in surface warming in the Arctic (north of 65°N) and in the rest of the Northern Hemisphere (south of 65°N), based upon the temperature dataset by *Cowtan and Way* [2013]. From *Coumou et al.* [2014]. *See electronic version for color representation.*

2014]. These findings are also consistent with the independent analysis as plotted in the lowest panel of Figure 4.1. This shows a moderate upward trend in the occurrence-frequency of a quasi-stationary wave 7 with an amplitude exceeding the seventieth percentile. This threshold is somewhat subjective and the upward trend is only marginally significant and thus this finding should be treated with caution. It nevertheless is consistent with the cluster analysis presented in *Coumou et al.* [2014]. We thus argue that the recent cluster of resonance events has led to a detectable increase in the overall frequency of high-amplitude quasi-stationary wave 7 since 2000. This occurred while transient wave activity, best captured by EKE, declined pronouncedly.

4.3.3. Statistics of Resonance Events

Based on the resonance periods identified by *Petoukhov et al.* [2013], *Coumou et al.* [2014] performed statistical analysis to quantify how anomalous upper-level wind fields and surface weather were during resonance events. To determine upper-level wave activity, spectral analysis of the 500 mb and 300 mb daily meridional wind field over a midlatitudinal belt stretching from 35°N to 65°N were performed using both the ERA-interim [*Dee et al.*, 2011] and NCEP-NCAR [*Kalnay et al.*, 1996] reanalyses. Since the meridional wind field was analyzed, wave amplitudes reflect actual north-south wind speeds. Further, also phase-speed and the daily zonal mean zonal wind in this midlatitudinal belt were extracted [*Coumou et al.*, 2014]. Figure 4.6 summarizes the key findings of these analyses presented as 2D probability density plots of phase-speed with respect to wave amplitude (top panels) and zonal mean wind (bottom panels) for waves 6 (left panels), 7 (middle panels), and 8 (right panels). The July-August climatological mean (solid curves) shows that the wave spectrum is dominated by eastward traveling waves (positive phase speed) with the speed increasing with wave number, as predicted by linear Rossby wave theory (equation [4.1]). Still, especially for waves 6 and 7, a sizeable proportion of the probability density distribution is associated with quasi-stationary waves (with an absolute phase speed c less than ~2 m/s) or with westward propagating waves. However, these slow moving waves

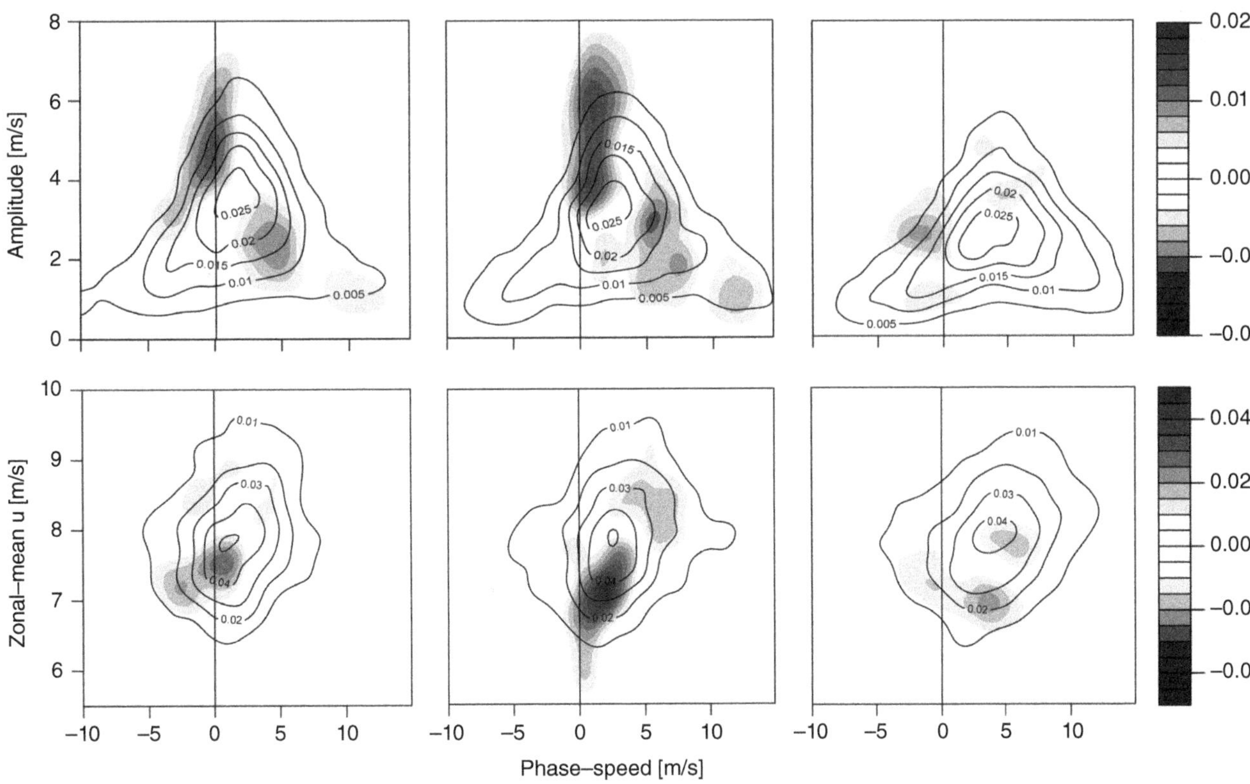

Figure 4.6 2D probability density distributions for daily values of phase speed (horizontal axis) against respectively (top) wave amplitude and (bottom) zonal mean zonal wind for (left) wave 6, (middle) wave 7, and (right) wave 8. Variables are based on ERA-interim reanalysis at 500 mb, averaged over 35°N–65°N for days in July–August (solid lines). Shading indicates anomalies from July-August climatology during resonance events showing increased frequencies of high-amplitude quasi-stationary waves. Adapted from *Coumou et al.* [2014]. *See electronic version for color representation.*

generally have low amplitudes. Shading in Figure 4.6 shows the anomaly in the distributions during resonance periods, showing distinct patterns. First of all, a pronounced increase in occurrence-frequency of high-amplitude quasi-stationary waves is observed, notably for waves 6 and 7. Next, the occurrence-frequency of fast-moving waves (i.e., faster than the July-August climatological mean phase speed) is substantially reduced, again most pronounced for waves 6 and 7. Further, *Coumou et al.* [2014] report that the mean phase speed during resonance periods reduces by more than a factor of two for wave 7 and by more than a factor of three for wave 6. Also distinct patterns are observed in the anomaly for the zonal mean zonal wind spectra (Fig. 4.6, bottom panels). Generally, a weakening of the zonal flow is observed, which is most pronounced for wave 7 (by roughly 5% in the mean). This weakening can thus explain some of the observed slowdown in wave propagation (following equation [4.1]) but the change is too small to fully account for the strong reduction in phase speed observed. Moreover, also the high-amplitude character of the quasi-stationary waves requires an amplification mechanism, not just a slowdown.

In a next step, it is analyzed whether surface weather conditions were also more extreme during these resonance months. To quantify this a simple midlatitudinal extreme (MEX) index is defined [*Coumou et al.*, 2014]:

$$MEX(x,t) = \frac{\left(\frac{1}{N}\sum_{i}^{N}\left(\frac{\Delta x_i(t)}{\sigma(x_i)}\right)^2 - \mu_{MEX}\right)}{\sigma_{MEX}} \quad (4.4)$$

Here x can in principle refer to any meteorological variable defined in the midlatitudes (35°N–65°N) but here we use near-surface temperature, with $\Delta x_i(t)$ its anomaly at time step t and $\sigma(x_i)$ its standard deviation at location i. The MEX index is normalized by subtracting its time-averaged mean (μ_{MEX}) and division by its standard deviation (σ_{MEX}) such that the climatological distribution centers around 0 and is defined in units of standard deviation [*Coumou et al.*, 2014]. If the MEX index has large positive values, this indicates that extreme temperatures (hot or cold) occur simultaneously in many locations throughout the midlatitudes. Combining hot and cold anomalies (by means of the square in equation [4.4]) makes sense as strong wave activity is expected to bring hot conditions in some regions and cold conditions in others [*Screen and Simmonds*, 2014].

Figure 4.7 shows probability density distributions for the MEX index based on daily and monthly mean surface temperatures. The July-August climatological distributions (black) are clearly asymmetric with a fat extreme tail and center around 0 (by definition). The daily heat index during resonance periods shows a statistically significant shift (p-value < 0.05) to more extremes with MEX indices beyond 2 standard deviations becoming much more likely. Moreover, the distribution of the monthly MEX index shows even more pronounced changes during resonance, with a significant shift (p-value < 0.05) toward more extremes. The very pronounced changes seen in monthly data, as compared to more moderate changes in daily data, indicates that it was especially the persistence in weather that led to extreme heat on longer, that is, monthly, timescales.

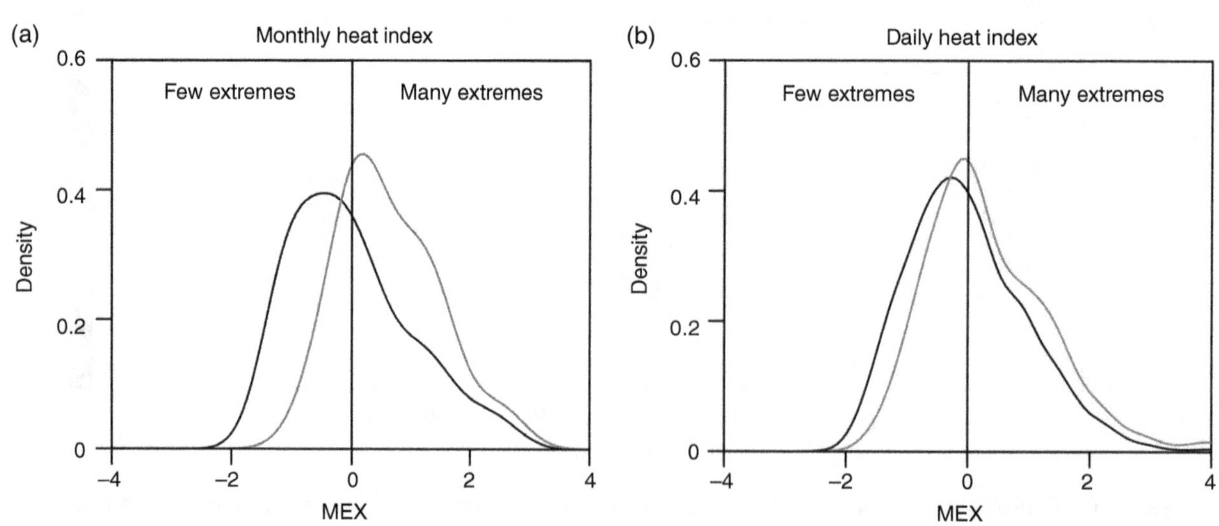

Figure 4.7 Midlatitudinal extreme index MEX (equation [4.4]) in units of standard deviation for (a) monthly heat extremes and (b) daily heat extremes for July-August climatology (black) and resonance events (gray). Adapted from *Coumou et al.* [2014]. *See electronic version for color representation.*

These statistical analyses thus reveal that, in support of the theoretical insights described above, resonance periods are characterized by (1) high-amplitude, slowly propagating waves and (2) persistent, and therefore extreme, surface weather.

4.4. CONCLUSIONS AND DISCUSSIONS

Boreal summer circulation has seen pronounced changes over the last decades, trends which seem to have amplified since the onset of rapid Arctic amplification around 2000. Especially the reduction in EKE, but also in zonal mean flow, have created conditions favorable for the buildup of heat and drought conditions over the continents. Moreover, a cluster of resonance events is observed since 2000, which has increased the occurrence-frequency of high-amplitude quasi-stationary waves with wave numbers close to 7. This generally implies a weakening of transient synoptic eddy activity and more-frequent states of quasi-stationary flow. The intimate links between quasi-stationary forced waves and transient synoptic eddies [*Chang et al.*, 2002] are not addressed here, but it seems reasonable to assume that a reduction in fast-wave activity will favor the occurrence of quasi-stationary flow. In fact, *Horton et al.* report significant increasing trends in anticyclonic summer circulation over the eastern United States, Europe, and western Asia since 1979 [*Horton et al.*, 2015]. Blocking high-pressure anticyclones are occurring more often and also persist longer in these regions [*Horton et al.*, 2015]. As stated above, these regions are directly influenced by the North Atlantic storm track and thus weakening EKE is likely to favor anticyclonic flow here. In any case, these different detected changes in observed large-scale summer circulation strongly point toward more-persistent flow patterns and therefore more extreme surface weather. This is also consistent with the dramatic increase in heat extremes in Europe and some other midlatitude regions [*Coumou et al.*, 2013; *Coumou and Robinson*, 2013; *Russo et al.*, 2014; *Christidis et al.*, 2014]. Also climate models project these trends in boreal summer circulation to continue under future high-emission scenarios [*Lehmann et al.*, 2014].

The underlying drivers behind these circulation changes are currently less well understood although they are generally consistent with theories and numerical model experiments involving Arctic change [*Francis and Vavrus*, 2012; *Cvijanovic and Caldeira,* 2015; *Schneider et al.*, 2014; *Petrie et al.*, 2015 ; also see Chapter 2]. The classical way to attribute drivers behind observed changes involves numerical modeling experiments with either in- or excluding certain forcing components [*Screen et al.*, 2013; *Trenberth et al.*, 2014; *Cvijanovic and Caldeira*, 2015]. Though certainly a lot can be learned this way, it has been shown that current GCMs appear deficient in simulating near stationary Rossby wave patterns during summer [*Schubert et al.*, 2011; *Overland et al.*, 2012]. A supplementary approach would be to apply recently developed causal-discovery algorithms [*Runge et al.*, 2012, 2014] to disentangle the contributions from different drivers in observational and climate model data. Such methods can detect and quantify causal interactions and they have successfully been applied to disentangle causality in ocean circulation [*Schleussner et al.*, 2014] and atmospheric circulation in the tropics [*Runge et al.*, 2014].

Most recent studies on trends in persistent circulation and extreme events have focused on boreal winter circulation [*Cohen et al.*, 2014; *Walsh*, 2014]. In this season, recent Arctic amplification has been strongest [*Cohen et al.*, 2014] and also possible tropical drivers are thought to be much more important [*Trenberth et al.*, 2014] (see Chapter 4 for tropical influences). Irrespective of these possible far-away drivers, the equator-to-pole temperature gradient in the midlatitudes has seen the most pronounced changes in boreal summer [*Coumou et al.*, 2015]. Further, due to its smaller year-to-year variability as compared to the other seasons, long-term trends in summer circulation are detectable at an earlier stage. But apart from that, the important physical mechanisms are likely to be quite different between warm and cold season. For example, in winter the variability in the polar night jet and sudden stratospheric warming events are thought to be important in forcing tropospheric circulation [*Baldwin and Dunkerton*, 2001; *Kim et al.*, 2014]. In summer such stratospheric interactions are unlikely to play a role. Clearly more research is needed to understand the differences between cold- and warm-season circulations including the drivers behind their variability. Much can be learned by analyzing possible resonance events in nonsummer seasons and also in the Southern Hemisphere. Due to its almost opposite nature in terms of land-ocean distribution, the Southern Hemisphere can be used to test different theories. For example, in boreal summer, double jet circulation patterns occur relatively frequently whereby the poleward jet forms around 70°N, that is, roughly at the boundary of the Arctic Ocean where a strong thermal contrast forms. Therefore, double jet patterns are somewhat unlikely in the Southern Hemisphere (although this remains to be tested) and this could thus provide insights in the relative role of double jets in triggering resonance events if the latter appear to exist in the Southern Hemisphere. Also long-term reanalysis, like the twentieth century reanalysis [*Compo et al.*, 2011], and climate models from the Coupled Climate Model Intercomparison Project [*Covey et al.*, 2003] should be analyzed for resonance events in both hemispheres. Key questions to address are whether climate models can actually reproduce such nonlinear wave dynamics and what the links are with the typical modes of variability like the Arctic Oscillation or ENSO.

Atmosphere dynamical changes associated with long-term climate change are a key contributor to uncertainty in future climate model projections [*Shepherd*, 2014; *Bony et al.*, 2015]. Here we have highlighted the importance of atmospheric circulation in causing devastating extreme surface weather events throughout the Northern Hemisphere midlatitudes. To reduce uncertainty in climate model projections and assess future impacts from extreme weather, the underlying processes and drivers will need to be understood in much more detail.

ACKNOWLEDGMENTS

JL was supported by the German Research Foundation (DFG, CO994/2-1) and DC and KK were supported by the German Federal Ministry of Education and Research (BMBF, 01LN1304A).

REFERENCES

Alexander, L. (2010), Extreme heat rooted in dry soils, *Nat. Geosci.*, *4*, 12–13.

Baldwin, M. P., and T. J. Dunkerton (2001), Stratospheric harbingers of anomalous weather regimes, *Science*, *294*, 581–584.

Barriopedro, D., E. M. Fischer, J. Luterbacher, R. M. Trigo, and R. García-Herrera (2011), The hot summer of 2010: Redrawing the temperature record map of Europe, *Science*, *332*, 220–224.

Bony, S., B. Stevens, D. M. W. Frierson, C. Jakob, M. Kageyama, R. Pincus, T. G. Shepherd, S. C. Sherwood, A. P. Siebesma, A. H. Sobel, M. Watanabe, and M. J. Webb (2015), Clouds, circulation and climate sensitivity, *Nat. Geosci.*, *8*, 261–268.

Chang, E. K. M., S. Lee, and K. Swanson (2002), Storm track dynamics, *J. Climate*, *15*, 2163–2183.

Christidis, N., G. S. Jones, and P. Stott (2014), Dramatically increasing chance of extremely hot summers since the 2003 European heatwave, *Nature Climate Change*, *5*, 46–50.

Cohen, J., J Screen, J C Furtado, M Barlow, D Whittleston, D Coumou, J Francis, K Dethloff, D Entekhabi, J Overland, and J. Jones (2014), Recent Arctic amplification and extreme mid-latitude weather, *Nat. Geosci.*, *7*, 627–637.

Compo, G. P., J. S. Whitaker, P. D. Sardeshmukh, N. Matsui, R. J. Allan, X. Yin, B. E. Gleason, R. S. Vose, G. Rutledge, P. Bessemoulin, S. Brönnimann, M. Brunet, R. I. Crouthamel, A. N. Grant, P. Y. Groisman, P. D. Jones, M. C. Kruk, A. C. Kruger, G. J. Marshall, M. Maugeri, H. Y. Mok, Ø. Nordli, T. F. Ross, R. M. Trigo, X. L. Wang, S. D. Woodruff, and S. J. Worley (2011), The Twentieth Century Reanalysis Project, *Quart. J. Roy. Meteor. Soc.*, *137*, 1–28.

Coumou, D., and A. Robinson (2013), Historic and future increase in the global land area affected by monthly heat extremes, *Environ. Res. Lett.*, *8*, 1–6.

Coumou, D., and S. Rahmstorf (2012), A decade of weather extremes, *Nature Climate Change*, *2*, 491–496.

Coumou, D., A. Robinson, and S. Rahmstorf (2013), Global increase in record-breaking monthly-mean temperatures, *Climate Change*, *118*, 771–782.

Coumou, D., J. Lehmann, and J. Beckmann (2015), The weakening summer circulation in the Northern Hemisphere mid-latitudes, *Science*, *348*, 324–327.

Coumou, D.,V. Petoukhov, and A. V. Eliseev (2011), Three-dimensional parameterizations of the synoptic scale kinetic energy and momentum flux in the Earth's atmosphere, *Nonlinear Process. Geophys.*, *18*, 807–827.

Coumou, D., V. Petoukhov, S. Rahmstorf, S. Petri, and H. J. Schellnhuber (2014), Quasi-resonant circulation regimes and hemispheric synchronization of extreme weather in boreal summer, *Proc. Natl. Acad. Sci. USA*, *111*, 12331–12336.

Covey, C., K. M. Achutarao, U. Cubasch, P. Jones, S. J. Lambert, M. E. Mann, T. J. Phillips, and K. E. Taylor (2003), An overview of results from the Coupled Model Intercomparison Project, *Global Planet. Change*, *37*, 103–133.

Cowtan, K., and R. G. Way (2013), Coverage bias in the HadCRUT4 temperature series and its impact on recent temperature trends, *Quart. J. Roy. Meteor. Soc.*, *133*, 459–477.

Cvijanovic, I., and K. Caldeira (2015), Atmospheric impacts of sea ice decline in CO2 induced global warming, *Climate Dyn.*, *44* (5–6); doi: 10.1007/s00382-015-2489-1.

Dee, D. P., S. M. Uppala, A. J. Simmons, P. Berrisford, P. Poli, S. Kobayashi, U. Andrae, M. A. Balmaseda, G. Balsamo, P. Bauer, P. Bechtold, A. C. M. Beljaars, L. van de Berg, J. Bidlot, N. Bormann, C. Delsol, R. Dragani, M. Fuentes, A. J. Geer, L. Haimberger, S. B. Healy, H. Hersbach, E. V.. Hólm, L. Isaksen, P. Kållberg, M. Köhler, M. Matricardi, A. P. McNally, B. M. Monge-Sanz, J-J. Morcrette, B-K. Park, C. Peubey, P. de Rosnay, C. Tavolato, J-N. Thépaut, and F. Vitart (2011), The ERA-interim reanalysis: Configuration and performance of the data assimilation system, *Quart. J. Roy. Meteor. Soc.*, *137*, 553–597.

Dole, R., M. Hoerling, J. Perlwitz, J. Eischeid, P. Pegion, T. Zhang, X. W. Quan, T. Xu, and D. Murray (2011), Was there a basis for anticipating the 2010 Russian heat wave?, *Geophys. Res. Lett.*, *38*.

Francis, J., and S. J. Vavrus (2015), Evidence for a wavier jet stream in response to rapid Arctic warming, *Environ. Res. Lett.*, *10* ; Online: http://stacks.iop.org/1748-9326/10/i=1/a=014005?key=crossref.74581076f734b2377ec8042d3aebe25d.

Francis, J. A., and S. J. Vavrus (2012), Evidence linking Arctic amplification to extreme weather in the mid-latitudes, *Geophys. Res. Lett.*, *39*, L06801.

Hong, C., H. Hsu, N. Lin, and H. Chiu (2011), Roles of European blocking and tropical-extratropical interaction in the 2010 Pakistan flooding, *Geophys. Res. Let.*, *38*, 6.

Horton, D. E., N. C. Johnson, D. Singh, D. L. Swain, B. Rajaratnam, and N. S. Diffenbaugh (2015), Contribution of changes in atmospheric circulation patterns to extreme temperature trends, *Nature*, *522*, 465–469.

Kalnay, E., M. Kanamitsu, R. Kistler, W. Collins, D. Deaven, L. Gandin, M. Iredell, S. Saha, G. White, J. Woollen, Y. Zhu, M. Chelliah, W. Ebisuzaki, W. Higgins, J. Janowiak, K. C. Mo, C. Ropelewski, J. Wang, A. Leetma, R. Reynolds, R. Jenne, and D. Joseph (1996), The NCEP/NCAR 40-year reanalysis project, *Bull. Amer. Meteor. Soc.*, 437–471.

Kim, B-M., S-W. Son, S-K. Min, J-H. Jeong, S-J. Kim, X. Zhang, T. Shim, and J-H. Yoon (2014), Weakening of the stratospheric polar vortex by Arctic sea-ice loss, *Nat. Commun.*, *5*, 4646.

Kyselý, J. (2008), Influence of the persistence of circulation patterns on warm and cold temperature anomalies in Europe: Analysis over the 20th century, *Global Planet. Change*, *62*, 147–63.

Kyselý, J., and R. Huth (2005), Changes in atmospheric circulation over Europe detected by objective and subjective methods, *Theor. App. Climatol.*, *85*, 19–36.

Lau, W. K. M., and K-M. Kim (2012), The 2010 Pakistan flood and Russian heat wave: Teleconnection of hydrometeorological extremes, *J. Hydrometeor.*, *13*, 392–403

Lehmann, J., and D. Coumou (2015), The influence of midlatitude storm tracks on hot, cold, dry, and wet extremes, *Sci. Rep.*, *5*, 1–9.

Lehmann, J., D. Coumou, K. Frieler, A. V. Eliseev, and A. Levermann (2014), Future changes in extratropical storm tracks and baroclinicity under climate change, *Environ. Res. Lett.*, *9*, 084002.

Mueller, B., and S. I. Seneviratne (2012), Hot days induced by precipitation deficits at the global scale, *Proc. Nat. Acad. Sci. USA*, *109*, 12398–12403.

O'Gorman, P. A. (2010), Understanding the varied response of the extratropical storm tracks to climate change, *Proc. Nat. Acad. Sci. USA*, *107*, 19176–19180.

Overland, J. E., J. A. Francis, E. Hanna, and M. Wang (2012), The recent shift in early summer Arctic atmospheric circulation, *Geophys. Res. Lett.*, *39*, L19804.

Palmer, T. (2014), Record-breaking winters and global climate change, *Science*, *344*, 803–804.

Pedlosky, J. (1979), *Geophysical Fluid Dynamics*, Springer, New York.

Petoukhov, V., S. Rahmstorf, S. Petri, and H. J. Schellnhuber (2013), Quasiresonant amplification of planetary waves and recent Northern Hemisphere weather extremes., *Proc. Nat. Acad. Sci. USA*, *110*, 5336–5341.

Petrie, R. E., L. C. Shaffrey, and R. T. Sutton (2015), Atmospheric response in summer linked to recent Arctic sea ice loss, *Quart. J. Roy. Meteor. Soc.*, *141*, 2070–2076.

Rahmstorf, S., and D. Coumou (2011), Increase of extreme events in a warming world, *Proc. Natl. Acad. Sci. USA.*, *108*, 17905–17909.

Runge, J., J. Heitzig, N. Marwan, and J. Kurths (2012), Quantifying causal coupling strength: A lag-specific measure for multivariate time series related to transfer entropy, *Phys. Rev. E*, *86*, 061121.

Runge, J., V. Petoukhov, and J. Kurths (2014), Quantifying the strength and delay of climatic interactions: The ambiguities of cross correlation and a novel measure based on graphical models, *J. Climate*, *27*, 720–739.

Russo, S., A. Dosio, R. G. Graversen, J. Sillmann, H. Carrao, M. B. Dunbar, A. Singleton, P. Montagna, P. Barbosa, and J. V. Vogt (2014), Magnitude of extreme heat waves in present climate and their projection in a warming world, *J. Geophys. Res. Atmos.*, *119*, 12,500–12,512.

Schär, C., P. L. Vidale, D. Lüthi, C. Frei, C. Häberli, M. A. Liniger, and C. Appenzeller (2004), The role of increasing temperature variability in European summer heatwaves, *Nature*, *427*, 332–336.

Schleussner, C. F., J. Runge, J. Lehmann, and A. Levermann (2014), The role of the North Atlantic overturning and deep ocean for multi-decadal global-mean-temperature variability, *Earth Syst. Dyn.*, *5*, 103–115.

Schneider, T., T. Bischoff, and H. Plotka (2014), Physics of changes in synoptic midlatitude temperature variability, *J. Climate*, *28* (6); doi: 10.1175/JCLI-D-14-00632.1.

Schneidereit, A., S. Schubert, P. Vargin, F. Lunkeit, X. Zhu, D. H. W. Peters, and K. Fraedrich (2012), Large-scale flow and the long-lasting blocking high over Russia: Summer 2010, *Monthly Weather Rev.*, *140*, 2967–2981.

Schubert, S., H. Wang, and M. Suarez (2011), Warm season subseasonal variability and climate extremes in the Northern Hemisphere: The role of stationary Rossby waves, *J. Climate*, *24*, 4773–4792.

Schubert, S. D., H. Wang, R. D. Koster, M. J. Suarez, and P. Y. Groisman (2014), Northern Eurasian heat waves and droughts, *J. Climate*, *27*, 3169–3207.

Screen, J. A., and I. Simmonds (2014), Amplified mid-latitude planetary waves favour particular regional weather extremes, *Nature Climate Change*, *4*, 704–709.

Screen, J. A., I. Simmonds, C. Deser, and R. Tomas (2013), The atmospheric response to three decades of observed Arctic sea ice loss, *J. Climate*, *26*, 1230–1248.

Shepherd, T. G. (2014), Atmospheric circulation as a source of uncertainty in climate change projections, *Nat. Geosci.*, *7*, 703–708.

Tachibana, Y., T. Nakamura, H. Komiya, and M. Takahashi (2010), Abrupt evolution of the summer Northern Hemisphere annular mode and its association with blocking, *J. Geophys. Res.*, *115*, 13.

Trenberth, K. E., J. T. Fasullo, G. Branstator, and A. S. Phillips (2014), Seasonal aspects of the recent pause in surface warming, *Nature Climate Change*, *4*, 911–916.

Walsh, J. E. (2014), Intensified warming of the Arctic: Causes and impacts on middle latitudes, *Glob. Planet. Change*, *117*, 52–63.

Wang, S. Y., L. E. Hipps, R. R. Gillies, X. Jiang, and A. L. Moller (2010), Circumglobal teleconnection and early summer rainfall in the US intermountain west, *Theor. App. Climatol.*, *102*, 245–252.

Wang, S. Y., R. E. Davies, and R. R. Gillies (2013), Identification of extreme precipitation threat across midlatitude regions based on short-wave circulations, *J. Geophys. Res. Atmos.*, *118*, 11059–11074.

Woollings, T., and M. Blackburn (2012), The North Atlantic jet stream under climate change and its relation to the NAO and EA patterns, *J. Climate*, *25*, 886–902.

5

Land Processes as the Forcing of Extremes: A Review

Min-Hui Lo, Tzu-Hsien Kuo, Hao-Wei Wey, Chia-Wei Lan, and Jen-Ping Chen

ABSTRACT

Climate extreme events are critical in the study of climate change due to their heightened societal and environmental impacts, such as water resource shortage, flooding, and agricultural loss. Under the global warming, extreme events could be more frequent and more intense. Since the supply of moisture over land is limited with large spatial inhomogeneity and highly influenced by climate variations, land-atmosphere interactions are regarded as a major source of uncertainty in climate models. Both positive and negative land-atmosphere feedbacks can coexist at different temporal and spatial scales. To what extent land-atmosphere interactions play a role in changing the pattern of climate extremes is of great importance. In this chapter, we review existing studies on the impact of land-atmosphere interactions on climate extremes, with a focus on soil moisture–rainfall feedbacks.

5.1. INTRODUCTION

Land process at the interface between land and atmosphere is a critical component to study in the climate system. The atmosphere can be heated by the sensible heat flux and moistened by evapotranspiration from the land surface below. For example, soil wetness affects the moisture supply for evaporation, and different properties of vegetation such as rooting depth and leaf area index can alter the amount of transpiration. The albedo of the different surfaces, which varies with land and vegetation types, can change the surface radiation budget. Changes in surface roughness can also affect wind speed and thus alter the heat flux released to the atmosphere.

The feedbacks of land to the atmosphere in different soil wetness states can be explained by the Budyko curve [*Budyko*, 1974]. The curve illustrates the ratio of actual evapotranspiration to maximum possible amount of evapotranspiration (i.e., potential evapotranspiration). When the land surface is saturated, the actual evapotranspiration is equal to potential evapotranspiration, which means that the local evapotranspiration is limited by the available input energy, the region is then defined as the wet regime or energy-limited regime. By contrast, if the land surface is not very wet (or dry) and actual evapotranspiration is smaller than the potential evapotranspiration, the region is defined as the dry regime or soil moisture-limited regime as shown in Figure 5.1 (Budyko-type curve). In soil moisture-limited regimes, adding moisture to the soil results in higher evapotranspiration, while in energy-limited regimes, extra moisture leads to no or minimal contributions to evapotranspiration [*Seneviratne et al.*, 2010]. A transition zone lies between wet and dry climates, where the atmosphere and land surface are strongly interacting or coupled [*Koster et al.*, 2004].

Changes in local evapotranspiration can have a strong impact on the atmosphere and affect land-atmosphere interactions. Enhanced knowledge of hydrological

Department of Atmospheric Sciences, National Taiwan University, Taipei, Taiwan

Climate Extremes: Patterns and Mechanisms, Geophysical Monograph 226, First Edition.
Edited by S.-Y. Simon Wang, Jin-Ho Yoon, Christopher C. Funk, and Robert R. Gillies.

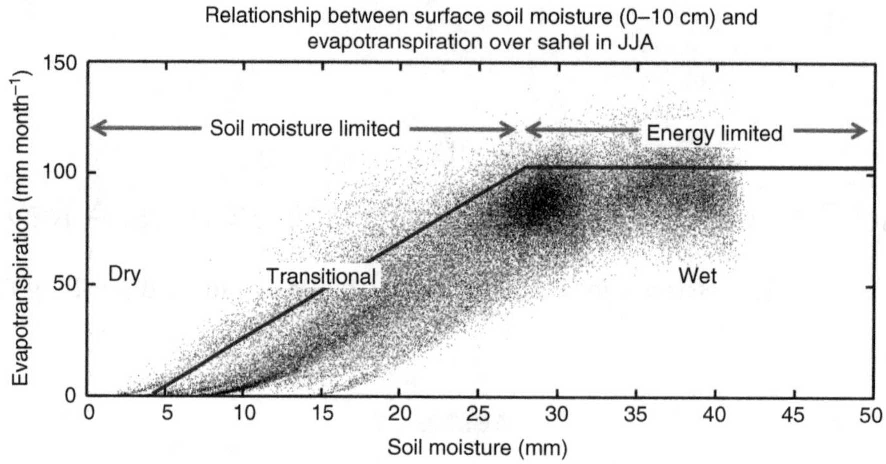

Figure 5.1 Schematic plot for the Budyko curve showing the nonlinear relationship between soil water and latent heat fluxes. The dots in the scatter plot are from Global Land Data Assimilation System (GLDAS) Noah model outputs over Sahel during boreal summer [*Rodell et al.*, 2004].

processes in land surface models can improve climate simulations on seasonal-to-interannual timescales. The Global Soil Wetness Project (GSWP) [*Dirmeyer et al.*, 1999] is an international land surface model intercomparison project, in which the land surface models were conducted in an off-line (uncoupled) mode, driven by prescribed atmospheric forcing datasets. The GSWP has provided a framework for comparisons and validations of land surface model simulations, and thus to improve and develop the model simulations [*Dirmeyer et al.*, 2011]. A follow-up international project, under the Global Energy and Water Cycle Experiment (GEWEX), the Global Land–Atmosphere Coupling Experiment (GLACE) [*Koster et al.*, 2004, 2006, 2010] have been used to study the land-atmosphere coupling strength and to explore hot spots over land, where soil moisture variations can significantly affect precipitation by ensemble experiments in coupled land-atmosphere models. The first phase of GLACE (GLACE-1) mainly addressed how the variations in precipitation and air temperature modulated by changes in land surface conditions [*Koster et al.*, 2006]. While the main contribution of GLACE-1 was to design an objective approach to isolate the land surface impacts on the atmosphere, the second phase of GLACE was targeted to quantify how realistic land surface initialization can improve seasonal forecast skill of precipitation and air temperature [*Koster et al.*, 2010]. In addition, under changing climatic conditions, improved understanding of the role of land processes and land-atmosphere interactions in climate and climate extremes becomes even more important [*Seneviratne et al.*, 2006; 2010]. Below we review existing studies on land surface processes, land subsurface processes, and climate extremes.

5.1.1. Land Surface Processes and Their Feedbacks to the Atmosphere

Precise and realistic climate modeling is essential for a clearer understanding of the ongoing and future global climate changes. Soil moisture is a major parameter influencing land-atmosphere interactions and thus is recognized as a major source of uncertainty in climate models [e.g., *Schär et al.*, 1999; *Koster et al.*, 2004; *Seneviratne et al.*, 2010; *Dirmeyer et al.*, 2013]. Soil moisture is highly spatially and temporally variable; therefore, knowledge of land hydrology is vital for determining moisture supply to the atmosphere through evapotranspiration. Soil moisture has a positive control on evapotranspiration, with generally negative feedback to air temperature [*Cheruy et al.*, 2013; *Lo and Famiglietti*, 2013]. However, the feedback of soil moisture to precipitation can be positive or negative, depending on local atmospheric stability and large-scale atmospheric circulation. For example, positive soil moisture–rainfall interaction has been observed in Illinois [*Findell and Eltahir*, 1997], in Kansas [*Eltahir*, 1998], and across the United States [*Koster et al.*, 2003], whereas negative feedbacks (wetter soil leading to less precipitation) have been observed in semiarid regions (e.g., over West Africa, Sahel, and Australia [*Taylor et al.*, 2012; *Guillod et al.*, 2015]), in which enhanced precipitation is generally accompanied by drier soils [*Hohenegger et al.*, 2009; *Taylor et al.*, 2012; *Guillod et al.*, 2015]. Negative feedbacks between soil moisture and the atmosphere precipitation have also been reported in the Amazon during the dry season [*Harper et al.*, 2014; *Lin et al.*, 2015].

Soil moisture can influence the aforementioned processes through delay and buffering effects, with consequences in the persistence of climate extremes. Land hydrological processes can record previous atmospheric forcing anomalies, and then manifest their effects in the following seasons or years. Land surface memory, defined as the lag response of soil moisture to precipitation [e.g., *Entekhabi et al.*, 1996; *Elthair,* 1998; *Lo and Famiglietti*, 2010], can record previous atmospheric forcing anomalies that propagate gradually to deeper soil layers and groundwater with a damping amplitude, and can later exert effects on the atmosphere at longer timescales (Fig. 5.2) [*Entekhabi et al.*, 1992, 1996]. The soil moisture–precipitation relationship depends on both precipitation frequency and the timescale of soil moisture retention [*Wei et al.*, 2008]. *Koster et al.* [2003] suggested that, during summer in midlatitude continental areas, soil moisture becomes more crucial than sea surface temperature in affecting atmospheric processes. An idealized simulation in a general-circulation model with intermediate complexity developed by *Chou et al.* [2001] showed that soil moisture variations could even affect the global monsoon system.

The effects of land-cover and land-use change (LCLUC) on the climate system have been a major concern for climate change assessments. LCLUC includes conversions of natural ecosystems into artificial ecosystems, mostly agricultural areas and some of urban areas [*Sala et al.*, 2000], having local and remote climate impacts. Soil moisture can affect precipitation directly and in situ (through changes in local evapotranspiration) or indirectly and ex situ (through changes in atmosphere circulations, which can affect precipitation patterns in remote areas) [*Kanamitsu and Mo*, 2003; *Mo*, 2008]. Therefore, LCLUC generally exerts a first-order effect (or direct effect) on local surface energy and water budget partitioning [*Cook et al.*, 2011] because modified soil wetness together with land and vegetation types cause changes in evapotranspiration and runoff [e.g., *Pielke*, 2001; *Adegoke et al.*, 2003; *Bonfils and Lobell*, 2007; *Kueppers et al.*, 2007; *Piao et al.*, 2007; *Lobell et al.*, 2009; *Sacks et al.*, 2009; *Sorooshian et al.*, 2011]. LCLUC can also have a second-order effect (or indirect effect) on atmospheric responses, for example, precipitation, which are more uncertain and affect larger and remote areas [*Yeh et al.*, 1984; *Puma and Cook*, 2010].

Converting forestland surface to agricultural and urban land surface modifies surface albedo, reduces soil moisture, and alters associated hydrological conditions [*Shepherd,* 2005; *Swann et al.*, 2012]. For example, increased or reduced water vapor flux because of irrigation, deforestation, and urbanization can modify convective available potential energy (CAPE), which has the potential to cause changes in spatiotemporal distributions and intensity of convection and precipitation [e.g., *Shepherd,* 2005; *Chen et al.*, 2007].

In addition, observational data also show an increase in the afternoon thunderstorm activity because of urbanization in Taipei [*Chen et al.*, 2007]. The urban heat island effect plays a crucial role in perturbing atmospheric thermodynamic and dynamic processes and has the potential of enhancing the rainfall system in downwind areas of northern Taiwan [*Lin et al.*, 2011].

5.1.1.1. Irrigation. Extensive irrigation and cultivation by humans result in changes in canopy resistance, soil albedo, and soil moisture that can modify surface evapotranspiration, particularly during the growing season

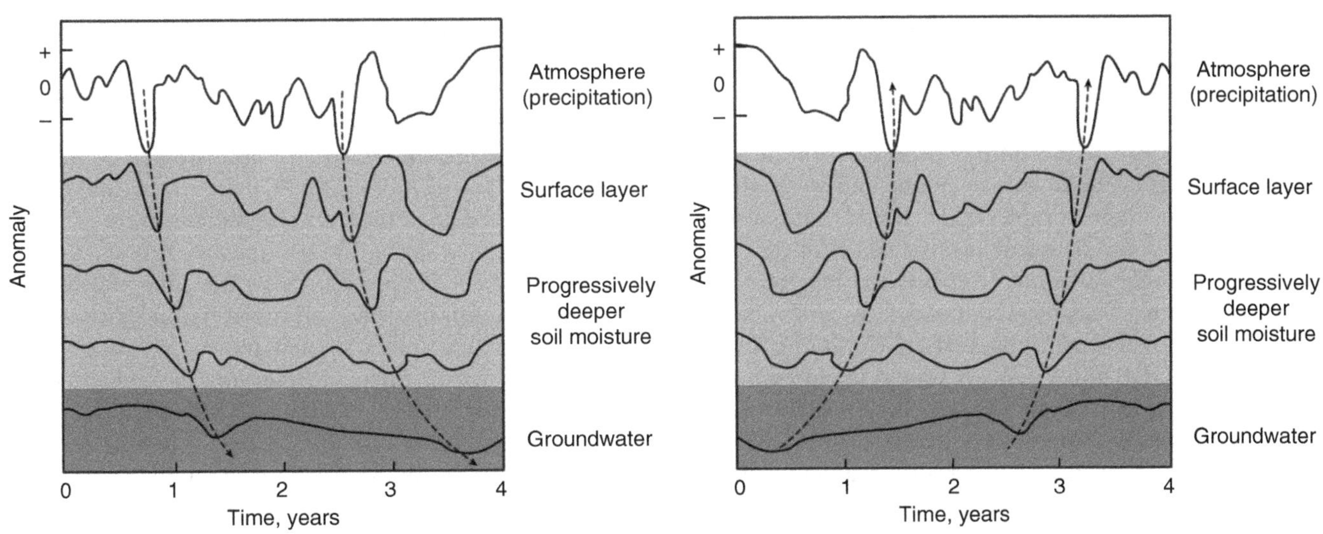

Figure 5.2 Schematic of propagation of precipitation anomalies in soil layers and groundwater (left), with reverse propagation the following season or year (right). Modified from *Changnon* [1987] and *Entekhabi et al.* [1992].

[*Chen et al.*, 2009; *Lo and Famiglietti*, 2013]. Irrigation can alter the partitioning of available land surface energy between sensible heat flux and latent heat flux by frequently increasing the latent heat flux and reducing the sensible heat flux, resulting in decreased surface temperature but increased air moisture [*Kueppers et al.*, 2008]. In addition, some research assessed irrigation impacts on local, regional, and global climate [e.g., *Gordon et al.*, 2005; *Douglas et al.*, 2006; *Bonfils and Lobell*, 2007; *Sacks et al.*, 2009; *Saeed et al.*, 2009; *Puma and Cook*, 2010; *Lee et al.*, 2011; *Lo and Famiglietti*, 2013; *Lo et al.*, 2013]. These studies indicated that agricultural irrigation could alter the regional climate by modifying water vapor and energy fluxes. Studies [e.g., *DeAngelis et al.*, 2010; *Puma and Cook*, 2010; *Lo and Famiglietti*, 2013] have shown that irrigation generally contributes significantly to increased precipitation in downwind regions. *Lee et al.* [2011] conducted a general circulation model (GCM) simulation to demonstrate that irrigation in the Middle East and central Asia in boreal spring and summer weakened the upper-level anticyclonic circulation, resulting in a weakened Asian jet during the summer.

5.1.1.2. Deforestation. *Negri et al.* [2004] found that, under weak synoptic-scale forcing, precipitation in deforested areas could be larger than that in high-density forests in the dry season. This conclusion was based on satellite observations showing less available evapotranspiration with larger convergence. In contrast, model simulations showed that in the dry season, deforestation could cause less evapotranspiration and higher temperature, which lead to reduced rainfall [*da Silva et al.*, 2008; *Lawrence and Vanecar*, 2015; *Lejeune et al.*, 2015]. Deforestation could also extend the dry season and delay the onset of monsoon [*Fu and Li*, 2004; *Costa et al.*, 2010]. Moreover, analysis of observational data showed that a deforested area presents a stable environment in the dry season, because CAPE is reduced in a drier atmosphere [*Wang et al.*, 2009; *Lejeune et al.*, 2015].

Irrigation and deforestation processes often exert opposite effects on land surface energy and water fluxes, which increase their complexity in the climate system. This excessive or deficient evapotranspiration from LCLUC can have substantial effects on the local and remote climate. The project Land-Use and Climate Identification of Robust Impacts (LUCID) [*Pitman et al.*, 2009] under IGBP and GEWEX was to explore impacts of LCLUC across multiple models. The LUCID project aimed to address the issues of (1) are there robust regional responses to LCLUC across different climate models and (2) are there remote impacts (teleconnections) commonly existing across multiple climate models? Answers to these questions will help in reducing the uncertainty and increasing the reliability of the current climate model simulations as well as for the future projections.

5.1.2. Land Subsurface Processes and Their Feedbacks to Soil Moisture and the Atmosphere

The effect of progressively deeper soil moisture and groundwater on soil moisture memory processes and the atmosphere has gathered considerable attention since the 1990s [e.g., *Famiglietti and Wood*, 1991, 1994; *Koster et al.*, 2000; *Liang et al.*, 2003; *Wu and Dickinson*, 2004; *Amenu et al.*, 2005; *Maxwell and Miller*, 2005; *Yeh and Eltahir*, 2005; *Fan et al.*, 2007; *Maxwell et al.*, 2007; *Miguez-Macho et al.*, 2007, 2008; *Niu et al.*, 2007; *Lo et al.*, 2008]. Groundwater can be stored in deep soil layers or aquifers, and thus has buffering effects on streamflow variability, surface soil moisture, and evapotranspiration [*Yeh and Eltahir*, 2005; *Kollet and Maxwell*, 2008; *Yeh and Famiglietti*, 2009]. Several studies have shown that deeper soil layers retain a longer memory of precipitation anomalies than surface layers do [e.g., *Liu and Avissar*, 1999a, b; *Wu et al.*, 2002; *Wu and Dickinson*, 2004; *Amenu et al.*, 2005]. These studies also pointed out the importance of representing shallow groundwater and its interaction with soil moisture in land surface hydrological simulations. For example, using model simulations, *Güntner et al.* [2007] showed that groundwater tends to contribute more to the interannual mode of total water storage variations than to seasonal variations because of a longer memory in groundwater aquifers. Thus, compared with surface soil moisture, deeper soil moisture and groundwater may play an even larger role in affecting climate (both the mean climate and climate variability) because of their longer memory timescale. Furthermore, *Liang et al.* [2014] showed that in the same warm phase of ENSO, changes in the location of El Niño (eastern Pacific and central Pacific El Niño) can cause asymmetrical responses in soil water storage and streamflow in the Mississippi River; however, without land subsurface hydrological storage, the impacts of EP and CP El Niño cannot persist from the winter season to subsequent seasons.

Based on the results from a simple stochastic model, *Bierkens and van den Hurk* [2007] suggested that long-term persistence in rainfall can be increased by groundwater convergence to shallow groundwater zones. *Lo and Famiglietti* [2010] further demonstrated that the hydrological connection between soil moisture and groundwater can affect the timing of soil moisture storage, and therefore play major roles in preserving the precipitation signals in terrestrial water storage and influencing streamflow in subsequent months or seasons. The authors also indicated that, depending on water table dynamics, groundwater can have nonlinear effects on the persistence of surface soil moisture. *Krakauer et al.* [2013] suggested that the addition of an aquifer to the Goddard Institute for Space Studies Model E general circulation model not only partially affected the mean climate, but also altered the seasonality and interannual persistence of soil moisture and climate. Using a coupled groundwater-land-atmosphere model for simulations over the Amazon River basin, *Lin et al.* [2015] showed that groundwater can increase latent heat fluxes and lower surface temperature, which increases the surface pressure gradient and, consequently, anomalous surface divergence. Therefore, dry season convection becomes even weaker when groundwater dynamics are included in the model. These findings indicate that subsurface and groundwater hydrological processes must be included in future numerical modeling studies to explore land-atmosphere interactions and the associated impacts on climate.

Understanding groundwater effects on soil moisture memory processes can help to more accurately quantify land-atmosphere interactions and climate feedback. Soil moisture generally increases after adding a groundwater component because of the additional supply of water to the root zone. However, land surface hydrological memory does not always increase when a groundwater component is incorporated, as demonstrated by *Lo and Famiglietti* [2010]. The authors showed that in regions where the water table level was intermediate, land surface hydrological memory could even decrease when soil moisture and capillary action from groundwater were not in phase with each other. Depending on water table dynamics, groundwater feedback to land surface memory can be positive, negative, or neutral (Fig. 5.3). Hence, in

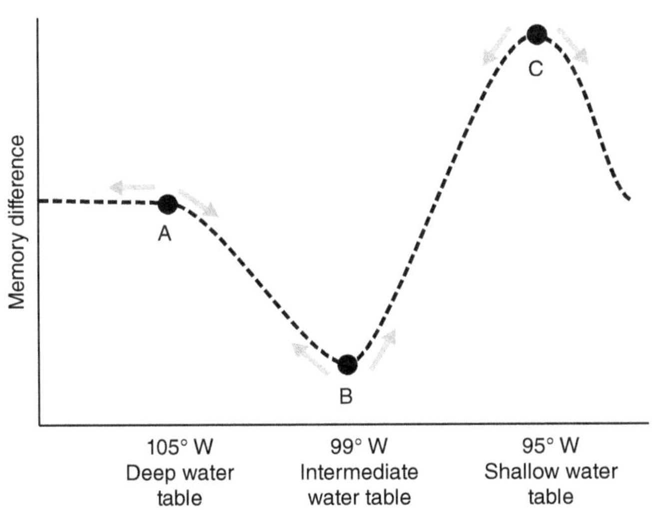

Figure 5.3 Schematic of memory difference variations with mean water table depth: point A represents the decoupled region; point B represents the region where groundwater has decreased land surface hydrologic memory the most; point C represents the regions where groundwater has increased land surface hydrologic memory the most. Gray arrows indicate how land surface hydrologic memory changes (increased, decreased, or not changed) with water table dynamics. Reprinted from *Lo and Famiglietti* [2010].

critical water table depth, the groundwater can significantly affect the surface energy budget and evapotranspiration [*Famiglietti and Wood*, 1994, 1995; *Kollet and Maxwell*, 2008; *Maxwell and Kollet*, 2008; *Lo and Famiglietti*, 2010]. Thus, a positive feedback of groundwater to land surface hydrological memory is observed in a transition zone between deep and shallow water tables [*Lo and Famiglietti*, 2011] because in the transition zone, capillary fluxes act as a buffer by reducing high-frequency soil moisture variations, thereby resulting in longer land surface hydrological memory. *Lo and Famiglietti* [2011] further indicated that, in the transition climatic zone, where land and atmosphere are strongly coupled, an increase in precipitation was observed in the GCM simulations when including groundwater dynamics. A positive feedback of land-atmosphere interaction was the major contributor to increased precipitation.

5.1.3. Climate Extremes

When studying the impact of global warming, it is even more essential to explore changes in climate extremes than changes in the climate mean state [*Trenberth et al.*, 2003], because the extreme events are expected to cause more societal damage [*Zhang et al.*, 2011; *IPCC*, 2012]. For example, heat wave, one of the climate extremes, is a prolonged period of higher than normal temperature, occasionally accompanied by humid weather. The European continent is occasionally subjected to heat wave events, which occur mostly in boreal summer. Therefore, many studies have been conducted to assess influence of land processes on heat waves for the present-day climate as well as in climate change scenarios [e.g., *Lorenz et al.*, 2010; *Hirschi et al.*, 2011; *Jaeger and Seneviratne*, 2011; *Mueller and Seneviratne*, 2012]. In addition, more frequent heavy precipitation events and less frequent light precipitation events have both been shown in observations [e.g., *Lau and Wu*, 2007; *Trenberth et al.*, 2007; *Shiu et al.*, 2009].

Some indices for climate extremes have been developed and defined in the past two decades [*Sillmann et al.*, 2013], particularly for temperature and precipitation. The tails of the probability density distribution (PDD) are essential from a statistical perspective. When the PDD is applied for temperature and precipitation, the tails (above or below a threshold value near the endpoints) of the PDD can generally be considered the extremes (such as maximum/minimum temperature, flood/drought, and heavy precipitation) that tend to be more relevant to society, environment, and the ecosystem [*Zhang et al.*, 2011] as shown in Figure 5.4.

To summarize, soil moisture memory effects have been demonstrated to be essential for the persistence of intraseasonal to interannual climate variability. The variance of soil moisture with regard to atmospheric precipitation shows a lag of 2–3 month [e.g., *Entekhabi et al.*, 1996; *Vinnikov et al.*, 1996; *Liang et al.*, 2014]. This characteristic of soil moisture persistence might potentially prolong and affect extreme climatic events. For variability at different timescales, both intraseasonal and interannual variability of soil moisture may contribute to climate extremes. Under global warming, climate extremes are projected to be more frequent and more intense according to GCM simulations [e.g., *Sun et al.*, 2007; *Chou et al.*, 2009; *Chen et al.*, 2012]; however, the extent to which land-atmosphere interactions can amplify or buffer such extremes remains unclear. On the basis of observations and modeling approaches, this chapter further reviews studies on how land processes affect climate extremes, particularly precipitation and temperature extremes.

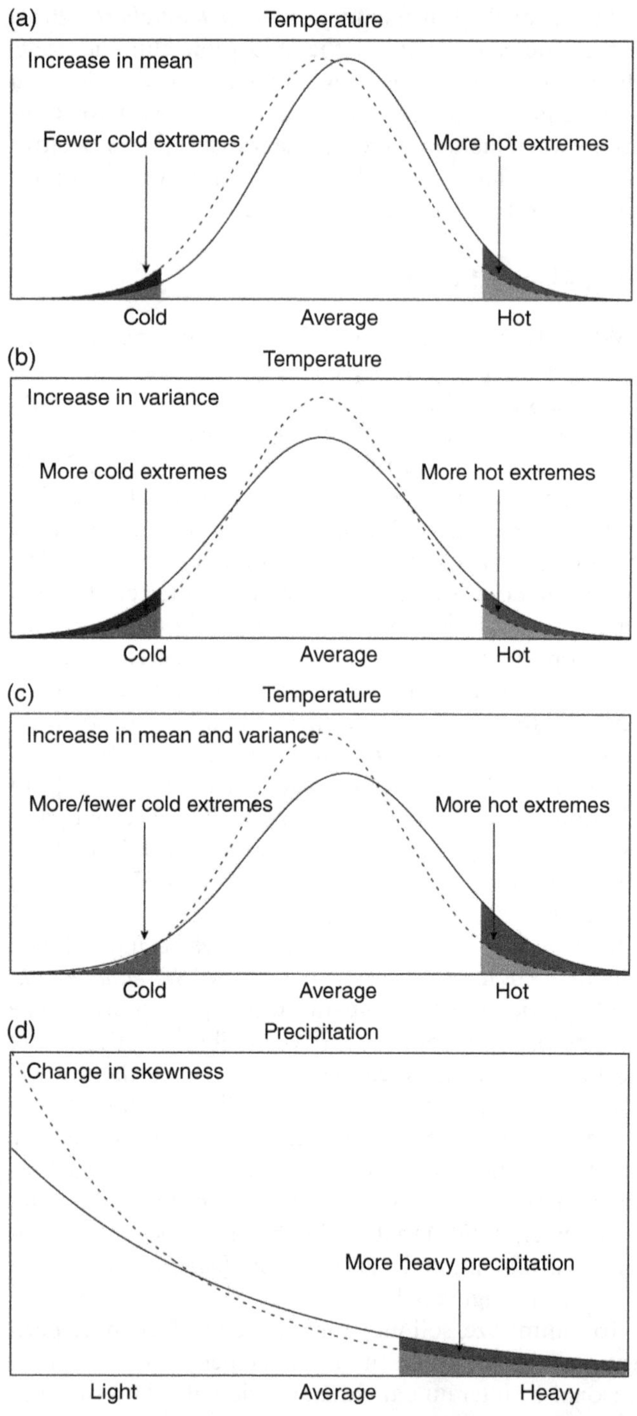

Figure 5.4 Schematic representations of the probability density function of daily temperature, which tends to be approximately Gaussian, and daily precipitation, which has a skewed distribution. Dashed lines represent a previous distribution and solid lines a changed distribution. The probability of occurrence, or frequency, of extremes is denoted by the shaded areas. In the case of temperature, changes in the frequencies of extremes are affected by changes (a) in the mean, (b) in the variance or shape, and (c) in both the mean and the variance. (d) In a skewed distribution such as that of precipitation, a change in the mean of the distribution generally affects its variability or spread, and thus an increase in mean precipitation would also imply an increase in heavy precipitation extremes, and vice versa. In addition, the shape of the right-hand tail could also change, affecting extremes. Furthermore, climate change may alter the frequency of precipitation and the duration of dry spells between precipitation events. (Parts a–c modified from *Folland et al.* [2001], and d modified from *Peterson et al.* [2008], as in *Zhang and Zwiers*, [2012].) Figure reprinted from IPCC AR5, Figure 1.8, 134, *Cubasch et al.* [2013], and same for the caption.

5.2. FORCINGS OF LAND PROCESSES ON CLIMATE EXTREMES

A land hydrological process plays a role as a low-pass filter, which can buffer the atmospheric forcing anomalies and delay the forcings-induced changes in soil hydrology [*Nicholson*, 2000]. However, land process can also prolong the effects of atmospheric drought or intensify climate extreme events because of interactions between the land surface and atmospheric boundary layer (BL). The combined effects of temperature and moisture, defined as moist static energy, can drive BL development and atmospheric convection [*Dirmeyer et al.*, 2013]. Under global warming, an increased surface temperature results in deeper BL depth and elevated lifting condensation level (LCL). The evapotranspiration flux can moisten BL and the sensible heat flux can lift the BL height. However, convective clouds are formed only when the BL top can penetrate the LCL; therefore, the probability of precipitation extreme events might increase because of delayed convection and accumulated moist static energy when the convection is finally triggered [*Dirmeyer et al.*, 2012]. *Dirmeyer et al.* [2012] further indicated that positive land-atmosphere interaction can broaden the probability distribution of precipitation intensity; hence, both extreme dry and extreme wet events are more frequent in a warmer climate.

Terrestrial soil moisture and atmosphere interactions potentially can affect climate extremes. The following is a list of studies on climate extremes, classified into three main themes, followed by a summary of the study findings.

1. Extreme precipitation and flood [e.g., *Clark*, 1987; *Pal and Eltahir*, 2002; *Bierkens and van der Hurk*, 2007; *Bradshaw et al.*, 2007; *Chen et al.*, 2007; *Laurance*, 2007; *Hurkmans et al.*, 2009; *Lin et al.*, 2011; *Miao et al.*, 2011; *Jothityangkoon et al.*, 2013; *Wei et al.*, 2015]
2. Drought [e.g., *Entekhabi et al.*, 1992; *Dirmeyer*, 1994; *Hong and Kalnay*, 2000, 2002; *D'Odorico and Porporato*, 2004; *Maxwell and Kollet*, 2008; *Teuling et al.*, 2013; *Zaitchik et al.*, 2013; *Müller et al.*, 2014]
3. Heat waves and extreme temperatures [e.g., *Bonan*, 1999, 2001; *Zhao and Pitman*, 2002; *Diffenbaugh*, 2005; *Mahmood et al.*, 2006; *Bonfils and Lobell*, 2007; *Fischer et al.*, 2007a, b; *Deo et al.*, 2009; *Grossman-Clarke et al.*, 2010; *Vautard et al.*, 2010; *Hirschi et al.*, 2011; *Avila et al.*, 2012; *Mueller and Seneviratne*, 2012; *Christidis et al.*, 2013; *Mahmood et al.*, 2013; *Stéfanon et al.*, 2014; *Mishra et al.*, 2015]

5.2.1. Extreme Precipitation and Flood

5.2.1.1. Wet Soil Feedback on Flood and Extreme Precipitation. *Pal and Eltahir* [2002] conducted a series of experiments with different soil moisture conditions in a regional climate model to explore the role of spring soil moisture distribution and anomalies during the 1993 summer flood in North America. The authors concluded that the anomalous wet soil moisture conditions in the Midwest of the United States were likely to cause the persistence and magnitude of the flood. In other words, wetter soil moisture conditions increased the convection, resulting in anomalous ascending motion with an anomalous low-pressure system, thereby strengthening the associated cyclonic flow. In addition, they suggested that soil moisture anomalies over small regions have impacts not only on rainfall in the local region but also the surrounding regions.

Wei et al. [2015] combined the Weather Research and Forecasting Model (WRF) simulations, precipitation observations, and reanalysis datasets to explore impacts of soil moisture and atmospheric moisture fluxes on precipitation mainly over the southern United States. The authors indicated that soil moisture contributes to precipitation directly through surface evapotranspiration, and can alter the precipitation efficiency. While the atmospheric moisture flux is more critical for the heavier rainfall occurrence, *Wei et al.* [2015] indicated that soil moisture has similar importance for intermediate to heavy rainfall (approximately 10 mm/day), because the role of soil moisture is mainly to trigger rainfall.

5.2.1.2. Deforestation Feedback on Flood. In addition to the increased flood risk associated with land-atmosphere interactions, deforestation can also increase the flood risk because of less transpiration from vegetation and less infiltration capacity from land use changes [*Clark*, 1987; *Bradshaw et al.*, 2007; *Laurance*, 2007]. Urbanization leads to less infiltration capacity with the creation of an impervious surface. *Hurkmans et al.* [2009] used the Variable Infiltration Capacity (VIC) model with land cover change scenarios to demonstrate that urbanization led to an increase in streamflow in the Rhine basin. *Jothityangkoon et al.* [2013] also demonstrated that deforestation could reduce the soil storage capacity, and thus increase the rainfall-runoff ratio. By using a distributed rainfall-runoff model simulation for the upper Ping River in northern Thailand, the authors showed the potential influence on the more extreme floods.

5.2.1.3. Urbanization Feedback on Flood. Urbanization can also affect the rainfall intensity and frequency [*Chen et al.*, 2007; *Lin et al.*, 2011]. *Miao et al.* [2011] used the WRF coupled with an urban canopy model (UCM) to study the urban heat island effect on a heavy rainfall event in Beijing, and demonstrated that effects of urbanization on heavy rainfall can vary depending on the intensity of urbanization. Strong urbanization effects could raise the BL height with strong vertical mixing resulting from surface heating, leading to an efficient development of convection and an increase in maximum rainfall. By contrast, weak urbanization was associated with surface drying, which potentially reduced maximum rainfall. The authors also indicated that urbanization-induced thermal fluxes (sensible heat and latent heat fluxes) play a larger role than the momentum fluxes in affecting the maximum rainfall. In addition, the authors considered urbanization effects and found that rainfall tends to be more locally concentrated; thus, heavy rainfall leads to potential flooding.

5.2.2. Drought

5.2.2.1. Dry Soil Feedback on Drought. *Entekhabi et al.* [1992] showed that in semiarid and semihumid climates, a drought can persist for several years because of the strong land-atmosphere coupling, although land surface effects might not directly induce drought. *Hong and Kalnay* [2000, 2002] studied the 1998 Oklahoma-Texas drought and showed that the large-scale circulation triggered the drought, and dry soil anomalies maintained the drought for several months. Dry soil induced positive feedback between soil moisture and the atmosphere, which led to less rainfall and recycling of moisture, and then led to drier soil. In addition to precipitation deficits, an enhancement in evapotranspiration can further reduce soil water during drought events, as demonstrated by *Teuling et al.* [2013] for the situation in Europe. The authors suggested that the amplification of soil water anomalies indicates the importance of considering the evapotranspiration processes in studying changes in the terrestrial water cycle under drought conditions.

Precipitation variation over the midlatitude continent is generally controlled by variations in soil moisture [e.g., *Koster et al.*, 2000]. A significant correlation between the lag response of precipitation and the soil moisture status can be shown in observations [*D'Odorico and Porporato*, 2004]. When soil is wetter, subsequent rainfall is enhanced by local moisture recycling. By contrast, a drier soil leads to less evapotranspiration, which then affects the development of BL dynamics by altering the partitioning of energy from latent heat fluxes to sensible heat fluxes. *D'Odorico and Porporato* [2004] indicated that the bimodal distribution of soil moisture dynamics often existed during the summer; in other words, soil tends to persist in either dry states or wet states. Therefore, the summer drought can be prolonged by earlier (spring) dry soil moisture.

Although the aforementioned studies showed that positive land-atmosphere interactions can intensify precipitation deficit and prolong drought events, *Zaitchik et al.* [2013] further indicated that in the 2006 southern Great Plains drought, precipitation tended to concentrate over the wet areas through local soil moisture feedback than over the dry areas. The authors also indicated the importance of including a dynamic surface albedo scheme, a function of soil moisture, in the NASA Unified WRF (NU-WRF) model used in the study. Their results showed that changes in soil moisture were associated with changes in albedo: brighter areas received less precipitation with dry soil moisture condition and darker areas received more precipitation with high soil moisture condition. However, although there are such spatial redistributions (heterogeneities) of precipitation, the total amount of precipitation in the simulated drought region and period is not significantly altered when considering a dynamic surface albedo scheme.

5.2.2.2. Vegetation Feedback on Drought. In an early study, *Dirmeyer* [1994] used a GCM with an idealized model configuration (aquaplanet with a rectangular land in the midlatitude) to explore effects of vegetation stress on midlatitude drought. He found that changes in vegetation can feed back to the atmosphere and thus affect the drought. In the case of low initial soil moisture combined with dormant vegetation, drought would become more severe, because the dormant vegetation could reduce the evapotranspiration by lowering the transpiration during the spring and early summer. Therefore, the result indicated that the temporary loss of vegetation might lead to a more severe drought due to a decrease in transpiration. For the western United States, *Diffenbaugh* [2005] suggested that global warming-induced vegetation feedbacks (including changes in surface albedo, leaf area index, and rooting depth) may further increase the frequency and magnitude of maximum temperature. Using model simulations, the authors showed that the magnitude of extreme climate responses from vegetation feedback could be the same as those from double CO_2 radiative forcing. Furthermore, they suggested that biological responses and feedback must be incorporated and considered in climate models to study extreme climate events under climate change in the future.

Müller et al. [2014] used a regional atmospheric model to examine vegetation feedback on the severe drought of 2008 in southern South America. By using a new and more realistic land-cover dataset and corresponding changes in the vegetation phenology (including greenness fraction, leaf area index, and stomatal resistance) and surface roughness, the model could more accurately simulate drought conditions. The authors showed that the regional atmospheric model was sensitive to changes in land cover and vegetation because of land-atmosphere interactions.

5.2.2.3. Groundwater Feedback on Rainfall Persistence. By using a land surface model coupled with an integrated groundwater flow model, *Maxwell and Kollet* [2008] and *Kollet and Maxwell* [2008] showed the strong dependence of atmospheric convergence and groundwater. Both studies further suggested that energy feedback from the land surface and groundwater are crucial for understanding the processes of recharge/discharge and drought in a changing climate. In addition, *Kollet and Maxwell* [2008] showed that if lateral groundwater flow is not considered, drought magnitude and length might be underestimated. Their results are consistent with those of *Bierkens and van den Hurk* [2007], who suggested that the long-term persistence of rainfall increases with an inclusion of groundwater convergence from a simple stochastic model analysis.

5.2.3. Heat Waves and Extreme Temperatures

5.2.3.1. Soil Moisture Feedbacks on Extreme Temperatures. Heat waves generally occur when there is a lack of precipitation in the beginning, and less latent heat cooling from the resulting drier soil further amplifies the magnitude of the heat wave and hence prolongs the event. In the European continent, because there is a high density of observation stations, the relation between soil moisture deficit and hot extremes can be explored by defining a drought index based on previous precipitation [*Hirschi et al.*, 2011]. The authors showed that occurrences of heat waves are correlated with the drought index in southeast Europe, which is in the soil moisture-limited regime. A distinct intensification of temperature extremes is found in the drier soil moisture condition. They further suggested that effects of soil moisture on temperature extremes are stronger in a dry soil moisture-limited than in a wet energy-limited regime.

Model experiments using prescribed and interactive soil moisture have been frequently conducted to explore impacts of land-atmosphere interactions on heat waves (including their strength and length) [e.g., *Fischer et al.* 2007a, b]. In the prescribed experiment, soil moisture is forced to maintain the climatology of either the control simulations or observations. In the interactive experiment, land hydrological processes are active; therefore, land surface fluxes not only balance the heat budget but also the water budget. Therefore, the impacts of land hydrological processes on heat waves can be identified through these two experiments. To quantify contributions of land-atmosphere interactions on heat waves, *Fischer et al.* [2007a] used the Climate High-Resolution Model (CHRM) [*Vidale et al.*, 2003] to conduct the aforementioned experiments for heat wave events in 1976, 1994, 2003, and 2005. The authors considered land-atmosphere coupling and observed that both the number of hot days and the duration of a maximum heat wave increased. This coupling resulted mainly from the drier soil condition that lead to low evapotranspiration (fewer latent heat fluxes) and a high surface temperature (higher sensible heat fluxes). In another study, *Fischer et al.* [2007b] specifically focused on the record-breaking heat wave event in 2003 and demonstrated that soil moisture change strongly amplified the extreme summer temperature anomalies in Europe. The changes in soil moisture altered the partitioning of available energy between latent heat fluxes and sensible heat fluxes, which play a crucial role in amplifying surface temperature extremes.

In addition to Europe, the association between precipitation deficits and hot extremes was also assessed for other places. By using global gridded datasets, *Mueller and Seneviratne* [2012] performed the same analysis at the global scale to discover areas where the precipitation deficit causes the maximum possibility of occurrence of hot extreme events. The authors found that many regions exhibit strong relationships between the number of hot days and preceding precipitation deficits in a hottest month, including South America, Iberian Peninsula, eastern Australia, most of North America, and Eastern Europe. The impacts were asymmetric toward the PDD tail for hot extremes but not the cold extremes, which is consistent with the finding of *Hirschi et al.* [2011].

5.2.3.2. Effects of Land Use Changes on Extreme Temperatures. On the basis of numerical experiments performed with CSIRO GCM, *Deo et al.* [2009] demonstrated that LCLUC affects the characteristics of climate extremes in eastern Australia. The model simulation results clearly indicated an increase in the dry and hot days after considering impacts of LCLUC. The major reason was the significant changes in the partitioning of available energy (shift from latent heat flux to sensible heat flux) because of land cover changes. *Avila et al.* [2012] used CSIRO Mk3L, coupled with the Community Atmosphere Biosphere Land Exchange model to explore the impact of LCLUC on climate extremes. In their study, two land-cover change experiments were performed: one for forest to represent natural vegetation and the other for cropland and pasture to represent anthropogenic deforestation. The authors used the Expert Team on Climate Change Detection and Indices [*Alexander et al.*, 2006], including coldest seasonal daily maximum/minimum temperature and warmest seasonal daily maximum/minimum temperature. Their results showed that the impact of LCLUC on coldest seasonal daily maximum/minimum temperature was minimal. By contrast, in many regions, LCLUC led to changes in the warmest seasonal daily maximum temperature by up to 4°C because the deforestation process reduced turbulent fluxes, which tended to minimize the effect of evaporative cooling. The other major reason was that the shallow rooting system in grasses and crops limited soil water access in the deeper soil layers, and thus latent heat fluxes decreased.

Zhao and Pitman [2002] used a numerical model to test impacts of land cover changes and increased carbon dioxide on the extreme temperature. Their simulations showed that land cover change can cause similar increases in the maximum temperature as does increasing carbon dioxide. Although most studies showed that LCLUC could cause an increase in the daily maximum temperature, asymmetric responses of the warmest day to deforestation in tropics and extratropics were shown by *Christidis et al.* [2013]. They used HadGEM2-ES with a dynamic vegetation component to explore alterations induced by land use change (mainly from forest to grassland) in the warmest days during 1951–2003. Their results indicated a significant cooling (warming) in warm extremes over the extratropical (tropical) regions. The albedo effect (an increase in albedo because of deforestation) was dominant in extratropics, and changes in evapotranspiration and surface turbulent flux were essential in the tropical region. These changes resulted in such asymmetric responses to global land use changes. Their study was a pioneering work to show that land use changes can regionally induce a significant cooling of daily temperature in warm extremes, which can mitigate the warming trend from greenhouse effects.

Because of different primary productivity (i.e., using atmospheric carbon dioxide to synthesize organic compounds mainly through photosynthesis processes) of forests and crops, regions with different vegetation types tend to have opposite effects on extreme temperature anomalies. *Stéfanon et al.* [2014] used numerical models (including the Model of the Regional Coupled Earth system, WRF, and Organizing Carbon and Hydrology in Dynamic Ecosystems land-surface model) to explore how vegetation affected the 2003 heat wave event in France. Their simulation results showed that the native vegetation type enhanced the positive temperature anomaly from March 2003 to June 2003, whereas it mitigated temperature extremes from July 2003 to October 2003. During the 2003 spring, agricultural plants had a higher photosynthetic efficiency and transpiration rates than trees did; therefore, stronger cooling effect existed over the cropland. In July, dry soil and shallow roots caused less transpiration from crops, but the forests could still transpire because of the deeper rooting system. Therefore, the forest area still had greater cooling effects than the cropland did. However, based on both model simulations and observational data analysis, *Bonan* [1999, 2001] showed that conversions of forest to croplands have resulted in decreases in surface temperature over the central and eastern United States (with a decrease in daily mean temperature by approximately 0.5 K during the summer). There is a large decrease in the daily maximum temperature by approximately 1.2 K over the north central United States. *Bonan* [1999] indicated that the cooling was mainly caused by increases in surface albedo and decreases in roughness length, which also reduces sensible heat fluxes. The opposite results from the aforementioned studies indicate the high uncertainties of the LCLUC effects on the surface extreme temperature, and warrant further investigations based on more observational datasets and more sophisticated land surface models.

Irrigation-induced cooling effects can alter the maximum near-surface daily temperature. *Bonfils and Lobell* [2007] used observational datasets to show that irrigation in California caused significant cooling of daily maximum temperature. *Mahmood et al.* [2006, 2013] also found the irrigation cooling effects on the mean maximum growing season temperature through analysis on observational datasets over the central United States.

In addition to changes in temperature and rainfall, *Vautard et al.* [2010] showed the impacts of land use changes on surface wind changes. Analysis from weather observation stations showed that, over the past few decades, surface wind speed decreased by 5%–15% in China, the Netherlands, the Czech Republic, the United States, and Australia. Strong winds decreased faster and more significantly than weak winds. However, upper atmospheric winds estimated both from sea level pressure gradients and from reanalysis datasets exhibited no such decreasing trend. Increase in surface roughness (mostly caused by land use changes) was mainly responsible for wind speed reduction and was also shown in the regional atmospheric model simulations [*Vautard et al.*, 2010].

To explore contributions from rapid urbanizations (1973–2005) to summer extreme heat wave events in Phoenix city, *Grossman-Clarke et al.* [2010] conducted experiments by using a regional atmospheric model with historical land-cover datasets. Their results indicated that nighttime temperatures in the urban region increased approximately 2 K with ongoing urbanization. In contrast, daytime temperatures increased by approximately 1 K, whereas maximum temperatures increased by approximately 3 K, when agricultural land was converted to urban land. The authors also mentioned that the reduction in cooler air advected from agricultural areas could partially contribute to the increase in temperature over the existing urban area. *Mishra et al.* [2015] used observational station data of precipitation, air temperature (mean, maximum, and minimum), and wind speed in urban areas for the period 1973–2012 for 217 urban regions worldwide are used to explore the impacts of urbanization on climate extremes. The results showed tremendous increases in the number of heat wave events, number of extremely hot days in 50% of urban regions, frequency of extremely hot nights in 66% of urban regions, and frequency of daily/annual maximum precipitation. In addition, the number of extremely windy days (days exceeded the 99th-percentile threshold of the daily mean wind speed) decreased in approximately 60% of urban regions, thus representing land process feedback to the atmosphere.

Finally, these studies of the aforementioned extreme events and their mechanism are summarized in Table 5.1. From observational evidence, soil moisture, as a major factor that controls land-atmosphere interactions, potentially affects climate extremes. Furthermore, modeling studies helped us to clarify the mechanism by which soil moisture affects climate. By modifying the mean and variability of soil moisture in models, we can alter the extent and strength of land-atmosphere coupling to investigate the roles of soil moisture in the climate extremes.

Table 5.1 Summary of Reviewed Studies on Models, Cases, and Land Forcings for Extremes

References	Models/Observations	Events/Regions	Forcings
Avila et al. [2012]	CSIRO Mk3L, CABLE	Extreme temperature/global	Vegetation cover change
Bierkens and van den Hurk [2007]	Stochastic soil moisture, evaporation, precipitation feedback model	Long-term persistence of precipitation	Groundwater convergence
Bonan [1999,2001]	CCM3, NCAR LSM	Extreme in temperature/central and eastern United States	Deforestation
Bonfils and Lobell [2007]	Stationary observations	Extreme in temperature/California United States	Irrigation
Bradshaw et al. [2007]	Dartmouth Flood Observatory	Flood/global	Deforestation
Chen et al. [2007]	Stationary observations	Increase of rainfall intensity/Taipei basin	Urbanization
Christidis et al. [2013]	HadGEM2-ES, DGVM	Extreme temperature/global	Deforestation
Clark [1987]	Stationary observations	Flood/East Somerset and North Dorset, England	Deforestation
D'Odorico and Porporato [2004]	Stationary observations	Summer drought/continental midlatitude regions	The lag response of precipitation to the soil moisture status
Deo et al. [2009]	Australia CSIRO GCM	Increase in number of dry and hot days/ eastern Australia	Land cover changes.
Diffenbaugh [2005]	MORCE, WRF, ORCHIDEE	Drought/Western United States	Vegetation feedbacks
Dirmeyer [1994]	GCM	Drought/midlatitude	Loss of vegetation decrease ET
Entekhabi et al. [1992]	Hydrologic balance coupled with land surface-atmosphere interaction model	Persistent drought	Strong coupling between land and the atmosphere
Fischer et al. [2007a, b]	CHRM	Heat waves/1976,1994,2003,2005 European summer heat wave	Drier soil lead to low evapotranspiration, higher surface temperature
Grossman-Clarke et al. [2010]	WRF, Noah UCM	Heat events/Phoenix	Urban development
Hirschi et al. [2011]	Stationary observations	Extreme temperature/southeastern Europe	Feedback of soil-moisture deficit to precipitation
Hong and Kalnay [2000;2002]	GSM/RSM;MRF	1998 Oklahoma-Texas drought	Soil moisture variability
Hurkmans et al. [2009]	VIC	Flood/Rhine River	Urbanization
Jaeger and Seneviratne [2011]	CLM RCM	Heat wave/Europe	Feedback of soil-moisture deficit to precipitation
Jothityangkoon et al. [2013]	Distributed rainfall, runoff model	Flood/the Upper Ping River in northern Thailand.	Deforestation
Laurance [2007]	Review paper	Flood	Deforestation
Lin et al. [2011]	WRF;UCM	Increase of rainfall intensity/northern Taiwan	Urbanization

(*Continued*)

Table 5.1 (Continued)

References	Models/Observations	Events/Regions	Forcings
Lorenz et al. [2010]	CLM RCM	Heat wave/Europe	Soil moisture memory
Mahmood et al. [2006, 2013]	Stationary observations	Extreme in temperature/central United States	Irrigation
Maxwell and Kollet [2008]	ParFlow, CLM	Drought/southern Great Plains, United States	Groundwater dynamics and land-energy feedbacks
Miao et al. [2011]	WRF, Noah-SLUCM	Heavy rainfall/Beijing	Urbanization
Mishra1 et al. [2015]	Stationary observations	Heat wave, wind stilling/global	Urban development
Mueller and Seneviratne [2012]	CRU, CPC;SPI	Hot days/global	Feedback of soil-moisture deficit to precipitation
Müller et al. [2014]	WRF	2008 southern South America drought	Land-cover changes and vegetation variability
Pal and Eltahir [2002]	RegCM	1993 summer flood/Midwest of the United States	Soil moisture variability
Stéfanon et al. [2014]	CCM, RegCM, BIOME4	Heat wave/Central France	Vegetation cover change
Teuling et al. [2013]	Stationary observations	Drought/Central and western Europe	Amplification of soil water anomalies
Vautard et al. [2010]	MM5	Surface wind speed declined/Northern Hemisphere	Deforestation, urbanization
Wei et al. [2015]	WRF	Heavier precipitation/Southern United States	Soil moisture changes surface evapotranspiration and the precipitation efficiency
Zaitchik et al. [2013]	NU-WRF	Drought/Southern Great Plains, United States	Precipitation tends to concentrate over the wet areas through local soil moisture feedbacks
Zhao and Pitman [2002]	CCM3, BATS	Extreme in temperature/global	Land cover change, increase in CO_2

5.3. SUMMARY

Weather and climate extreme events are expected to occur more often under global warming and climate changes. Flood and drought both directly affect food and water security as well as plant growth. Land processes and land-atmosphere interactions have been shown to affect the occurrence of climate extremes. Regionally and globally, dry soil moisture conditions and reduced river streamflow during drought could cause water shortages and threaten agricultural water supplies. In the framework of land-atmosphere interactions, wet soil is associated with a lower near-surface temperature but land-atmosphere feedback can lead to either enhancement or suppression of precipitation. Complicating the matter, model-dependency is also a concern as the soil moisture–rainfall feedbacks had been found to be opposite in different simulations (e.g., between a convection permitting model signifying a negative feedback and parameterized convection schemes representing a positive feedback [*Hohenegger et al.*, 2009]). Most IPCC models lack good representation of groundwater/aquifer components and this deficiency hinders the exploration of soil moisture–related and groundwater-related feedback on climate change. Some aforementioned studies suggest that after incorporating groundwater dynamics into climate models, extreme events may further increase, suggesting that current predictions may underestimate potential increases in hydrologic extremes.

Further knowledge is required for this land-atmosphere interaction from a collection of observation analysis, model development, and model intercomparisons. The latter saw light in the tunnel, such as the Local-Coupled Land-Atmospheric Modeling Project and GLACE-type experiments designed to address how land processes affect climate extremes. The GLACE-CMIP5 (Coupled Model Intercomparison Project Phase 5)-type multimodel experiment can be utilized for investigation of the impact of soil moisture on the climate in the future projections [*Kumar et al.*, 2013; *Seneviratne et al.*, 2013]. The extent to which humans influence land-atmosphere interactions is a fundamental and critical question. For instance, *Mahmood et al.* [2010] and *Pielke et al.* [2011] highlighted the importance of land-cover and land-use change (LCLUC) in the changing climate and also indicated the need to explore impacts of LCLUC on climate extremes. The gained knowledge in LCLUC will help in agriculture and water management planning processes in the essential food-producing region of the world.

ACKNOWLEDGMENTS

This study was supported by the Ministry of Science and Technology, Taiwan (MOST 103-2111-M-002-006 and MOST 104-2111-M-002-008). Figure 5.4 in this chapter is reprinted from IPCC AR5, Figure 1.8, 134, *Climate Change 2013: The Physical Science Basis.*

REFERENCES

Adegoke, J., R. Pielke Sr., J. Eastman, R. Mahmood, and K. Hubbard (2003), Impact of irrigation on midsummer surface fluxes and temperature under dry synoptic conditions: A regional atmospheric model study of the U.S. High Plains, *Monthly Weather Rev.*, *131*, 556–564; doi:10.1175/520-0493(2003)131<0556:IOIOMS>2.0.CO;2.

Alexander, L. V., et al. (2006), Global observed changes in daily climate extremes of temperature and precipitation, *J. Geophys. Res.*, *111*, D05109; doi:10.1029/2005JD006290.

Amenu, G. G., P. Kumar, and X. Z. Liang (2005), Interannual variability of deep-layer hydrologic memory and mechanisms of its influence on surface energy fluxes, *J. Climate*, *18*, 5024–5045.

Avila, F. B., A. J. Pitman, M. G. Donat, L. V. Alexander, and G. Abramowitz (2012), Climate model simulated changes in temperature extremes due to land cover change, *J. Geophys. Res.*, *117*, D04108.

Bierkens, M. F. P., and B. J. J. M. van den Hurk (2007), Groundwater convergence as a possible mechanism for multi-year persistence in rainfall, *Geophys. Res. Lett.*, *34*, L02402; doi:10.1029/2006GL028396.

Bonan, G. B. (1999), Frost followed the plow: Impacts of deforestation on the climate of the United States, *Ecol. Appl.*, *9*(4), 1305–1315; doi :10.1890/1051- 0761(1999)009[1305:FFTPIO] 2.0.CO;2.

Bonan, G. B. (2001), Observational evidence for reduction of daily maximum temperature by croplands in the Midwest United States, *J. Climate*, *14*, 2430–2442; doi :10.1175/1520-0442(2001)014 < 2430:OEFROD > 2.0.CO;2.

Bonfils, C., and D. Lobell (2007), Empirical evidence for a recent slowdown in irrigation-induced cooling, *Proc. Natl. Acad. Sci.*, *104*, 13,582–13,587; doi:10.1073/pnas.0700144104.

Bradshaw, C. J. A., N. S. Sodhi, K. S.-H. Peh, and B. W. Brook (2007), Global evidence that deforestation amplifies flood risk and severity in the developing world, *Glob. Change Biol.*, *13*, 1–17.

Brovkin, V., et al. (2013), Effect of anthropogenic land-use and land-cover changes on climate and land carbon storage in CMIP5 projections for the twenty-first century, *J. Climate*, *26*(18), 6859–6881; doi :10.1175/JCLI-D-12-00623.1.

Budyko, M. I. (1974), *Climate and Life*, Academic Press, Cambridge, MA.

Changnon, S. A., Jr.(1987), *Detecting Drought Conditions in Illinois*, ISWS/CIR-169-87, Illinois State Water Survey, Champaign, IL.

Chen, C. A., C. Chou, and C. T. Chen (2012), Regional perspective on mechanisms for tropical precipitation frequency and intensity under global warming, *J. Climate*, *25* (8), 8487–8501.

Chen, S., J. Chen, G. Lin, W. Zhang, H. Miao, L. Wei, J. Huang, and X. Han (2009), Energy balance and partition in Inner Mongolia steppe ecosystems with different land use types, *Agric. For. Meteor.*, *149*, 1800–1809.

Chen, T.-C., S.-Y. Wang, and M.-C. Yen (2007), Enhancement of afternoon thunderstorm activity by urbanization in a valley: Taipei, *J. Appl. Meteor. Clim.*, *46*, 1324–1340.

Cheruy, F., A. Campoy, J. C. Dupont, A. Ducharne, F. Hourdin, M. Haeffelin, et al. (2013), Combined influence of atmospheric physics and soil hydrology on the simulated meteorology at the SIRTA atmospheric observatory, *Climate Dyn.*, *40*, 2251–2269.

Chou, C., J. D Neelin, and H. Su (2001), Ocean-atmosphere-land feedbacks in an idealized monsoon, *Quart. J. Roy. Meteor. Soc.*, *127*, 1869–1891; doi:10.1002/qj.49712757602.

Chou, C., J. D. Neelin, C.-A. Chen, and J.-Y. Tu (2009), Evaluating the "rich-get-richer" mechanism in tropical precipitation change under global warming, *J. Climate*, *22*, 1982–2005.

Christidis, N., P. A. Stott, G. C. Hegerl, and R. A. Betts (2013), The role of land use change in the recent warming of daily extreme temperatures, *Geophys. Res. Lett.*, *40*, 589–594.

Clark, C. (1987), Deforestation and floods, *Environ. Conserv.*, *14* (1), 67–69.

Cook, B. I., M. J. Puma, and N.Y. Krakauer (2011), Irrigation induced surface cooling in the context of modern and increased greenhouse gas forcing, *Climate Dyn.*, *37*, 1587–1600; doi :10.1007/s00382-010-0932-x.

Costa, M., and G. Pires (2010), Effects of Amazon and Central Brazil deforestation scenarios on the duration of the dry season in the arc of deforestation, *Int. J. Climatol.*, *30*(13), 1970–1979.

Cubasch, U., D. Wuebbles, D. Chen, M. C. Facchini, D. Frame, N. Mahowald, and J.-G. Winther (2013), Introduction, in *Climate Change 2013: The Physical Science Basis, Contribution of Working Group I to the Fifth Assessment Report of the Intergovernmental Panel on Climate Change*, edited by T. F. Stocker, D. Qin, G.-K. Plattner, M. Tignor, S. K. Allen, J. Boschung, A. Nauels, Y. Xia, V. Bex, and P. M. Midgley, 119–158, Cambridge University Press, Cambridge, UK and New York ; doi :10.1017/CBO9781107415324.007.

da Silva, R., R. D. Werth, and R. Avissa (2008), Regional impacts of future land-cover changes on the Amazon basin wet-season climate, *J. Climate*, *21* (6), 1153–1170.

DeAngelis, A., et al. (2010), Evidence of enhanced precipitation due to irrigation over the Great Plains of the United States, *J. Geophys. Res.*, 10.1029/2010JD013892.

Deo, R. C., J. I. Syktus, C. A. McAlpine, P. J. Lawrence, H. A. McGowan, and S. R. Phinn (2009), Impact of historical land cover change on daily indices of climate extremes including droughts in eastern Australia, *Geophys. Res. Lett.*, *36*, L08705; doi :10.1029/2009GL037666

Diffenbaugh, N. S. (2005), Sensitivity of extreme climate events to CO2-induced biophysical atmospherevegetation feedbacks in the western United States, *Geophys. Res. Lett.*, *32*, L07702; doi :10.1029/2004GL022184.

Dirmeyer, P., Y. Jin, B. Singh, and X. Yan (2013), Trends in land–atmosphere interactions from CMIP5 simulations, *J. Hydrometeor.*, *14*, 829–849; doi: http://dx.doi.org/10.1175/JHM-D-12-0107.1.

Dirmeyer, P. A. (1994), Vegetation as a feedback mechanism in mid-latitude drought, *J. Climate*, *7*, 1463–1483.

Dirmeyer, P. A. (2011), A history and review of the Global Soil Wetness Project (GSWP), *J. Hydrometeor*, *12*, 729–749; doi: http://dx.doi.org/10.1175/JHM-D-10-05010.1.

Dirmeyer, P. A., A. J. Dolman, and N. Sato (1999), The pilot phase of the Global Soil Wetness Project, *Bull. Amer. Meteor. Soc.*, *80*, 851–878.

Dirmeyer, P. A., et al. (2012), Evidence for enhanced land-atmosphere feedback in a warming climate, *J. Hydrometeor.*, *13*, 981–995.

D'Odorico, P., and A. Porporato (2004), Preferential states in soil moisture and climate dynamics, *Proc. Nat. Acad. Sci.*, *101*, 8848–8851.

Douglas, E. M., D. Niyogi, S. Frolking, J. B. Yeluripati, R. A. Pielke Sr., N. Niyogi, C. J. Vörösmarty, and U. C. Mohanty (2006), Changes in moisture and energy fluxes due to agricultural land use and irrigation in the Indian monsoon belt, *Geophys. Res. Lett.*, *33*, L14403; doi :10.1029/2006GL026550.

Eltahir, E. A. B. (1998), A soil moisture rainfall feedback mechanism 1, theory and observations, *Water Resour. Res.*, *34*, 765–776.

Entekhabi, D., I. Rodriguez-Iturbe, and F. Castelli (1996), Mutual interaction of soil moisture state and atmospheric processes, *J. Hydrol.*, *184*, 3–17.

Entekhabi, D., I. Rodriguez-Iturbe, and R. L. Bras (1992), Variability in large-scale water balance with land surface–atmosphere interaction, *J. Climate*, *5*, 798–813.

Famiglietti, J. S., and E. F. Wood (1991), Evapotranspiration and runoff from large land areas: Land surface hydrology for atmospheric general circulation models, *Surv. Geophys.*, *12*, 179–204; doi :10.1007/BF01903418.

Famiglietti, J. S., and E. F. Wood (1994), Multiscale modeling of spatially variable water and energy balance processes, *Water Resour. Res.*, *30*, 3061–3078.

Famiglietti, J. S., and E. F. Wood (1995), Effects of spatial variability and scale on areally averaged evapotranspiration, *Water Resour. Res.*, *31*(3), 699–712.

Fan, Y., G. Miguez-Macho, C. Weaver, R. Walko, and A. Robock (2007), Incorporating water table dynamics in climate modeling, part I: Water table observations and the equilibrium water table, *J. Geophys. Res.*; doi:10.1029/2006JD008111.

Findell, K. L., and E. A. B. Eltahir (1997), An analysis of the soil moisture-rainfall feedback, based on direct observations from Illinois, *Water Resour. Res.*, *33*, 725–735.

Fischer, E. M., S. I. Seneviratne, D. Lüthi, and C. Schär (2007b), The contribution of land–atmosphere coupling to recent European summer heat waves, *Geophys. Res. Lett.*, *34*, L06707, http://dx.doi.org/10.1029/2006GL029068.

Fischer, E. M., S. I. Seneviratne, P. L. Vidale, D. Lüthi, and C. Schär (2007a), Soil moisture–atmosphere interactions during the 2003 European summer heat wave, *J. Climate*, *20*, 5081–5099.

Folland, C. K., et al., (2001) Observed climate variability and change. *In Climate Change 2001: The Scientific Basis. Contribution of Working Group I to the Third Assessment Report of the Intergovernmental Panel on Climate Change* by J. T. Houghton, Y. Ding, D. J. Griggs, M. Noquer, P. J. van der Linden, X. Dai, K. Maskell and C. A. Johnson (eds.).

Cambridge University Press, Cambridge, United Kingdom and New York, NY, USA, 101–181.

Fu, R., and W. Li (2004), The influence of the land surface on the transition from dry to wet season in Amazonia, *Environ. Conserv.*, *78*, 97–110, http://dx.doi.org/10.1007/s00704-004-0046-7.

Gordon, L. J., W. Steffen, B. F. Jonsson, C. Folke, M. Falkenmark, and A. Johannessen (2005), Human modifications of global water vapor flows from the land surface, *Proc. Nat. Acad. Sci.*, *102*, 7612–7617.

Grossman-Clarke, S., J. A. Zehnder, T. Loridan, and C. S. B. Grimmond (2010), Contribution of land use changes to near-surface air temperatures during recent summer extreme heat events in the Phoenix metropolitan area, *J. Appl. Meteor. Clim.*, *49*, 1649–1664.

Guillod, B. P., B. Orlowsky, D. G. Miralles, A. J. Teuling, and S. I. Seneviratne (2015), Reconciling spatial and temporal soil moisture effects on afternoon rainfall, *Nature Communications*, *6*, 6443.

Güntner, A., J. Stuck, S. Werth, P. Döll, K. Verzano, and B. Merz (2007), A global analysis of temporal and spatial variations in continental water storage, *Water Resour. Res.*, *43*, W05416; doi:10.1029/2006WR005247.

Harper, A., I. T. Baker, A. S. Denning, D. A. Randall, D. Dazlich, and M. Branson (2014), Impact of evapotranspiration on dry season climate in the Amazon forest, *J. Climate*, *27*, 574–591; doi: http://dx.doi.org/10.1175/JCLI-D-13-00074.1.

Hirschi, M., S. I. Seneviratne, V. Alexandrov, F. Boberg, C. Boroneant, O. B. Christensen, H. Formayer, B. Orlowsky, and P. Stepanek (2011), Observational evidence for soil-moisture impact on hot extremes in southeastern Europe, *Nat. Geosci.*, *4*, 17–21.

Hohenegger, C., P. Brockhaus, C. S. Bretherton, and C. Schär (2009), The soil moisture–precipitation feedback in simulations with explicit and parameterized convection, *J. Climate*, *22*, 5003–5020; doi:10.1175/2009JCLI2604.1.

Hong, S.-Y., and E. Kalnay (2000), Role of sea surface temperature and soil-moisture feedback in the 1998 Oklahoma-Texas drought, *Nature*, *408*, 842–844.

Hong, S.-Y., and E. Kalnay (2002), The 1998 Oklahoma-Texas drought: Mechanistic experiments with NCEP global and regional models, *J. Climate*, *15*, 945–963; doi :10.1175/1520-0442(2002)015 < 0945:TOTDME > 2.0.CO;2.

Hurkmans, R. T. W. L., W. Terink, R. Uijlenhoet, E. J. Moors, P. A. Troch, and P. H. Verburg (2009), Effects of land use changes on streamflow generation in the Rhine basin, *Water Resour. Res.*, *45*, W06405.

IPCC (2012), *Managing the Risks of Extreme Events and Disasters to Advance Climate Change Adaptation. A Special Report of Working Groups I and II of the Intergovernmental Panel on Climate Change*, Cambridge University Press, Cambridge, UK and New York, USA.

Jaeger, E., and S. Seneviratne (2011), Impact of soil moisture–atmosphere coupling on European climate extremes and trends in a regional climate model, *Climate Dyn.*, *36*, 1919–1939.

Jothityangkoon, C., C. Hirunteeyakul, K. Boonrawd, and M. Sivapalan (2013), Assessing the impact of climate and land use changes on extreme floods in a large tropical catchment, *J. Hydrol.*, *490*, 88–105.

Kanamitsu, M., and K. C. Mo (2003), Dynamical effect of land surface processes on summer precipitation over the Southwestern United States, *J. Climate*, *16*, 496–509, http://dx.doi.org/10.1175/1520-0442(2003)016<0496:DEOLSP>2.0.CO;2.

Kollet, S. J., and R. M. Maxwell (2008), Capturing the influence of groundwater dynamics on land surface processes using an integrated, distributed watershed model, *Water Resour. Res.*, *44*, W02402.

Koster, R. D., et al. (2000), A catchment-based approach to modeling land surface processes in a general circulation model, 1. Model structure, *J. Geophys. Res.*, *105*, D20, 24,809–24,822.

Koster, R. D., et al. (2004), Regions of strong coupling between soil moisture and precipitation, *Science*, *305*, 1138–1140.

Koster, R. D., et al. (2006), GLACE: The Global Land-Atmosphere Coupling Experiment. Part 1: Overview, *J. Hydrometeor.*, *7*, 590–610.

Koster, R. D., et al. (2010), Contribution of land surface initialization to subseasonal forecast skill: First results from a multi-model experiment, *Geophys. Res. Lett.*, *37*, L02402; doi: 10.1029/2009GL041677.

Koster, R. D., M. J. Suarez, R. W. Higgins, and H. M. Van den Dool (2003), Observational evidence that soil moisture variations affect precipitation, *Geophys. Res. Lett.*, *30*(5), 1241; doi: 10.1029/2002GL016571.

Krakauer, N. Y., M. J. Puma, and B. I. Cook (2013), Impacts of soil-aquifer heat and water fluxes on simulated global climate, *Hydrol. Earth Syst. Sci.*, *17*, 1963–1974.

Kueppers, L., et al. (2008), Seasonal temperature response to land-use change in the western United States, *Global Planet. Change*, *60*, 250–264; doi:10.1016/j.gloplacha.2007.03.005.

Kueppers, L. M., M. A. Snyder, and L. C. Sloan (2007), Irrigation cooling effect: Regional climate forcing by land-use change, *Geophys. Res. Lett.*, *34*, L03703; doi:10.1029/2006GL028679.

Kumar, S., P. A. Dirmeyer, V. Merwade, T. DelSole, J. M. Adams, and D. Niyogi (2013), Land use/cover change impacts in CMIP5 climate simulations: A new methodology and 21st century challenges, *J. Geophys. Res. Atmos.*, *118*, 6337–6353; doi:10.1002/jgrd.50463.

Lau, K. M., and H. T. Wu (2007), Detecting trends in tropical rainfall characteristics, 1979 – 2003, *Int. J. Climatol.*, *27*, 979–988; doi:10.1002/joc.1454.

Laurance, W. F. (2007), Forests and floods, *Nature*, *449*, 409– 410.

Lawrence, D., and K. Vandecar (2015), Effects of tropical deforestation on climate and agriculture, *Nature Climate Change*, *5*, 27–36; doi:10.1038/nclimate2430.

Lee, E., W. J. Sacks, T. N. Chase, and J. A. Foley (2011), Simulated impacts of irrigation on the atmospheric circulation over Asia, *J. Geophys. Res.*, *116*, D08114; doi:10.1029/2010JD014740.

Lejeune, Q., E. Davin, B. Guillod, and S. Seneviratne (2015), Influence of Amazonian deforestation on the future evolution of regional surface fluxes, circulation, surface temperature and precipitation, *Climate Dyn.*, *44*, 2769–2786.

Liang, X., Z. Xie, and M. Huang (2003), A new parameterization for surface and groundwater interactions and its impact on water budgets with the variable infiltration capacity (VIC) land surface model, *J. Geophys. Res.*, *108*, 8613; doi: 19.1029/2002JD003090.

Liang, Y.-C., M.-H. Lo, and J.-Y. Yu (2014), Asymmetric responses of land hydroclimatology to two types of El Niño in the Mississippi River Basin, *Geophys. Res. Lett.*, *41*, 582–588; doi:10.1002/2013GL058828

Lin, C. Y., W. C. Chen, P. L. Chang, and Y. F. Sheng (2011), Impact of the urban heat island effect on precipitation over a complex geographic environment in northern Taiwan, *J. Appl. Meteor. Clim.*, *50*, 339–353; doi:10.1175/2010JAMC2504.1.

Lin, Y.-H., M.-H. Lo, and C. Chou (2015), Potential negative effects of groundwater dynamics on dry season convection in the Amazon River basin (in review).

Liu, Y., and R. Avissar (1999a), A study of persistence in the land–atmosphere system using a general circulation model and observations, *J. Climate*, *12*, 2139–2153.

Liu, Y., and R. Avissar (1999b), A study of persistence in the land-atmosphere system with a fourth order analytical model, *J. Climate*, *12*, 2154–2168.

Lo, M.-H., P. J.-F. Yeh, and J. S. Famiglietti (2008), Constraining water table depth simulations in a land surface model using estimated baseflow, *Adv. Water Resour.;* doi:10.1016/j.advwatres.2008.06.007.

Lo, M.-H., and J. S. Famiglietti (2010), The effect of water table dynamics on land surface hydrologic memory, *J. Geophys. Res.*, *115*, D22118; doi:10.1029/2010JD014191.

Lo, M-H., and J. S. Famiglietti (2011), Precipitation response to land subsurface hydrologic processes in atmospheric general circulation model simulations, *J. Geophys. Res.*, *116*, D05107; doi:10.1029/2010JD015134.

Lo, M.-H., and J. S. Famiglietti (2013), Irrigation in California's Central Valley strengthens the southwestern U.S. water cycle, *Geophys. Res. Lett.*, *40*; doi:10.1002/grl.50108.

Lo, M.-H., C.-M. Wu, H.-Y. Ma, and J. S. Famiglietti (2013), The response of coastal stratocumulus clouds to agricultural irrigation in California, *J. Geophys. Res. Atmos.*, *118*; doi:10.1002/jgrd.50516.

Lobell, D., G. Bala, A. Mirin, T. Philips, R. Maxwell, and D. Rotman (2009), Regional differences in the influence of irrigation on climate, *J. Climate*, *22*, 2248–2255; doi:10.1175/2008JCLI2703.1.

Lorenz, R., E. B. Jaeger, and S. I. Seneviratne (2010), Persistence of heat waves and its link to soil moisture memory, *Geophys. Res. Lett.*, *37*, L09703.

Mahmood, R., S. A. Foster, T. Keeling, K. G. Hubbard, C. Carlson, and R. Leeper (2006). Impacts of irrigation on 20th-century temperatures in the northern Great Plains, *Global Planet. Change*, *54*, 1–18.

Mahmood, R., R. A. Pielke Sr., K. G. Hubbard, D. Niyogi, G. Bonan, P. Lawrence, et al. (2010), Impacts of land use land cover change on climate and future research priorities, *Bull. Amer. Meteor. Soc.*, *91*, 37–46.

Mahmood, R., T. Keeling, S. A. Foster, and K. G. Hubbard (2013a), Did irrigation impact 20th century air temperature in the High Plains aquifer region?, *Appl. Geogr.*, *38*, 11–21; doi:10.1016/j.apgeog.2012.11.002

Maxwell, R. M., and N. L. Miller (2005), Development of a coupled land surface and groundwater model, *J. Hydrometeor.*, *6*, 233–247.

Maxwell, R. M., and S. J. Kollet (2008), Interdependence of groundwater dynamics and land-energy feedbacks under climate change, *Nat. Geosci.*, *1*, 665–669.

Maxwell, R. M., F. K. Chow, and S. J. Kollet (2007), The groundwater land-surface-atmosphere connection: Soil moisture effects on the atmospheric boundary layer in fully-coupled simulations, *Adv. Water Resour.*, *30*, 2447–2466.

Miao, S., F. Chen, Q. Li, and S. Fan (2011), Impacts of urban processes and urbanization on summer precipitation: A case study of heavy rainfall in Beijing on 1 August 2006, *J. Appl. Meteor. Climatol.*, *50*, 806–825.

Miguez-Macho, G.., H. Li, and Y. Fan (2008), Simulated water table and soil moisture climatology over North America: Does the water table matter? *Bull. Amer. Meteor. Soc.*, *89*(5), 663–672; doi:10.1175/BAMS-89-5-663.

Miguez-Macho, G., Y. Fan, C. P. Weaver, R. Walko, and A. Robock (2007), Incorporating water table dynamics in climate modeling: 2. Formulation, validation, and simulations of soil moisture fields, *J. Geophys. Res.*, *112*, D13108; doi:10.1029/2006JD008112.

Mishra1, V., et al. (2015), Changes in observed climate extremes in global urban areas, *Environ. Res. Lett.*, *10*, 024005.

Mo, K. C. (2008), Influence of sea surface temperature on soil moisture and precipitation interactions over the southwest, *J. Geophys. Res.*, *113*, D12116; doi:10.1029/2007JD009221.

Mueller, B., and S. I. Seneviratne (2012), Hot days induced by precipitation deficits at the global scale, *Proc. Nat. Acad. Sci.*, *109*, 12398–12403.

Müller, O. V., E. H. Berbery, D. Alcaraz-Segura, and M. B. Ek (2014), Regional model simulations of the 2008 drought in southern South America using a consistent set of land surface properties, *J. Climate*, *27*, 6754–6778.

Negri, A. J., R. F. Adler, L. Xu, and J. Surratt (2004), The impact of Amazonian deforestation on dry season rainfall, *J. Climate*, *17*, 1306–1319; doi:10.1175/1520-0442(2004)017<1306:TIOADO>2.0.CO;2.

Nicholson, S. (2000), Land surface processes and Sahel climate, *Rev. Geophys.*, *38*, 117–139.

Niu, G.-Y., Z.-L. Yang, R. E. Dickinson, L. E. Gulden, and H. Su (2007), Development of a simple groundwater model for use in climate models and evaluation with Gravity Recovery and Climate Experiment data, *J. Geophys. Res.*, *112*, D07103; doi:10.1029/2006/JD007522.

Pal, J. S., and E. A. B. Eltahir (2002), Teleconnections of soil moisture and rainfall during the 1993 midwest summer flood, *Geophys. Res. Lett.*, *29*(18), 1865; doi:10.1029/2002GL014815.

Peterson, T. C., et al. (2008) Why weather and climate extremes matter. *In Weather and Climate Extremes in a Changing Climate.* Regions of Focus: North America, Hawaii, Caribbean, and U.S. Pacific Islands by Karl, T. R., G. A.

Meehl, C. D. Miller, S. J. Hassol, A. M. Waple, and W. L. Murray (eds.). A Report by the U.S.Climate Change Science Program and the Subcommittee on Global Change Research, Washington, DC., USA, 11–33.

Piao, S., P. Friedlingstein, P. Ciasis, N. de Noblet-Ducoudré, D. Labata, and S. Zaehle, (2007), Changes in climate and land use have a larger direct impact than rising CO2 on global river runoff trends, *Proc. Nat. Acad. Sci.*, *104*, 15242–15247.

Pielke, R. A. (2001), Influence of the spatial distribution of vegetation and soils on the prediction of cumulus convective rainfall, *Rev. Geophys.*, *39*, 151–177.

Pielke, R. A., Sr., A. Pitman, D. Niyogi, R. Mahmood, C. McAlpine, F. Hossain, K. K. Goldewijk, U. Nair, R. Betts, S. Fall, M. Reichstein, P. Kabat, and N. De Noblet (2011), Land use/land cover changes and climate: Modeling analysis and observational evidences, *WIRE's Clim. Change*, *2*, 828–850; doi:10.1002/wcc.144.

Pitman, A. J., et al. (2009), Uncertainties in climate responses to past land cover change: First results from the LUCID intercomparison study, *Geophys. Res. Lett.*, *36*, L14814; doi:10.1029/2009GL039076.

Puma, M., and B. Cook (2010), Effects of irrigation on global climate during the 20th century, *J. Geophys. Res.*, *115*, 1–15; doi:10.1029/2010JD014122.

Rodell, M., et al. (2004), The Global Land Data Assimilation system, *Bull. Amer. Meteor. Soc.*, *85*, 381–394; doi:10.1175/BAMS-85-3-381.

Sacks, W. J., B. I. Cook, N. Buenning, S. Levis, and J. H. Helkowski (2009), Effects of global irrigation on the near-surface climate, *Climate Dyn.*, *33*, 159–175.

Saeed, F., S. Hagemann, and D. Jacob (2009), Impact of irrigation on the South Asian summer monsoon, *Geophys. Res. Lett.*, *36*, L20711; doi:10.1029/2009GL040625.

Sala, O. E., et al. (2000), Global biodiversity scenarios for the year 2100, *Science*, *287*, 1770–1776.

Schär, C., D. Lüthi, U. Beyerle, and E. Heise (1999), The soil-precipitation feedback: A process study with a regional climate model, *J. Climate*, *12*, 722–741. doi: http://dx.doi.org/10.1175/1520-0442(1999)012<0722:TSPFAP>2.0.CO;2.

Seneviratne, S. I., D. Lüthi, M. Litschi, and C. Schär (2006), Land-atmosphere coupling and climate change in Europe, *Nature*, *443*, 205–209; doi:10.1038/nature05095.

Seneviratne, S. I., et al. (2010), Investigating soil moisture-climate interactions in a changing climate: A review, *Earth-Science Rev.*, *99*, 125–161.

Seneviratne, S. I., et al. (2013), Impact of soil moisture-climate feedbacks on CMIP5 projections: First results from the GLACE-CMIP5 experiment, *Geophys. Res. Lett.*, *40*, 5212–5217; doi:10.1002/grl.50956.

Shepherd, J. M. (2005), A review of current investigations of urban-induced rainfall and recommendations for the future, *Earth Interact.*, *9*, 1–27; doi: http://dx.doi.org/10.1175/EI156.1.

Shiu, C.-J., S. C. Liu, and J.-P. Chen (2009), Diurnally asymmetric trends of temperature, humidity and precipitation in Taiwan, *J. Climate*, *22*, 5635–5649.

Sillmann, J.,V. V. Kharin, F. W. Zwiers,X. Zhang, and D. Bronaugh (2013), Climate extremes indices in the CMIP5 multimodel ensemble, Part 2, Future climate projections, *J. Geophys. Res. Atmos.*, *118*, 2473–2493; doi:10.1002/jgrd.50188.

Sorooshian, S., J. Li, K-L. Hsu, and X. Gao (2011), How significant is the impact of irrigation on the local hydroclimate in California's Central Valley? Comparison of model results with ground and remote-sensing data, *J. Geophys. Res.*, *116*, D06102; doi:10.1029/2010JD014775.

Stéfanon, M., S. Schindler, P. Drobinski, N. de Noblet-Ducoudré, and F. D'Andrea (2014), Simulating the effect of anthropogenic vegetation land cover on heat wave temperatures over central France, *Climate Res.*, *60*, 133–146.

Sun, Y., S. Solomon, A. Dai, and R. W. Portmann (2007), How often will it rain?, *J. Climate*, *20*, 4801–4818.

Swann, A. L. S., I. Y. Fung, and J. C. H. Chiang (2012), Midlatitude afforestation shifts general circulation and tropical precipitation, *Proc. Nat. Acad. Sci.*, *109*, 712–716.

Taylor, C. M., R. A. M. de Jeu, F. Guichard, P. P. Harris, and W. A. Dorigo (2012), Afternoon rain more likely over drier soils, *Nature*, *489*, 423–426.

Teuling, A. J., A. F. Van Loon, S. I. Seneviratne, I. Lehner, M. Aubinet, B. Heinesch, C. Bernhofer, T. Gruünwald, H. Prasse, and U. Spank (2013), Evapotranspiration amplifies European summer drought, *Geophys. Res. Lett.*, *40*, 2071–2075; doi:10.1002/grl.50495.

Trenberth, K. E., A. Dai, R. M. Rasmussen, and D. B. Parsons (2003), The changing character of precipitation, *Bull. Amer. Meteor. Soc.*, *84*, 1205–1217.

Trenberth, K. E., et al. (2007), Observations: Surface and atmospheric climate change, in *Climate Change 2007: The Physical Science Basis*, edited by S. Solomon et al., 235–336, Cambridge University Press.

Vautard, R., J. Cattiaux, P. Yiou, J-N. Thépaut, P. Ciais (2010), Northern Hemisphere atmospheric stilling partly attributed to an increase in surface roughness, *Nat. Geosci.*, *3*, 756–761.

Vidale, P. L., D. Lüthi, C. Frei, S. I. Seneviratne and C. Schär (2003),Predictability and uncertainty in a regional climate model, *J. Geophys. Res.*, *108*, D18, 4586; doi:10.1029/2002JD002810.

Vinnikov, K., A. Robock, N. A. Speranskaya, and C. A. Schlosser (1996), Scales of temporal and spatial variability of midlatitude soil moisture, *J. Geophys. Res.*, *101*(D3), 7163–7174.

Wang, J., F. J. F. Chagnon, E. R. Williams, A. K. Betts, N. O. Renno, L. A. T. Machado, G. Bisht, R. Knox, R. L. Brase (2009), Impact of deforestation in the Amazon basin on cloud climatology, *Proc. Nat. Acad. Sci.*, *106* (10), 3670–3674.

Wei, J., R. E. Dickinson, and H. Chen (2008), A negative soil moisture–precipitation relationship and its causes, *J. Hydrometeor.*, *9*(6), 1364–1376.

Wei, J. F., H. Su, and Z. L. Yang (2015), Impact of moisture flux convergence and soil moisture on precipitation: A case study for the southern United States with implications for the globe, *Climate Dyn.*; doi:10.1007/s00382-00015-02593-00382.

Wu, W., and R. E. Dickinson (2004), Time scales of layered soil moisture memory in the contest of land-atmosphere interaction, *J. Climate*, *17*, 2752–2764.

Wu, W., M. A. Geller, and R. E. Dickinson (2002), Soil moisture profile variability in response to long-term precipitation, *J. Hydrometeor.*, *3*(5), 604–613.

Yeh, P. J.-F., and E. A. B. Eltahir (2005), Representation of water table dynamics in a land surface scheme, Part I: Model development, *J. Climate*, *18*, 1861–1880.

Yeh, P. J.-F., and J. S. Famiglietti (2009), Regional groundwater evapotranspiration in Illinois, *J. Hydrometeor.*, *10*, 464–478.

Yeh, T. C., R. T. Wetherald, and S. Manabe (1984), Effect of soil moisture on the short term climate and hydrology change: A numerical experiment, *Monthly Weather Rev.*, *112*, 474–490.

Zaitchik, B. F., J. A. Santanello, S. V. Kumar, and C. D. Peters-Lidard (2013), Representation of soil moisture feedbacks during drought in NASA Unified WRF (NU-WRF), *J. Hydrometeor.*, *14*, 360–367.

Zhang, X., L. Alexander, G. C. Hegerl, P. Jones, A. K. Tank, T. C. Peterson, B. Trewin, and F. W. Zwiers (2011), Indices for monitoring changes in extremes based on daily temperature and precipitation data, *WIRE's Clim. Change*, *2*, 851–870; doi:10.1002/wcc.147.

Zhang, X., and F. Zwiers (2012) Statistical indices for the diagnosing and detecting changes in extremes. *In Extremes in a Changing Climate: Detection, Analysis and Uncertainty* by A. AghaKouchak, D. Easterling, K. Hsu, S. Schubert, and S. Sorooshian (eds.). Springer Science+Business Media, Heidelberg, Germany and New York, NY, USA, 1–14.

Zhao, M., and A. J. Pitman (2002),The impact of land cover change and increasing carbon dioxide on the extreme and frequency of maximum temperature and convective precipitation, *Geophys. Res. Lett.*, *29*, 1078; doi:10.1029/2001GL013476.

Part II
Processes of Climate Extremes

6

Timing of Anthropogenic Emergence in Climate Extremes

Andrew D. King[1], Markus G. Donat[2], Ed Hawkins[3], and David J. Karoly[1]

ABSTRACT

In many areas of the world, the observed trends of warming of temperature extremes and intensification of precipitation extremes have been attributed to human-induced climate change. The topic of calculating the "time of emergence" has expanded in recent years with a growing body of literature in this field. So when did (or when will) an anthropogenic signal in these climate extremes emerge? In this chapter, we first describe some of the established approaches taken to tackling this problem in previous studies. Approaches that have been taken to calculate the time of emergence have been varied and are generally based on the statistical distributions of a climate variable for a given location or region. Here we apply one methodology to observational and CMIP5 model data. We find that simulated emergence of temperature indices occurs earlier in the tropics than at higher latitudes. There are fewer signs of emergence in precipitation extremes in the model data. Our observational analysis points to some initial signs of emergence in temperature indices, which are likely to become clearer in the next few decades.

6.1. INTRODUCTION

As shown in other chapters of this book, climate extremes are changing and are expected to continue to change into the future. In many areas of the world, the observed trends of warming of temperature extremes and intensification of precipitation extremes have been attributed to human-induced climate change [e.g., *Hegerl et al.*, 2004; *Min et al.*, 2011]. In this chapter, we take a different approach in studying these climate changes by asking the question: when did (or when will) an anthropogenic signal in these climate extremes emerge? This is not a straightforward question to answer, so we first describe some of the established approaches taken to tackling this problem used in previous studies. The topic of calculating the "time of emergence" has expanded in recent years with a growing body of literature in this field. Approaches that have been taken to calculate the time of emergence have been varied and are generally based on the statistical distributions of a variable (e.g., mean monthly temperatures) for a given location or region.

[1]*ARC Centre of Excellence for Climate System Science, School of Earth Science, University of Melbourne, Melbourne, Australia*

[2]*ARC Centre of Excellence for Climate System Science, Climate Change Research Centre, University of New South Wales, Sydney, NSW, Australia*

[3]*National Centre for Atmospheric Science, Department of Meteorology, University of Reading, Reading, UK*

Climate Extremes: Patterns and Mechanisms, Geophysical Monograph 226, First Edition.
Edited by S.-Y. Simon Wang, Jin-Ho Yoon, Christopher C. Funk, and Robert R. Gillies.

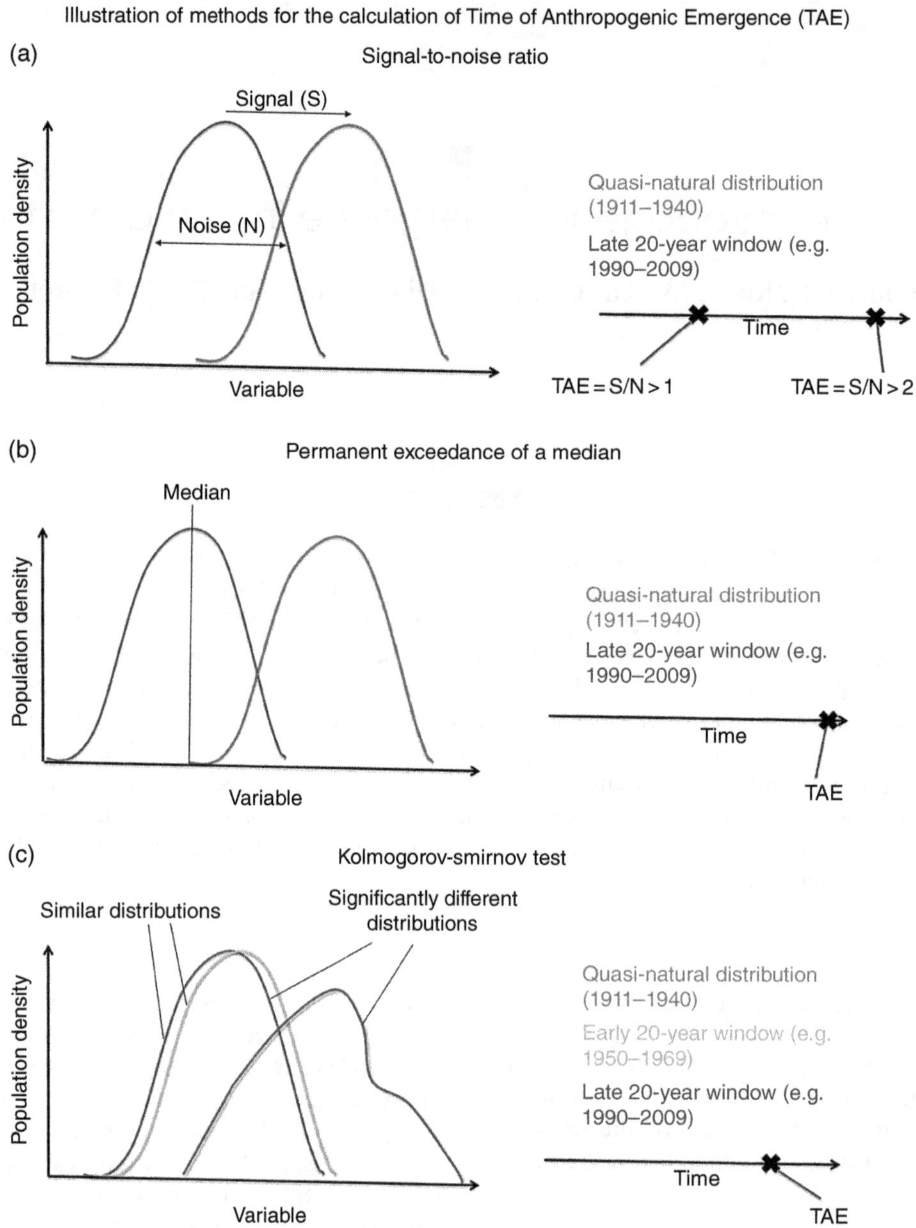

Figure 6.1 An illustration of the calculation of the time of anthropogenic emergence. As time passes, the climate warms further such that it is more different from its preindustrial state (approximated by the blue curve). The time of anthropogenic emergence can be defined as the point in time when (a) the signal-to-noise ratio exceeds a given threshold, (b) the statistical distribution exceeds the previous average (median), or (c) the statistical distributions for the baseline period and a later period are significantly different (see section 6.2, Defining Time of Emergence, for more details). (*See insert for color representation of the figure.*)

6.2. DEFINING TIME OF EMERGENCE

There are three principal methodologies used to calculate time of emergence, which are described here and through the illustration in Figure 6.1. Here we describe each method and discuss advantages and disadvantages.

The first method is the calculation of signal-to-noise ratios. For a given time series, the signal may be calculated as the change in the mean of statistical distributions between the baseline period and a later period (as in Fig. 6.1a). It could also be estimated using a regression of the variable of interest (e.g., surface air temperature) onto the global mean trend or projection for that variable [*Hawkins and Sutton*, 2012]. The noise may be defined as the standard deviation of the variable over a baseline period or long model simulations. The time of emergence is then set as the time when the signal-to-noise ratio reaches a given threshold and does not return to values below that threshold. *Hawkins and Sutton* [2012] used thresholds of one and two to define when the time of emergence had occurred.

The signal-to-noise ratio calculation for the time of emergence is particularly useful for studying the emergence of mean monthly, seasonal, and annual temperature distributions as these tend to have near-Gaussian distributions. The signal-to-noise methodology may be adapted for the study of mean and extreme precipitation indices by investigating changes in the properties of generalized extreme value (GEV) fits to distributions [*Maraun,* 2013].

Another popular method for emergence calculations is the exceedance of a previous median value (illustrated in Fig. 6.1b). The median is calculated for a given baseline period and moving windows over later periods are used to calculate when permanent emergence above the median threshold can be found. This methodology has been applied to temperature extremes in the United States [*Scherer and Diffenbaugh*, 2014]. This method makes no assumptions about the shape of a distribution and is a relatively strict criterion as even one cold year or season late in the period of study will lead to no emergence being found. For larger regions and climate variables averaged over longer periods, emergence is more likely as interannual variability is reduced [*Fischer et al.*, 2013]. Using this method, it is unlikely emergence would be found for precipitation variables, such as monthly total precipitation, either, because of the large interannual variability in precipitation. The climate may change substantially without emergence being detected using the exceedance of a median method.

The third method we describe is the use of a statistical test to describe whether distributions are significantly different. Popular examples include the Kolmogorov-Smirnov statistic (KS-test; illustrated in Fig. 6.1c) and the Anderson-Darling test. In this case, a baseline distribution is compared with moving window distributions of the same variable to test whether the distributions are likely to have come from the same population. If the distributions are significantly different (and the distributions in subsequent moving windows remain significantly different from the baseline), then emergence has occurred. Again, this methodology makes no assumptions about the shapes of the statistical distributions. It may also be applied to rainfall indices as well as temperature. One potential disadvantage of this method is that a significant difference between the moving windows and the baseline may take very different forms (either through shifts in the distributions or changes in higher moments of the distributions). The KS-test statistic alone does not give information on the nature of these differences, so further analysis may be necessary.

All these methods require choices to be made on the threshold used to define emergence. For example, the signal-to-noise ratio used to define when emergence has occurred could be set at one or two (as in *Hawkins and Sutton* [2012]) or another number. Similarly, the KS-test statistic may be defined as significant at the 5% level (as in *Mahlstein et al.* [2011, 2012]; *King et al.* [2015]) or another level. Also, permanent exceedance over a previous median could be altered to permanent exceedance over any percentile threshold. While the use of different significance levels or thresholds allows for sensitivity tests, it also can lead to confusion when interpreting results.

Another issue to consider when defining the time of emergence is the timing of the baseline period. The majority of previous studies use a recent baseline period (such as 1986–2005 in *Hawkins and Sutton* [2012]) and study when emergence occurs into the future using projections from model simulations. Other studies such as *Mahlstein et al.* [2012] and King *et al.* [2015] use a baseline period from the late nineteenth or early twentieth centuries and calculate when the time of emergence has occurred in the past or will occur into the future. Both approaches to the choice of baseline have their merits. The use of a recent baseline period is particularly relevant to those interested in adaptation to future climate change. It is based on the assumption that society and the environment have fully adapted to the climatic changes that have already occurred up to now. This may be a reasonable assumption to make in some parts of the world and not others. The use of a preindustrial baseline means that changes that have occurred may be called anthropogenic, hence the use of the term *Time of Anthropogenic Emergence* in *King et al.* [2015]. Changes from a preindustrial baseline may also be linked to detection and attribution, as regions for which there has already been a significant change in the climate (i.e., the time of anthropogenic emergence has already occurred) would be more likely to have an attributable climate change signal (assuming that the other requirements for an attribution are met, such as the existence of high-quality observations).

Here, we use a Kolmogorov-Smirnov test ($p < 0.05$) and apply it to an early twentieth century baseline to calculate the time of anthropogenic emergence of climate means and extremes in CMIP5 simulations and observations. The KS-test does not make assumptions about the properties of statistical distributions (which are more likely to be nonnormal in the case of extreme indices and precipitation variables than for mean temperature) and this method allows for changes in higher moments of the distributions (such as variance and skewness) to be detected.

6.3. DATA AND METHODS

The data and methods used here follow those of *King et al.* [2015] closely except for one major difference. In this study, the baseline period is later (1911–1940 as opposed to 1860–1910 in *King et al.* [2015]). The use of a later baseline allows for greater comparison between results based on model simulations and station observations as many long-running observational time series start in the early twentieth century or are more reliable from this period onward.

The model analysis is based on 23 model simulations from the Fifth Phase of the Coupled Model Intercomparison Project (CMIP5) [*Taylor et al.*, 2012] combining historical (1860–2005) and RCP8.5 scenario (2006–2099) simulations from six climate models. These six models all have mean temperature and precipitation and extreme indices available for at least three historical and RCP8.5 runs. The mean climate was represented by the seasonal mean surface air temperature (tas) and precipitation (pr) variables. The climate extremes indices [*Zhang et al*, 2011] used in this study (shown in Table 6.1) were the seasonal maximum and minimum values of daily maximum temperature (TXx and TXn, respectively) and daily minimum temperature (TNx and TNn, respectively), and the seasonal maximum 1 day and consecutive 5 day precipitation (RX1D and RX5D, respectively). These climate extremes indices, calculated for the different CMIP5 runs, were obtained from the Environment Canada CLIMDEX website [http://www.cccma.ec.gc.ca/data/climdex/] and have been used in many studies as measures of climate extremes [e.g., *Zhang et al*, 2013]. Monthly values of these mean and extreme variables were regridded onto a common 2.5° x 2.5° grid. Seasonal values (June to August and December to February) were calculated at each gridbox and variable for each model run for 1860–2099, such that the extreme indices represent the maximum (in the case of TXx, TNx, RX1D, and RX5D) or minimum values (in the case of TXn and TNn) for the season.

Table 6.1 Definitions of the Mean and Extreme Climate Indices Used in This Analysis

Mean and extreme indices	Definition of index
Tas	Seasonal mean temperature
Pr	Seasonal total precipitation
TXx	Highest daily maximum temperature per season
TXn	Lowest daily maximum temperature per season
TNx	Highest daily minimum temperature per season
TNn	Lowest daily minimum temperature per season
RX1D	Maximum 1 day precipitation per season
RX5D	Maximum consecutive 5 day precipitation per season

Comparable time series were also calculated for three long-running station observational series (Jena in Germany, Sydney in Australia, and Toronto in Canada) and the Central England Temperature series [*Parker et al.*, 1992]. The mean temperature and total precipitation data were sourced from the National Centers for Environmental Information (NCEI) Global Historical Climatology Network (GHCN) dataset (http://www.ncdc.noaa.gov/cdo-web/search). The extremes indices, calculated from the same dataset, were obtained from the CLIMDEX website (http://www.climdex.org/; *Donat et al.*, [2013]). These observational time series were selected for their long-running complete (or near-complete) records.

Within each of the 23 model runs and the four observational time series, quasi-natural distributions of each variable were calculated, for each gridbox and season, over the 1911–1940 period when anthropogenic forcing on the climate was much smaller than it is today. Moving windows of 20 yr periods were compiled from 1941 onward and similarity with the quasi-natural distributions was tested using a Kolmogorov-Smirnov (KS-) test with statistical significance defined at the 5% level [*Mahlstein et al*, 2012; *King et al.*, 2015]. The use of a 20 yr window is a compromise between having a large enough sample to estimate the distribution, and not having substantial warming during the window period (as discussed in *King et al.* [2015]). The KS-test method was chosen for its ability to detect changes across the different properties of statistical distributions (i.e., location, shape, scale, etc.) and its widespread use in climate science, often in model evaluation [e.g., *von Storch and Zwiers*, 1999; *Lewis and Karoly*, 2013a]. The time of anthropogenic emergence (TAE) is defined as the year when the KS-test implies that the variable, defined seasonally for the 20 yr windowed period, is drawn from a different distribution than the quasi-natural period, and that all subsequent 20 yr windowed periods are also drawn from different distributions than the quasi-natural period. The TAE is indexed to the first year of the 20 yr moving windows. This method was applied to variables at individual land-based gridboxes. Median TAE values were calculated at each land gridbox across all 23 model runs. Comparisons were made between the observational TAE values and the TAE for the gridbox in which the stations are located.

6.4. RESULTS

The median modeled TAE values (Fig. 6.2) show earlier emergence of temperature indices in equatorial regions than in the extratropics. This result is due to there being

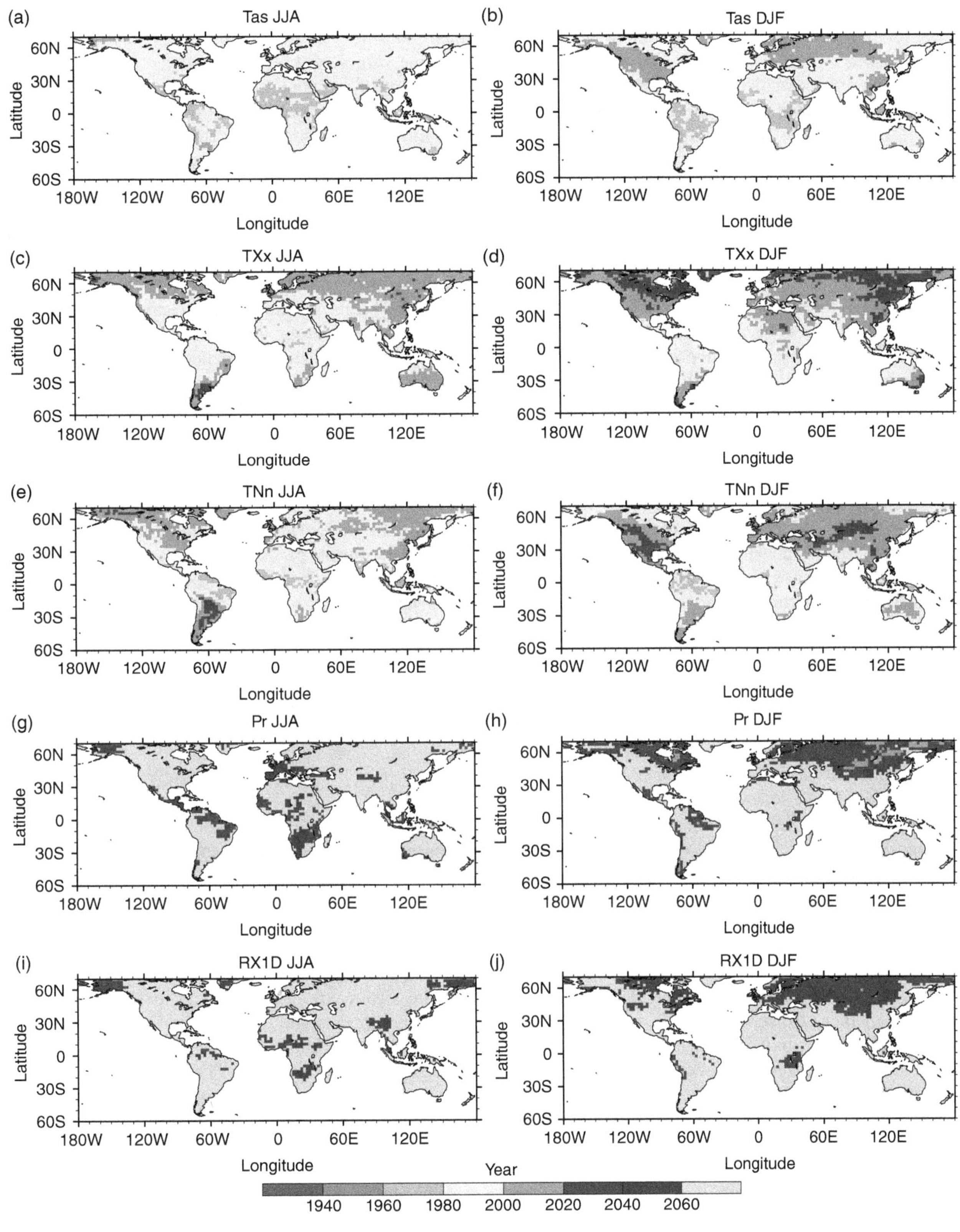

Figure 6.2 Median simulated time of anthropogenic emergence for climate means and extremes. Maps of median TAE averaged across 23 model simulations for (a, b) mean surface air temperature, (c, d) highest daily maximum temperature, (e, f) lowest daily minimum temperature, (g, h) total precipitation, and (i, j) maximum 1 day precipitation for (a, c, e, g, i) June–August and (b, d, f, h, j) December–February.

lower interannual variability in the tropics than at higher latitudes meaning that a smaller amount of warming is needed for emergence to occur. This spatial variability in timing of emergence can also be seen in previous studies [e.g., *Hawkins and Sutton*, 2012; *King et al.*, 2015]. In many regions of the world, median simulated emergence has already occurred in mean summer and winter temperatures. This does not necessarily mean that observed emergence values would be similar to the median due to the effect of interannual variability being reduced by showing the median. Rather, these maps give a sense of the expected emergence timing through the forced signal alone.

Temperature extremes tend to emerge later than mean temperatures and this is because they have greater interannual variability. This is unsurprising given that the mean seasonal temperatures are the average of roughly 90 values and these extremes are based on single values. Indeed, it is remarkable that the extremes show emergence and that, across the low latitudes, median simulated emergence has occurred prior to 2015.

The precipitation indices do not tend to show emergence prior to the end of the 21st century across most land regions of the globe. However, there are indications of widespread emergence in mean and extreme precipitation indices across northern high latitudes in wintertime. This is associated with a consistent signal of increased total precipitation and more intense heavy precipitation events [*King et al.*, 2015].

A limited comparison to observational results is made with three long-running station time series and the Central England Temperature series. As the observed time series finish in 2014 (2002 in the case of Toronto) and the simulated time series extend into the future, permanent emergence is less likely in the observed time series than in the model simulations. Figure 6.3 shows that in the majority of indices there is no emergence at all in the observations (indicated by the black crosses). The unfilled circles show emergence before the end of the time series, but by less than 20 yr. For emergence values less than 20 yr before the end of the observational time series, we cannot have confidence that the emergence is permanent. In fact, *Hawkins et al.* [2014] found emergence in models more than 50 yr prior to the end of simulations was in some cases not associated with permanent emergence. Given the short length of our observational time series such a strict criterion for emergence is not applied here.

Multidecadal variability, associated with climate modes, such as the Interdecadal Pacific Oscillation and the Atlantic Multidecadal Oscillation, may be the cause of temporary emergence from the baseline period. In Figure 6.3, where there are filled circles there is emergence more than 20 yr prior to the end of the observational time series and we can have some confidence that the emergence may be permanent. This can only be seen in mean summer (DJF) temperature in Sydney and summertime values of TNx at all three stations. There is no indication of permanent emergence seen in the Central England Temperature indices or in any of the precipitation indices. It is interesting that the clearest signals for observed emergence are in an extreme temperature index measuring the warmest minimum temperatures in summer. Some previous studies have indicated that there has been enhanced warming of minimum temperatures, when compared with maximum temperatures, leading to a reduction in the diurnal temperature range [e.g., *Karl et al.*, 1991]. Climate models fail to capture the magnitude of the observed trend [e.g., *Lewis and Karoly*, 2013b], which may be why the emergence in the indices based on maximum and minimum temperature indices tend to be similar in the model simulations.

For several indices at different locations, there are signs of potential emergence that are consistent with emergence before now in some model simulations. This is the case for mean summer temperature in the CET series and several extreme temperature indices at Jena and Sydney. While we cannot be confident that this is permanent emergence, the fact that the model simulated emergence times are similar suggests that in the next decade or two more robust signs of emergence may be detectable. At Jena, there is potential emergence in both the heaviest 1 day and consecutive 5 day summer precipitation indices, however, this is substantially earlier than in any of the 23 model simulations.

6.5. DISCUSSION

In this chapter we have taken a different approach to examining the changing nature of climate extremes by investigating the time of anthropogenic emergence (TAE) from a quasi-natural baseline period. We considered different approaches that may be used to calculate the time of emergence and chose to use a Kolmogorov-Smirnov (KS-) test to calculate the point in time when an anthropogenic signal emerges in climate extremes in a set of model simulations and four observational time series.

The simulated time of emergence is earliest in the tropics for temperature extremes. These maps are in contrast with the observed trends in climate extremes, which tend to show the strongest warming signals in polar regions (Chapter 2) and, for temperature extremes, midlatitudes [e.g., *Donat et al.*, 2013b]. This difference is due to the stronger interannual climate variability in high latitudes causing the emergence of an anthropogenic warming signal from background noise to be delayed in these regions compared with lower latitudes.

Consistent with *King et al.* [2015], there is later emergence of extreme temperature indices than means and

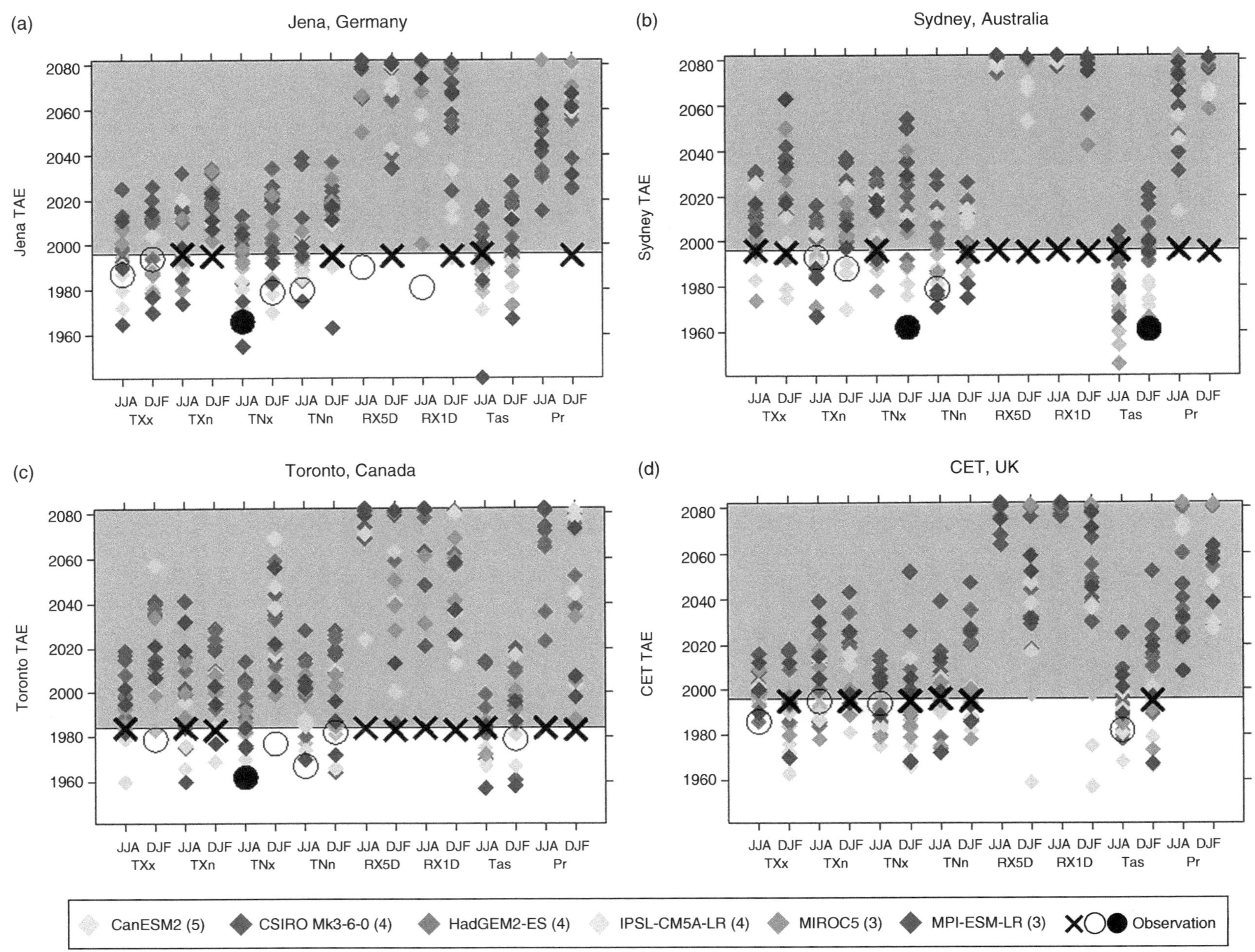

Figure 6.3 Time of anthropogenic emergence and model uncertainty. Graphs of TAE values for (a) Jena, (b) Sydney, (c) Toronto, and (d) Central England Temperature (CET) calculated for each station and corresponding model gridbox extreme and mean index in June–August and December–February shown for CanESM2 (light blue), CSIRO-Mk3-6-0 (dark blue), HadGEM2-ES (green), IPSL-CM5A-LR (yellow), MIROC5 (orange), and MPI-ESM-LR (red). The observed TAE is shown in black. Crosses indicate no emergence in the observations prior to the end of the time series, empty circles indicate emergence up to 20 yr prior to the end of the time series and filled black circles indicate emergence at least 20 yr prior to the end of the series. In observed time series with missing values, no symbol is shown. (*See insert for color representation of the figure.*)

limited signs of emergence in precipitation indices except at high latitudes in winter in our model simulations. There is remarkable agreement in the modeled TAE values with *King et al.* [2015] despite the use of different baseline periods. While there are slightly later emergence times for temperature indices in the tropics than in *King et al.* [2015], due to emergence there occurring prior to 1940 in that paper, which is during the baseline period used here, there are only small differences in the emergence time outside the tropics and in the precipitation indices. The strong agreement in results increases confidence in these simulated TAE values.

It could be argued that if a significant change in one variable (e.g., temperature) is detected, then the climate, a combination of many variables, has been noticeably altered. Therefore, any other related variables have also altered in the new warmer climate. For example, if temperatures become warmer, but no change in precipitation is detectable, then even though the frequency of precipitation deficits may not have changed, drought impacts could be worse, as warmer temperatures are associated with increased evaporation.

While it is true that temperature and precipitation changes are not independent, it is more pragmatic to separate variables of interest and study their changes separately. The information gained by doing so can inform decision makers on the changes in risks to a particular area that are posed due to the changes in individual variables. The later emergence of precipitation indices compared with temperature indices is unsurprising given that warming is the primary consequence of the anthropogenic influence on the climate and precipitation changes are a secondary effect, altered through complex interactions between the changing ability of the atmosphere to hold moisture [e.g., *Held and Soden*, 2006; *Trenberth et al.*, 2015] and dynamical changes.

The observed TAE values tend to show a lack of permanent emergence, as in *King et al.* [2015], except in the TNx index in summertime. As discussed previously, the earlier emergence in TNx may be related to stronger observed warming of minimum temperatures seen across much of the globe. The scope for validation of the simulated TAE values using these observations is limited, however, there is broad agreement in the emergence timing between the observations and the model simulations at the same locations. Unfortunately, the locations of earliest emergence in the model simulations (i.e., the tropics) are also areas of the world where few long-running station observations exist. The sensitivity of the KS-test to changes in statistical distributions also means that changes in observation techniques, related to instrumentation and human practices, can cause spurious emergence timings that are not associated with human influences on the climate.

6.6. CONCLUSIONS

This chapter illustrates some of the possibilities that investigating the timing of emergence offers. Analysis of the time of emergence gives us a different perspective on anthropogenic climate change and allows us to examine where the effects of climate change are already being felt. This is especially important for our environment, which is highly tuned to the climate variability we experience, and especially at low latitudes where that window of variability is narrow [e.g., *Mahlstein et al.*, 2011]. The information provided from these studies may be of particular use in the adaptation and mitigation of climate change and this is especially important given that the regions experiencing the earliest significant climatic changes in the tropics tend to be less well equipped to adapt [*Mahlstein et al.*, 2011]. While we have illustrated the utility of examining the time of emergence, we have also shown that there are difficulties in its calculation, particularly in observational datasets. In decades to come, as the climate continues to change, a fuller investigation and calculation of the time of emergence in observational datasets may become possible.

ACKNOWLEDGMENTS

We thank the reviewers for their constructive comments and the editorial board for handling our submission. Andrew King, Markus Donat, and David Karoly were supported by Australian Research Council grant CE110001028. Markus Donat was also supported by Australian Research Council grant DE150100456. Ed Hawkins was supported by the UK National Environmental Research Council. We acknowledge the support of the NCI facility in Australia. We acknowledge the World Climate Research Programme's Working Group on Coupled Modeling, which is responsible for CMIP, and we thank the climate modeling groups for producing and making available their model output. For CMIP, the U.S. Department of Energy's Program for Climate Model Diagnosis and Intercomparison provides coordinating support and led development of software infrastructure in partnership with the Global Organization for Earth System Science Portals.

REFERENCES

Donat, M. G., et al. (2013b), Updated analyses of temperature and precipitation extreme indices since the beginning of the twentieth century, *J. Geophys. Res.*, *118*, 2098–2118.

Donat, M. G., L. V. Alexander, H. Yang, I. Durre, R. Vose, and J. Caesar (2013a), Global land-based datasets for monitoring climatic extremes, *Bull. Amer. Meteor. Soc.*, *94*, 997–1006; http://dx.doi.org/10.1175/BAMS-D-12-00109.1.

Fischer, E. M., U. Beyerle, and R. Knutti (2013), Robust spatially aggregated projections of climate extremes, *Nat. Climate Change*, *3*, 1033–1038.

Hawkins, E., and R. Sutton (2012), Time of emergence of climate signals, *Geophys. Res. Lett.*, *39*, L01702.

Hawkins, E., et al. (2014), Uncertainties in the timing of unprecedented climates, *Nature*, *511*, E3–5.

Hegerl, G. C., F. W. Zwiers, P. A. Stott, and V. V. Kharin (2004), Detectability of anthropogenic changes in annual temperature and precipitation extremes, *J. Climate*, *17*, 3683–3700.

Held,, I. M., and B. J. Soden (2006), Robust responses of the hydrological cycle to global warming., *J. Climate*, *19*, 5686–5699; doi: 10.1175/JCLI3990.1.

Karl, T. R., G. Kukla, V. N. Razuvayev, M. J. Changery, R. G. Quayle, R. T. Heim Jr., D. R. Easterling, and C. B. Fu (1991), Global warming: Evidence for asymmetric diurnal temperature change, *Geophys. Res. Lett.*, *18*, 2253–2256.

King, A. D., M. G. Donat, E. M. Fischer, E. Hawkins, L. V. Alexander, D. J. Karoly, A. J. Dittus, S. C. Lewis, and S. E. Perkins (2015), The timing of anthropogenic emergence in climate extremes, *Environ. Res. Lett.*, *10*, 094015; doi: 10.1088/1748-9326/10/9/094015.

Lewis, S. C., and D. J. Karoly (2013a), Anthropogenic contributions to Australia's record summer temperatures of 2013, *Geophys. Res. Lett.*, *40*, 3705–3709.

Lewis, S. C., and D. J. Karoly (2013b), Evaluation of historical diurnal temperature range trends in CMIP5 models, *J. Climate*, *26*, 9077–9089; doi: http://dx.doi.org/10.1175/JCLI-D-13-00032.1.

Mahlstein, I., G. Hegerl, and S. Solomon (2012), Emerging local warming signals in observational data, *Geophys. Res. Lett.*, *39*, L21711.

Mahlstein, I., R. Knutti, S. Solomon, and R.W. Portmann (2011), Early onset of significant local warming in low latitude countries, *Environ. Res. Lett.*, *6* (034009).

Maraun, D. (2013), When will trends in European mean and heavy daily precipitation emerge?, *Environ. Res. Lett.*, *8*, 014004.

Min, S-K., X. Zhang, F. W. Zwiers, and G. C. Hegerl (2011), Human contribution to more-intense precipitation extremes, *Nature*, *470*, 378–381.

Parker, D. E., T. P. Legg, and C. K. Folland (1992), A new daily central England temperature series, *Int. J. Climatol.*, *12*, 317–342.

Scherer, M., and N. S. Diffenbaugh (2014), Transient twenty-first century changes in daily-scale temperature extremes in the United States, *Climate Dyn.*, *42*, 1383–1404.

Taylor, K. E., R. J. Stouffer, and G. A. Meehl (2012), An overview of CMIP5 and the experiment design, *Bull. Amer. Meteor. Soc.*, *93*, 485–498.

Trenberth, K. E., J. T. Fasullo, and T. G. Shepherd (2015), Attribution of climate extreme events., *Nature Climate Change*, *5*, 725–730.

von Storch, H., and F. W. Zwiers (1999), *Statistical Analysis in Climate Research*, Cambridge University Press.

Zhang, X., H. Wan, F. W. Zwiers, G. C. Hegerl, and S-K. Min (2013), Attributing intensification of precipitation extremes to human influence, *Geophys. Res. Lett.*, *40*, 5252–5257.

Zhang, X., L. Alexander, G. C. Hegerl, P. Jones, A. Klein Tank, T. C. Peterson, B. Trewin, and F. W. Zwiers (2011), Indices for monitoring changes in extremes based on daily temperature and precipitation data, *Wiley Interdiscip. Rev. Climate Change*, *2*, 851–870.

7

Recent Increases in Extreme Temperature Occurrence over Land

Scott J. Weaver[1], Arun Kumar[2], and Mingyue Chen[2]

ABSTRACT

Recently observed global and US temperature increases are probed from the perspective of several hundred climate realizations afforded by the availability of reforecast climate model runs from the NCEP Climate Forecast System Version 2 (CFSv2). The large number of seasonal realizations with the observed time varying CO_2 affords a unique opportunity to explore the role of greenhouse gas changes on 3 mo seasonal mean temperature increases and, specifically, whether they are the result of a shift in the mean temperature distribution or an increase in its variability. It is found that significant positive shifts in the temperature probability density function (PDF) occurs primarily as the result of the time varying CO_2 included in the historical model runs, although a contribution from natural climate variability modes cannot be categorically excluded. The temperature PDF comparison further indicates that the increasing global and US temperatures over the last 30 yr are predominantly the result of shifts in the mean temperature distribution and not increasing temperature variability. As such, the likelihood of increases in the occurrence of warm temperature extremes will continue to increase worldwide, leading to significant impacts on many socioeconomic sectors such as agriculture and public health.

7.1. INTRODUCTION

The IPCC AR5 report states, "it is extremely likely that human influence has been the dominant cause of the observed warming since the mid twentieth century," and "1983–2012 was likely the warmest 30-year period of the last 1400 years" [*IPCC*, 2013]. This anthropogenic influence is primarily through the continued increase in atmospheric greenhouse gas (GHG) concentrations. One consequence of the rapid increase in global temperatures is that the occurrence of extreme warm temperature events has been steadily rising in recent decades [*Meehl et al.*, 2007]. Indeed, the early 21st century has seen a number of extreme heat waves that have caused significant societal and economic impacts, including the European heat wave of 2003, the Russian heat wave of 2010, and heat waves over the southern Great Plains and Midwest of the United States in 2011 and spring 2012. Furthermore, mean wintertime temperatures have been increasing, although without an observed decrease in cold air outbreaks [*Vavrus et al.*, 2006], while the length of growing season has been expanding [*Xu et al.*, 2013].

These events have brought to the forefront questions regarding the role of anthropogenic climate change in increasing the probability of temperature extremes and, specifically, whether the increase in these extreme temperatures is a reflection of a shift in the mean temperature to the warmer side, or due to a change in its variability, or both. Confounding understanding is that while observed evidence indicates temperatures are increasing worldwide, the interpretation of the statistics (i.e., mean and variability) are not in agreement, with some studies

[1] *Environmental Defense Fund, Washington, DC, USA*
[2] *NOAA/NWS/NCEP Climate Prediction Center, College Park, Maryland, USA*

Climate Extremes: Patterns and Mechanisms, Geophysical Monograph 226, First Edition.
Edited by S.-Y. Simon Wang, Jin-Ho Yoon, Christopher C. Funk, and Robert R. Gillies.

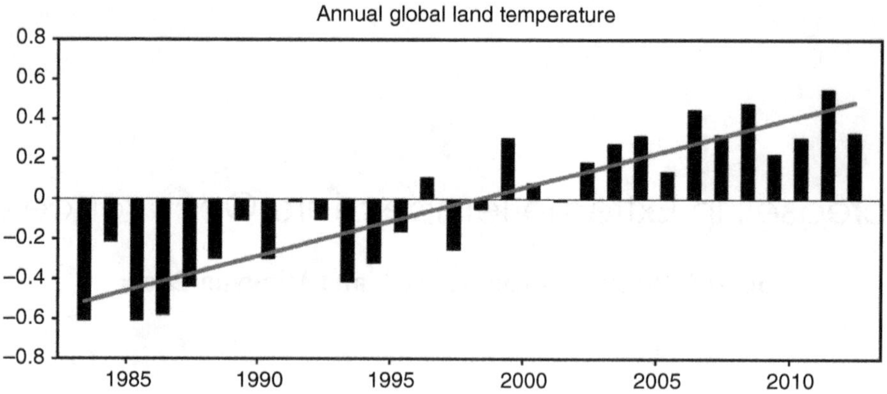

Figure 7.1 Observed annual global land mean 2 m temperature anomalies (bars) for 1983–2012 from the GHCN/CAMS dataset. The line indicates the linear trend. *See electronic version for color representation.*

indicating increases in extreme temperatures are a result of a shift in the mean temperature [*Huntingford et al.*, 2013; *Rhines and Huybers* 2013], while others suggest there has been increased temperature variability in recent decades as the result of anthropogenic climate change [*Donat and Alexander* 2012; *Hansen et al.*, 2012].

To illustrate these recent temperature changes, Figure 7.1 shows the observed annual globally averaged land-only temperature anomalies for 1983–2012 from the GHCN-CAMS dataset [*Fan and van den Dool,* 2008]. It is striking that in just the most recent three decades, globally averaged land temperatures have increased approximately 1 K. However, given that the observational record includes the totality of forcing mechanisms (i.e., natural and anthropogenic influences), it is not clear as to the dominant cause of such rapid increases in global land surface temperature, and whether they are mostly forced by anthropogenic influences or by multidecadal natural climate variability as expressed through global SST variability. Fortunately, global climate models can be used to isolate these internal (natural) and external (anthropogenic) components of the climate system. To understand the former influence, climate model experiments typically include prescribing observed sea surface temperatures (SST) as a boundary condition to multi-ensembles of global atmospheric models (i.e., AMIP style), while the latter involve free running simulations in which only forcing external to the climate system is prescribed, such as GHG concentrations, solar variability, and volcanic forcings (i.e., CMIP style).

Nevertheless, both the observational and model (i.e., AMIP/CMIP) approaches are important for unraveling the interconnectedness between natural variability and anthropogenic forcing, although, both suffer from some limitations. The observed record is only one realization of the climate state and convolves interannual-to-decadal natural variability with that from anthropogenic causes. Indeed, the large-scale natural climate variability can exert a strong influence when examining regional changes in temperatures [*Weaver,* 2013]. As such, it is difficult to tease out the anthropogenic signal from the large natural interannual-to-decadal variability. AMIP simulations clarify the role of the SST-forced component of natural variability, however do not have realistic air-sea coupling, which may be important for the evolution of the climate system in nature. On the other hand, while the CMIP approach includes air-sea coupling and is useful for isolating the externally forced signal (i.e., GHG emissions) from the natural variability, the simulated climate cannot be constrained by observations internal to the climate system. Given the inherent characteristics of the CMIP (i.e., small ensemble size) and AMIP (more ensemble members but uncoupled climate state evolution) simulations, neither approach is adequate to fully address the question of whether recently observed increases in land-surface temperature are a result of a shift in the mean temperature distribution or an increase in its variability.

A method to alleviate some of these issues is to use the large ensembles of seasonal reforecasts produced by operational climate prediction centers to increase the number of climate realizations that can be analyzed. For example, the National Centers for Environmental Prediction (NCEP) has recently completed 30 yr (1983–2012) of coupled atmosphere and ocean reforecasts for the Climate Forecast System Version 2 [*Saha et al.*, 2013]. These reforecasts include time varying CO_2 in their initialization history. Accordingly, the large numbers of initializations, lead times, and multiple ensemble members per initialization provides several thousand climate realizations for each season over the 30 yr period, allowing one to better investigate changes in climate statistics.

Figure 7.2 demonstrates this concept by showing a hypothetical comparison between a CMIP model free run and that from the operational reforecast process currently in use at many climate prediction centers worldwide. Suppose one is interested in analyzing the climate

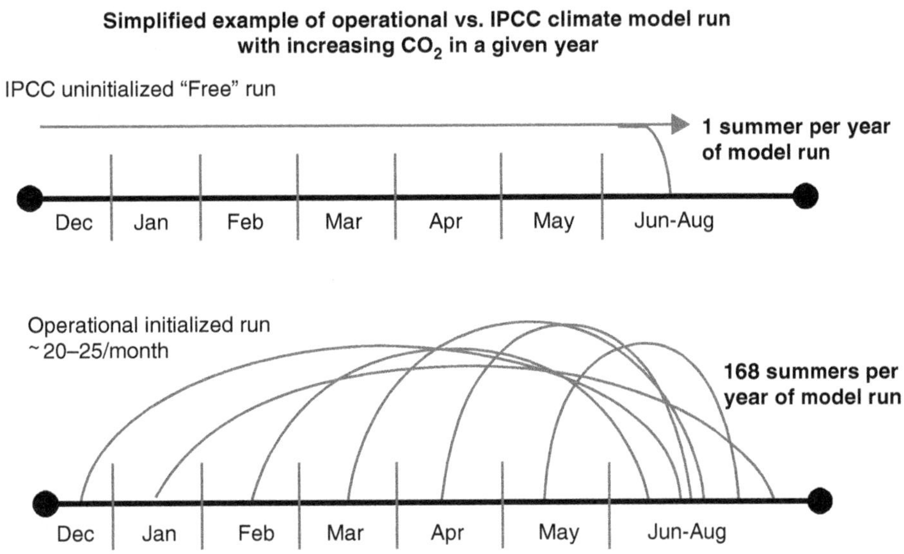

Figure 7.2 Schematic showing the comparison of climate realizations between a CMIP-style free run and an operational climate model reforecast run producing a summer (JJA) projection.

response for a particular variable (e.g., temperature) during the summer (JJA) of a given year. In the CMIP representation, it is possible to cull only one value for the given climate variable, whereas, in the operational reforecast run, numerous values of a given variable can be subset for a particular summer, is this case 168 (see section 7.2 for details). Moreover, given the varying lead times in the reforecast initialization, a diversity of outcomes are generated that are all equally plausible in that year's climate.

Despite the relatively short 30 yr record in the CFSv2 reforecasts, this approach has unique value when compared to AMIP- and CMIP-style climate model runs. In the CFSv2 reforecast paradigm, SST predictions are relatively close to the observed patterns, thus retaining the natural variability component of SST, while at the same time, air-sea coupling is retained. (This does not necessarily imply near perfect SST throughout the long 9 mo integrations when compared to observations.) To be sure, CMIP runs can similarly be used to characterize externally forced climate variability [*Taylor et al.*, 2012], however, they often suffer from a lack of significantly large ensembles (at least compared to several thousand that can be gleaned from the CFSv2 reforecasts) and may have large biases in SST variability [*Li and Xie,* 2012] and so are unable to reproduce natural variability in SST in the single realization of the observed history. As such, the strategy employed here provides an opportunity to use reforecasts in a novel way, leveraging the CFSv2 reforecast dataset to understand recent shifts in the statistics of temperature as a result of only the observed changes in greenhouse gas concentrations and in the absence of significant SST biases.

7.2. DATA AND METHODOLOGY

Our analysis is based on the reforecasts from CFSv2, which consists of fully coupled components of the ocean, atmosphere, and land [*Saha et al.*, 2013]. The atmospheric component is the NCEP global forecast system (GFS) at a horizontal resolution of T126 (~100 km) with 64 vertical levels extending from the surface to 0.26 hPa. The oceanic component is the Geophysical Fluid Dynamics Laboratory (GFDL) Modular Ocean Model version 4 (MOM4), which uses 40 levels in the vertical, a zonal resolution of 0.5°, and a meridional resolution of 0.25° between 10°S to 10°N, gradually coarsening through the tropics until becoming fixed at 0.5° poleward of 30°S and 30°N.

The CFSv2 reforecasts were initialized from the Climate Forecast System Reanalysis (CFSR) [*Saha et al.*, 2010]. The CFSR is the latest version of the NCEP climate reanalysis with the first guess from a coupled atmosphere-ocean model that is similar to CFSv2 model with a higher horizontal resolution of T382 (~35 km). In this study, we analyze the CFSv2 reforecasts initialized four times per day every fifth day with lead times out to 9 mo from 1983–2012. For each target season in a given year, a total of 168 realizations can be mined for surface temperature.

To analyze the potential influence of time varying CO_2, the probability density functions (PDF) of 3 mo averaged seasonal temperature anomalies in histogram format are computed for both the summer (JJA) and winter (DJF) for the first four full years (1983–1986), hereafter denoted early period, and the last four full years (2009–2012), hereafter denoted late period. Using reforecasts with a maximum 10 mo lead allows a 7 mo lead for the seasonal realizations due to the required 3 mo truncation necessary to create 3 mo averages, providing 168 seasonal realizations per year (7 mo x 6 initialization days per month x 4 initializations per day) and 672 over each 4 yr analysis period. The displayed PDFs are for the global mean and continental United States (CONUS) and do not include Alaska and Hawaii.

Observed land-surface temperature data are from the GHCN-CAMS dataset [*Fan and van den Dool,* 2008], while SST observations are from the Extended Reconstructed Sea Surface Temperature Version 3 (ERSSTv3) [*Smith et al.,* 2008]

7.3. RESULTS

Figure 7.3 shows the mean global land-surface temperature anomaly differences between the late and early periods in the CFSv2 for all 12 rolling seasons (upper), winter (middle), and summer (lower). There are regionally varying positive temperature anomaly differences over most of the global land surface. The warming is strongest in the higher latitude regions of the Northern Hemisphere, especially during boreal summer. In boreal winter, the strongest warming is mostly confined to the interior continental regions of Asia and North America. These spatial characteristics are generally consistent with recent studies using CMIP climate model projections of global warming, including the higher amplitude in the Arctic, and the southwest to northeast orientation over North America during DJF [*IPCC,* 2007]. It is interesting that these temperature difference patterns do not resemble those from the primary modes of natural climate variability (i.e., ENSO). Sensitivity to longer time periods (i.e., 8 yr) is shown in Figure 7.4. Similar patterns emerge to those in the 4 yr sampling albeit with a weaker temperature difference (please note contour interval difference between Figs. 7.3 and 7.4 and that all shading for both figures indicates all positive values).

To analyze the characteristics of the period shifts in the distribution of global land-surface temperature, Figure 7.5 shows the histogram of globally and 3 mo averaged land-surface temperature for winter (upper) and summer (lower) for the early (blue bars) and late (red bars) periods binned by 0.1 K intervals. The 672 model seasons show a distinct positive shift of 0.8 K in the mean of the PDF for both winter and summer from the early to late periods. Additionally, there is a strong seasonal difference in the modeled mean global land-surface temperatures during both the early and late sampling periods and it is likely due to the larger percentage of land cover in the Northern Hemisphere as compared to the Southern Hemisphere. Overall, the PDF shifts are quite distinct with more overlap during the boreal winter. Shifts in the temperature variability as diagnosed from the standard deviation are small relative to the overall magnitude in the global average from one period to another. This would suggest that global land-surface temperature changes in the CFSv2 over the last three decades are due primarily to a shift in the mean temperature, and not an increase in temperature variability. Nevertheless, it is obvious that relative to the earlier period, extreme seasonal temperature anomalies are far more likely to occur in the present as a result of a mean shift in the temperature distribution.

The shifts in the distribution of global land temperatures are noteworthy. However, given the recent extreme temperature events over the CONUS, it is also necessary to focus on regional changes to the temperature distributions. Unlike the global temperature distribution shifts, there are much stronger seasonal differences in the characteristics of the land-surface temperature PDFs over the CONUS (Fig. 7.6). Despite the similar shifts in the mean when comparing the global and US land-surface temperature differences, there is substantial overlap and less separation between the early and late periods during DJF on account of the larger temperature variability and north-south temperature gradient during boreal winter. As in the global representation, the standard deviation shifts from the early to late periods are quite weak in the US average, although in DJF the strongest change in standard deviation is noted out of the four distributions that are shown between Figures 7.5 and 7.6. Using JJA as an example, it can be seen that over the United States the probability of exceeding a summer seasonal mean temperature of 297 K (75.3 °F) as a result of only changes in greenhouse gas forcing has shifted from 0% (0 out of 672 cases) in the early period to approximately 29% (195 out of 672 cases) in the later period. Mean seasonal temperature increases such as these significantly enhance the potential for extreme heat events that impact socioeconomic sectors such as agriculture and public health.

7.3.1. Sensitivity to Initialized SST

Given the use of initialized predictions in this analysis, a question arises as to what extent natural variability of SST may be influencing the shifts in the temperature distributions, or if the period difference of predicted SST reflects a pattern similar to the observed or CMIP global warming trend. Figure 7.7 shows the SST anomaly period difference as diagnosed from the CFSv2 (upper) and ERSST data (lower). It is not surprising that there is substantial disagreement in the sign of SST patterns over portions of the tropical and eastern North Pacific, given that the methodology extracts all lead time predictions for a given season. This reduces the influence of modes related to internal SST variability by ensemble averaging,

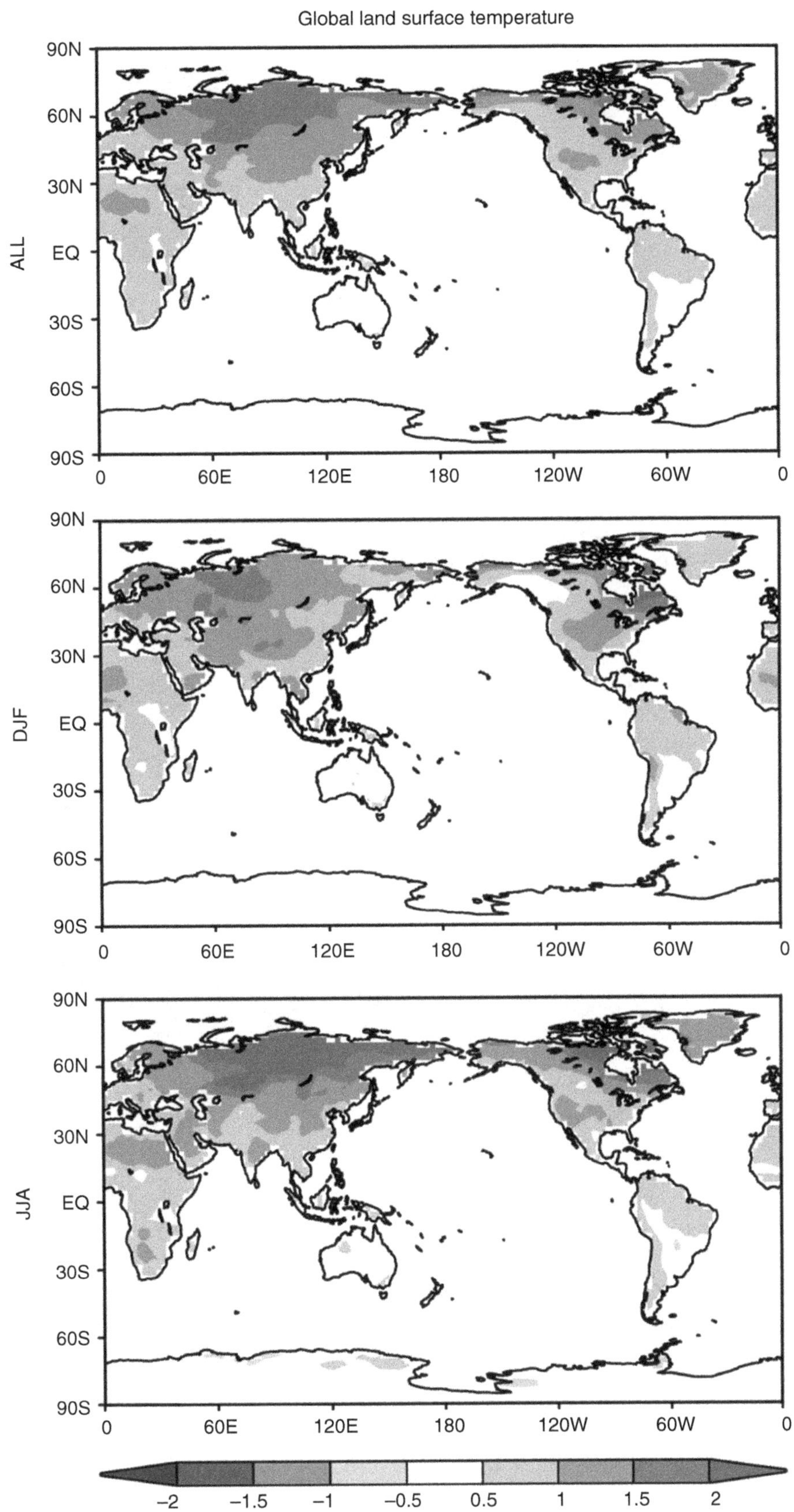

Figure 7.3 Global land surface temperature anomaly differences in the CFSv2 for the periods of 2009–2012 minus 1983–1986 for all 12 rolling seasons (upper), DJF (middle), and JJA (lower). Each seasonal period has 672 seasonal realizations. Contour interval is 0.5 K and results indicate that all shaded values are positive.

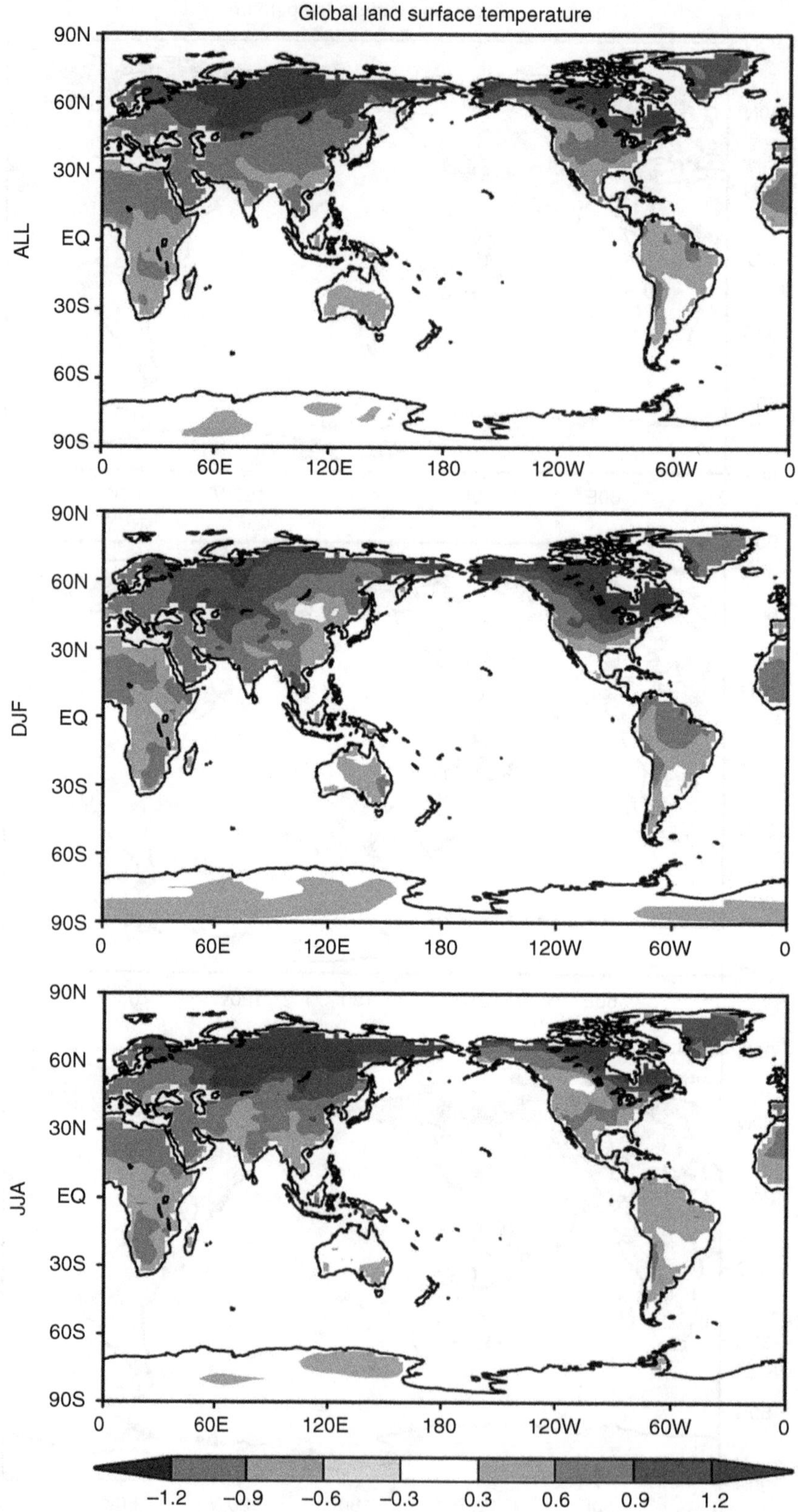

Figure 7.4 Global land surface temperature anomaly differences in the CFSv2 for the periods of 2005–2012 minus 1983–1990 for all 12 rolling seasons (upper), DJF (middle), and JJA (lower). Each seasonal period has 1344 seasonal realizations. Contour interval is 0.3 K and results indicate that all shaded values are positive.

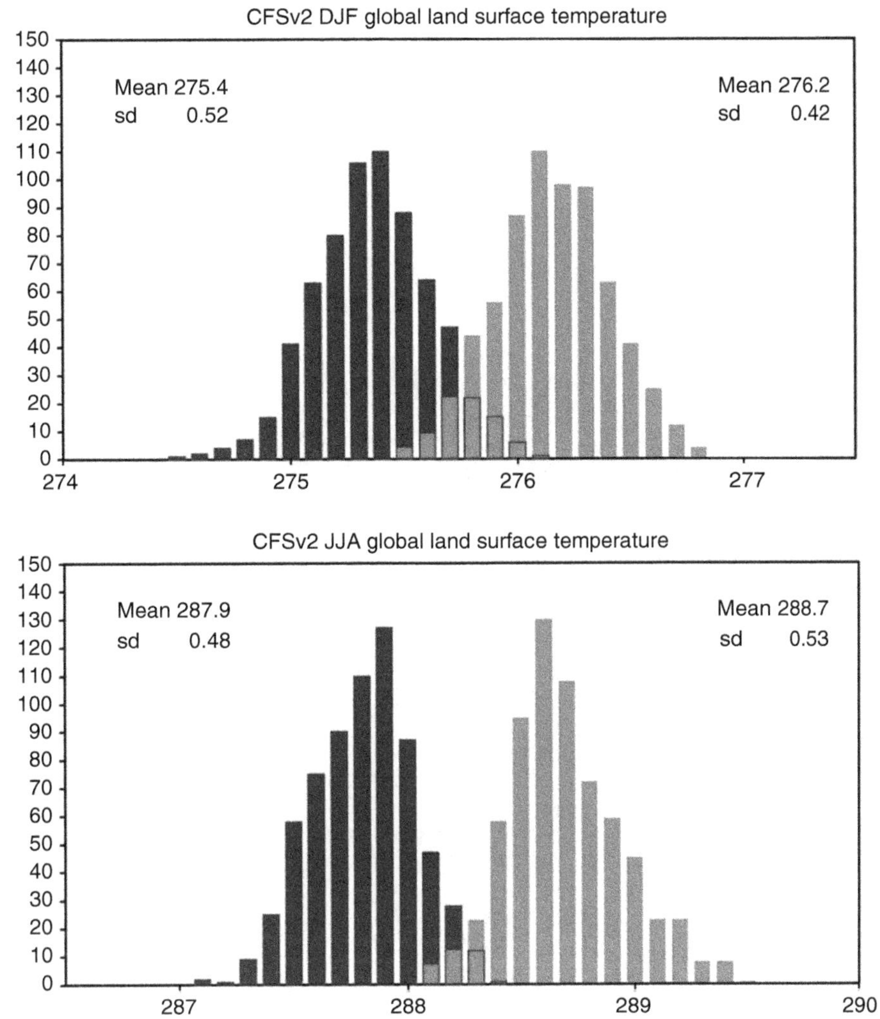

Figure 7.5 Global land surface temperature anomaly histograms as a function of the early (1983–1986 dark gray) and late (2009–2012 light gray) periods in the CFSv2 for DJF (upper) and JJA (lower). Each period has 672 seasonal realizations. Temperature units are in K and sd is standard deviation. *See electronic version for color representation.*

and by doing so, enhances the influence of time-varying changes in CO_2. A weak contribution of natural SST variability is evident in the eastern North Pacific that shows a weak warming pattern in the CFSv2 period differences, and, further, there is a hint of the observed decadal-like horseshoe-like pattern as seen by the weaker warming when compared to the surrounding areas.

Conversely, the eastern tropical Pacific shows much stronger warming in the CFSv2 than in its observed counterpart. It is unclear as to the cause of this warming, however, this pattern is consistent with numerous CMIP model projections as a response to enhanced greenhouse gas forcing [*Li and Xie,* 2012]. This suggests that the CFSv2 prediction of SST lies between the SST response to CO_2 (as inferred from CMIP5) and observations (which includes natural and anthropogenic responses), however is closer to that of CMIP5.

To be sure, it is possible that the warming in the eastern Pacific is also partially related to SST model biases in that region and that this bias may influence regional temperature changes. Despite the conventional wisdom in ascribing changes in the eastern Pacific SST (i.e., ENSO) to the climate of North America, that teleconnection is primarily forced by changes in SST much farther to the west over the central tropical Pacific [*Barsugli and Sardeshmukh,* 2002]. As such, even if a portion of the eastern Pacific warming is related to model bias, it is likely not a significant contributor to the shifts in global land surface or US temperature distributions and is further confirmed by the spatial patterns of land-surface temperature changes in Figure 7.1, which do not resemble the observed ENSO temperature-regression patterns (http://www.cpc.ncep.noaa.gov/products/precip/CWlink/ENSO/regressions/).

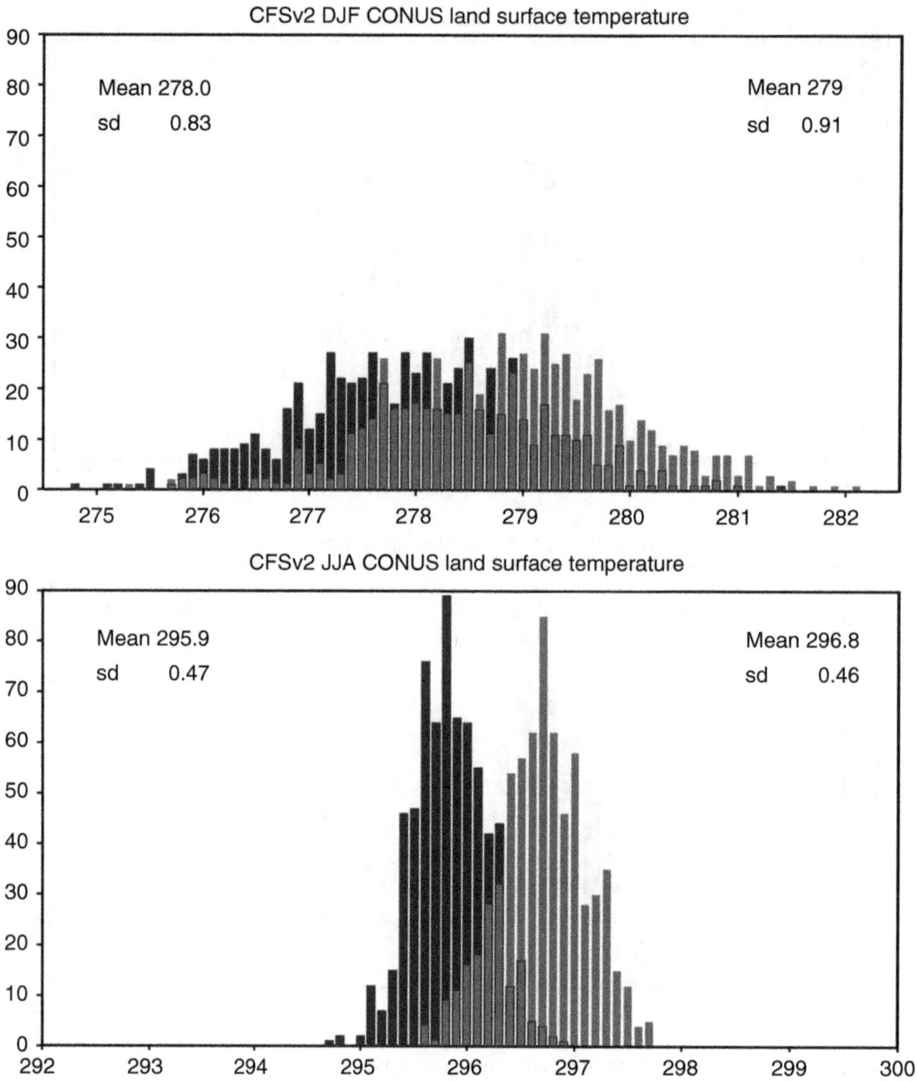

Figure 7.6 US land surface temperature anomaly histograms as a function of the early (1983–1986 dark gray) and late (2009–2012 light gray) periods in the CFSv2 for DJF (upper) and JJA (lower). Each period has 672 seasonal realizations. Temperature units are in K and sd is standard deviation.

7.4. CONCLUSIONS

Recent increases in extreme temperature events across many regions of the world have brought to the forefront questions regarding the role of GHG-induced increases in temperatures and whether this augmentation is the result of a shift in mean temperatures or an indication of increasing temperature variability. More than one thousand climate realizations from the NCEP Climate Forecast System are mined to answer this question.

While operational climate models are typically used for seasonal forecasting, here a strategy is developed that takes advantage of their extremely large number of ensemble members to analyze the statistics of more than one thousand climate states for any particular season. The method can be used to supplement and even improve upon both the AMIP- and CMIP-style climate model analysis, as the former does not have air-sea coupling, while the latter includes air-sea coupling, however employs a low number of ensemble members. Using the reforecast archives ameliorates aspects of both of these weaknesses and is thus instrumental in adding to the portfolio of climate change analyses.

Analyzing temperature anomalies from the large number of seasonal climate realizations in the NCEP CFSv2 reforecast archives shows that the increase in both global and US seasonal temperature anomalies over the last three decades is primarily a result of shifts in the mean temperature distributions in response to increasing GHG concentrations. This suggests that the occurrence

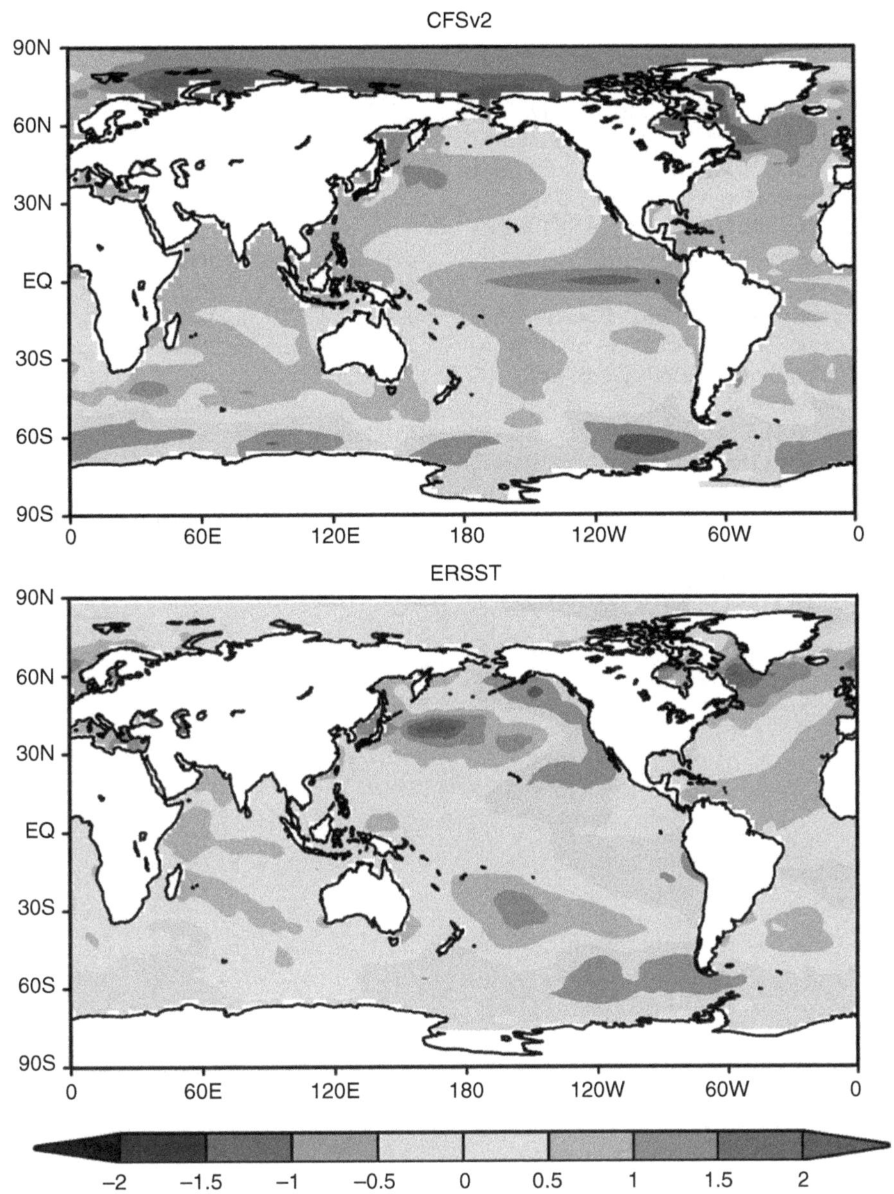

Figure 7.7 Global sea surface temperature anomaly differences for the periods of 2009–2012 minus 1983–1986 in the CFSv2 (upper) and the ERSSTv3 (lower). (*See insert for color representation of the figure.*)

of extreme temperature events is far more likely now than just 30 yr ago, as the result of anthropogenic climate change.

REFERENCES

Barsugli, J. J., and P. D. Sardeshmukh (2002), Global atmospheric sensitivity to tropical SST anomalies throughout the Indo-Pacific basin, *J. Climate*, *15*, 3427–3442.

Donat, M. G., and L. V. Alexander (2012), The shifting probability distribution of global daytime and night-time temperatures, *Geophys. Res. Lett.*, *39*, L14707; doi:10.1029/2012GL052459.

Fan, Y., and H. van den Dool (2008), A global monthly land surface temperature analysis for 1948-present, *J. Geophys. Res.*, *113*, DO1103; doi:10.1029/2007JD008470.

Hansen, J., M. Sato, and R. Ruedy (2012), Perception of climate change, *PNAS;* doi:10.1073pnas.1205276109.

Huntingford, C., and coauthors (2013), No increase in global temperature variability despite changing regional patterns, *Nature*, *500*, 327–331.

IPCC (2007), *Climate Change 2007: The Physical Science Basis, Contribution of Working Group I to the Fourth Assessment Report of the Intergovernmental Panel on Climate Change*, edited by S. Solomon, D. Qin, M. Manning, Z. Chen, M. Marquis, K. B. Averyt, M.,Tignor and H. L. Miller, Cambridge University Press, Cambridge, UK and New York, USA.

IPCC (2013), Summary for policy makers, in *Climate Change 2013: The Physical Science Basis. Contribution of Working Group I to the Fifth Assessment Report of the Intergovernmental Panel on Climate Change*, edited by T. F. Stocker,, D. Qin, G.-K. Plattner, M. Tignor, S. K. Allen, J. Boschung, A. Nauels, Y. Xia, V. Bex and P. M. Midgley, Cambridge University Press, Cambridge, UK and New York, USA.

Li, G., and S.-P. Xie (2012), Origins of tropical-wide SST biases in CMIP multi-model ensembles, *Geophys. Res. Lett.*, *39*; doi: 10.1029/2012GL053777.

Meehl, G. A., J. M. Arblaster, and C. Tebaldi (2007), Contributions of natural and anthropogenic forcing to changes in temperature extremes over the United States, *Geophys. Res. Lett.*, *34*, L19709; doi:10.1029/2007GL030948.

Rhines, A., and P. Huybers (2013), Frequent summer temperature extremes reflect changes in the mean, not the variance, *PNAS;* doi:10.1073pnas.121874110.

Saha, S., and coauthors (2010), The NCEP climate forecast system reanalysis, *Bull. Amer. Meteor. Soc.*, *91*, 1015–1057.

Saha., S., and coauthors (2013), The NCEP climate forecast system version 2, *J. Climate,* in press; doi: 10.1175JCLI-D-12-00823.1.

Smith, T. M., R. W. Reynolds, T. C. Peterson, and J. Lawrimore (2008), Improvements to NOAA's historical merged land-ocean surface temperature analysis, *J. Climate*, *21*, 2283–2296.

Taylor, K. E., R. J. Stouffer, and G. A. Meehl (2012), An overview of CMIP5 and the experiment design, *Bull. Amer. Meteor. Soc.*, *93*, 485–498.

Vavrus, S., J. E. Walsh, W. L. Chapman, and D. Portis (2006), The behavior of extreme cold air outbreaks under greenhouse warming, *Int. J. Climatol.*, *26*, 1133–1147.

Weaver, S. J. (2013), Factors associated with decadal variability in Great Plains summer surface temperatures, *J. Climate*, *26*, 343–350.

Xie, S-P., and coauthors (2010), Global warming pattern formation: Sea surface temperature and rainfall, *J. Climate*, *23*, 966–986.

Xu, L., and coauthors (2013), Temperature and vegetation seasonality diminishment over northern lands, *Nature Climate Change*; doi: 10.1038/NCLIMATE1836.

8

Why Future Shifts in Tropical Precipitation Will Likely Be Small: The Location of the Tropical Rain Belt and the Hemispheric Contrast of Energy Input to the Atmosphere

Aaron Donohoe[1] and Aiko Voigt[2]

ABSTRACT

The region of tropical convective precipitation (the ITCZ) migrates by approximately 12° of latitude over the seasonal cycle with an annual mean position slightly north of the equator. We demonstrate that the ITCZ is located in the hemisphere in which the atmosphere is heated more strongly. A 3° latitude ITCZ shift requires approximately 1 PW of energy input into the atmosphere in one hemisphere with equal cooling in the other hemisphere. This relationship is shown to apply to the climatological seasonal cycle, the interannual variability and the response to both paleoclimatic and anthropogenic forcing. Future changes in the ITCZ position will likely conform the 3° latitude shift per PW of hemispherically asymmetric atmospheric heating. Since hemispherically asymmetric forcings and climate feedbacks are small (<< 1 PW), future changes in ITCZ position will likely be very small (<1° latitude). Indeed, ensembles of coupled climate models simulate ITCZ shifts that are not significantly different from zero in response to an instantaneous quadrupling of carbon dioxide.

8.1. INTRODUCTION

The contrast between the moist tropics and dry subtropics is a fundamental feature of the climate system and is a consequence of the mass overturning circulation of the atmosphere. Solar heating of the low latitudes causes upward atmospheric motion near the equator resulting in a region of convective precipitation (Fig. 8.1a). The upward mass flux does not penetrate above the stable tropopause but, instead, diverges horizontally giving rise to poleward flow aloft. Eastward deflection of the poleward wind by the Coriolis force in the upper branch of the circulation results in the westerly subtropical jets and also demands a meridional temperature gradient via thermal wind balance [*Held and Hou*, 1980]. The latter becomes unstable to atmospheric weather systems (baroclinic instability) for a meridionally expansive circulation and, thus, the meridional extent of the poleward flow aloft is limited by the simultaneous constraints of conservation of angular momentum and thermal wind balance [*Walker and Schneider*, 2006; *Lu and Vecchi*, 2014]. Mass converges aloft on the poleward side of the Hadley circulation resulting in subsidence and dry surface conditions associated with the subtropics. Thus, the contrast between the dry subtropics and moist tropics is intimately related to the strength, position. and width of the atmospheric mass overturning circulation in the tropics, commonly referred to as the Hadley cell [*Hadley*, 1735].

Patterns of tropical precipitation and dryness have widespread biological and socioeconomic impacts. The moist tropics are home to some of the most rich and

[1] *Karlsruhe Institute of Technology, Institute of Meteorology and Climate Research – Department Troposphere Research, Karlsruhe, Germany; Lamont-Doherty Earth Observatory, Columbia University, New York, NY, USA*

[2] *Lamont Doherty Earth Observatory, Columbia University, New York, NY, USA*

Climate Extremes: Patterns and Mechanisms, Geophysical Monograph 226, First Edition.
Edited by S.-Y. Simon Wang, Jin-Ho Yoon, Christopher C. Funk, and Robert R. Gillies.

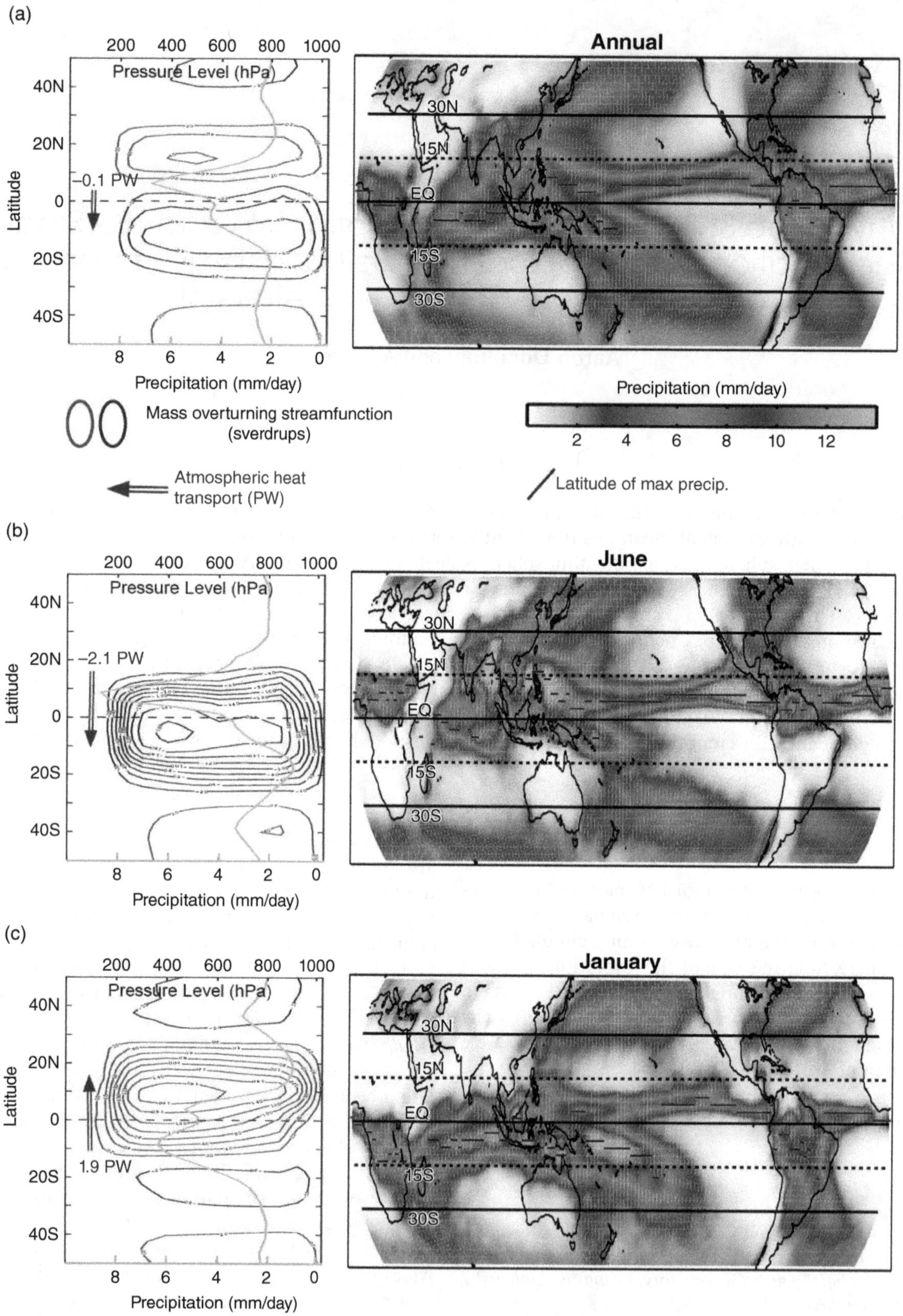

Figure 8.1 (Right panels) The colors show the climatological precipitation from the NOAA merged analysis [*Xie and Arkin*, 1996] and blue lines denote the latitude of maximum precipitation at each longitude. (Left panels) The purple line is the zonal mean precipitation and the colors show the mass overturning stream function in Sverdrups from the NCEP reanalysis [*Kalnay et al.*, 1996]. The stream function is defined as positive (red) when the circulation is clockwise and negative (blue) when the circulation is counterclockwise. The vertically and zonally integrated atmospheric energy transport at the equator (AHT_{EQ}) is denoted by the green arrow. (a) Annual mean; (b) June; (c) January. *See electronic version for color representation.*

diverse species of flora and fauna on the planet [*Koppen*, 1936], whereas the subtropical deserts are among the most sparsely populated regions of the planet due to their lack of biodiversity and water resources. Interannual variations in tropical precipitation have large implications for agricultural production, especially in the semiarid regions where water resources are marginal [*Sivakumar et al.*, 2005]. Thus, understanding future changes in the position, intensity, and width of the Hadley cell and its associated impact on large-scale patterns of precipitation climatology and variability is of critical importance to society. Moreover, the Hadley cell is part of the global circulation of the atmosphere and changes in the strength and location of the Hadley cell have the potential to change tropical cyclones [*Vimont and Kossin*, 2007; *Merlis et al.*, 2013] and extratropical jet streams [*Ceppi et al.*, 2013]

Here we focus on the processes that determine the location of intense tropical precipitation in the tropical rain belt on seasonal and longer timescales. In the Northern Hemisphere (NH) summer, the latitude of most intense precipitation (blue lines in Fig. 8.1b) is north of the equator with an intense band of precipitation around 6–10° N in the Pacific and extending out to beyond 15°N over the land domain, especially in the Indian monsoon. During the Southern Hemisphere (SH) summer, a zonal band of precipitation in the tropical Pacific remains north of the equator but has deintensified significantly. A north-west to south-east tilted band of precipitation in the south tropical Pacific, the South Pacific convergence zone (SPCZ), has emerged as the region of most intense precipitation (Fig. 8.1c). The annual mean precipitation map represents a smoothed and smeared version of the seasonally migrating regions of intense precipitation that follow the sun.

While zonal inhomogeneities in precipitation, including the monsoonal systems, are prevalent and important from both societal and meteorological perspectives, we will focus here on the pattern of zonal mean precipitation that defines the tropical-subtropical contrast and its relationship to the Hadley cell. In the boreal summer, the zonal mean precipitation maximum (left panel of Fig. 8.1b) is clearly in the Northern Hemisphere, with its magnitude of 8 mm/day exceeding the subtropical precipitation minimum by a factor of 4. During the austral summer, the zonal mean precipitation clearly peaks south of the equator although there is a remnant of the precipitation maximum north of the equator seen during the boreal summer. While the literature sometimes distinguishes between the zonal band of precipitation in tropical North Pacific and the SPCZ precipitation, here we treat the precipitation in the two regions as a single entity: the seasonally migrating convective precipitation associated with the upwelling branch of the Hadley circulation (left panels of Fig. 8.1). We refer to the location of the zonal mean precipitation maximum as the intertropical convergence zone (ITCZ), so named for the horizontal convergence of winds in the lower atmosphere that give rise to the upward atmospheric motion and convective precipitation. We note that the region of convective precipitation that we use to define the ITCZ here is inseparable from the region where the meridional winds converge at the surface since convection (upward motion in the atmosphere) demands low-level convergence for an incompressible fluid (Boussinesq continuity equation) subject to the boundary condition of no vertical flow through the surface. As a result, the ITCZ is colocated with the ascending branch of the Hadley cell and any shift in the tropical precipitation requires a concurrent shift in the Hadley circulation.

The annual mean precipitation is best thought of as a statistical average of a seasonally bimodal spatial distribution of precipitation, with regions of intense precipitation prevalent north of the equator in the Pacific, in the SPCZ region and also in the monsoons of India, Southeast Asia, Australia, and northern Africa. In this regard, the annual mean precipitation and Hadley cell (Fig. 8.1a) represents a physical state that is seldom realized, just as the mean of a sine curve is unlikely to occur and merely represents the average of a bimodal distribution (which happens to be symmetric about the mean in the case of a pure sine curve). While the annual mean precipitation is definitely skewed northward off the equator, we demonstrate below that the annual mean displacement of the ITCZ off the equator represents the small residual of seasonal ITCZ migrations off the equator that are an order of magnitude larger than the annual mean.

The colocation of the ITCZ and the ascending branch of the Hadley circulation demands that the atmosphere move energy away from the hemisphere in which the ITCZ is located with the energy export proportional to ITCZ displacement off the equator. This relationship results from the following properties of the tropical atmospheric circulation as summarized in Figure 8.2a:

1. Because the upper branch of the Hadley circulation flows from the ascending to descending branches, the upper level mass flux at the equator is away from the ITCZ.

2. Conservation of mass requires that there is equal mass flux in the upper and lower branch of the Hadley circulation and, therefore, the net moisture and energy fluxes at the equator will be in the direction of the mass flux at the vertical level of the moisture and energy maxima.

3. The total energy of an air parcel is equal to the moist static energy (MSE).

$$MSE = Lq + C_P T + gZ \tag{8.1}$$

where L is the latent heat of vaporization of water vapor, q is the specific humidity, C_p is the heat capacity of air at

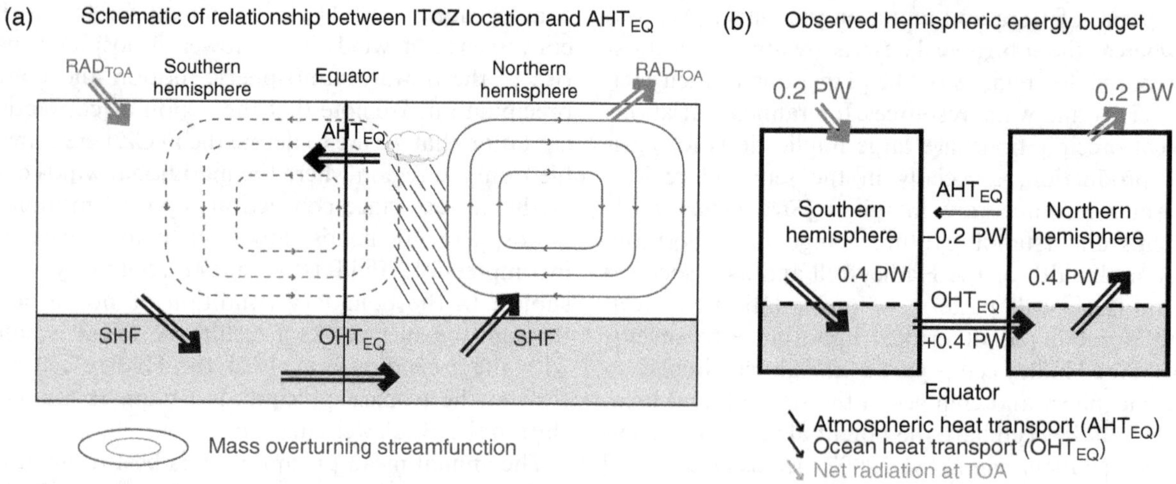

Figure 8.2 (a) Cartoon of the relationship between ITCZ location, the Hadley cell, AHT_{EQ}, and the terms in the hemispheric energy budget of the atmosphere. (b) Observational estimates of the terms in the annual mean hemispheric energy budget. Values are derived in *Marshall et al.* [2013] and *Frierson et al.* [2013].

constant pressure, T is the temperature, g is the acceleration of gravity, and Z is the height above sea level. While both moisture (q) and temperature (T) achieve their (tropospheric) maxima at or near the surface, the increase in geopotential (gZ) with altitude generally exceeds the decrease in the sum of latent (Lq) and sensible energies (C_pT), resulting in MSE increasing with altitude. If this were not true, the atmospheric column as a whole would be unstable to moist convection. As a result of this combined with point (2) above, the net energy flux across the equator is away from the ITCZ. This logic neglects the role of transient and stationary eddy energy transport which are generally an order of magnitude smaller than that associated with the mass overturning circulation in the deep tropics [*Held*, 2001].

4. In contrast to energy, moisture content maximizes at the surface. Thus the net moisture flux is in the sense of the surface mass flux, is toward the ITCZ [*Frierson et al.*, 2013] and provides the moisture convergence required for precipitation exceeding evaporation within the ITCZ region.

5. The stream function magnitude at the equator will increase as the ITCZ moves farther off the equator. Provided that changes in the bulk moist stability do not change appreciably with the ITCZ shift, the energy flux across the equator is away from the ITCZ and is proportional to the ITCZ displacement off the equator.

A negative correlation between atmospheric energy transport across the equator (hereafter AHT_{EQ}) and ITCZ location is therefore expected based on the prominent dynamics in the deep tropics. But is this relationship useful or merely diagnostic? Since the energy storage in the atmospheric column is small, AHT_{EQ} must be balanced by a hemispheric contrast of energy input to the atmosphere by way of either the radiation at the top of atmosphere (TOA) or surface energy fluxes. Therefore, the relationship between ITCZ location and AHT_{EQ} allows one to relate the ITCZ location to the hemispheric-scale atmospheric energy budget of the climate system. Specifically, we argue in this chapter that ITCZ location and hemispheric energy budget are connected via the approximate relationship:

$$ITCZ_{LOC} \approx -3\left(°PW^{-1}\right)AHT_{EQ}. \tag{8.2}$$

The negative sign in equation (9.2) means that an $ITCZ_{LOC}$ in the NH (+) is associated with southward (-) energy transport across the equator and vice versa. The factor of 3° PW^{-1} suggests that, if the ITCZ shifted 3° of latitude, the associated atmospheric circulation would export 1 PW (= 10^{15} W) away from the hemisphere to which the ITCZ has shifted. This translates to a heating of 4 W m^{-2} averaged over a single hemisphere with equal cooling in the opposite hemisphere. To put that number in perspective, one would have to simultaneously double CO_2 in one hemisphere and halve CO_2 in the other hemisphere in order to achieve the same climate forcing. We emphasize that the factor of 3° PW^{-1} is an approximate relationship that seems to hold for many model simulations and applications but there is no theory that allows one to derive this number from first principles and no a priori reason why this number appears to hold across different climate states and timescales. It is intended more as a guide for understanding the magnitude of forcing and energetic feedback processes needed for a given ITCZ shift. We will present evidence that this quantitative relationship is fundamentally constrained by the seasonal relationship between the ITCZ location and AHT_{EQ} because the seasonal migration of the ITCZ is so large compared to that forced by external forcing and the forced change simply represents a small modification of the seasonal cycle. We further argue that it is difficult to achieve a hemispheric asymmetry of climate forcing and/or climate response of this magnitude (~1PW) and, thus, the $ITCZ_{LOC}$ appears to be a relatively stable feature of the climate system, and ITCZ shifts greater than a couple degrees of latitude will be unlikely even under rather extreme climate forcing such as those associated with glacial cycles.

We will demonstrate that the 3° latitude ITCZ shift per PW of energy input into a single hemisphere relationship is robust and consistent across timescales in both models and observation-based estimates including (1) the seasonal cycle, (2) the annual mean climatology, (3) the interannual variability, and (4) the response to external forcing, including both paleoclimatic and anthropogenic forcing. This theory for the ITCZ location and the hemispheric energy budget allows us to understand the magnitude and direction of future ITCZ shifts from the large-scale patterns of climate forcings and feedbacks and their uncertainties. As such we will need a quantitative relationship between ITCZ location and AHT_{EQ}. In section 8.2, we present empirical estimates of this relationship across timescales and also discuss the energetic processes that give rise to a hemispheric contrast of energy input to the atmosphere. In section 8.3, we address the physical processes that control the value of the $ITCZ_{LOC}/AHT_{EQ}$ relationship and relate the amplitude of seasonal ITCZ migrations to the annual mean ITCZ location and the meridional extent of tropical convective precipitation. We then extend these results to the ITCZ shifts expected under global warming (section 8.4).

8.2. THE RELATIONSHIP BETWEEN ITCZ POSITION AND HEMISPHERIC CONTRAST OF ATMOSPHERIC HEATING

In this section, we present evidence for the of 3° latitude ITCZ shift per PW of AHT_{EQ} in both observational datasets and in model simulations across a myriad of timescales ranging from the seasonal cycle to the long-term response to external forcing. We begin by developing a framework for relating the hemispheric contrast of TOA radiation and surface energy fluxes to AHT_{EQ},which will allow us to analyze the magnitude of ITCZ shifts demanded by energetic processes given the 3° PW^{-1} relationship.

8.2.1. The Hemispheric Contrast of Atmospheric Heating Leading to AHT_{EQ}

The atmospheric energy transport across the equator is fundamentally a consequence of the hemispheric contrast of atmospheric heating; in order to move energy from the NH to the SH, there must be a source of atmospheric heating in the NH. Here, we derive the relationship between AHT_{EQ} and the hemispheric energy budget for the annual mean climatology, which we will later extend to other timescales. By way of equation (8.2), we can then quantitatively relate the ITCZ location, including its variability and future changes, to the hemispheric energy budget. This is the goal of this chapter.

In the annual mean, the ascending branch of the Hadley cell and ITCZ are located in the NH (Fig. 8.1a), which results in southward AHT_{EQ} in the Hadley circulation. The hemispheric energy budget demands a spatially integrated atmospheric heating equal to AHT_{EQ} and an equal magnitude atmospheric cooling in the SH. This heating can be provided either by a net radiative heating at the TOA or by way of the surface energy fluxes from the ocean to the atmosphere:

$$-AHT_{EQ} = \langle NET_{RAD} \rangle + \langle SHF \rangle \quad (8.3)$$

where brackets denote the spatial integrals over the NH, NET_{RAD} is the net radiative heating at the TOA, and SHF is the surface energy flux defined as positive to the atmosphere. Global energy balance requires that the global integral of NET_{RAD} and SHF be zero in order for the energy budget of the climate system as a whole and surface to be in steady state. Therefore, equation (8.3) could also be defined from the spatial integrals of NET_{RAD} and SHF over the SH with signs reversed. (Note that for the case of a global mean disequilibrium, the global mean imbalances can be removed prior to the calculation of the hemispheric mean or, equivalently, half the difference between the SH and NH integrals can be taken. For the purposes here, we have also neglected the atmospheric energy storage, which is, in general, small but nonnegligible on seasonal timescales.) In simple terms, the net heating of the atmosphere in either hemisphere must be balanced by energy export across the equator and an equal magnitude of cooling in the opposite hemisphere.

The annual mean net energy flux from the surface to the atmosphere is a consequence of ocean heat transport convergence provided the ocean energy content is stable (i.e., at the climatological timescale). Therefore, the climatological hemispheric energy budget in equation (8.3) can be rewritten as:

$$-AHT_{EQ} = \langle NET_{RAD} \rangle + OHT_{EQ} \quad (8.4)$$

since the ocean heat transport convergence spatially integrated over either hemisphere is equal to the heat transport across the border, in this case the equator (Fig. 8.2a). Moving the OHT_{EQ} term on the right-hand side of the equator to the left results in the statement that the coupled (atmosphere plus ocean) energy export across the equator is balanced by net radiative heating/cooling of the hemispheres.

In this section, we explore the terms in the hemispheric energy budget (equations [8.3] and [8.4]) and the implication for the $ITCZ_{LOC}$ via equation (8.2).

8.2.2. Seasonal Cycle of ITCZ Location and AHT_{EQ}

To quantify the ITCZ location, we use the centroid of zonal mean precipitation (P_{CENT}) equatorward of 20° defined as the latitude for which there are equal amounts of precipitation between the regions from P_{CENT} to 20N and from P_{CENT} to 20S [*Frierson and Hwang*, 2012]. This metric coincides with the precipitation maximum if the precipitation is sharply peaked (i.e., a delta function), whereas it represents a precipitation weighted average between two peaks for a bimodal precipitation distribution such as the annual and zonal average precipitation (Fig. 8.1a). The zonal mean ITCZ (P_{CENT}) is displaced farthest north (7.3°N) in August and farthest south (5.3°S) in February (Fig. 8.3b) with an annual mean position (1.6°N) that reflects the small asymmetry between the northward and southward seasonal migration off the equator. In all seasons, the maximum precipitation is colocated with the ascending branch of the Hadley circulation (left panels of Fig. 8.1), where the meridional gradient of the mass overturning circulation is greatest. In the annual mean, the upward motion and convective precipitation are colocated with the zero in the stream function. However, during both solstices, there is a prevalent magnitude asymmetry between the winter and summer Hadley cells, with the winter cell intensifying by approximately a factor of 3 relative to its annual mean (cf. the left panels of Fig. 8.1a to those in 8.1b-c) while the summer Hadley cell all but disappears. As a result of this amplitude asymmetry between the winter and summer Hadley cell, the maximum upward velocities and precipitation are located within the winter Hadley cell and equatorward of the zero stream function. Thus, previous attempts in the literature to connect the ITCZ location with the location of the zero mass overturning stream function (i.e., the energy flux equator) are limited in providing quantitative insights into the relationship between the ITCZ location and AHT_{EQ} for a system with a seasonal cycle [*Schneider et al.,* 2014].

The atmosphere moves 2.1 PW of energy from north to south across the (green arrow and labeling in the left panel of Fig. 8.1b) equator during the boreal summer, which is associated with a counterclockwise mass overturning circulation of approximately 170 Sv (1 Sv is a Sverdrup mass flux equivalent of 10^9 kg S^{-1}). In the opposite season, the mass overturning circulation and atmospheric energy flux at the equator have reversed direction with comparable but slightly smaller amplitude; the atmosphere moves 1.9 PW of energy south to north across the equator associated with a 140 Sv, clockwise mass overturning circulation. Overall, the seasonal variations in energy transport across the equator are incredibly highly correlated with the mass overturning stream function at the equator ($R^2=0.99$) with a regression coefficient that implies a 14K contrast in the moist potential temperature ($\approx MSE/C_p$) between the northward and southward flowing air in the Hadley cell [*Donohoe et al.,* 2013b]. Given the observed moist stability of the tropics, the above 14K contrast suggests an average vertical separation of 600 hPa between the equatorward and poleward mass flux, which is qualitatively consistent with the meridional velocity weighted MSE contrast between the equatorward and poleward flow in the ~800 hPa deep circulation. This result implies that seasonal variations on the moist stability of the tropics and atmospheric energy transport exert a secondary influence on the interhemispheric energy transport as compared with variations in the strength and location of the Hadley cell.

Seasonal variations in the ITCZ location and AHT_{EQ} are incredibly well correlated ($R^2=0.99$) with a regression coefficient of -2.7 ± 0.6 ° PW^{-1} (Fig. 8.3b). We will demonstrate that the approximate quantitative relationship between $ITCZ_{LOC}$ and AHT_{EQ} of -3 ° PW^{-1} holds not only for the seasonal cycle but also for a myriad of timescales to be discussed below. This relationship suggests that a 3° latitude ITCZ shift is accompanied by 1 PW of energy exchange between the two hemispheres, which must be accompanied by the same quantity of atmospheric heating/cooling in each hemisphere. Coupled climate models show remarkably similar relationships between the seasonal cycle of ITCZ location and AHT_{EQ} both in terms of the order 3° latitude ITCZ shift per PW of AHT_{EQ} (cf. the slopes of the thin dashed lines and thick solid blue line in Fig. 8.3b) and the amplitude of the seasonal cycle of ITCZ location [*Donohoe et al.,* 2013b] suggesting that the physical processes relating the mutual dependence of ITCZ location and AHT_{EQ} to the Hadley cell are robust in models and nature alike. We note that a nearly identical (-2.9° PW^{-1}) seasonal relationship between the ITCZ location and AHT_{EQ} is found if the ITCZ location is defined as the centroid of precipitation over the ocean domain only, suggesting that the amplitude of the seasonal migration of the ITCZ is unaffected by the alternating monsoons of Asia and Australia.

Seasonal variations in the hemispheric contrast of energy input to the atmosphere (and therefore AHT_{EQ} and $ITCZ_{LOC}$) are driven by changes in the meridional distribution of insolation. The latter causes the ASR contrast between the two hemispheres to vary by 22 PW seasonally as measured by the amplitude of the annual harmonic. Less than 10% of the seasonal variations in energy input into each hemisphere is fluxed to the

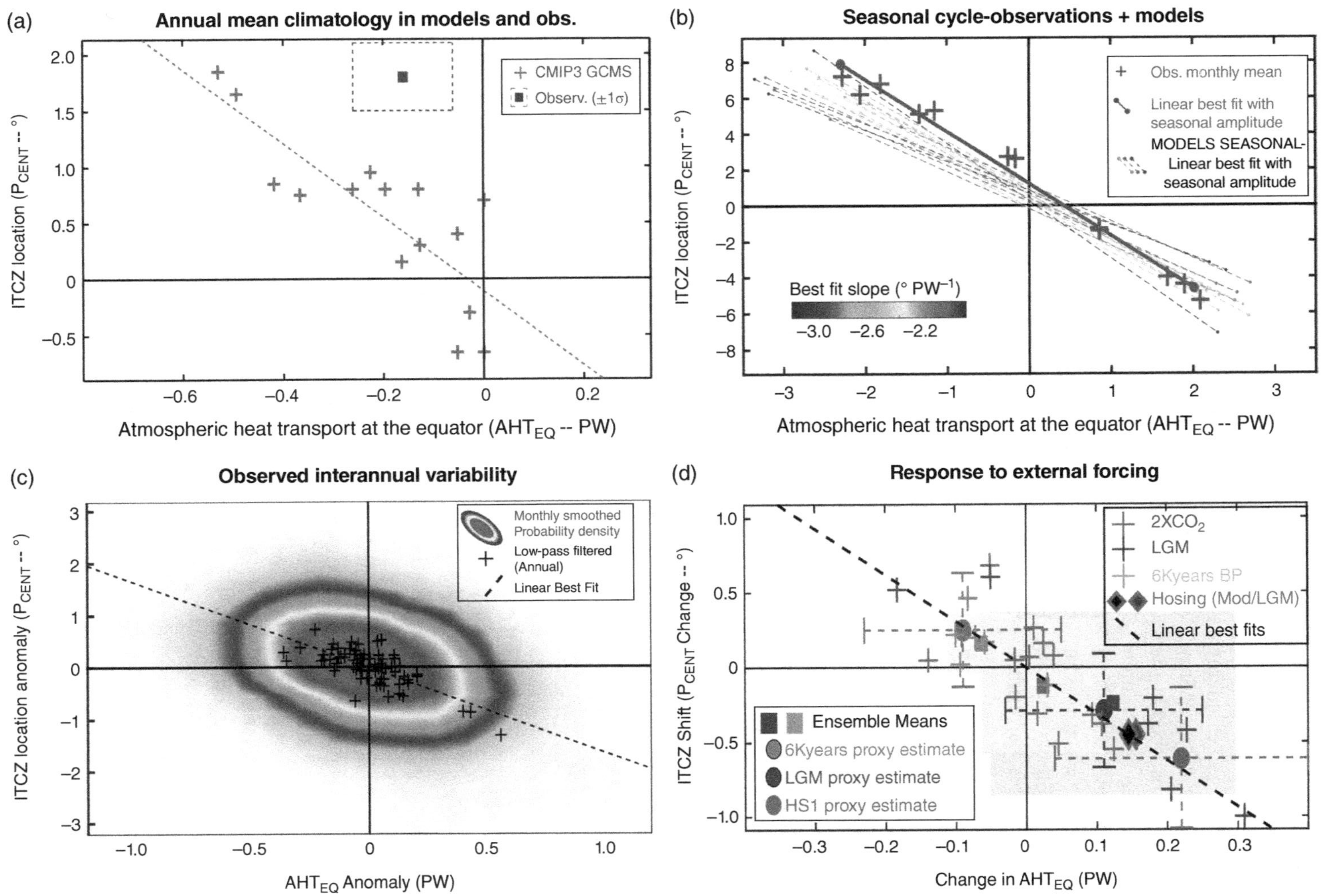

Figure 8.3 The relationship between ITCZ location and the atmospheric energy transport across the equator spanning different timescales in observation-based estimates, models, and paleoclimate reconstructions. In all cases, the centroid of tropical precipitation (P_{CENT} ; between 20N and 20S) is used as a metric for the ITCZ location. Although the domain and range differ substantially between the panels, the relative scaling of the axes is constant throughout the panels.(a) Annual mean climatology in (red crosses) coupled CMIP3 preindustrial simulations and (blue box) observational estimates with uncertainty (1σ) indicated by the dashed blue rectangle, based on interannual variability and spread between the different sets of reanalyses [*Marshall et al.*, 2013]. (b) Observational estimate of seasonal cycle based on NCEP reanalysis energy fluxes and NOAA CPC merged analysis precipitation [*Donohoe et al.*, 2013b]. Monthly means are the blue crosses and the thick dashed blue line represents the linear best fit with length equal to the amplitude (first harmonic) of the seasonal cycle. Each of the dashed lines represents the slope and amplitude of a CMIP3 model with the color representing the slope value as given by the color bar inset in the figure. (c) Observational estimates of interannual variability [*Donohoe et al.*, 2014a] of the ITCZ (from NOAA CPC merged analysis) and AHT_{EQ} (from NCEP reanalysis). The colors are the smoothed probability distribution function of monthly anomalies, the crosses are the annual data (low pass filtered with cutoff period of 1 yr) and the dashed line is the linear best fit. (d) Response to anthropogenic and paleoclimate forcing [*Donohoe et al.*, 2013b] alongside proxy estimates based on sea surface temperature reconstructions [*Mcgee et al.*, 2014]. Each cross is a different simulation, color coded by experiment type (red = $2XCO_2$, blue = LGM, and green = mid-Holocene) and the solid squares and shaded areas represent the ensemble average and one standard deviation. Circles and dashed crosses are the paleo estimates and uncertainty. (*See insert for color representation of the figure.*)

opposite hemisphere while approximately 75% is stored within the ocean, 8% is stored within the atmosphere, and 7% is emitted back to space as OLR. It is unsurprising that the majority of seasonal variations in radiation at the TOA are stored in the ocean since the majority of the ASR passes through the atmosphere and is absorbed by the surface. The ocean has a large heat capacity and therefore takes time to heat up before it fluxes energy upward to the atmosphere via latent and sensible energy fluxes. Thus, most of the solar insolation absorbed at the surface never finds its way into the atmospheric column on seasonal timescales. But why only 10% of the seasonal variations in insolation are fluxed across the equator is an outstanding question in climate science. Surely the climate system would be very different if this percentage were altered significantly.

The seasonal migration of the Hadley cell and ITCZ off the equator is the climate system's response to hemispherical asymmetric insolation. Because the seasonal insolation forcing is large compared to any external forcing (i.e., anthropogenic or paleoclimatic), the robust relationship between ITCZ location and the hemispheric contrast of atmospheric heating over the seasonal cycle (in both observational estimates and models) provides a constraint to how the system responds to internal variability and external forcing. We will further argue that the same quantitative relationship of 3° latitude ITCZ shift per PW of AHT_{EQ} discussed here applies to all timescales because ITCZ shifts at longer timescales can be viewed as small modifications of the climatological seasonal cycle and, therefore, are well approximated by linearization about the seasonal cycle.

8.2.3. Annual Mean ITCZ Location

The annual and zonal mean precipitation has peaks in the tropics of both hemispheres at approximately 5° (Fig. 8.1a) and, we argue, is a physical state that is seldom realized but, rather, represents the (spatially smeared and smoothed) statistical average of a bimodal system. The NH peak is stronger in magnitude, which reflects the ITCZ migrating farther off the equator with a stronger precipitation maximum during boreal summer as compared to austral summer (cf. Fig. 8.1b and c). The annual mean P_{CENT} is 1.9°N and represents a weighted average location between the bimodal peaks. We note that this metric places the ITCZ closer to the equator than simply picking out the precipitation maximum at approximately 5°N [*Frierson et al.,* 2013].

The annual mean ITCZ location north of the equator is associated with a Hadley cell that is shifted into the NH (Fig. 8.1a) and 0.2 ± 0.1 PW of southward AHT_{EQ} where error bounds reflect the spread across the different reanalysis products (NCEP, ERA, and JRA) as analyzed by *Marshall et al.* [2013] and *Fasullo and Trenberth* [2008]. The latter demands an equivalent magnitude of atmospheric heating in the NH via equation 8.4, which is accomplished by the surface energy fluxes by way of the large-scale oceanic circulation (OHT_{EQ}) transporting 0.4 ± 0.2 PW northward across the equator and an opposing net radiative cooling ($\langle NET_{RAD} \rangle$) of the NH that is smaller in magnitude (-0.2 ± 0.1 PW – Fig. 8.2b). The atmospheric and oceanic energy transports across the equator oppose each other and the ITCZ position in the NH would not be energetically possible without the oceanic circulation and, more specifically, the energy transport in the Atlantic Meridional Overturning Circulation (AMOC) [*Frierson et al.,* 2013]. This conclusion is based on the fact that the net radiation at the TOA heats the SH and cools the NH and, acting alone, would favor an ITCZ in the SH. We explore the processes leading to the hemispheric contrast of radiative heating below.

The annual mean insolation is completely symmetric about the equator. Therefore, any hemispheric contrast in absorbed shortwave radiation (ASR), the net shortwave radiation at the TOA, is a consequence of differences in the planetary albedo of the two hemispheres. *Voigt et al.* [2013] recently demonstrated that the two hemispheres have identical planetary albedos. This result is perhaps surprising given that the predominance of land in the NH is associated with a larger surface albedo in the NH than the SH. However, the Southern Ocean is very cloudy, which results in more shortwave reflection in the midlatitudes of the SH as compared with the same latitudes in the NH [*Donohoe and Battisti,* 2011]. The NH is slightly (1.24 K) warmer than the SH [*Kang et al.,* 2014]. Consistent with the temperature dependence of longwave emission by the Stefan Boltzmann law, the NH emits more (0.8 W m^{-2}) outgoing longwave radiation (OLR) than the SH resulting in a net radiative cooling of the NH of 0.2 PW.

How well do coupled climate models simulate the hemispheric contrast of energy input into the climate system and ITCZ location? Figure 8.3a shows that the ensemble of CMIP3 coupled preindustrial simulations [*Meehl et al.,* 2007] simulate a wide range of ITCZ locations and AHT_{EQ}. Despite the wide range of ITCZ locations simulated in the models, including three in the SH, the intermodel spread in ITCZ location is well correlated with the AHT_{EQ} as would be expected based on the mutual dependence of $ITCZ_{LOC}$ and AHT_{EQ} on the Hadley cell location. Thus, the model bias and spread in ITCZ location appears to be a consequence of the failure of the models to simulate the hemispheric scale energy budget discussed in the previous paragraph. In particular, model differences and an ensemble average bias in the planetary albedo over the Southern Ocean explain the intermodel spread and model bias toward a more southward positioned ITCZ [*Hwang and Frierson*, 2013]; this result suggests that the well-known double ITCZ bias in climate models is primarily a consequence of too much ASR in the SH resulting in a net heating of the SH atmosphere and, thus, an ITCZ in the SH with associated northward AHT_{EQ}.

We note that the slope of the linear best fit to the intermodel spread in Figure 8.3a is -3.2 ± 0.7 ° PW^{-1}. This relationship between ITCZ location and AHT_{EQ} is similar to that already seen for the seasonal cycle (section 8.2.2) and we will demonstrate also applies to the ITCZ shift due to external forcing (section 8.2.4). We note that observational past estimates from satellite data and atmospheric reanalysis (blue square in Fig. 8.3a) fail to fall on this line within the observational uncertainty. We believe, however, that the "failure" of the observed annual mean climatology to fit on this line does not necessarily contradict the hypothesis of a universal $ITCZ_{LOC}/AHT_{EQ}$ slope because the observed annual mean AHT_{EQ} is complicated by several issues including:

1. There is a significant energy flux in the stratosphere in the reanalyses [*Marshall et al.*, 2013] that is poorly constrained by measurements.
2. Stationary waves make a significant contribution to the annual mean AHT_{EQ} in nature [*Heaviside and Czaja*, 2013]. The stationary eddy contribution to AHT_{EQ} is nearly seasonally invariant and the vast majority of seasonal changes in AHT_{EQ} are due to variations in the Hadley cell strength [*Donohoe et al.*, 2013b]
3. While the AHT_{EQ} in the Hadley cell is dominated by the deep circulation cell between the surface and upper troposphere, at the equator there is smaller magnitude shallow surface cell with southward flow at the top of the tropical boundary layer (700 hPa, where MSE is a vertical minimum) that results in anomalous northward AHT_{EQ} [*Marshall et al.*,, 2013].

It is unclear which of the above three processes, which lead to a breakdown between the annual mean ITCZ location and AHT_{EQ} across the equator via the mutual dependence on the Hadley cell, contribute a significant amount of AHT_{EQ} in climate models.

8.2.4. Interannual Variability of ITCZ Location

The ITCZ location varies interannually by of order 1° (2σ of annual mean variability = 0.7°) with approximately twice as much variability at the monthly timescale. We focus here on the annual mean anomalies, which have a more robust connection to Hadley cell anomalies. The interannual variability of the ITCZ location and AHT_{EQ} over the satellite era show the expected negative correlation (Fig. 8.3c). A northward ITCZ displacement is associated with an intensification of the climatological Hadley cell with anomalous precipitation north of the equator in the upwelling branch of the Hadley cell and decreased precipitation south of the equator in the region of anomalous subsidence [*Donohoe et al.*, 2014a]. The regression coefficient between ITCZ location and AHT_{EQ} is 1.8 ± 0.8° PW^{-1}. This result indicates smaller ITCZ sensitivity to interhemispheric heating than seen over the seasonal cycle but is not statistically different from the approximately 3° latitude ITCZ shift per PW of AHT_{EQ} seen in the previous two sections. We note that the correlation between El Niño, the prominent mode of east-west sea surface temperature variability in the tropical Pacific, and both the ITCZ location and AHT_{EQ} is not significant.

What processes contribute to interannual variations in the heating of the NH atmosphere at the expense of the SH leading to a northward ITCZ shift? The lack of a consistent observing network precludes an evaluation of the observed hemispheric energy budget but we can evaluate the processes at play in a coupled climate model displaying similar interannual statistics between $ITCZ_{LOC}$ and AHT_{EQ}. In the GFDL climate model (AM2.1), anomalous heating of the NH atmosphere is associated with positive net radiative anomalies in the northern tropics, which appear to be a positive feedback from the ITCZ shift itself. The northward shifted deep-convective clouds reflect additional shortwave radiation but also emit less OLR since the cloud tops are higher and colder. The reduced ASR and OLR in the ITCZ region partially cancel each other but the OLR reduction is larger in magnitude leading to a radiative gain in the region the ITCZ has shifted to. This mechanism provides a positive feedback to any initial ITCZ shift since the radiative gain must be balanced by more AHT_{EQ} away from the ITCZ and, hence, a further ITCZ shift in the same direction. *Voigt et al.* [2014] found this mechanisms to be important in other climate models as well, but also showed that its magnitude and even its sign (positive or negative feedback) depends on how models represent the coupling between tropical clouds and tropical circulation. It also remains unclear how important this mechanism is in the observed climate system.

Alternative mechanisms of hemispherically asymmetric climate variability (i.e., those leading to ITCZ shifts) include (1) variations in the OHT_{EQ} including decadal variations in the Atlantic Meridional Overturning Circulation (AMOC), (2) extratropical cloud radiative feedbacks, and (3) the expansion of sea (or land) ice in a single hemisphere. These mechanisms have yet to emerge as a source of significant ITCZ variability on interannual timescales in either climate model simulations or the observed record. Our quantitative scaling between ITCZ shifts and the hemispheric energy budget allows us to assess the impact some of these mechanisms might have on the ITCZ shift. For example, the basic state OHT_{EQ} is approximately 0.5 PW and is primarily associated with the AMOC. Some models produce interannual variations in the AMOC that are approximately 20% of the climatological mean circulation. Less than half of that variability (< 10% of total) persists at timescales of variability longer than a year. If the entirety of that variability was realized as 0.05 PW of compensating southward AHT_{EQ}, the associated ITCZ shift would be 0.15° or less than 20 km ITCZ shift, likely in the noise of an observation network and well below the horizontal resolution of a global climate model. This is likely an upper bound of the impact of the AMOC energy transport on the ITCZ since much of the AMOC energy transport is stored in the deep ocean, even at the decadal scale, and some portion of the AMOC heating anomaly will be radiated to space as an OLR anomaly. Thus, the AHT_{EQ} will compensate for but be significantly smaller in magnitude than the initial OHT_{EQ}. Similarly, the impact of cloud and surface albedo anomalies on the interannual variability of the ITCZ can be estimated from the impact of each on the variability of the hemispheric average planetary albedo over the satellite era. Monthly anomalies in the hemispheric average planetary albedo from the satellite observations [*Kato et al.*, 2006; *Voigt et al.*, 2013] lead to single-hemisphere heating anomalies that are smaller in magnitude (1σ) than 0.1 PW with atmospheric contributions to planetary albedo contributing <0.09 PW and surface contributions to planetary albedo contributing <0.03 PW. Together, these estimates suggest that the interannual variability of planetary albedo has a small hemispheric asymmetry that would contribute to ITCZ migrations of less than 50 km if acting in isolation with no other feedbacks [*Voigt et al.*, 2013].

8.2.5. ITCZ Shift due to External Forcing

We now consider the annual mean ITCZ shift due to external forcing. We argue that the approximate relationship of 3° latitude ITCZ shift per PW of AHT_{EQ}, which we saw for the annual mean climatology, seasonal cycle and interannual variability, also applies to externally forced annual mean ITCZ shifts. Therefore, understanding how forcing mechanisms and climate feedbacks project onto the hemispheric energy budget allows one to predict the magnitude and direction of the ITCZ shift. Here we consider different coupled climate model forcings in conjunction to emphasize that the aforementioned scaling of ITCZ shifts can be generalized to many climate states: (1) 1% CO_2 increase per year to doubling (hereafter $2XCO_2$), (2) the last glacial maximum (LGM), (3) the mid-Holocene, (4) freshwater hosing experiments to shut down the AMOC circulation, and (5) a transient climate simulation running from the LGM to present day. We will see that the ITCZ shift is small (~1° latitude) for all climate forcing seen here because it is hard to realize a hemispheric contrast of atmospheric heating of more than a couple tenths of a PW even when considering fairly large climate forcings such as the introduction of the Laurentide ice sheet to the NH.

The change in ITCZ location and AHT_{EQ} in all experiments is shown in Figure 8.3d, with each cross representing a different model simulation color coded by the experiment type (see labeling inset). The ITCZ response to each forcing experiment differs drastically between the different climate models. The ensemble average shift in the LGM and $2XCO_2$ simulations is not significantly different from zero while the northward shift in the mid-Holocene is marginally significant (95% confidence interval). The intermodel spread in ITCZ shift is strongly correlated with the change in AHT_{EQ} ($R = -0.84$) with a regression coefficient of $-3.2°\ PW^{-1}$ when considering all forcing experiments collectively. This result again suggests that both the ITCZ location and AHT_{EQ} are mutually dependent on the Hadley cell location for all climate states. Furthermore, the order 3° latitude ITCZ shift per PW of energy transport suggests that the same dependence between the ITCZ, AHT_{EQ}, and Hadley cell found over the seasonal cycle, annual mean climatology and interannual variability also determine the response to external forcing, implying that the bulk stability of the tropics and the relationship between the location of the upwelling branch of the Hadley circulation and the value of the mass overturning stream function at the equator is nearly climate state invariant. Most importantly, this result suggests that if we had a theory for how climate forcing and feedbacks project onto the hemispheric energy budget, we could make a quantitative prediction for the change in the ITCZ location. (Some portion of the hemispheric asymmetry in energy input to the atmosphere may be a consequence of the ITCZ shift itself, either due to radiative changes associated with shifting convection [*Voigt et al.*, 2014] or changes in surface fluxes associated with ocean circulation anomalies initiated by changing trade winds. These effects can be incorporated into a predictive model for ITCZ shifts, in response to an energy input external to the ITCZ shift, by introducing a feedback gain associated with the energetic changes due to the ITCZ shift as discussed in *Donohoe et al.* [2014a].) Furthermore, the ITCZ shift is fairly small (< 1°) in all simulations because the shift is constrained to the approximately $3°\ PW^{-1}$ relationship and the change in the hemispheric contrast of energy input to the atmosphere does not exceed 0.3 PW even for the highly hemispherically asymmetric forcing associated with the Laurentide ice sheet in the LGM. We briefly discuss the change in the hemispheric atmospheric energy budget in the LGM experiments, the processes leading to the intermodel spread below, and a comparison of the model results to paleoclimatic proxy records below. A more extensive analysis of the simulated ITCZ due to CO_2 forcing is provided in section 8.4.

The LGM simulations are forced by a prescribed land ice topography that features a large Laurentide ice sheet in the NH [*Peltier,* 2004] with more modest ice sheet enhancement in the SH. Overall, the surface albedo of the NH increases by approximately 0.03 relative to that of the SH in the ensemble average. Surprisingly, the ensemble average response to LGM boundary conditions and forcing is not significantly different from zero despite the large and hemispherically asymmetric forcing. If the change in surface albedo was realized at the TOA, the energy input to the NH would decrease by 1.3 PW, which we would extrapolate to a 4° southward ITCZ shift provided that the energy deficit was balanced by a compensating northward AHT_{EQ}.We explain below why the ensemble average ITCZ shift is so much smaller in these simulations and why the ITCZ even shifts northward in two of the models below.

In order for the surface albedo to make an impact in the TOA radiation budget (and therefore the atmospheric energy budget), solar radiation must be transmitted downward through the atmosphere, be reflected by the surface albedo, and then get transmitted back to the TOA. Therefore, the surface contribution to planetary albedo should *scale* as the atmospheric transmissivity squared [*Donohoe and Battisti*, 2011]. Approximately 20% of insolation is absorbed within the atmospheric column, primarily by ozone and water vapor. Also, 25% of the insolation is reflected back to space, primarily off the top of clouds. Therefore, the atmospheric shortwave transmissivity is approximately 0.55, which means that a three-unit change in surface albedo will translate to a one-unit change in planetary albedo and the atmospheric energy budget. In the LGM case, the atmosphere is slightly more shortwave transparent than the present day, and the anticipated deficit of energy to the NH 1.3 PW based on changes in surface albedo alone is reduced to 0.6 PW when accounting for the atmospheric transmissivity squared. Furthermore, cloud cover is reduced over the Laurentide ice sheet, which causes a 0.17 PW heating of the NH atmosphere resulting in a net shortwave deficit of 0.43 PW when added to the surface albedo impact. The NH cools relative to the SH, which causes a reduction of OLR by 0.32 PW via the Planck feedback. Thus, the ensemble average total radiative deficit of the NH is 0.11 PW and the resulting change in the AHT_{EQ} is accomplished by a small (~0.2°) southward ITCZ shift. The key point of the above analysis is that the anticipated energy deficit of the NH induced by the ice sheet albedo (of order 1.3 PW) is ameliorated by the basic state atmospheric transmissivity, cloud feedbacks and the Planck feedback to produce a very modest (of order 0.1 PW) change in the energy input to the NH, a modest AHT_{EQ} anomaly and, thus, a small ITCZ shift. We see that radiative feedbacks act to partially balance any hemispherically asymmetric forcing. This results in an ITCZ position that is a fairly stable property of the climate system.

In the two LGM simulations where the ITCZ shifts northward, radiative feedbacks and changes in ocean circulation result in an ITCZ shift that differs in direction from that anticipated from the forcing alone. In one simulation (MPI ECHAM5), the Planck feedback associated with the cooling of the NH exceeds the decrease in ASR in the NH. As a result, the introduction of the Laurentide ice sheet acts as a source of energy to the NH in part because the high topography (upwards of 5km) causes the surface to cool and emit less OLR. The other model with a northward ITCZ shift during the LGM simulation (IAP-FGOALS) has nearly compensating changes in the SW cooling and longwave warming due to the introduction of the Laurentide ice sheet to the NH. However, the model shows a significant increase in the northward OHT_{EQ} that results in a compensating southward AHT_{EQ} and northward ITCZ shift. These cases emphasize that the hemispheric asymmetry of radiative feedbacks can result in an ITCZ shift whose sign is questionable even in the presence of a hemispherically asymmetric forcing.

8.2.5.1. ITCZ Shifts from Paleoproxy Records. Much effort has gone into reconstructing past changes in the position and intensity of the ITCZ using proxy data thought to reflect changes in hydrology or wind strength. These include records of vegetation, lake water balance, precipitation isotopes, river sediment discharge, surface ocean isotopes, and marine primary productivity. Together, these records suggest the following ITCZ shifts over the paleohistory of the Earth:

1. a southward shift of the ITCZ relative to its modern position during the LGM [*Arbuszewski et al.,* 2013];
2. a further, rapid southward shift of the ITCZ associated with a proposed AMOC shutdown during Heinrich events when there is widespread evidence of North Atlantic cooling [*Peterson et al.,* 2000] and an equally abrupt northward shift during the Bølling Allerød when there is evidence that the AMOC circulation reinvigorated;
3. a northward ITCZ shift in the mid-Holocene [*Haug et al.,* 2001] especially during the boreal summer as deduced from the presence of lakes in the now arid Sahara [*deMenocal et al.,* 2000];
4. a southward ITCZ shift during the little ice age (1400–1850) in the tropical Pacific [*Sachs et al.,,* 2009].

Multiproxy reconstructions suggest a simultaneous moistening of sites north of the modern-day ITCZ location and drying of sites south of the modern-day ITCZ during northward ITCZ shifts, which greatly strengthens the interpretation of the results. However, most studies can give only a qualitative ITCZ shift. More specifically, because the climatological precipitation decreases between the ITCZ location and the subtropical precipitation minimum at approximately 20°, any meridional translation of the ITCZ of less than 20° latitude will result in concurrent precipitation changes of opposite signs north and south of the ITCZ. Do the ITCZ shifts deduced above correspond to small (< 1°) ITCZ shifts similar to those seen in the climate models or to shifts of order 10° as have been claimed in the literature based on the notion that the ITCZ itself must migrate over a location in order to moisten the region?

Recent work has used paleo-SST reconstructions alongside observational estimates of the relationship between tropical SST gradients to begin to answer the above question. Similar to the relationship between AHT_{EQ} and the ITCZ location, the ITCZ position is also strongly correlated with SST contrast (ΔSST) between the northern tropics (0–20°N) and southern tropics (0–20°S) with the ITCZ moving toward the warmer SSTs. The regression coefficient between ITCZ position and SST contrast is approximately 2° per K [*Donohoe et al.*, 2013b] and is nearly invariant with timescale (i.e., similar statistics hold for the seasonal cycle in observations and models, the interannual variability and the response to external forcing). Paleo reconstructions of tropical SSTs from isotopic analysis allow a calculation of ΔSST during the LGM (23–19 kyr before present), Heinrich event 1 (HS1, 17–15 kyr before present), and the mid-Holocene (7–5 kyr before present). The latter can be used alongside the relationships between ΔSST, ITCZ location, and AHT_{EQ} to place the modeling results presented here in the context of the observational estimates (shaded circles and hatched uncertainties in Fig. 8.3d). While there is considerable uncertainty in the proxy reconstructions (the dashed lines extend across the zero axis suggesting the estimates are not significantly different than zero), the climate model ensemble average ITCZ shift and change in AHT_{EQ} for the LGM and mid-Holocene are in excellent agreement with the proxy estimates (cf. the shaded green and blue squares with the shaded circles of the same color). We emphasize here not the correspondence of the numbers from the modeling simulations and the proxy records but, rather, that the small magnitude of the ITCZ shifts seen in the modeling studies is validated by the proxy records. More specifically, no model simulation produces an ITCZ shift of greater than 1° latitude and the proxy evidence is inconsistent with ITCZ shifts of greater than 1° for all time periods considered here even when accounting for the uncertainties in the proxy SST reconstructions and the relationships between ITCZ location and ΔSST. Models and observational evidence thus suggest that the ITCZ location is fairly stable in time. We emphasize, however, that the above discussion is limited to the annual zonal-mean ITCZ location and that regional changes to which proxy records are sensitive to (especially those associated with monsoons) are not subject to the same constraints.

The energetic framework for ITCZ shifts suggests that a complete shutdown of the AMOC would cause a fairly modest (~1°) southward ITCZ shift. The AMOC carries approximately 0.5 PW of energy northward across the equator [*Frierson et al.,*, 2013]. Therefore, a complete AMOC shutdown would reduce the energy input to the NH by 0.5 PW demanding an equal quantity of combined increase in northward AHT_{EQ} and OLR reduction in the NH (via the Planck feedback and surface cooling); 0.5 PW is an upper bound of the AHT_{EQ} change resulting from a complete AMOC shutdown. This anomalous AHT_{EQ} would demand a 1.5° southward ITCZ shift given our scaling of 3° PW^{-1} for a complete AMOC shutdown. Indeed, model simulations of both the modern-day and LGM systems forced by (unrealistically large) freshwater input to the Atlantic [*Chiang and Bitz* 2005] result in an approximate 50% weakening of the AMOC and simulate a 0.5° southward ITCZ (purple outlined diamonds in Fig. 8.3d) shift, with a 0.15 PW increase in AHT_{EQ} that is in very good agreement with the logic above for a halving of AMOC strength. Additionally, the proxy evidence from HS1, a time when the AMOC strength was thought to be severely reduced due to iceberg freshwater input to the North Atlantic, indicates a southward ITCZ shift of order 0.6° (purple circle in Fig. 8.3d), which also agrees with the magnitude of the changes anticipated from models and theory. Overall, the balance of evidence suggests that a complete AMOC shutdown would result in an ITCZ shift of around 1° latitude.

8.2.5.2. ITCZ Response to Climate Forcing over the Last 23,000 Years. *He et al.* [2010] recently simulated the transient climate evolution (TraCE) of the last 23,000 yr in a coupled simulation with the NCAR CCSM3 model. This simulation is forced by time-evolving orbital parameters, greenhouse gases, and prescribed land ice topography from reconstructions. Additionally, freshwater forcing of the Atlantic is applied during times when paleoproxy data suggest iceberg discharges and a slowdown of the AMOC circulation. This simulation gives a plausible reconstruction of the LGM climate, the deglaciation, abrupt NH cooling associated with HS1 and the Younger Dryas, the rapid warming into the Bølling Allerød,and the rather quiescent and stable climate of the Holocene in a single, continuous simulation [*Liu et al.,*, 2009]. We use this simulation to illustrate the magnitude and mechanisms of ITCZ shifts over the recent paleo history of the Earth.

The ITCZ location varies by approximately 2° latitude (red lines in Fig. 8.4a) from its southernmost location

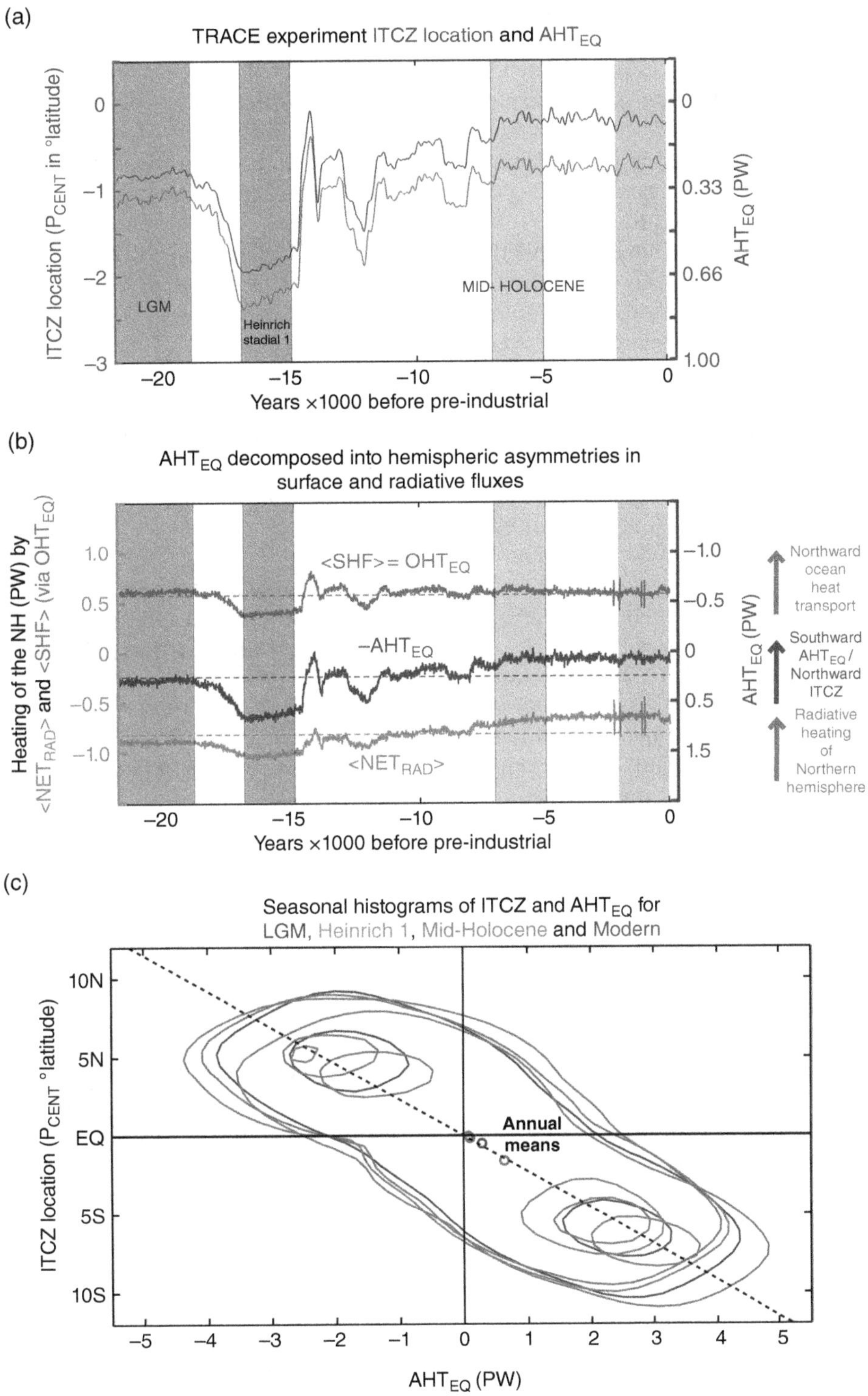

Figure 8.4 (a) Time series of 10 yr low pass filtered (red) ITCZ location (as measured by P_{CENT}) and (blue) AHT_{EQ} in the CCSM3 transient climate evolution (TraCE) experiment. The AHT_{EQ} has been negated and scaled by a factor of 3 as shown on the *y*-axis to the right. The vertical shading shows the time periods used to define the (blue) LGM, (gray) Heinrich Stadial 1, and (green) mid-Holocene. (b) The partitioning of AHT_{EQ} into (green) surface and radiative (orange) heating of the NH via equations (8.3) and (8.4). The AHT_{EQ} has been negated (right *y*-axis) so that the positive axis corresponds to radiative heating of the NH and a northward ITCZ shift. (c) The contours show smoothed probability distribution functions in the $AHT_{EQ}/ITCZ_{LOC}$ plane for the (blue) LGM, (gray) Heinrich Stadial 1, (green) mid-Holocene, and (red) modern. Annual mean values are shown with small circles. The dashed black line shows the linear best fit to all the data, which fall on the seasonal best fit. (*See insert for color representation of the figure.*)

during HS1 (17–15 kyr ago) to its northernmost location during the early Bølling Allerød (14,500 yr ago). (We note the CCSM3 has a basic state ITCZ bias with an ITCZ slightly south of the equator. This bias is a consequence of enhanced ASR in the Southern Ocean due to reduced cloud reflection relative to the observational estimates.) The ITCZ location and AHT_{EQ} changes are nearly locked to each other for all time periods and transitions (cf. the red and blue lines in Fig. 8.4a). We have multiplied the AHT_{EQ} axis (right, blue axis in Fig. 8.4a) by the factor of -3° latitude PW^{-1} for ease of comparison. We see that the same relationship between ITCZ location and AHT_{EQ} that was found over the observed seasonal cycle, interannual variability, the model spread in climatological mean state, and response to external forcing also applies to the transient transitions in the TraCE simulation. We now ask: what physical processes influence the direction and magnitude of the hemispheric contrast in energy input to the atmosphere and therefore dictate the ITCZ shift via changes in AHT_{EQ}?

The southward AHT_{EQ} is decomposed into changes in the net radiative heating of the NH ($<NET_{RAD}>$) and heating of the NH by way of the OHT_{EQ} (and subsequent surface energy fluxes) via equation (8.4). Decadal averages are shown in Figure 8.4b, which are sufficiently long to allow changes in atmospheric and oceanic energy content to be ignored and, hence, the near equality of <SHF> and OHT_{EQ}. As expected, the OHT_{EQ} is reduced by approximately 0.3 PW after the freshwater pulse heading into HS1 (around 17,500 yr ago) followed by a recovery and overshoot beyond the modern-day value during the Bølling Allerød after the freshwater forcing has been removed (14,500 yr ago). A similar reduction and subsequent recovery of smaller magnitude is observed during the Younger Dryas (13 kyr ago) when the applied freshwater forcing is smaller in magnitude. The OHT_{EQ} reduction is consistent with an approximately 50% reduction in the strength of the AMOC circulation. The partial shutdown of the AMOC triggers Arctic sea-ice expansion that reduces $<NET_{RAD}>$ by approximately half the magnitude of the change in OHT_{EQ} via the positive ice albedo feedback. The longer term upward trend in NET_{RAD} is associated with the poleward retreat of snow and ice cover in the NH moving toward the present, including the retreat of the Laurentide ice sheet. Over the entirety of the simulation, there is more variance in the radiative input of energy into the NH than that in OHT_{EQ} despite the visually apparent abrupt transitions in the AMOC circulation; the decadal average range (2σ) of $<NET_{RAD}>$ is 0.28 PW compared with 0.16 PW for OHT_{EQ}.

The TraCE results emphasize that the 3° ITCZ per PW of (differential) energy input into the NH atmosphere holds across timescale and climate state and that understanding the magnitude and sign of ITCZ shifts relies on understanding how climate forcing and feedbacks impact the hemispheric contrast of energy input to the atmosphere. The very extreme freshwater forcing of HS1 leads to approximately a 0.3 PW reduction of AMOC strength (slightly amplified by the ice albedo feedback), whereas the melting of the Laurentide ice sheet results in an enhancement of the radiative input into the NH of 0.4 PW at the 10,000 yr timescale. All in all, the total forcing and feedbacks on the hemispheric scale are of order 0.7 PW, which results in ITCZ variability of approximately 2°. This variability is associated with the deglaciation, arguably one of the most extreme climate transitions our planet has seen in its more recent past. These numbers thus likely set an upper bound on the anticipated future changes in the zonal-mean ITCZ position due to anthropogenic forcing.

8.3. RELATIONSHIP BETWEEN THE SEASONAL CYCLE OF ITCZ MIGRATION AND THE ANNUAL MEAN PRECIPITATION DISTRIBUTION

What processes determine the quantitative relationship between ITCZ location and AHT_{EQ} of 3° latitude per PW? If the annual mean Hadley cell were simply meridionally translated alongside the ITCZ, the change in AHT_{EQ} would be approximately equal to the meridional gradient of AHT evaluated at the equator (a first-order Taylor expansion) times the meridional ITCZ shift. The annual mean AHT has a gradient of approximately 0.1 PW per degree at the equator. Therefore this conceptual model projects that the ITCZ should shift by 10° latitude per PW of AHT_{EQ}, which is a sensitivity approximately three times greater than what we found from observations and models in the previous section. A similar formulation for ITCZ sensitivity to energy input based on the net energy input into the tropical atmosphere was recently expanded on by *Schneider et al.* [2014].

Why is the ITCZ three times less sensitive to the hemispheric contrast in forcing and feedbacks than the conceptual model based on the meridional translation of the annual mean climatology above suggest? We argue that the large seasonal migrations of the Hadley cell and ITCZ render the annual mean a less meaningful physical state that is seldom realized in the real (or modeled) climate system. As a result, a linearization about the annual mean state provides unrealistic ITCZ sensitivity to external forcing due to the exclusion of a seasonal cycle. Specifically, the magnitude asymmetry between the winter and summer Hadley cells that is present in the seasonal extremes but not in the annual mean (compare Fig. 8.1 b and c with Fig. 8.1a) causes the quantitative relationship between the ITCZ and AHT_{EQ} to differ fundamentally between the annual mean and seasonal extremes for two reasons:

1. In the annual mean, the ITCZ and maximum upward velocities are colocated with the zero stream function of the Hadley circulation due to the near symmetry of the Hadley circulation about the ITCZ. In contrast, the magnitude asymmetry between the winter and summer Hadley cells during the seasonal extremes results in maximum upward velocity and ITCZ location where the meridional gradient of the stream function is greatest. Thus, during the course of the seasonal cycle, the ITCZ is located within the winter Hadley cell and equatorward of the zero stream function.

2. The intensification of the winter Hadley cell as the ITCZ moves away from the equator toward the seasonal extremes results in a stronger mass overturning circulation at the equator than would be expected if the Hadley cell simply translated meridionally [*Lindzen and Hou,* 1988].

As a result of the above mechanisms, the ITCZ shift off the equator is smaller than the Hadley cell shift (via 1) and the AHT_{EQ} changes more than would be expected from the meridional translation of the annual mean Hadley cell (via 2). Both mechanisms cause the ITCZ location to be less sensitive to a hemispheric contrast of energy input relative to the linearization about the annual mean state. Why these mechanisms reduce the ITCZ sensitivity by a factor of 3 is an outstanding question in climate dynamics. If the meridional pattern of AHT (from NCEP reanalysis) is linearized about the location where the AHT is zero (the energy flux equator) during the extreme seasons (JJA and DJF), the implied ITCZ sensitivity is 4.2 and 4.3° PW^{-1}, respectively (not shown), in closer agreement to the empirical result of 3° PW^{-1} demonstrated here but still a higher sensitivity due to the energy flux equator moving farther off the equator than the ITCZ during the extreme seasons.

In this section, we expand on the relationship between the seasonal cycle of the ITCZ location and the annual mean response to external forcing (section 8.3.1) and how the amplitude of the seasonal cycle of the ITCZ determines the meridional extent of tropical convective precipitation (section 8.3.2). We argue that, because the annual mean state is seldom realized in nature, the seasonal relationship between ITCZ location and AHT_{EQ} determines the annual mean statistics in response to external forcing.

8.3.1. Annual Mean ITCZ Shifts Are Constrained by the Seasonal Relationship Between ITCZ Location and AHT_{EQ}

The annual mean ITCZ location, Hadley cell, and AHT_{EQ} are the small residual of large seasonal migrations of the Hadley cell off the equator. Here we demonstrate that annual mean changes in the ITCZ location represent small changes in the duration or amplitude of the seasonal migrations of the ITCZ off the equator and, as a result, the seasonal relationship between ITCZ location and AHT_{EQ} determines the same relationship for the annual mean ITCZ shift. Figure 8.4c shows smoothed seasonal histograms of the ITCZ location and AHT_{EQ} for the two-thousand-year periods centered on the (blue) LGM, (gray) HS1, (green) mid-Holocene, and (red) modern (shaded bars of the same color in Fig. 8.4a-b) in the TraCe simulations. As we noted for the observed seasonal cycle (section 8.2b), the ITCZ and AHT_{EQ} are most often found in a seasonal extreme position with the ITCZ located around 5° N (S) and with a corresponding southward (northward) AHT_{EQ} of – (+) 2 PW. The annual mean ITCZ location and AHT_{EQ} (open circles) are seldom realized and represent the statistical average of the seasonal modes. As such, the annual mean $ITCZ_{LOC}$ and AHT_{EQ} lies on the line connecting the seasonal modes, which has the slope of approximately 3° PW^{-1}.

The differences in the seasonal histograms of $ITCZ_{LOC}$ and AHT_{EQ} across the different paleoclimate epochs in TraCE are fairly modest. More specifically, in all states the system is most likely to be in a seasonal mode and the relationship between $ITCZ_{LOC}$ and AHT_{EQ} over the seasonal cycle is consistently 3° PW^{-1} in all climate states. We speculate that this result holds because the seasonal heating of the atmosphere induced by the seasonal migration of the sun is much larger in magnitude than any external forcing such that the perturbations to the seasonal cycle are small compared to the amplitude of the basic state seasonal cycle. We do see small variations in the duration of the seasonal modes and the magnitude of the ITCZ displacement off the equator during the seasonal modes across the different paleoclimate epochs. Changes in the seasonal ITCZ modes can cause annual mean ITCZ shifts via three possible mechanisms:

1. *Equatorward/poleward shifts during a single season.* As example: the most poleward migration of the ITCZ in the boreal summer during the LGM is nearly identical to that of the modern but the ITCZ migrates farther poleward during the austral summer (cf. the blue and red contours in the lower right quadrant of Fig. 8.4c). As a result, there is an annual mean southward ITCZ shift during the LGM compared with the modern.

2. *Opposing equatorward/poleward shifts during both seasons.* As example: during HS1 the ITCZ migrates less far off the equator during the boreal summer as compared with the LGM, and the ITCZ migrates farther off the equator during the austral summer (cf. the gray and blue contours in both the lower-right and upper-left quadrants). As a result, there is an annual mean southward ITCZ shift during HS1 compared with the modern.

3. *Changes in the duration of the season.* As example, during the mid-Holocene, the ITCZ spends less time in the boreal summer mode and more time in the austral summer mode as compared with the modern (cf. the relative areas of the red and green contours in the upper-left and lower-right quadrants). This mechanism shifts the annual mean ITCZ toward the more frequently occupied mode, to the south in this case. In the mid-Holocene, there is a simultaneous poleward ITCZ shift of the boreal summer ITCZ mode, which nearly cancels the contribution above. We note that the concurrent intensification and shortening of the duration of the boreal summer with near zero net effect matches the precessional forcing during the mid-Holocene, which has zero annual mean due to Kepler's second law.

We note that simultaneous poleward or equatorward ITCZ shifts during both seasons (symmetric stretching/compression) cancel each other in the annual mean and have little impact on the annual mean ITCZ location. We return to the implication of this mode of variability on the meridional extent of the tropics in the next subsection.

In all cases, because the annual mean ITCZ shift is a consequence of small variations in the seasonality of the ITCZ migration off the equator, *the change in annual mean ITCZ location and AHT_{EQ} must follow the seasonal relationship between the ITCZ and AHT_{EQ}*. For the reasons discussed in the introduction to this section, the 3° ITCZ shift per PW of AHT_{EQ} over the seasonal cycle is approximately a factor of 3 less than would be expected from a linearization about the annual mean climatology. Given the robustness of the mutual dependence of ITCZ location and AHT_{EQ} on the seasonal migration of the Hadley cell off the equator, we would anticipate future changes in the ITCZ location to be governed by the hemispheric contrast of climate forcing and feedbacks according to the 3° latitude ITCZ shift per PW AHT_{EQ}.

8.3.2. The Meridional Extent of Tropical Convective Precipitation and the Amplitude of the Seasonal Cycle

The distinction between the moist tropics and dry subtropics is fundamentally set by the prevailing vertical motion in the atmosphere associated with the upwelling and downwelling branches of the Hadley circulation, respectively. The annual mean meridional profile of precipitation represents the smearing out of many instantaneous realizations of intense and localized precipitation that migrate in time over the seasonal cycle. Therefore, the annual mean meridional extent of the region of tropical convective precipitation is a consequence of how far off the equator the ITCZ moves over the course of the seasonal cycle. Here, we think of the width of the tropical precipitation as roughly corresponding to the region of ascending atmospheric motion that extends from the mass overturning stream function minimum at the equatorward edge of the SH subtropics to the stream function maximum at the equatorward edge of the NH subtropics. We note that this metric of the width of tropical precipitation is independent of the mean ITCZ position.

As an illustration of the connection between the seasonal migration of the ITCZ and the width of tropical convective precipitation, we examine the impact of seasonal migrations of the ITCZ off the equator on the meridional profile of annual mean precipitation in an ensemble of slab ocean aquaplanet simulations, an atmosphere coupled to a motionless ocean. We modify the amplitude of the seasonal cycle by changing the ocean depth from 2.4 m to 50 m. In the deeper runs, more of the seasonal variations in insolation are stored within the ocean and never enter the atmospheric column to impact the atmospheric temperature and circulation [*Donohoe and Battisti,* 2013]. In the shallower ocean simulations, the intensification of the insolation in the summer hemisphere rapidly heats up the ocean, which subsequently fluxes energy upward to the atmosphere to drive a large seasonal cycle of atmospheric temperature and energy input to the atmosphere. As a result, there is a much larger hemispheric contrast of energy input to the atmosphere and, hence, larger AHT_{EQ} during the solstices in the shallow ocean runs than the deep ocean runs (cf. the orange lines in Fig. 8.5c–d). The larger AHT_{EQ} is accomplished by strongly displacing the Hadley cell off the equator, resulting in a large mass overturning circulation at the equator (gray contours in Fig. 8.5d) and an ITCZ in the summer hemisphere (blue lines). In contrast, in the deep ocean simulations, the Hadley cell and ITCZ barely move off the equator during the solstice seasons. In these runs, hemispheric symmetry demands that the annual mean ITCZ and Hadley cells be centered on the equator.

The impact of the amplitude of seasonal migrations of the ITCZ off the equator on the annual mean precipitation is truly profound. In the 50 m ocean depth simulation, the ITCZ hardly moves seasonally. Thus, the annual mean precipitation looks much like the seasonal extremes with a well-defined region of intense precipitation confined to within 10° of the equator and an intense Hadley cell comparable in magnitude to the circulation during the solstices (Fig. 8.5a). In contrast, the large seasonal migration of the Hadley cell in the 2.4 m simulations results in the meridional smearing out of the convective precipitation over a meridional broad region; there is almost no precipitation contrast between the tropics and extratropics (Fig. 8.5b). Similarly, the time averaging of large seasonal variations in the Hadley circulation results in a small time averaged stream function at any location; the regions of annual mean upward and downward atmospheric motion are not well defined. The meridional

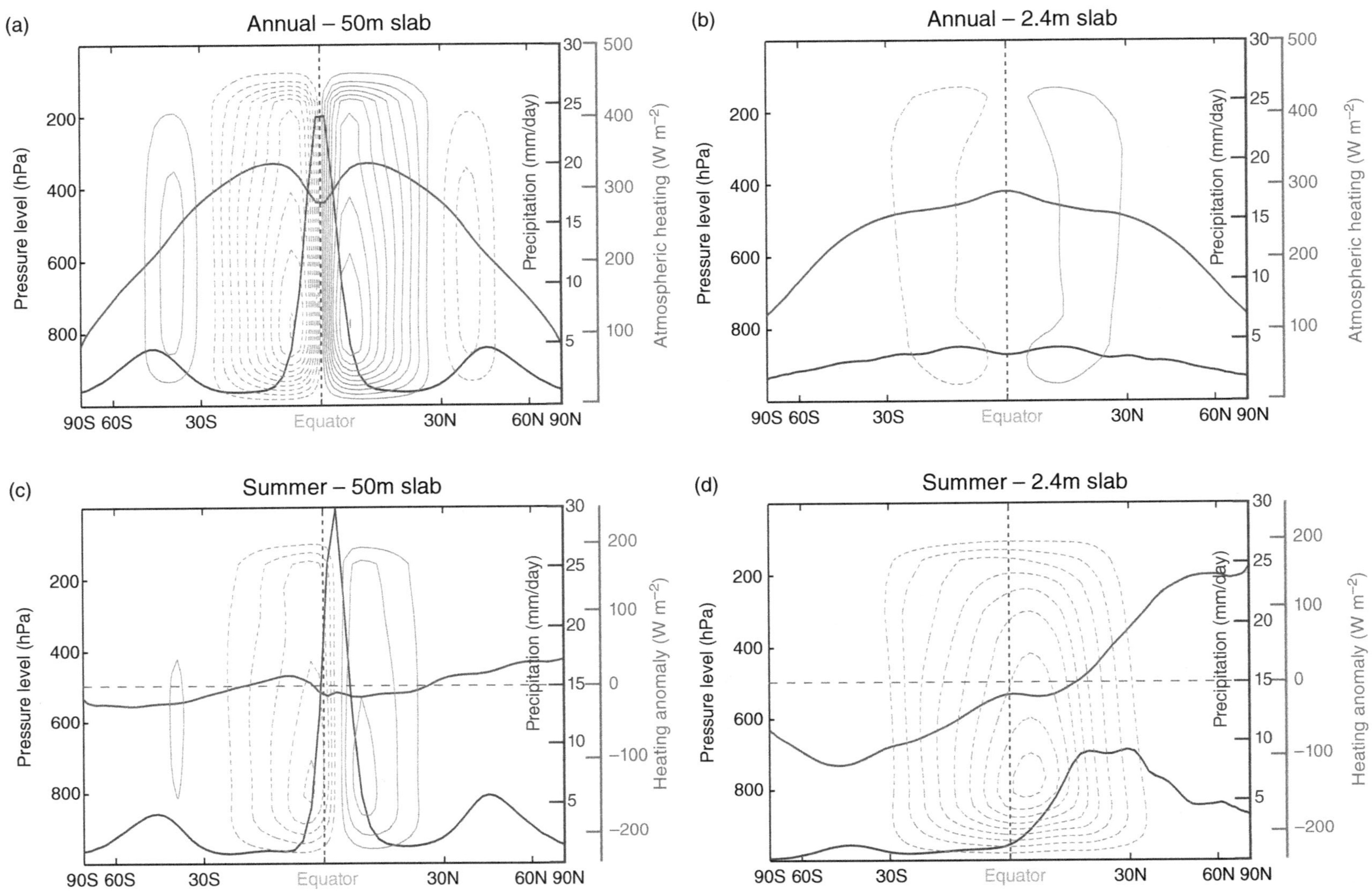

Figure 8.5 (a and b) Annual mean and (c and d) NH summer mass overturning stream function (gray contours) and precipitation (blue lines) for the (a and c) 50 meter and (b and d) 2.4 meter slab ocean aquaplanet simulations. The orange line is the energy input to the atmosphere (defined as the shortwave absorption in the atmospheric column plus the upward turbulent and longwave flux at the surface) with annual mean shown in the top panels and the seasonal anomaly shown in the lower panels. The mass overturning stream function is shown in gray contours with solid lines denoting clockwise rotation and dashed line denoting counterclockwise rotation. The contour interval is 20 Sv (10^9 kg s^{-1}) for the top panels and 50 Sv for the bottom panels. *See electronic version for color representation.*

extent of tropical convective precipitation is fundamentally set by the seasonal migration of the ITCZ off the equator, which is demanded by the seasonal energy input into the atmosphere.

As discussed in section 8.2b, only 10% of the seasonal variations in the hemispheric contrast of ASR are realized as AHT_{EQ}. Therefore, the potential for climate forcing and feedbacks to change the amplitude of the seasonal cycle of atmospheric heating, and hence the meridional width of tropical convective precipitation, is immense. In general, the moistening of the atmosphere tends to amplify the seasonal cycle of atmospheric heating [*Donohoe and Battisti* 2013], the melting of sea ice diminishes the seasonal energy input to the atmosphere [*Dwyer et al.,* 2012] while emission of sulfate aerosols diminish [*Hwang et al.,* 2013] and those of black carbon enhance the seasonal cycle of atmospheric heating. Overall, understanding how climate forcing and feedbacks impact the seasonal amplitude of energy input to the atmosphere and thus the meridional width of the Hadley cell [*Johanson and Fu,* 2009] and tropical convective precipitation is a fundamental challenge for global climate change forecasts.

8.4. IMPLICATIONS FOR FUTURE ITCZ SHIFTS UNDER GLOBAL WARMING

How does anthropogenic forcing lead to a hemispheric contrast of energy input to the atmosphere and, thus, meridional ITCZ shifts? Climate forcing by atmospheric constituents with residence times that are shorter than the interhemispheric transport timescale (1–2 yr) [*Bowman and Cohen,* 1997] are localized to the region (hemisphere) of emission. For example, sulfate aerosol emissions are overwhelmingly in the NH and, thus, lead to a reduction in ASR in the NH. Therefore, we would expect the ITCZ to shift southward in response to sulfate aerosol emissions. Indeed, the observational record and climate model simulations of the past century both show a southward ITCZ shift during the 1980s when sulfate aerosol emissions from India peaked [*Hwang et al.,*, 2013]. Other anthropogenic forcing agents that are spatially localized (e.g., land use change, black carbon accumulation on ice, etc.) are also expected to directly force ITCZ shifts and we can use the approximate 3° latitude PW^{-1} of spatially integrated forcing as a rough (neglecting feedback processes) guide to the magnitude of the associated ITCZ shifts.

Here, we focus on the ITCZ shifts induced by well-mixed (i.e., long-lived) forcing agents such as CO_2 that exert nearly spatially uniform forcing on the climate system [*Feldl et al.,* 2014] and, therefore, do not directly force ITCZ shifts. Rather, these forcing agents cause the planet as a whole to warm and this surface warming has a characteristic hemispheric asymmetry due to the predominance of ocean heat uptake in the Southern Ocean in the transient case and the spatial pattern of climate feedbacks in the equilibrium case. From an energetic perspective, global warming can lead to a hemispheric contrast of energy input to the atmosphere (and, thus, an ITCZ shift) by the following mechanisms:

1. Transient ocean heat uptake in one hemisphere, which acts as a (temporary) energy sink. For example, heat uptake in the Southern Ocean removes energy from the atmosphere locally and is expected to "pull" southward energy flux across the equator, which is accomplished by a northward ITCZ shift.

2. Shortwave feedbacks that differ between the two hemispheres. For example, sea and land ice has and is projected to decline more rapidly in the NH than the SH resulting in more ASR in the NH through the ice albedo feedback. This leads to a northward ITCZ shift. Additionally, contrasting cloud shortwave feedbacks between the Southern Ocean and northern extratropics in simulations of global warming [*Zelinka and Hartmann,* 2012] have been associated with substantial ITCZ shifts that differ in both direction and magnitude between models [*Frierson and Hwang,* 2012].

3. Differential surface warming between the two hemispheres leading to energy loss via the emission of longwave radiation. Although the Planck feedback (the change in OLR per unit change in surface temperature) does differ in magnitude spatially [*Soden and Held,* 2006], the spatial variations are smaller than the mean magnitude of the Planck feedback. Therefore, areas that warm more lose more radiation to space. As a result, the Planck feedback acts as a damping of spatial gradients in temperature, which themselves must be sustained by local forcing, feedback processes, and/or (up the temperature gradient) energy fluxes in the atmosphere or ocean. Specifically, the preferential surface warming of the NH (Arctic amplification) leads to energy loss from the NH via OLR, which, if acting in isolation, would lead to a southward ITCZ shift.

4. Changes in the large-scale ocean circulation. The melting of land ice in the NH in a warming planet results in freshwater input to the North Atlantic, which is projected to decrease the AMOC strength by as much as 50% [*Swingedouw et al.,* 2006] resulting in a southward ITCZ shift.

The simulated changes in zonal mean atmospheric heating 150 yr after an instantaneous CO_2 quadrupling (hereafter $4XCO_2$) in the CMIP5 multimodel ensemble are shown in Figure 8.6a. There is significant intermodel spread (differences between the thin dashed lines along the *y*-axis) in the energy flux changes. The solid lines show the ensemble average energy flux change and the shaded region indicates the 95% confidence interval of the mean, the ensemble average change is significantly different from zero where the shaded region does not intercept the x-axis. There are several notable and statistically robust regional changes in both the TOA (red and green) and surface energy fluxes including:

1. Increased energy input to the ocean in Southern Ocean where the surface temperature increase is significantly smaller than the global mean in all models. In this region, ocean upwelling associated with Ekman suction by the westerly surface winds keeps the ocean cool. As a result, the increased downwelling longwave flux associated with greenhouse forcing and, additionally, the warming aloft (due to atmospheric heat transport) exceeds the increase in upward energy flux (primarily latent and sensible heat fluxes) since the sea surface has warmed less than it would in the absence of the upwelling of cold ocean waters [*Armour et al.,* 2015].

2. Increased ocean heat uptake around the NH sea-ice edge (60° N). The loss of sea ice in this region leads to the absorption of solar radiation by the ocean surface that would have otherwise been reflected off the ice. This mechanism primarily acts to increase ocean energy input in the summer, whereas the loss of wintertime sea ice leads to ocean energy loss since the ocean is warmer than the local atmosphere in the winter, which results in anomalous upward latent and sensible energy fluxes to the atmosphere if the ocean is open.

3. The increase in both ASR and OLR in the northern extratropics. Both OLR and ASR increase in response to radiative feedbacks associated with substantial surface warming. ASR increases for two reasons: (a) the sea ice and snow on land have melted resulting in a darker surface that reflects less solar radiation and (b) the increased moisture content of the atmosphere absorbs more shortwave radiation that would have otherwise been incident on the reflective cloud tops and surface resulting in less SW radiation reflected to space. Changes in solar reflection off of clouds vary drastically between the models and result in intermodel differences in ASR changes. However, the robust increase in ASR can be explained entirely by the robust aforementioned ice albedo and shortwave water vapor feedbacks [*Donohoe et al.,* 2014b] in the absence of any ensemble average cloud feedback. The increase in OLR in the NH is associated with and expected from the increase in surface temperature, which achieves a spatial maxima in the northern extratropics, via the Planck feedback.

Despite the previously noted robust regional changes in the energy fluxes at the TOA and surface under global warming, even the sign of the change in the hemispheric contrast (defined as half the spatial integral over the NH minus that in the SH) is ambiguous between the CMIP5 models. Eight models show increased and nine models show decreased energy input into the NH atmosphere in response to $4XCO_2$. Because the hemispheric contrast of energy input to the atmosphere is equal to the AHT_{EQ} (equation [8.3]) and, as shown and discussed throughout this chapter, the ITCZ shift is proportional to the change in AHT_{EQ}, there is no robust and consistent ITCZ shift in the ensemble average due to CO_2 forcing. The lack of a consistent change in the hemispheric energy budget is a consequence of the intermodel differences in the hemispheric contrast of TOA radiation changes ($\sigma = 0.14$ PW and mean = 0.06 PW) despite the robust ensemble average atmospheric energy loss in the SH associated with ocean heat uptake (mean = 0.12 PW and $\sigma = 0.08$). This point can be visualized in Figure 8.6c where the total change in the hemispheric contrast of atmospheric heating (= AHT_{EQ}) is represented by the colored lines with slope -1 as the sum of the contributions from net radiation at the TOA (abscissa = $\Delta <NET_{RAD}>$) and surface energy fluxes (ordinate = $\Delta <SHF>$) via equation (8.3). While the models overwhelmingly occupy the upper two quadrants due to a robust increase in $<SHF>$ associated with atmospheric energy loss to the Southern Ocean, the models have wide spread in $<NET_{RAD}>$ along the abscissa with no robust ensemble average change. As a result, the total change in AHT_{EQ} spreads across the zero contour (gray line through the origin) with a nearly equal number of models predicting increases in energy input to the NH atmosphere (red contours) as those predicting decreases (blue contours) with the total change being dictated by the simulated change in $<NET_{RAD}>$ (dashed purple line).

Why is the change in $<NET_{RAD}>$ under global warming ambiguous in sign despite the robust regional changes in ASR and OLR in the northern extratropics associated with polar amplification, the ice albedo feedback of melting ice and snow, and atmospheric moistening? The net radiative heating of the atmosphere in that region is the difference between the increased ASR and increased OLR in that region and is, therefore, much smaller in magnitude and less statistically robust than the changes in either ASR or OLR alone (black line in Fig. 8.6b; note the shaded black line intersects the x-axis in most regions). We can visualize the compensation between changes in the hemispheric contrast in ASR and OLR by decomposing the total change in $<NET_{RAD}>$ into component contributions from ASR and OLR in Figure 8.6d. The models distinctly occupy the upper-right quadrant due to a robust increase in both ASR ($\Delta <ASR> = 0.29$ PW) and OLR ($\Delta <OLR> = 0.23$ PW) in the NH. However, the changes in $<ASR>$ and $<OLR>$ are positively correlated (dashed purple line) and, thus, the total change on $<NET_{RAD}>$ is ambiguous in sign (models spread across the black, zero contour line into both the red and blues) and dictated primarily by the change in $<ASR>$.

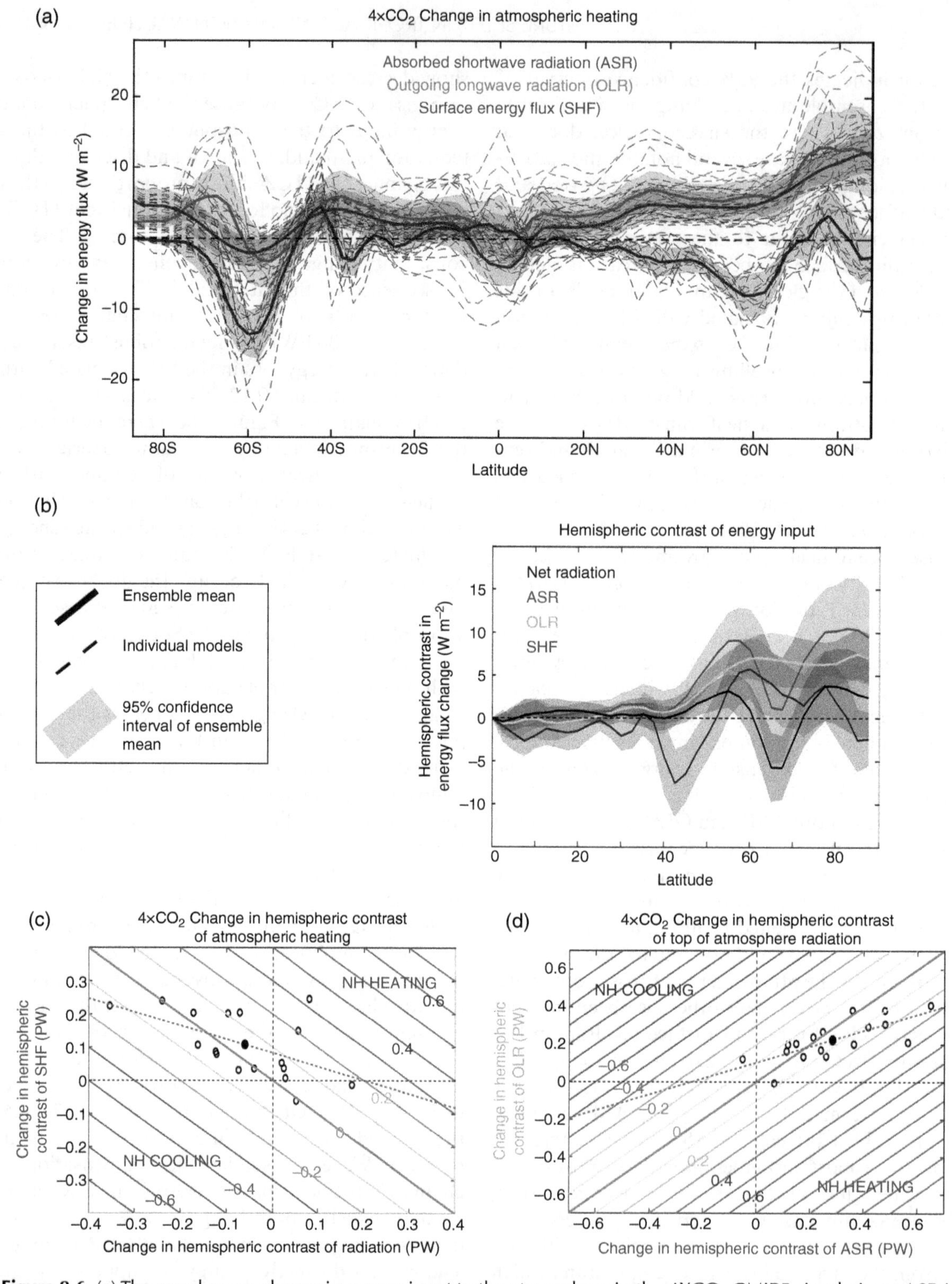

Figure 8.6 (a) The zonal mean change in energy input to the atmosphere in the $4XCO_2$ CMIP5 simulations. ASR is shown in red, OLR in green, and the surface energy flux in blue. The dashed lines are the individual models, the solid line is the ensemble mean, and the shaded region is the 95% confidence interval of the ensemble mean. (b) As in (a) except the hemispheric contrast of the energy flux change (half the difference between the NH and SH) with total change in energy input to the atmosphere in black. (c) The spatially integrated hemispheric contrast of energy input to the NH atmosphere by SHF (abscissa) and TOA radiation (ordinate). The total change in energy input is the sum of the abscissa and ordinate and is indicated by the diagonal colored contours with slope negative unity. The zero contour is gray. The open circles are individual models and the solid circle is the ensemble mean. The dashed purple line is the linear best fit. (d) As in (c) but for the decomposition of the change in the hemispheric contrast of TOA radiation into ASR and OLR components. In this case, the total TOA change is the difference between ASR and OLR and is indicated by the slanted color lines with positive unity slope. (*See insert for color representation of the figure.*)

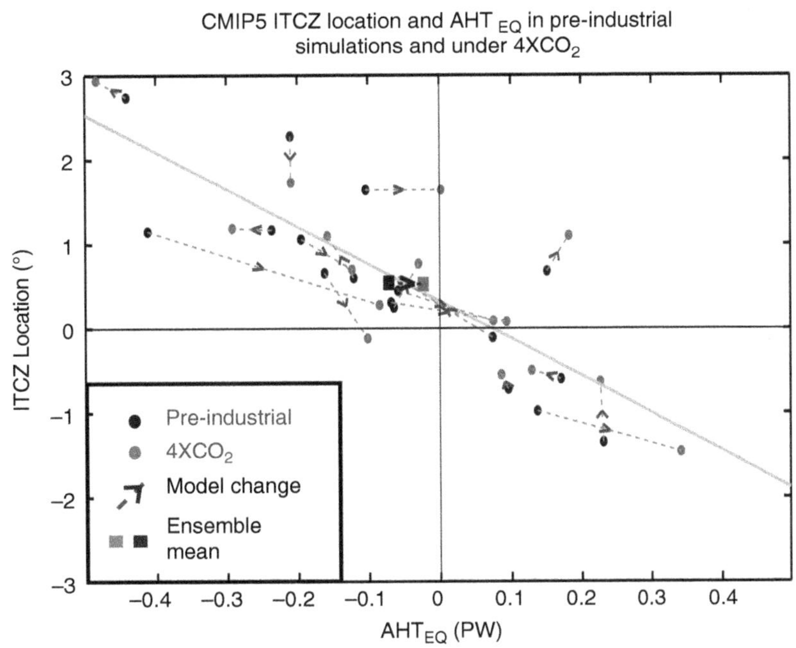

Figure 8.7 Scatter plot of ITCZ position (ordinate) and AHT_{EQ} (abscissa) in the CMIP5 simulations. The blue dots show the preindustrial simulation in individual models and the red dots are for the $4XCO_2$ simulations. The arrows indicate the change due to global warming. The ensemble average is shown by the shaded squares and the purple line is the linear best fit to all simulations. *See electronic version for color representation.*

As has been demonstrated throughout this chapter, meridional shifts in the ITCZ location demand a change in the AHT_{EQ} and, thus, the hemispheric contrast in energy input to the atmosphere. As discussed previously, some models simulate increased energy input into the NH due to $4XCO_2$, while a nearly equal number of models simulate more cooling of the NH atmosphere. As a result, the CMIP5 ITCZ shift due to global warming is not significantly different than zero; the ensemble mean ITCZ shift is 0.02° and the intermodel spread (2σ) is 0.50°. In the CMIP5 models, intermodel differences in ITCZ location and AHT_{EQ} are well correlated in the preindustrial simulations (blue circles in Fig. 8.7) and the $4XCO_2$ simulations (red circles) with an approximate relationship of 3° PW^{-1}. Therefore, the intermodel differences in the ITCZ shift and change in AHT_{EQ} are also well correlated ($R = 0.71$), the changes in the ITCZ/AHT_{EQ} plane indicated by the black arrows in Figure 8.7 are primarily along the approximate 3° PW^{-1} regression slope indicated by the purple line. Because the direction of the ITCZ shift and AHT_{EQ} vary widely (the arrows point in different directions), the ensemble average changes (dark black arrow) are small and not significantly different from zero. The essential point is that any ITCZ shift must be accompanied by a change in the hemispheric energy budget of the atmosphere and (despite several prominent regional energetic changes) the uncertainties in the projected changes to the hemispheric contrast of atmospheric heating due to global warming are larger than any projected changes.

Therefore, it is unclear if we should expect the ITCZ to shift northward or southward under global warming. However, the approximate quantitative relationship between the ITCZ location and the hemispheric contrast of atmospheric heating (~3° PW^{-1}), which seems to capture past ITCZ shifts on timescales ranging from the seasonal cycle, to the interannual variability, to the shift due to external forcing, suggests that ITCZ shifts will be small in the annual zonal mean (< 1°). Larger magnitude shifts would demand substantial changes in the large-scale energy budget of the climate system, which are unrealized in climate models, even under extreme forcing such as CO_2 quadrupling and the forcing of the last glacial maximum.

REFERENCES

Arbuszewski, J., P. deMenocal, C. Clroux, L. Bradtmiller, and A. Mix (2013), Meridional shifts of the Atlantic intertropical convergence zone since the last glacial maximum, *J. Atmos. Sci.*, *6*, 959962.

Armour, K.C., J. Marshall, J. Scott, A. Donohoe, and E. R. Newsom (2015), Southern Ocean warming delayed by circumpolar upwelling and equatorward transport, submitted to *Nat. Geosci.*

Bowman, K. P., and P. J. Cohen (1997), Interhemispheric exchange by seasonal modulation of the Hadley circulation, *J. Atmos. Sci.*, *54*, 2045–2059.

Ceppi, P., T.-T. Hwang, X. Liu, D. M. W. Frierson, and D. L. Hartmann (2013), The relationship between the ITCZ and the Southern Hemispheric eddy-driven jet, *J. Geophys. Res. Atmos.*, *118*, 5136–5146.

Chiang, J., and C. Bitz (2005), The influence of high latitude ice on the position of the marine intertropical convergence zone, *Climate Dyn.;* doi: 10.1007/s00 382–005–0040–5.

deMenocal, P., et al. (2000), Abrupt onset and termination of the African Humid Period: Rapid climate responses to gradual insolation forcing, *Quat. Sci. Rev.*, *19*, 347–361.

Donohoe, A., and D. Battisti (2011), Atmospheric and surface contributions to planetary albedo, *J. Climate*, *24* (16), 4401–4417.

Donohoe, A., and D. Battisti (2013), The seasonal cycle of atmospheric heating and temperature, *J. Climate*, *26* (14), 4962–4980.

Donohoe, A., D. Frierson, and D. Battisti (2013a), The effect of ocean mixed layer depth on climate in slab ocean aquaplanet experiments, *Climate Dyn.*, *26*; doi:10.1007/s00382-013-1843-4.

Donohoe, A., J. Marshall, D. Ferreira, and D. McGee (2013b), The relationship between ITCZ location and atmospheric heat transport across the equator: from the seasonal cycle to the last glacial maximum, *J. Climate*, *26* (11), 3597–3618.

Donohoe, A., J. Marshall, D. Ferreira, K. Armour, and D. McGee (2014a), The inter-annual variability of tropical precipitation and inter-hemispheric energy transport, *J. Climate*, *27* (9), 3377–3392.

Donohoe, A., K. C. Armour, A.G. Pendergrass, and D. S. Battisti (2014b), Shortwave and longwave contributions to global warming under increasing CO_2, *PNAS* (10 Nov); doi:10.1073/pnas.141219.

Dwyer, J., M. Biasutti, and A. Sobel (2012), Projected changes in the seasonal cycle of surface temperature, *J. Climate*, *25*, 6359–6374.

Fasullo, J. T., and K. E. Trenberth (2008), The annual cycle of the energy budget: part 2, Meridional structures and poleward transports, *J. Climate*, *21*, 2313–2325.

Feldl, N., D. M. W. Frierson, and G. H. Roe (2014), The influence of regional feedbacks on circulation sensitivity, *Geophys. Res. Lett.*, *41*; doi:10.1002/2014GL059336.

Frierson, D., Y. T. Hwang, N. S. Fuckar, R. Seager, S. M. Kang, A. Donohoe, E. Maroon, X. Liu, and D. S. Battisti (2013), Why does tropical rainfall peak in the northern hemisphere? the role of the oceans meridional overturning circulation, *Nat. Geosci.*, *6*, 940–944.

Frierson, D. M. W., and Y.-T. Hwang (2012), Extratropical influence on itcz shifts in slab ocean simulations of global warming, *J. Climate*, *25*, 720–733.

Hadley, G. (1735), Concerning the cause of the general tradewinds, *Philos. Trans. R. Soc.*, *29*, 58–62.

Haug, G., K. Hughen, D. Sigman, L. Peterson, and U. Rohl (2001), Southward migration of the intertropical convergence zone through the Holocene, *Science*, *293*, 1304–1308.

He, F., Z. Liu, O. Bette, P. Clark, A. Carlson, and E. Brady (2010), North Atlantic melting water forcing of early deglacial warming in the southern ocean and two-step north pacific ventilation: A transient GCM study with CCSM3.

Heaviside, C., and A. Czaja (2013), Deconstructing the Hadley cell heat transport, *Quart. J. Roy. Meteor. Soc.*, *139* (677), 2181–2189.

Held, I. (2001), The partitioning of the poleward energy transport between the tropical ocean and atmosphere, *J. Atmos. Sci.*, *58*, 943–948.

Held, I., and A. Y. Hou (1980), Nonlinear axially symmetric circulations in a nearly inviscid atmosphere, *J. Atmos. Sci.*, *37*, 515–533.

Hwang, Y., and D. Frierson (2013), Link between the double-intertropical convergence zone problem and cloud bias over southern ocean, *110,* 4935–4940.

Hwang, Y.-T., D. M. W. Frierson, and S. M. Kang (2013), Anthropogenic sulfate aerosol and the southward shift of tropical precipitation in the 20th century, *Geophys. Res. Lett.*, *40*, 1–6; doi: 10.1002/grl50502.

Johanson, C., and Q. Fu (2009), Hadley cell widening: Model simulations versus observations, *J. Climate*, *22*, 2713–2725.

Kalnay, E., et al. (1996), The NCEP/NCAR 40-year reanalysis project, *Bull. Amer. Meteor. Soc.*

Kang, S., R. Seager, D. Frierson, and X. Liu (2014), Croll revisited: Why is the Northern Hemisphere warmer than the Southern Hemisphere?, *Climate Dyn.*, *16*; doi:10.1007/s00382-014-2147-z.

Kato, S., N. Loeb, P. Minnis, J. Francis, T. Charlock, D. Rutan, E. Clothiaux, and S. Sun-Mack (2006), Seasonal and interannual variations of top-of-atmosphere irradiance and cloud cover over polar regions derived from ceres data set, *Geophys. Res. Lett.*, *33*; doi:10.1029/2006GL026 685.

Koppen, W. (1936), Das geographisca system der klimate, in *Handbuch der Klimatologie*, edited by W. Koppen and G. Geiger, 1. C. Gebr, Borntraeger, 1–44.

Lindzen, R. S., and A. Y. Hou (1988), Hadley circulations of zonally averaged heating centered off the equator, *J. Atmos. Sci.*, *45*, 2416–2427.

Liu, Z., B. Otto-Bliesner, F. He, E. Brady, R. Tomas, P. Clark, and A. Carlson (2009), Transient simulation of last deglaciation with a new mechanism for Bølling-Allerød warming, *Science*, *325* (5938), 310–314.

Lu, J., and G. A. Vecchi (2014), Hadley circulation, *Encyclopedia of Atmospheric Sciences*, 2nd ed., edited by G. North, J. Pyle, and F. Zhang, Elsevier.

Marshall, J., A. Donohoe, D. Ferreira, and D. McGee (2013), The oceans role in setting the mean position of the intertropical convergence zone, *Climate Dyn.*, *14*.

Meehl, G. A., C. Covey, T. Delworth, M. Latif, B. McAvaney, J. F. B. Mitchell, R. J. Stouffer, and K. E. Taylor (2007), The WCRP CMIP3 multi-model dataset: A new era in climate change research, *Bull. Amer. Meteor. Soc.*, *88*, 1383–1394.

Merlis, T. M., M. Zhao, and I. M. Held (2013), The sensitivity of hurricane frequency to ITCZ changes and radiatively forced warming in aquaplanet simulations, *Geophys. Res. Lett.*, *40*, 4109–4114.

Peltier, W. (2004), Global glacial isostasy and the surface of the ice-age earth: The ice-5g (vm2) model and grace, *Ann. Rev. Earth Planet. Sci.*, *22*, 111–149.

Peterson, L., G. Haug, K. Hughen, and U. Rohl (2000), Rapid changes in the hydrologic cycle of the tropical atlantic during the last glacial, *Science*, *290*, 1947–1951.

Sachs, J., D. Sachse, R. Smittenberg, Z. Zhang, D. Battisti, and S. Golubic (2009), Southward movement of the pacific intertropical convergence zone ad 1400–1850, *natgeo*, *2*, 519–525.

Schneider, T., T. Bischoff, and G. Haug (2014), Migrations and dynamics of the intertropical convergence zone, *Nature*, *513*, 45–53.

Sivakumar, M. V. K., H. P. Das, and O. Brunini (2005), Impacts of present and future climate variability and change on agriculture and forestry in the arid and semi-arid tropics, *Climate Change*, *70*, 31–72.

Soden, B., and I. Held (2006), An assessment of climate feedbacks in coupled ocean-atmosphere models, *J. Climate*, *19*, 3354–3360.

Swingedouw, D., P. Braconnot, and O. Marti (2006), Sensitivity of the Atlantic meridional overturning circulation to the melting from northern glaciers in climate change experiments, *Geophys. Res. Lett.*, *33*, L07711; doi:10.1029/2006GL025765.

Vimont, D. J., and J. P. Kossin (2007), The Atlantic meridional mode and hurricane activity, *Geophys. Res. Lett.*, *34*, L07709.

Voigt, A., B. Stevens, J. Bader, and T. Mauritsen (2013), The observed hemispheric symmetry in reflected shortwave irradiance, *J. Climate*, *26*, 468–477.

Voigt, A., S. Bony, J. Dufresne, and B. Stevens (2014), The radiative impact of clouds on the shift of the intertropical convergence zone, *Geophys. Res. Lett.*, *41* (12), 4308–4315.

Walker, C. C., and T. Schneider (2006), Eddy influences on Hadley circulations: Simulations with an idealized GCM, *J. Atmos. Sci.*, *63*, 3333–3350.

Xie, P., and P. Arkin (1996), Analyses of global monthly precipitation using gauge observations, satellite estimates, and numerical model predictions, *J. Climate*, *9*, 840–858.

Zelinka, M., and D. Hartmann (2012), Climate feedbacks, and their implications for poleward energy flux changes in a warming climate, *J. Climate*, *25*, 608–624.

9

Weather-Climate Interactions and MJO Influences

Paul E. Roundy[1], Naoko Sakaeda[2], Kyle MacRitchie[3], and Lawrence Gloeckler[1]

ABSTRACT

Weather events acting on timescales of days to weeks interact with the evolutions of events acting on longer timescales. Extreme weather events interact with other extreme events elsewhere in the world. This chapter describes how intraseasonal convective variability in the tropics, such as the Madden-Julian Oscillation (MJO) interacts with the seasonal cycle, the El Niño/Southern Oscillation (ENSO), and synoptic weather events including breaking Rossby waves and tropical cyclones, and how these interactions influence the evolution of subsequent outcomes. An overview of the interaction between the MJO, the seasonal cycle, and ENSO is followed by a case study of the interaction of the MJO with developing El Niño conditions during 1997 and 2014. Although the sea-surface temperature patterns of these two events were similar during early spring, this interaction apparently led to amplification of the 1997 event and the relative absence of strong El Niño conditions later in 2014. Analysis of the association between the MJO and Rossby wave breaking is then presented. Finally, assessment of the associations between the MJO and tropical cyclones over the western North Pacific Ocean is presented, demonstrating their combined interactions with global atmospheric circulation that can change weather outcomes around the world.

9.1. INTRODUCTION

The most pronounced local impacts of climate change would occur in association with extreme weather events superimposed on the altered climate. Thus, a major thrust of recent efforts in the climate community has been to assess how extreme regional events such as cold air outbreaks, heat waves, tropical cyclones, floods, droughts, and severe weather might change with the climate. Many of these types of events are poorly simulated in climate models because of insufficient spatial resolution and insufficient quality parameterization of sub-grid-scale convection and radiation processes [e.g., *Kerr*, 2011]. This chapter presents examples of how weather and climate events across the world can be interconnected so that the physics of natural climate and weather phenomena depend on each other, thereby complicating our ability to simulate and predict extreme events. A major focus of this chapter is the Madden-Julian Oscillation (MJO) [*Madden and Julian*, 1972; *Zhang*, 2005], which is associated with alternating eastward-moving planetary scale regions of enhanced and suppressed moist deep convection favoring warm pool regions in the tropics. The MJO modulates weather events around the world and influences the evolution of interannual climate variability. We first discuss how the MJO evolves together with the seasonal cycle, the El Niño/Southern Oscillation (ENSO), and the extratropical circulation, then continue with a case study illustration of how El Niño is intrinsically

[1] *Department of Atmospheric and Environmental Sciences, University at Albany, Albany, New York, USA*

[2] *NOAA Earth System Research Laboratory, Physical Sciences Division, Boulder, Colorado, USA*

[3] *NASA Goddard Space Flight Center, GES-DISC, Greenbelt, Maryland, USA*

Climate Extremes: Patterns and Mechanisms, Geophysical Monograph 226, First Edition.
Edited by S.-Y. Simon Wang, Jin-Ho Yoon, Christopher C. Funk, and Robert R. Gillies.

coupled to intraseasonal and synoptic weather events such as the MJO and westerly wind bursts [e.g., *Seiki and Takayabu*, 2007]. Transitions in ENSO states can occur across specific intraseasonal events (as illustrated in the 1997 case study). We show that parts of the extratropical circulation response to MJO convection and the subsequent evolution of the convection itself depend on its interaction with breaking extratropical Rossby waves and on its loose modulation of tropical cyclone formation and movement. Rossby wave breaking can lead to precipitation extremes of either sign, and tropical cyclones can lead to midlatitude Rossby wave dispersion events that sometimes cause cold air outbreaks over eastern North America and other significant weather outcomes downstream. Rossby waves can also be guided back into the tropics, where they may influence the evolution of subsequent MJO events (e.g., *Sakaeda and Roundy,* 2015). This interconnectedness in the system implies that modeling many types of regional extreme weather events requires more than simply downscaling coarse climate model signals to nested regional models [e.g., *Pielke,* 2012] because extreme outcomes in a region can depend on poorly simulated extreme weather in distant parts of the world.

The authors hope that an improved understanding of these types of interactions between signals across scales of time and space will ultimately yield conversation about how to better simulate these signals in climate models to yield better understanding and more informed prediction of the broader climate system and regional climate change. In order to enhance such discussion, this chapter includes both summaries of established science as well as new science that might prove controversial.

9.2. THE INTERACTIONS BETWEEN THE MJO, BACKGROUND STATE, AND SYNOPTIC WEATHER

9.2.1. The MJO and the Seasonal Background State

The MJO exists on the intraseasonal bridge between weather and climate. The circulation and moist convection associated with the MJO evolve within a background state set by all signals with timescales longer than intraseasonal. This background state influences the behavior of MJO convection and circulation including its response in the extratropical circulation. The MJO is also coupled to and loosely depends on its interactions with shorter timescale weather events. Understanding how the MJO interacts with its background state and the weather is crucial to explaining the observed variability of the MJO and understanding how it might change in a future climate state.

The structure of convective and circulation anomalies of the MJO evolve with the seasons [*Roundy and Frank*, 2004; *Zhang and Dong*, 2004]. Figure 9.1 shows maps of variance of intraseasonal precipitable water and outgoing longwave radiation (OLR) anomalies filtered in the zonal wavenumber frequency domain for the MJO (wavenumber 0–6 and periods 30–100 days) during four seasons. OLR anomalies are applied as a rough tracer of deep convection in the tropics. These results show that the latitude of MJO convective activity shifts with the seasons, generally following high sea-surface temperature and the coinciding regions of moisture convergence in the lower troposphere [*Zhang and Dong*, 2004]. MJO convection is most symmetric about the equator during boreal spring (Fig. 9.1). During boreal winter, MJO convection favors the Southern Hemisphere intertropical convergence zone (ITCZ) or monsoon regions, while the boreal summer MJO convection favors monsoon regions and tends to follow the Northern Hemisphere ITCZ as it crosses the Western Hemisphere. Convective anomalies in these regions of enhanced activity tend to move eastward and southward during southern summer and eastward and northward during northern summer.

The background circulation and moisture distribution help to drive the northward propagation of individual boreal summer MJO events [*Jiang et al.*, 2004; *Demott et al.*, 2013]. Figure 9.2 (from *Jiang et al.* [2004]) suggests how the northward propagation of MJO convection during boreal summer is driven by background easterly vertical wind shear. This background wind shear generates barotropic vorticity on the north side of MJO convection, which induces low-level convergence of moisture. *Jiang et al.* [2004] and *Demott et al.* [2013] also suggest that the positive meridional gradient of background moisture and the intraseasonal converging wind in the lower troposphere induces northward advection of moisture on the north side of MJO convection, leading to a northward advance of the convection. The advection of background moisture by intraseasonal wind anomalies helps to drive the MJO propagation and its convective onset in other seasons as well [e.g., *Kim et al.*, 2014; *Zhao et al.*, 2013]. These studies suggest that the seasonally varying background state including circulation and moisture help drive the seasonal variability of MJO characteristics.

9.2.2. The MJO and the El Niño/Southern Oscillation

Interannual variability such as ENSO also influences the geographical evolution of MJO convective activity and the evolution of the structures of its individual events [*Kessler*, 2001; *Pohl and Matthews*, 2007; *McPhaden et al.*, 2006]. For example, during the warm phase of ENSO, MJO convection tends to propagate farther eastward over the central Pacific basin as the region of high sea-surface temperature also extends eastward. Figure 9.3

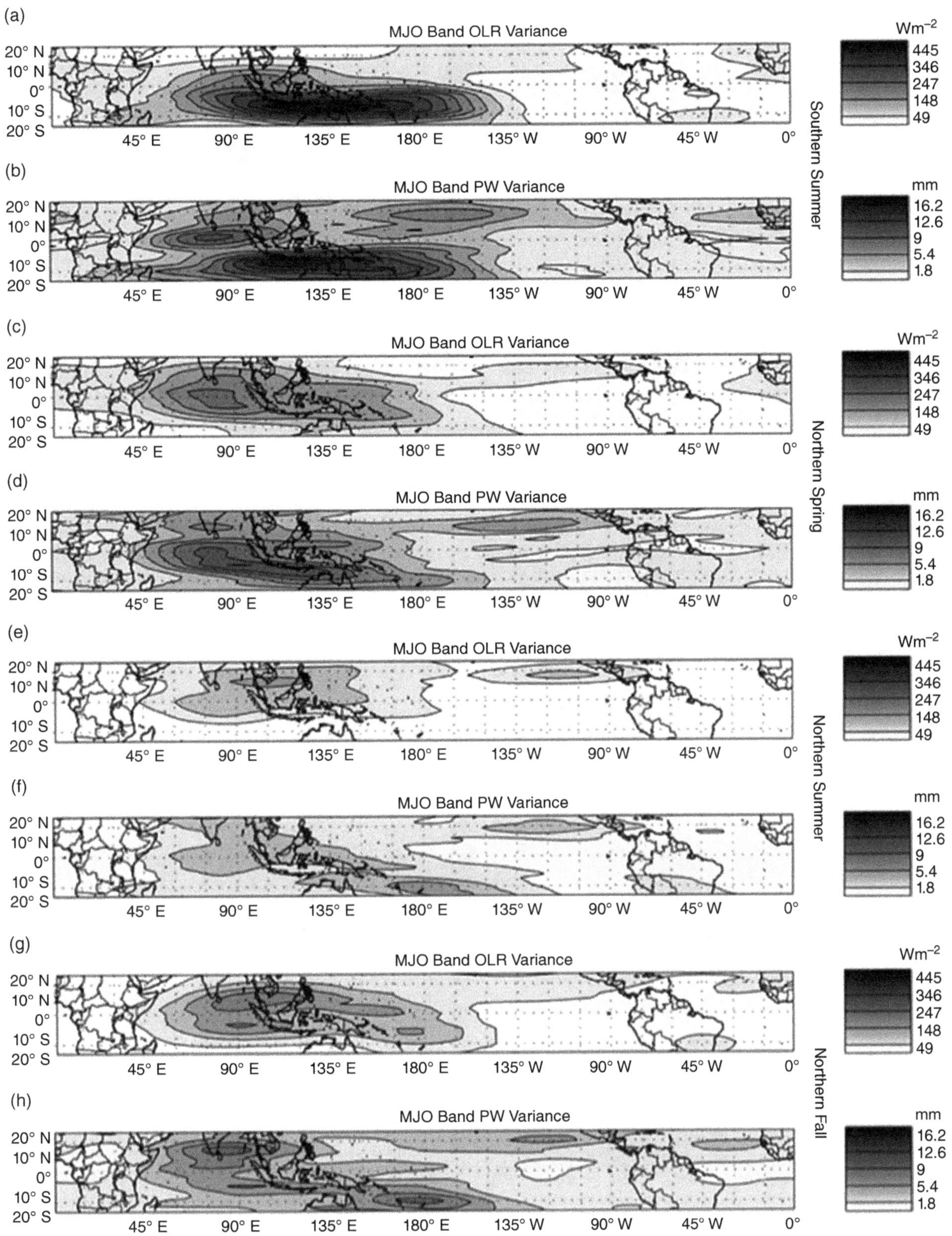

Figure 9.1 Seasonal maps of OLR and precipitable water (PW) variance in the eastward-propagating intraseasonal band. (a) and (b), the southern summer results for OLR and for PW, respectively; (c) and (d), the same for northern spring; (e) and (f) the result for northern summer; (g) and (h) the northern fall. Adapted from *Roundy and Frank* [2004]. ©American Meteorological Society. Used with permission.

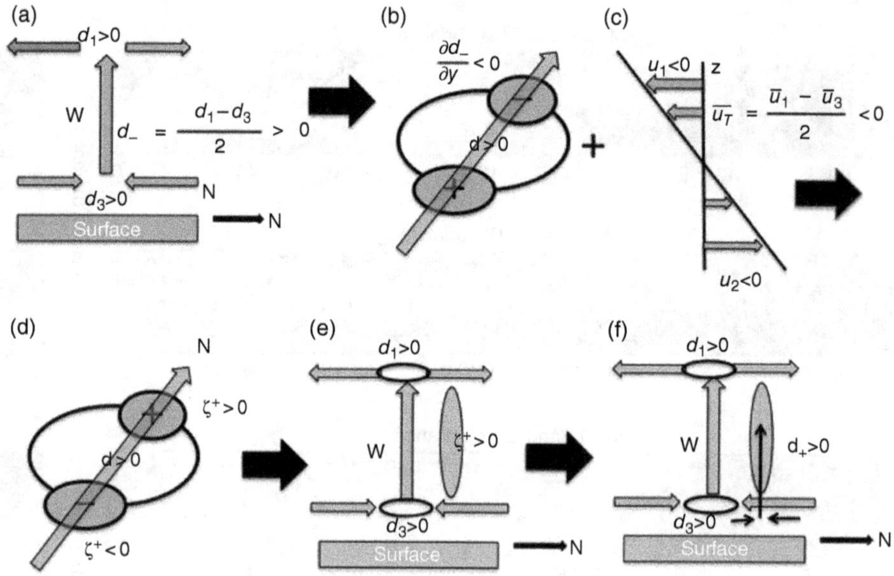

Figure 9.2 Schematic diagram for the vertical shear mechanism. (a) Consider initially an intraseasonal convection with a baroclinic structure. (b) This leads to mass divergence north of the convection center where d_ marks divergence. (c) In the presence of the easterly shear of the mean flow. (d and e) A positive barotropic vorticity is induced north of the convection, leading to (f) a barotropic divergence in situ. The latter further leads to a planetary boundary layer convergence and thus a northward shift of convective heating. Adapted from *Jiang et al.* [2004]. ©American Meteorological Society. Used with permission.

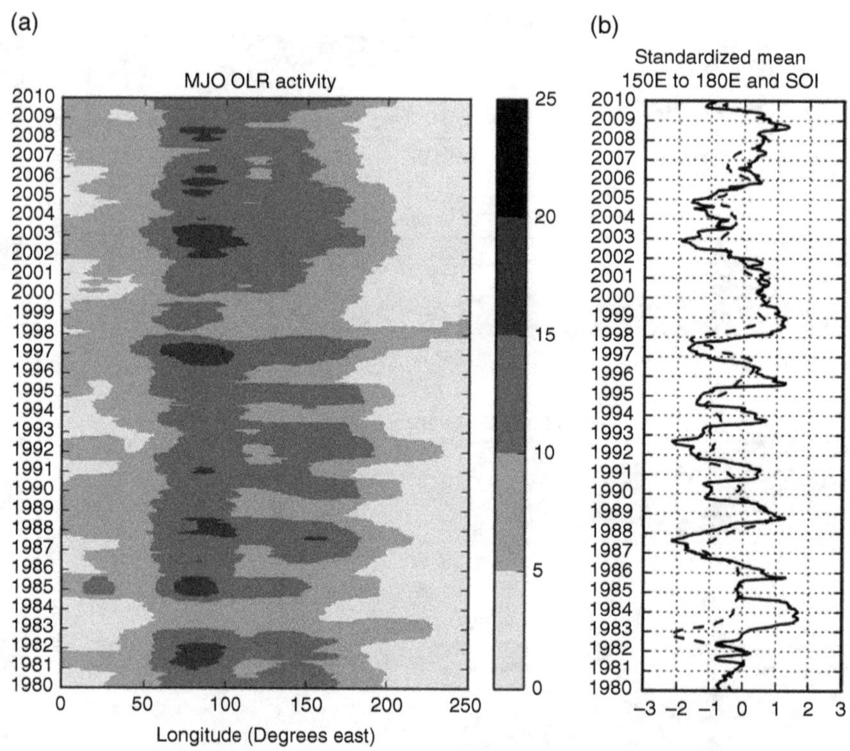

Figure 9.3 Interannual amplitude of intraseasonal OLR (5°N–5°S) (W m-2) constructed by squaring the bandpassed data, smoothing the squares with a 1 yr running mean, and then taking the square root. (Left) Amplitude in the global tropic strip, centered on the major region of variance (abscissa extends around the globe, broken at the southern American coast at 80°W). (Right) Time series of OLR interannual amplitude averaged over the western Pacific (150°E–180°) (solid line, scale at top) in comparison with the Southern Oscillation Index (dotted line, scale at bottom). Year ticks on each panel are at 1 January of each year, with year labels centered at midyear. Replotted for a longer period following *Kessler* [2001].

shows that the interannual variability of MJO convection over the central Pacific is well correlated with an ENSO index. While ENSO influences the activity of the MJO, changes in MJO activity can feed back onto the development of ENSO. Surface westerly wind bursts often associated with the MJO can advect warm sea-surface waters eastward and induce oceanic Kelvin waves that can help develop the warm phase of ENSO [*McPhaden et al.*, 2006; *Roundy and Kravitz*, 2009]. The relationship between MJO activity during the boreal spring and El Niño occurrence during the following winter is stronger than during other seasons [*Pohl and Matthews*, 2007; *McPhaden et al.*, 2006], though this relationship extends beyond the spring when El Niño events are already in progress [*Roundy and Kiladis*, 2006]. Furthermore, *Roundy and Kravitz* [2009] suggest that the differences in the structural evolution of the MJO at different stages of ENSO can help maintain trends toward or away from El Niño. For example, during trends toward El Niño, downwelling oceanic Kelvin waves forced by westerly wind bursts over the western Pacific Ocean are less likely to encounter extensive periods of easterly wind anomalies along their trajectories farther east, resulting in less attenuation of the wave signal and stronger warming outcomes. The westerly wind bursts and oceanic Kelvin waves associated with the MJO are stronger during the El Niño development phase, which advances El Niño [*Hendon et al.*, 1998; *Bergman et al.*, 2001; *Roundy and Kravitz*, 2009; *Seiki et al.*, 2009].

The structures of El Niño events vary continuously between extremes that have become known as central Pacific or Modoki El Niño and east Pacific El Niño (e.g., *Ashok et al.*, 2007; *Johnson*, 2013; *Roundy*, 2015; *Capotondi et al.*, 2015). Although debate remains vigorous with respect to the extent to which the modes at the extremes behave independently, recent research has shown that the relationship between the MJO and ENSO varies between ENSO events at these extremes (*Gushchina and Dewitte*, 2012; *Feng et al.*, 2015). The evolution of the MJO and its associated westerly wind bursts influences the evolving structure of SST anomalies during El Niño events (e.g., *Chen et al.*, 2015).

The impact of the background atmospheric circulation on the MJO extends globally, including its mid- and high-latitude circulation response. The MJO induces planetary scale heating anomalies in the tropical atmosphere that drive circulations in the tropics and around the world. The anomalous positive heating associated with MJO convection often couples with lower-tropospheric convergence and upper-tropospheric divergence. The Coriolis torque by the meridionally converging and diverging flow and the deep convective heating profile in the tropics induces a pair of upper-tropospheric anticyclonic and a pair of lower-tropospheric cyclonic Rossby gyres about the equator [e.g., *Rui and Wang*, 1990; *Kiladis et al.*, 2005]. These planetary-scale Rossby gyres induce midlatitude wave trains and modulate the structure and strength of subtropical jet streams [*Matthews and Kiladis*, 1999; *Moore et al.*, 2010]. Wave activity originates from the tropics in these regions and propagates poleward. Consistent with this view, *Sardeshmukh and Hoskins* [1988] showed that advection of background potential vorticity by divergent winds induces Rossby waves. They suggested that the location and strength of Rossby wave generation by tropical convection depends on the relative positioning of divergent circulations induced by anomalous heating and on the gradients of background potential vorticity nearby. As an illustration of the wave train response, Figure 9.4 shows a series of time-lag composites of 200 hPa circulation over the Northern Hemisphere following times when the enhanced envelope of MJO convection is over the eastern Indian Ocean and the maritime continent. This wave train response propagates around the world, where it modulates weather outcomes from extremes in temperature (e.g., *Matsueda and Takaya*, 2015), to extremes in rainfall (e.g., *Jones*, 2000; *Jones et al.*, 2004), to the distribution of severe weather events (e.g., *Thompson and Roundy*, 2013). The midlatitude wave trains generated by the MJO modulate teleconnection patterns such as the Pacific North American teleconnection (PNA), the North Atlantic Oscillation (NAO), and the Arctic Oscillation [*Mori and Watanabe*, 2008; *Lin et al.*, 2009; *L'Heureux and Higgins*, 2008]. In turn, these high-latitude patterns apparently influence the evolution of subsequent MJO convection [e.g., *Lin et al.*, 2009; *Roundy*, 2014; *Sakaeda and Roundy*, 2015].

9.2.3. Interaction Between the MJO and the Midlatitude Atmospheric Circulation

The anomalous heating and circulation associated with the MJO induces midlatitude wave trains, but the background flow helps determine the subsequent evolution and structure of the wave trains. An idealized model experiment by *Bao and Hartmann* [2014] suggests that the midlatitude circulation response to the MJO depends on the longitudinal structure of the subtropical jet streams, suggesting that accurate representation of the subtropical jets is critical for realistic simulation of the extratropical circulation response to the MJO. The MJO also induces anomalous precipitation in the subtropics that is not directly associated with its convective envelope within the tropics [*Kim et al.*, 2006; *Jeong et al.*, 2008]. The anomalous precipitation in the subtropics results from vertical motions analogous to quasigeostrophic forcing at higher latitudes, through the interaction of a time-mean subtropical westerly jet stream and the anomalous circulation induced by the MJO heat source [*Kim et al.*, 2006; Jeong *et al.*, 2008]. Other observational and modeling

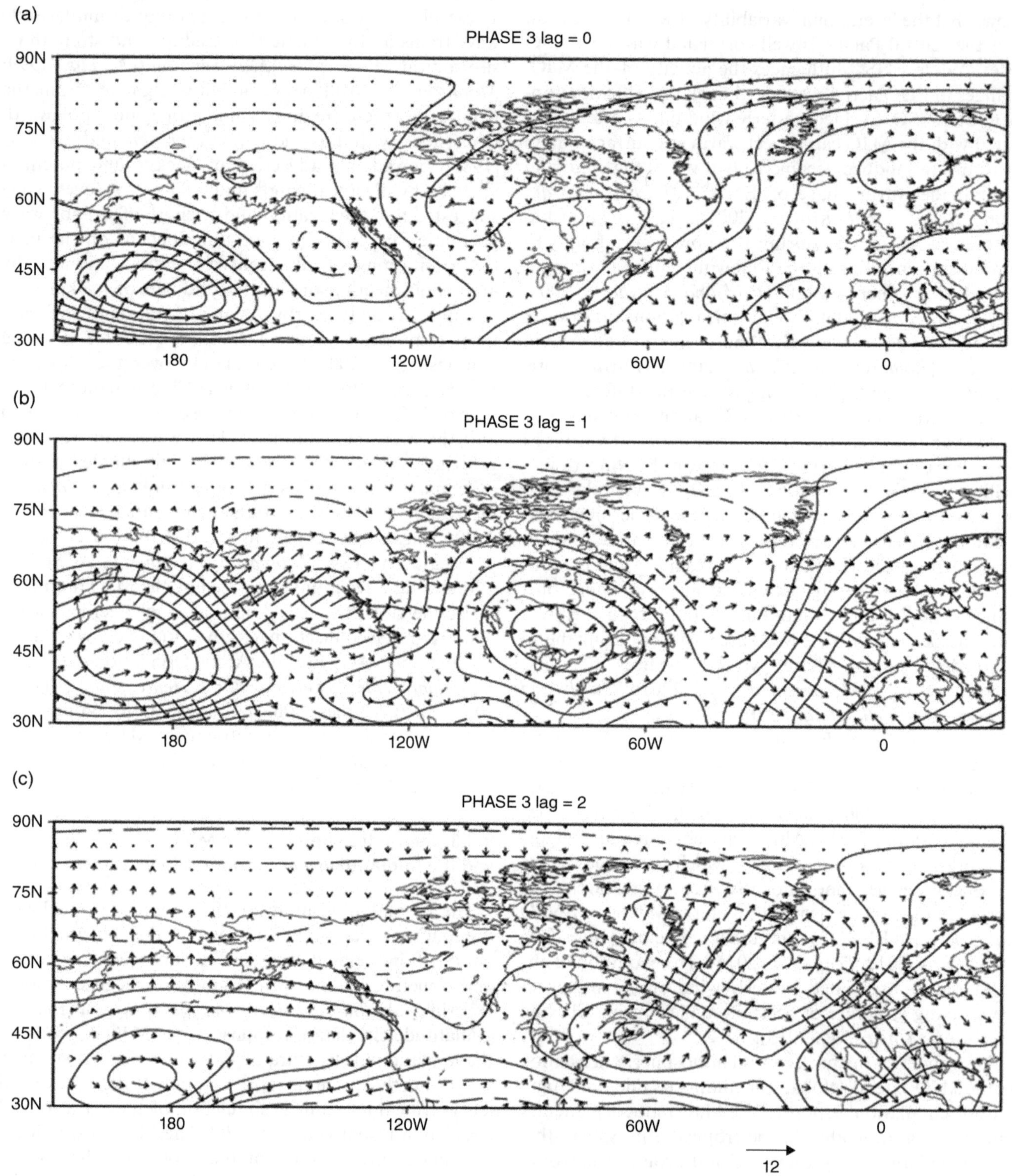

Figure 9.4 The 200 hPa wave activity flux with respect to MJO phase 3 for (a) lag 0, (b) lag 1, and (c) lag 2 pentads. The arrows are the horizontal wave flux vectors, and the contours the 200 hPa streamfunction anomalies. Contour interval is 1 × 106 $m^2\ s^{-1}$. Contours with negative values are dashed. Scaling for arrows is given below (c) (unit: $m^2\ s^{-2}$). Wave activity flux with magnitude smaller than 0.5 $m^2\ s^{-2}$ is not plotted. From *Lin et al.* [2009]. ©American Meteorological Society. Used with permission.

studies have also shown that the background state helps drive the observed extratropical circulation anomalies associated with the MJO [e.g., *Tyrell et al.*, 1996; *Frederiksen and Lin*, 2013; *Sakaeda and Roundy*, 2014].

The intensity and geographical distribution of MJO convection varies seasonally, contributing to changes in the structure and evolution of the extratropical circulation response to the MJO, but *Newman and Sardeshmukh* [1998] showed that the extratropical circulation response to anomalous heating in the tropics also changes with the seasonally varying extratropical background state (Fig. 9.5). Figure 9.5 shows that the same forcing by tropical convection can induce different extratropical circulation responses depending on the extratropical background state.

Roundy et al. [2010] showed that the extratropical circulation patterns associated with the MJO depend on the ENSO state. ENSO modulates the activity of the MJO by

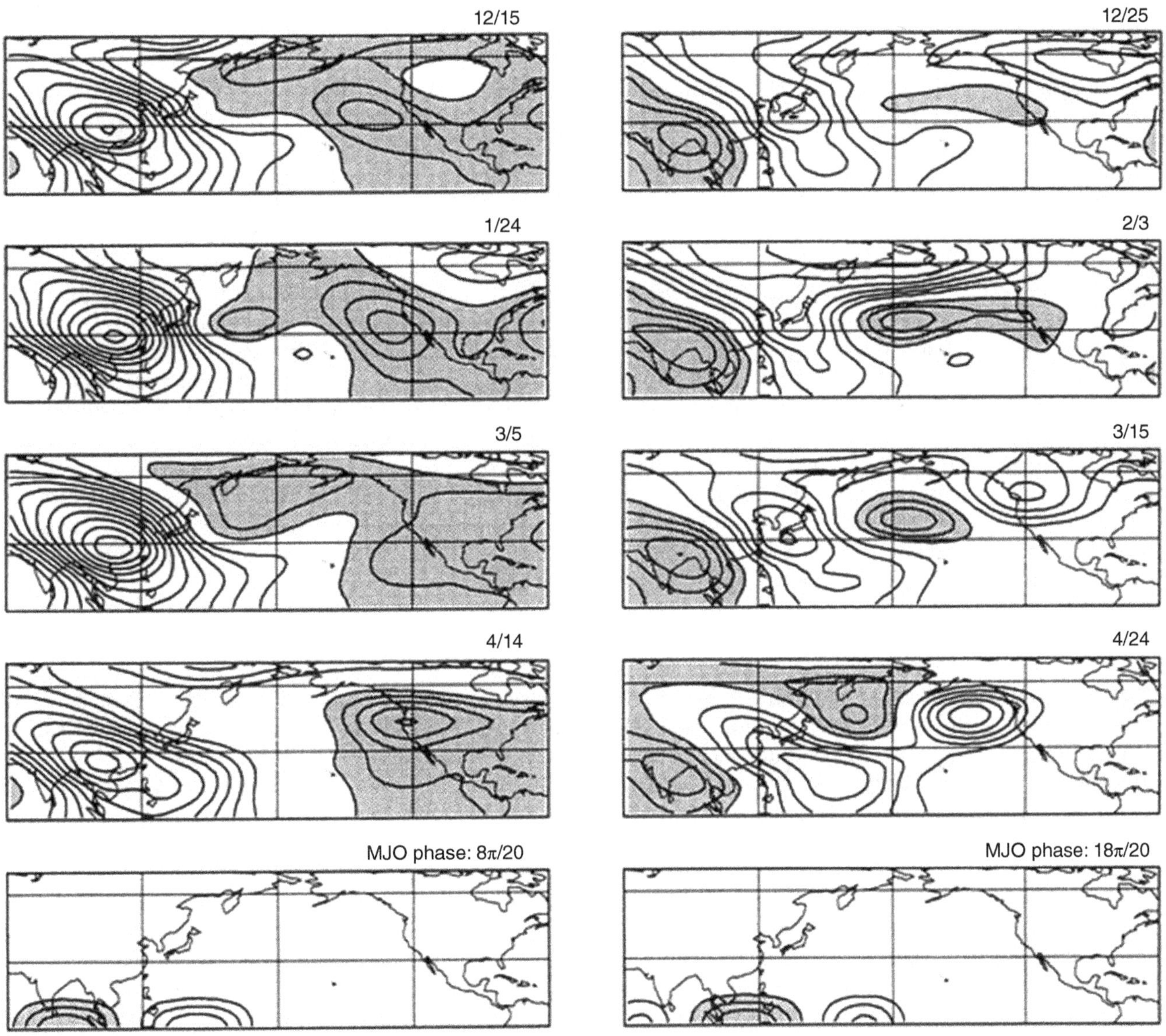

Figure 9.5 Maps of 300 mb height anomaly response obtained in the propagation forcing run on specified dates corresponding to the phases of the MJO indicated at the bottom of each column. The plots are shown 40 days apart, that is, at consecutive occurrences of the same phase of the MJO. Note that height anomalies on the right occur 10 days after the height anomalies on the left. Contour interval is 5 m. The zero contour has been removed for clarity, and negative values less than −5 m are dashed. From *Newman and Sardeshmukh* [1998]. ©American Meteorological Society. Used with permission.

altering the background oceanic and atmospheric state in the tropics (thereby influencing MJO convection), but ENSO also impacts the circulation response to the MJO by redistributing background potential vorticity across the global atmosphere, thereby influencing wave propagation and blocking. Furthermore, ENSO modulates the structure and strength of subtropical jet streams. The modulated characteristics of the MJO convection and the altered background extratropical circulation interact nonlinearly to induce different midlatitude response patterns to the MJO during the warm and cold phases of ENSO. Figure 9.6 shows absolute values of 300 hPa geopotential height anomalies composited over all MJO phases during different phases of ENSO. The differences in the patterns show that the strong anomalous flow associated with the MJO occurs over different regions during different ENSO states. *Seiki et al.* [2009] suggest that the interaction between the MJO, ENSO, and extratropical circulation during periods of El Niño development induces circulation patterns that help moisture to converge over the western Pacific basin, thereby enhancing formation of twin cyclones and westerly wind bursts on the equator, which further contribute toward the development of El Niño.

The background flow helps to determine the activity and characteristics of the MJO and the extratropical circulation induced by the MJO. Therefore, in order to improve the quality of simulations of the global circulation response to the MJO, we do not only need to improve the simulation skill of MJO convection but we must also accurately represent the background state and its interaction with the MJO.

In addition to the sensitivities of MJO activity to its background state, the MJO also actively interacts with weather phenomena characterized by shorter timescales both in and out of the tropics. For example, the MJO modulates the characteristics of midlatitude transient circulations and the frequency and location of Rossby wave breaking events [*Moore et al.*, 2010; *Matthews and Kiladis*, 1999]. Figure 9.7, from *Sakaeda and Roundy* [2014], illustrates that the intraseasonal cyclonic circulation imposed on the subtropical jet stream over the eastern North Pacific Ocean shifts the storm tracks of synoptic eddies equatorward. The anomalously cyclonically sheared synoptic eddies also shift their trough and ridge axes to be more meridionally oriented. The resultant changes in the characteristics of the transient eddies feed back onto the intraseasonal zonal wind by reducing the northward flux of zonal momentum by the eddies, which induces anomalous westerly wind acceleration to the south of the subtropical jets. This nonlinear interaction between the intraseasonal and synoptic circulations helps drive the extratropical circulation associated with the MJO.

The extratropical response to the MJO involves complicated dynamics of interactions with the background state and higher-frequency transient circulations. Thus, further understanding of these interactions across scales is a key to improve extended forecast skill and climate projections. The following sections discuss case studies of the interactions between the MJO and ENSO and two-way interactions between the MJO and midlatitude Rossby wave breaking events and tropical cyclones.

9.3. A CASE STUDY ON INTERACTIONS BETWEEN THE MADDEN-JULIAN OSCILLATION AND EL NIÑO

During early February 1997, after a relative drought of nearly two months, heavy rainfall developed over the equatorial Indian Ocean (Fig. 9.8a). This moist convection began to evolve eastward at roughly 4–8 ms^{-1} as an active convective phase of the MJO. During the second half of February, the MJO convection spread eastward across the maritime continent and entered the West Pacific basin, where it continued to move eastward before its decline in mid March. Within and to the west of this advancing convection, daily mean westerly wind anomalies of 2–8 ms^{-1} developed. (Red contours on Fig. 9.8a represent westerly wind anomalies.) The corresponding wind stress (taux) anomalies are plotted in Figure 9.9a, from *Menkes et al.* [2014]. Westerly wind stress on the equatorial ocean surface accelerated warm water eastward (fine black contours in Fig. 9.9 b, d, and e represent surface currents) and forced a sea-surface ridge anomaly and a thermocline depression, which then emerged as the downwelling phase of an oceanic Kelvin wave, propagating the eastward and downward current anomaly eastward at 2–3 ms^{-1} across the Pacific basin (solid red contours March through April on Fig. 9.8b represent the associated dynamic height anomalies), arriving near the coast of South America by early May. Sea-level gauge data from Christmas Island of Kiribati (1.9°N 157.4°W) for the first half of 1997 (not shown) has a minimum around 1 March, followed by the Kelvin wave, which was associated with roughly a 20 cm increase in sea level by its crest at the island on 1 April. Figures 9.9b and 9.10b show that sea-surface temperature anomalies increased roughly 1°C across this wave event, and increased sea-surface temperatures facilitated enhanced deep convection and westerly wind anomalies from the dateline eastward, both on seasonal and intraseasonal timescales. Further MJO events arrived over the western Pacific Ocean and became associated with westerly wind bursts in April, May, and July, ultimately leading to what may have been the strongest El Niño event of the twentieth century. *Menkes et al.* [2014] analyzed different evolutions in an ocean model forced by different westerly wind burst configurations. The dashed light gray curve in Figure 9.9d shows the eastern

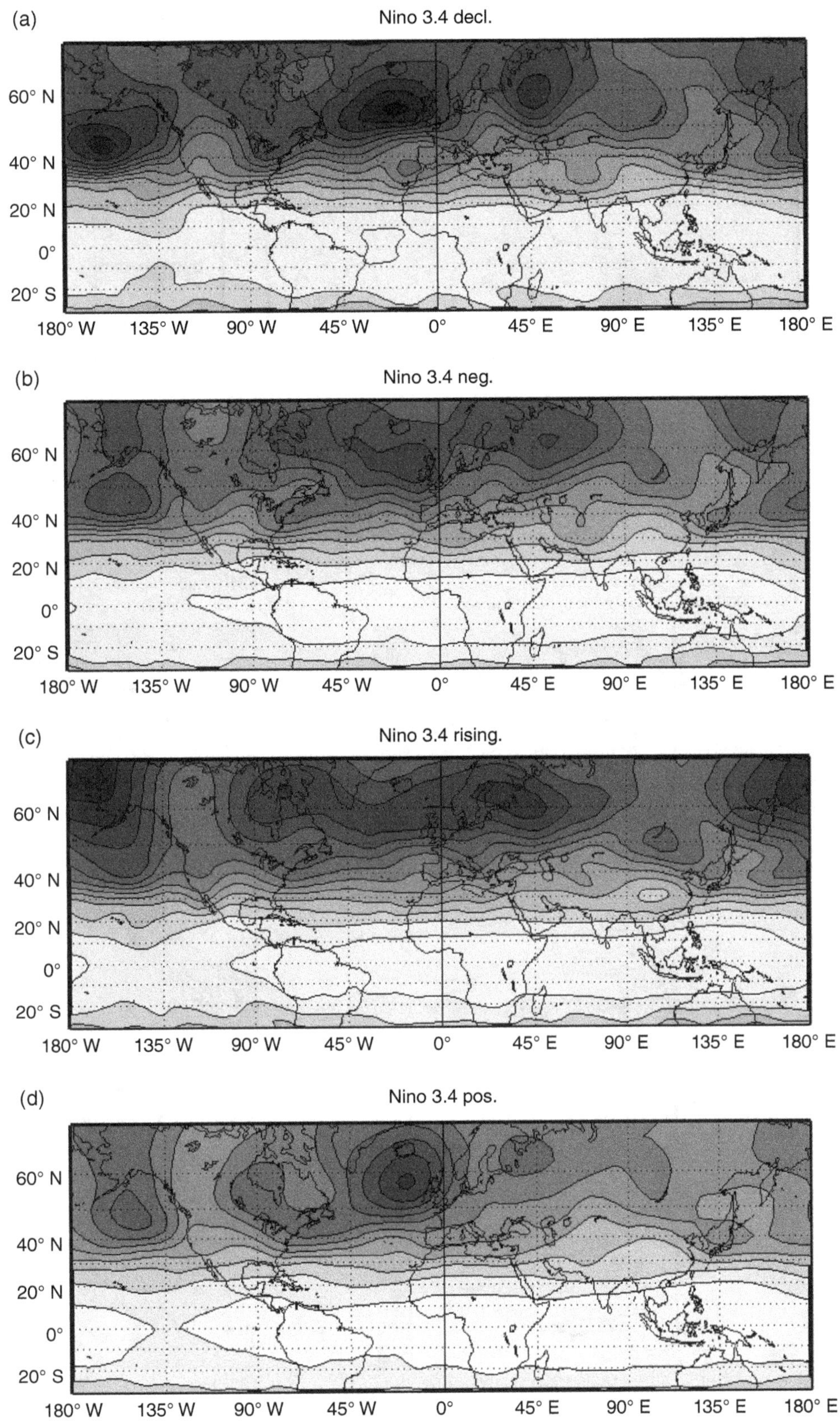

Figure 9.6 Absolute values of composite 300 hPa geopotential height anomalies averaged over each phase of the MJO for (a) cooling, (b) cold, (c) warming, and (d) warm ENSO. The contour interval is 5 m, and maximum values exceed 50 m. From *Roundy et al.* [2010]. ©American Meteorological Society. Used with permission.

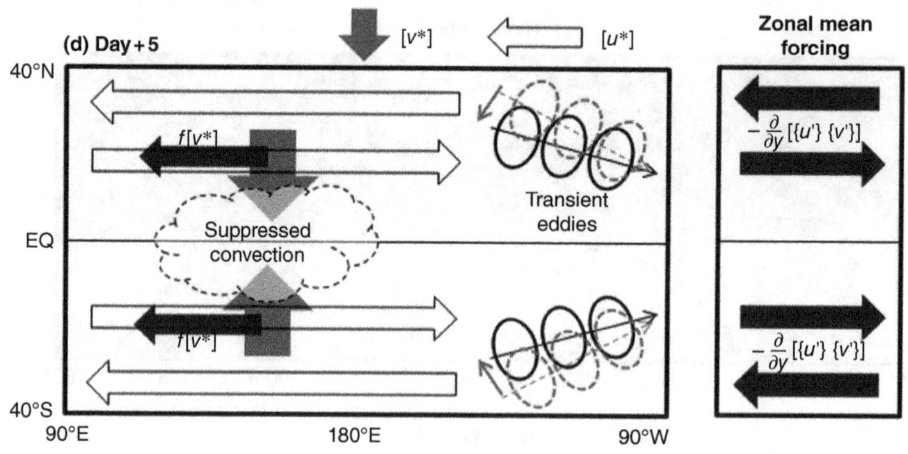

Figure 9.7 Schematics of the mechanism of the intraseasonal modulation of synoptic eddies feeding back onto the upper-tropospheric intraseasonal zonal-mean zonal wind. Circulations are drawn in black and intraseasonal zonal wind accelerations are drawn in dark gray. From *Sakaeda and Roundy* [2014]. ©American Meteorological Society. Used with permission.

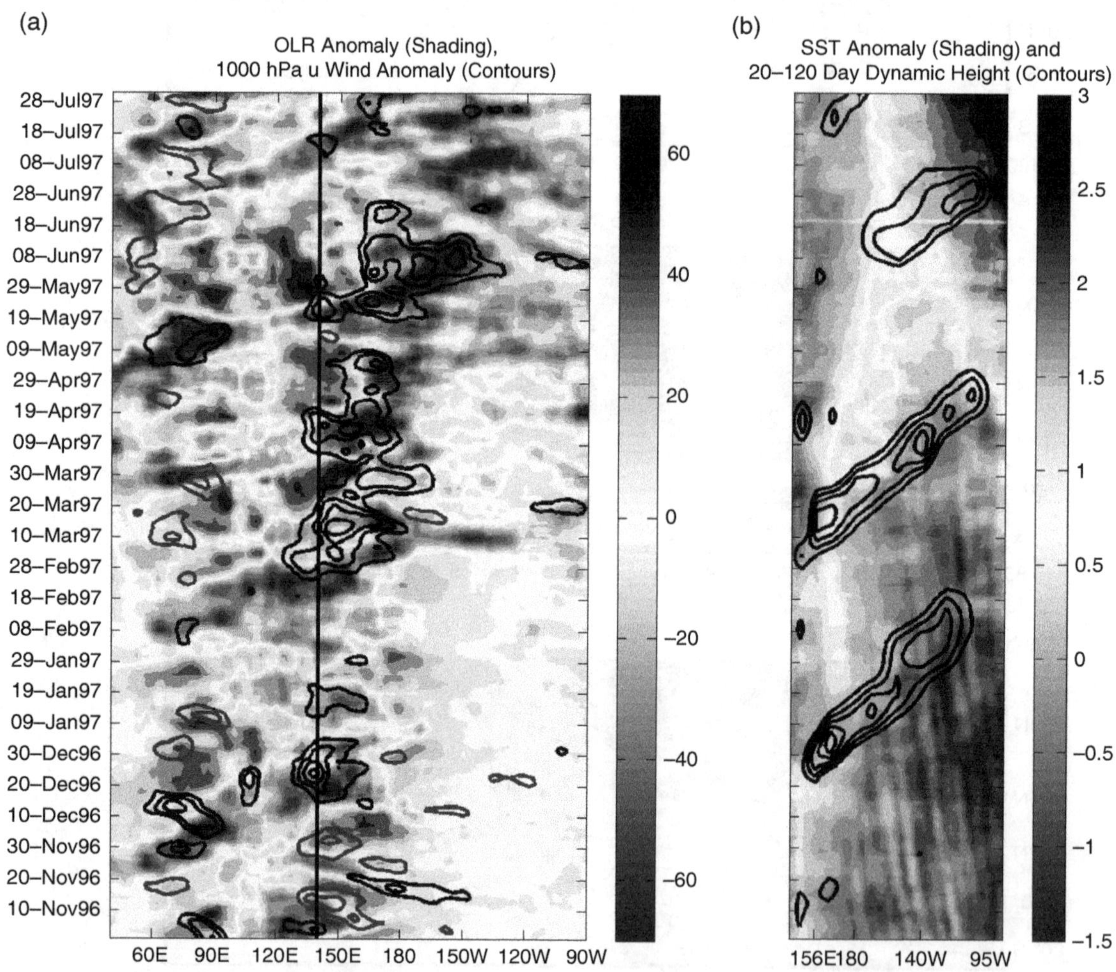

Figure 9.8 (a) Longitude-time diagram of unfiltered OLR anomalies (shading, Wm^{-2}) and zonal wind anomalies (contours, plotted every 2 ms^{-1} with the zero line omitted, westerly anomalies in dark gray), both averaged from 2.5°N to 2.5°S for the development period of the 1997–1998 El Niño. (b) SST anomalies (shading) and 20–120 day band filtered dynamic height anomalies (contours every 2 cm, with only positive anomalies plotted for simplicity) averaged from 2°N to 2°S. *See electronic version for color representation.*

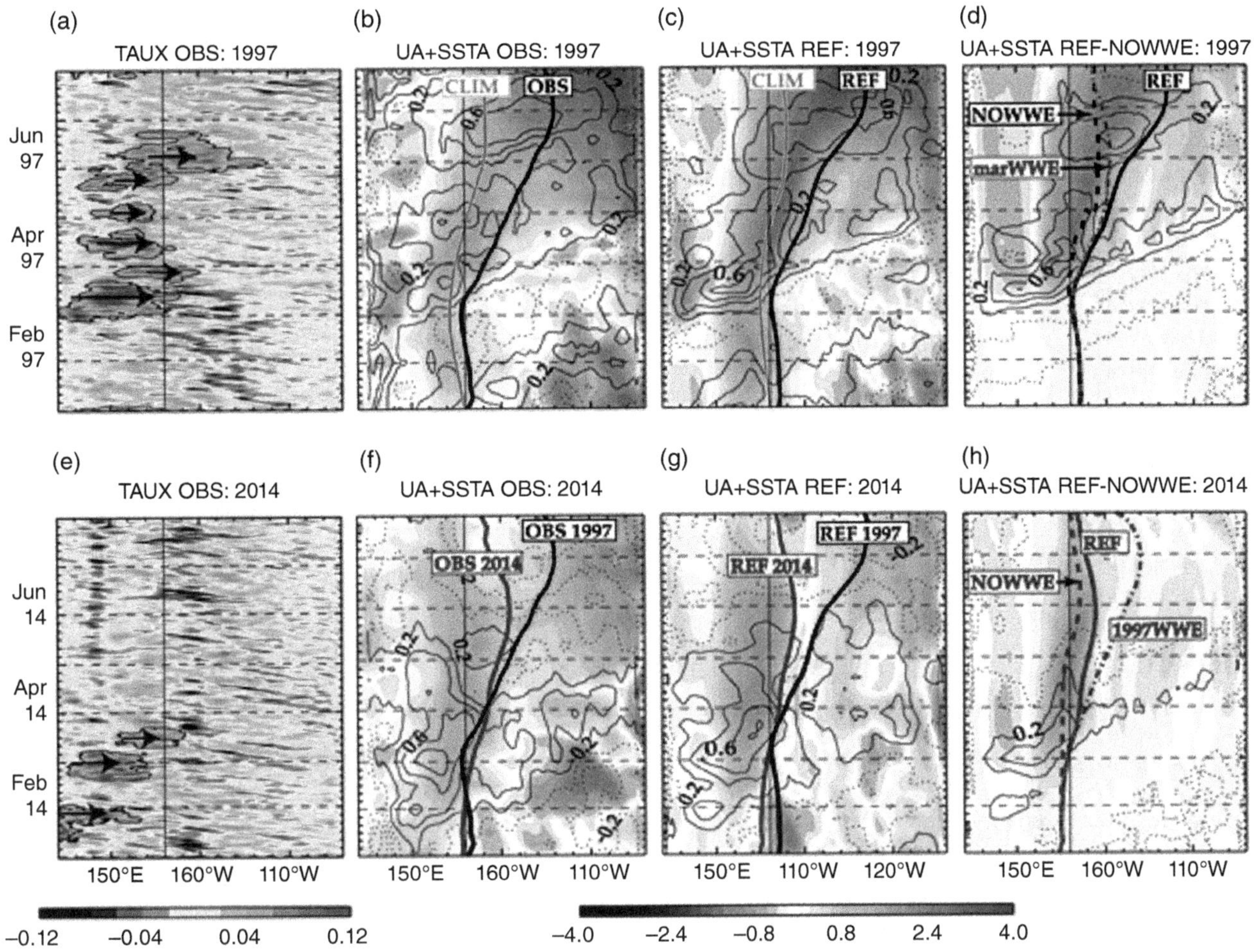

Figure 9.9 January–July 1997 3°N–3°S zonal time section of (a) zonal wind stress referred to as TAUX anomalies (colors with observations [OBS] referring to the NCEP2 model forcing) and WWEs (contours) and SST (color), surface currents (thin black contours) anomalies for (b) observations, (c) REF experiment, and (d) REF minus NO WWE experiments. (e–h) Same as (a–d) but for 2014. Plain light gray lines indicate the EEWP position defined as the longitudinal position of the 28.5°C isotherm in the 3°S–3°N band. Plain grey lines indicate the climatological position of the EEWP. In (d), the dashed light gray (respectively, dash-dotted dark gray) line indicates the EEWP position for the NOWWE (respectively, MarWWE) experiments. In (f) and (g), the EEWP is indicated in blue for 1997 and in red for 2014. In (h), the dashed red (respectively, dash-dotted dark gray) line indicates the EEWP position for the NO WWE (respectively, 1997WWE) experiments. Labels of the experiment EEWP are added for clarity on each panel. Reproduced from *Menkes et al.* [2014]. (*See insert for color representation of the figure.*)

extent of the 28.5C isotherm (as a measure of the eastern edge of the warm pool) if the March westerly wind event did not occur. The "ref" line in the same figure represents the actual movement of the same isotherm as it responded to the March wind event, subsequent wind events, and other terms in the ocean surface heat budget.

Figures 9.9 e and 9.10b show that during late January, late February, March, and mid April 2014, a similar series of intense outbreaks of moist convection and westerly wind bursts developed across the western and central Pacific basin. Atmospheric convection indicated by negative OLR anomalies in gray shading in Figure 9.10a advanced toward the dateline, with each successive event attaining high amplitude east of the previous event. Central Pacific SST anomalies increased markedly following the February event, and east Pacific SST anomalies increased following the March event, yielding an SST anomaly pattern in April similar to that at the same time of year during 1997. Conditions appeared at the time to be consistent with the evolution of the super El Niño event of 1997, but after April, westerly wind events largely ceased, and although SST anomalies and atmospheric convection did develop consistent with weak to moderate El Niño conditions, strong El Niño conditions did not develop. The dashed dark gray line in Figure 9.9h shows the easternmost extent of 28.5°C water in the model experiment of *Menkes et al.* [2014] excluding the westerly wind event in late February and early March.

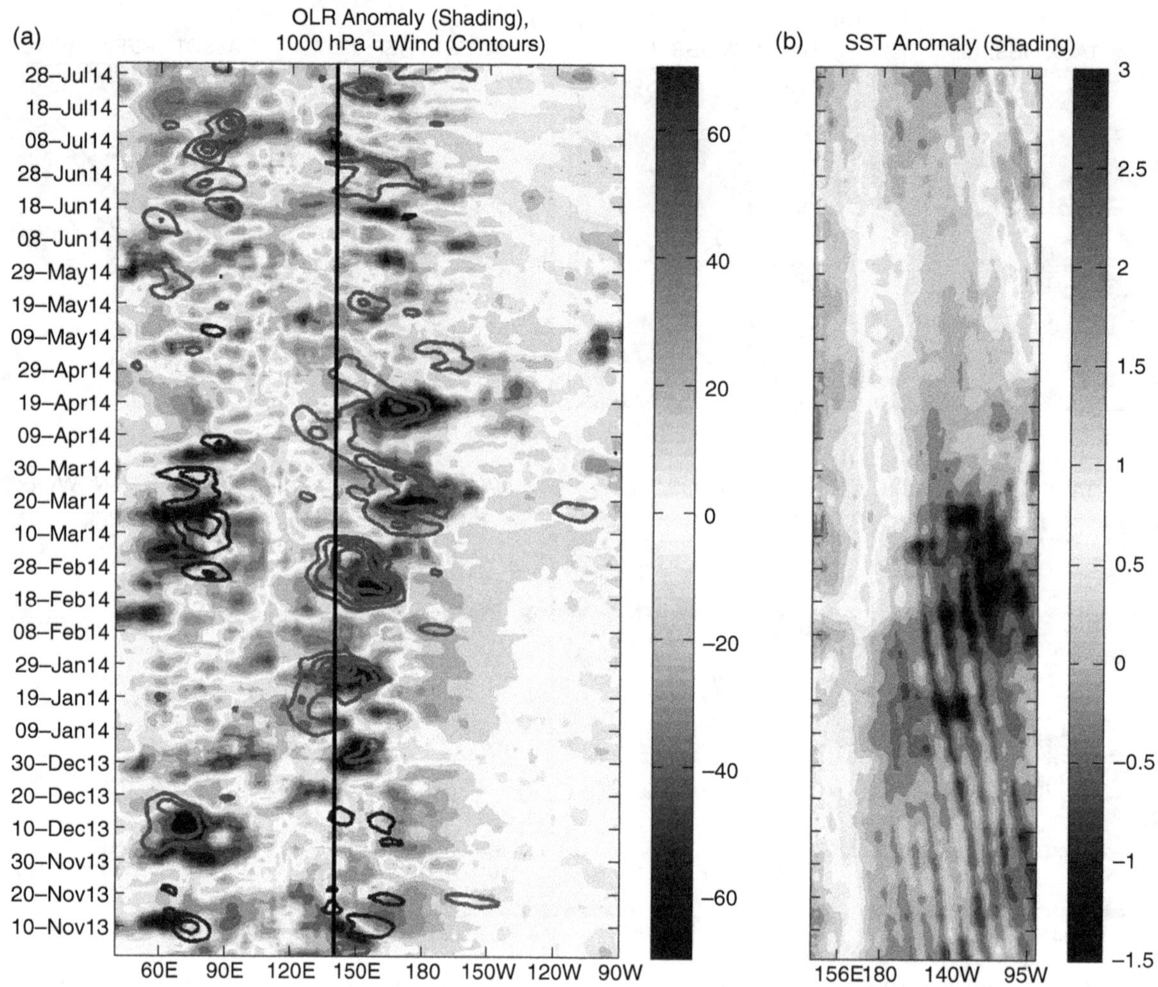

Figure 9.10 Same as Figure 9.8, except for the weaker 2014 ENSO warm event. Dynamic height anomalies are omitted from panel (b) because fewer buoys were maintained during 2014 than during 1997, leading to data gaps.

Although this result suggests that the wind event positively impacted SST, in the absence of further westerly wind events after April, the ocean did not achieve warming outcomes consistent with 1997. Although ENSO specialists are likely to uncover many differences between the events of 1997 and 2014, previous works have shown that continued westerly wind events during 1997 allowed that event to grow [*Lengaigne et al.*, 2004; *Menkes et al.*, 2014], whereas almost complete loss of westerly wind burst activity after April 2014 did not support similar growth [*Menkes et al.*, 2014].

The structures of the SST anomaly patterns during February to April 1997 and February to April 2014 were similar, and most operational coupled ocean atmosphere forecast models developed strong El Niño conditions across spring 2014 when initialized in early spring. To the extent that the comparison between 1997 and 2014 was valid for February through April, and to the extent that these coupled ocean atmosphere models simulate well the observed climate system, these points suggest that the coupled system supported growth of strong El Niño conditions in 2014. Yet, westerly wind stress events in 2014 largely ended in early April. *Vecchi et al.* [2006] and *Eisenman et al.* [2005] argued that westerly wind events might be part of the coupled ocean atmospheric climate system, forced by continuously evolving sea-surface temperature. *Vecchi et al.* [2006] showed that an atmospheric model forced by observed SST during early 1997 without initializing MJO signals arriving from the west developed westerly wind bursts with similar timing to those observed. Yet, each of the westerly wind events was also observed to be associated with an MJO convective event that began roughly a month earlier over the western Indian basin. The proposed forcing of western and central Pacific wind stress anomalies by increasing sea surface temperature clearly could not have worked backward in time to excite the Indian basin MJO events, so the only reasonable conclusion is that these MJO events influenced the observed SST signal, which then fed back on the model atmospheric circulations to extend westerly wind bursts in the coupled system with the correct general timing. Forcing of the ocean in a primed condition by internal atmospheric variability not controlled, but modulated, by ENSO apparently initiates a coupled response that behaves differently in different ENSO states [e.g., *Roundy and Kiladis*, 2006; *Roundy and Kravitz*, 2009]. In other words, whenever atmospheric internal variability leads to MJO events, ENSO influences their behaviors in ways that can feed back on ENSO, but the sea-surface temperature state of ENSO alone does not guarantee their occurrence, suggesting that in spite of its influence on the weather, ENSO responds to semistochastic variability in the atmosphere in ways that would ultimately reduce ENSO predictability. Initialization of ocean models with atmospheric forcing up to 1 January 1997 produced an El Niño event in the subsequent months, but not as strong as other models that included the MJO/WWB events of February through May 1996 [e.g., *Shi et al.*, 2010]. The trajectory of individual El Niño events must therefore depend on the population of individual MJO and westerly wind events that interact with it. Apparently, two MJO cycles lacking substantial westerly wind bursts over the western and central Pacific Ocean during May and June 2014 led to a weak El Niño outcome in 2014, but continued extensive intraseasonal forcing during 1997 led to the strongest El Niño event of the last century (*Menkes et al.*, 2014).

So, why did intraseasonal westerly wind events nearly cease in May 2014 and yet continue in 1997? Most authors suggest that the MJO originates as internal variability of the atmosphere, but since it interacts with weather events (such as tropical cyclones and breaking Rossby waves) and with the evolving oceanic and atmospheric climate base states through which it propagates, many factors might explain changes in its behavior. We suggest one possible explanation to encourage further research. Recently, *Roundy* [2014] and *Sakaeda and Roundy* [2015] demonstrated that MJO convection and circulation become stronger over the Indian Ocean following periods of upper tropospheric easterly wind on the equator over the Western Hemisphere and Africa. Weaker than average MJO events tend to follow periods of westerly winds. The intraseasonal zonal wind dominated the assessments of *Roundy* [2014] and *Sakaeda and Roundy* [2015]. Yet, the different structures in the evolving base state winds associated with the adjustment toward El Niño might have impacted intraseasonal variability differently during 2014 than during 1997. We begin with the hypotheses that background winds associated with the evolving El Niño events enhanced MJO activity over the Pacific Ocean during 1997 (thereby increasing westerly wind burst activity) and reduced similar activity during 2014. Analysis of an index of the MJO along with upper tropospheric zonal wind data reveals the level of consistency of observations and reanalysis data with this hypothesis. Although this type of observational analysis cannot prove a hypothesis, if the observations are inconsistent without clear explanation, it must be rejected.

The seasonal cycle, including the primary and its first four harmonics, was removed from daily 200 hPa zonal wind data from the NCEP/NCAR reanalysis (based on the 1974–2012 climatology). These data were averaged from 5°N to 5°S and from 90°W to 50°E to generate an index. Results are not sensitive to changes in the width of this region. A 61 day centered moving average was applied to remove intraseasonal variability. Averaging this index from May through June 1997 yields 2 m s^{-1} and averaging over the same months in 2014 yields 3.8 ms^{-1}. Only 7% of the dates in this time series exceed 3.4 ms^{-1}, and 22% of values land between 1 and 3 ms^{-1}. Observations would not disprove this hypothesis if the set of past MJO events when the wind index was near the values in 1997 and 2014 behaved in similar manners to the events those years.

The real time multivariate MJO (RMM) index of *Wheeler and Hendon* [2004] was applied as an approximate metric of the amplitude and location of MJO convection and lower tropospheric westerly wind. Other indices such as the OLR-MJO index (OMI) of *Kiladis et al.* [2014] yield similar results, though the RMM index explicitly includes wind signals that are relevant to this analysis. The RMM index approximates the state of the MJO by breaking it into eight phases that loosely relate to the longitudes of the centers of regions of active MJO convection and associated circulation. For example, phases 2–3 suggest convection and equatorial westerly wind in the lower troposphere over the Indian Ocean, 4–5 over the maritime continent, 6–7 over the Western and Central Pacific basin, and phases 8–1 over the Western Hemisphere and Africa. Phases 2–5 favor easterly wind stress anomalies on the equatorial central Pacific Ocean and phases 6–8 favor westerly stress anomalies. In general, the amplitude of the RMM index varies across the phases, and the amount of time spent in each phase also tends to change with the base state through which the MJO propagates. A measure of the relative cumulative impacts of different RMM phases on the environment during a given period of time would therefore depend on the amplitude of the signal and on the amount of time spent in the different phases. Thus, during a period of time sufficiently long to sample a given phase, a rough metric of the cumulative impact of the MJO on its environment in that phase is the product of average amplitude in that phase and the number of days spent in that phase divided by the total number of days in the sample. Any period of time substantially longer than one MJO cycle would provide comparison of the relative integrated impacts of each phase.

By this metric, during May–June 2014, phases 6–8 were 40% as active as phases 2–4. In contrast, during the same period in 1997, phases 6–8 were 50% more active than phases 2–4. This distribution is consistent with the relative lack of westerly wind events during 2014 and with enhanced westerly wind during 1997. How did MJO activity during these periods compare with activity during past events with upper tropospheric wind index values in similar ranges? To construct a large sample, data from all dates April through July from June 1974 through 2012 were analyzed, and the above MJO-impact metric was calculated from the RMM index for periods of time when the smoothed wind index was between 1.5 and 2.5 ms^{-1} (similar to 1997) and from 3–4 ms^{-1} (similar to 2014). Inclusion of 1997 does not significantly change results, and 2014 was excluded. Variations on these ranges were tested, leading to similar general conclusions. The solid red curve on Figure 9.11 shows the result for wind states similar to 2014. Activity during phases 3–4 is substantially and statistically significantly greater than for phases 7–8. In contrast, results for states similar to 1997 (blue curve) show enhanced activity in phases 7–8 and reduced activity during phases 3–4. Further analysis of zonal wind anomalies during MJO events with RMM amplitude in excess of 1 sigma for May through June suggests that phases 3–4 tend to be associated with slightly negative surface zonal wind anomalies averaged from 2.5°N to 2.5°S and from 160°E to 160°W during periods of time when Western Hemisphere upper tropospheric wind conditions were similar to those of 1997 and 2014, while conditions like 2014 were associated with stronger westerly wind anomalies during phase 7 than for conditions similar to those of 1997. Yet, such phase 7 events in conditions similar to 2014 tend to occur less than half as often as the weak anomalies of phase 3 or phase 4, implying that past periods of time during May and June with strong background upper tropospheric Western Hemisphere westerly wind anomalies like those in 2014 do not favor any more than brief periods of strong surface westerly winds over the central Pacific Ocean with more than twice as much time spent in weak westerly wind states or in easterly wind states. Results therefore do not refute the hypothesis that the background upper tropospheric equatorial westerly wind from near South America east across Africa influenced the evolution of the MJO differently during 2014 than during 1997 in a way that would have helped amplify the 1997 El Niño event but not the 2014 event. Further testing, including

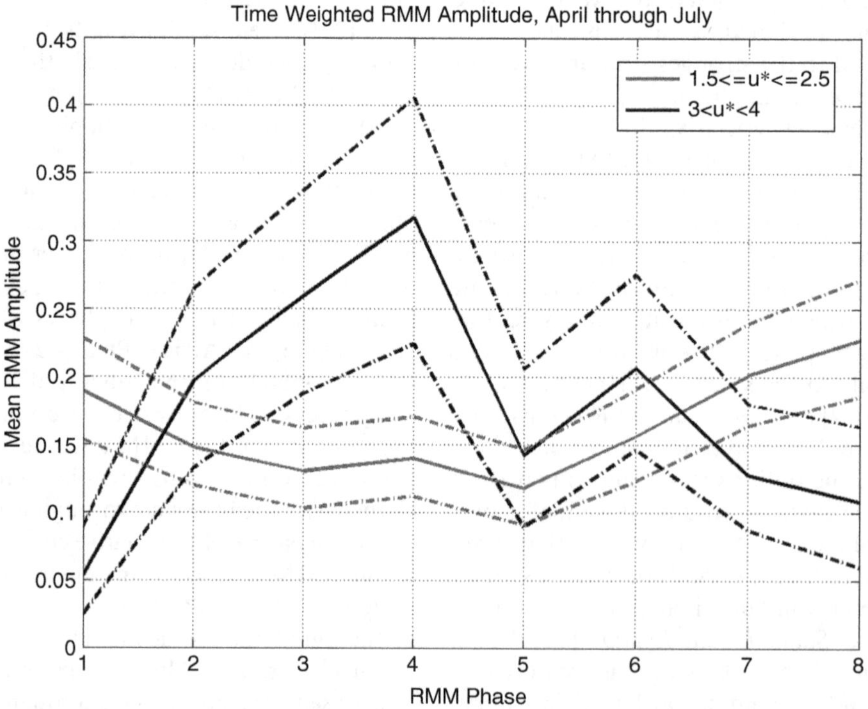

Figure 9.11 Time integrated RMM index amplitude as a function of phase number, weighted by the number of days spent in each phase divided by the number of days in the April through July sample from June 1974–2012. The solid light gray and dark gray curves denote results for periods of time when the smoothed Western Hemisphere/Africa zonal wind index was between 1.5 and 2.5 ms^{-1} and between 2.5 and 3.4 ms^{-1}. Dashed curves outline the bootstrap 90% confidence intervals based on the same calculation obtained from random resampling, with replacement, from the entire list of dates without regard for phase or the wind index.

model experiments, is necessary to better understand these findings. Nevertheless, these results suggest that the different atmospheric circulation structures associated with the developing 2014 and 1997 El Niño events, including their impacts on upper tropospheric zonal wind, may have altered the behavior of the MJO in ways that would have amplified the 1997 event to extreme levels and resisted growth of the 2014 event.

This relationship between ENSO and the weather suggests that simulation of the statistics of ENSO including its extreme events requires that coupled ocean atmosphere global climate models simulate the evolving spatial structures and punctuated variance of the types of weather events that disrupt or advance ENSO. Other authors suggest that these interactions between ENSO and the weather influence not only the amplitudes of El Niño events, but also how their structures evolve with time, including whether they favor the central or eastern Pacific basin [*Shi et al.*, 2010; *Gushchina and Dewitte*, 2012].

9.4. INTERACTIONS BETWEEN THE MJO AND BREAKING WAVES

The relationship between the MJO and Rossby wave breaking is an important component of intraseasonal variability that connects the midlatitude atmospheric circulation with convection and circulation in the tropics. Previous studies suggest that the variability of the subtropical jets and the generation of the NAO teleconnection patterns associated with the MJO are partly driven by the changes in the location and frequency of midlatitude Rossby wave breaking associated with the MJO (*Cassau*, 2008; *Lin et al.*, 2009; *Moore et al.*, 2010). The modulation of the location and frequency of midlatitude Rossby wave breaking associated with the MJO is an important nonlinear phenomenon that helps determine how the MJO interacts with the midlatitude circulation and extreme weather events. *McIntyre and Palmer* [1984] describe wave breaking as an "irreversible deformation of material contours," a definition that has permeated modern literature [e.g., *Holton*, 2004; *Moore et al.*, 2010, etc.]. Rossby waves, which exist because of the meridional gradient of Ertel's potential vorticity (PV), often break at the ends of jet streams where PV gradients relax. The background horizontal wind shear then influences whether the wave breaks anticyclonically or cyclonically [e.g., *Thorncroft et al.*, 1993], which correspond with the lifecycles 1 and 2 definitions, respectively, of *Thorncroft et al.* [1993], in patterns like those labeled LC1 and LC2 on the maps in Figure 9.12.

The relationship between the MJO and AWB events is documented by *Moore et al.* [2010] and by *MacRitchie and Roundy* [2015], hereafter MR15. *Moore et al.* found that an upper-tropospheric anticyclone develops to the northwest of the area of tropical convection associated with the MJO as it approaches the maritime continent. They further suggested that this ridge strengthens and shifts the Pacific subtropical jet stream poleward, which primes the atmosphere for an AWB event farther east over the central Pacific Ocean. This event often leads to a subsequent cyclonic wave break over the northeast Pacific (Fig. 9.12) that also extends the subtropical jet eastward.

MR15 explored this idea further. Using the OMI MJO index from *Kiladis et al.* [2014], MR15 determined the anomalous percentage of AWB events per day in each OMI phase during the extended winters of October–March from 1979 to 2010. The results are shown in Figure 9.13.

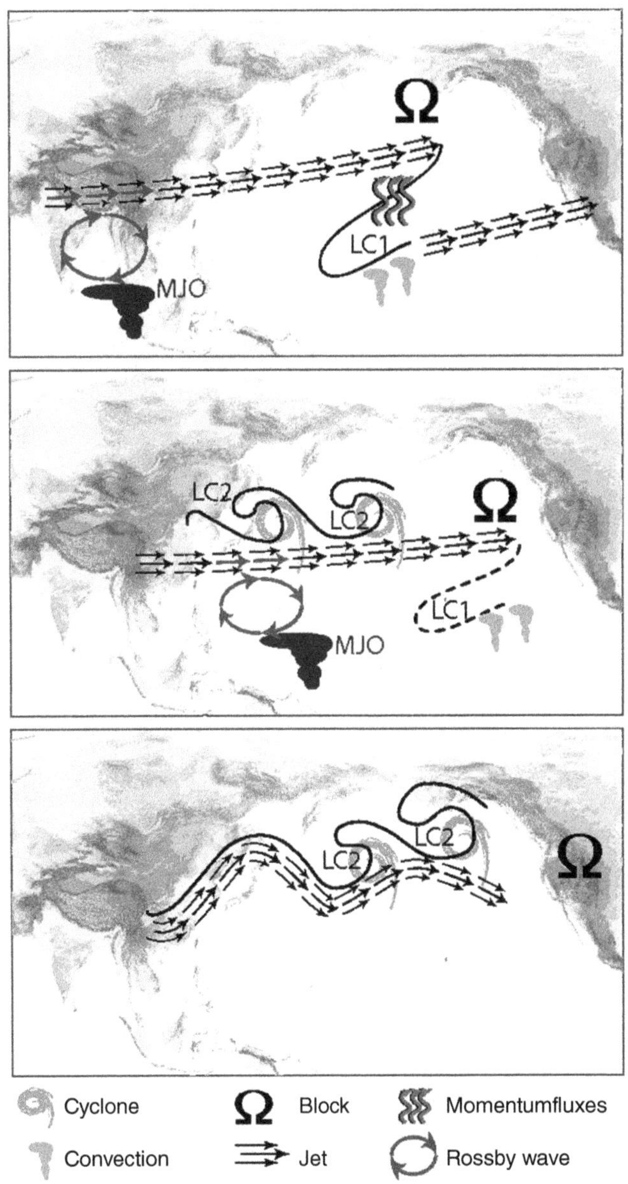

Figure 9.12 Figure 9 from *Moore et al.* [2010] showing a schematic of the way in which the MJO induces AWB. ©American Meteorological Society. Used with permission.

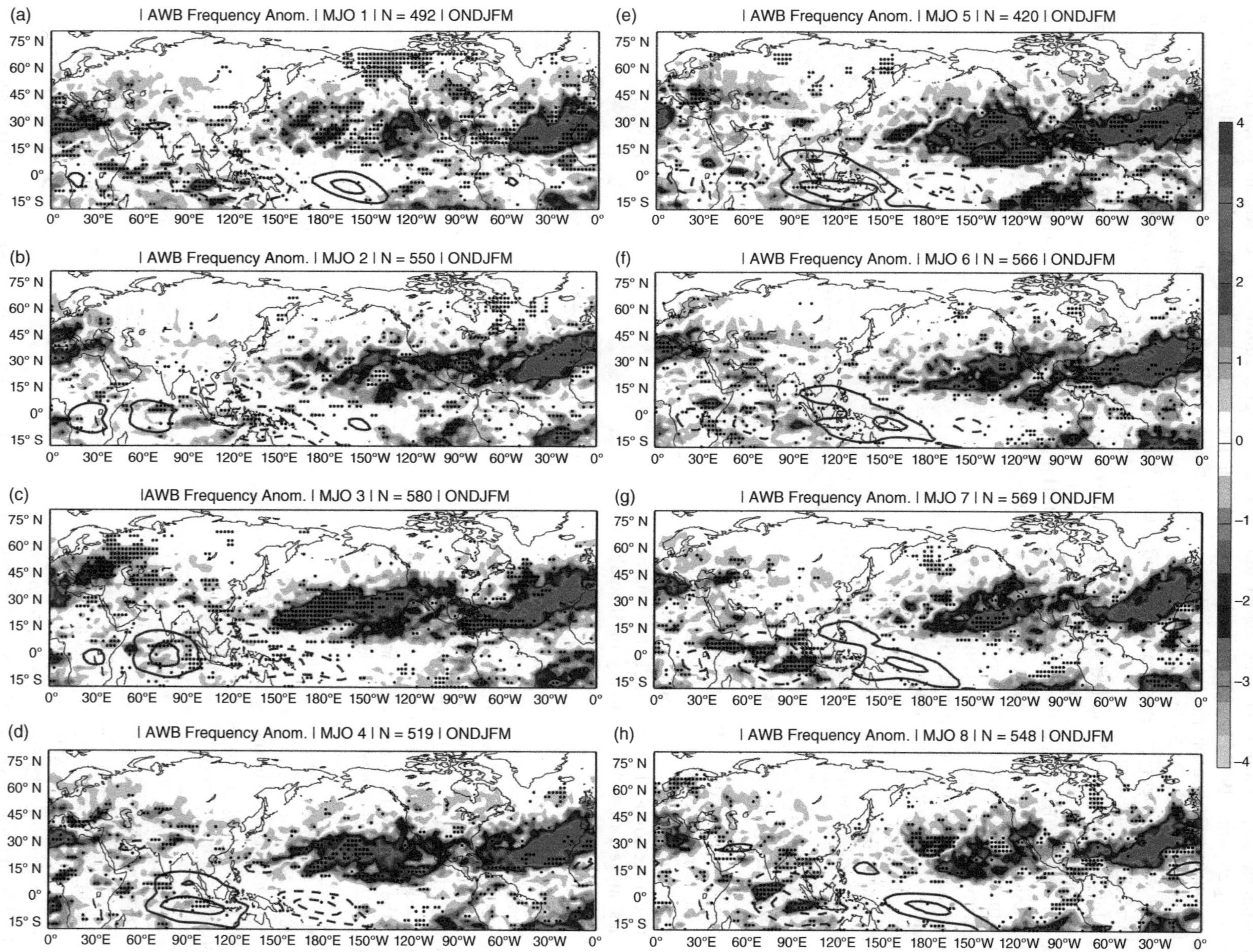

Figure 9.13 Composite during specified MJO phase and events that occur from October–March. Combined 330K and 350K AWB frequency departure from climatology (shaded) according to color bar in units of % day-1. Solid (dashed) red lines indicate areas of negative (positive) MJO filtered OLR starting at negative (positive) 6 W m-2 and contoured by 6 W m-2. Black dots indicate areas where the shaded values are statistically significantly different from climatology at the 90% level using a t-test. (*See insert for color representation of the figure.*)

Key results from this figure include largely suppressed AWB activity over much of the central Pacific, the US/Canada border, and the northern Atlantic during phases 1 and 8. During the middle phases, there is, broadly, enhanced AWB activity over the central and eastern Pacific as well as the North Atlantic oceans. The enhanced AWB activity during the middle phases, when MJO convection is most active approaching the maritime continent, suggests a link between the two features.

To explore this link further, MR15 partitioned the aforementioned MJO climatology into events in which AWB occurred during phase 3 over the central Pacific in a box with bounds 10°N–30°N and 120°E–120°W and events in which AWB did not occur there during phase 3. This analysis only included events that began in phases 1 or 2 and maintained amplitude of 0.75 continuously through phase 8. Composites over the two sets of MJO events are shown in Figure 9.14.

The left column of Figure 9.14 shows the progression of MJO events with AWB in the specified box through the odd numbered phases. The right column shows the same progression for the MJO events without AWB in the box. The events in the left-hand column begin (panel a) with weak active convection over the Indian Ocean and an area of suppressed convection over the maritime continent. The active convection intensifies as the OMI index progresses into phase 3 and an anomalous ridge (denoted by the letter A) occurs to the north of the active convection. Downstream, an anomalous ridge and an anomalous trough to its south occur over the central Pacific, which is the signature of the beginnings of a wave break. The wave continues to break through phase 5, leaving behind an anomalous trough over the north central Pacific, just east of the dateline (the breaking wave is indicated by the diagonal arrow). This trough reinforces or becomes the upper tropospheric cyclonic gyre commonly associated with the suppressed convective phase of the MJO. This gyre (denoted by the letter C) is absent during phase 5 in the events without wave breaking. Anomalous equatorial westerly wind anomalies south of the trough strengthen through phase 7. This enhancement in upper-level winds is collocated with a region of active convection on the east side of the trough anomaly that propagates across the central and eastern Pacific.

Weaker convection develops in MJO events without AWB (right-hand column) in all phases except 1. Indeed, MR15 showed that the amplitude of the MJO convective events in the left-hand column is statistically significantly greater than the amplitude of the events in the right-hand column in all phases except 1. This result suggests that already strong MJO convective events over the eastern Indian basin are more likely to initiate AWB over the north central Pacific basin through ridge amplification early in their life cycles, which later apparently contributes constructively to the MJO convection (including both its suppressed and active convective phases) and the associated atmospheric circulation. These results suggest coupling between synoptic and intraseasonal weather in the midlatitudes and intraseasonal variability of the tropics.

9.5. INTERACTIONS BETWEEN THE MJO, TROPICAL CYCLONES, AND THE EXTRATROPICAL CIRCULATION

The MJO and tropical cyclones are important contributors to the evolution of anomalous global circulation patterns. Numerous studies have sought to address relationships between organized intraseasonal (e.g., MJO) convection and the modulation of large-scale circulation patterns, particularly over the Northern Hemisphere extratropics [e.g., *Feranti et al.*, 1990; *Vecchi and Bond,* 2004; *Lin and Brunet*, 2009; *Vitart and Molenti*, 2010; *Gloeckler and Roundy*, 2013; etc.]. Similarly, associations between recurving tropical cyclones and the subsequent evolution of extratropical circulation patterns have been the subjects of several publications [e.g., *Anwender et al.*, 2008; *Anwender et al.*, 2010; *Cordeira and Bosart,* 2010; *Archambault et al.*, 2013; etc.]. Tropical cyclone activity is generally not independent from the MJO, as the MJO modulates the large-scale environment in which tropical cyclones organize and mature [e.g., *Hall et al.*, 2001; *Frank and Roundy*, 2006; *Barrett and Leslie*, 2009; *Camargo et al.*, 2009; *Klotzbach and Oliver*, 2014]. Thus, large-scale circulation outcomes associated with the MJO are often influenced by the presence or absence of tropical cyclones at particular geographical locations. Recent work suggests that tropical cyclones occasionally modulate not only the extratropical response to MJO convection, but also the structure of the circulation associated with the MJO itself.

To illustrate the relationships between the MJO, tropical cyclones, and the extratropical circulation, the RMM index discussed in section 9.3 was utilized to identify days in which the MJO was likely to be active (i.e., days in which the index exceeded 1 standard deviation) in each of eight phases during boreal fall (SON) between the years 1979 and 2010. Phase 6 (hereafter RMM 6) was selected for further investigation because tropical cyclones tend to be most active during SON over the West Pacific warm pool [e.g., *Frank and Roundy*, 2006], and MJO convection is also most active over this region during RMM 6. Tropical cyclone density for all active RMM 6 days during SON 1979–2010 (not shown) includes a maximum over the western Pacific Ocean, extending eastward from the South China Sea to about 150°E. A box bounded by 10°N and 25°N, and 120°E and 150°E (hereafter the West Pacific domain) was used to identify days in which

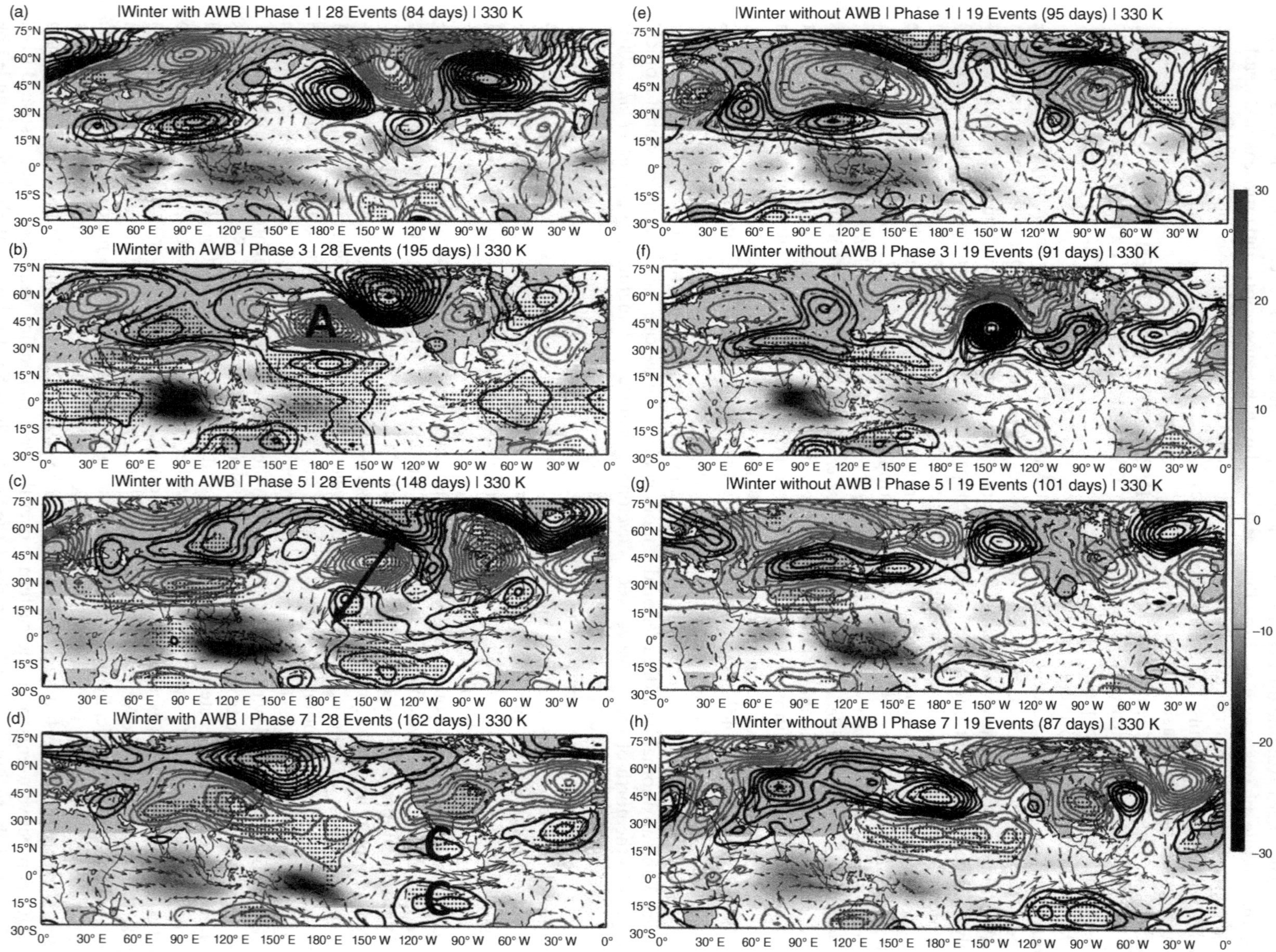

Figure 9.14 (Left column) Composites of MJO events that have AWB within a box bounded by 10°N – 30°N and 120°E – 120°W during phase 3. (Right column) Composites of MJO events that do not have AWB within the aforementioned box. Shading represents MJO filtered OLR in W m-2 according to the color bar. Positive (negative) 250 hPa geopotential height anomalies, contoured with solid (dashed) lines, are contoured every 15 m beginning at positive (negative) 15 m. Vectors are 250 hPa anomalous winds and are scaled to be comparable between panels. The maximum winds in each panel are around 10 ms-1. Dotted regions, only plotted north of 15°N, indicate areas in which the geopotential height pattern between each pair of figures is statistically significantly different at the 90% level.

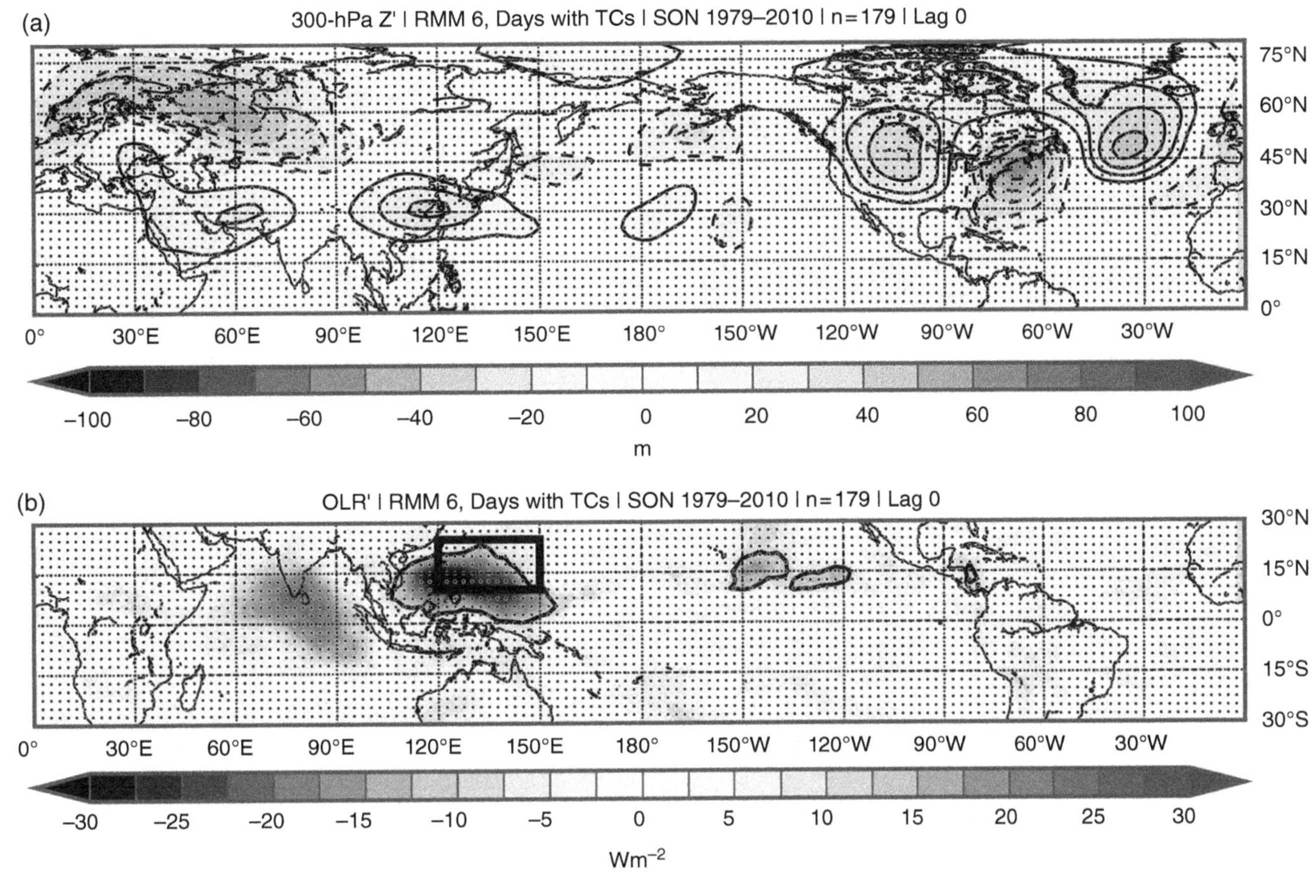

Figure 9.15 Composite 300 hPa geopotential height (top) and OLR (bottom) anomalies for all days during RMM 6 in which a tropical cyclone was present in the West Pacific domain (boxed region in the bottom panel). Contours of negative geopotential height anomalies are dashed. Negative OLR anomalies are enclosed in a solid contour. Anomalies not stippled are statistically significantly different from those composited over the entire RMM 6 sample at the 95% level.

tropical cyclones were present or absent in the vicinity of the RMM 6 tropical cyclone density maximum. These lists, identified as the tropical cyclone and non- tropical-cyclone composite subsets, include 178 days and 53 days, respectively.

Figures 9.15 and 9.16 show 300 hPa geopotential height anomalies (top panel) and OLR anomalies (bottom panel) composited over tropical cyclone and non- tropical-cyclone days, respectively, corresponding to the West Pacific domain. Anomalies that are not stippled are statistically significantly different at the 95% level from those composited over the entire sample of active RMM 6 days (not shown) using a two-sample *t*-test. Figure 9.16 suggests that neither OLR nor 300 hPa geopotential height anomalies belonging to the tropical cyclone composite subset are statistically distinguishable from anomalies composited over the entire sample of active RMM 6 days. On the other hand, the absence of negative OLR anomalies east of the Philippines in the non- tropical-cyclone composite subset (Fig. 9.16, bottom panel) represents a statistically significant difference in the convective pattern relative to the full sample composite. The relative minimum in convection suggested by the shading extends northwestward into the West Pacific domain. Thus, the lack of tropical cyclones east of the Philippines during RMM 6 is associated with a change in the structure of the MJO convective envelope. Moreover, the change in the structure of MJO convection contributes to changes in the evolution of extratropical circulation anomaly patterns, as evidenced in the top panel in Figure 9.16.

So, what might contribute to tropical cyclone activity, or lack thereof, coincident with MJO convection during RMM 6? Figure 9.17 shows remnant OLR and 850 hPa wind anomalies following the removal of MJO-filtered OLR and 850 hPa wind anomaly fields. The top and bottom panels correspond to the tropical cyclone and non-tropical-cyclone composite subsets, respectively, and occur 4 days prior to the arrival of the MJO in RMM 6. The top panel elucidates a broad cyclonic gyre centered on the West Pacific domain, accompanied by active

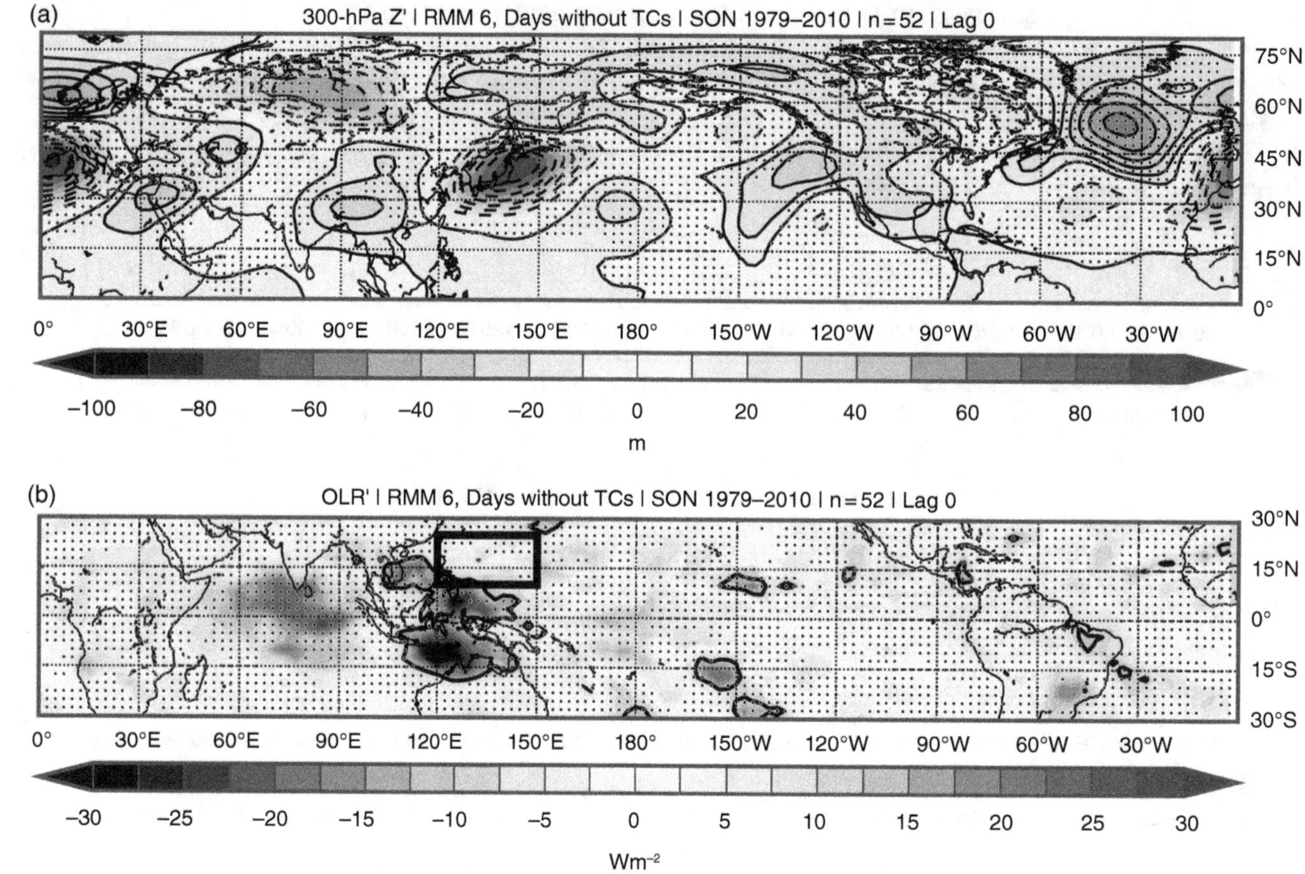

Figure 9.16 As in Figure 9.15, but for non-tropical-cyclone days during RMM 6.

convection along the southern and eastern flanks of the gyre, in the vicinity of enhanced low-level convergent flow. While not shown, this gyre formed near 150°E 6 days earlier and exhibited eastward and poleward propagation over the course of 15–20 days. Additionally, a less coherent westward-propagating cyclonic circulation south of the equator accompanied its Northern Hemisphere counterpart, suggesting the presence of an equatorial Rossby or mixed Rossby-gravity wave [e.g., Kiladis *et al.*, 2009]. This westward-propagating wave likely enhanced tropical cyclone activity over the West Pacific domain as it intersected the MJO convective envelope [e.g., *Frank and Roundy*, 2006; *Schreck et al.*, 2011; etc.].

In contrast to the top panel in Figure 9.17, the bottom panel suggests an opposite-signed circulation anomaly accompanying suppressed convection (yellow shades) over the West Pacific domain. These OLR and wind anomaly patterns suggest the intersection of MJO convection with the convectively suppressed phase of an equatorial Rossby or mixed Rossby gravity wave, which likely inhibited tropical cyclone activity during RMM 6. While the presence or absence of these waves affects the structure of the MJO convective envelope, differences in the evolution of the extratropical circulation related to MJO convection that preceded the organization of these waves might have contributed to timing differences in their intersection with MJO convection, therefore modulating tropical cyclones activity over the West Pacific domain. Thus, the MJO and the extratropical circulation work in tandem to influence the evolution of the other.

As suggested throughout this section, differences in the structure of the MJO convective envelope contribute to differences in the evolution of extratropical circulation patterns. Figure 9.18 shows time-longitude sections of 300 hPa geopotential height anomalies corresponding to the tropical cyclone (left) and non- tropical-cyclone (right) composite subsets. Anomalies in both panels are averaged between 30°N and 50°N, and the vertical black lines mark the longitudinal boundaries of a box drawn over the eastern North Pacific Ocean, North America, and the western North Atlantic Ocean (hereafter the North America domain). This box will be used to compare the evolution of circulation patterns associated with the two composite subsets. Prior to lag −10 days (ten days before the arrival of the MJO in RMM 6), height

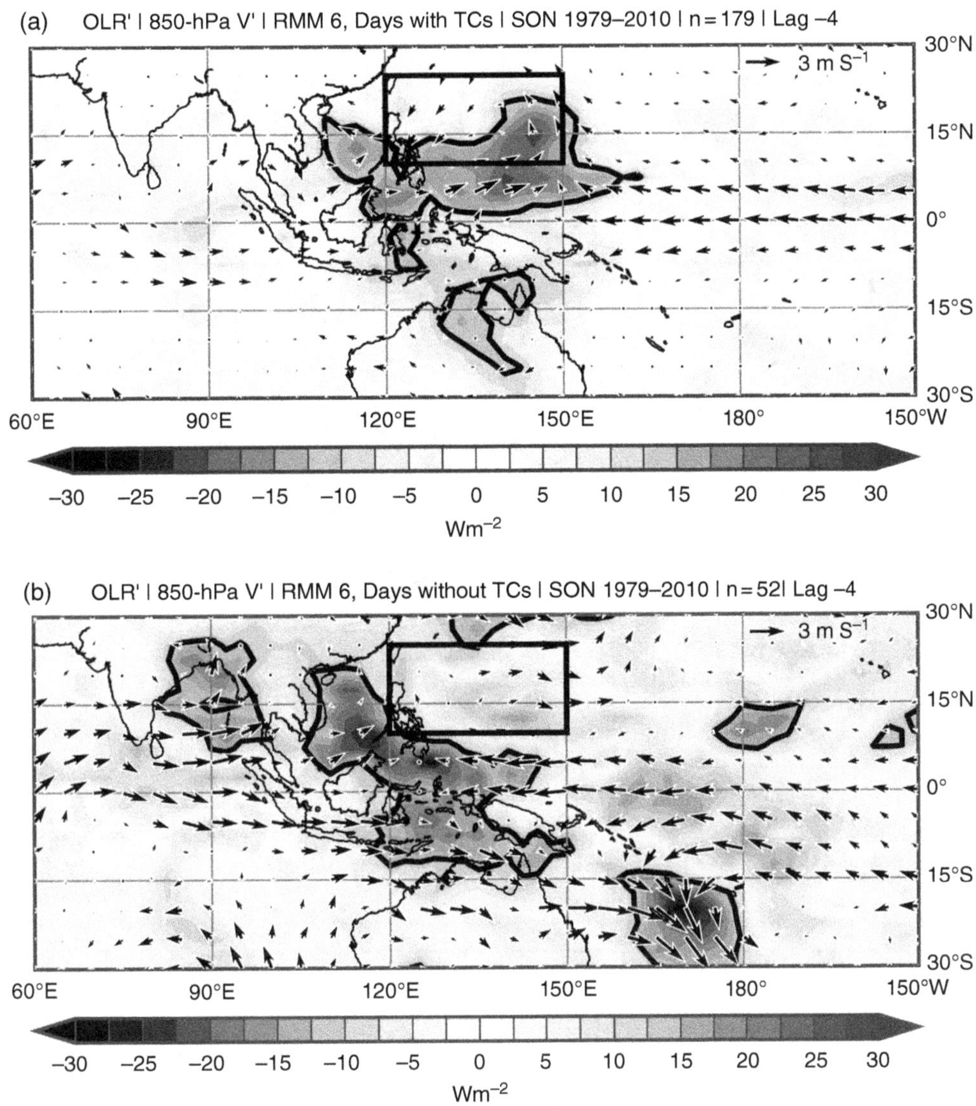

Figure 9.17 OLR and 850 hPa wind anomalies corresponding to (top) tropical cyclone and (bottom) non-tropical-cyclone composite subsets during RMM 6. A solid contour encloses negative OLR anomalies. Anomalies were computed by subtracting MJO-filtered OLR, u- and v-wind anomalies from their respective unfiltered anomaly fields. A reference vector is included in the top right of each panel.

anomalies in the tropical cyclone composite subset are up to 90° out of phase with those in the non- tropical-cyclone composite subset across parts of the North America domain, but a well-defined Rossby wave train (alternating red and blue shades) is apparent in both panels. Between lags −10 days and −5 days, coincident with the eastward propagation of MJO convection across the eastern Indian Ocean and maritime continent (not shown), height anomalies in the tropical cyclone composite subset become more in phase with those in the non-tropical-cyclone composite subset, although amplitude differences exist between the two composite subsets. It is at these lags that the composite height anomalies achieve their greatest statistically significant spatial correlations of 0.5 to 0.8 across the North America domain, followed by a rapid decorrelation of the height anomaly fields at subsequent lags.

Differences in the evolution of height anomaly patterns across the North America domain following lag −5 days occur in conjunction with increasing differences in the structure of MJO convection in both composite subsets. Organized convection over the western North Pacific Ocean in the tropical cyclone composite subset contributes to the amplification of a Rossby wave train that extends across the North America domain through lag +5 days. This Rossby wave response can be maintained in part because of the presence of large-scale, organized convection over the western Pacific Ocean. In the

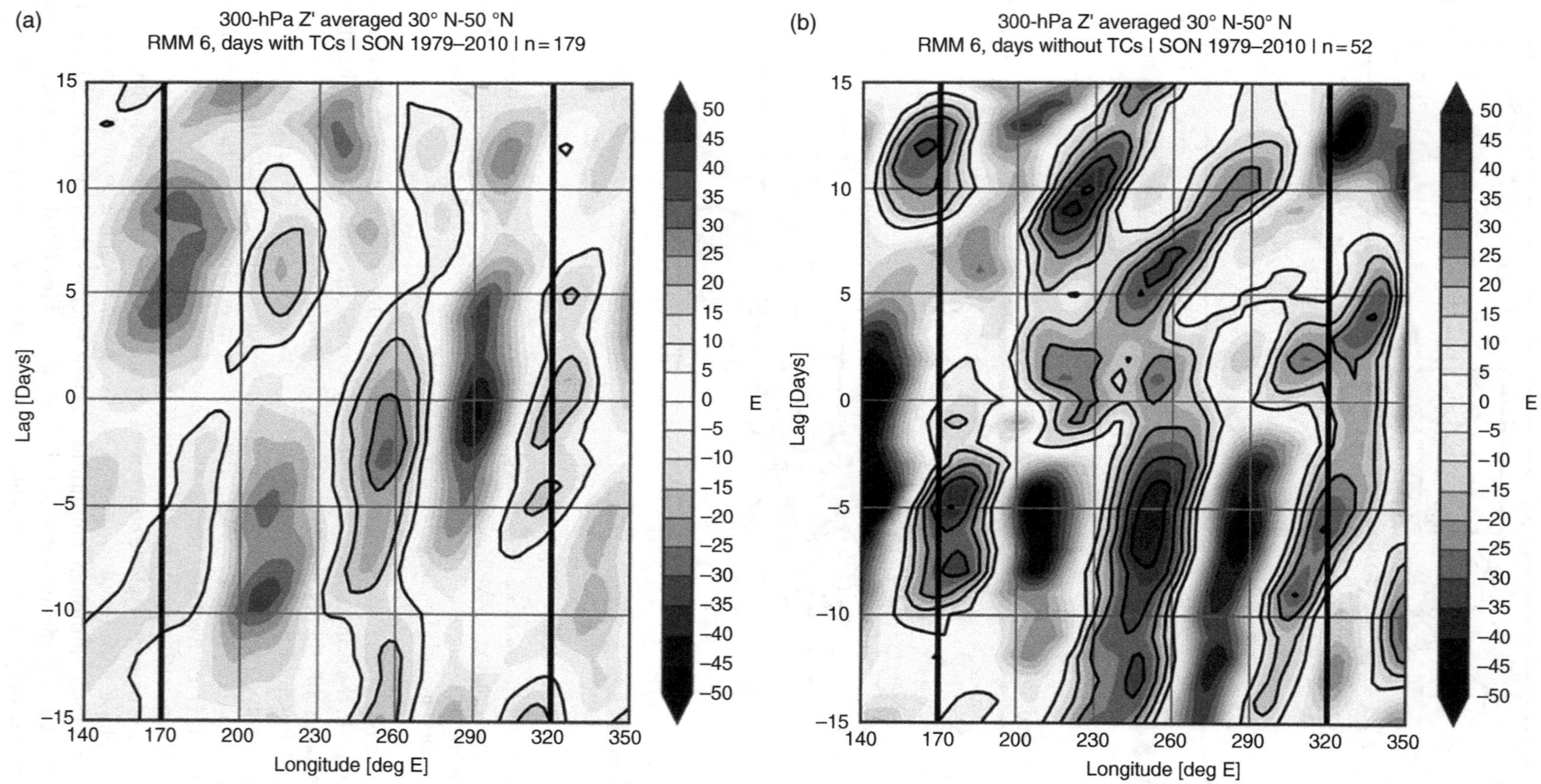

Figure 9.18 The 300 hPa geopotential height anomalies averaged between 30°N and 50°N. Vertical black lines mark the longitudinal boundaries of the North America domain. Anomalies with solid contours are positive.

non- tropical-cyclone composite subset, a wave train resembling the one observed in the tropical cyclone composite subset breaks down prior to lag 0 days as convection across the western Pacific Ocean becomes suppressed. A second, more quickly evolving wave train is evident in the tropical cyclone composite subset after lag 0 days, with origins over the western North Pacific Ocean. It is possible that this wave train formed in response to recurving tropical cyclones over the western North Pacific Ocean, especially when considering the absence of a similar wave train in the non- tropical-cyclone composite subset. Thus, the presence or absence of tropical cyclones in conjunction with MJO convection during RMM 6 (and other RMM phases not discussed) yields substantial differences in large-scale extratropical circulation patterns, which in turn would feed back on the MJO and equatorial waves. Results show that, during RMM 6 in boreal fall, the extratropical circulation response pattern associated with the MJO with tropical cyclones in the region of greatest tropical cyclone activity does not differ significantly from the more general composite based on the MJO only. Yet, other composites based on other phases with and without tropical cyclones (not shown) suggest that sometimes circulation responses to tropical cyclones interfere constructively and other times destructively with extratropical circulation outcomes otherwise associated with the MJO. Results support the view that the MJO modulates tropical cyclone activity and that the MJO and tropical cyclones can interact to produce circulation outcomes downstream.

9.6. SUMMARY

This chapter demonstrates that intraseasonal and synoptic weather events interact with the background interannual climate state to alter weather events and climate outcomes. The case studies in section 9.3 demonstrate the two-way interaction between ENSO and the MJO that dramatically change the behaviors of both phenomena. The tropical circulation associated with the evolving ENSO state can modulate MJO activity, which can then feed back on the development of ENSO amplitude and structure. Beyond the tropics, significant variability of midlatitude Rossby wave breaking events in concert with the MJO is demonstrated in section 9.4. The circulation associated with active MJO convection modulates the structure of the subtropical North Pacific jet and influences the structural evolution of midlatitude Rossby waves. This Rossby wave breaking can further feed back onto the MJO circulation by enhancing upper-level gyres that are characteristic of the MJO. Section 9.5 further highlights the associations of tropical cyclones with some of the observed extratropical circulation patterns associated with an average MJO event. Similar MJO events become associated with vastly different extratropical circulations depending on the presence of tropical cyclones, suggesting that the observed average intraseasonal circulation associated with the MJO is a nonlinear cumulative result of both the MJO convection and the synoptic-scale weather events that evolve with it, including tropical cyclones. These studies demonstrate the complexity of interactions between climate and weather and suggest that understanding those interactions might improve the simulation and prediction of extreme weather events.

REFERENCES

Anwender, D., P. A. Harr, and S. C. Jones (2008), Predictability associated with the downstream impacts of the extratropical transition of tropical cyclones: Case studies, *Monthly Weather Rev.*, *136*, 3226–3247.

Anwender, D., S. C. Jones, M. Leutbecher, and P. A. Harr (2010), Sensitivity experiments for ensemble forecasts of the extratropical transition of Typhoon Tokage (2004), *Quart. J. Roy. Meteor. Soc.*, *136*, 183–200.

Archambault, H. M., L. F. Bosart, D. Keyser, and J. M. Cordeira (2013), A climatological analysis of the extratropical flow response to recurving western North Pacific tropical cyclones, *Monthly Weather Rev.*, *141*, 2325–2346.

Ashok, K., S. K. Behera, S. A. Rao, H. Weng, and T. Yamagata (2007), El Niño Modoki and its possible teleconnection, *J. Geophys. Res.*, *112*, C11007.

Bao, M., and D. L. Hartmann (2014), The response to MJO-like forcing in a nonlinear shallow-water model, *Geophys. Res. Lett.*, *41*, 1322–1328.

Barrett, B. S., and L. M. Leslie (2009), Links between tropical cyclone activity and Madden-Julian Oscillation phase in the North Atlantic and Northeast Pacific basins, *Monthly Weather Rev.*, *137*, 727–744.

Bergman, J. W., H. H. Hendon, and K. M. Weickmann (2001), Intraseasonal air–sea interactions at the onset of El Niño, *J. Climate*, *14*, 1702–1719. doi: http://dx.doi.org/10.1175/1520-0442(2001)014<1702:IASIAT>2.0.CO;2.

Camargo, S. J., M. Wheeler, and A. H. Sobel (2009), Diagnosis of MJO modulation of tropical cyclogenesis using an empirical index, *J. Atmos. Sci.*, *66*, 3061–3074.

Capotondi, A., A. T. Wittenberg, M. Newman, E. D. Lorenzo, J.-Y. Yu, P. Braconnot, J. Cole, B. Dewitte, B. Giese, E. Guilyardi, F.-F. Jin, K. Karnauskas, B. Kirtman, T. Lee, N. Schneider, Y. Xue, and S.-W. Yeh (2015), Understanding ENSO Diversity, *Bull. Amer. Meteor. Soc.*, *96*, 921–938; doi: http://dx.doi.org/10.1175/BAMS-D-13-00117.1.

Cassau, C (2008), Intraseasonal interaction between the Madden–Julian Oscillation and the North Atlantic Oscillation. *Nature.* *455*, 523–527 doi:10.1038/nature07286

Chen, D, T. Lian, C. Fu, M. A. Cane, Y. Tang, R. Murtugudde, X. Song, Q. Wu, and L. Zhou (2015), Strong influence of westerly wind bursts on El Niño diversity, *Nat. Geosci.*, *8*, 339–345; doi:10.1038/ngeo2399

Cordeira, J. M., and L. F. Bosart (2010), The antecedent large-scale conditions of the "perfect storms" of late October and early November 1991, *Monthly Weather Rev.*, *138*, 2546–2569.

DeMott, C. A., C. Stan, and D. A. Randall (2013), Northward propagation mechanisms of the boreal summer intraseasonal oscillation in the ERA-interim and SP-CCSM, *J. Climate*, *26*, 1973–1992.

Eisenman, I., L. S. Yu, and E. Tziperman (2005), Westerly wind bursts: ENSO's tail rather than the dog?, *J. Climate*, *18*, 5224–5238.

Feng, J., P. Liu, W. Chen, and X. Wang (2015), Contrasting Madden-Julian Oscillation activity during various stages of EP and CP El Niños, *Atmos. Sci. Lett.*, *16*, 32–37; doi: 10.1002/asl2.516

Ferranti, L., T. N. Palmer, F. Molenti, and K. Klinker (1990), Tropical-extratropical interaction associated with the 30–60-day oscillation and its impact on medium and extended range prediction, *J. Atmos. Sci.*, *47*, 2177–2199.

Frank, W. M., and P. E. Roundy (2006), The role of tropical waves in tropical cyclogenesis, *Monthly Weather Rev.*, *134*, 2397–2417.

Frederiksen, J. S., and H. Lin (2013), Tropical-extratropical interactions of intraseasonal oscillation, *J. Atmos. Sci.*, *70*, 3180–3197.

Gloeckler, L. C., and P. E. Roundy (2013), Modulation of the extratropical circulation by combined activity of the Madden-Julian oscillation and equatorial Rossby waves during boreal winter, *Monthly Weather Rev.*, *141*, 1347–1357.

Gushchina, D., and B. Dewitte (2012), Intraseasonal tropical atmospheric variability associated with the two flavors of El Niño, *Monthly Weather Rev.*, *140*, 3669–3681; doi: http://dx.doi.org/10.1175/MWR-D-11-00267.1.

Hall, J. D., A. J. Matthews, and D. J. Karoly (2001), The modulation of tropical cyclone activity in the Australian region by the Madden-Julian oscillation, *Monthly Weather Rev.*, *129*, 2970–2982.

Hendon, H. H., B. Liebmann, and J. D. Glick (1998) Oceanic Kelvin waves and the Madden-Julian Oscillation, *J. Atmos. Sci.*, *55*, 88–101; doi: http://dx.doi.org/10.1175/1520-0469(1998)055<0088:OKWATM>2.0.CO;2.

Holton, J. R. (2004), *An Introduction to Dynamic Meteorology*, 4th ed., Elsevier Academic Press.

Jeong, J.-H., B.-M. Kim, C.-H. Ho, and Y.-H. Noh (2008), Systematic variation in wintertime precipitation in East Asia by MJO-induced extratropical vertical motion, *J. Climate*, *21*, 788–801.

Jiang, X., T. Li, and B. Wang (2004), Structures and mechanisms of the northward propagating boreal summer intraseasonal oscillation, *J. Climate*, *17*, 1022–1039.

Johnson, N. C. (2013), How many ENSO flavors can we distinguish?, *J. Climate*, *26* (13), 4816–27.

Jones, C. (2000), Occurrence of extreme precipitation events in California and relationships with the Madden-Julian Oscillation, *J. Climate*, *13*, 3576–3587.

Jones, C., D. E. Waliser, K. M. Lau, W. Stern (2004), Global occurrences of extreme precipitation and the Madden-Julian Oscillation: Observations and predictability, *J. Climate*, *17*, 4575–4589.

Kerr, R. A. (2011), Vital details of global warming are eluding forecasters, *Science*, *14*, 173–174; doi: 10.1126/science.334.6053.173

Kessler, W. S. (2001), EOF representations of the Madden-Julian Oscillation and its connection with ENSO, *J. Climate*, *14*, 3055–3061.

Kiladis, G. N., J. Dias, K. H. Straub, M. C. Wheeler, S. N. Tulich, K. Kikuchi, K. M. Weickmann, and M. J. Ventrice (2014), A comparison of OLR and circulation-based indices for tracking the MJO, *Monthly Weather Rev.*, *142*, 1697–1715.

Kiladis, G. N., K. H. Straub, P. T. Haertel (2005), Zonal and vertical structure of the Madden-Julian Oscillation, *J. Atmos. Sci.*, *62*, 2790–2809.

Kiladis, G. N., M. C. Wheeler, P. T. Haertel, K. H. Straub, and P. E. Roundy (2009), Convectively coupled equatorial waves, *Rev. Geophys.*, *47*, RG2003; doi:10.1029/2008RG000266.

Kim, B.-M., G.-H. Lim, and K.-Y. Kim (2006), A new look at the middle-MJO teleconnection in the northern hemisphere winter, *Quart. J. Roy. Meteor. Soc.*, *132*, 485–503.

Kim, D., J.-S. Kug, and A. H. Sobel. (2014), Propagating versus nonpropagating Madden-Julian Oscillation events, *J. Climate*, *27*(1), 111–125.

Klotzbach, P. J., and E. Oliver (2014), The Madden-Julian Oscillation's impacts on worldwide tropical cyclone activity, *J. Climate*, *27*, 2317–2330.

Lengaigne, M., E. Guilyardi, J.-P. Boulanger, C. Menkes, P. Delecluse, P. Inness, J. Cole, J. Slingo (2004), Triggering of El Niño by westerly wind events in a coupled general circulation model. *Climate Dynamics*, *23*, 601–620. DOI 10.1007/s00382-004-0457-2

L'Heureux, M. L., and R. W. Higgins (2008), Boreal winter links between the Madden-Julian Oscillation and the Arctic Oscillation, *J. Climate*, *21*, 3040–3050.

Lin, H. L., G. Brunet, and J. Derome (2009), An observed connection between the North Atlantic Oscillation and the Madden-Julian Oscillation, *J. Climate*, *22*, 364–380, doi: 10.1175/2008JCLI2515.1.

MacRitchie, K., and P. E. Roundy (2016), The two-way relationship between the Madden-Julian Oscillation and anticyclonic wave breaking, *Quart. J. Roy. Meteor. Soc.*, *142*, 2159–2167; doi:10.1002/qj.2809.

Madden, R. A., and P. R. Julian (1972) Description of global-scale circulation cells in the tropics with a 40–50 day period. *J. Atmos. Sci.*, *29*, 1109–1123.

Matsueda, S., and Y. Takaya (2015), The global influence of the Madden-Julian Oscillation on extreme temperature events, *J. Climate*, *28*, 4141–4151; doi: http://dx.doi.org/10.1175/JCLI-D-14-00625.1.

Matthews, A. J., and G. N. Kiladis (1999), The tropical-extratropical interaction between high-frequency transients and the Madden-Julian Oscillation, *Monthly Weather Rev.*, *127*, 661–677.

McIntyre, M. E., and T. N. Palmer (1984) The "surf zone" in the stratosphere, *J. Atmos. Terr. Phys.*, *51*, 29–43.

McPhaden, M. J., X. Zhang, H. H. Hendon, and M. C. Wheeler (2006), Large scale dynamics and MJO forcing of ENSO variability, *Geophys. Res. Lett.*, *33*, L167072; doi: 10.1029/2006GL026776.

Menkes, C. E., M. Lengaigne, J. Vialard, M. Puy, P. Marchesiello, S. Cravatte, and G. Cambon (2014), About the role of westerly wind events in the possible development of an El Niño in 2014, *Geophys. Res. Lett.*, *41*, 6476–6483, doi:10.1002/2014GL061186.

Moore, R. W., O. Martius, and T. Spengler (2010) The modulation of the subtropical and extratropical atmosphere in the Pacific basin in response to the Madden-Julian Oscillation, *Monthly Weather Rev.*, *138*, 2761–2779.

Mori, M., and M. Watanabe (2008), The growth and triggering mechaisms of the PNA: A MJO-PNA coherence, *J. Meteor. Soc. Japan*, *86*, 213–236.

Newman, M., and P. D. Sardeshmukh (1998), The impact of the annual cycle on the North Pacific/North American response to remote low-frequency forcing, *J. Atmos. Sci.*, *55*, 1336–1353.

Pielke, R. A., Sr. (2012), Regional climate downscaling: What's the point?, *Eos*, *93*, 52–53.

Pohl, B. and A.J. Matthews (2007) Observed Changes in the Lifetime and Amplitude of the Madden–Julian Oscillation Associated with Interannual ENSO Sea Surface Temperature Anomalies. *J. Climate*, *20*, 2659–2674, doi: 10.1175/JCLI4230.1.

Roundy, P.E., (2014) Some Aspects of Western Hemisphere Circulation and the Madden–Julian Oscillation. *J. Atmos. Sci.*, *71*, 2027–2039, doi: 10.1175/JAS-D-13-0210.1.

Roundy, P. E., (2015), On the interpretation of EOF analysis of ENSO, atmospheric Kelvin waves, and the MJO, *J. Climate*, *28*, 1148–1165; doi: http://dx.doi.org/10.1175/JCLI-D-14-00398.1.

Roundy, P. E., and G.N. Kiladis (2006), Observed Relationships between Oceanic Kelvin Waves and Atmospheric Forcing. *J. Climate*, *19*, 5253–5272; doi: 10.1175/JCLI3893.1.

Roundy, P. E., and J. R. Kravitz (2009), The association of the evolution of intraseasonal oscillations to ENSO phase, *J. Climate*, *22*, 381–395.

Roundy, P. E., and W. M. Frank (2004), A climatology of waves in the equatorial region, *J. Atmos. Sci.*, *61*, 2105–2132.

Roundy, P. E., K. MacRitchie, J. Asuma, and T. Melino (2010), Modulation of the global atmospheric circulation by combined activity in the Madden-Julian Oscillation and the El Niño–Southern Oscillation during boreal winter, *J. Climate*, *23*, 4045–4059.

Rui, H., and B. Wang (1990), Development characteristics and dynamic structure of tropical intraseasonal convection anomalies, *J. Atmos. Sci.*, *47*, 357–379.

Sakaeda, N., and P. E. Roundy (2014), The role of interactions between multiscale circulations on the observed zonally averaged zonal wind variability associated with the Madden-Julian Oscillation, *J. Atmos. Sci.*, *71*, 3816–3836.

Sakaeda, N. and P.E. Roundy (2015) The Development of Upper-Tropospheric Wind over the Western Hemisphere in Association with MJO Convective Initiation. *J. Atmos. Sci.*, *72*, 3138–3160, doi: 10.1175/JAS-D-14-0293.1.

Sardeshmukh, P. D., and B. J. Hoskins (1988), The generation of global rotational flow by steady idealized tropical divergence, *J. Atmos. Sci.*, *45*, 1228–1251.

Schreck, C. J., J. Molinari, and K. I. Mohr (2011), Attributing tropical cyclogenesis to equatorial waves in the western North Pacific, *J. Atmos. Sci.*, *68*, 195–209.

Seiki, A., and Y. N. Takayabu (2007) Westerly wind bursts and their relationship with intraseasonal variations and ENSO. Part II: Energetics over the western and central Pacific. *Mon. Wea. Rev.*, *135*, 3346–3361.

Seiki, A. Y., N. Takayabu, K. Yoneyama, N. Sato, and M. Yoshizaki (2009), The oceanic response to the Madden-Julian Oscillation and ENSO, *SOLA*, *5*, 93–96.

Shi, L., H. H. Hendon, O. Alves, M. C. Wheeler, D. Anderson, and G. Wang (2010), On the importance of initializing the stochastic part of the atmosphere for forecasting the 1997/1998 El Niño, *Climate. Dyn.*; doi:10.1007/s00382-010-0933-9.

Thompson, D. B., and P. E. Roundy (2013), The relationship between the Madden-Julian Oscillation and U.S. violent tornado outbreaks in the spring, *Monthly Weather Rev.*, *141*, 2087–2095; doi: http://dx.doi.org/10.1175/MWR-D-12-00173.1.

Thorncroft, C. D., B. J. Hoskins, and M. McIntyre (1993) Two paradigms of baroclinic-wave life-cycle behavior, *Quart. J. Roy. Meteor. Soc.*, *119*, 17–55.

Tyrrell, G. C., D. J. Karoly, and J. L. McBride (1996) Links between tropical convection and variations of the extratropical circulation during TOGA COARE, *J. Atmos. Sci.*, *53*, 2735–2748.

Vecchi, G. A., and N. A. Bond (2004), The Madden-Julian Oscillation (MJO) and northern high latitude wintertime surface air temperatures, *Geophys. Res. Lett.*, *31*, L04104, doi:10.1029/2003GL018645.

Vecchi, G. A., A. T. Wittenberg, and A. Rosati (2006), Reassessing the role of stochastic forcing in the 1997–1998 El Niño, *Geophys. Res. Lett.*, *33*, L01706, doi:10.1029/2005GL024738.

Vitart, F., and F. Molenti (2010), Simulation of the Madden-Julian Oscillation and its teleconnections in the ECMWF forecast system, *Quart. J. Roy. Meteor. Soc.*, *136*, 842–855.

Wheeler, M.C. and H.H. Hendon (2004), An All-Season Real-Time Multivariate MJO Index: Development of an Index for Monitoring and Prediction. *Mon. Wea. Rev.*, *132*,1917–1932, doi: 10.1175/1520-0493(2004)132<1917:AARMMI>2.0.CO;2.

Zhang, C., and M. Dong (2004), Seasonality in the Madden-Julian Oscillation, *J. Climate*, *17*, 3169–3180.

Zhao, C., T. Li, and T. Zhou (2013), Precursor signals and processes associated with MJO initiation over the tropical Indian Ocean, *J. Climate*, *26*, 291–307.

10

Recent Climate Extremes Associated with the West Pacific Warming Mode

Chris Funk[1] and Andrew Hoell[2]

ABSTRACT

Here we analyze empirical orthogonal functions (EOFs) of observations and a 30 member ensemble of Community Earth System Model version 1 (CESM1) simulations, and suggest that precipitation declines in the Greater Horn of Africa (GHA) and the northern Middle East/Southwestern Asia (NME/SWE: Iran, Iraq, Kuwait, Syria, Saudi Arabia north of 25°N, Israel, Jordan, and Lebanon) may be interpreted as an interaction between La Niña-like decadal variability and the West Pacific Warming Mode (WPWM). While they exhibit different SST patterns, warming of the Pacific cold tongue (ENSO) and warming of the western Pacific (WPWM) produce similar warm pool diabatic forcing, Walker circulation anomalies, and terrestrial teleconnections. CESM1 SST EOFs indicate that *both* La Niña-like WPWM warming and El Niño-like east Pacific warming will be produced by climate change. The temporal frequency of these changes, however, are distinct. WPWM varies decadally, while ENSO is dominated by interannual variability. Future WPWM and ENSO warming may manifest as a tendency toward warm West Pacific SST, punctuated by extreme warm East Pacific events. WPWM EOFs from Global Precipitation Climatology Project (GPCP) precipitation also identify dramatic WPWM-related declines in the Greater Horn of Africa and NME/SWE.

10.1. INTRODUCTION

The physicist John Archibald Wheeler [*Wheeler and Zurek*, 1990] suggested that "time is nature's way to keep everything from happening at once." The analog of this conception in climate science might be multimodal analysis, which identifies characteristic space-time patterns, typically associated with specific frequencies, like the El Niño Southern Oscillation (ENSO) [*Madden and Julian*, 1972; *Walker and Bliss*, 1932; *Walker*, 1923], the Madden-Julien Oscillation (MJO) [*Madden and Julian*, 1972], or the Pacific Decadal Oscillation (PDO) [*Mantua and Hare*, 2002]. This type of analysis often informs atmospheric global circulation model experiments, which often explore various combinations of the dominant global sea surface temperature (SST) empirical orthogonal functions (EOFs) [*Schubert et al.*, 2009] with the three dominant EOFs typically corresponding to ENSO, a low frequency "trend" mode, and a decadal PDO-like mode of variability. Here, building on a recent paper focused on the West Pacific Warming Mode [*Funk and Hoell*, 2015], the first ENSO-residual global EOF of SSTs, we suggest that this type of multimodal analysis, applied to observations or coupled atmosphere-ocean general circulation model (AOGCM) climate change simulations, indicates that modes of variability are nature's way of keeping everything from warming at once. More specifically, using observed SSTs and precipitation, and SSTs and precipitation from a 30 member climate change ensemble from the

[1] *US Geological Survey and University of California Santa Barbara Geography, Santa Barbara, California, USA*

[2] *NOAA Earth System Research Laboratory, Physical Sciences Division, Boulder, Colorado, USA*

Climate Extremes: Patterns and Mechanisms, Geophysical Monograph 226, First Edition.
Edited by S.-Y. Simon Wang, Jin-Ho Yoon, Christopher C. Funk, and Robert R. Gillies.

Community Earth System Model version 1 (CESM1) [*Kay et al.*, 2015], we show that the two dominant, but opposed, modes of SST variability (and warming) are ENSO and WPWM. While ENSO operates on a characteristic 3–7 yr time scale, the WPWM corresponds closely with SST trend modes of variability [*Compo and Sardeshmukh*, 2010; *Hoell and Funk*, 2013a; *Newman*, 2013; *Williams and Funk*, 2011] and ENSO-residual SST trends [*Cane et al.*, 1997; *Compo and Sardeshmukh*, 2010; *Solomon and Newman*, 2012]. Rather than conceptualizing climate change as a one size fits all ubiquitous simultaneous warming of SSTs, the analysis of observations and simulations presented below suggests that a modal conception of change and variability may provide a more informative and useful framework for discussing climate change.

10.2. BACKGROUND

This study builds on a recent analysis of *Funk and Hoell* [2015], which examined the WPWM in the following datasets: observed SSTs [*Smith et al.*, 2008], SSTs from a multimodel climate change ensemble of Historical Phase 5 Coupled Model Intercomparison Project (CMIP5) [*Taylor et al.*, 2011], Simple Ocean Data Assimilation (SODA) ocean temperatures, sea surface heights (SSHs) and currents, and twentieth century reanalysis near surface zonal winds and precipitation [*Compo et al.*, 2011] and GPCP precipitation [*Adler et al.*, 2003]. Following recent ENSO-residual trend analyses *Cane et al.* [1997], *Compo and Sardeshmukh* [2010], *Solomon and Newman*, [2012], and *Funk and Hoell* [2015] analyzed each of these datasets, examining the ENSO-related component, and the first EOF after ENSO influences were removed by regression. This global ENSO-residual EOF mode was termed the West Pacific Warming Mode (WPWM) because of a strong warm pool warming response in the SST EOFs. The WPWM EOFs from observed SSTs and climate change simulations were very similar, indicating a strong radiative influence in the "western V" region stretching from Indonesia into the northern and southern subtropical Pacific. Conversely, the ENSO-component of the climate change simulations indicated rapid East Pacific warming, a result at odds with the observed cooling in that region. *Funk and Hoell* [2015] interpreted this result as a combination of natural decadal variability combined with a tendency for the models to overestimate the strength of El Niños.

Dynamically, the WPWM is associated with a stronger central-to-western Pacific SST gradient, and an intensification of the Walker circulation similar to that described by *Williams and Funk* [2011]. *Funk and Hoell* [2015] shows that this intensification is associated with stronger central Pacific easterly trade winds, an enhanced western/central Pacific precipitation dipole, and a similar west-versus-east dipole in sea surface heights and ocean temperatures at 25 m, 70 m, and 171 m. *Funk and Hoell* [2015] performs Community Atmosphere Model version 5 (CAM5) simulations driven with a WPWM SST pattern, and shows boreal spring precipitation decreases across East Africa, the Middle East, the southern United States, Mexico, and southern South America.

Funk and Hoell [2015] applied a common WPWM analysis to these based on the global ENSO-residual EOF. *Funk and Hoell* [2015] concluded:

1. Observation-based and historical CMIP5 SST WPWM eigenvectors and principal components (PCs) match closely, hence the observed western V-like warming over the past few decades has been largely produced by radiative forcing and not internal Pacific Decadal Variability (PDV).
2. The ENSO mode of the observed and historical CMIP5 SST track fairly closely, but the CMIP5 eigenvectors are about twice as large as the observed eigenvectors across the eastern Pacific. As a result, eastern Pacific SST have been cooler than predicted.
3. The interaction of internal and external radiative forcing produced rapid West Pacific (WP)/North Pacific (NP) warming in the 1990s and 2000s, with a concomitant jump in the west Pacific SST gradient. Running 30 yr PC cross correlations also rise steeply, indicating increasing radiative control. The observed 1978–2005 warming of the WP/NP was very unlikely without greenhouse gas (GHG) forcing.
4. The pattern of the WPWM circulation responses in global zonal surface winds and precipitation is consistent with a Matsuno-Gill-like response to diabatic forcing over the maritime continent. An equatorial acceleration of central Pacific trade winds appears associated with an east west-dipole in Pacific basin ocean temperatures and SSH.
5. Correlations between the WPWM PCs themselves and between the WPWM PCs and WP tidal gauge heights, WP-EP precipitation, and dateline zonal winds are all strong ($> |0.81|$) and significant. Low-frequency variations in these fields, and the recent rise in western Pacific SSH, appear to be strongly influenced by radiative forcing. The recent increase in the Walker circulation is related to the WPWM and radiative warming in the Western Pacific.
6. The west Pacific SST gradient, dateline zonal surface winds, and the WP-EP precipitation gradient modulate weather on global scales. The WPWM of boreal spring GPCP precipitation indicates significant drying across parts of East Africa, the Middle East, western Asia, southern North America, and southern South America.

Increases are identified in northern Australia, southern Africa, Southeast Asia, and northern South America. An atmospheric GCM experiment recreates many of these changes. The associated precipitation changes can be substantial, more than half a standardized deviation.

7. The combination of results 1 through 7 imply that radiative forcing in the western V region has contributed to SST variations associated with substantial changes in precipitation.

Here we build on the analysis of *Funk and Hoell* [2015] and extend the WPWM analysis by (1) applying WPWM analysis to a new 30 member ensemble of CESM1 climate change simulations, forced with observed greenhouse gas concentrations, aerosols, and variations in solar insolation and by (2) looking beyond the March–June season used in *Funk and Hoell* [2015], examining GPCP WPWM changes by season. This analysis shows that two regions/seasons are particularly sensitive to WPWM-forced drying: northern Middle East/Southwestern Asia (NME/SWE) during October–March and the Greater Horn of Africa (GHA) during April–June. These teleconnections, we propose, are the result of, respectively, off-equatorial and equatorially trapped Rossby wavelike responses to Indo-Pacific diabatic forcing. Seminal analyses by *Gill* [1980] and *Rodwell and Hoskins* [1996] have linked Indo-Pacific diabatic heating to the climatological average conditions associated with these dry climates. More recent research has suggested that above normal Indo-Pacific heating and SST warming may be associated with seasonal rainfall deficits or drying trends in SWA [*Mathew Barlow et al.*, 2002; *M. Barlow et al.*, 2007; *Hoell et al.*, 2014a] and the GHA [*Funk et al.*, 2008; *Funk et al.*, 2014; *Hoell and Funk*, 2013b; *Liebmann et al.*, 2014; *Lyon*, 2014; *Lyon and DeWitt*, 2012; *Williams and Funk*, 2011; *Yang et al.*, 2014]. The multimodal interpretation that we present here, consistent with our recent studies [*Funk et al.*, 2014; *Hoell and Funk*, 2013a, b; *Hoell et al.*, 2013; *Hoell et al.*, 2014a; *Hoell et al.*, 2014b], suggests that the recent increases in NME/SWE and GHA drought frequencies have arisen through the interaction of a trendlike WPWM mode [*Funk and Hoell*, 2015] and ENSO-like decadal variability [*Lyon et al.*, 2013; *Meehl et al.*, 2013].

10.3. DATA AND METHODS

This analysis uses six datasets: 1900–2014 National Oceanic and Atmospheric Administration (NOAA) Extended Reconstruction SST version 4 (ERv4) [*Smith et al.*, 2008], 1900–2014 CESM1 SSTs, precipitation and 2 m air temperatures from a 1900–2014 30 member ensemble of 0.94 x 1.25° AOGCM simulations [*Kay et al.*, 2015], 1901–2014 GPCC precipitation observations [*Becker et al.*, 2013; *Schneider et al.*, 2013], 1901–2014 Goddard Institute for Space Studies (GISS) land/ocean surface temperatures [*Hansen et al.*, 2010], and 1979–2014 GPCP satellite-station land-ocean precipitation [*Adler et al.*, 2003].

We begin by examining WPWM and ENSO SST EOFs, as applied in *Funk and Hoell* [2015]. ENSO variations are represented by the first PC of tropical Pacific 1900–2014 SST (15°S–15°N, 125°E–80°W). This PC is regressed against seasonal SST grids, and the global residuals from this regression are used to estimate the first EOF, which we refer to as the WPWM. The WPWM is the first EOF of ENSO-residual global 1900–2014 SST (cosine-weighted from 70°S–70°N). We assess these modes of variability for the ERv4 and CESM1 SSTs for two seasons: boreal winter/spring (October–March) and boreal summer/fall (June–September). All analyses of the CESM1 simulations were carried out on individual simulations, and then composited by averaging across the 30 simulations. We then briefly examine the ERv4 and CESM1 SSTs' standard deviations and correlations with globally averaged SST.

We next examine regressions between the WPWM and ENSO PCs and CESM1 precipitation and 2 m air temperatures. We then expand the March–June WPWM GPCP analysis presented in *Funk and Hoell* [2015] to January–March, April–June, July–August, and September–December compositing periods. This study reveals strong WPWM sensitivities in NME/SWE and the Greater Horn of Africa. We conclude with an evaluation of 1901–2014 GPCP and GISS air temperatures for these regions.

10.3.1. WPWM and ENSO PCs and EOFs

In Figure 10.1 we present 1900–2014 PC time series for the WPWM and ENSO modes of variability from the CESM1 and ERv4 SST. By construction, these modes are independent at seasonal timescales. Both modes, however, exhibit warming (increasing SST) over the 1900–2014 time period. The variance of the WPWM, in the observations and simulations, is much lower than the ENSO mode, and since the mid-1970s the observed WPWM has increased in step with predictions from the CESM1 climate change simulations. This warming is large (~+2 standardized deviations) compared to interannual variations of the WPWM.

The ENSO ERv4 and CESM1 also indicate warming, but the magnitude of this warming compared to the variance of this ENSO SSTs is smaller; 1970–2014 warming is about one standard deviation in these CESM1 simulations and about one-half of a deviation in the ERv4. In the observations, PDV-related warming in the eastern

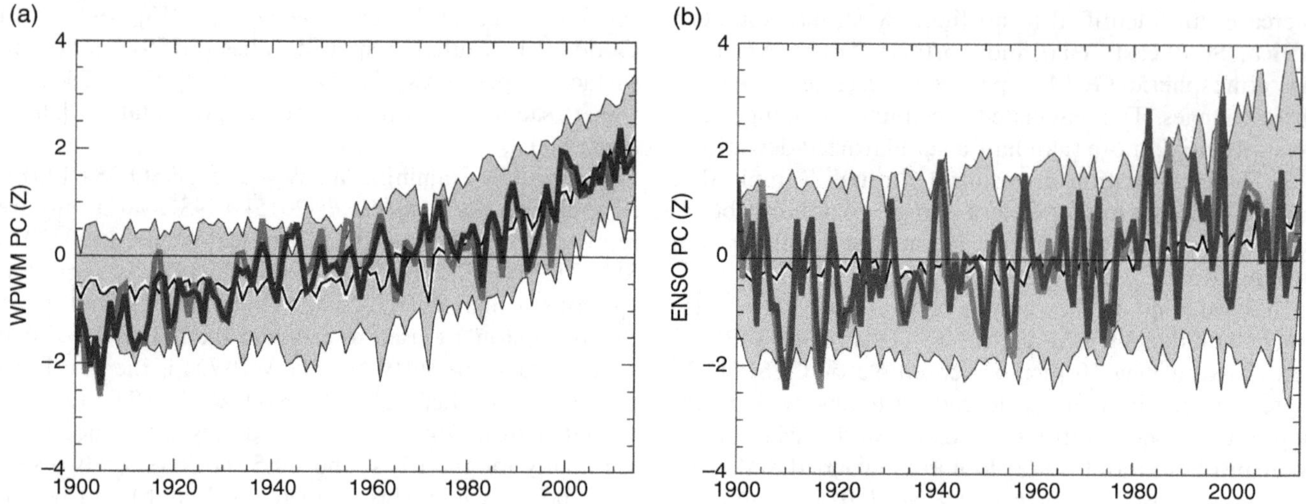

Figure 10.1 (a) WPWM PCs from ERv4 DJFMAM SST (black) and JJASON SST (gray) and the mean CESM1 ensemble WPWM PC for DJFMAM (white) and JJASON (black). The gray polygon shows 95% CESM1 confidence intervals. (b) Same but for the ENSO PC. *See electronic version for color representation.*

Pacific can be seen in the late 1970s [*Graham*, 1994]. This is then followed by cooling in the eastern Pacific in the late 1990s [*Meehl et al.*, 2013], which helped produce a hiatus in global warming [*England et al.*, 2014; *Trenberth et al.*, 2014].

Figure 10.1 indicates that a global warming signal will involve the interaction of these two dominant SST modes. With much higher interannual variances, and large decadal excursions, ENSO-related variations will produce both El Niño and La Niña conditions, which will then interact with much more persistent WPWM warming patterns. Estimates of these WPWM patterns are shown in Figure 10.2. These maps are produced by multiplying the 1999–2014 versus 1979–1998 change in the WPWM PC by the WPWM eigenvector. Note strong similarity between the WPWM changes estimated from the observed SSM and CESM1 SST. These warming patterns are characterized by a western V stretching from near the maritime continent into the subtropical Pacific of both hemispheres. Similar structures have been identified with long-term trends in ENSO-residual SST [*Cane et al.*, 1997; *Compo and Sardeshmukh*, 2010; *Solomon and Newman*, 2012]. The cooling along the tropical central Pacific may be associated with upwelling caused by increased equatorial trade winds and Indo-Pacific diabatic heating; such a response would be consistent with a Matsuno-Gill-like equatorial atmospheric Kelvin wave responses [*Funk and Hoell*, 2015] and recent increases in west Pacific SSHs [*Funk and Hoell*, 2015; *Merrifield*, 2011; *Merrifield and Maltrud*, 2011]. Note the strong correspondence between both the CESM1 and ERv4 PCs (Fig. 10.1) and change maps (Fig. 10.2). We suggest that this arises because the trend mode of observed SST has a strong radiative component associated with GHGs and aerosols. The characteristic western V warming structure in the Pacific appears related to the SST gradient between the central and western tropical Pacific [*Hoell and Funk*, 2013a], stronger central Pacific trade winds, and increased subtropical convergence zones [*Funk and Hoell*, 2015].

Note also that, in the tropics, WPWM SST increases are generally associated with regions with high water vapor content, such as the tropical Atlantic and Indian oceans. This water vapor perhaps acts as a potent, but localized, GHG, amplifying the effects of anthropogenic warming by well-mixed gases like CO_2 and methane.

Figure 10.3 shows similar change plots for the ENSO EOF. Each locations' SST were regressed against the 1900–2014 ENSO PC, and the change in this PC between 1999–2014 and 1979–1998 used to estimate ENSO-related SST changes. The observed SSTs exhibit eastern Pacific cooling, thought to be forced by a PDV-related hiatus in global temperature increases [*Meehl et al.*, 2013]. Increased diabatic forcing in the warm pool helped drive changes in midlatitude circulations that produced cooler Northern Hemisphere temperatures, while stronger trade winds helped cool the eastern Pacific [*England et al.*, 2014; *Trenberth et al.*, 2014]. Note in Figure 10.2a,b and in *Meehl et al.* [2013] Figure 10.5 that the estimates of ENSO (Fig. 10.2) and PDV [*Meehl et al.*, 2013] focus on the eastern Pacific, with little loading on the western V region.

Consistent with recent studies analyzing other models [*Kent et al.*, 2015; *Sandeep et al.*, 2014], the CESM1 models show a fairly vigorous (~+0.4°C) increase in eastern Pacific SST (Fig. 10.3c,d). This translates into a more El Niño-like climate in model simulations than has been

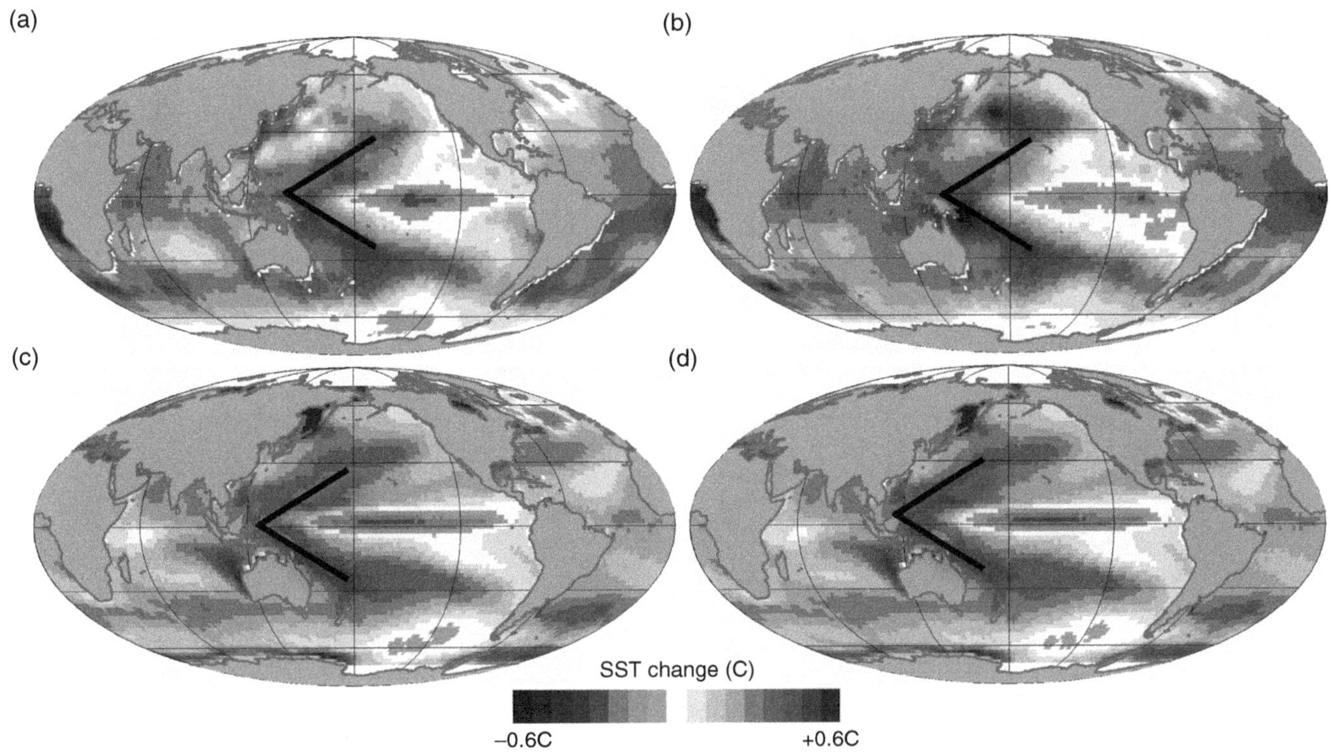

Figure 10.2 WPWM changes between 1999–2014 and 1979–1998 for (a) observed ERv4 DJFMAM SSTs and (b) JJASON SSTs and simulated (CESM1) SSTs for (c, d) DJFMAM and JJASON. Black lines denote schematic representation of the western V region.

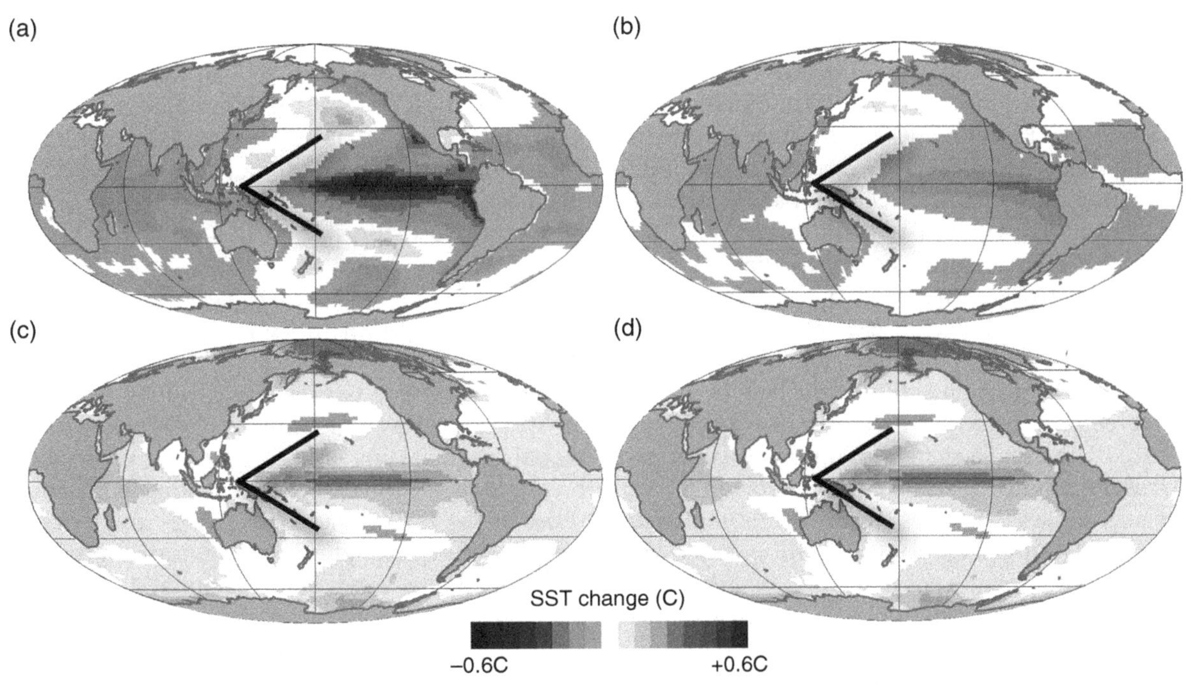

Figure 10.3 ENSO-related changes between 1999–2014 and 1979–1998 for (a) observed ERv4 DJFMAM SSTs and (b) JJASON SSTs and simulated (CESM1) SSTs for (c, d) DJFMAM and JJASON. Black lines denote schematic representation of the western V region. (*See insert for color representation of the figure.*)

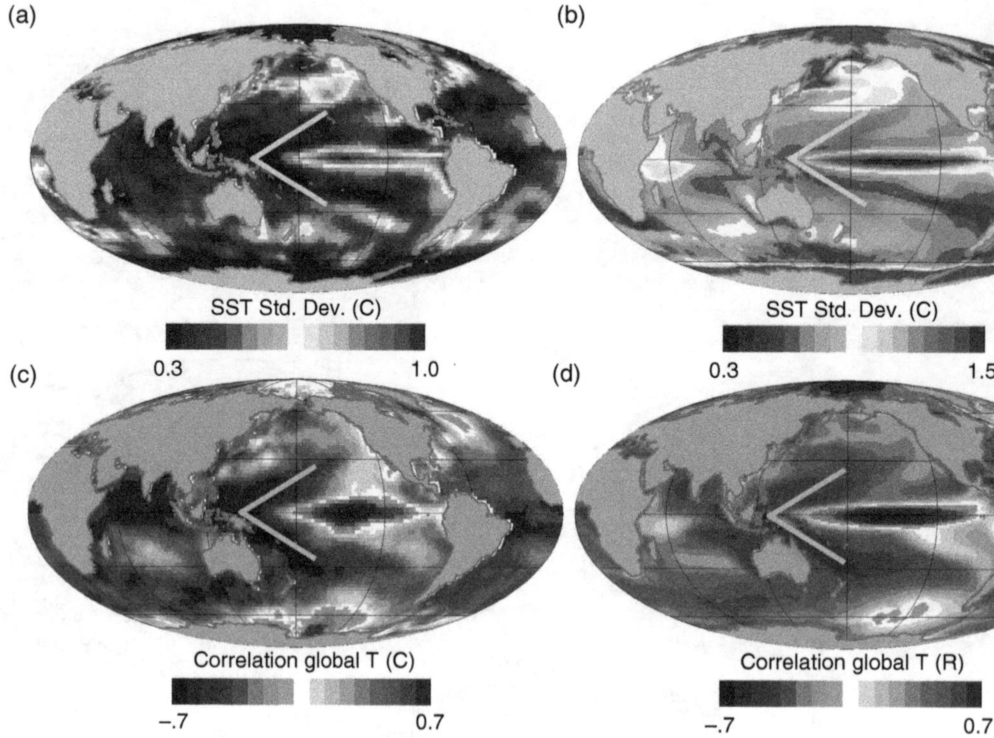

Figure 10.4 1900–2014 standard deviation of (a) ERv4 and (b) CESM1 DJFMAM SST. (c, d) 1900–2014 correlation between ENSO-residual DJFMAM SST and global average SSTs for ERv4 and CESM1.

observed during 1999–2014. This bias may also have a role in the El Niño-like climate projected by models. The CESM1 model appears to overestimate ENSO variability (Fig. 10.4a,b), and the magnitude of the ENSO eigenvectors and ENSO-related warming [*Funk and Hoell,* 2015]. Note, however, the variance of the ENSO mode (Fig. 10.1) and ENSO-SSTs (Fig. 10.4) is much higher than the WPWM and western V SSTs. This variance makes it difficult to identify the true rate of eastern Pacific warming with just a few decades of SST observations.

Of particular interest to us, however, from a climate extremes perspective, is the potential for interaction between the WPWM and cool eastern Pacific ENSO/PDV states. If the CESM1 model represents a good estimate of ENSO-warming, there will still be strong La Niña events, due to the high variance of ENSO (Figs. 10.1b, 10.4b), and a chance for internally generated PDV [*Meehl et al.*, 2013]. When eastern Pacific SST are cool, for a given season like 2010/2011 or a multiyear period like 1999–2014, these cool SST will interact with warming in the low variance ocean regions. Figure 10.4c,d shows the 1900–2014 correlation between ENSO-residual SST and global ocean SST averages. The CESM1 results were calculated independently for each simulation, and then averaged. Again, note the similarity between the correlations based on the observed and modeled SSTs. These correlation patterns also resemble the EOF loadings shown in Figure 10.2, reinforcing the interpretation of the WPWM as a trend mode of variability that keys in on low-variance ocean regions (Fig. 10.4a,b) that are well correlated with global temperature increases (Fig. 10.4c,d). We refer to this as the West Pacific Warming Mode because the dynamic response associated with this mode tends to focus on a west Pacific/east Pacific precipitation, SSH, and subsurface ocean dipole, mediated by an increase in central Pacific trade winds [*Funk and Hoell,* 2015].

10.3.2. WPWM and ENSO Relationships to Precipitation and Air Temperatures

We next explore the relationship between boreal winter/spring and summer/fall WPWM and ENSO and global precipitation and 2 m air temperatures using bivariate regressions. In each location, standardized time series of the WPWM and ENSO PCs are used to predict standardized seasonal rainfall totals or nonstandardized air temperature anomalies. Precipitation was standardized to help represent meaningful variations across large changes in mean precipitation. For observed precipitation, we use 1950–2014 GPCC precipitation data; data prior to 1950 were not used because of sparse observation networks in many regions [*Funk et al.*, 2015]. Figure 10.5 shows the ERv4/GPCC and CESM1 WPWM

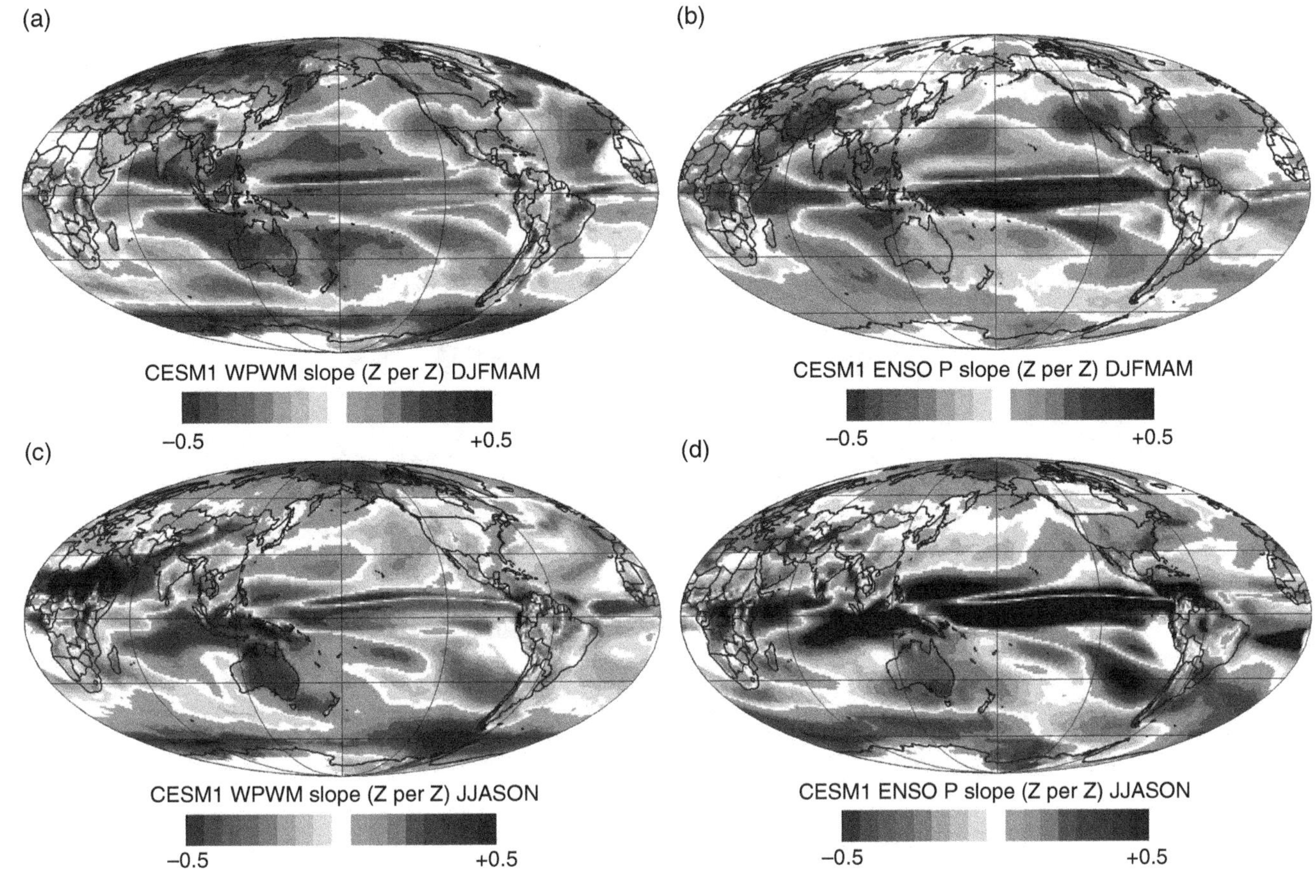

Figure 10.5 (a, b) Regressions between DJFMAM 1900–2014 CESM1 standardized precipitation and WPWM and ENSO SST PCs. (c, d) Same but for JJASON.

and ENSO regression results. For the CESM1 results, it is quite striking how opposed in sign the regression coefficients are. Despite the fact that the WPWM and ENSO SST EOF patterns are very different (Figs. 10.2 and 10.3), the CESM1 precipitation responses to these forcing patterns are very similar, with most of the same areas responding in an inverted way to warm WPWM and ENSO conditions. The ultimate source of this similarity, we would suggest, is likely due to the almost identical (but inverted) Indo-Pacific precipitation responses to WPWM and ENSO warming. Different patterns of SST warming create similar changes in Indo-Pacific diabatic warming, producing changes in the equatorial precipitation structure and off-equatorial Rossby wave responses over southwest Asia [*Mathew Barlow et al.*, 2002; *M. Barlow et al.*, 2007; *Hoell et al.*, 2014a] and North America [*Held et al.*, 2002].

The implication of Figure 10.5, from a multimodal climate perspective, is quite interesting. It suggests that precipitation in many regions of the world responds in similar ways to WPWM warming and La Niña-like eastern Pacific warming. Convolved within the CESM1 simulations we find two distinct signals, a low frequency trend mode that has La Niña-like teleconnections and consequences, and a higher frequency ENSO mode with the opposite impacts. In practice, this could indicate a transition toward a more Niña-like new normal, likely to be punctuated by very warm El Niño events.

Of course, such changes will also occur in the context of warming air temperatures, and we present in Figure 10.6 similar regressions with CESM1 air temperatures. The WPWM warming appears associated with increasing air temperatures in northern Eurasia, the Arctic, and North America. Note that since the WPWM has increased about two standardized deviations since the 1970s (Fig. 10.1a) these images denote substantial temperature increases. ENSO temperature influences focus mostly on warming in Australia, South America, and Africa.

10.3.3. GPCP WPWM Changes

We next present seasonal GPCP WPWM EOF results, as presented for MAMJ in *Funk and Hoell* [2015]. These results are based on the first EOF of ENSO-residual 1979–2014 GPCP precipitation. Changes between the

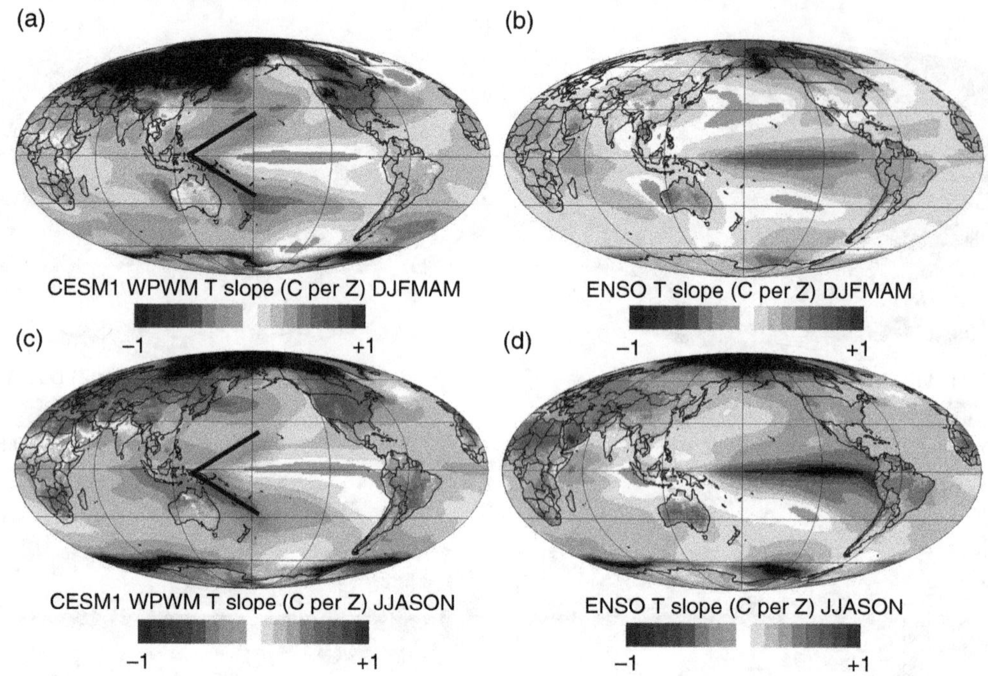

Figure 10.6 (a, b) Regressions between 1900–2014 DJFMAM CESM1 2 m air temperatures and WPWM and ENSO SST PCs. (c, d) Same but for JJASON.

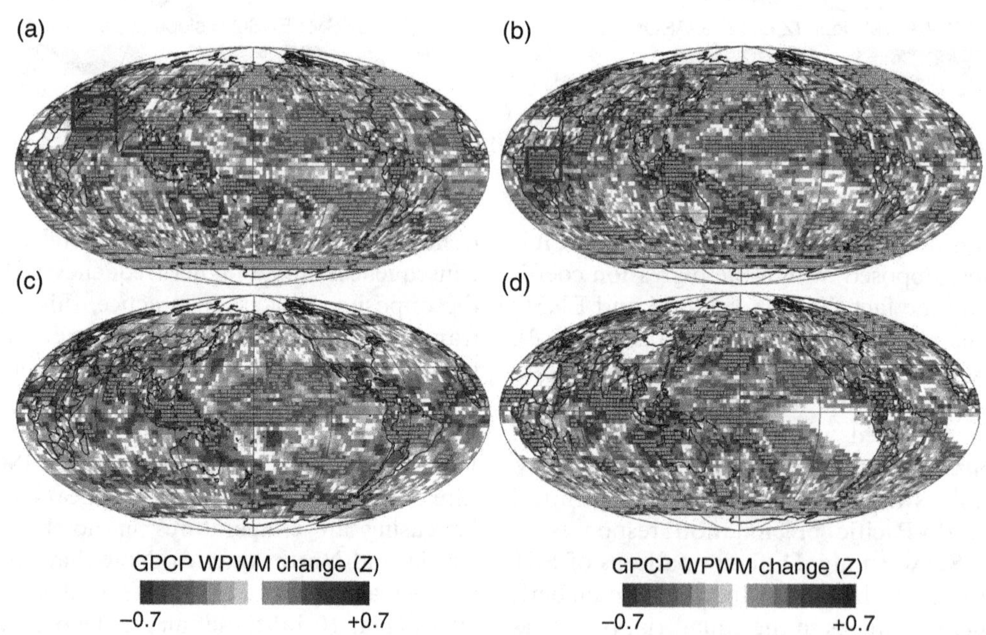

Figure 10.7 1999–2014 versus 1979–1998 GPCP changes associated with the WPWM EOF for (a) JFM, (b) AMJ, (c) JAS, and (d) OND. (*See insert for color representation of the figure.*)

1999–2014 and 1979–1998 WPWM PC were multiplied by the corresponding EOF and plotted in Figure 10.7. Values with mean seasonal precipitation of less than 10 mm have been screened. In each season one finds a strong dipole between the western and eastern Pacific precipitation responses, with another precipitation increase located in or near Central America. This wet-dry-wet tropical Pacific pattern also appears in the CESM1 precipitation responses (Fig. 10.5a,c), although the CESM1 responses appear more zonal.

The GPCP WPWM impacts on North and South America shift substantially by season. In JFM and AMJ, drying associated with subtropical high-pressure cell over the northeast Pacific extends into the southern United States and northern Mexico (Fig. 10.7). This drying relates to an intensification and drying of the subtropical highs in the eastern Pacific, and seems broadly consistent with the CESM1 WPWM teleconnections for DJFMAM (Fig. 10.5b). During boreal summer (JAS) significant drying impacts disappear in the United States, and areas of South America exhibit negative precipitation impacts. The GPCP WPWM results (Fig. 10.7) appear quite similar to Figure 10.1 of a recent study [*Gao et al.*, 2014], which finds a robust spring drying signal across the southern United States and Mexico, and a boreal summer drying across the northern United States. The boreal spring drying pattern, but not the summer signal, appears clearly in the WPWM results (Fig. 10.5).

For eastern Africa and NME/SWE, impacts also vary by season. During JFM, there is a large coherent region of drying stretching from Ethiopia to northeastern Iran. This same region appears drier during OND as well (Fig. 10.7d), but fewer locations indicated significant precipitation reductions. During AMJ there are substantial precipitation decreases identified in the Greater Horn of Africa region (Fig. 10.7c). While the CESM1 WPWM results actually suggest some precipitation increases over the Greater Horn of Africa (Fig. 10.7a), these increases appear amid drying signals to the east and west; it seems likely that CESM1 is getting a reasonable picture of large scale zonal Walker circulation anomaly, but missing some aspects of the complex regional circulation. During JAS, the WPWM Indo-Pacific precipitation increases appear to merge with Indian monsoon rainfall increases. Precipitation increases are seen across the Sahel and north tropical Atlantic as well. These latter changes are certainly heavily influenced by decadal SST variations in the Atlantic.

The PC time series for the seasonal GPCP WPWM analysis are shown in Figure 10.8a. All four time series exhibit substantial decadal variations, with substantial increases beginning in the mid-1990s. The JFM and AMJ time series, more influenced perhaps by Southern Ocean SST, exhibit less interannual variability than JAS and OND. All seasons indicate that a large transition in global precipitation accompanied the western V-like warming identified in the SST WPWM analysis.

10.3.4. GPCC Precipitation and GISS Air Temperature Time Series for NME/SWE and Greater Horn of Africa

We conclude by plotting long (1901–2014) time series of 15 yr running mean GPCC precipitation and air temperatures anomalies for the NME/SWE (October–May) and the Greater Horn of Africa (April–June) (Fig. 10.8). These areas were selected because they are water insecure

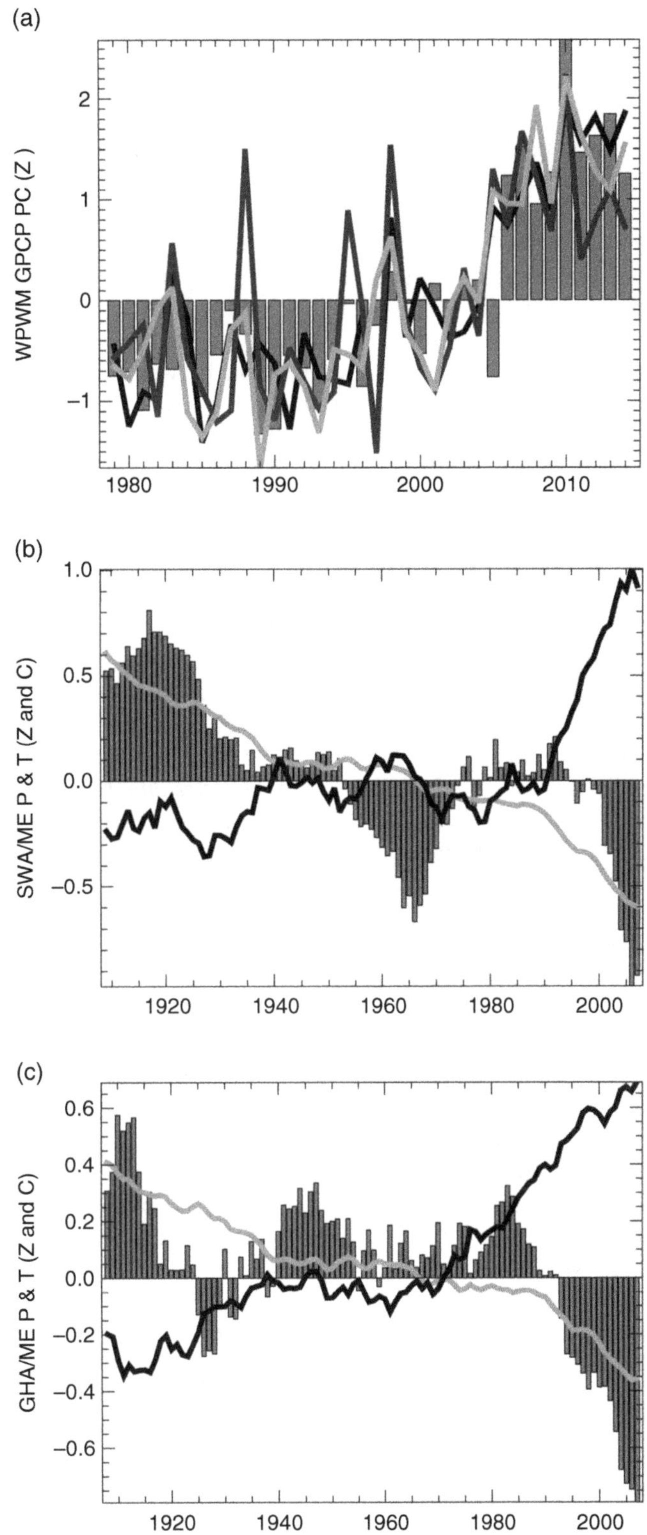

Figure 10.8 (a) GPCP WPWM PCs for JFM (black bars), AMJ (black), JAS (dark gray), and OND (light gray). (b) SWA/ME GPCC precipitation (black bars), estimated precipitation (light gray), and GISS air temperatures (dark gray). (c) Same but for GHA. *See electronic version for color representation.*

regions with rapidly growing populations, and exhibited coherent and significant GPCP WPWM drying signals. In the NME/SWE (Fig. 10.8b), we see a dramatic downward shift in precipitation beginning in the later 1990s, with the past 15 yr being apparently the driest on record. Air temperatures have also increased by about 1°C since the early 1990s. The rapid transition from relatively stable 1940–1980 air temperatures to rapidly increasing air temperatures may indicate a dynamic forcing mechanism.

In the Greater Horn of Africa (Fig. 10.8c), we find a fairly linear decrease in precipitation between the early 1980s and 2014, with the most recent 15 yr period being identified as the lowest on record. Air temperatures appear to have increased in a fairly linear manner since the early 1960s, with the total amount of warming being on the order of + 0.6°C.

10.4. SUMMARY AND DISCUSSION

The past 15 yr (1999–2014) have presented a complex tableau of climate change, variability, and extremes. On one hand, total emissions and greenhouse gasses continue to increase [*Le Quéré et al.*, 2013] with a total 2014 radiative forcing of almost 4 Wm^{-2} [*Butler and Montzka*, 2015]. While this radiative forcing will certainly increase the Earth's absorption of longwave radiation, this increase will be modulated by changes in the distribution of water vapor; areas experiencing water vapor increases (decreases) will experience more (less) local radiative forcing. In the oceans, patterns of warming will be further complicated by ocean-atmosphere dynamics, associated upwelling and downwelling, and changes in the depth of the mixed layer. What one sees from climate change simulations, such as the CESM1 simulations analyzed here, is that warming within a given simulation resembles warming in the observed ocean. As radiative forcing increases, there are more regions exhibiting exceptional warmth. The pattern of warming, however, shifts from season to season, depending on the state of the climate. There are important (and separate) modes of climate variability that help produce the warming trends. To understand the historical and future drivers of global change, these need to be closely examined.

In this brief study, we represented the state of the climate in one of the simplest ways possible, by the first two dominant EOFs – ENSO and the WPWM. The CESM1 indicates increases (warming) in both these modes. This modal perspective has important implications for identifying, and predicting, seasonal climate extremes associated with very warm SST.

As a thought experiment, assume that these modes perfectly encompass the observed SST variability, and that the combination of these modes, given enough samples, at any one time, would reveal a perfectly even (equitropic) warming pattern. This equitropic warming pattern would not be the climate that we experienced. Rather, during any given season, nature would provide a sample from the WPWM mode and ENSO mode. One year (soon) we will likely experience a very strong El Niño event. More mixed greenhouse gasses, more water vapor, and more heat energy in the Pacific Ocean will create very warm SSTs in the high-variance eastern Pacific (Fig. 10.4a,b).

The next year, La Niña-like conditions in the eastern Pacific might interact in a positive way with the WPWM, creating a stronger wet Pacific SST gradient [*Hoell and Funk*, 2013a]. Our work has suggested that the combination of warm west Pacific SST and La Niña can amplify drying along the northwestern rim of the Indian Ocean [*Hoell et al.*, 2014b], leading to drying over NME/SWE and the Greater Horn of Africa. Our research has also suggested that the gradient between the central and western equatorial Pacific SST helps drive seasonal precipitation anomalies over Africa [*Funk et al.*, 2014; *Liebmann et al.*, 2014] and Asia [*Hoell et al.*, 2014a]. This SST gradient is certainly influenced by anthropogenic WPWM-like forcing [*Funk and Hoell [2015]*] and ENSO/PDV variability [*Lyon et al.*, 2013].

One possibility, suggested by the analysis here, might be that we may transition into a period dominated by a more Niña-like new normal that is periodically punctuated by very warm El Niño events. When PDV/ENSO variability results in neutral ENSO conditions, we may experience modest La Niña-like impacts, possibly similar to Figure 10.5a,c. When we experience strong La Niña events, the CESM1 simulations suggest that the associated climate responses to this east Pacific cooling may combine with WPWM influences. This may be a source of the exceptional recent drying experienced in the NME/SWE and Greater Horn of Africa (Fig. 10.8).

REFERENCES

Adler, R. F., et al. (2003), The version 2 Global Precipitation Climatology Project (GPCP) monthly precipitation analysis (1979–present), *J. Hydrometeor.*, *4*, 1147–1167.

Barlow, M., A. Hoell, and F. Colby (2007), Examining the wintertime response to tropical convection over the Indian Ocean by modifying convective heating in a full atmospheric model, *Geophys. Res. Lett.*, *34*(19), L19702.

Barlow, M., H. Cullen, and B. Lyon (2002), Drought in central and Southwest Asia: La Niña, the warm pool, and Indian Ocean precipitation, *J. Climate*, *15*(7), 697–700.

Becker, A., P. Finger, A. Meyer-Christoffer, B. Rudolf, K. Schamm, U. Schneider, and M. Ziese (2013), A description of the global land-surface precipitation data products of the global precipitation climatology centre with sample applications including centennial (trend) analysis from 1901–present, *Earth Syst. Sci. Data*, *5*, 71–99.

Butler, J. H., and S. A. Montzka (2015), The NOAA annual greenhouse gas index (AGGI), edited, NOAA, NOAA Earth System Research Laboratory, R/GMD, Boulder, CO.

Cane, M. A., A. C. Clement, A. Kaplan, Y. Kushnir, D. Pozdnyakov, R. Seager, S. E. Zebiak, and R. Murtugudde (1997), Twentieth-century sea surface temperature trends, *Science*, *275*(5302), 957–960.

Compo, G. P., and P. D. Sardeshmukh (2010), Removing ENSO-related variations from the climate record, *J. Climate*, *8*(23), 1957–1978.

Compo, G. P., et al. (2011), The twentieth century reanalysis project, *Quart. J. Roy. Meteor. Soc.*, *137*(654), 1–28.

England, M. H., S. McGregor, P. Spence, G. A. Meehl, A. Timmermann, W. Cai, A. S. Gupta, M. J. McPhaden, A. Purich, and A. Santoso (2014), Recent intensification of wind-driven circulation in the Pacific and the ongoing warming hiatus, *Nature Climate Change*, *4*(3), 222–227.

Funk, C., and A. Hoell (2015), The leading mode of observed and CMIP5 ENSO-residual sea surface temperatures and associated changes in Indo-Pacific climate, *J. Climate*, *28*(11), 4309–4329.

Funk, C., A. Hoell, S. Shukla, I. Bladé, B. Liebmann, J. B. Roberts, and G. Husak (2014), Predicting East African spring droughts using Pacific and Indian Ocean sea surface temperature indices, *Hydrol. Earth Syst. Sci. Discuss*, *11*(3), 3111–3136.

Funk, C., M. D. Dettinger, J. C. Michaelsen, J. P. Verdin, M. E. Brown, M. Barlow, and A. Hoell (2008), Warming of the Indian Ocean threatens eastern and southern African food security but could be mitigated by agricultural development, *Proc. Nat. Acad. Sci. USA*, *105*(31), 11081–11086.

Funk, C., S. Nicholson, D. Klotter, M. Landsfeld, and P. Peterson (2015), The centennial trends greater horn of Africa precipitation dataset, *Nat. Sci. Data*, *2*; doi: 10.1038/sdata.2015.50.

Gao, Y., L. R. Leung, J. Lu, Y. Liu, M. Huang, and Y. Qian (2014), Robust spring drying in the southwestern US and seasonal migration of wet/dry patterns in a warmer climate, *Geophys. Res. Lett.*, *41*(5), 1745–1751.

Gill, A. E. (1980), Some simple solutions for heat-induced tropical circulation, *Quart. J. Roy. Meteor. Soc.*, *106*, 447–462.

Graham, N. E. (1994), Decadal-scale climate variability in the tropical and North Pacific during the 1970s and 1980s: Observations and model results, *Climate Dyn.*, *10*(3), 135–162.

Hansen, J., R. Ruedy, M. Sato, and K. Lo (2010), Global surface temperature change, *Rev. Geophys.*, *48*(4).

Held, I. M., M. Ting, and H. Wang (2002), Northern winter stationary waves: Theory and modeling, *J. Climate*, *15*(16), 2125–2144.

Hoell, A., and C. Funk (2013a), The ENSO-related West Pacific sea surface temperature gradient, *J. Climate*, *26*(23), 9545–9562.

Hoell, A., and C. Funk (2013b), Indo-Pacific sea surface temperature influences on failed consecutive rainy seasons over eastern Africa, *Climate Dyn.*, *43* (5–6), 1645–1660.

Hoell, A., C. Funk, and M. Barlow (2013), The regional forcing of Northern Hemisphere drought during recent warm tropical west Pacific Ocean La Niña events, *Climate Dyn.*, *42*(11–12), 3289–3311.

Hoell, A., C. Funk, and M. Barlow (2014b), La Niña diversity and the forcing of northwest Indian Ocean rim teleconnections, *Climate Dyn.*, *42* (11–12), 3289–3311.

Hoell, A., F. C. Barlow, and N. Barlow (2014a), The forcing of Southwest Asia teleconnections by low frequency Indo-Pacific Sea surface temperature variability during boreal winter, *J. Climate* (in press).

Kay, J., C. Deser, A. Phillips, A. Mai, C. Hannay, G. Strand, J. Arblaster, S. Bates, G. Danabasoglu, and J. Edwards (2015), The Community Earth System Model (CESM) large ensemble project: A community resource for studying climate change in the presence of internal climate variability, *Bull. Amer. Meteor. Soc.*, *96*(8), 1333–1349.

Kent, C., R. Chadwick, and D. P. Rowell (2015), Understanding uncertainties in future projections of seasonal tropical precipitation, *J. Climate*, *28*(11), 4390–4413.

Le Quéré, C., et al. (2013), Global carbon budget 2013, *Earth Syst. Sci. Data Discuss.*, *6*(2), 689–760.

Liebmann, B., M. P. Hoerling, C. Funk, I. Bladé, R. M. Dole, D. Allured, X. Quan, P. Pegion, and J. K. Eischeid (2014), Understanding recent Eastern Horn of Africa rainfall variability and change, *J. Climate*, *27*(23), 8630–8645.

Lyon, B. (2014), Seasonal drought in the Greater Horn of Africa and its recent increase during the March–May long rains, *J. Climate*, *27*(2014), 7953–7975.

Lyon, B., and D. G. DeWitt (2012), A recent and abrupt decline in the East African long rains, *Geophys. Res. Lett.*, *39*(L02702).

Lyon, B., A. G. Barnston, and D. G. DeWitt (2013), Tropical pacific forcing of a 1998–1999 climate shift: Observational analysis and climate model results for the boreal spring season, *Climate Dyn.*, *43* (3–4), 893–909.

Madden, R. A., and P. R. Julian (1972), Description of global-scale circulation cells in the tropics with a 40–50 day period, *J. Atmos. Sci.*, *29*(6), 1109–1123.

Mantua, N., and S. Hare (2002), The Pacific decadal oscillation, *J. Oceanogr.*, *58*(1), 35–44.

Meehl, G. A., A. Hu, J. M. Arblaster, J. Fasullo, and K. E. Trenberth (2013), Externally forced and internally generated decadal climate variability associated with the interdecadal Pacific oscillation, *J. Climate*, *26*(18), 7298–7310.

Merrifield, M. A. (2011), A Shift in Western tropical Pacific sea level trends during the 1990s, *J. Climate*, *24*(15), 4126–4138.

Merrifield, M. A., and M. E. Maltrud (2011), Regional sea level trends due to a Pacific trade wind intensification, *Geophys. Res. Lett.*, *38*(21), L21605.

Newman, M. (2013), An empirical benchmark for decadal forecasts of global surface temperature anomalies, *J. Climate*, *26*(14), 5260–5269.

Rodwell, M. J., and B. J. Hoskins (1996), Monsoons and the dynamics of deserts, *Quart. J. Roy. Meteor. Soc.*, *122*(534), 1385–1404.

Sandeep, S., F. Stordal, P. Sardeshmukh, and G. Compo (2014), Pacific Walker circulation variability in coupled and uncoupled climate models, *Climate Dyn.*, *43*(1–2), 103–117.

Schneider, U., A. Becker, P. Finger, A. Meyer-Christoffer, M. Ziese, and B. Rudolf (2013), GPCC's new land surface precipitation climatology based on quality-controlled in situ data and its role in quantifying the global water cycle, *Theor. App. Climatol.*, *115*(1–2), 15–40.

Schubert, S., et al. (2009), A U.S. CLIVAR project to assess and compare the responses of global climate models to drought-related sst forcing patterns: Overview and results, *J. Climate*, *22*(19), 5251–5272.

Smith, T. M., R. W. Reynolds, T. C. Peterson, and J. Lawrimore (2008), Improvements to NOAA's historical merged land–ocean surface temperature analysis (1880–2006), *J. Climate*, *21*(10), 2283–2296.

Solomon, A., and M. Newman (2012), Reconciling disparate twentieth-century Indo-Pacific ocean temperature trends in the instrumental record, *Nature Climate Change*, *2*(9), 691–699.

Taylor, K. E., R. J. Stouffer, and G. A. Meehl (2011), An overview of CMIP5 and the experiment design, *Bull. Amer. Meteor. Soc.*, *93*(4), 485–498.

Trenberth, K. E., J. T. Fasullo, G. Branstator, and A. S. Phillips (2014), Seasonal aspects of the recent pause in surface warming, *Nature Climate Change*, *4*(10), 911–916.

Walker, G., and E. Bliss (1932), World weather V, *J. Roy. Meteor. Soc.*, *4*, 53–84.

Walker, G. T. (1923), Correlation in seasonal variations of weather, VIII: A preliminary study of world weather, in *Memoirs of the Indian Meteorological Department*, edited, Indian Meteorological Department, Calcutta.

Wheeler, J., and W. Zurek (1990), Complexity, entropy and the physics of information, *Information, Physics, Quantum: The Search for Links*.

Williams, P., and C. Funk (2011), A westward extension of the warm pool leads to a westward extension of the Walker circulation, drying eastern Africa, *Climate Dyn.*, *37*(11–12), 2417–2435.

Yang, W., R. Seager, M. A. Cane, and B. Lyon (2014), The East African long rains in observations and models, *J. Climate*, *27*(19), 7185–7202.

11

Connections Between Heat Waves and Circumglobal Teleconnection Patterns in the Northern Hemisphere Summer

Haiyan Teng and Grant Branstator

ABSTRACT

Based on a 12,000 year integration of a climate model free of year-to-year variations in both sea surface temperature and external forcing, we have investigated linkages between heat waves in the Northern Hemisphere summer and quasi-stationary planetary wave anomalies produced by atmospheric internal variability. We find heat waves are more frequent in the subtropics and at higher latitudes, where intraseasonal variability is more pronounced relative to synoptic variability thus making persistent extreme events more likely. The intraseasonal variability of the upper-troposphere meridional wind is led by a pair of zonal wavenumber-5 circumglobal teleconnection (CGT) patterns that resulted from the mean jet's waveguide effect. When circulation anomalies have unusually high projections onto the CGT patterns, the probability of heat waves can be increased/decreased over much of the hemisphere by a factor of 4 both concurrently and in the following two weeks. Because these CGT patterns are quasi-stationary, through their influence on the likelihood of heat waves, they provide a source of predictability for heat waves on subseasonal timescales. Yet, because heat waves happen so infrequently, even if a circulation anomaly makes a fourfold increase in the likelihood of a heat wave occurring, the probability of a heat wave is still very low.

11.1. INTRODUCTION

With more intense and frequent heat waves in recent decades [*Easterling et al.*, 2000; *Meehl and Tebaldi*, 2004], there has been a growing demand for understanding key processes and feedbacks contributing to such events. In this study, we aim to investigate one potentially important mechanism, namely, to what extent *anomalous quasi-stationary planetary waves,* due to atmospheric internal dynamics, may affect the probability of heat waves. A better understanding of this relationship should also assist in assessing the potential impacts on heat waves of changes in planetary wave characteristics that occur in a future warmer climate.

We focus on heat waves that last for about a week or two in the Northern Hemisphere during the June–August (JJA) season. The emphasis on the intraseasonal timescale sets the heat waves we study apart from extreme weather events caused by synoptic eddies. Although both phenomena may exhibit similar regional features, such as high-temperature anomalies at the surface and anticyclonic circulation anomalies aloft, there is evidence that heat waves tend to be associated with nearly global-scale anomalous quasi-stationary planetary wave patterns [*Namias,* 1982; *Chang and Wallace*, 1987; *Chen and Newman*, 1998; *Lyon and Dole*, 1995; *Schubert et al.*, 2011], which appear to be conducive to the production of long-lasting events characterized by high surface temperature anomalies. The connection between such planetary wave anomalies and prolonged extreme weather conditions in midlatitudes was statistically supported by *Screen and Simmonds* [2014] and the high amplitude and persistent

National Center for Atmospheric Research, Boulder, Colorado, USA

Climate Extremes: Patterns and Mechanisms, Geophysical Monograph 226, First Edition.
Edited by S.-Y. Simon Wang, Jin-Ho Yoon, Christopher C. Funk, and Robert R. Gillies.

planetary wave anomalies are further explained by Rossby wave resonance [*Petoukhov et al.*, 2013; *Coumou et al.*, 2014].

Throughout this paper, the term *quasi-stationary planetary wave anomalies* refers to zonally asymmetric features of *intraseasonal variability* of the atmospheric circulation. That is, here we are studying the relationship of anomalous heat wave probabilities to anomalous planetary wave patterns rather than to the climatological mean waves. The most famous form of quasi-stationary planetary wave anomalies is probably that of atmospheric teleconnection patterns that have strong variations in the meridional direction, for example, the Pacific–North American pattern (PNA) [*Wallace and Gutzler*, 1981] and the North Atlantic Oscillation [*Hurrell*, 1995]. A second type of planetary wave anomaly emphasizes the waveguide effect of the mean jet [*Branstator*, 1983; *Hoskins and Ambrizzi*, 1993]. They are often referred to as circumglobal teleconnection (CGT) patterns following *Branstator* [2002]. Unlike those canonical teleconnection patterns that have a distinct meridional component to their group velocities [*Hoskins and Karoly*, 1981], the waveguide modes propagate along latitude circles because they tend to be meridionally confined to the vicinity of the jets due to the strong meridional gradients that exist on the flanks of the jets. Despite much weaker mean jets and consequently a less pronounced waveguide effect in the summertime, a wavenumber-5 pattern somewhat like the wintertime leading CGT pattern is also found in the summer [*Ding and Wang*, 2007]. Studies indicate it has a strong impact on local extremes [*Wang et al.*, 2010; *Teng et al.*, 2013].

Although internal atmospheric dynamics are sufficient to produce CGT patterns [*Ding et al.*, 2011], external forcing can also induce them on subseasonal to decadal timescales. For example, *Ding and Wang* [2005] suggest that monsoon heating anomalies can amplify CGT anomalies, and model simulated climate responses to increasing greenhouse gas forcing sometimes show a strong resemblance to the intrinsic wavenumber-5 CGT pattern. Just as internally generated planetary wave anomalies have the potential to affect the probability of heat waves, so can externally forced planetary wave variations. This adds to the difficulty of assessing to what degree an observed heat wave event was made more likely because of natural processes and to what extent it was affected by climate change (e.g., the debate over the cause of the Russian heat wave; *Dole et al.* [2011]).

Whereas observations are key to describing characteristics of individual extreme events, it is difficult to rigorously test the degree to which quasi-stationary wave anomalies resulting from natural variability influence heat waves. This difficulty arises from the scarcity of these events in the observational record and the inherent noisiness of midlatitude weather fluctuations. To avoid these problems, recently *Teng et al.* [2013] employed a 12,000 year integration of an atmospheric general circulation model. They found that summertime heat waves over the continental United States tend to be accompanied by a wavenumber-5 CGT pattern in the upper troposphere. More important from a prediction perspective was their finding that heat wave events are often preceded by the CGT pattern 15–20 days before the heat waves occur.

While the above study, together with those of others, highlighted the importance of the relationship between heat waves and quasi-stationary planetary wave anomalies, it was incomplete in several respects. One particular limitation is that *Teng et al.* [2013] only focused on heat waves in the US region. The purpose of this study is to address whether planetary wave events may also play an important role in heat waves in other places in the Northern Hemisphere. Motivated by *Teng et al.* [2013], we will continue to focus on the CGT patterns and their connections with heat waves. Since these patterns span most of the Northern Hemisphere, they have the potential to affect the likelihood of heat waves in many regions. As we will demonstrate, concentrating on CGT patterns does capture substantial linkages between heat waves and quasi-stationary planetary wave anomalies.

The chapter is organized as follows: Section 11.2 describes the datasets and methodology. Section 11.3 provides an overview of the distribution of heat waves in the Northern Hemisphere summer. In section 11.4, we investigate the relationship between heat waves and planetary wave anomalies from two different perspectives. We first select four domains at widely separated locations along the summer mean jet to identify anomalous planetary wave patterns associated with heat waves in those regions. Those findings steer us to our second approach, which is to investigate connections between the leading CGT patterns and heat waves throughout the entire Northern Hemisphere. Implications of this model-based study for the real world are discussed in section 11.5.

11.2. DATA AND METHODS

The general circulation model that *Teng et al.* [2013] considered and that we analyze here is CAM3. CAM3 is a particular configuration of a comprehensive atmosphere-ocean-land-sea ice general circulation model, namely National Center for Atmospheric Research Community Climate System Model version 3 (CCSM3) [*Collins et al.*, 2006]. In this configuration of CCSM3, only the atmosphere and land components are actively coupled. We ran CAM3 for 12,000 years at a T42 horizontal resolution, which corresponds to approximately a 2.8° latitude/longitude grid spacing. Monthly varying SST and sea ice concentrations as well as external forcing (solar, anthropogenic) were fixed at present-day climatological levels. We have also employed the reanalysis data from National Centers for Environmental Prediction–National Center for Atmospheric Research (NCEP-NCAR) [*Kalnay et al.*, 1996]. Although it is short for studying probabilities of extreme events, this dataset provides a benchmark for basic characteristics of intraseasonal variability in both surface air temperature and planetary waves. Daily NCEP-NCAR reanalysis data during 1948–2013 have been regridded to CAM3s T42 grids to facilitate the comparison.

There is no generally agreed upon definition of a heat wave. Although relative humidity is an important factor when considering the impacts of heat waves, here we devote our attention solely to events that distinguish themselves by having persistent high temperature anomalies. Specifically, at each grid point, if a day's averaged surface air temperature (TAS) along with averaged TAS during each of the three preceding and three succeeding days all exceed a corresponding threshold value, we call the day a heat wave day. The purpose of imposing the 7 day criterion is to help separate long-lasting, and thus more catastrophic, events from local short-lived events. (Changing 7 days to 5 days does not affect the major conclusions of this study.) The threshold value varies with both grid point and time of the year. For a particular day of the year and grid point, it corresponds to the 97.5 percentile of TAS anomalies for all historical values within a centered 30 day running window. To facilitate computation, for the CAM3 dataset, we divided the total 12,000 years into 12 approximately equal length segments, and applied the heat wave criteria in each segment separately.

Based on this definition, the threshold values exhibit a distinct temporal evolution within the JJA season. In order to get a rough sense of their amplitude, we plot the seasonally averaged threshold values for both CAM3 and nature in Figure 11.1. In general, the threshold values are much higher over land than over ocean, due to the large heat capacity of the ocean mixed layer (and fixed SST in CAM3). Hereafter, we will focus only on heat waves over land in the Northern Hemisphere. Over the continents, the values increase with latitude from about 3–4°C in the subtropical area to above 7–8°C close to the polar region. In many regions, the model slightly overestimates the threshold values. The discrepancies can be caused by model bias, an insufficient amount of observational data, and the atmospheric stand-alone configuration of the

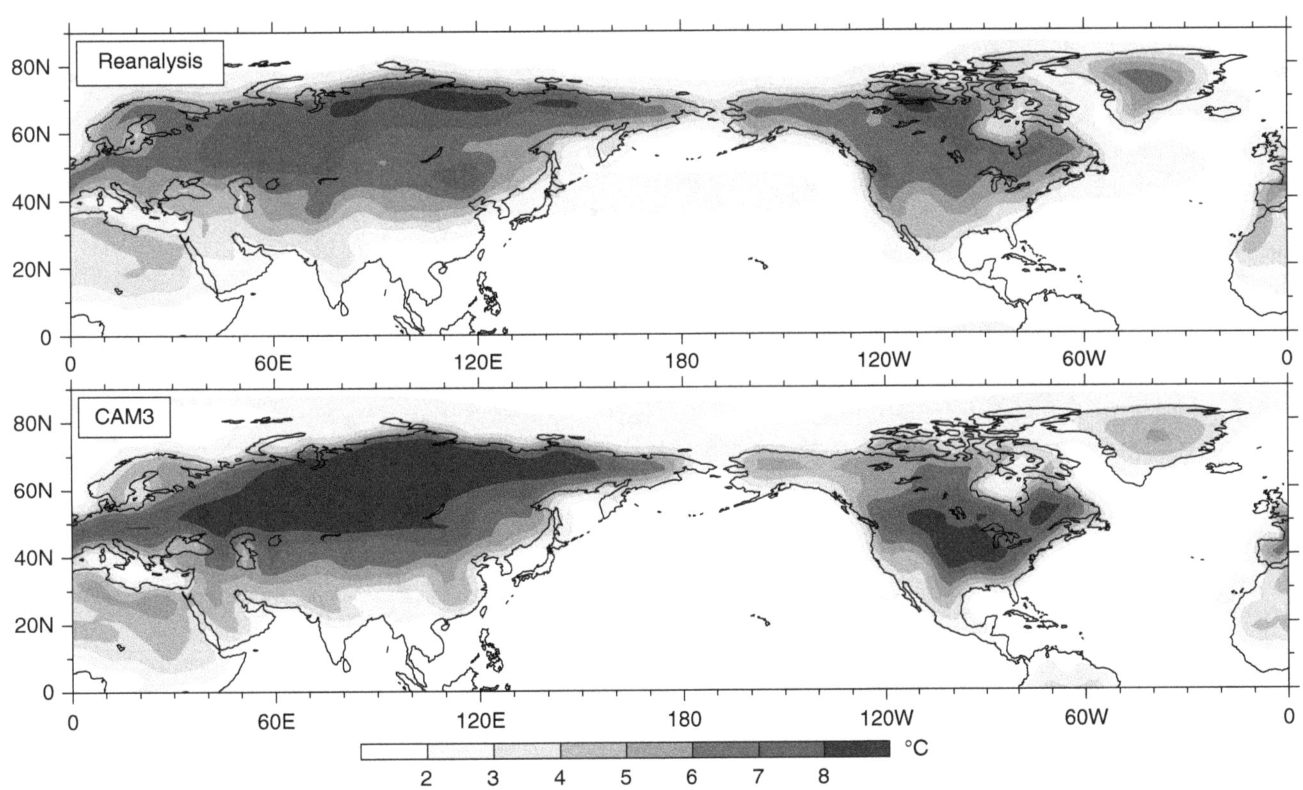

Figure 11.1 June to August averaged heat wave threshold values in the NCEP-NCAR reanalysis during years 1948–2013 (top) and in a 1000 year CAM3 simulation (bottom). For a particular day of the year and grid point, the threshold value corresponds to the 97.5th percentile of daily averaged surface air temperature anomalies for all historical values within a centered 30 day running window.

model. A comprehensive comparison of the model and the observations is difficult due to the brevity of the observations. Nevertheless, there are qualitative similarities in the threshold values between the model and the reanalysis data indicating our use of the CAM3 experiment is not unreasonable.

11.3. DISTRIBUTION OF HEAT WAVES

The occurrence rate of heat wave days during JJA is presented in Figure 11.2. Higher rates, which can be caused by both more events and longer duration, are mainly located in two latitude bands in CAM3, with one band at high latitudes close to the polar region and the second band in the subtropics (Fig. 11.2, bottom). Within these two bands, there are roughly 5–20 heat wave days during JJA per 100 years at any given location. These numbers roughly double or triple if we lower the 7 day criterion to 5 days. Both the pattern and the magnitude of the occurrence rate are consistent among each of the twelve 1000 year segments of the CAM3 simulation, indicating robustness of the statistics. On the other hand, the two-band feature is unclear in the reanalysis (Fig. 11.2, top), which could be caused by the brevity of the data. Consistent with this possibility, when we randomly discard one or two years' reanalysis data, important features are affected substantially. For example, we find the high occurrence rate area over western Russia disappears if 2010, which was the year of an unprecedented Russian heat wave event, is left out. The atmosphere stand-alone configuration of the CAM3 experiment may also contribute to the discrepancy, since slowly varying tropical SST and associated diabatic heating can remotely force intraseasonal and interannual variability in the extratropics, possibly inducing heat waves in regions where their likelihood would be scarce if the atmospheric variability were subject only to midlatitude intrinsic fluctuations.

Why are there more heat wave days in these two latitude bands in CAM3? A possible explanation is related to our definition of heat waves, which requires prolonged events. A consequence of this requirement is that for heat waves to be prevalent in a region, there needs to be pronounced variability in TAS on intraseasonal timescales. Indeed, if we reduce the 7 day criterion for heat waves to 1 day, frequency of the heat wave days becomes almost uniformly distributed around the globe. The two-band structure starts to appear after we increase the criterion to longer than 3 days. On the other hand, when we plot the standard deviation of 20–90 day bandpass filtered TAS anomalies during northern summer, it generally increases with

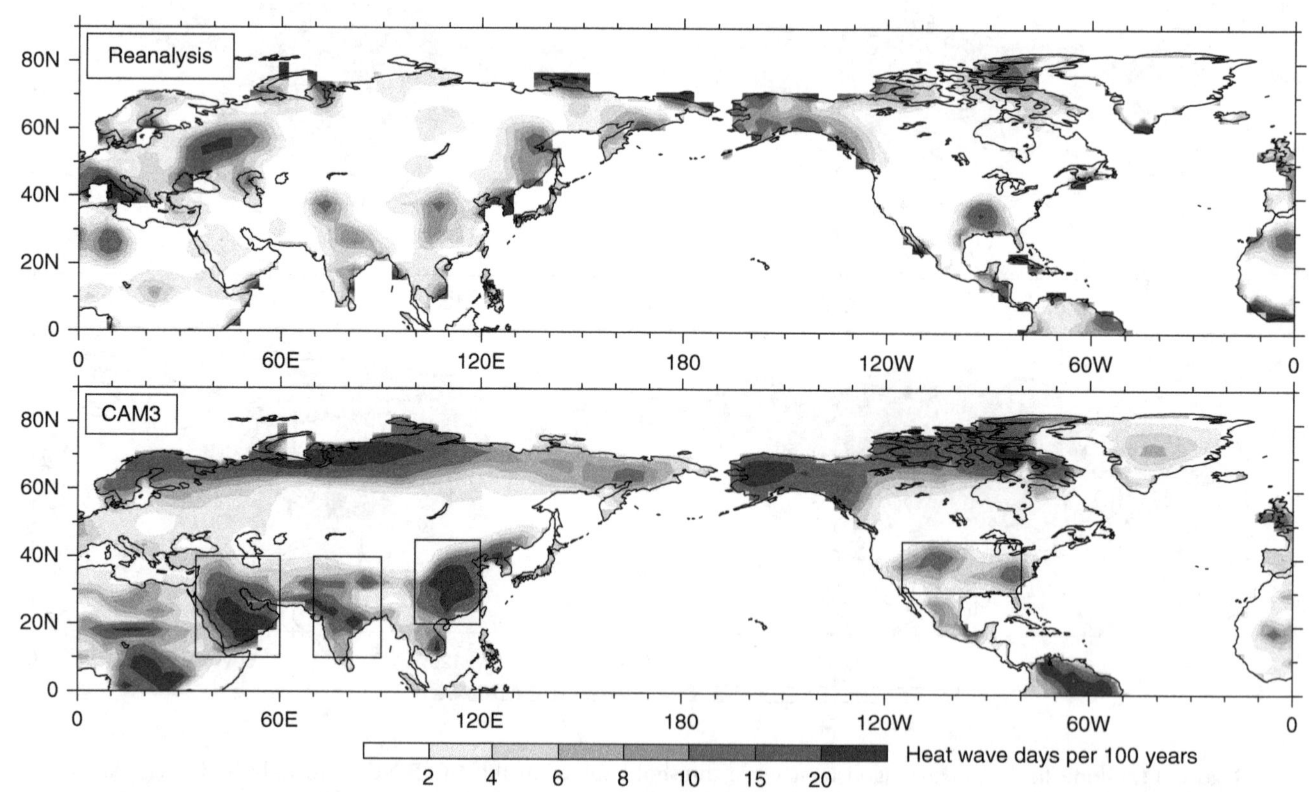

Figure 11.2 Heat wave occurrence rate during Jun-Aug in the reanalysis (top) and the CAM3 12,000 year simulation (bottom).

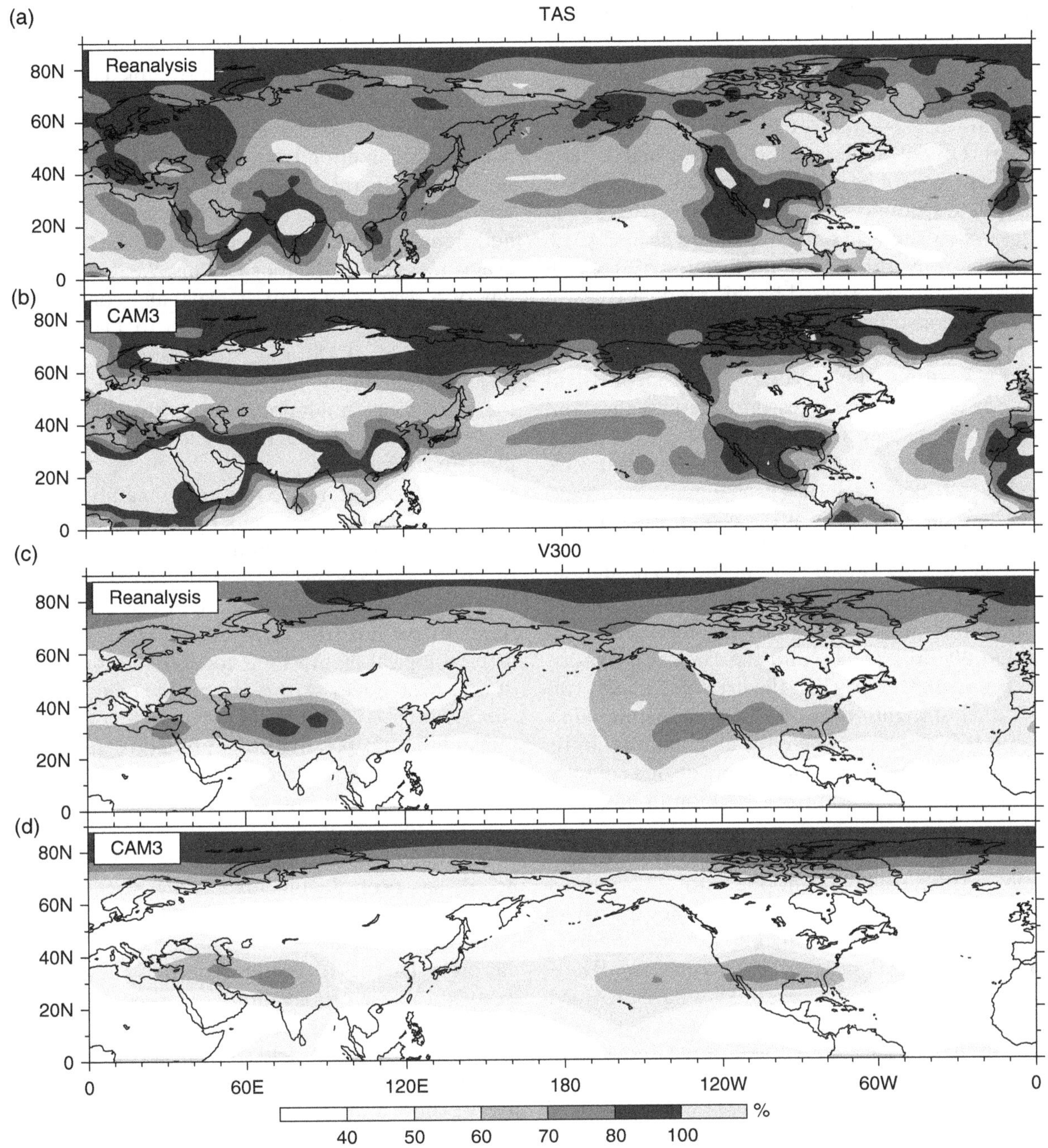

Figure 11.3 Ratio of standard deviation of 20–90 day band-pass filtered and 2–8 day high-pass filtered surface air temperature (TAS, a, b) and 300 hPa meridional wind (v300, c, d) during June to August from the reanalysis (a, c) and the CAM3 simulation (b, d).

latitude rather than having two maximum latitude bands. Hence, the presence of high-amplitude intraseasonal variability is not sufficient for a region to have a high likelihood of heat waves.

To understand why high-amplitude intraseasonal variability does not necessarily lead to more heat wave days, it is helpful to recognize that our definition incorporates a locally defined threshold. What is needed for the threshold to be met at some location is not that there be substantial low frequency variability relative to other locations but rather that the low frequency variability be strong compared to high-frequency variability at that location. To find locations where this criterion is met, we have calculated the ratio of 20–90 day intraseasonal variability to 2–8 day synoptic timescale variability. Figure 11.3b shows this ratio for TAS in CAM3. When compared to Figure 11.2b, this figure confirms that in CAM3, it is regions with a higher ratio that have more heat wave days. Not only is the ratio higher in the same latitude bands where heat wave days are more frequent, but even locally (e.g., southeast China, India, Arabia, and the southern United States) regions with a higher ratio are more likely to have heat waves. In nature, the ratio of intraseasonal- to synoptic-scale TAS variability (Fig. 11.3a) maximizes in two bands in similar locations as in CAM3. Hence, under our hypothesis that the location of frequent heat waves is largely determined by this ratio, heat waves in nature should also favor the same two bands seen for CAM3 in Figure 11.2b. The placement of heat wave locations for nature in Figure 11.2a is not inconsistent with this possibility if one takes into account the sampling problems noted previously.

If the basic premise of our study, namely that heat wave likelihood is affected by intraseasonal planetary wave anomalies, is valid, the higher intraseasonal-to-synoptic-variability ratio for TAS in the two latitude bands is possibly tied to the upper troposphere circulation. Figures 11.3c and 11.3d appear to support this idea. They display the same ratio of variability but for the meridional wind at 300 hPa (v300), and the same two-banded structure is apparent. For v300, the existence of the two bands is rather straightforward to interpret as the minimum in 40–60°N is exactly where the maximum synoptic variability exists in the upper troposphere during summer as revealed by plots of storm track location in both reanalysis and CAM3 (not shown).

The subtropical band of pronounced heat wave activity and large intraseasonal-to-synoptic ratio is located slightly to the south of the mean jet (not shown), where CGT patterns play an important role in regulating low-frequency variability [*Branstator,* 2002; *Ding and Wang,* 2005]. While a number of previous studies have provided evidence of the connection between the CGT patterns and temperature and precipitation in Europe, India, East Asia, and North America [*Ding and Wang*, 2005; *Wang et al.,* 2010], in the following section we will look more carefully for the connection between the CGT patterns and heat waves by using our ultralong CAM3 dataset.

11.4. PLANETARY WAVES ASSOCIATED WITH THE HEAT WAVES

11.4.1. Regions

In order to get some sense of how heat waves in different regions are connected to the temporal evolution of planetary wave anomalies, we concentrate on four regions within the subtropical latitude band with more heat wave days. They include the United States (115°–80°W, 30°–45°N), East China (100°–120°E, 20°–45°N), South Asia (70°–90°E, 10°–40°N), and the Middle East (35°–60°E, 10°–40°N) (outlined by the boxes in Fig. 11.2). We consider only heat wave events that occupied more than 10% of the total area of a region thus filtering out some isolated events possibly resulting from local processes. To avoid contamination from previous heat wave events, we only use events that have no heat wave days in the preceding 15 days. After filtering the heat waves with these two additional criteria, we get approximately 1500–2000 events for each of the four regions. We designate the first day of a heat wave event as Day0 and constructed composites on each day from Day-15 to Day5 using daily anomalies of TAS and streamfunction at 300 hPa (psi300). Due to the large sample size made available by the extra long simulation, we are able to detect signals associated with these extreme events with a simple composite technique.

For the US region, the results are mainly a replica of a similar calculation reported by *Teng et al.* [2013]. On Day0, the entire continental United States is dominated by an anticyclonic circulation aloft and warm temperature anomalies at the surface that exceed 4°C near the center of the box (third row of Fig. 11.4). These regional meteorological features are consistent with observations during heat waves in this region [*Namias*, 1982; *Chang and Wallace*, 1987]. Outside North America, there are four additional positive streamfunction anomalies indicating anticyclonic circulations over the Middle East, East Asia, the Aleutians, and the North Atlantic. All together, the five anticyclonic circulation centers form a wavenumber-5 CGT pattern, which, as the introduction noted, has been documented in the scientific literature [*Branstator,* 2002; *Ding and Wang*, 2005].

A similar wavenumber-5 pattern also appears on each day from Day-15 to Day15. Together they show a slow westward phase propagation at a speed of roughly 2–3° longitude per day. As described in *Teng et al.* [2013], both Student's t-test and a field significance test indicate these wavenumber-5 patterns are statistically robust. For brevity, we show only the progression of these features with composites on Days -10, -5, 0, and 5 in Figure 11.4. By comparing with circulation anomalies in the lower troposphere (figure not shown), we find that overall the wavenumber-5 structure is barotropic, although the lower tropospheric circulation anomalies over the Asian summer monsoon region are fairly weak.

The Plumb flux [*Plumb,* 1985; *Karoly et al.,* 1989] has been frequently used as a diagnostic of the energy propagation of stationary atmospheric Rossby waves. When applied to each of the composites in Figure 11.4, it indicates (arrows in Fig. 11.4) that although the phase of the wavenumber-5 Rossby wave that is associated with US heat waves propagates slightly westward from Day-10 to Day0, energy moves consistently eastward. This is

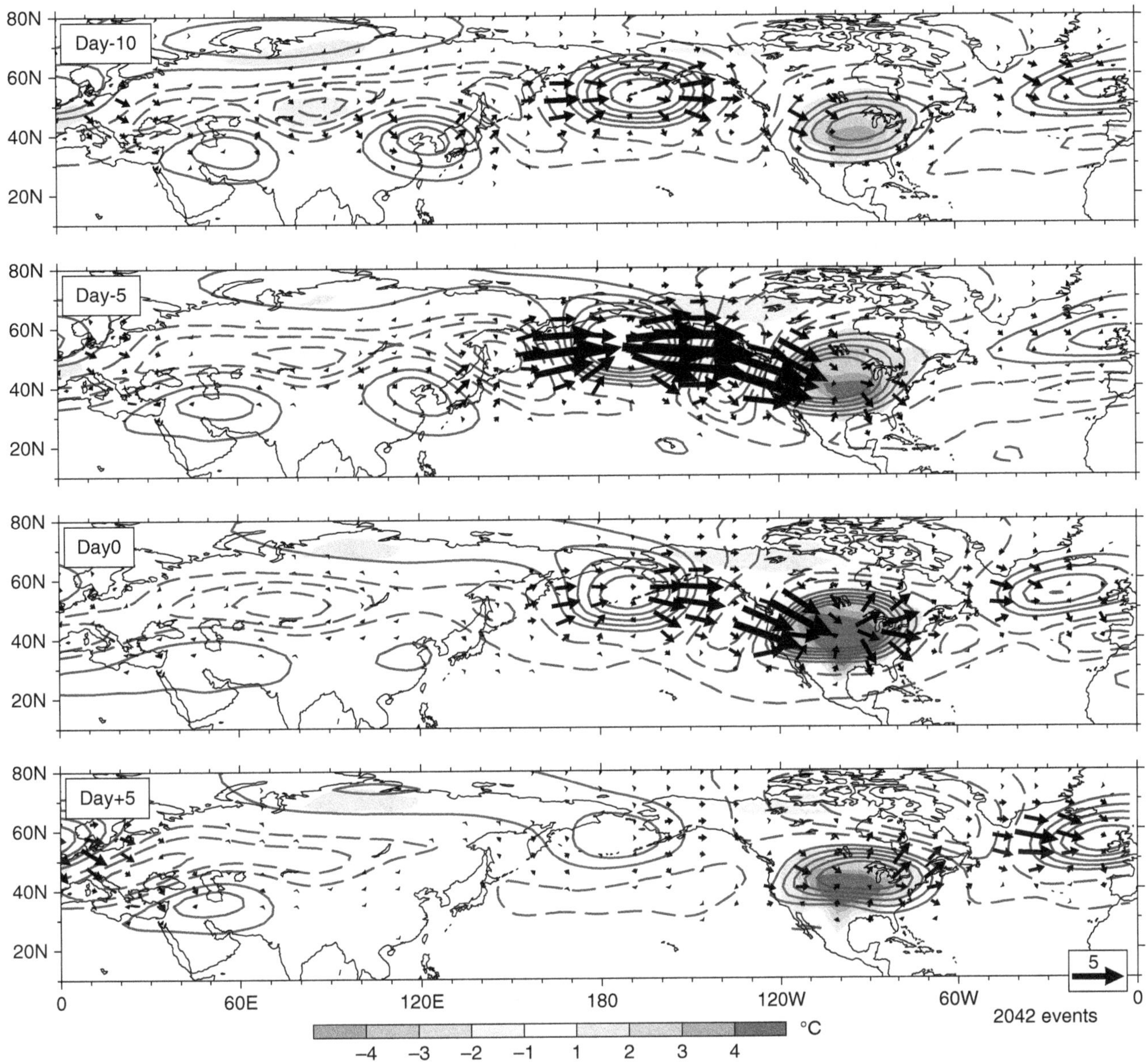

Figure 11.4 Heat wave composite for the US domain (outlined in Fig. 11.2) on Day-10, Day-5, Day0, and Day5. Contours represent 300 hPa stream function anomalies for ±1, 2, 3, 4, 5 × 10^6 m^2s^{-1}. Shading represents surface air temperature anomalies and arrows denote the Plumb flux vectors with magnitudes larger than 0.1 m^2s^{-2}. The total number of heat wave events is labeled at the bottom right corner.

consistent with the eastward propagation of the amplitude envelope seen in the progression of composite maps. Composite plots from Day-10 to Day10 (not shown) suggest that energy circles the globe during that time, which is roughly consistent with the group velocity of a stationary zonal wavenumber-5 disturbance trapped in the tropospheric jet [*Hoskins and Karoly*, 1981]. Over the Pacific, there are also substantial similarities in both the Plumb fluxes and the Rossby wave with some observed heat wave events (for example, the 1980 heat wave/drought studied by *Lyon and Dole* [1995]).

The Plumb fluxes have a substantial divergent component in the Aleutian region, which can be an indication of a local energy source. *Teng et al.* [2013] showed that the synoptic eddies in this location help to maintain the circulation anomalies associated with US heat waves in CAM3 through vorticity fluxes. *Teng et al.* [2013] also pointed out that energy resulting from mean state gradients in this region could also contribute to their maintenance.

Similarly, we construct composites of TAS and psi300 for heat waves in the other three boxes. Figures 11.5 and 11.6 show these composites for the China and South Asia heat wave cases. As expected, high temperature anomalies on Day0 correspond to anticyclonic circulation anomalies aloft in each case. On the global scale, interestingly, heat waves in these domains all tend to be led by a similar wavenumber-5 CGT pattern that exists 10 days earlier. This result is robust in both t-tests and field significance tests. However, though the same wavenumber-5 pattern contributes to Day-10 psi300 anomalies in each case, there are also noticeable regional differences among the cases on Day-10 and on succeeding days. First, the regions with strongest features are different for each case. On Day0 this is simply true because the highest amplitude circulation features are the anticyclones over the heat wave regions. On days preceding Day0, the regions with strongest features are in different locations because, as the Plumb vectors in Figures 11.4–11.6 indicate, there is eastward energy propagation in each case, which means the largest amplitudes are somewhat upstream of each heat wave box. Second, the dominant phase speed and even direction are not the same in each case. The most distinct example of this can be seen by comparing the United States (Fig. 11.4) and South Asia (Fig. 11.6) cases. As we have mentioned, for the former situation there is primarily slow westward phase propagation over the North Pacific, while for the latter there is definite eastward phase propagation across Asia. This difference may

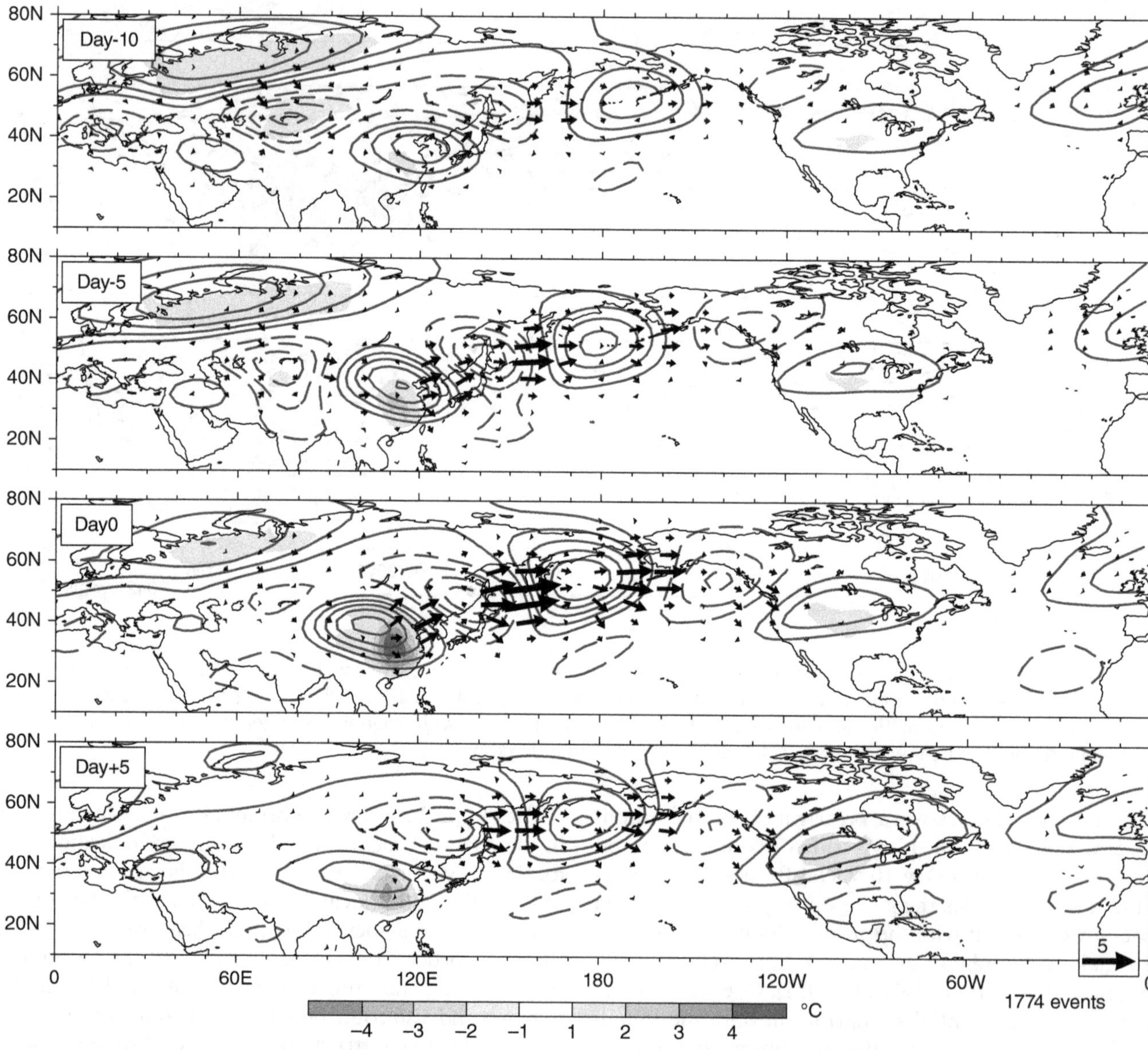

Figure 11.5 Same as Figure 11.4 but for heat waves in East China, and the contour intervals are ±0.5, 1, 2, 3, 4 × 10^6 m^2s^{-1}.

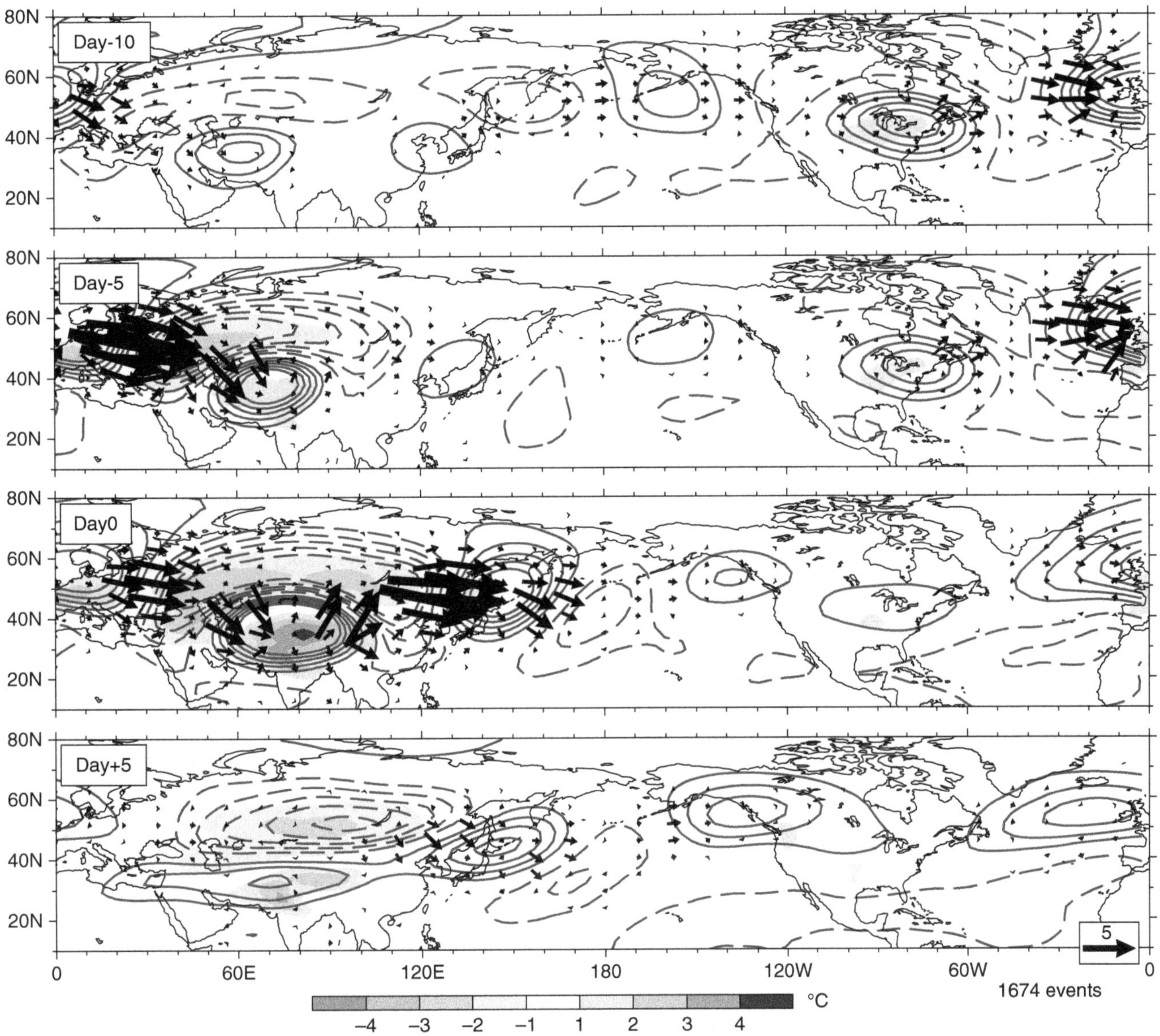

Figure 11.6 Same as Figure 11.5 but for heat waves in the South Asia domain.

possibly be related to the contrasting amount of Doppler shifting produced by the stronger westerly jet that exists across Asia than across the North Atlantic. Third, the Plumb fluxes are not as suggestive of dynamical sources of energy that are confined to a particular region in the China and South Asia cases as they are in the US case. One reason for this could be that during northern summer the storm tracks are north of 40°N, so synoptic eddy fluxes are not likely to make an important contribution to the maintenance of the features that reside in the Asian jet in these two cases. On the other hand, *Kosaka et al.* [2009] have argued that intraseasonal disturbances in the summertime Asian jet are maintained by interactions with both horizontal and vertical shears in the mean state so those sources may act to maintain all three cases in Figures 11.4–11.6.

The Middle East case is rather different from the other three. Although there is a very weak wavenumber-5 structure on Day-10 (figure not shown), over all the composite maps lack resemblance to patterns that have been previously documented by literature. One possibility is that the planetary anomalies associated with heat waves in the Middle East are less circumglobal. It may also be a situation where the composite technique is inadequate. For example, if the Middle East heat waves are associated with more than one planetary wave pattern, simple compositing will combine these patterns leading to features that are difficult to interpret.

11.4.2. Two Leading Waveguide Patterns

The above composite maps generally support the hypothesis that heat waves in the subtropical regions are connected to planetary wave events. In particular, in the three selected regions, heat waves all tend to be preceded by 10 days by a similar anomalous circulation pattern that is reminiscent of the CGT patterns that we discussed in the introduction. This similarity has motivated us to perform a second type of analysis of heat wave/planetary wave relationships that keys on the CGT patterns rather than on heat waves. In this approach, we look at the most prominent CGT patterns in CAM3 and systematically investigate if there are any regions in the Northern Hemisphere where heat waves are associated with these patterns.

As in *Branstator* [2002], we use v300 to depict CGT patterns and they are defined by empirical orthogonal function analysis (EOF) of 20–90 day filtered v300 in 20°–90°N during June to August. The spatial structure of the leading two EOFs is displayed as shading in Figure 11.7, while the regressed psi300 anomalies are superimposed as contours. The two patterns account for 9% and 8% of the total intraseasonal variance in v300 at 20–90°N (or 11% and 10% of the variance at 30°–70°N). Both EOFs have a distinct zonal wavenumber-5 structure along the summer mean jet (stippling). They extend more or less completely around the North Hemisphere. The lead-lag relationship between the two principal components (PC) suggests that EOF2 lags EOF1 by 7 days, but the 7 day lag correlation coefficient is weak, which implies this slowly westward propagating feature moves irregularly. EOF3 is also a wavenumber-5 pattern trapped in the jet, and it has its strongest features between the North Atlantic and West Asia (not shown). Since the first two EOFs are better separated from EOF3 than EOF3 is separated from EOF4 based on the rule of thumb of *North et al.* [1982], we mainly focus on the first two patterns in the study.

It turns out that the psi300 anomalies appearing in the heat wave composite maps for the United States, East China, South Asia, and the Middle East, and the corresponding v300 composites, are all in some way connected to some CGT patterns. To quantify the similarity, we use pattern correlation coefficients to compare the v300 composite maps (not shown) to the two leading v300 EOFs (Fig. 11.7). For comparison, we also find pattern correlations between the v300 composites and v300 EOFs 3–10. For the composites associated with US heat waves, we find that for Day-10 through Day-8, the highest pattern correlations among all 10 EOFs is with EOF1. Then from Day-7 through Day0, the composites are most like EOF2. Following heat wave Day0, the composites are most similar to EOF3. For the circulation anomalies associated with the China box, it is EOF2 that is most prominent during the buildup phase and minus EOF1 that becomes the most important pattern after the heat waves peak. Note that for

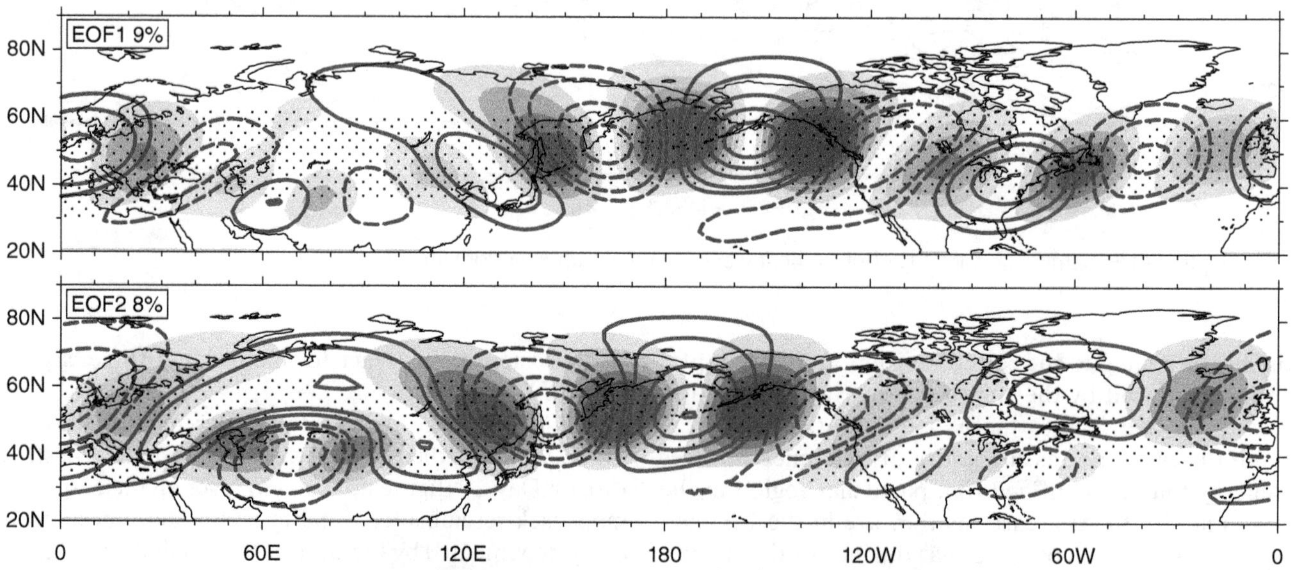

Figure 11.7 The leading two EOFs of 20–90 day band-pass filtered 300 hPa meridional wind (shading) and the corresponding regressed 300 hPa streamfunction anomalies (contour) during June to August from a 1000 year CAM3 integration. The contour intervals for the streamfunction anomalies are ±0.5, 1, 2, 3, 4 × 10^6 m^2s^{-1}. Stippling denotes regions with 300 hPa seasonal mean zonal wind exceeding 10 ms^{-1}. (*See insert for color representation of the figure.*)

both of these cases, the evolution from EOF1 to EOF2 or from EOF2 to minus EOF1 corresponds to the gradual westward phase propagation we described above. For South Asia heat waves, the opposite is true. There pattern correlations are strongest with EOF1 during Day-10 through Day-5 and then minus EOF2 dominates from Day-4 through Day5 consistent with eastward phase propagation. In all of these comparisons, the strongest pattern correlation for a given composite is at least 0.47 while values as large as 0.71 are seen in some cases. The story for the circulations that accompany Arabian heat waves is somewhat different. There no pattern correlation with EOF1 or EOF2 between Day-9 and Day 5 is greater than 0.37. However, pattern correlations with EOF3 range between 0.49 and 0.68 during Day-10 and Day-1, so even this location may be influenced by waveguide circulation anomalies.

Based on these results, we find that heat waves in three out of the four selected regions (Figs. 11.4–11.6) tend to be accompanied or preceded by EOF1 or EOF2. Thus, times when the circulation resembles EOF1 and EOF2 are times when heat waves are more likely to occur than they normally are in these three regions. To be more quantitative and to determine whether there are other regions where heat wave likelihood is influenced by EOF1 and EOF2 circulation anomalies, we explore how heat wave probabilities are affected by the degree to which psi300 anomalies resemble the EOF1 and EOF2 psi300 regression patterns of Figure 11.7. To do this we first need to quantify the strength of these CGT patterns. We take the area weighted dot product of 20–90 day filtered psi300 anomalies during June to August from all 12,000 years of our simulation with the two regressed patterns derived from a 1000 year segment of the run (Fig. 11.7). We are particularly interested in CGT events with absolute amplitude equal to and larger than two standard deviations. Those events for which either of the CGT psi300 patterns exceeds this threshold are referred to as extreme CGT1 or CGT2 events.

Next we inspect heat wave days at each grid point, and calculate the percentage of heat waves that have at least one extreme CGT event, either CGT1 or CGT2, during 0–15 days before the heat wave occurrence. Consistent with the composite analysis, subtropical South Asia, East Asia, and North America all show relatively high percentages, indicating heat waves over those regions are more closely connected with the extreme CGT events (Fig. 11.8). Because we search a wide 16 day window for the extreme CGT events and both positive and negative 2 standard deviation EOF1 and EOF2 events are counted as extreme CGT events, the chance for a random day to meet these criteria is not rare. In order to find the chance for a random day *d* to meet the criteria, we pick 2000 random days (the number 2000 roughly corresponds to the number of heat wave days in 12,000 years), then calculate the percentage of days having at least one extreme CGT event 0–16 days before day *d*. When we repeat this calculation 10,000 times, the 95th percentile of such percentages is 28.5%. Nevertheless, the percentages in Figure 11.8 over subtropical South Asia, East Asia, and North America are much higher than this 95th percentile value, further supporting the conclusion that heat waves over those regions are connected to unusually high amplitude CGT events. The ratio over the Middle East, on the other hand, is smaller than over South Asia and China, which is also consistent with our composite analysis finding that heat waves in the Middle East lack connections with the two leading CGT patterns.

Figure 11.8 also shows regions outside the subtropics, for example, western Europe, eastern Russia, and large parts of midlatitude North America, where heat waves are associated with extreme CGT events. This motivates us to quantify systematically how the extreme CGT events may affect heat wave probability hemisphere-wide. To do this, we calculate the probability of a heat wave occurring at each grid point 0 days, 5 days, 10 days, and

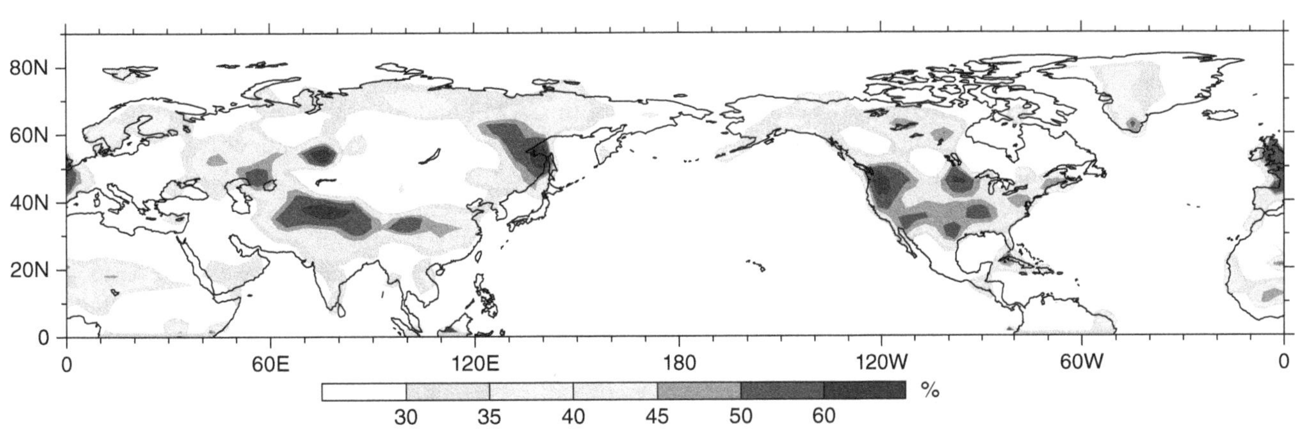

Figure 11.8 Percentage of heat waves in CAM3 that are preceded by an extreme CGT1 or CGT2 event during 0–15 days before the heat wave occurrence.

15 days after an extreme CGT event. These probabilities are then normalized with the probability of a heat wave occurring at the same location without any preconditions. The resulting ratio, which we refer to as the probability factor, reflects how an extreme CGT event influences the chance of a heat wave occurring at each grid point a given number of days after the circulation extreme. When doing this, unlike for Figure 11.8, we consider extreme CGT1 and CGT2 events separately.

We first focus on extreme CGT1 events (Fig. 11.7, top). The heat wave probability factor for those events is presented as shading, and psi300 composites constructed from daily anomalies are superimposed as contours (Fig. 11.9). With Day0 designated as the day when the EOF1 projection exceeds 2 standard deviations, we calculate both the heat wave probability factor and psi300 composite for Day0, Day5, Day10, and Day15. In all four panels, the sign of the probability factor matches that of the psi300 composite over most of the plotted area, and both variables have a distinct wavenumber-5 structure. This provides strong evidence that heat wave occurrence is connected to the CGT1 events.

Although we cannot conclude anything about causality between the two variables based solely on the map on Day0, the clear wave propagation feature in the four maps is very suggestive that the planetary wave anomalies are causing the anomalous heat wave probabilities and not vice versa. For while we are unaware of any mechanism for heat wave propagation independent of the influence of planetary waves, the planetary wave propagation is consistent with well-understood dynamical mechanisms. From Day0 to Day15, the two anticyclones over the east coast of North America and western Europe both move slowly westward, while the anticyclone over Iran moves eastward and the anticyclone over Korea and Japan moves northeastward. This slow propagation can be easily reproduced with a lag regression analysis based on a 1000 year segment of the simulation, which suggests that the CGT1 pattern is highly predictable on the intraseasonal timescale, with or without involvement of heat waves. Similarly, we plot the probability factors and psi300 composites for the extreme CGT2 events (Fig. 11.10). For this CGT pattern, too, there is a clear relationship between circulation and heat wave probability with propagation indicating the relationship exists because the circulation is affecting heat wave likelihood.

11.5. SUMMARY AND DISCUSSION

The overall objective of this study is to investigate possible connections between heat waves in the Northern Hemisphere summer and quasi-stationary planetary wave anomalies produced by atmospheric internal dynamics. It is difficult to determine such relationships for nature due to the limited number of observed heat waves. For nature it is also a challenge to distinguish the role of internal dynamics from external forcing for individual events. Instead, we have taken advantage of a 12,000 year AGCM experiment for which robust relationships can be identified. The most important outcomes of the investigation are that in this AGCM simulation, (1) there is a pair of distinctive anomalous planetary wave patterns consisting of five highs and lows that more or less encompass the Northern Hemisphere that precede and/or coincide with heat waves in many regions in both Asia and North America; (2) the wavenumber-5 patterns appear to be largely a consequence of internal midlatitude processes due to the waveguide effect of the summer mean jet; and (3) these patterns retain some predictability at the 2 week range and consequently may provide useful information on the likelihood of heat waves at this range. While a previous report that analyzed the same AGCM experiment [*Teng et al.*, 2013] served as the primary motivation for this study, here we have expanded the focus from heat waves in the United States to the entire Northern Hemisphere and have explicitly documented the effect of each member of the wavenumber-5 pair of patterns.

It is natural to wonder whether our model-based results apply to nature, for CAM3 has marked biases [*Collins et al.*, 2006]. For our study, one important bias concerns the possibility that wavenumber-5 is too dominant in CAM3. When *Teng et al.* [2013] compared midlatitude wavenumber-frequency spectra for CAM3 and nature, they found that CAM3 has a strong maximum in variance for zonal wavenumber-5 while in nature there is approximately equal variance in wavenumbers 5 and 6. This discrepancy can also be seen if one performs the same EOF analysis of intraseasonal v300 fluctuations for reanalysis fields that we displayed for CAM3 fields in Figure 11.7. For nature, the leading two EOFs (Fig. 11.11) have a smaller zonal scale than CAM3 EOF1 and 2. They are also weaker in that together they explain 13% of nature's v300 variance in 20°–90°N while the two explain 16% of the variance in CAM3. It may also be noteworthy that *Teng et al.* [2013] found that wavenumber-5 disturbances in both CAM3 and nature on average tend to have slow westward phase propagation while wavenumber-6 disturbances tend to propagate eastward. Since nature appears to have more wavenumber-6 variance, the westward phase propagation that we found in some regions for circulation anomalies associated with heat waves in CAM3 may be replaced by eastward propagation in nature. In fact, some case studies have reported that US extreme climate events are triggered by eastward propagating Rossby waves [*Chen and Newman*, 1998; *Schubert et al.*, 2011]. Some other studies linked extreme weather to amplified zonal wavenumbers 6–8 [*Petouhkov et al.*, 2013; *Coumou et al.*, 2014] instead of wavenumber-5. It may be that the difference

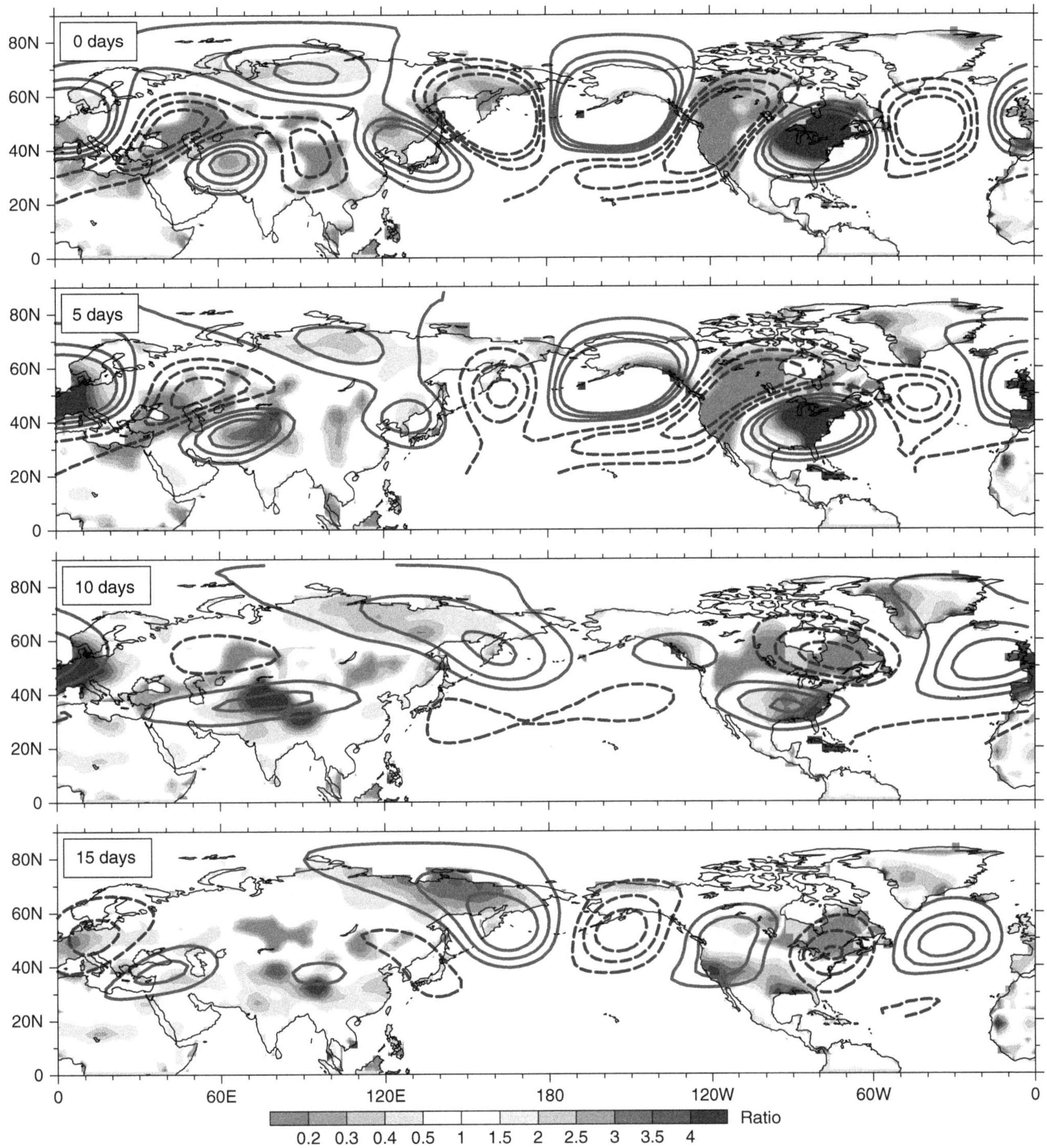

Figure 11.9 Shading represents ratio of probability of a heat wave occurring 0 days (top), 5 days (second row), 10 days (third row), and 15 days (bottom) after an extreme CGT1 event relative to probability of a heat wave without any precondition. Contour represents composite of 300 hPa streamfunction anomalies of those extreme CGT1 events on the corresponding days at $\pm 1, 2, 3 \times 10^6\ m^2s^{-1}$ intervals. (*See insert for color representation of the figure.*)

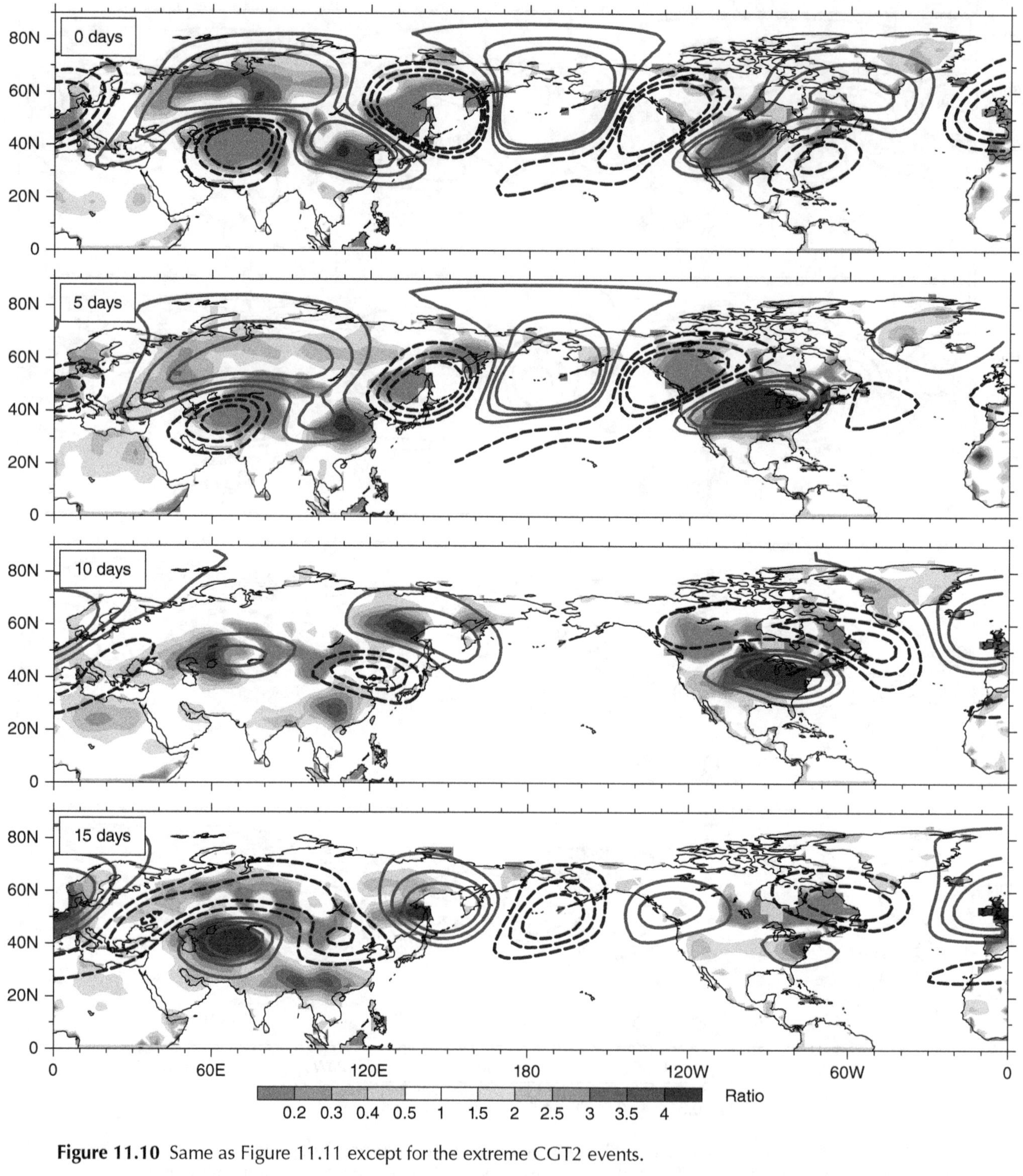

Figure 11.10 Same as Figure 11.11 except for the extreme CGT2 events.

in preferred scales and direction of phase propagation between CAM3 and nature can be traced to differences in the mean jet, which can be seen by comparing the stippled regions in Figures 11.7 and 11.11. But to verify that this is the reason for the difference in scale and direction of phase propagation goes beyond the scope of our study.

Despite the CAM3 shortcomings, there are reasons that our results are relevant for nature. One reason is the resemblance of the leading CGT pattern in the model and nature. Over much of the North Pacific and North America, the cyclones and anticyclones occur at similar longitudes and their latitudinal position is guided by the

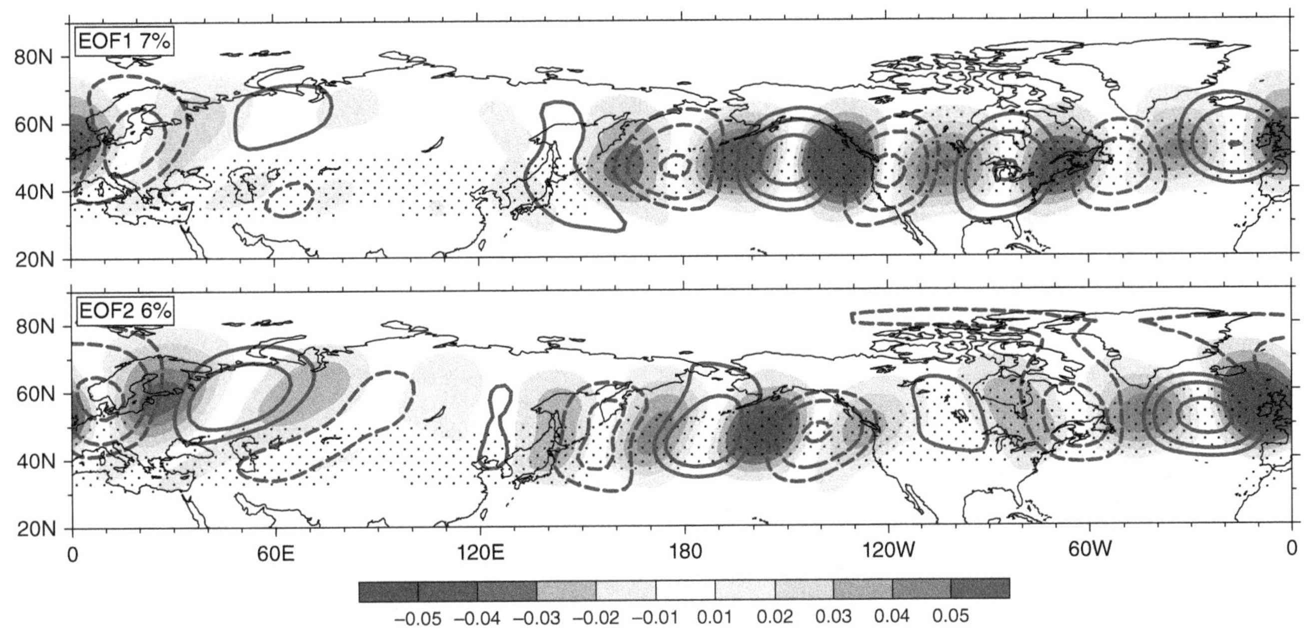

Figure 11.11 Same as Figure 11.7 but for the reanalysis data during 1948–2013.

meandering of the mean jet in both systems (Figs. 11.7, 11.11). As mentioned above, without sufficient data, it is difficult to detect the connection between planetary wave events and heat waves in nature. Even so, when we include all events rather than just extremes and regress the 20–90 day filtered psi300 anomalies onto surface air temperature anomalies averaged over the US or East China domains, we find similar regressed circulation patterns in the model and the reanalysis (figure not shown). More important, in this calculation the model regressed psi300 pattern resembles the composite maps on Day0 of the heat waves (Figs. 11.4, 11.5). Together, these results suggest that the model is capable of simulating planetary circulation features associated with heat waves.

While the present work could surely benefit from employment of models with a more realistic representation of nature, we believe the basic concepts in this study are not affected by the model bias. Its findings open a new source of predictability for heat waves on subseasonal timescales that has not been fully comprehended and quantified before. Until now, the search for sources of intraseasonal predictability has primarily focused on atmospheric phenomena driven by tropical intraseasonal variability associated with the Madden-Julian Oscillation (MJO) [*Madden and Julian*, 1971; *Waliser et al.*, 2003] or Asian summer monsoons [*Ding and Wang*, 2007], or by slowly varying boundary conditions including sea surface temperature [*Fu et al.*, 2008] and soil moisture [*Koster et al.*, 2010]. Without studies like ours, one may not expect any predictive information of heat waves from intrinsic midlatitude intraseasonal planetary wave variations. After all, typical instantaneous states are unpredictable after about 10 days because the evolution of atmospheric features is chaotic. This fact, however, does not preclude the possibility that time-averaged or unusually predictable patterns could be predictable on longer timescales [*Tribbia and Baumhefner*, 1988; *Branstator et al.*, 1993]. Our results, whose robustness relies on use of a long integration of a climate model, suggest CGT patterns may be such highly predictable time averaged patterns. And together with the apparent impact they have on the likelihood of heat waves in the subtropical band, this may make the likelihood of heat waves in certain regions, if not individual heat wave events, predictable on subseasonal timescales.

On the less promising side, even after taking into account the predictability associated with CGT patterns, the probability of heat waves is still far from being adequate to assure skillful forecasts of individual extreme events. For our results suggest that during a 2 standard deviation planetary wave event, the likelihood of a heat wave in even the most highly affected regions is around only 1%. Furthermore, the implications of this study may go beyond forecasts of heat waves. Recently, there is a growing perception by the general public that extreme events are occurring more frequently due to climate change [*Trenberth and Fasullo*, 2012]. Contrary to the well-accepted thermodynamic arguments [*Meehl and Tebaldi*, 2004], how the dynamic processes influence such extremes remains poorly understood. While global warming might indirectly affect extreme events by changing the amplitude and structure of planetary wave fluctuations [*Petouhkov et al.*, 2013; *Coumou et al.*, 2014], one needs to exercise extreme caution when attributing individual heat wave and associated anomalous planetary wave events to climate change, for studies like ours indicate that prolonged planetary wave anomalies that influence the likelihood of heat waves can be produced simply by natural variability.

ACKNOWLEDGMENTS

The GCM experiments were run and postprocessed by A. Mai. Portions of this study were supported by the Regional and Global Climate Modeling Program (RGCM) of the US Department of Energy's, Office of Science (BER), Cooperative Agreement DE-FC02-97ER62402, by NOAA NA14OAR4310224 and NASA NEWS via NA09OAR4310187, and by the National Science Foundation. Computing resources were provided by the National Energy Research Scientific Computing Center, which is supported by BER of the US Department of Energy. The National Center for Atmospheric Research is sponsored by the National Science Foundation.

REFERENCES

Branstator, G. (1983), Horizontal energy propagation in a barotropic atmosphere with meridional and zonal structure, *J. Atmos. Sci.*, *40*, 1689–1708.

Branstator, G. (2002), Circumglobal teleconnections, the jet stream waveguide, and the North Atlantic Oscillation, *J. Climate*, *15*, 1893–1910.

Branstator, G., A. Mai, and D. Baumhefner (1993), Identification of highly predictable flow elements for spatial filtering of medium- and extended-range numerical forecasts, *Monthly Weather Rev.*, *121*, 1786–1802.

Chang, F.-C., and J. M. Wallace (1987), Meteorological conditions during heat waves and droughts in the United States Great Plains, *Monthly Weather Rev.*, *115*, 1253–1269.

Chen, P., and M. Newman (1998), Rossby wave propagation and the rapid development of upper-level anomalous anticyclones during the 1988 U.S. drought, *J. Climate*, *11*, 2491–2504.

Collins, W. D., and coauthors (2006), The Community Climate System Model version 3 (CCSM3), *J. Climate*, *19*, 2122–2143.

Coumou, D., V. Petoukhov, S. Rahmstorf, S. Petri, and H. J. Schellnhuber (2014), Quasi-resonant circulation regimes and hemispheric synchronization of extreme weather in boreal summer, *Proc. Nat. Acad. Sci. USA*, *111*, 12331–12336.

Ding, Q., and B. Wang (2005), Circumglobal teleconnection in the Northern Hemisphere summer, *J. Climate*, *18*, 3483–3505.

Ding, Q., and B. Wang (2007), intraseasonal teleconnection between the summer Eurasian wave train and the Indian monsoon, *J. Climate*, *20*, 3751–3767.

Ding, Q., B. Wang, J. M. Wallace, and G. Branstator (2011), Tropical-extratropical teleconnections in boreal summer: Observed interannual variability, *J Climate*, *24*, 1878–1896.

Dole, R., et al. (2011),Was there a basis for anticipating the 2010 Russian heat wave? *Geophys. Res. Lett.*; doi:10.1029/2010GL046582.

Easterling, D. R., et al. (2000), Climate extremes: Observations, modeling, and impacts, *Science*, *289*, 2068–2074,.

Fu, X., B. Yang, Q. Bao, and B. Wang (2008), Seas surface temperature feedback extends the predictability of tropical intraseasonal oscillation, *Monthly Weather Rev.*, *136*, 577–597.

Hoskins, B. J., and D. J. Karoly (1981), The steady linear response of a spherical atmosphere to thermal and orographic forcing, *J. Atmos. Sci.*, *38*, 1179–1196.

Hoskins, B. J., and T. Ambrizzi (1993), Rossby wave propagation on a realistic longitudinally varying flow, *J. Atmos. Sci.*, *50*, 1661–1671.

Hurrell, J. W. (1995), Decadal trends in the North Atlantic Oscillation: Regional temperatures and precipitation, *Science*, *269*, 676–679.

Kalnay, E., and coauthors (1996), The NCEP/NCAR 40-year reanalysis project, *Bull. Amer. Meteor. Soc.*, *77*, 437–471.

Karoly, D. J., R. A. Plumb, and M. Ting (1989), Example of the horizontal propagation of quasi-stationary waves, *J. Atmos. Sci. 46*, 2802–2811.

Kosaka, Y., H. Nakamura, M. Watanabe, and M. Kimoto (2009), Analysis on the dynamics of a wave-like teleconnection pattern along the summertime Asian jet based on a reanalysis dataset and climate model simulations, *J. Meteor. Soc. Japan*, *87*, 561–580.

Koster, R. D., and coauthors (2010), The contribution of land surface initialization to subseasonal forecast skill: First results from the GLACE-2 project, *Geophys. Res. Lett.*, *37*; doi:10.1029/2009GL041677.

Lyon, B., and R. M. Dole (1995), A diagnostic comparison of the 1980 and 1988 US summer heat wave-droughts, *J. Climate*, *8*, 1658–1675.

Madden, R. A., and P. R. Julian (1971), Detection of a 40–50 day oscillation in the zonal wind in the tropical Pacific, *J. Atmos. Sci.*, *28*, 702–708.

Meehl, G. A., and C. Tebaldi (2004), More intense, more frequent, and longer lasting heat waves in the 21st century, *Science*, *305*, 994–997.

Namias, J. (1982), Anatomy of Great Plains protracted heat waves (especially the 1980 U.S. summer drought), *Monthly Weather Rev.*, *110*, 824–838.

North, G. R., and coauthors (1982), Sampling errors in the estimation of empirical orthogonal functions, *Monthly Weather Rev.*, *110*, 699–706.

Petoukhov, V., and coauthors (2013), Quasi-resonant amplication of planetary waves and recent Northern Hemisphere weather extremes, *Proc. Nat. Acad. Sci. USA*, *110*, 5336–5341.

Plumb, R. A. (1985), On the three-dimensional propagation of stationary waves, *J. Atmos. Sci.*, *42*, 217–229.

Schubert, S., H. Wang, and M. Suarez (2011), Warm seasonal subseasonal variability and climate extremes in the Northern Hemisphere: The role of stationary Rossby waves, *J. Climate*; doi:10.1175/JCLI-D-10-05035.1.

Screen, J. A., and I. Simmonds (2014), Amplified mid-latitude planetary waves favor particular regional weather extremes, *Nature Climate Change*; doi:10.1038/NCLIMTE2271.

Teng, H., G. Branstator, H. Wang, G. A. Meehl, and W. M. Washington (2013), Probability of US heat waves affected by a subseasonal planetary wave pattern, *Nat. Geosci.*; doi:10.1038/ngeo1988.

Trenberth, K. E., and J. T. Fasullo (2012), Climate extremes and climate change: The Russian heat wave and other climate extremes of 2010, *J. Geophys. Res.*; doi:10.1029/2012JD018020.

Tribbia, J. J., and D. P. Baumhefner (1988), The reliability of improvements in deterministic short-range forecasts in the presence of initial state and modeling deficiencies, *Monthly Weather Rev.*, *116*, 2276–2288.

Waliser, D. E., K. M. Lau, W. Stern, and C. Jones (2003), Potential predictability of the Madden-Julian oscillation, *Bull. Amer. Meteor. Soc.*, *84*, 33–50.

Wallace, J. M., and D. S. Gutzler (1981), Teleconnections in the geopotential height field during the Northern Hemisphere winter, *Monthly Weather Rev.*, *109*, 784–812.

Wang, S.-Y., L. E. Hipps, R. R. Gillies, X. Jiang, and A. L. Moller (2010), Circumglobal teleconnection and early summer rainfall in the U.S. intermountain west, *Theor. App. Climatol.*, *102*, 245–252.

Part III
Regional Climate Extremes

12

North American Drought and Links to Northern Eurasia: The Role of Stationary Rossby Waves

Hailan Wang, Siegfried D. Schubert, and Randal D. Koster

ABSTRACT

This chapter provides an overview of the role of stationary Rossby waves in the subseasonal development of warm season drought over North America and subsequent downstream development of climate anomalies over northern Eurasia. The results are based on a case study of a stationary Rossby wave event that developed during 20 May–15 June 1988. Simulations with the NASA Goddard Earth Observing System version 5 (GEOS-5) atmospheric general circulation model highlight the importance of the mean jet streams in guiding and constraining the path and speed of wave energy propagation. In particular, convective anomalies that developed over the western Pacific in late May (in the presence of the strong North Pacific jet) produce a predilection for persistent upper-level high anomalies over central North America about 10 days later, leading to the rapid development of severe dry conditions there. There are indications of continued downstream wave energy propagation that reaches northern Eurasia about 2 weeks later, leading to the development of dry conditions over eastern Europe and western Russia, and cool and wet conditions over western Europe and central northern Eurasia. The results suggest that stationary Rossby waves can serve as a source of predictability for subseasonal development of droughts over North America and northern Eurasia.

12.1. INTRODUCTION

A drought usually results from a prolonged period of below-normal precipitation in a given region. It can cause soil dryness and groundwater shortage, and may contribute to wildfire, fatality, and even population migration. Because of its substantial impact on the agriculture and ecosystem of the affected regions, drought is one of the most expensive recurring natural disasters to affect North America and northern Eurasia. Within the United States alone, major droughts and heat waves during the last three decades have resulted in costs exceeding 100 billion dollars [*Lott and Ross*, 2000; *Ross and Lott*, 2003; *Smith and Katz*, 2013]. Take the 1988 North American drought, for example [*Ross and Lott*, 2003]. It not only caused severe damage to agriculture and related industries but also led to prolonged dust storms; it is also well known for extensive wildfires over western North America. Furthermore, the 1988 drought and its associated record-setting heat waves accounted for an estimated 5,000 to 10,000 deaths. Ranked as the costliest drought in US history, the severe 1988 drought caused $60 billion (1988 US dollars) in damage in the United States, and an additional $1.8 billion (1988 Canadian dollars) in Canada. The societal impact of droughts thus requires an advanced scientific understanding of processes that drive them, and a proper incorporation of these processes in forecast models in order to provide reliable drought prediction and drought early warning system to the society and policy makers.

NASA Goddard Space Flight Center, Global Modeling and Assimilation Office, Greenbelt, Maryland, USA

Climate Extremes: Patterns and Mechanisms, Geophysical Monograph 226, First Edition.
Edited by S.-Y. Simon Wang, Jin-Ho Yoon, Christopher C. Funk, and Robert R. Gillies.

The last few decades have seen significant advances in our understanding of the large-scale controls on droughts. For North America, it is now well established that certain spatial patterns of sea surface temperature (SST) anomalies are conducive to generating precipitation deficits or meteorological droughts there, achieved by exerting changes in atmospheric circulation there through atmospheric teleconnection. Examples of such SST patterns include those associated with the El Niño Southern Oscillation (ENSO) on seasonal-to-interannual timescales and those associated with the Pacific Decadal Oscillation (PDO) and the Atlantic Multidecadal Oscillation (AMO) on decadal and longer timescales [e.g., *Schubert et al.*, 2009]. Extensive observational [e.g., *Ting and Wang,* 1997; *Nigam et al.*, 2011; *Hu and Feng*, 2012; *Dai*, 2013] and AGCM modeling [e.g., *Hoerling and Kumar*, 2003; *Schubert et al.*, 2004; *Seager et al.*, 2005; *Schubert et al.*, 2009; *Wang et al.*, 2010] studies show that ENSO and PDO in their cold phases and the AMO in its warm phase produce a tendency for drought conditions over North America, with the Pacific playing the dominant role. It is also known that the impact of SST anomalies varies substantially from region to region. Droughts over the southern plains and southwestern North America are significantly forced by certain tropical SST anomalies and appear to have some predictability on seasonal timescales, whereas droughts over the northern plains and substantial areas of Canada are more strongly determined by atmospheric internal variability and appear less predictable [*Hoerling et al.*, 2009; *Seager and Hoerling*, 2014; *Schubert et al.*, 2015]. In addition, while certain SST patterns and random atmospheric internal variability can be important in initiating drought, sustaining and/or amplifying droughts over North America often involves feedback from local soil moisture [e.g., *Koster et al.*, 2003; *Ferguson et al.*, 2010]. For example, a month with low precipitation leads to a drier-than-average soil, which in turn can cause lower-than-average evaporation, helping to encourage low precipitation. Such feedback between land and atmosphere plays a particularly important role in sustaining droughts over the central United States during the warm season [*Koster et al.*, 2006]. See *Seager and Hoerling* [2014] for a comprehensive overview of the atmosphere and ocean origins of North American droughts.

In northern Eurasia, precipitation and surface temperature variations on interannual timescales are not as strongly influenced by SST patterns; they are instead dominated by the effects of atmospheric internal variability [*Schubert et al.*, 2015]. *Schubert et al.* [2014], in a review article on northern Eurasian droughts and heat waves, found that some of the most severe summer heat waves and droughts are linked to distinct Rossby wave trains spanning the continent that, while producing severe heat in one location, cause a juxtaposition of wet and cool conditions in regions thousands of miles to the east or west. They note that longer-term droughts (lasting multiple years) do occur but are largely confined to the southern parts of northern Eurasia, where there appears to be a weak link to SST and an important control from soil moisture.

Despite significant progress made in understanding the processes that lead to droughts over North America and northern Eurasia, accurate drought predictions from weeks to a season in advance remain a challenge. While the role of SST anomalies for North American droughts has been extensively studied (as noted above), the SST impacts are often embedded in a substantial amount of atmospheric internal variability that is unforced by SST, limiting the potential predictability of precipitation over North America. The problem is even more significant in northern Eurasia where the connection to SST is much weaker, and thus drought prediction there poses an even greater challenge.

Summertime quasi-stationary Rossby waves are known to play a key role in Northern Hemisphere (NH) atmospheric circulation and surface meteorology variability on subseasonal timescales [*Schubert et al.*, 2011; *Schubert et al.*, 2014]. In particular, such waves have been crucial to the development of recent short-term warm season heat waves and droughts over North America (e.g. the 1988, 1998, and 2012 summer droughts) and northern Eurasia (e.g., the 2003 summer heat wave over Europe and the 2010 summer drought and heat wave over Russia). For example, past studies [e.g., *Namias,* 1991; *Chen and Newman*, 1998; *Schubert et al.*, 2011] suggested that the intense anomalous upper-level anticyclonic flow anomalies associated with the 1988 drought are linked to quasi-stationary Rossby waves that originate from the western Pacific.

The Rossby waves are guided by the mean jet streams [*Branstator*, 2002; *Schubert et al.*, 2011], as evidenced by the fact that the meridional gradients and nearly circumpolar extent of the warm season mean jet streams can provide important guides for Rossby waves [e.g., *Ambrizzi et al.*, 1995]. Stationary wave modeling diagnosis of leading patterns of summertime subseasonal atmospheric circulation variations shows that these waves are dominantly maintained by submonthly vorticity transient flux convergences [*Schubert et al.*, 2011]. *Koster et al.* [2014] further revealed that the leading subseasonal atmospheric circulation patterns over North America can be phase-locked and amplified by certain spatial patterns of soil moisture anomalies. The nature of these drought-inducing stationary Rossby waves, including physical processes that act to initiate and sustain them, is as yet not well understood.

This chapter provides an overview of the current understanding of the role of stationary Rossby waves in the development and maintenance of warm season drought over North America and northern Eurasia on subseasonal timescales. The study is also an extension of our previous work [*Schubert et al.*, 2011; *Koster et al.*, 2014] in that it further explores (via a case study) the processes by which stationary Rossby waves affect the subseasonal development of North American droughts, and their subsequent effects over northern Eurasia.

The chapter is organized as follows. Section 12.2 describes the reanalysis data and AGCM experiments. Section 12.3 presents our main results. Section 12.3.1 reviews the leading modes of subseasonal atmospheric circulation variability and their associated surface variations. Here we update the results in *Schubert et al.* [2011] to include more recent years using NASA's Modern-Era Retrospective Analysis for Research and Applications (MERRA) [*Rienecker et al.*, 2011] and MERRA-Land [*Reichle et al.*, 2011] reanalysis data through 2014. Section 12.3.2.1 focuses on the 1988 North American drought, describing the key processes through which the stationary Rossby waves lead to the rapid buildup of severe drought conditions over central North America. Section 12.3.2.2 examines the connection between the subseasonal development of North American drought and the subsequent development of surface anomalies over northern Eurasia via stationary Rossby waves. The summary and conclusions are given in section 12.4.

12.2. REANALYSIS DATA AND THE GEOS-5 AGCM EXPERIMENTS

This study makes extensive use of daily and monthly atmospheric circulation and surface fields from recent reanalysis data: MERRA [*Rienecker et al.*, 2011] and MERRA-Land [*Reichle et al.*, 2011]. MERRA was produced with the Goddard Earth Observing System Version 5 (GEOS-5) [*Rienecker et al.*, 2008] consisting of the GEOS-5 Atmospheric General Circulation Model (AGCM) and the Grid-point Statistical Interpolation (GSI) analysis system. It was run at a resolution of ½° latitude × 2/3° longitude with 72 levels extending to 0.01 hPa. MERRA-Land is an off-line land-only assimilation in which the precipitation forcing consists of a gauge-based data product from the NOAA Climate Prediction Center (CPC) merged with MERRA precipitation, and the land model is an updated version of the NASA GEOS-5 Catchment Land Surface Model. MERRA-Land has been shown to be more accurate than MERRA in terms of surface hydrological fields and is recommended for land surface hydrological studies [*Reichle et al.*, 2011]. More information about MERRA and MERRA-Land data can be found at https://gmao.gsfc.nasa.gov/research/merra/.

Our investigation focuses on the boreal summer months with much of the analysis based on monthly data. The subseasonal (30–90 day) component of variability is obtained by first removing the annual cycle and then computing the departures of summer monthly anomalies from June-July-August (JJA) seasonal mean anomaly. In order to study the detailed evolution of stationary Rossby waves and the subseasonal development of regional climate extremes, we also make use of daily atmospheric circulation and surface temperature fields from MERRA and daily precipitation and surface soil wetness from MERRA-Land.

New results from this study are based on experiments performed with the NASA GEOS-5 AGCM [*Rienecker et al.*, 2008; *Molod et al.*, 2012], combined with a data assimilation technology called "replay" (described below). The GEOS-5 AGCM employs the finite-volume dynamics of *Lin* [2004] and various moist physics packages described in *Bacmeister et al.* [2006] including a modified form of the Relaxed Arakawa-Schubert convection scheme [*Moorthi and Suarez*, 1992], a prognostic cloud microphysics scheme [*Bacmeister et al.*, 2006], and the Catchment Land Surface Model [*Koster et al.*, 2000]. In our experiments, the model is run with 72 hybrid-sigma vertical levels extending to 0.01 hPa and with 1° horizontal resolution on a latitude/longitude grid.

The GEOS-5 data assimilation system employed in MERRA was developed so that the analysis increments produced during the assimilation cycle are inserted gradually over a 6 hour interval into the model using the Incremental Analysis Updates (IAU) approach described in *Bloom et al.* [1996]. This approach has been modified so that the model can be "replayed" against an existing analysis. The replay methodology reads in an existing analysis field (in this case MERRA) and mimics the IAU approach to provide a continuous model run that is forced by analysis increments that change every 6 hours, thereby constraining the model to remain close to the original analysis. This approach has been generalized so that it allows us to constrain only certain components of the atmosphere (e.g., a limited region, or a subset of model variables).

Table 12.1 lists the AGCM experiments performed, focusing on the stationary Rossby wave event that occurred during 20 May – 15 June 1988. The experiments consist of two free-running AGCM hindcasts and two sets of AGCM regional replay experiments in which the constraint to MERRA is applied either over eastern Asia and the western Pacific (90°E–140°E, 0°–70°N), or over North America (130°W–70°W; 0°–70°N), while letting the model atmosphere evolve freely elsewhere.

Table 12.1 Experiments Performed for a Stationary Rossby Wave Event That Developed During 20 May–15 June 1988 Using the NASA GEOS-5 AGCM and MERRA

Experiments	Replay region	Replay variables	Atmospheric and land initial conditions	Land feedback	Processes
A1	N/A	N/A	*Control*: 21z, 20 May of 1980–2010 *Anomaly*: 21z, 20 May 1988 plus small atmospheric perturbations	Yes	A
A2	N/A	N/A	*Control:* same as A1 *Anomaly:* 21z, 20 May of 1980–2009 for atmosphere; 21z, 20 May 1988 for land	Yes	B
B1	90° E–140° E; 0° –70° N	U,V,T,Q,Ps	Same as A1	Yes	A, C, D, E
B2	90° E–140° E; 0° –70° N	U,V,T,Q,Ps	*Control*: 21z, 2 May of 1980–2010 *Anomaly*: 21z, 2 May of 1980–2009	Yes	C, D, E
B3	90° E–140° E; 0° –70° N	U,V,T,Q,Ps	Same as B2	No	C, E
B4	90° E–140° E; 0° –70° N	U,V	Same as B2	Yes	C, D
C1	130° W–70° W; 0° –70° N	U,V,T,Q,Ps	Same as B2	Yes	C, D, E

Note: The experiments A1–A2 and B1–B4 are used to study the processes that affect stationary Rossby waves and their effects over North America. These processes include (A) the 21z 20 May 1988 atmospheric and land initial conditions, (B) the 21z 20 May 1988 land initial conditions, (C) observed stationary wave sources over the western Pacific, (D) soil moisture feedback, and (E) corrected model mean jet over the North Pacific. The experiment C1 is used to investigate the connections between the 1988 North American drought and surface anomalies developed over northern Eurasia about 2 to 3 weeks later. Each experiment consists of two sets of ensembles, a control ensemble and an anomaly ensemble. The control ensemble consists of 31 members for years 1980–2010 forced with observed SST, initialized using MERRA conditions for years 1980–2010. The anomaly ensemble consists of 30 members for year 1988 forced with the 1988 observed SST. The atmospheric and land initial conditions used vary in the experiments. All experiments are run through the end of July, at a horizontal resolution of 1° latitude/longitude.

Each of the experiments consists of two sets of ensembles: a control ensemble and an "anomaly" ensemble. The control ensemble consists of 31 hindcast runs forced with observed SST for the summers of 1980 through 2010 (one hindcast for each summer). The ensemble average of those runs is used to establish a hindcast climatology. The anomaly ensemble consists of 30 hindcast runs for the summer of 1988 forced with the 1988 observed SST. The atmospheric and land initial conditions used vary from experiment to experiment and are detailed below.

In the free-running AGCM hindcast experiment A1, the 31 members of the control ensemble are initialized at 21z 20 May in each of the years 1980–2010, using the atmospheric and land conditions for that date and time from MERRA. The 30 anomaly simulations are initialized at 21z 20 May 1988, just before the stationary Rossby waves develop in the observations, and run through the end of July 1988. The atmospheric initial conditions for the 30 anomaly ensemble members are obtained by adding small perturbations to the MERRA atmospheric state at 21z 20 May 1988; the land variables in all of the anomaly ensemble members are initialized with MERRA land conditions at 21z 20 May 1988. Experiment A2 is the same as experiment A1 except that it uses MERRA atmospheric state at 21z 20 May of years 1980–2009 to initialize the atmosphere of its 30 anomaly members. The regional replay experiment B1 is the same as experiment A1 except that it constrains all of the model basic atmospheric variables to be close to those of MERRA over eastern Asia and the western Pacific (90°E–140°E; 0°–70°N). Experiment B2 is the same as experiment B1 except that the initialization occurs in early May (21z 2 May) instead of 21z 20 May, and it uses MERRA atmospheric and land conditions for 21z 2 May of years 1980 through 2009 to initialize its 30 anomaly ensemble members for the summer of 1988. This is to remove most, if not all, of the effects of late May atmospheric and land initial conditions on the results, allowing the ensemble mean of B2 to highlight the effects of the stationary Rossby waves that originate from the western Pacific. Experiment B3 is the same as experiment B2 except that it disables soil moisture feedback by continually resetting the soil moisture prognostic variables in the land model to seasonally varying climatological values. Experiment B4 is identical to experiment B2 except that it replays only the zonal and meridional wind over eastern Asia and western Pacific (90°E–140°E; 0°–70°N); air temperature, atmospheric moisture, and surface pressure are allowed to evolve freely. Finally, experiment C1 is identical to B2 except that the MERRA-based replay of the atmospheric variables is only over North America (130°W–70°W; 0°–70°N). We note that for each of the experiments, the control ensemble strictly follows the experimental design of the anomaly ensemble (e.g., dates of atmospheric and land initial conditions, replay region and variables, soil moisture feedback). Thus, among all the experiments performed, only A1 and A2 have identical control ensembles. The rest of the experiments differ from one another and also differ from A1/A2 in their control ensembles.

Experiment A1 is used to assess the GEOS-5 AGCM's ability to reproduce the stationary Rossby wave event of interest, and Experiment A2 is used to isolate the effect of the late May 1988 land initial conditions. The comparison between A1 and A2 thus highlights the contribution from the late May 1988 atmospheric initial conditions. Among the regional replay experiments, B1 offers the greatest potential to reproduce the stationary Rossby waves and their effects over North America, since it includes the most ingredients: the contributions from the late May 1988 atmospheric and land initial conditions, the observed stationary wave sources over the western Pacific, the (mostly) corrected model mean jet over the North Pacific (see section 12.3.2), and soil moisture feedback over land. The examination of and comparison between experiments B1 and B4 allow us to isolate and assess the relative contributions of these different ingredients to the evolution of the wave and its effects over North America. The experiment C1 is used to study the connections between surface anomalies over North America and northern Eurasia via stationary Rossby waves.

12.3. RESULTS

12.3.1. The Leading Modes of Subseasonal Atmospheric Variability

To study subseasonal atmospheric circulation variations, we focus on 250 hPa meridional wind, a quantity that highlights wave activity in the upper troposphere. The leading patterns of subseasonal atmospheric circulation variability are extracted by performing rotated empirical orthogonal function (REOF) analysis in which varimax rotation [e.g., *Richman*, 1986] is used to separate the leading wave structures geographically. Figure 12.1 shows the top five leading patterns of subseasonal atmospheric circulation variability during boreal summer, obtained using a REOF analysis of MERRA 250 hPa intraseasonal monthly meridional wind anomalies over the NH for June, July, and August of 1980–2014. The first REOF explains 8% of the total variance and consists of a wave train pattern that appears to emanate from the northeastern coast of North America. The wave train pattern spans the North Atlantic and much of northern Eurasia north of the mean south Asian jet. The second REOF is characterized by a wave train embedded in the North Atlantic jet. Over North America, it consists of (for the sign displayed here) a northerly flow anomaly over the west and a southerly flow anomaly over the east, corresponding to an upper-level anticyclonic flow anomaly over central North America. (Note: Here and in the following, our descriptions of the REOF patterns and associated surface anomalies are based on the sign of the patterns displayed in Fig. 12.1. The reader should keep in mind that the sign is arbitrary.) Further downstream at the North Atlantic jet exit, there are indications that the wave train (while weak) splits into two branches, with the northern branch spanning much of northern Eurasia and the southern branch embedded in the south Asian jet. The fourth REOF is similar overall to the second REOF, except for a 90° phase shift and for being more confined to North America and the North Atlantic. The third and fifth REOFs show similar wave trains confined to the North Pacific jet stream, distinguished primarily by a 90° phase shift. The third REOF in particular has a notable component over North America where it shows a northerly flow anomaly to the west and a southerly flow anomaly to the east, corresponding to an anomalous upper-level high over the central United States. Figure 12.1 clearly illustrates the importance of the mean summertime NH jet streams in guiding these leading subseasonal atmospheric circulation patterns.

These leading modes of subseasonal atmospheric circulation variability are associated with considerable subseasonal variations of surface meteorological fields (Fig. 12.2). The REOF1 is associated with notable surface temperature and precipitation anomalies over much of northern Eurasia, consisting of surface cooling and precipitation surpluses over southern Europe and central Russia, accompanied by surface warming and precipitation deficits over eastern Europe. The surface anomalies associated with the REOF2 exhibit a wave-train-like pattern spanning North America and northern Eurasia, with considerable surface warming and precipitation deficits over the central and eastern United States, surface cooling and precipitation increases over much of the Europe, and surface warming and precipitation deficits near 60°E. The wave-train-like pattern (Fig. 12.2, second panels from top) suggests that these subseasonal surface anomalies over North America and northern Eurasia could be linked through the action of stationary Rossby waves, which we will explore in section 12.3.2.2. By comparison, REOF4 primarily affects North America; it is associated with surface warming over western North America and cooling to the east, with notable precipitation deficits over central North America. Both REOF3 and REOF5 are associated with surface anomalies spanning the North Pacific. The REOF3 exhibits notable effects over North America: surface cooling over the Pacific Northwest and surface warming centered over the southern United States, along with precipitation surpluses over the northern United States and deficits over the

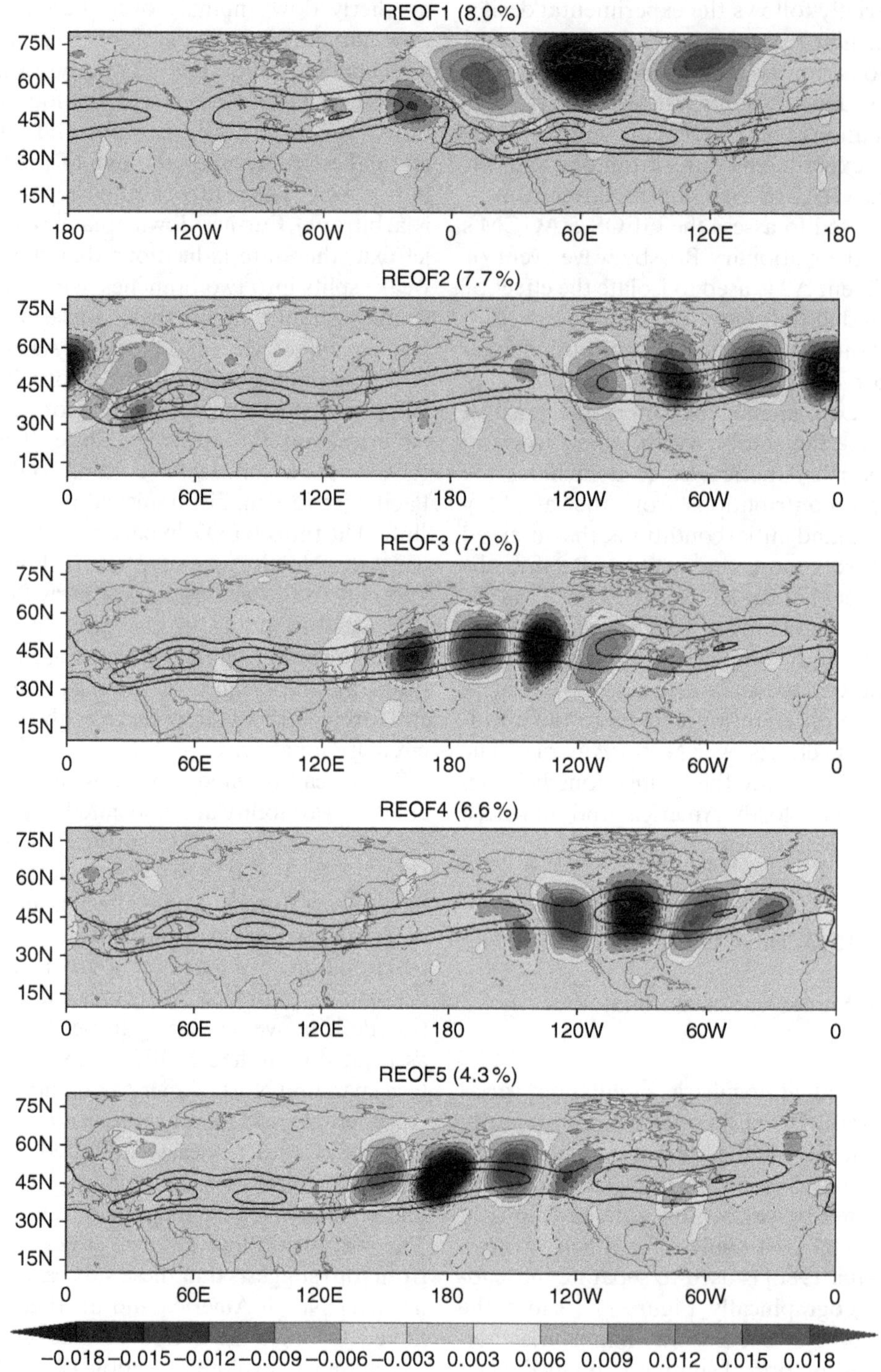

Figure 12.1 The leading REOFs of MERRA 250 hPa intraseasonal monthly mean meridional wind anomalies for JJA of 1980–2014. Units are arbitrary. The thick black contours are the long-term JJA mean (1980–2010) 250 hPa zonal wind at 15, 20, 25 m s^{-1}. Note that the central longitude in the panel for REOF1 is 180°E, whereas it is 0°E for the other REOFs.

southern United States. The REOF5 displays a moderate surface warming over the Pacific Northwest and moderate surface cooling and precipitation increases over the southern United States. We note that these surface anomalies, which are based on MERRA and MERRA-Land data, are very similar to those based on the CRU TS3.0 [*Harris et al.*, 2014] and GPCP [*Adler et al.*, 2003] observations (not shown).

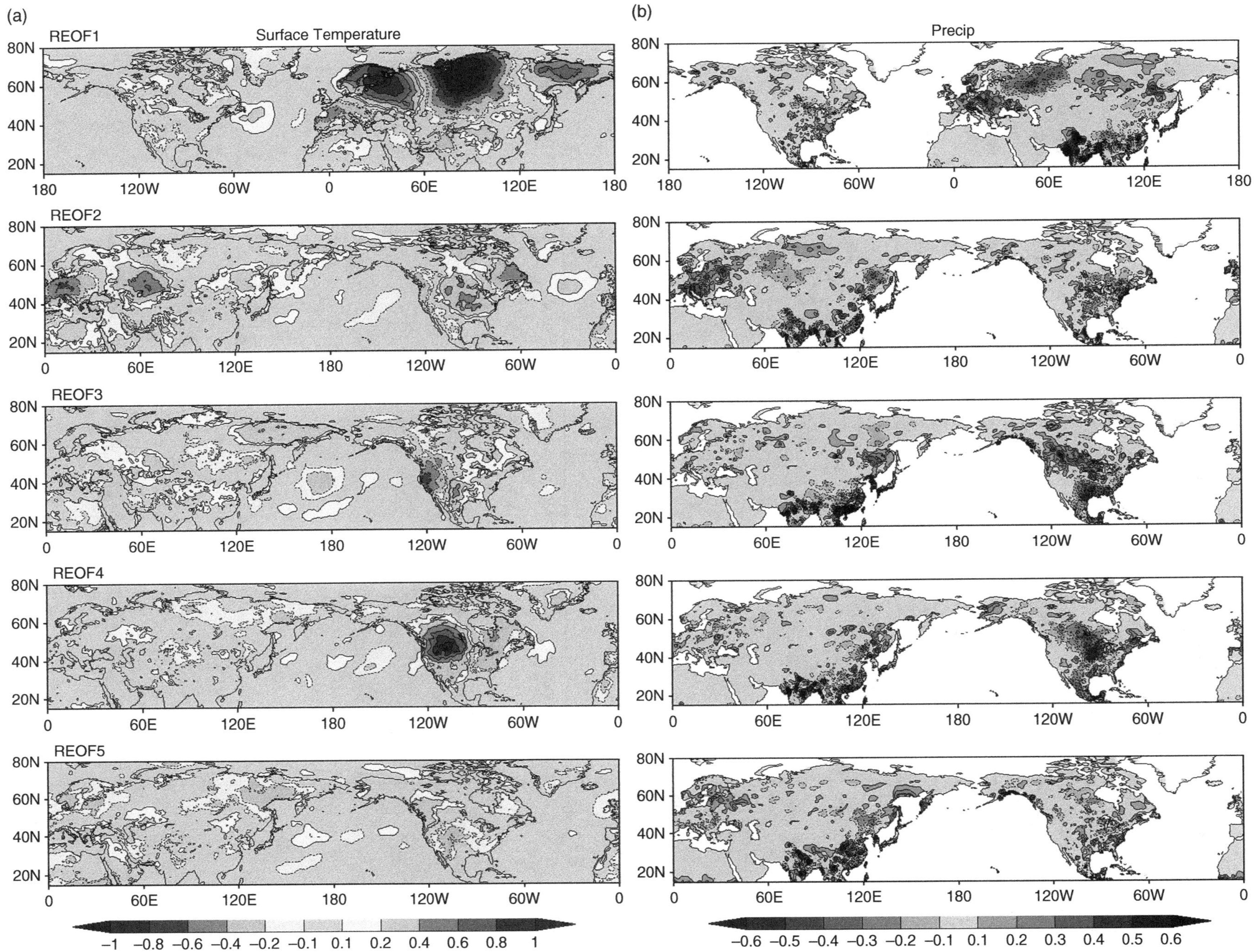

Figure 12.2 The subseasonal anomalies of (a) MERRA surface temperature (K), and (b) MERRA-Land precipitation (mm day-1) associated with the five leading REOFs of MERRA 250 hPa subseasonal meridional wind anomalies, based on a linear regression against the normalized RPCs of these leading REOFs for JJA of 1980–2014.

The subseasonal SST anomalies associated with the leading REOFs are vertically aligned such that warm (cold) SST anomalies lie beneath upper-level high (low) anomalies. This appears to reflect warm season air-sea interaction in the NH extratropical oceans where the rather shallow mixed layer allows a modest surface heat flux forcing from the overlying atmospheric circulation anomaly to exert considerable changes in mixed layer temperature, which in turn could feed back onto the atmosphere, reinforcing the subseasonal atmospheric circulation anomaly. While not investigated here, there is observational evidence suggesting the importance of extratropical air-sea interaction for sustaining subseasonal atmospheric circulation anomalies [*Peña et al.*, 2004; *Wang and Lupo*, 2009].

In order to quantify the importance of the leading REOFs in accounting for subseasonal surface meteorological variations, we show in Figure 12.3 the fraction of the subseasonal variability of surface temperature and precipitation explained by the top 10 leading REOFs. The leading REOFs primarily affect subseasonal surface variations over North America and northern Eurasia. They explain more than 60% of subseasonal surface temperature variations over much of the central and northern United States, Europe, and western Russia, and more than 30% of subseasonal SST variations over the North Pacific

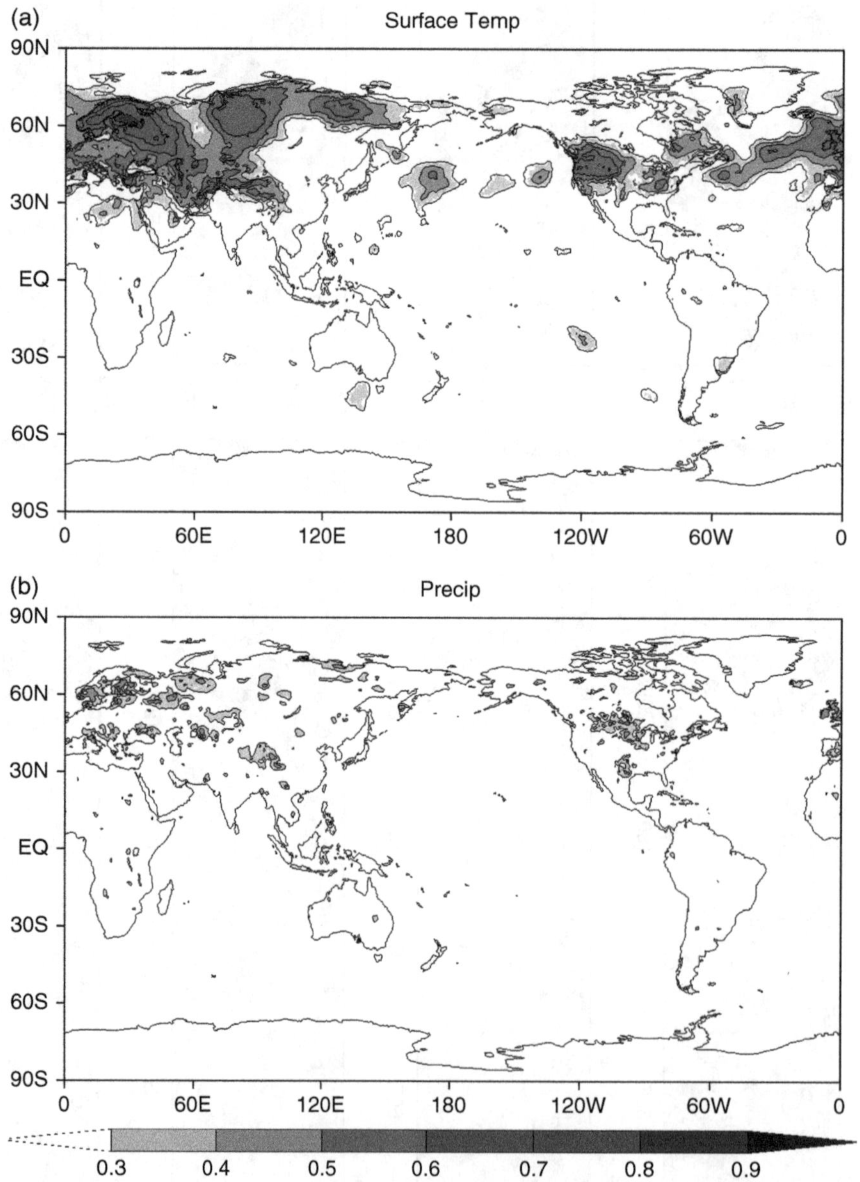

Figure 12.3 The spatial distribution of the fraction of monthly (JJA1980–2014) subseasonal variance of (a) surface temperature and (b) precipitation explained by the first ten 250 hPa subseasonal meridional wind REOFs, based on a linear regression. Contours start at 0.3 with a 0.1 interval. Shaded values are significant at the 5% (10%) level for surface temperature (precipitation) based on a Monte Carlo test.

and North Atlantic. The fraction of explained variance for precipitation is considerably smaller than that for surface temperature, with values exceeding 30% occurring primarily over the northern plains and Texas in North America as well as over northern Europe and western Russia.

12.3.2. The Influence of Stationary Rossby Waves on Drought

It is now well established that the leading modes of subseasonal atmospheric circulation variability play an important role in accounting for the subseasonal development of many regional climate extremes [*Schubert et al.*, 2011]. A better understanding of the causes and predictability of these climate extremes requires an exploration of the origin and predictability of stationary Rossby waves, as well as a study of the key processes through which these stationary waves lead to the climate extremes.

Here we focus on North America and northern Eurasia, the regions that appear to be most affected by the leading REOFs (Fig. 12.3), by performing a case study of a drought event during which stationary Rossby waves play an important role. To select the drought event, we examine in Figure 12.4 the rotated principal components (RPCs) of the top five leading REOFs and focus our attention on the modes that strongly affect North

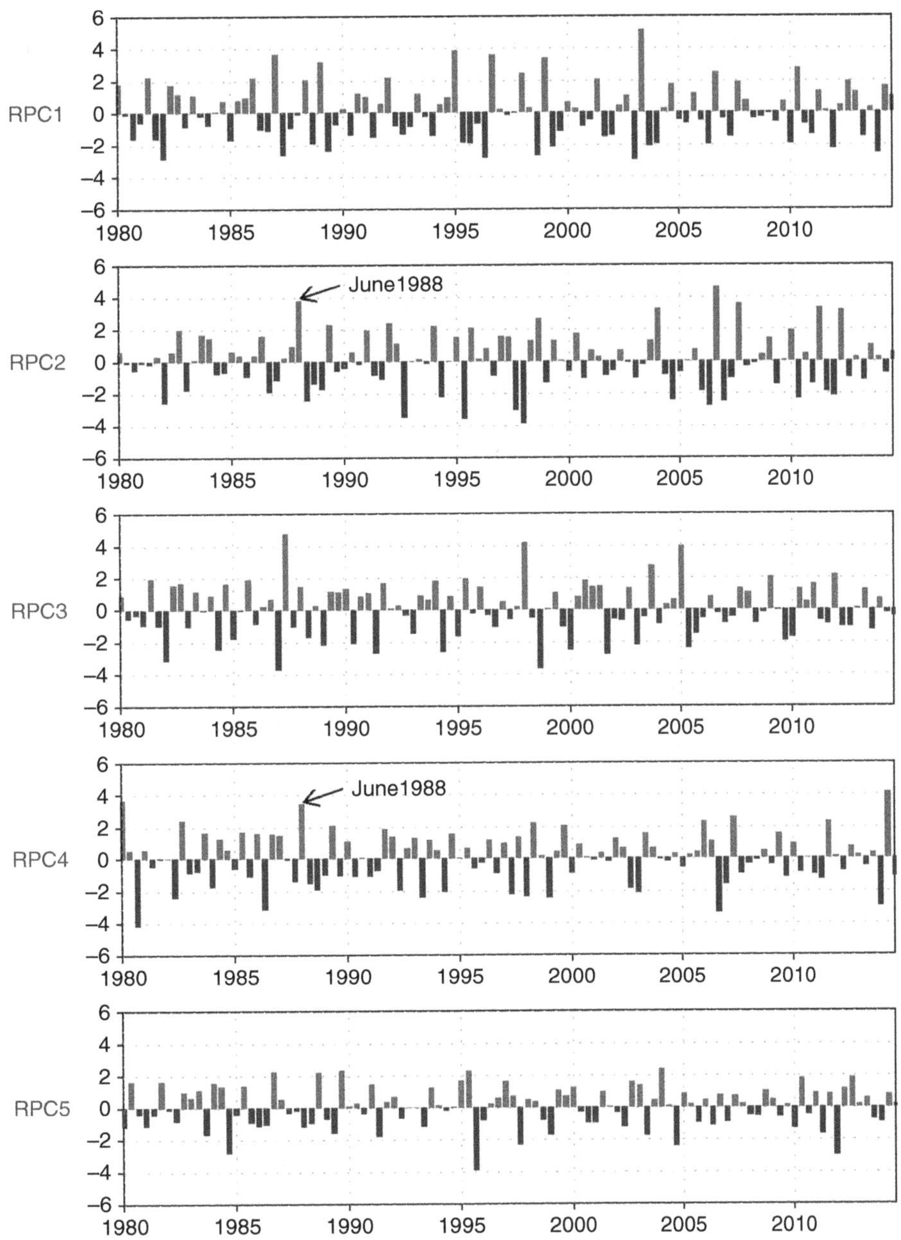

Figure 12.4 The leading RPCs of MERRA 250 hPa subseasonal monthly mean meridional wind anomalies for June, July, and August of 1980–2014.

America, that is, REOF2 and REOF4. Among the drought events that occurred during the time period 1980–2014, the month of June 1988 stands out: it shows strong projections on both REOF2 and REOF4.

Figure 12.5 shows the monthly anomalies for June 1988. The 250 hPa meridional wind anomaly (Fig. 12.5a) displays a wavelike pattern that appears to emanate from North America, crosses the North Atlantic, and then splits at the North Atlantic jet exit, with the northern branch extending eastward across much of northern Eurasia, and the southern branch confined to the south Asia jet. The anomalies have strong projections on both REOF2 and REOF4 over North America, and they resemble REOF2 over regions farther downstream. The surface temperature anomalies (Fig. 12.5b) over North America show strong surface warming of 8K over central North America with cooling to the northeast. This is accompanied by a substantial precipitation deficit of -3 mm day^{-1} (Fig. 12.5c) and surface soil dryness of -25% (Fig. 12.5d) over the eastern portion of the United States. The surface anomalies over northern Eurasia exhibit a wave-train-like pattern, with surface cooling, precipitation surpluses, and soil wetness over southern Europe and central Russia, and surface warming and precipitation deficits along with surface soil dryness over western Russia. Overall, the surface temperature and precipitation anomalies for June 1988 (Fig. 12.5b,c) show the combined contributions from REOF2 and REOF4 over North America as well as a prominent contribution from REOF2 over northern Eurasia (Fig. 12.2). In addition, there are regional SST anomalies of more than 1K over the North Pacific and North Atlantic, apparently reflecting the effects of extratropical air-sea interaction.

In order to better appreciate the stationary Rossby wave event(s) that occurred during June 1988 and the associated surface anomalies, we examine in Figure 12.6 the time evolution of the daily 250 hPa meridional wind and surface anomalies. Figure 12.6a clearly shows two stationary Rossby wave events that developed during 20 May–15 June and 16 June–8 July 1988. The energy associated with both waves propagates eastward from the North Pacific to western Eurasia, with the peak magnitude reached over North America. The wave event during 20 May–15 June is particularly prominent. It appears to originate from the western Pacific, with the energy propagating across the North Pacific to North America where it reaches its peak magnitude and persists for about 3–4 weeks. While the stationary Rossby waves remain quasi-stationary over North America and affect surface conditions there, the wave energy slowly propagates farther downstream across the North Atlantic, reaching Europe and western Russia about 2 weeks later; its magnitude meanwhile gradually weakens. The projections of daily 250 hPa meridional wind anomalies onto REOF2 and REOF4 (not shown) show that the first stationary Rossby wave event is dominated by REOF2, whereas the second is dominated by REOF4. This explains the aforementioned strong projections of the monthly mean anomalies for June 1988 on both REOF2 and REOF4 (Fig. 12.4). This also emphasizes the necessity of using daily data to study the waves and their effects on the development of regional surface anomalies.

Strong local surface anomalies develop rather quickly (Fig. 12.6b,c,d) as soon as the stationary Rossby waves are established over North America and northern Eurasia. In particular, surface warming, precipitation deficit, and surface soil dryness anomalies over both continents develop within 1–2 days under the overlying upper-level anticyclonic flow associated with the stationary Rossby waves. The amplitudes of the surface anomalies over northern Eurasia are overall weaker than those over North America. Figure 12.6c, d shows that while central North America is already experiencing precipitation deficits and surface soil dryness in early and mid-May, the sustained extreme precipitation deficits and surface dryness from late May through much of June 1988, presumably linked to the stationary Rossby waves, are much more distinct. In comparison, the surface anomalies over northern Eurasia appear to develop only after the stationary Rossby waves arrive, implying that these anomalies are initiated and then maintained by the waves. Furthermore, associated with the eastward wave energy propagation from North America to western north Eurasia, the surface anomalies over northern Eurasia occur about 2 weeks after those over North America. This suggests that surface anomalies in North America and Eurasia may be linked via stationary Rossby waves. We will explore this further in section 12.3.2.2.

Based on the examination of Figure 12.6, we choose the 20 May–15 June 1988 stationary Rossby wave event as the case study for our investigation. It is used to study the processes by which stationary Rossby waves are established over North America and lead to the quick buildup of severe drought conditions there. This case study also allows us to examine potential connections, via stationary Rossby waves, between North American drought and subsequent surface anomalies over northern Eurasia.

We use AGCM modeling in our analyses because it allows us to isolate and assess the key processes that affect stationary Rossby waves and associated regional climate extremes. An inspection of the warm season mean NH jet in the GEOS-5 AGCM, however, shows that while the simulation of the North Atlantic jet is generally reasonable, the model is deficient in simulating the North Pacific jet (Fig. 12.7b). The North Pacific mean jet in the GEOS-5 AGCM (Fig. 12.7b) is considerably weaker and less organized than the observed as estimated from MERRA (Fig. 12.7a), a deficiency also seen in many other current

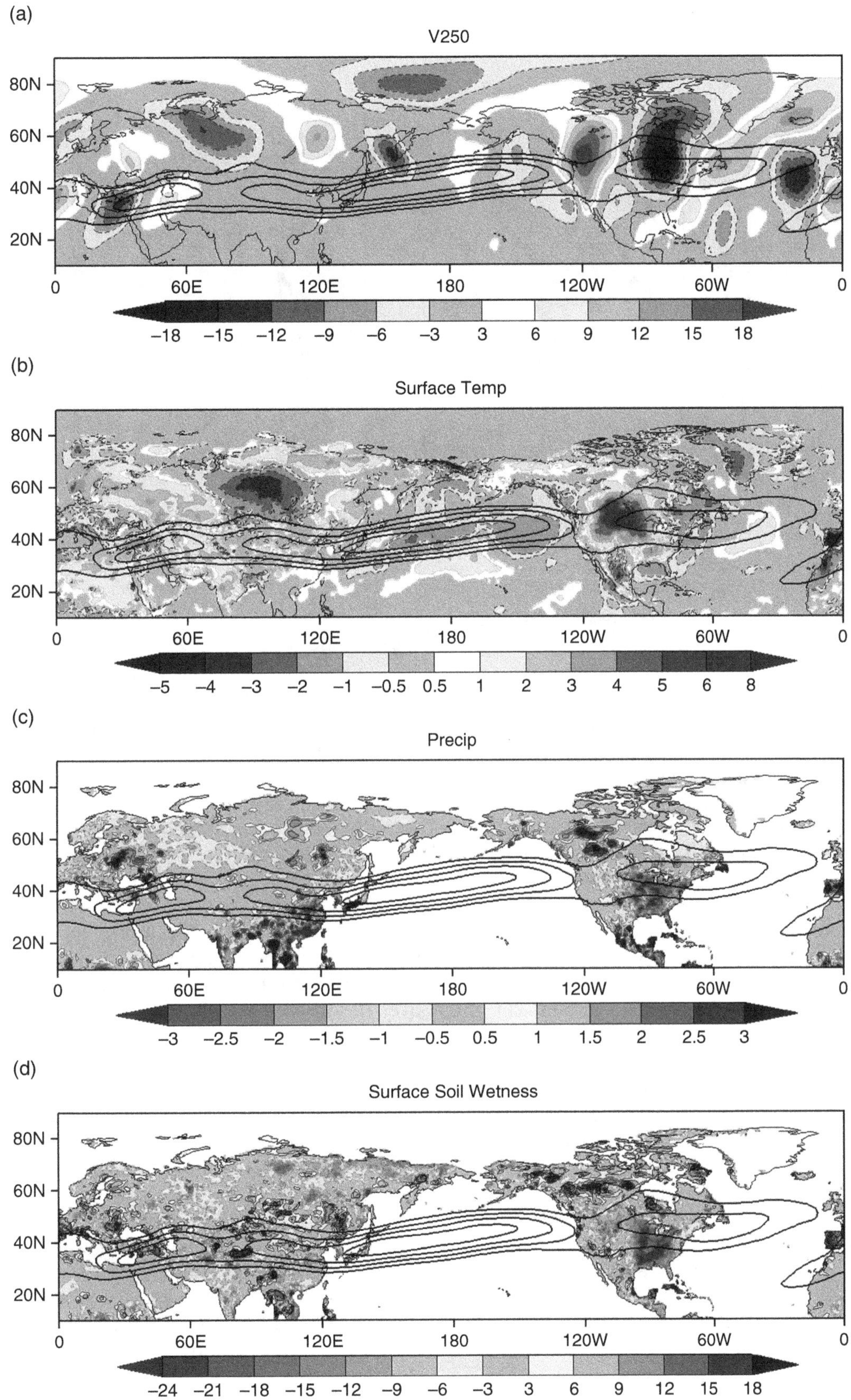

Figure 12.5 The June 1988 monthly anomalies for (a) MERRA 250 hPa meridional wind (m s^{-1}), (b) MERRA surface temperature (K), (c) MERRA-Land precipitation (mm day-1), and (d) MERRA-Land surface soil wetness (%). These anomalies are obtained as deviations from their 1980–2010 monthly means. The contours show the long-term mean (1980–2010) of June 250 hPa zonal wind, starting at 15 m s^{-1} with a 5 m s^{-1} interval. (*See insert for color representation of the figure.*)

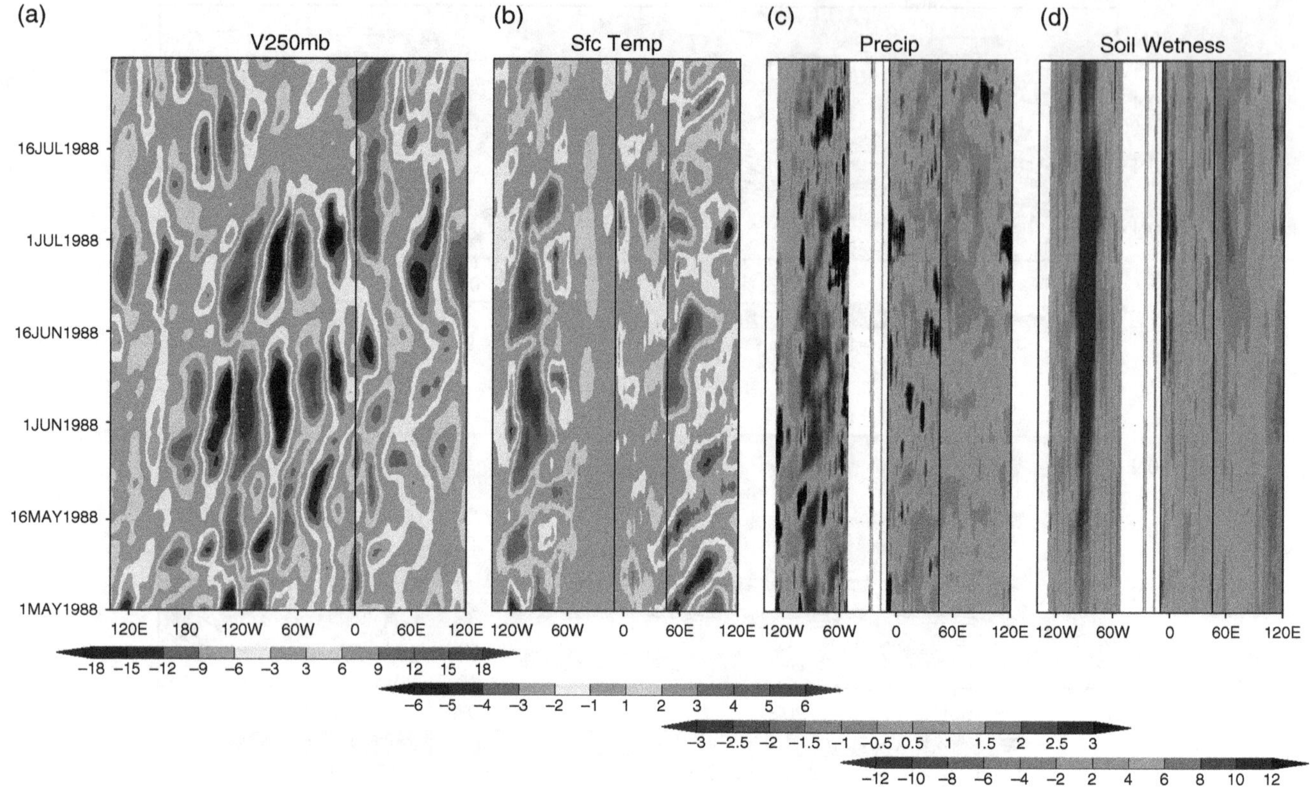

Figure 12.6 Evolution of (a) MERRA daily 250 hPa meridional wind ($m\ s^{-1}$) averaged between 40°N–60°N west of 0° and 50°N–70°N east of 0°, with a 10 day running mean smoother applied, (b) MERRA daily surface temperature anomalies (K), (c) MERRA-Land daily precipitation anomalies (mm day-1), and (d) MERRA-Land daily surface soil wetness anomalies (%), averaged between 30°N–50°N west of 10°W, between 35°N–55°N for 10°W–45°E, and between 50°N–70°N east of 45°E, with a 5 day running mean smoother applied, during the period 1 May–31 July 1998. The color shadings displayed from left to right are for panels (a) through (d), respectively. (*See insert for color representation of the figure.*)

AGCMs (not shown). The poor simulation of the North Pacific jet limits the usefulness of the free-running GEOS-5 AGCM for this study.

We use the "replay" technology to circumvent this model bias. We first determine the region that provides the forcing for the mean North Pacific jet by performing several regional replay experiments in which the model atmosphere is constrained over various regions upstream of the jet (not shown). The results show that the air temperature over central and north Asia and the diabatic heating over eastern Asia and the western Pacific (90°E–140°E; 0°–70°N; Fig. 12.8) play a key role in maintaining the North Pacific jet, in part through thermal wind balance. Constraining the GEOS-5 model atmosphere to be close to MERRA over this region (experiment B2) brings the simulated jet (ensemble average of 250 hPa zonal wind in the control ensemble of B2) close to the observed jet (compare Fig. 12.7a,c). In experiment B2, the simulated mean jet is in the correct location, with the correct shape and magnitude over much of the North Pacific; only minor biases are still seen over the northeastern Pacific, where the model jet is still somewhat weak. Furthermore, since the stationary Rossby wave event of interest originates from convective anomalies over the western Pacific [*Chen and Newman*, 1998], the locally replayed region (Fig. 12.8) in experiment B2 is appropriate. Given the improved jet and the utilization in replay mode of the observed locality of the Rossby wave sources, experiment B2 provides a useful framework for a study of wave propagation across the North Pacific and the subsequent effects on drought development over North America.

We also evaluate the mean North Pacific jet in the other regional replay experiments. The jet produced in experiment B1 (not shown) is similar to that in experiment B2, which is expected because the atmospheric and land initial conditions at 21z 20 May only affect the following 2 weeks and thus have little impact on the JJA mean jet. The mean NH jet in B3 (not shown) also resembles that in B2, suggesting a minor impact of soil moisture feedback on the mean jet. In contrast, constraining only the zonal and meridional wind fields over the region

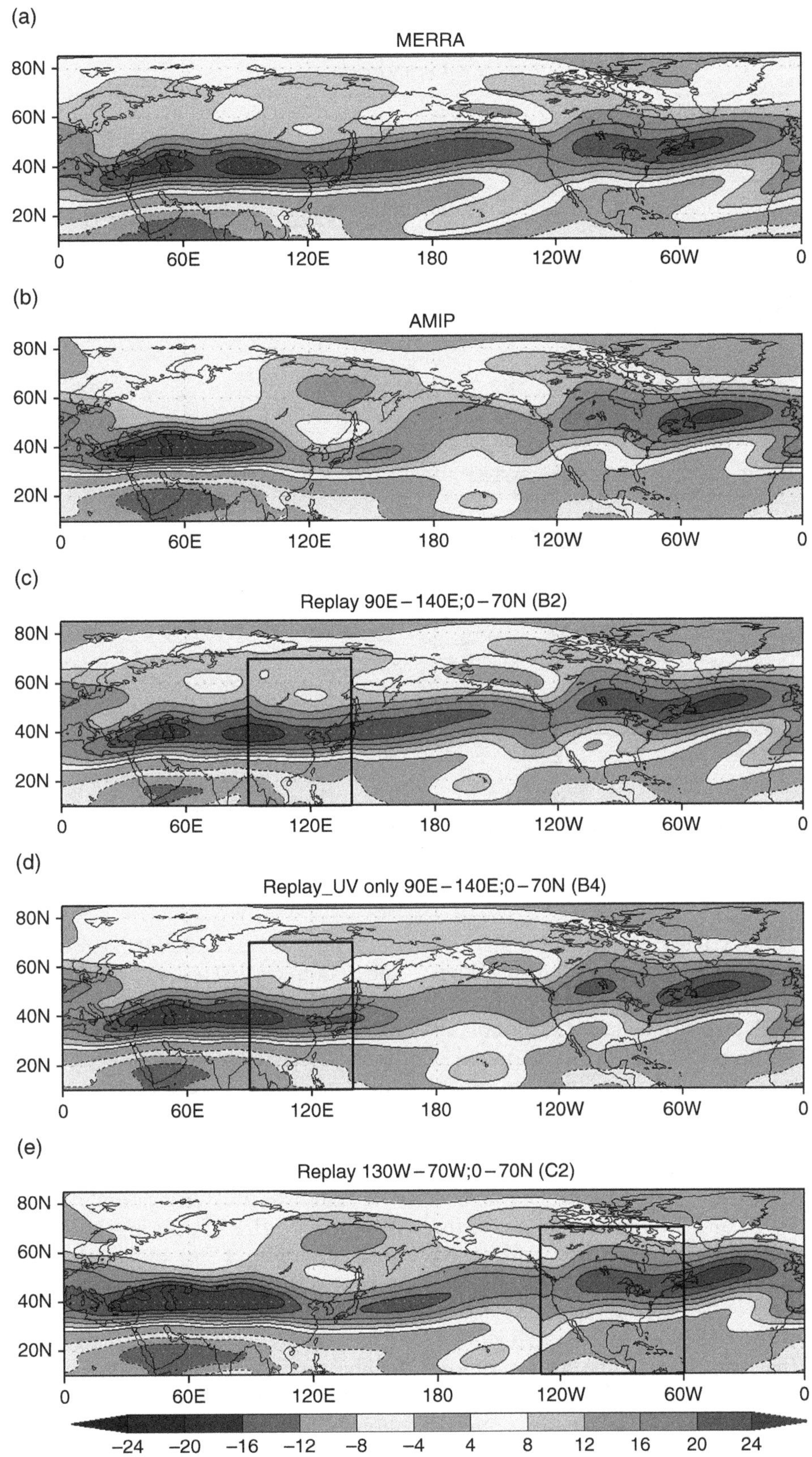

Figure 12.7 The long-term mean (1980–2010) of JJA 250 hPa zonal wind (m s^{-1}) in (a) MERRA, (b) the GEOS-5 AMIP simulations, (c) the GEOS-5 AGCM regional replay experiment in which the model atmosphere is constrained to be close to MERRA over eastern Asia and the western Pacific (90°E–140°E; 0°–70°N) (experiment B2), (d) same as (c) except that only zonal and meridional wind fields are replayed (experiment B4), (e) same as (c) except for the AGCM regional replay experiment over North America (130°W–70°W; 0°–70°N) (experiment C1). The portions of the replay regions in the plotting domain are indicated using black boxes.

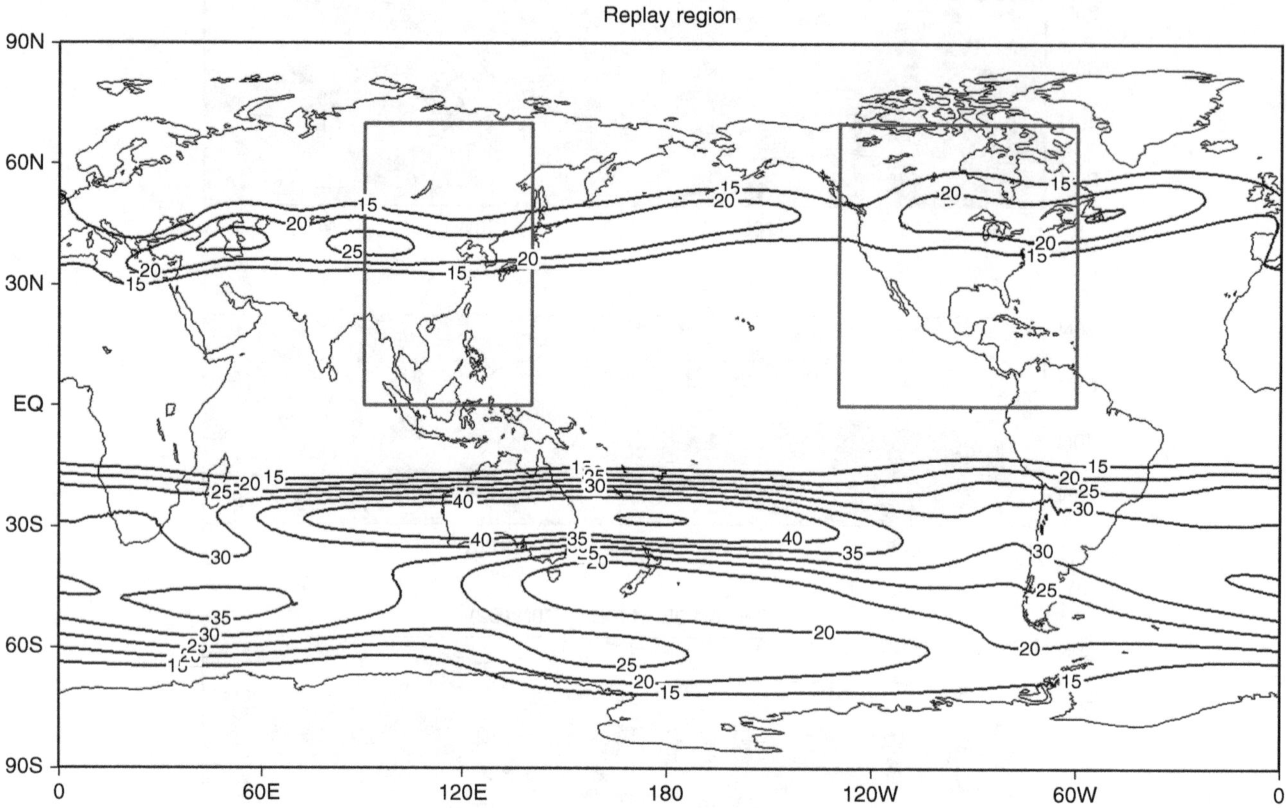

Figure 12.8 The regions of eastern Asia and the western Pacific (90°E–140°E, 0°–70°N) and North America (230°E–290°E, 0°–70°N) used in the GEOS-5 AGCM regional replay experiments. The contours show the long-term mean (1980–2010) of JJA 250 hPa zonal wind from MERRA, starting from 15 m s^{-1} with a 5 m s^{-1} interval.

90°E–140°E; 0°–70°N (experiment B4) produces a North Pacific mean jet with the correct location and shape but with considerably weaker amplitude over the central and eastern North Pacific (Fig. 12.7d); deficiencies and biases associated with atmospheric variables not constrained in B4 (e.g., air temperature and specific humidity) are apparently able to feed back onto the zonal and meridional winds, leading to considerably larger analysis increments for zonal wind in B4 than in B2, which subsequently affects the mean North Pacific jet downstream. Note that a comparison between B2 and B4 allows us to assess the impact of the magnitude of the mean North Pacific jet on wave energy propagation across the North Pacific and subsequent conditions over North America.

The examination of the NH mean jet in experiment C1 (Fig. 12.7e) shows that constraining the model atmosphere to be close to MERRA over North America (Fig. 12.8) leads to moderate improvements over North America, where the model jet strengthens by about 5 m s^{-1} and is placed slightly southward. There are, however, still moderate differences over western north Eurasia, where the model underestimates the mean zonal wind by about 5 m s^{-1}.

12.3.2.1. The 1988 North American Drought. In this subsection, we first assess the skill of the GEOS-5 AGCM in reproducing the stationary Rossby wave event of interest; we then investigate the key factors that influence stationary Rossby waves and their effects over North America. Figures 12.9 and 12.10 show the stationary Rossby waves produced in the two free-running GEOS-5 AGCM hindcast experiments (A1 and A2), the regional replay experiments B1 through B4, and nature (MERRA). The atmospheric circulation anomalies over North America in MERRA during 20 May–15 June are overall quasi-stationary (Fig. 12.9a); they consist of a distinct northerly flow anomaly over western North America and a southerly flow anomaly over eastern North America (Fig. 12.10a), corresponding to an upper-level positive height anomaly over central North America. The free-running GEOS-5 AGCM hindcast from A1 (Fig. 12.9b) shows rather limited skill beyond 1 week, after which the circulation anomalies show notable retrograde (westward) motion rather than remaining quasi-stationary as in MERRA (Fig. 12.9a), a reflection of the weak mean zonal jet in the free-running model (Fig. 12.7b). Consequently, the circulation anomalies in experiment

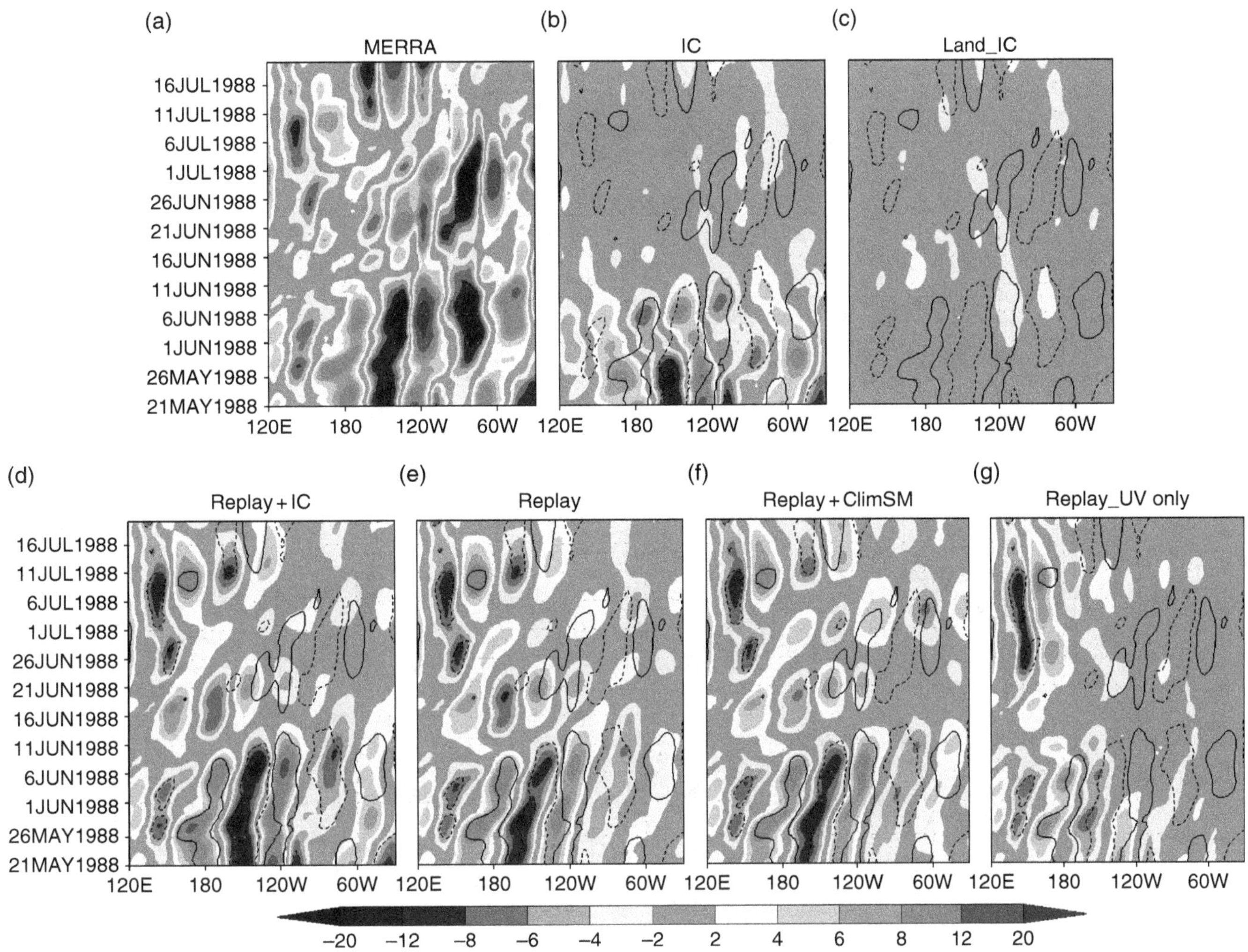

Figure 12.9 Evolution of daily 250 hPa meridional wind component (m s^{-1}) averaged between 30°N–60°N with a 10 day running mean smoother applied in (a) MERRA, (b) the ensemble mean of the GEOS-5 AGCM hindcast initialized using MERRA atmospheric and land conditions at 21z 20 May 1988 (experiment A1), (c) same as (b) but for the AGCM hindcast that uses MERRA atmospheric conditions at 21z 20 May of years 1980–2009 to initialize the atmosphere of its anomaly ensemble (experiment A2), (d) the ensemble mean of the GEOS-5 AGCM regional replay over eastern Asia and the western Pacific (90°E–140°E, 0°–70°N) initialized using MERRA atmospheric and land conditions at 21z 20 May, 1988 (experiment B1), (e) same as (d) but initialized using MERRA atmospheric and land conditions at 21z 2 May of years 1980–2009 (experiment B2), (f) same as (e) but have land feedback disabled (experiment B3), and (g) same as (e) but replay only zonal and meridional wind fields (experiment B4). The black contours in (b)–(g) indicate the −6 m s^{-1} and 6 m s^{-1} contours of MERRA in (a).

A1 are opposite to those in MERRA in early June (Fig. 12.9b). The mean atmospheric circulation anomalies over North America for experiment A1 averaged over 26 May–10 June 1988 are displaced westward relative to those in MERRA, with a northerly flow anomaly over northwest North America and a southerly flow anomaly over central North America (Fig. 12.10b). The spread between the ensemble members is modest overall (Fig. 12.10c), due to the imposition of the 21z 20 May 1988 atmospheric and land initial conditions. The free-running GEOS-5 AGCM hindcast that makes use only of MERRA land initial conditions at 21z 20 May 1988 (A2) essentially shows no skill in producing the observed stationary Rossby waves (Figs. 12.9c, 12.10d), with rather large intramember spread (Fig. 12.10e). The contrast between A1 and A2 highlights the importance of the late May atmospheric initial state for the AGCM hindcast; the contribution from the late May dry land initial condition is minor.

While not shown here, we also examined the contribution of SST anomalies to the AGCM hindcast by performing two other AGCM hindcast experiments that

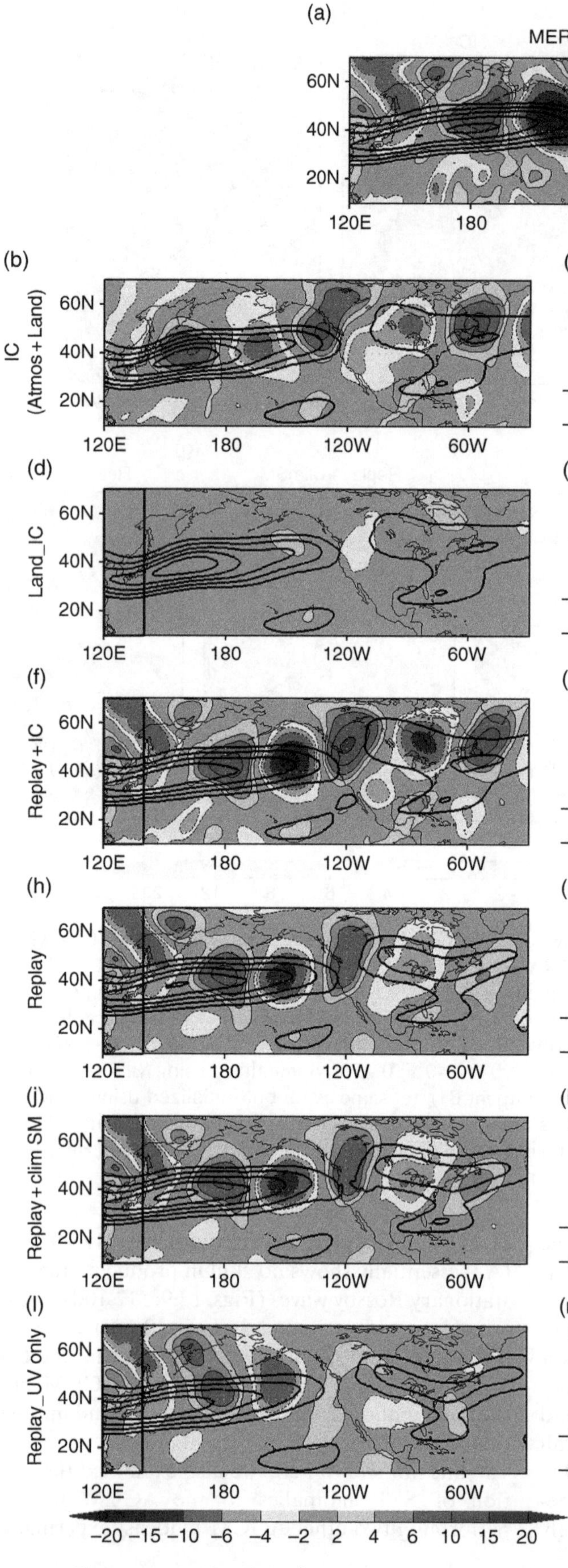

Figure 12.10 (Continued)

are the same as A1 and A2 but using climatological SST (1980–2010). The results from these hindcast experiments are similar overall to those from A1 and A2, indicating that the SST anomalies played a very minor (if any) role in forcing the anomalies that occurred over North America during the period 20 May–15 June 1988 (not shown).

As discussed earlier, experiment B1 includes all of the key factors considered in our experiments: the observed stationary Rossby wave sources over the western Pacific, a mostly corrected North Pacific mean jet, the atmospheric and land initial conditions in late May, and soil moisture feedback over land. This combination results in a significantly improved reproduction of the Rossby wave event of interest (Figs. 12.9d, 12.10f,g; focus on the case study period of 20 May–15 June). In fact, among all the regional replay experiments, B1 shows the strongest agreement with MERRA. The circulation anomalies in the ensemble mean of B1 strongly resemble those of MERRA over the North Pacific as well as North America (Figs. 12.9d, 12.10f). The ensemble spread among the 30 members gradually increases as the wave energy propagates away from the replay region (Fig. 12.10g). Over North America, most of the 30 ensemble members agree with their ensemble mean in producing a northerly flow over western North America and a southerly flow over eastern North America. There are indications of a slight lagging of the waves in B1 relative to MERRA over the northeast Pacific and western North America (Fig. 12.10g), which could be due to the modestly weaker simulated jet there (Fig. 12.7c). The considerable improvement in B1 over A1 confirms that the stationary wave sources reside in the western Pacific. By comparison, the stationary Rossby event during 16 June–8 July 1988 is not captured in the ensemble mean of B1, suggesting that these waves could originate from processes outside the replay region (90°E–140°E; 0°–70°N).

The results from B2 are similar overall to those from B1 except that they have moderately smaller amplitudes due to the modestly larger ensemble spread (Figs. 12.9e, 12.10h,i), and that they displace the circulation anomalies westward relative to MERRA in late May, due to the lack of constraints from late May initialization. It is worth noting that despite the diverse atmospheric and land initial conditions used in the anomaly runs, most of the 30 ensemble members agree with MERRA in showing a northerly flow anomaly over western North America and a southerly flow anomaly over eastern North America (Fig. 12.10i). This suggests that with a mostly corrected North Pacific mean jet, the observed stationary wave sources in the western Pacific do provide a predilection for persistent upper-level anticyclonic flow anomalies over central North America about 1 to 2 weeks later.

The results from B3 (Figs. 12.9f, 12.10j, k) are in strong agreement with those from B2, with only a modestly smaller ensemble spread over North America and regions farther downstream. The overall good agreement between B3 and B2 suggests that during the period (3–4 weeks) in which the stationary Rossby waves existed over North America, local soil moisture variations induced by the wave did not significantly feed back on the overlying upper-level atmospheric circulation anomalies.

The results from B4 (Figs. 12.9g, 12.10 l, m), the experiment that constrains only the winds through the replay approach, distinctly differ from those of B2 (Figs. 12.9e, 12.10h,i). While B4 is similar to B2 in prescribing the observed stationary wave sources over the western Pacific, the mean jet over the North Pacific (Fig. 12.7d) is considerably weaker than that in B2 (Fig. 12.7c) east of the dateline, thus imposing a weaker constraint on wave propagation there. As a result, B4 shows a slower eastward propagation and a larger ensemble spread east of dateline than B2 (compare Fig. 12.10m and Fig. 12.10i); the ensemble mean of B4 over North America not only shows a rather weak magnitude but also places the responses in the wrong locations (Fig. 12.10 l).

Upon their establishment over North America, the stationary Rossby waves exert a strong impact on the

Figure 12.10 (Continued) The 250 hPa meridional wind anomalies (m s^{-1}) averaged over the period 26 May–10 June 1988 in (a) MERRA; (b) the ensemble mean of the GEOS-5 AGCM hindcast initialized using MERRA atmospheric and land conditions at 21z 20 May 1988 (experiment A1); (c) the mean 250 hPa meridional wind response over the period 26 May–10 June 1988 averaged over 30°N–60°N in the 30 ensemble members (thin gray lines) and their ensemble mean (thick black line) in the AGCM hindcast (experiment A1), and the corresponding anomalies in MERRA (thick gray line); (d)–(e) same as (b)–(c) but for the AGCM hindcast initialized using MERRA atmospheric conditions at 21z 20 May of years 1980–2009, and MERRA land conditions at 21z 20 May 1988 (experiment A2); (f)–(g) same as (b)–(c) but for the AGCM regional replay experiment over eastern Asia and the western Pacific (90°E–140°E, 0°–70°N) initialized using MERRA atmospheric and land conditions at 21z 20 May 1988 (experiment B1); (h)–(i) same as (f)–(g) but initialized using MERRA atmospheric and land conditions at 21z 2 May of years 1980–2009 (experiment B2); (j)–(k) same as (h)–(i) but have land feedback disabled (experiment B3); (l)–(m) same as (h)–(i) but replay only zonal and meridional wind (experiment B4). The black contours in (a), (b), (d), (f), (h), (j), and (l) indicate the 250 hPa zonal wind (m s^{-1}) averaged over 28 May–15 June 1988 in MERRA and the AGCM experiments. They start at 15 m s^{-1} with a 5 m s^{-1} interval. *See electronic version for color representation.*

local fields. MERRA (Fig. 12.11a–c) shows that during the late May–middle June stationary Rossby wave event, there are distinct surface warming anomalies over central North America and strong precipitation deficits and surface dryness over central and eastern America. The 26 May–10 June mean surface anomalies in the AGCM hindcast (experiment A1) (Fig. 12.11d–f) are somewhat similar to MERRA except that the anomalies are weaker in magnitude and are also shifted westward due to the westward displacement of upper-level atmospheric circulation anomalies (Fig. 12.9b). The isolated impact of the late May land initial conditions on the surface temperature and precipitation anomalies over North America is rather weak (Fig. 12.11g,h), consistent with the weak atmospheric circulation anomalies in A2 (Figs. 12.9c, 12.10d,e). The ensemble mean surface anomalies in experiment B1 (Fig. 12.11j–l) show the best agreement with those in MERRA (Fig. 12.11a–c), benefiting from a good reproduction of the observed stationary Rossby waves over North America (Figs. 12.9d, 12.10f,g). The strong surface soil dryness anomalies in A1, A2, and B1 mainly reflect the late May 1988 dry land initial condition.

The ensemble mean surface anomalies in B2 (Fig. 12.11m–o) are similar overall to those in B1, except that magnitudes are reduced by 50% for surface temperature and precipitation and by considerably more for surface soil wetness, since unlike B1, B2 does not benefit from a dry land initialization in late May. The surface anomalies in B2 essentially reflect the effects of the overlying quasi-stationary upper-level high anomalies that originate from the stationary Rossby wave sources over the western Pacific.

Disabling soil moisture feedback in B3 only modestly affects the precipitation anomalies over the central and eastern United States (compare Fig. 12.11n and q), but does have moderate impact on surface temperature anomalies. The magnitude of surface warming response over central North America in B3 (Fig. 12.11p) is about 60% of that in B2 (Fig. 12.11m), suggesting soil moisture feedback contributes to about 40% of surface warming in B2.

The results of experiment B3 (Figs. 12.9f, 12.10j,k, 12.11 p–r) may at first glance seem counter to the findings of *Koster et al.* [2014], who found a significant impact of soil moisture anomalies on such waves. Note, however, that *Koster et al.* [2014] studied a very different problem: how the imposition of large soil moisture anomalies may affect the atmospheric circulation. In B2, initial soil moisture conditions were not extreme on average, and while some soil moisture anomalies are generated in B2 over the course of the simulations, they are not especially large (compare Fig. 12.11c and o). Comparing B3 with B2, therefore, does not address the question of how large soil moisture anomalies may influence wave structure. More germane to the *Koster et al.* [2014] study are the results of A2, which does impose a strong soil moisture anomaly. Here we can only speculate on why the wave response is smaller in A2 than in *Koster et al.* [2014]. One potential explanation involves the spatial breadth of the imposed soil moisture anomaly in A2, as indicated in a map of the initial soil moisture anomaly (not shown) and as implied in Figure 12.11i, the initial soil moistures in both the southern Great Plains and the northwest United States were anomalously dry in A2, which would lead (according to the results of *Koster et al.* [2014]) to conflicting impacts on the nature of the overlying wave. That is, the impacts of the southeastern and northwestern dry soil moisture anomalies in A2 may, in this case, have partially canceled each other out.

While both B2 and B4 prescribe the observed stationary Rossby wave sources over the western Pacific, the surface anomalies in the latter (Fig. 12.11s–u) are weak overall and show little resemblance to MERRA, due to the poor simulation in B4 of atmospheric circulation anomalies over North America (Figs. 12.9g, 12.10 l,m) stemming from the simulation of a weak North Pacific mean jet east of the dateline (Fig. 12.7d). The contrast between B2 and B4 illustrates the importance for a model to correctly simulate the North Pacific mean jet, not only its latitudinal location and shape but also its magnitude, so as to let the wave energy propagate across the North Pacific with the correct speed and direction, and subsequently exert the correct regional effects over North America. Since it usually takes 1 to 2 weeks for the wave energy to propagate across the North Pacific, knowledge of western North Pacific sources for stationary Rossby waves can serve as a potential source of predictability for North American droughts.

12.3.2.2. North American Drought and Surface Anomalies over Northern Eurasia. Observations (Figs. 12.2, 12.5, 12.6, 12.12a) imply that subseasonal surface anomalies over North America and those over northern Eurasia may be connected to each other via stationary Rossby waves. Observed surface anomalies in northern Eurasia, however, are often strongly influenced by local processes that may mask out the effects of such waves [e.g., *Schubert et al.*, 2014]. Here we address these issues by performing a regional replay experiment in which the GEOS-5 model atmosphere is constrained to be close to MERRA over North America (experiment C1) and then studying what occurs downstream.

The ensemble mean 250 hPa meridional wind in C1 (Fig. 12.12b) shows that while the strong quasi-stationary atmospheric circulation anomalies remain over North America (as necessitated by the replay design) leading to the severe drought conditions there, there are indications

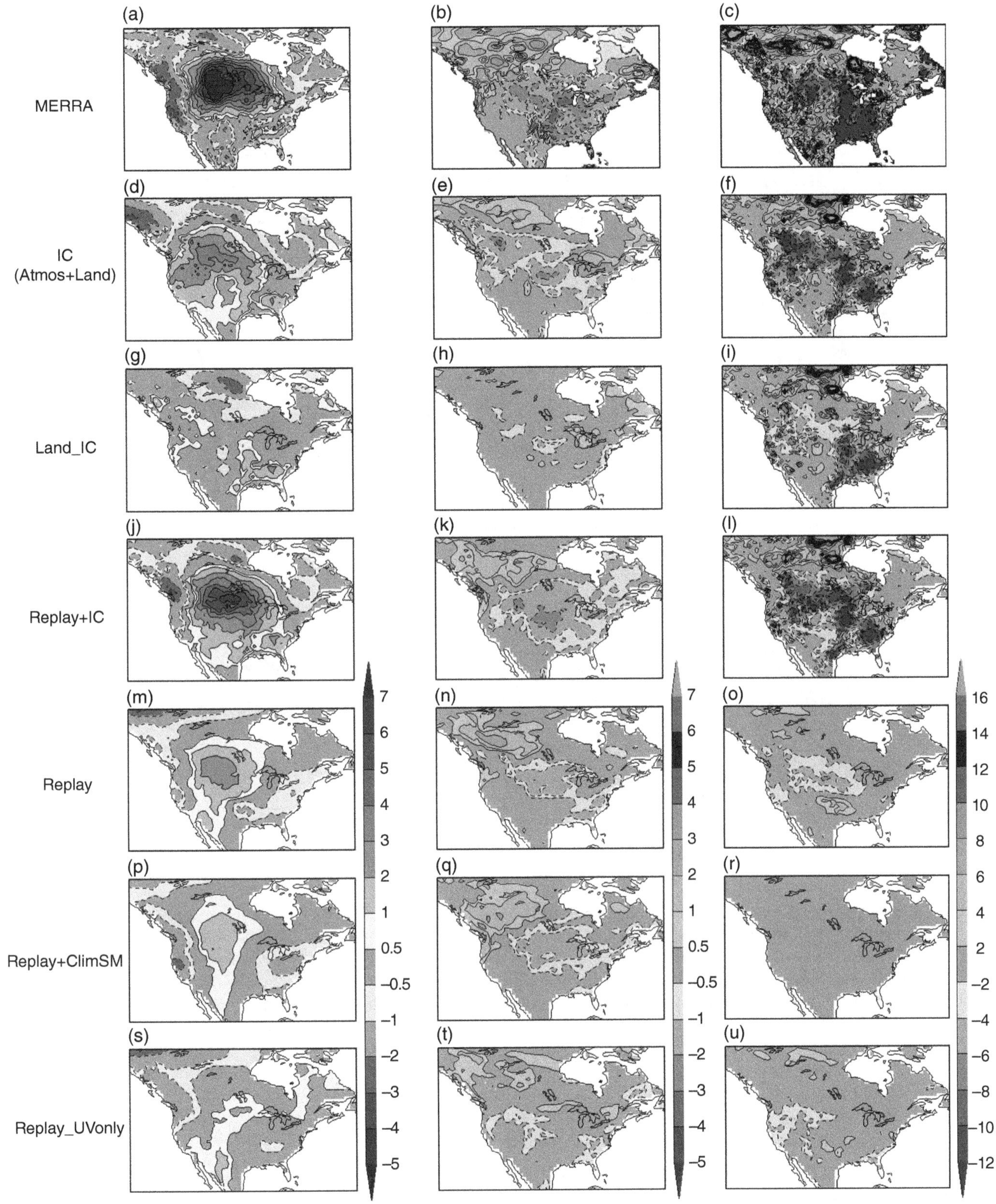

Figure 12.11 The anomalies of (a) MERRA surface temperature (K); (b) MERRA-Land precipitation (mm day-1); and (c) MERRA-Land surface soil wetness (%) averaged over the period 26 May–10 June 1988; (d)–(f) same as (a)–(c) but for the GEOS-5 AGCM hindcast initialized using MERRA atmospheric and land conditions at 21z 20 May 1988 (experiment A1); (g)–(i) same as (a)–(c) but for the GEOS-5 AGCM hindcast initialized using MERRA atmospheric conditions at 21z 20 May of years 1980–2009, and MERRA land conditions at 21z 20 May 1988 (experiment A2); (j)–(l) same as (a)–(c) but for the GEOS-5 AGCM regional replay over eastern Asia and the western Pacific (90°E–140°E;0°–70°N) initialized using MERRA atmospheric and land conditions at 21z 20 May 1988 (experiment B1); (m)–(o) same as (j)–(l) but initialized using MERRA atmospheric and land conditions at 21z 2 May of years 1980–2009 (experiment B2); (p)–(r) same as (m)–(o) but have land feedback disabled (experiment B3); (s)–(u) same as (m)–(o) but replay only zonal and meridional wind over eastern Asia and the western Pacific (90°E–140°E;0°–70°N) (experiment B4).

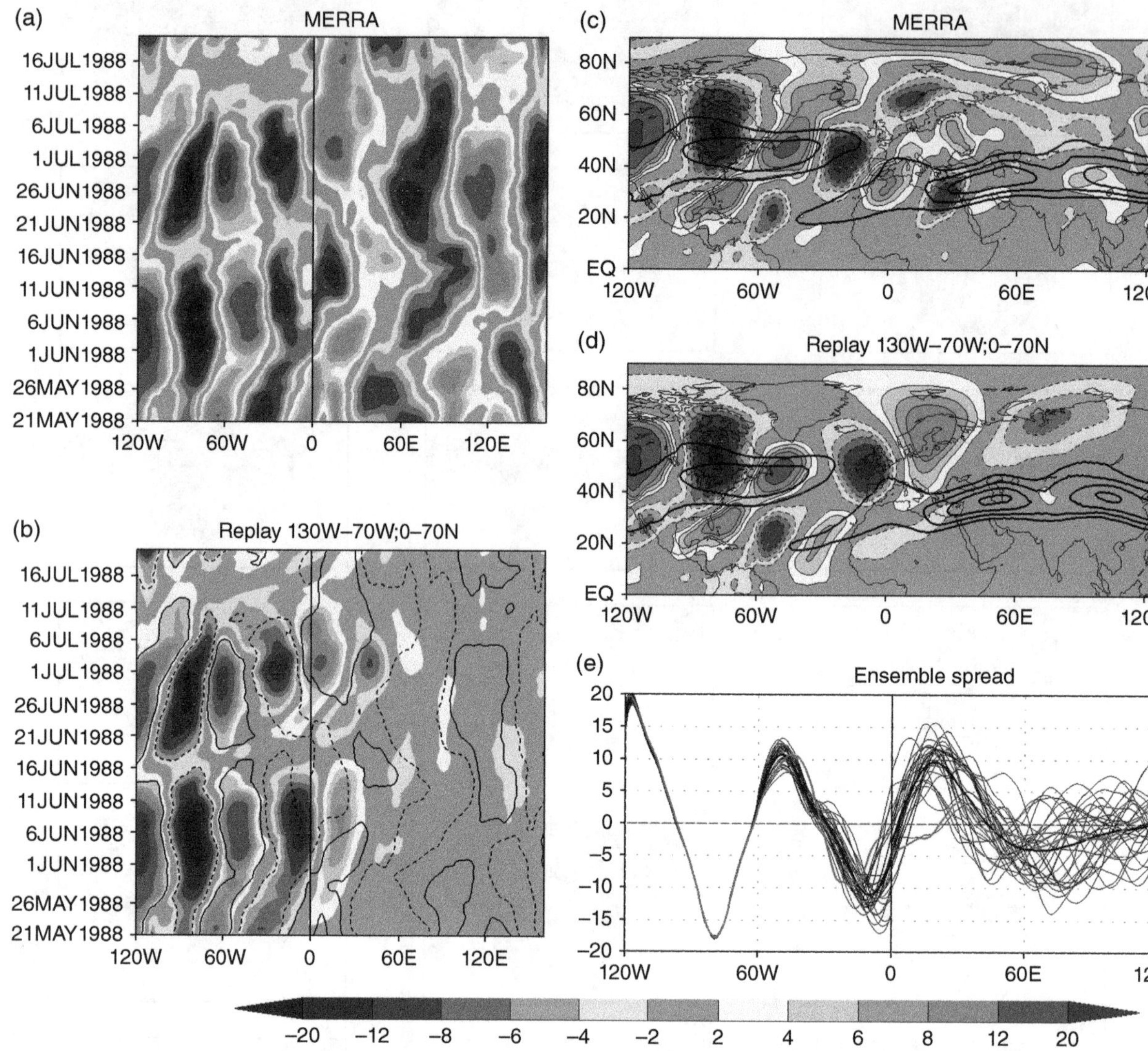

Figure 12.12 The 250 hPa meridional wind anomalies (m s^{-1}) averaged over 40°N–60°N west of 0° and 50°N– 70°N east of 0°, with 10 day running mean applied in (a) MERRA and (b) the ensemble mean of the regional replay experiment in which the model atmosphere is constrained to be close to MERRA over North America (130°W–70°W; 0°–70°N) (experiment C1); the 250 hPa meridional wind anomalies averaged over 28 May–15 June 1988 in (c) MERRA, (d) the ensemble mean of the regional replay experiment over North America (experiment C1); (e) the mean 250 hPa meridional wind anomalies over 28 May–15 June 1988 averaged over 40°N–60°N west of 0° and 50°N–70°N east of 0°, in the 30 ensemble members (thin gray lines), the 30 member ensemble mean (thick black line), and MERRA (thick gray line). The black contours in (b) indicate the –6 m s^{-1} and 6 m s^{-1} contours of MERRA in (a). The black contours in (c) and (d) indicate the 250 hPa zonal wind (m s^{-1}) averaged over 28 May–15 June 1988 in MERRA and experiment C1, starting at 15 m s^{-1} with a 5 m s^{-1} interval. *See electronic version for color representation.*

in C1 of wave energy propagation downstream under the guiding effect of the North Atlantic jet, reaching Europe and western Russia about 2 weeks later. In particular, off the North Atlantic jet exit, the ensemble mean shows a distinct wave train propagating eastward along the northern edge of northern Eurasia (Fig. 12.12d). The model ensemble spread (Fig. 12.12e) over the North Atlantic is strongly constrained by the strong North Atlantic jet and then increases notably off the jet exit where the upper zonal wind weakens markedly. The majority of the ensemble members consistently show a southerly flow anomaly off the west coast of Europe, a northerly flow anomaly over central Europe, and a southerly flow anomaly farther downstream over central northern Russia. The wave-train-like responses over Europe and northern Russia in the ensemble mean are displaced somewhat north of those of MERRA (Fig. 12.12c), possibly due to the underestimation of the mean zonal wind there (by about 5 m s^{-1}).

The stationary Rossby waves over northern Eurasia affect the local surface conditions considerably (Fig. 12.13). The upper-level anticyclonic flow anomalies associated with the stationary Rossby waves act to both warm and dry the atmosphere and land surface, leading to local surface warming, precipitation deficits, and soil moisture dryness, whereas the upper-level cyclonic flow anomalies exert surface anomalies of the opposite sign (Fig. 12.13b, d, f). Relative to the observations (Fig. 12.13a, c, e), the model surface anomalies are displaced northward, consistent with the northward displacement of the model-produced stationary Rossby waves (Fig. 12.12d).

Despite the noted model biases, experiment C1 provides evidence that stationary Rossby waves (resembling REOF2) that lead to drought over central North America can subsequently affect northern Eurasia. They can provide a predilection for dry conditions to develop over eastern Europe and western Russia and for cool and wet conditions to develop over both western Europe and central northern Russia.

The fact that the surface anomalies over northern Eurasia lag those over North America by about 2 weeks has important implications for subseasonal forecasts in Europe and western Russia. Improved prediction skill for weeks three to four can potentially be realized if the forecast model used properly represents stationary Rossby wave generation over North America, the wave energy propagation across the North Atlantic (requiring a realistic simulation of the North Atlantic jet), and the subsequent stationary Rossby wave impacts over northern Eurasia.

While our investigation here focuses on the 20 May–15 June 1988 stationary Rossby wave event, which appears to originate from convective processes over the western Pacific, we note that any stationary Rossby waves over North America, regardless of how they are initiated and sustained, can subsequently lead to subseasonal surface anomalies over Europe and western Russia. For example, using a series of idealized AGCM experiments, *Koster et al.* [2014] obtained an REOF2-like atmospheric circulation pattern over North America by imposing a persistent dry soil moisture anomaly over the central United States and a persistent wet soil moisture anomaly over the northwestern United States. As the wave energy is propagated downstream, the model ensemble mean exhibits a wave-train-like pattern over Europe and western Russia (Fig. 12.14) that is somewhat similar to that in experiment C1 (Fig. 12.12d).

12.4. SUMMARY AND CONCLUSIONS

The leading modes of subseasonal atmospheric circulation variability play an important role in determining subseasonal surface meteorological variations during the warm season, and they have been crucial in the development of many recent short-term warm season climate extremes over North America and northern Eurasia. These leading modes are guided by the NH mean jet streams and are maintained by submonthly transients [*Schubert et al.*, 2011]. There is evidence that subseasonal atmospheric circulation anomalies over land can be phase locked and amplified by certain spatial patterns of soil moisture anomalies [*Koster et al.*, 2014].

Using the NASA GEOS-5 AGCM and a data assimilation technology called "replay," we perform a case study of the 20 May–15 June 1988 stationary Rossby wave event. Through this case study, we systematically investigate the processes through which stationary Rossby waves lead to the development of the severe 1988 drought conditions over central North America, which subsequently influence surface anomalies over northern Eurasia. In particular, we perform a series of AGCM experiments designed to isolate the contributions to this event from (1) atmosphere and land initial conditions in late May 1988, (2) observed stationary wave sources over the western Pacific, (3) the correct simulation of the model mean jet over the North Pacific, and (4) soil moisture feedback over land.

While the free-running GEOS-5 AGCM provides a generally reasonable simulation of the warm season North Atlantic jet, it is deficient in simulating the North Pacific mean jet; the model North Pacific jet is considerably weaker and less organized than the observed. This model deficiency significantly affects the simulation and hindcasts of stationary Rossby waves and their effects over North America and northern Eurasia, and thus limits the usefulness of the free-running GEOS-5 AGCM for this study. To circumvent this deficiency, we perform a series of regional replay experiments in which the GEOS-5 model atmosphere is constrained to be close to MERRA over eastern Asia and the western Pacific. This constraint indeed allows the GEOS-5 AGCM to produce a greatly corrected North Pacific mean jet, providing us with a useful modeling framework for our analysis.

Our modeling results show the crucial importance of the strong mean jet streams over the North Pacific and North Atlantic for guiding and constraining wave energy propagation path and speed and thereby providing a potential source of predictability. For the 20 May–15 June 1988 stationary Rossby wave event investigated here, the free-running model hindcast skill is limited to only about 1 week, beyond which the hindcasted waves show westward retrograde displacement instead of remaining quasi-stationary as in the observations, due to the weak mean jet over the North Pacific. With the mostly corrected (through replay) North Pacific mean jet, our results show that convective anomalies over the western

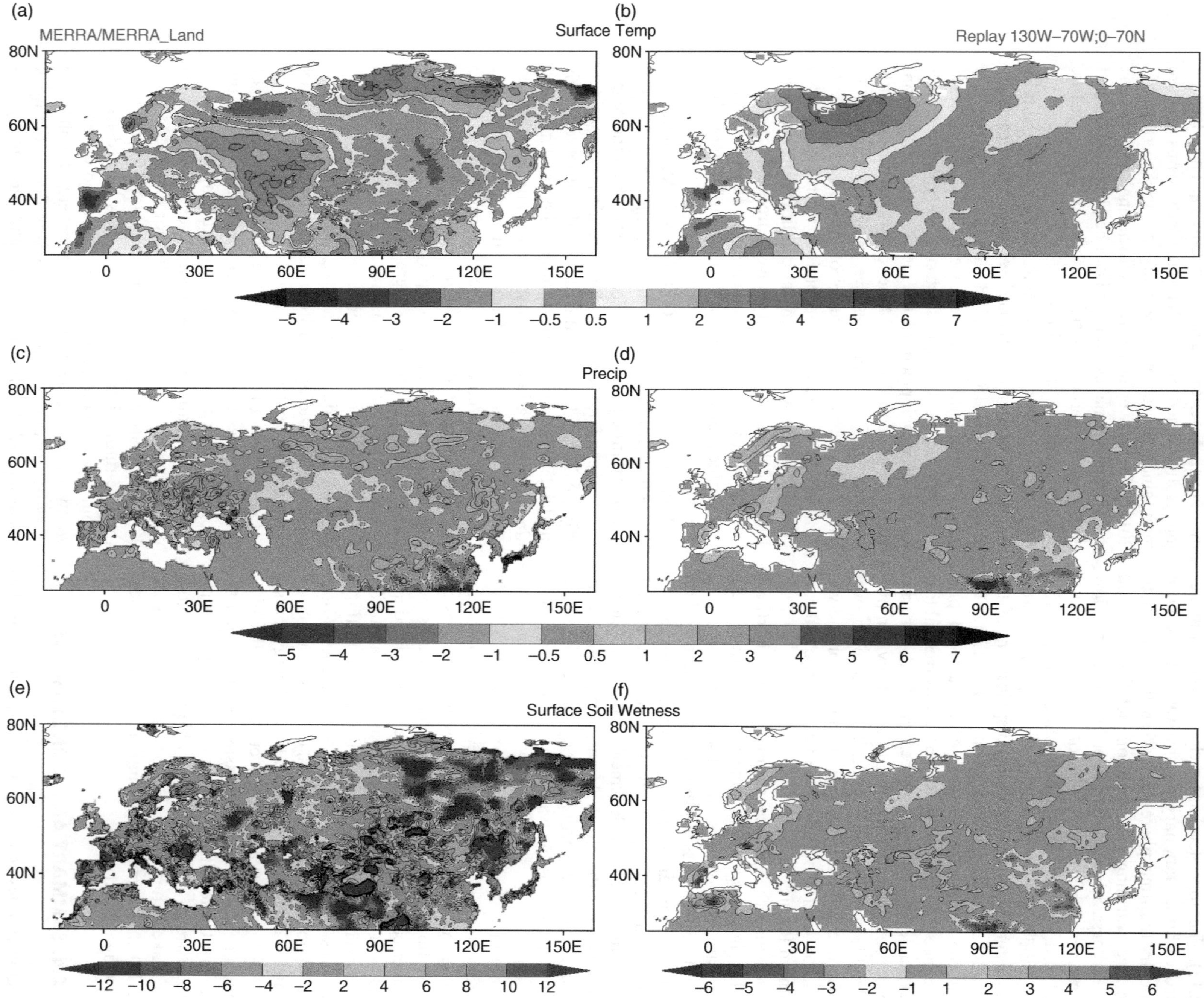

Figure 12.13 The surface temperature anomalies (K) averaged over the period 28 May–15 June 1988 in (a) MERRA, and (b) the ensemble mean of the AGCM regional replay experiment over North America (130°W–70°W;0°–70°N) (experiment C1); (c) same as (a) but for MERRA-Land precipitation (mm day^{-1}); (d) same as (b) but for precipitation (mm day^{-1}); (e)–(f) same as (c)–(d) but for surface soil wetness (%).

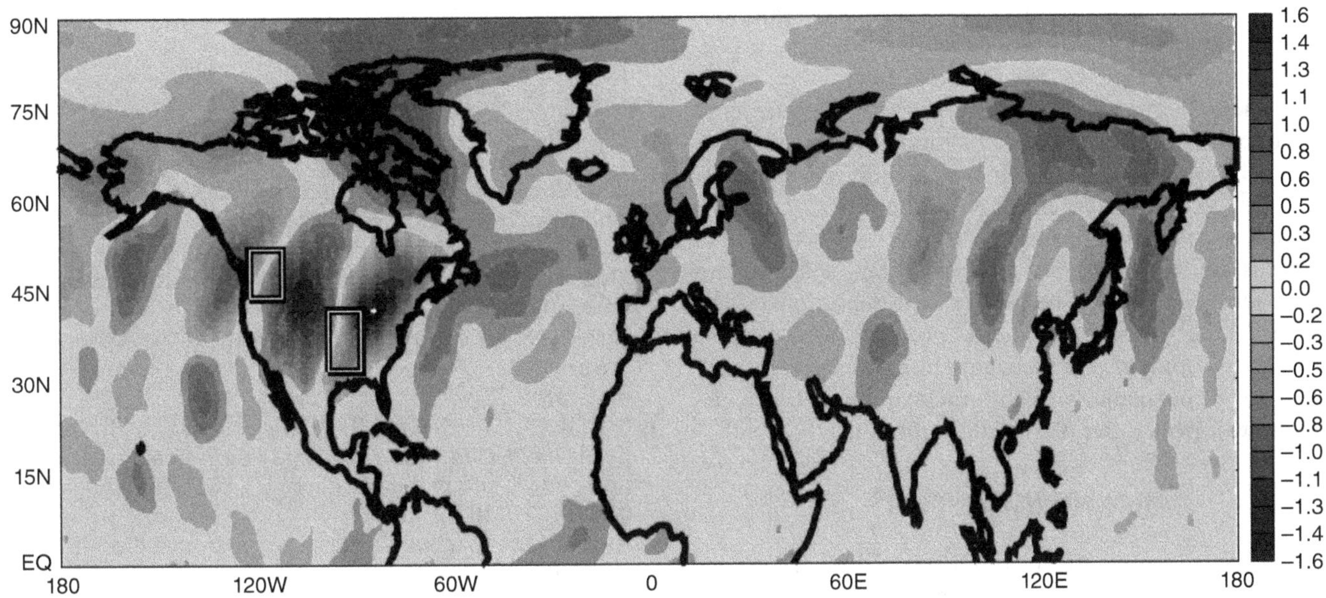

Figure 12.14 The July 250 hPa meridional wind modifications (relative to a control) produced by an experiment in which wet and dry soil moisture anomalies were imposed in the northwest United States and southern Great Plains, respectively (see indicated rectangles). Results from *Koster et al.* [2014].

Pacific do produce a predilection for sustained upper-level high anomalies over central North America about 1 to 2 weeks later. Such high anomalies subsequently lead to strong local precipitation deficits, soil dryness, and surface warming, and thereby to severe drought conditions there. For the period that the stationary Rossby waves remain over North America, local soil moisture anomalies induced by the waves amplify the temperature effect somewhat but do not appear to feed back noticeably on the waves and the precipitation.

While the stationary Rossby waves over North America lead to local drought conditions, there are indications of wave energy propagation farther downstream under the wave guiding effect of the North Atlantic jet. This results in the establishment of stationary Rossby waves over Europe and western Russia about 2 weeks later, facilitating (for this case) drought conditions over eastern Europe and western Russia, and cool and wet conditions over western Europe and central northern Eurasia.

Since it usually takes about 2 weeks for the wave energy to propagate across the North Pacific (North Atlantic) to reach North America (northern Eurasia), these waves can potentially serve as a source of subseasonal forecast skill in these downstream locations. To access this skill, however, it would be critically important for a forecast model to correctly simulate the mean NH jet streams, not only their location and shape but also their magnitude, in order to allow the correct simulation of wave propagation (path and speed) across the northern oceans.

While this study focuses on the summer 1988 drought, the modeling approach employed here can similarly be applied to other climate extremes involving stationary Rossby waves. In addition, since the weak mean jet bias over the North Pacific in the GEOS-5 AGCM is also seen in many other current models, drought studies using these other AGCMs may find our replay approach useful.

A number of key research challenges remain. First, the predictability of stationary Rossby waves is unclear and needs to be assessed. This requires an identification of the processes that initiate stationary Rossby waves and an assessment of their predictability. The origins of stationary Rossby waves could vary from event to event. While some stationary Rossby waves originate from regional subseasonal SST variations and thus possess certain predictability, others could be generated by pure atmospheric internal variability and are hence unpredictable. The origins of stationary Rossby waves need to be further investigated by analyzing past events using observations and reanalyses. In addition, the processes that potentially sustain the stationary Rossby waves, including extratropical air-sea interaction and local land soil moisture feedback, also need to be better understood. Second, possible interaction between the short-term stationary Rossby waves and climate variations on decadal and longer timescales needs focused study [*Schubert et al.*, 2014]. There is, for example, considerable evidence that PDV, AMV, and global warming can play important roles in the occurrence of long-term drought [e.g., *McCabe et al.*, 2004; *Cook et al.*, 2015; *Schubert et al.*, 2015], yet the possibility that such long-term changes might impact stationary Rossby waves and short-term drought development (e.g., through changing the mean jet streams) is yet to be addressed. Third, to incorporate properly the effects of stationary Rossby waves, current models need to be considerably improved, or at least artificially corrected, to provide a correct representation of the NH mean jet, as well as to represent properly the processes that initiate and sustain stationary Rossby waves (e.g., soil moisture feedback over land). The proper representation of stationary Rossby waves in a model provides hope for considerably improved subseasonal drought forecast skill. This is of particularly high practical importance, for example, for operational drought outlooks.

ACKNOWLEDGMENTS

This study is supported by the NOAA Climate Program Office Modeling, Analysis, Prediction, and Projections (MAPP) program.

REFERENCES

Adler, R. F., and coauthors (2003), The version 2 Global Precipitation Climatology Project (GPCP) monthly precipitation analysis (1979–present), *J. Hydrometeor.*, *4*, 1147–1167.

Ambrizzi, T., B. J. Hoskins, and H.-H. Hsu (1995), Rossby wave propagation and teleconnection patterns in the austral winter, *J. Atmos. Sci.*, *52*, 3661–3672.

Bacmeister, J. T., M. J. Suarez, and F. R. Robertson (2006), Rain reevaporation, boundary layer–convection interactions, and Pacific rainfall patterns in an AGCM, *J. Atmos. Sci.*, *63*, 3383–3403.

Bloom, S., L. Takacs, A. DaSilva, and D. Ledvina (1996), Data assimilation using incremental analysis updates, *Monthly Weather Rev.*, *124*, 1256–1271.

Branstator, G. (2002), Circumglobal teleconnections, the jet stream waveguide, and the North Atlantic Oscillation, *J. Climate*, *15*, 1893–1910.

Chen, P., and M. Newman (1998), Rossby wave propagation and the rapid development of upper-level anomalous anticyclones during the 1988 U.S. drought, *J. Climate*, *11*, 2491–2504.

Cook, B. I., T. R. Ault, and J. E. Smerdon (2015), Unprecedented 21st-century drought risk in the American Southwest and Central Plains, *Sci. Adv.*, *1*(1), e1400082; doi:10.1126/sciadv.1400082.

Dai, A. (2013), The influence of the Inter-decadal Pacific Oscillation on U.S. precipitation during 1923–2010, *Climate Dyn.*, *41*, 633–646.

Ferguson, I. M., J. A. Dracup, P. B. Duffy, P. Pegion, and S. Schubert (2010), Influence of SST forcing on stochastic characteristics of simulated precipitation and drought, *J. Hydrometeor.*, *11*, 754–769.

Harris, I., P. D. Jones, T. J. Osborn, and D. H. Lister (2014), Updated high-resolution grids of monthly climatic observations–the CRU TS3.10 Dataset., *Int. J. Climatol.*, *34*, 623–642.

Hoerling, M., X. W. Quan, and J. Eischeid (2009), Distinct causes for two principal U.S. droughts of the 20th century, *Geophys. Res. Lett.*, *36*, L19708; doi:10.1029/2009GL039860.

Hoerling, M. P., and A. Kumar (2003), The perfect ocean for drought, *Science*, *299*, 691–699.

Hu, Q., and S. Feng (2012), AMO- and ENSO-driven summertime circulation and precipitation variations in North America, *J. Climate*, *25*, 6477–6495.

Koster, D. R., Y. Chang, and S. D. Schubert (2014), A mechanism for land–Atmosphere feedback involving planetary wave structures, *J. Climate*, *27*, 9290–9301.

Koster, R. D., M. J. Suarez, and M. Heiser (2000), Variance and predictability of precipitation at seasonal-to-interannual timescales, *J. Hydrometeor.*, *1*, 26–46.

Koster, R. D., M. J. Suarez, R. W. Higgins, and H. M. Van den Dool (2003), Observational evidence that soil moisture variations affect precipitation, *Geophys. Res. Lett.*, *30*, 1241; doi:10.1029/2002GL016571.

Koster, R. D., Z. Guo, P. Dirmeyer, and coauthors (2006), GLACE, The global land-atmosphere coupling experiment, 1, Overview, *J. Hydrometeor.*, *7*, 590–610.

Lin, S.-J. (2004), A "vertically Lagrangian" finite-volume dynamical core for global models, *Monthly Weather Rev.*, *132*, 2293–2307.

Lott, N., and T. Ross (2000), *A Climatology of Recent Extreme Weather and Climate Events,* Technical Report No. 2000-02, NOAA/NESDIS, National Climatic Data Center, Ashville, NC.

McCabe, G. J., M. A. Palecki, and J. L. Betancourt (2004), Pacific and Atlantic Ocean influences on multidecadal drought frequency in the United States, *Proc. Nat. Acad. Sci.*, *101*, 4136–4141.

Molod, A., L. Takacs, M. Suarez, J. Bacmeister, I.-S. Song, and A. Eichmann (2012), The GEOS-5 atmospheric general circulation model: Mean climate and development from MERRA to Fortuna, NASA Tech. Rep. Series on Global Modeling and Data Assimilation, 28, NASA TM-2012-104606.

Moorthi, S., and M. J. Suarez (1992), Relaxed Arakawa-Schubert: A parameterization of moist convection for general circulation models, *Monthly Weather Rev.*, *120*, 978–1002.

Namias, J. (1991), Spring and summer 1988 drought over the contiguous United States, causes and prediction, *J. Climate*, *4*, 54–65.

Nigam, S., B. Guan, and A. Ruiz-Barradas (2011), Key role of the Atlantic Multidecadal Oscillation in 20th century drought and wet periods over the Great Plains, *Geophys. Res. Lett.*, *38*, L16713; doi:10.1029/2011GL048650.

Peña, M., M. Cai, and E. Kalnay (2004), Life span of subseasonal coupled anomalies, *J. Climate*, *17*, 1597–1604.

Reichle, R. H., R. D. Koster, G. J. M. De Lannoy, B. A. Forman, Q. Liu, S. P. P. Mahanama, and A. Toure (2011), Assessment and enhancement of MERRA land surface hydrology estimates, *J. Climate*, *24*, 6322–6338.

Richman, M. B. (1986), Rotation of principal components, *Int. J. Climatol.*, *6*, 293–335.

Rienecker, M. M., and coauthors (2008), The GEOS-5 Data Assimilation System, Documentation of versions 5.0.1, 5.1.0, and 5.2.0, Tech. Rep. Series on Global Modeling and Data Assimilation, *27,* NASA/TM-2008-104606.

Rienecker, M. M., and coauthors (2011), MERRA-NASA's modern-era retrospective analysis for research and applications, *J. Climate*, *24*, 3624–3648.

Ross, T., and N. Lott (2003), *A Climatology of 1980–2003 Extreme Weather and Climate Events,* Technical Report No. 2003-01, NOAA/NESDIS, National Climate Data Center, Ashville, NC.

Schubert, S., and coauthors (2015), Global drought: A synthesis of current understanding with a focus on precipitation, *J. Climate*, Submitted.

Schubert, S., H. Wang, and coauthors (2009), A USCLIVAR project to assess and compare the responses of global climate models to drought-related SST forcing patterns: Overview and results, *J. Climate*, *22*, 5251–5272.

Schubert, S., H. Wang, and M. Suarez (2011), Warm season subseasonal variability and climate extremes in the Northern Hemisphere: The role of stationary Rossby waves, *J. Climate*, *24*, 4773–4792.

Schubert, S. D., H. Wang, R. D. Koster, M. J. Suarez, and P. Y. Groisman (2014), Northern Eurasian heat waves and droughts, *J. Climate*, *27*, 3169–3207.

Schubert, S. D., M. J. Suarez, P. J. Pegion, R. D. Koster, and J. T. Bacmeister (2004), On the cause of the 1930s Dust Bowl, *Science*, *33*, 1855–1859.

Seager, R., and M. Hoerling (2014), Atmosphere and ocean origins of North American drought, *J. Climate*, *27*, 4581–4606.

Seager, R., Y. Kushnir, C. Herweijer, N. Naik, and J. Velez (2005), Modeling the tropical forcing of persistent droughts and pluvials over western North America: 1856–2000, *J. Climate*, *18*, 4068–4091.

Smith, A., and R. Katz (2013), U.S. Billion-dollar weather and climate disasters: data sources, trends, accuracy and biases, *Nat. Hazards*.; doi: 10.1007/s11069-013-0566-5.

Ting, M., and H. Wang (1997), Summertime U.S. precipitation variability and its relation to Pacific sea surface temperature, *J. Climate*, *10*, 1853–1873.

Wang, H., S. Schubert, M. Suarez, and R. Koster (2010), The physical mechanisms by which the leading patterns of SST variability impact U.S. precipitation, *J. Climate*, *23*, 1815–1836.

Wang, Y., and A. R. Lupo (2009), An extratropical air-sea interaction over the North Pacific in association with a preceding El Niño episode in early summer, *Monthly Weather Rev.*, *137*, 3771–3785.

13

The California Drought: Trends and Impacts

Shih-Yu (Simon) Wang[1], Jinho Yoon[2], Robert R. Gillies[1], and Huang-Hsiung Hsu[3]

ABSTRACT

California has experienced a prolonged drought lasting approximately five consecutive years from 2012 to 2016, covering both a La Niña (in 2011–2012) and an El Niño (in 2015–2016) events. Studies have attributed the cause of the drought to a variety of natural variabilities and different pathways of teleconnection with tropical and arctic origins. The diversity of theories explaining the mechanisms leading to this prolonged drought event is multivariate and can be confusing. In this chapter, the circulation and sea surface temperature (SST) patterns during this prolonged drought are analyzed. Some dynamic aspects of these circulation and SST anomalies are reviewed, based on both diagnostic studies and modeling scenarios that suggest attribution in the context of current model sophistication and representation. Possible mechanisms and interactions include an amplified cycle and teleconnection impacts of the El Niño–Southern Oscillation (ENSO) combined with the influence of the reduced Arctic sea ice, moreover these processes contain both natural variability and anthropogenic greenhouse gas effect. The amplified ENSO-cycle impacts on the water cycle extremes, fire risks, and groundwater depletion are discussed.

13.1. INTRODUCTION

The prolonged drought situation in California (2012–2016) led its governor to declare a state of emergency and launch a water conservation mandate. The severe drought caused statewide water stress, unprecedented economic loss, and an extraordinary increase in wildfires. In 2014 alone, the drought is estimated to have caused the state's economy $2.2 billion with a concomitant loss of 17,100 jobs [*Howitt et al.*, 2014]. Moreover, the early onset of the 2014 dry season fueled an extraordinary jump in wildfires; by August a state of emergency was declared just for a single wildfire that had burned 32,000 acres [*Yoon et al.*, 2015b]. Paleoclimate reconstructions indicate that the 2013–2014 drought was the most severe in the last 1200 years, with a resulting moisture deficit worse than any previous continuous span of dry years [*Griffin and Anchukaitis*, 2014]. Given the extraordinary severity and duration of the drought over California, a pressing question posed was whether the state will experience continued shortfalls in natural water resources. The answer to this question required a comprehensive exploration by applying climate dynamics in order to assess attribution. In addition, a surge of recent studies provided supplementary analysis and explanation.

While warming in the lower troposphere alone can cause widespread drought due to evaporative demand

[1] *Utah Climate Center, Utah State University, Logan, Utah, USA; and Department of Plants, Soils, and Climate, Utah State University, Logan, Utah, USA*

[2] *School of Earth and Environmental Engineering, Gwangju Institute of Science and Technology, Gwangju, South Korea*

[3] *Research Center for Environmental Changes, Academia Sinica, Taipei, Taiwan*

Climate Extremes: Patterns and Mechanisms, Geophysical Monograph 226, First Edition.
Edited by S.-Y. Simon Wang, Jin-Ho Yoon, Christopher C. Funk, and Robert R. Gillies.

[*Dai*, 2011], the majority of the emerging studies have shown that, as anthropogenic warming continues, natural climate variability has become increasingly unable to compensate for the drying effect of warming. For instance, *Shukla et al.* [2015] showed that if temperature during the 2013–2014 water year resembled the 1916–2012 climatological distribution, there would be a > 86% probability that winter snowpack, spring soil moisture, and runoff percentiles would have been less severe than those observed in 2013–2014. *Cheng et al.* [2015] suggested that long-term increases in radiative forcing induced both increases in annual precipitation and surface temperature over California; these are consistent with prior climate change model studies. *Diffenbaugh et al.* [2015] found that continued global warming can cause a transition in a regime in which the seasonal, annual, and multiannual precipitation deficit will coincide with increasingly warm conditions, and the recent warm-dry event in California reflects such a regime. Furthermore, *AghaKouchak et al.* [2014] argued that global warming and the associated increase in extreme temperatures substantially increased the "return period" (occurrence probability) of concurrent droughts and heat waves. *Funk, Chris et al.* [2014] and *Williams et al.* [2015] both observed that anthropogenic warming intensified the drought as part of a chronic drying trend that is becoming increasingly detectable and is projected to continue growing throughout the rest of this century. All the aforementioned exist because temperature plays an important role in exacerbating drought, even though low precipitation was the main driver of the recent California drought.

A recent study using large ensemble analysis [*Yoon et al.*, 2015a] suggested an increase in the fluctuation of precipitation over California in association with the El Niño–Southern Oscillation (ENSO) cycle. *Funk et al.* [2014] noted that while coupled climate models indicated an overall increase in California precipitation, this trend was actually composed of competing tendencies, an ENSO-related component associated with more precipitation and an ENSO-residual component related to drier California winters. Extensions of this analysis identify both warming in the West Pacific Warming Mode (WPWM), and the ENSO region, but with differing patterns and temporal characteristics (see Chapter 10 of this volume).

Other studies emphasize natural variability in the making of the California drought. *Mao et al.* [2015] argued that the extent to which climate change is implicated in the drought is not clear, as they found only a modest warming effect exacerbating the 2013–2014 drought and likewise the 1976–1977 drought. *Mao et al.* proposed that these drought events were mainly the result of natural precipitation variability. In conducting climate simulations using prescribed sea surface temperature (SST) as the forcing, *Wang and Schubert* [2014] showed that SST warming occurring over most of the global ocean can induce dry conditions over California, the implication being that atmospheric internal variability accounted for the extreme magnitude of the 2013–2014 drought (also see Chapter 11). The 2013–2014 California drought also accompanied a very cold winter in central and eastern North America, and a later study [*Hartmann*, 2015] supported this role of SST in causing the 2013–2014 cold winter. Using a different climate model, *Lee et al.* [2015] also found that SST-induced anomalous circulations led to extreme weather in the United States during the 2013–2014 winter, once again suggestive of a predominant role of natural variability. However, *Lee et al.* linked the event intensity to the anomalously warm SST and low Arctic sea ice concentration in the Bering Sea, an phenomenon that is connected with the Arctic warming [*Vihma*, 2014] (see Chapter 2).

These seemingly contradictory findings and associated theories cloud the issue, especially for readers outside the fields of dynamic meteorology. Understanding the mechanisms that govern changes in extreme events is difficult, especially at the regional scale, as they are likely a combination of both anthropogenic and natural causes, with both remote and local climate influences. In this chapter, we summarize the results from a selection of studies and growing evidence that affords an overview with regard to the mechanistic processes of the recent California drought, as well as the drought impacts on fire risk and groundwater depletion.

13.2. THE PROLONGED DROUGHT OF 2012–2016

Both the research community and the media [*NBC*, 2016] made statements about the recent California drought lasting for at least 4 years (i.e., 2012–2015) [*Mann and Gleick*, 2015; *Williams et al.*, 2015]. However, the climate anomalies observed in the oceans and the atmosphere were noticeably dissimilar for year to year during this time period. It therefore begs the question as to how each year's unique climate pattern exacerbated drought conditions. To answer this question, we first examined 250 hPa geopotential height anomalies during November–February (i.e., rainy season for California) from 2011–2012 to 2015–2016, as outlined in Figure 13.1. The February distribution of the Palmer Drought Severity Index (PDSI) [*Dai*, 2011] in California was plotted alongside to depict the corresponding drought patterns. Some features are noteworthy: (1) the 2013–2014 drought pattern in California underwent a distinctive intensification; (2) the 2014–2015 geopotential height anomalies resemble those of 2013–2014 and exhibit a comparable drought pattern; these two similar circumstances are distinct from

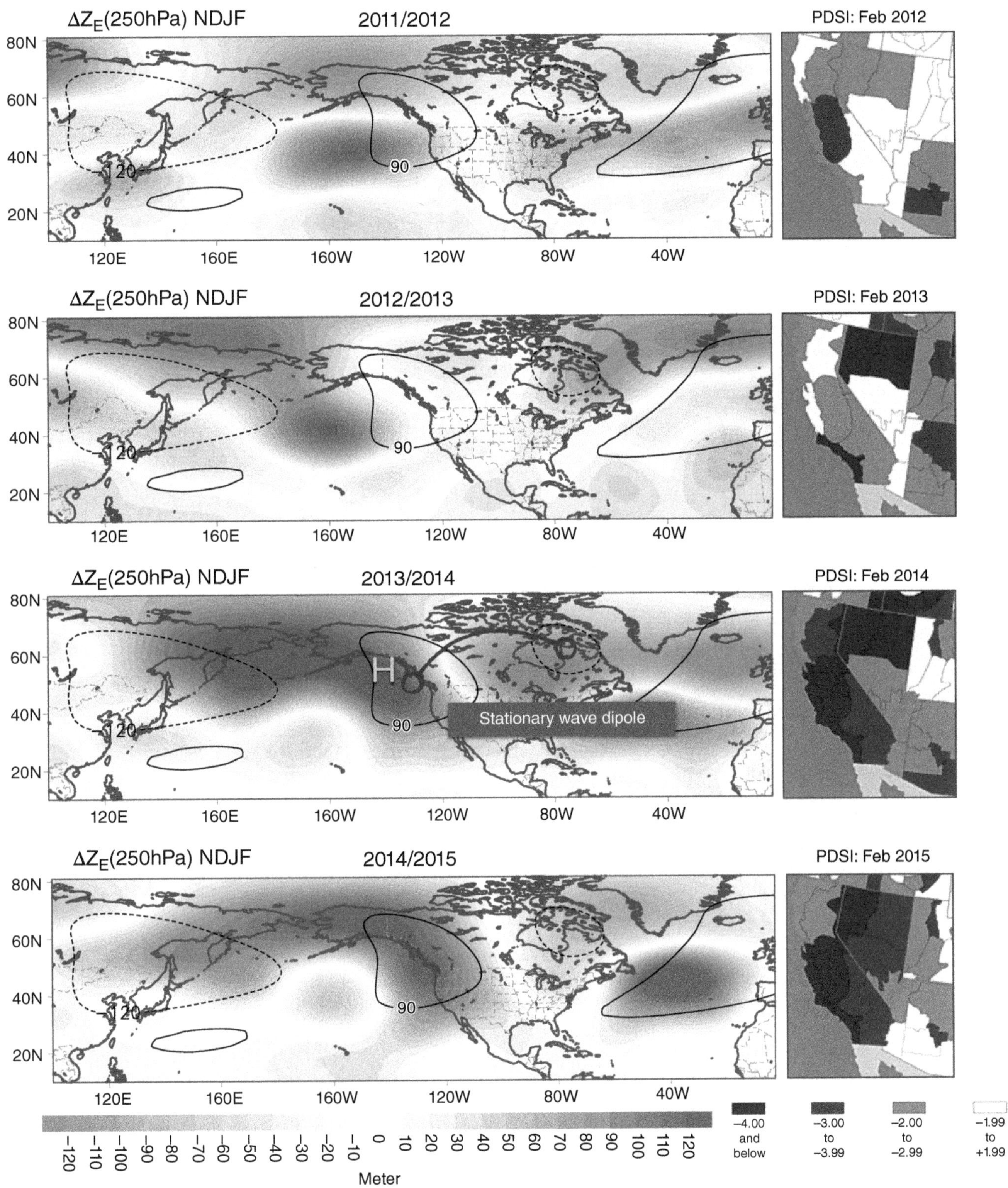

Figure 13.1 The 250 hPa geopotential height eddies (ΔZE; shadings) during November–February from (top to bottom) 2011–2012 through 2014–2015, overlaid with the long-term seasonal mean ZE (contours of 90 and −120 m). Data source: NCEP-NCAR Reanalysis. The dipole referred to in the text is indicated by the red line connecting the western ridge and eastern trough in 2013–2014. Panels on the right-hand side show the February PDSI adopted from the US Drought Portal.

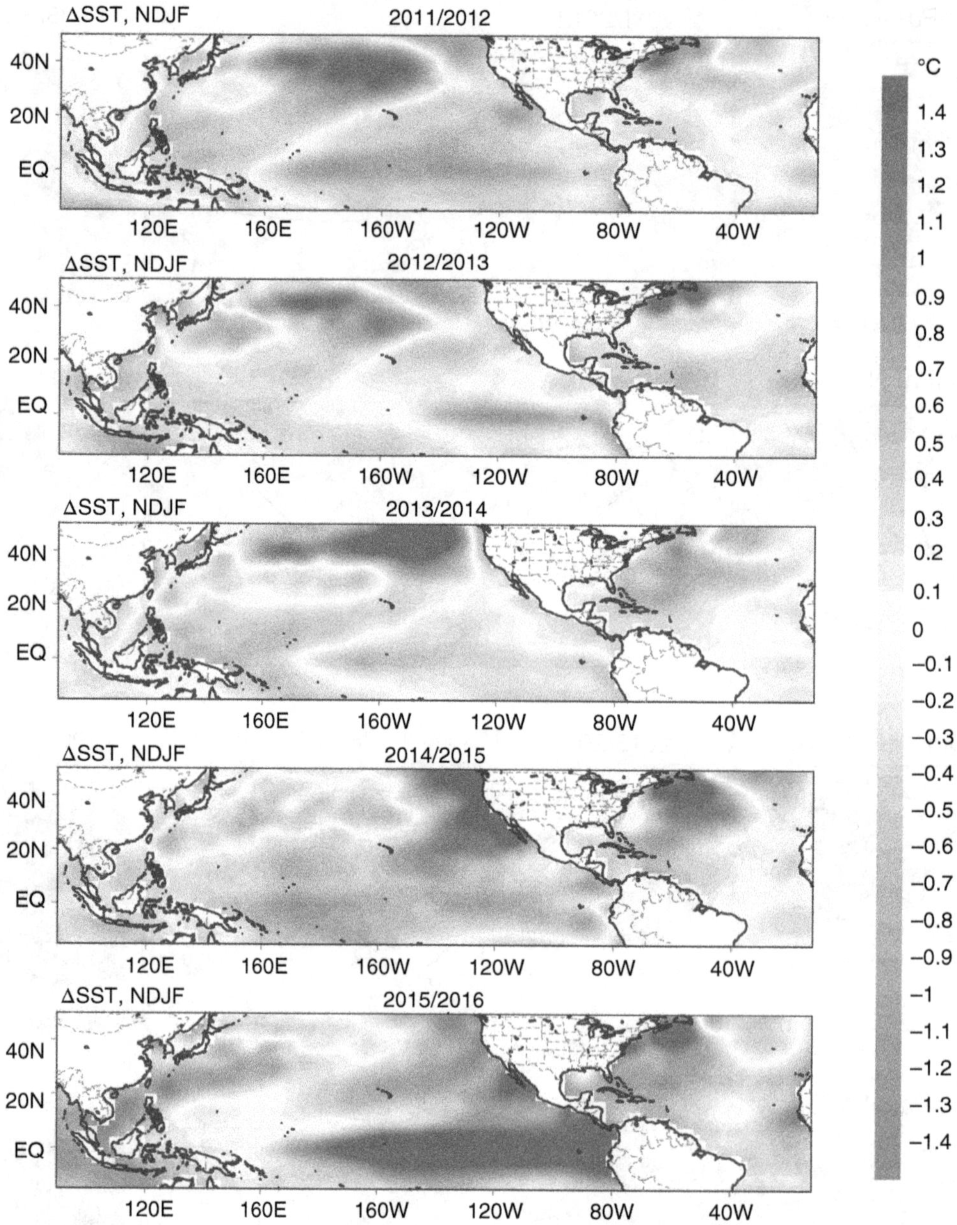

Figure 13.2 The sea surface temperature anomalies (ΔSST) corresponding to Figure 13.1 with the addition of the 2015–2016 pattern. Data source: ERSSTv3.

the prior two winters that exhibit relatively weak circulation anomalies. (3) The 2011–2012 geopotential height anomalies over the North Pacific are essentially opposite to that of 2014–2015 and this latter feature corresponds to the sequence of events from the 2011–2012 La Niña to transition to the 2014–2015 El Niño. The corresponding sea surface temperature (SST) anomalies, as shown in Figure 13.2, illustrate the La Niña-to-El Niño transition. The initiation of this prolonged drought was a response to La Niña lasted into 2015–2016 conditions in the presence of a strong El Niño.

The contradictory situations of drought condition manifesting as a result of El Niño and La Niña is puzzling. To resolve the paradox, we overlaid the winter-mean geopotential height in Figure 13.1 with the zonal mean removed (i.e., eddies). It is obvious that a winter mean stationary wave across North America resulted in a standing ridge in the northwest and likewise a trough in the northeast, forming a dipole as indicated by the red line. The circulation anomalies during 2013–2014 and 2014–2015 are unique in that the anomalous ridges and troughs over North America are phase-coincident with the stationary-wave dipole. In particular, the anomalous ridge accompanying the 2013–2014 California drought exceeded the 95th percentile of the entire period of record [*Lee et al.*, 2015]. Thus, it would seem that the 2013–2015 circulation anomalies signify an amplified winter stationary wave, which made the climatologically warmer western continent warmer (and drier) and the colder eastern continent colder (and wetter). These anomalous synoptic features are different from the ones associated with a La Niña as depicted by previous studies [*Ropelewski and Halpert*, 1986; *Schonher and Nicholson*, 1989; *Mo and Higgins*, 1998].

13.3. ROLE OF ENSO CYCLE

The aforementioned analyses and some prior studies suggest that the 2012–2015 circulation anomalies that could result from a combination of tropical and higher-latitude forcings are *not* directly linked to the extreme phases of ENSO. The SST and wind anomalies that emerged in late 2013 led some researchers to predict a fully developed 2014 El Niño [*Ludescher et al.*, 2014]; however, the anticipated El Niño did not happen due to a lack of western Pacific wind forcing and other oceanic factors [*McPhaden*, 2015]. An El Niño did finally develop in late 2015, as shown in Figure 13.2, yet the drought persisted. Thus, the consecutive drought conditions in California in the absence of a major ENSO event suggest a linkage within ENSO's evolution cycle rather than its extreme phases.

Using the Large Ensemble Project data produced by the Community Earth System Model (CESM), which consists of 30 members of long simulations, *Yoon et al.* [2015a] projected that both intense drought and excessive flooding will increase by at least 50% toward the end of the 21st century. This projection echoes the analysis of *Funk et al.* [2014], which relates the opposing influences of ENSO and the ENSO-residual SST forcing to the occurrence of California drought (see also Chapter 10). *Yoon et al.* [2015a] affiliated the increase in drought and flood conditions with a strengthened relation to ENSO, through not only its extreme warm and cold phases but also its precursor patterns. Here, ENSO precursor refers to optimal SST and surface wind conditions formed approximately 9–15 months prior to a fully developed El Niño or La Niña event [*Capotondi and Sardeshmukh*, 2015]. For instance, the literature has shown that wintertime variability in the atmospheric circulation related to the North Pacific Oscillation (NPO) significantly impacts the SSTA anomalies in the central North Pacific [*Chang et al.*, 2007; *Alexander et al.*, 2010]. The NPO is a leading pattern of wintertime middle- to high-latitude sea level pressure variability [*Linkin and Nigam*, 2008],

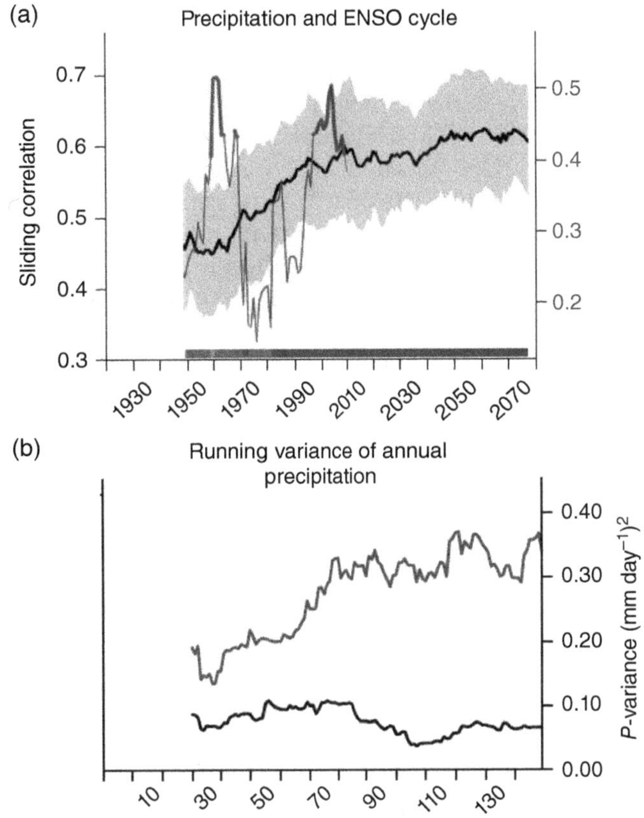

Figure 13.3 (a) The multivariate correlations between the Niño-3.4, Niño-3.4(Y-1) and annual precipitation averaged for water year in California, defined as October to September in the following year, within a running 30 year window during the November–January season. (b) 30 year running variance of annual precipitation averaged over California from the 1% of CO^2 ramping experiment (red) and 1% CO^2 ramping experiment without global ocean feedback (black). Modified from *Yoon et al.* [2015a]. *See electronic version for color representation.*

which interacts with tropical SST and wind anomalies across the equatorial Pacific and facilitates the development of El Niño in the following year [*Vimont et al.*, 2001; *Vimont et al.*, 2003]. The NPO pattern was particularly strong in the 2013–2015 winters and, as suggested by *Wang et al.* [2014], its circulation anomalies may be linked to the NPO and hence to ENSO precursor in the warming climate.

To illustrate this linkage, Figure 13.3a shows the sliding multivariate correlations between the California precipitation, the ENSO index, and the ENSO precursor index (1 year before ENSO; adopted from *Yoon et al.*, [2015a]). The correlations reveal a persistent increase in the association between the ENSO cycle and California winter precipitation after 1995, with an even more pronounced increase revealed in reanalysis data (red line). The result shown in Figure 13.3a echoes the increase in the correspondence between the El Niño precursor and California drought under the greenhouse gas (GHG) forcing as was observed by *Wang et al.* [2015], who analyzed the Coupled Model Intercomparison Project Phase 5 (CMIP5) models. As further validation, *Yoon et al.* [2015b] conducted further CESM experiments through two sets of sensitivity experiments with a 1% CO_2 increase per year, one with oceanic feedbacks and the other without. This is shown in Figure 13.3b. It was found that the increased occurrence and intensity of water cycle extremes in California are linked to the increase in anthropogenic greenhouse gases (GHG), and such a connection is only possible through coupling the atmosphere with the ocean to enable the ENSO cycle-induced teleconnections.

Previous studies [*Lau and Weng*, 2002; *DeWeaver and Nigam*, 2004] have found that the SST and convective activity anomalies in the far western tropical Pacific can induce cross-Pacific atmospheric wave trains that affect North America, even in the absence of ENSO. Furthermore, the Pacific warm pool has become increasingly warm over the past decade [*Funk and Hoell*, 2015] resulting in an increase of teleconnection strength [*Kumar et al.*, 2010; *Palmer*, 2014]. The existence of cross-Pacific wave trains supports the process by which an enhanced ENSO precursor in the warming western Pacific [*Wang et al.*, 2013] can link to the circulation dipole over North America. Subsequent studies [*Hartmann*, 2015; *Wang et al.*, 2015] also found a similar connection between the ENSO precursor (i.e., the NPO) and the dipole that accompanied the 2013–2014 California drought. Regarding the linkage of increased water cycle extremes with the ENSO cycle, which has long been recognized as the prime climate modulator of precipitation and snowpack in California [*Ropelewski and Halpert*, 1986; *Schonher and Nicholson*, 1989; *Mo and Higgins*, 1998], *Yoon et al.* [2015a] combined the ENSO index, which includes both its peak (current year) and precursor (previous year) values, and found a persistent increase in the association between the ENSO cycle and water cycle extremes in California after 1995.

13.4. ARCTIC INFLUENCES

An increasing body of research has indicated that low sea ice concentration (SIC) in the Arctic Ocean and positive SST anomaly in the western tropical Pacific can induce anomalous circulation in the extratropics [*Francis and Vavrus*, 2012; *Ding et al.*, 2014; *Frankignoul et al.*, 2014], an example of which is the role low SIC played in the 2014 extreme arctic air outbursts [*Kim et al.*, 2014]. Regarding the cause of 2014's extreme circulation anomalies in western North America, *Wang and Schubert* [2014] and *Funk et al.* [2014] pointed out the importance of the exceptionally warm SST in the extratropical North Pacific, which is likely connected to Arctic SIC anomalies through an atmospheric feedback. Chapter 2 of this book provides a thorough review of the Arctic connection with midlatitude extreme events.

Based upon these observations, *Lee et al.* [2015] showed that anomalously warm SST in the western tropical Pacific and the extratropical North Pacific, as well as extremely low SIC in the Arctic region, as noted in *Kim et al.* [2014], jointly led to the extreme atmospheric responses in the North Pacific, North America, and the North Atlantic during the winter of 2013–2014. *Lee et al.* [2015] conducted numerical experiments using the European Centre Hamburg Atmosphere Model version 5 (ECHAM5) [*Roeckner et al.*, 2003] and simulated the atmospheric circulation during December 2013 to February 2014, using the observed SST and Arctic SIC anomalies as external forcings. For prescribed SST anomalies, *Lee et al.*'s experiments consisted of the Extratropical North Pacific (ExtraNP), tropical western Pacific (TWP), tropical eastern Pacific (TEP), and the entire Pacific. (Observed climatological monthly SST was prescribed outside these domains.) The observed SIC north of 45°N was prescribed in Pacific + SIC and TWP + SIC runs. An additional experiment to reveal the influence of local SIC in the Bering Sea and Sea of Okhotsk was also performed (designated as BO SIC).

The simulated 500 hPa geopotential height anomalies of *Lee et al.* [2015] are shown in Figures 13.4b–h and compared with the observed 500 hPa geopotential height (Fig. 13.4a). The Pacific + SIC simulation produced the most realistic response to the prescribed forcing among all experiments (Fig. 13.4b). This includes forcings from all Pacific SST anomalies and the entire Arctic SIC anomalies. The Pacific + SIC simulation captured most of the 500 hPa geopotential height features over the North Pacific and North America, and even some features over Europe. The next closest simulations came from TWP + SIC (Fig. 13.4c) and TWP (Fig. 13.4d) but with weaker amplitudes over the Eurasian continent and the Arctic. The ExtraNP simulation (Fig. 13.4e) enhanced the geopotential height over the Sea of Okhotsk while producing a circulation dipole pattern across the eastern North Pacific and the North Atlantic. This represents a positive feedback of warm SST in the North Pacific (which might be initiated by atmospheric circulation anomalies earlier) to reinforce the high-amplitude ridge that impacted California. The BO SIC simulation (Fig. 13.4g) suggests that low SIC in the Bering Sea and the Sea of Okhotsk contributed the most to the anomalous ridge over the Bering Sea, though this ridge did not directly impact western North America. These results echo those of *Kim et al.* [2014] that decreased sea-ice cover in the Barents-Kara seas during the early winter

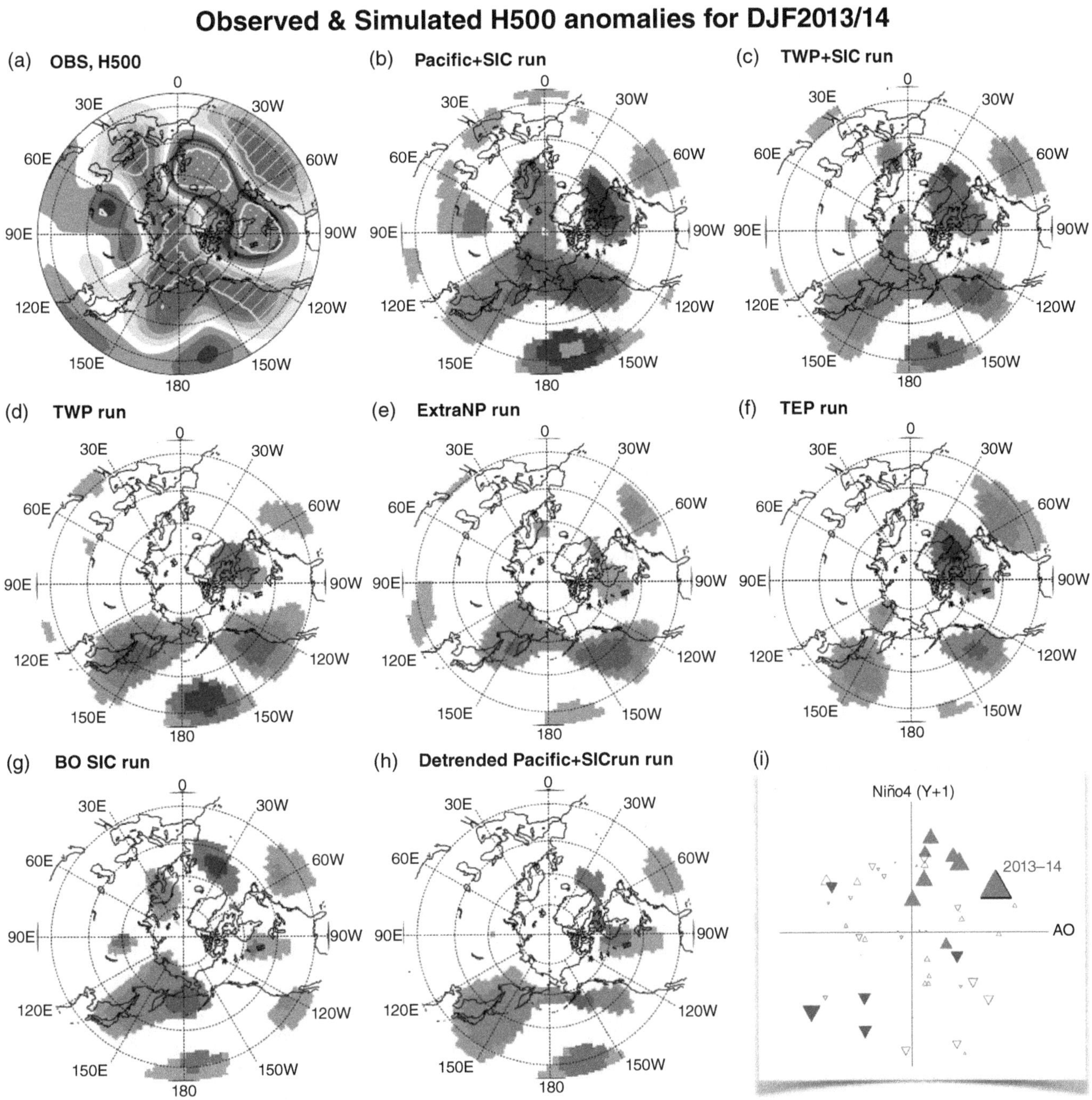

Figure 13.4 Observed and simulated anomalous fields for DJF 2013–2014. (a) Observed H500 anomaly. (b–h) Same as in Figure 13.3a but for the simulated H500 in the Pacific + SICrun, TWP + SICrun, TWPrun, ExtraNPrun, TEPrun, BO SICrun, and detrended Pacific + SIC run. The hatching and dots in (a) denote values exceeding the 95th and 5th percentiles, respectively. The shading in the simulated fields represents anomalies above the 90% confidence level from the 10 member ensemble simulations (adopted from *Lee et al.* [2015]). (i) Scatterplot for the AO and Niño4 (Y +1) indices for 1960–2013 with the scatters represented by values of the dipole index; all three indices are standardized (between 1 and 1). Red and blue triangles indicate positive and negative values, while those exceeding 2 standard deviations are color filled (adopted from *Wang et al.* [2014]). *See electronic version for color representation.*

months resulted in upward propagation of planetary-scale waves and subsequent weakening of the stratospheric polar vortex in midwinter.

The Arctic Oscillation (AO) reached a strong positive phase in the winter of 2013–2014. Noting a possible Arctic effect, *Wang et al.* [2014] compared the indices of AO and Niño4 of the following year (Y + 1); this is shown in Figure 13.4i as scatter plots. These two indices are not correlated but when overlapped with the dipole index *Wang et al.* [2014] noted that, while strong positive phases of the dipole are concentrated in the first quadrant, strong negative phases are clustered in the third quadrant. These two apparent groups suggest an AO modulation on the dipole in addition to tropical forcing. Moreover, since anthropogenic warming is partially responsible for the warming in the Pacific and the melting of the Arctic sea ice [*Bindoff et al.*, 2013], it is possible that the rising SST and reducing SIC over the past few decades have contributed to amplifying the atmospheric anomalies observed in the winter of 2013–2014. Indeed, *Lee et al.* [2015] conducted another experiment by removing the long-term (warming) trends in the Pacific + SIC (Figure 13.4h) and found that the geopotential height responses were much weaker, revealing a compounding effect of anthropogenic warming on the recent (post-2013) intensification of the atmospheric circulation anomalies that accompanied the severe drought in California.

13.5. DROUGHT IMPACTS ON CALIFORNIA

13.5.1. Wildfire Risk

Based upon the aforementioned idea that the amplified teleconnections can enhance California's water cycle extremes [*Yoon et al.*, 2015a], *Yoon et al.* [2015b] found an increase in fire risk in California in relation to human-induced climate change. They computed wildfire probabilities directly within the CESM simulations, which are primarily driven by the dryness of the surface soil and vegetation, representing the availability of fuel load [*Thonicke et al.*, 2001]. *Yoon et al.* [2015b] computed extreme fire risk measures in terms of the fractional area and the extreme fire danger days for the preindustrial and RCP8.5 runs. As shown in Figure 13.5a,b, their results indicate a steady increase starting from the early 1990s and 2000s. Analyzing the 1800 year preindustrial simulation of the CESM, the preindustrial simulations envelop entirely the fluctuation in extreme fire risk measures for the period spanning 1920–1980. Beginning in the 1990s, however, a clear separation emerges between the extreme fire risks driven by the anthropogenic GHG and natural climate variability. The period of 2013–2015 lies in a time

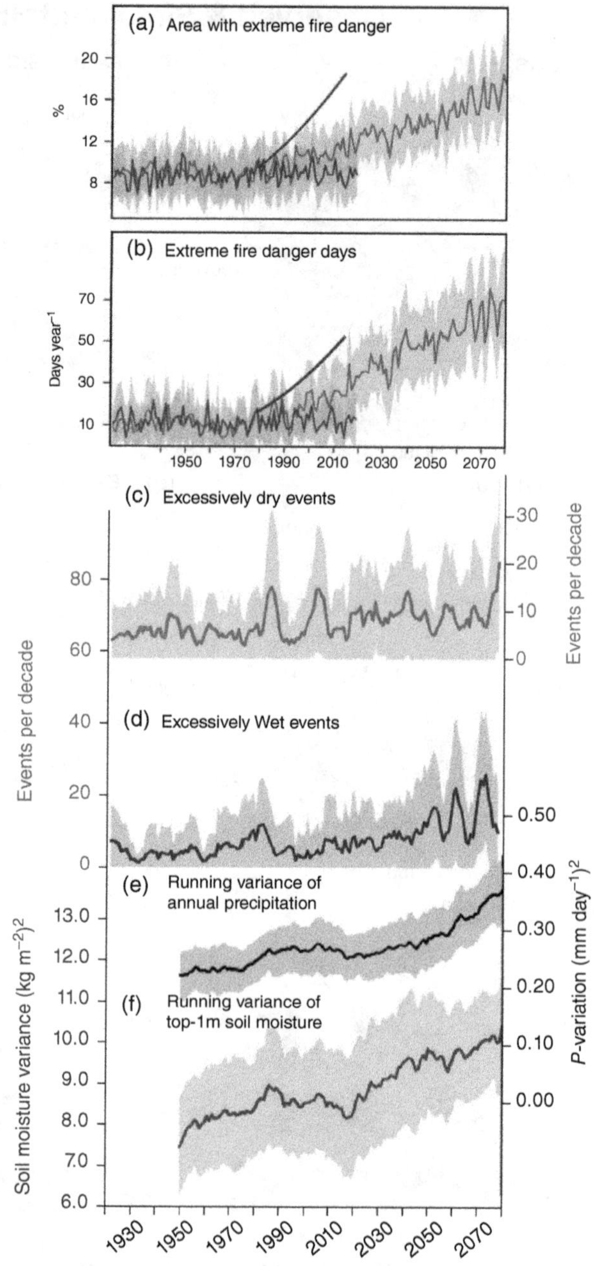

Figure 13.5 (a) Fractional area (%) under the extreme fire risk, and (b) the extreme fire danger (days per year) over California from the large ensemble simulation of the CESM1. Red (blue) indicates the historical and RCP8.5 (preindustrial) runs. Gray lines indicate 50% of the second order trend of the extreme fire risk measures based on the NLDAS2 data. (c) Excessively dry events per year with 5 year running average (see text and Methods section), (d) excessively wet events per year with 5 year running average, (e) 30 year running variances of annual precipitation, and (f) 30 year running variances of annual soil moisture in the top 1 m of soil layers averaged over California. Solid lines indicate a 30 member ensemble mean of CESM1, while shaded areas indicate model spread; roughly 50% of the ensemble members are within the shaded areas. Modified from *Yoon et al.* [2015a,b]. *See electronic version for color representation.*

of accelerated increase in extreme fire risk that is expected to surpass the range of natural climate variability in the early 21st century. The observationally derived extreme fire risk shows a much faster increases (gray lines in Fig. 13.5a,b).

To understand the extent to which the increase in extreme fire risk will exceed that of natural variability within the coming decade, *Yoon et al.* [2015a] derived a metric of "water cycle extremes" as the occurrence of both excessively wet and dry events per year in which the standardized precipitation index (SPI) exceeds ± 2 in magnitude; this quantifies the degree to which California experiences the alternation between extreme wet and extreme dry events. Figure 13.5c shows that the occurrence of excessively dry events in California increases considerably, and Figure 13.5d indicates an even more pronounced increase in the occurrence of excessively wet events. The results of *Yoon et al.* [2015a,b] suggest that future water cycle extremes will fluctuate more drastically than those in the present climate, for example, extreme dry event occurrence increases twofold from the period of 1930–1940 to 2070–2080, while extreme wet event occurrence increases from about 4 during 1930–1940 to roughly 15 during 2070–2080 (roughly threefold). This result is also consistent as realized from the CMIP5 ensemble [*Yoon et al.*, 2015a].

How are the projected increases in water cycle extremes and the increase in extreme fire risk connected? To increase extreme fire risk, two basic requirements are needed: (1) abundant fuel load (i.e., surface vegetation coverage enhanced through precipitation), and (2) the occurrence of a hot-and-dry climate regime (i.e., drought) to dry the vegetation. By computing the 30 year running variances of both annual precipitation (Fig. 13.5e) and the top 1 m soil moisture content (Fig. 13.5f) in California simulated by the CESM, the results mirror the water cycle extremes index that continually increases toward the end of the 21st century. Observed rainfall in California also shows an increase in the 30 year running variance (not shown).

13.5.2. Groundwater Depletion

Since 2013, the intense drought in California has exposed the longstanding problem of groundwater depletion in the Central Valley, home to one of the world's most productive agricultural regions. Historically, drought and reduced groundwater storage occurred almost hand in hand in the Central Valley. When drought conditions ended, groundwater storage would normally rebound. This is the relationship we see in records from about 1960–2000. However, a recent study found that this relationship has changed over the last decade and a half. Analyzing groundwater level derived from about 500 wells in the Central Valley (Fig. 13.6a), *Wang et al.* [2016] found a continued depletion of groundwater lasting a full year after drought has ended, a phenomenon that was not observed in earlier records before the 21st century. Whereas previously when drought ended, groundwater resources would begin to recover, now groundwater continues to decline, even through a wet period.

Figure 13.6b shows the September–December PDSI alongside the groundwater well levels and liquid water equivalent thickness (LWET) derived from the Gravity Recovery and Climate Experiment (GRACE) satellites [*Famiglietti et al.*, 2011]. The low-frequency variability in all these variables is discernable; in particular, the tendency for any drought to last longer than 2–3 years has become more pronounced after the 20th century. Figure 13.6c presents sliding correlations with a trailing 15 year window between PDSI (year 0) and groundwater level (year +1). The lagged (year +1) correlations increase drastically and become significant ($p < 0.01$) in the beginning of the 21st century. Prior to that, the year +1 correlations are insignificant so that the situation of protracted groundwater decline a full year after drought was not the case. This result suggests that the effects of drought in California on groundwater depletion have become enduring. By comparison, the no-lag correlations are rather stable and remain marginally significant throughout the analysis period. The LWET-PDSI correlations over the 2002–2014 period (indicated by open circles) align with the well data analysis (Fig. 13.6c).

As further analysis, a wavelet transform coherence was applied on the PDSI and groundwater level time series and the result is shown in Figure 13.6d. There, significant coherency between the two variables in the 4–6 year frequency appears after 1995, with a phase difference of 75°, which amounts to a time lag of 1 year within a 4–6 year "cycle"; this lends support to the increased year +1 correlations presented in Figure 13.6c. We note that the 50° phase difference within the significant 10–16 year coherency indicates a time lag of about 2–3 years. Consequently, the year +1 correlations during this time period do not reveal as strongly the prolonging effect of drought on groundwater depletion.

Changes to surface water deliveries may not be the leading cause for the increase in year +1 correlations. For example, in the Colorado River basin, *Famiglietti* [2014] noted a disconnect between reservoir storage and groundwater level, which leads to the conclusion that "the steepest rate of groundwater storage decline (in the upper Basin in 2013) follows exceptional drought conditions in 2012 and record low Rocky Mountain

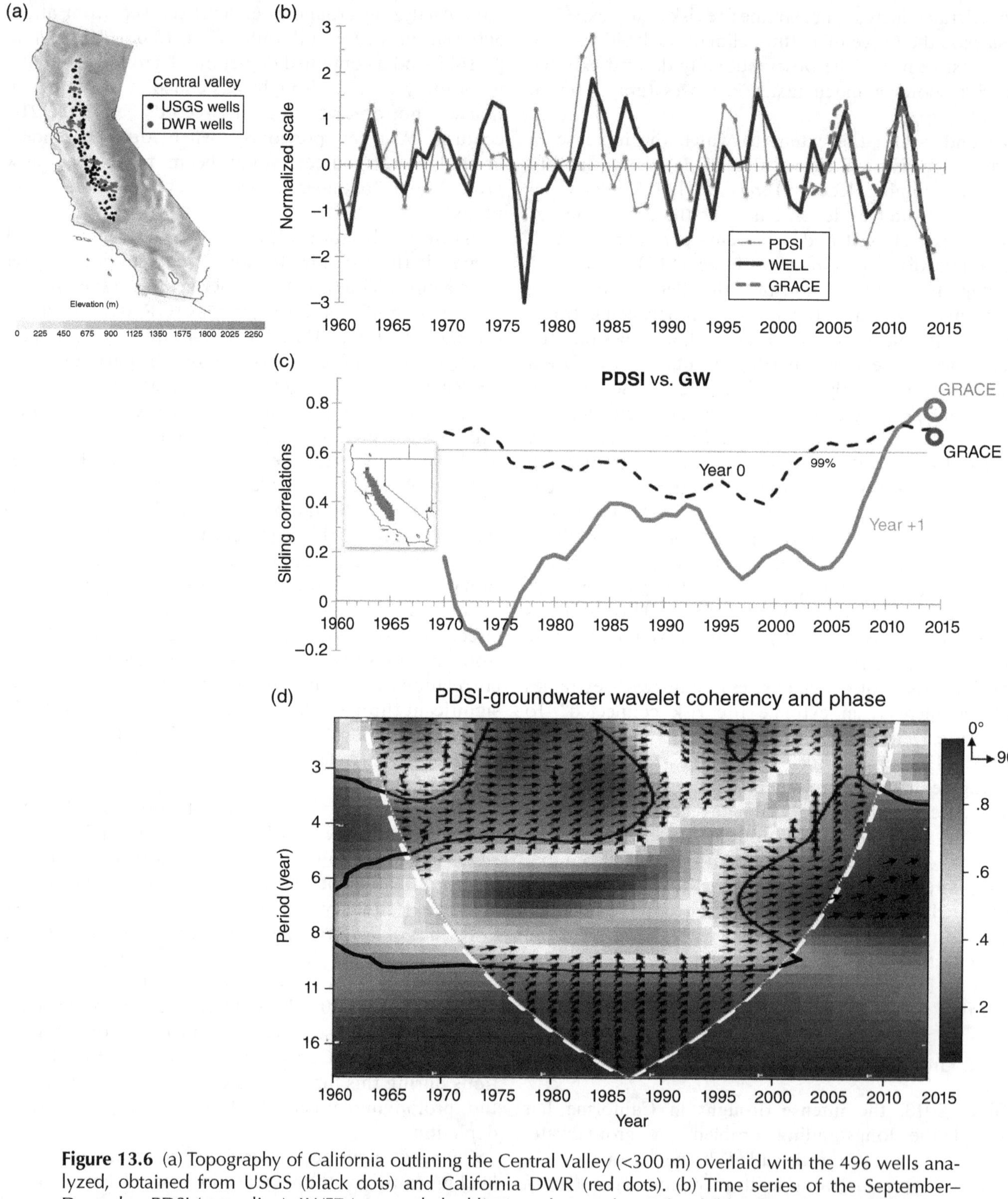

Figure 13.6 (a) Topography of California outlining the Central Valley (<300 m) overlaid with the 496 wells analyzed, obtained from USGS (black dots) and California DWR (red dots). (b) Time series of the September–December PDSI (green line), LWET (orange dashed line), and groundwater level (blue line) from 1960 to 2014. (c) Sliding correlations between the Central Valley PDSI and the groundwater level (GW) in the following year (year +1; solid line) and in the same year (year 0; dashed line), computed with a 15 year running window (one-sided). The LWET correlations with PDSI are indicated by thick circles for 2002–2014. Gray horizontal lines indicate the 99% confidence interval. (d) Wavelet coherency (shading) and phase (vectors) between the PDSI and groundwater. Vectors pointing to the right indicate a quarter-phase. The cone of influence and the 95% confidence interval based on red noise are hatched/contoured. Modified from *Wang et al.* [2016]. *See electronic version for color representation.*

snowpack." While low snowpack does affect surface water availability and thus tends to increase pumping of groundwater, the implication from the *Famiglietti* [2014] analysis (alongside his Fig. 13.3) suggests that climatic factor is the leading contributor to large-scale groundwater behavior, rather than changes in reservoir storage. Nonetheless, according to the observations by *Bertoldi et al.* [1991] and *Anderson et al.* [2015], extended drought periods did prompt further withdrawal from the aquifer thereby further prolonging the depletion of groundwater.

13.6. CONCLUDING REMARKS

California's modern drought history reached a milestone in the winter of 2013–2014 as the severe drought continued into 2016. Synoptically, a high-amplitude and persistent ridge occupied the Gulf of Alaska. The formation mechanism of this anomalous ridge is manifold and has been attributed to low Arctic sea ice, extremely high SST anomalies over the North Pacific, and persistent sources of Rossby wave energy emanating from the western tropical Pacific, as well as some forms of anthropogenically induced local climate change. The ridge's subsequent intensification generated a surge of wave activity fluxes downwind and deepened the trough over the northeast United States, forming a dipole while causing consecutive cold winter weather events over eastern North America. A number of studies that investigated the California drought have indicated that the abnormally strong ridge (and associated dipole) is projected to intensify. Other studies further cautioned future episodes of extremely wet seasons to occur between the periods of drought. Moreover, the amplified alternation between excessively wet and extremely dry seasons can increase fire risks. This observed and projected tendency for California to experience an increase in the water cycle extremes is linked to the lifecycle of ENSO and associated teleconnections during its transition phases, and how it may amplify in the future.

Putting these together, the reported findings suggest that predicting future drought and/or wet episodes for California might be feasible at long lead times, provided that the reported integral improvement toward long-lead predictions for ENSO and its lifecycle [*Barnston et al.*, 2012] will continue to advance (see Chapter 18). By improving ENSO prediction and utilization of the prediction information, as well as monitoring of Arctic sea ice, one ultimately can drive more informed management practices that pertain to water resource management and supervision practices in California as well as the western United States.

REFERENCES

AghaKouchak, A., L. Cheng, O. Mazdiyasni, and A. Farahmand (2014), Global warming and changes in risk of concurrent climate extremes: Insights from the 2014 California drought, *Geophys. Res. Lett.*, *41*(24), 8847–8852.

Alexander, M. A., D. J. Vimont, P. Chang, and J. D. Scott (2010), The impact of extratropical atmospheric variability on ENSO: Testing the seasonal footprinting mechanism using coupled model experiments, *J. Climate*, *23*(11), 2885–2901.

Anderson, R. G., M. H. Lo, S. Swenson, J. S. Famiglietti, Q. Tang, T. H. Skaggs, Y. H. Lin, and R. J. Wu (2015), Using satellite-based estimates of evapotranspiration and groundwater changes to determine anthropogenic water fluxes in land surface models, *Geosci. Mod. Dev. Discuss.*, *8*(4), 3565–3592.

Barnston, A. G., M. K. Tippett, M. L. L'Heureux, S. Li, and D. G. DeWitt (2012), Skill of real-time seasonal ENSO model predictions during 2002–2011: Is our capability increasing?, *Bull. Amer. Meteor. Soc.*, *93*(5), 631–651.

Bertoldi, G. L., R. H. Johnston, and L. D. Evenson (1991), Ground water in the Central Valley, California: A summary report., U.S. Geological Survey Professional Paper 1401-A; http://pubs.usgs.gov/pp/1401a/report.pdf.

Bindoff, N. L., P. A. Stott, M. AchutaRao, M. R. Allen, N. Gillett, D. Gutzler, K. Hansingo, G. Hegerl, Y. Hu, and S. Jain (2013), Detection and attribution of climate change: From global to regional, in *Climate Change 2013: The Physical Science Basis, Contribution of Working Group I to the Fifth Assessment Report of the Intergovernmental Panel on Climate Change*, edited by T. F. Stocker, D. Qin, G.-K. Plattner, M. Tignor, S. K. Allen, J. Boschung, A. Nauels, Y. Xia, V. Bex, and P. M. Midgley, Cambridge University Press, Cambridge, UK, and New York, USA.

Capotondi, A., and P. D. Sardeshmukh (2015), Optimal precursors of different types of ENSO events, *Geophys. Res. Lett.*, *42*(22), 9952–9960.

Chang, P., L. Zhang, R. Saravanan, D. J. Vimont, J. C. Chiang, L. Ji, H. Seidel, and M. K. Tippett (2007), Pacific meridional mode and El Nino–Southern oscillation, *Geophys. Res. Lett.*, *34*(16), L16608.

Cheng, L., M. Hoerling, A. AghaKouchak, B. Livneh, X.-W. Quan, and J. Eischeid (2015), How has human-induced climate change affected California drought risk?, *J. Climate*, *29*(1), 111–120.

Dai, A. (2011), Drought under global warming: A review, *Wiley Interdisc. Rev. Climate Change*, *2*(1), 45–65.

DeWeaver, E., and S. Nigam (2004), On the forcing of ENSO teleconnections by anomalous heating and cooling, *J. Climate*, *17*(16), 3225–3235.

Diffenbaugh, N. S., D. L. Swain, and D. Touma (2015), Anthropogenic warming has increased drought risk in California, *Proc. Nat. Acad. Sci.*, *112*(13), 3931–3936.

Ding, Q., J. M. Wallace, D. S. Battisti, E. J. Steig, A. J. Gallant, H.-J. Kim, and L. Geng (2014), Tropical forcing of the recent rapid Arctic warming in northeastern Canada and Greenland, *Nature*, *509*(7499), 209–212.

Famiglietti, J. S. (2014), The global groundwater crisis, *Nature Climate Change*, *4*(11), 945–948.

Famiglietti, J. S., M. Lo, S. L. Ho, J. Bethune, K. J. Anderson, T. H. Syed, S. C. Swenson, C. R. de Linage, and M. Rodell (2011), Satellites measure recent rates of groundwater depletion in California's Central Valley, *Geophys. Res. Lett.*, *38*(3), L03403.

Francis, J. A., and S. J. Vavrus (2012), Evidence linking Arctic amplification to extreme weather in midlatitudes, *Geophys. Res. Lett.*, *39*(6), L06801.

Frankignoul, C., N. Sennéchael, and P. Cauchy (2014), Observed atmospheric response to cold season sea ice variability in the Arctic, *J. Climate*, *27*(3), 1243–1254.

Funk, C., and A. Hoell (2015), The leading mode of observed and CMIP5 ENSO-residual sea surface temperatures and associated changes in Indo-Pacific climate, *J. Climate*, *28*(11), 4309–4329.

Funk, C., A. Hoell, and D. Stone (2014), Examining the contribution of the observed global warming trend to the California droughts of 2012/13 and 2013/14, *Bull. Amer. Meteor. Soc.*, *95*(9), S11.

Griffin, D., and K. J. Anchukaitis (2014), How unusual is the 2012–2014 California drought?, *Geophys. Res. Lett.*, *41*(24), 9017–9023.

Hartmann, D. L. (2015), Pacific sea surface temperature and the winter of 2014, *Geophys. Res. Lett.*, *42*(6), 1894–1902.

Howitt, R., J. Medellín-Azuara, D. MacEwan, J. Lund, and D. Sumner (2014), Economic analysis of the 2014 drought for California agriculture, prepared for California Department of Food and Agriculture by UC Davis Center for Watershed Sciences and ERA Economics.

Kim, B.-M., S.-W. Son, S.-K. Min, J.-H. Jeong, S.-J. Kim, X. Zhang, T. Shim, and J.-H. Yoon (2014), Weakening of the stratospheric polar vortex by Arctic sea-ice loss, *Nat. Commun.*, *5*, 4646; doi:10.1038/ncomms5646.

Kumar, A., B. Jha, and M. L'Heureux (2010), Are tropical SST trends changing the global teleconnection during La Niña?, *Geophys. Res. Lett.*, *37*(12).

Lau, K.-M., and H. Weng (2002), Recurrent Teleconnection Patterns Linking Summertime Precipitation Variability over East Asia and North America, *J. Meteor. Soc. Japan*, *80*(6), 1309–1324.

Lee, M. Y., C. C. Hong, and H. H. Hsu (2015), Compounding effects of warm sea surface temperature and reduced sea ice on the extreme circulation over the extratropical North Pacific and North America during the 2013–2014 boreal winter, *Geophys. Res. Lett.*, *42*(5), 1612–1618.

Linkin, M. E., and S. Nigam (2008), The north pacific oscillation-west Pacific teleconnection pattern: Mature-phase structure and winter impacts, *J. Climate*, *21*(9), 1979–1997.

Ludescher, J., A. Gozolchiani, M. I. Bogachev, A. Bunde, S. Havlin, and H. J. Schellnhuber (2014), Very early warning of next El Niño, *Proc. Nat. Acad. Sci.*, *111*(6), 2064–2066.

Mann, M. E., and P. H. Gleick (2015), Climate change and California drought in the 21st century, *Proc. Nat. Acad. Sci.*, *112*(13), 3858–3859.

Mao, Y., B. Nijssen, and D. P. Lettenmaier (2015), Is climate change implicated in the 2013–2014 California drought? A hydrologic perspective, *Geophys. Res. Lett.*, *42*(8), 2805–2813.

McPhaden, M. (2015), Playing hide and seek with El Niño, *Nature Climate Change*, *5*(9), 791–795.

Mo, K. C., and R. W. Higgins (1998), Tropical influences on California precipitation, *J. Climate*, *11*(3), 412–430.

NBC News (2016), Northern California Winter Storms Bring Minor Drought Improvement; http://www.nbclosangeles.com/news/local/California-Drought-El-Nino-Update-Storms-Rain-Snow-366064501.html.

Palmer, T. (2014), Record-breaking winters and global climate change, *Science*, *344*(6186), 803–804.

Roeckner, E., G. Bauml, L. Bonaventura, R. Brokopf, M. Esch, M. Giorgetta, S. Hagemann, I. Kirchner, L. Kornblueh, and E. Manzini (2003), The atmospheric general circulation model EHAM5. Part I: model description.

Ropelewski, C. F., and M. S. Halpert (1986), North American precipitation and temperature patterns associated with the El Niño/Southern Oscillation (ENSO), *Monthly Weather Rev.*, *114*(12), 2352–2362.

Schonher, T., and S. Nicholson (1989), The relationship between California rainfall and ENSO events, *J. Climate*, *2*(11), 1258–1269.

Shukla, S., A. Steinemann, S. F. Iacobellis, and D. R. Cayan (2015), Annual Drought in California: Association with Monthly Precipitation and Climate Phases, *J. Appl. Meteor. Clim.*, *54*(11), 2273–2281.

Thonicke, K., S. Venevsky, S. Sitch, and W. Cramer (2001), The role of fire disturbance for global vegetation dynamics: coupling fire into a Dynamic Global Vegetation Model, *Glob. Ecol. Biogeog.*, *10*(6), 661–677.

Vihma, T. (2014), Effects of Arctic sea ice decline on weather and climate: A review, *Surv. Geophys.*, *35* (5), 1175–1214.

Vimont, D. J., D. S. Battisti, and A. C. Hirst (2001), Footprinting: A seasonal connection between the tropics and mid-latitudes, *Geophys. Res. Lett.*, *28*(20), 3923–3926.

Vimont, D. J., D. S. Battisti, and A. C. Hirst (2003), The Seasonal Footprinting Mechanism in the CSIRO General Circulation Models, *J. Climate*, *16*(16), 2653–2667.

Wang, H., and S. Schubert (2014), Causes of the extreme dry conditions over California during early 2013, *Bull. Amer. Meteor. Soc.*, *95*(7), S7–S11.

Wang, S.-Y., M. L'Heureux, and J.-H. Yoon (2013), Are greenhouse gases changing ENSO precursors in the western North Pacific?, *J. Climate*, *26*(17), 6309–6322.

Wang, S.-Y., W.-R. Huang, and J.-H. Yoon (2015), The North American winter "dipole" and extremes activity: A CMIP5 assessment, *Atmos. Sci. Lett.*, in press.

Wang, S.-Y., L. Hipps, R. R. Gillies, and J.-H. Yoon (2014), Probable causes of the abnormal ridge accompanying the 2013–2014 California drought: ENSO precursor and anthropogenic warming footprint, *Geophys. Res. Lett.*, *41*(9), 2014GL059748.

Wang, S.-Y., Y.-H. Lin, R. R. Gillies, and K. Hakala (2016), Indications for protracted groundwater depletion after drought over the Central Valley of California, *J. Hydrometeor.*, 947–955.

Williams, A. P., R. Seager, J. T. Abatzoglou, B. I. Cook, J. E. Smerdon, and E. R. Cook (2015), Contribution of anthropogenic warming to California drought during 2012–2014, *Geophys. Res. Lett.*, *42*(16), 6819–6828.

Yoon, J.-H., S.-Y. Wang, R. R. Gillies, B. Kravitz, L. E. Hipps, and P. J. Rasch (2015a), Increasing water cycle extremes in California and in relation to ENSO cycle under global warming, *Nature Communications*; doi: 10.1038/ncomms9657.

Yoon, J.-H., S.-Y. S. Wang, R. R. Gillies, L. Hipps, B. Kravitz, and P. J. Rasch (2015b), 2. Extreme fire season in California: A glimpse into the future?, *Bull Amer. Meteor. Soc*, *96*, S5–9.

14

Observed Trends in US Tornado Frequency

Adam J. Clark

ABSTRACT

Tornadoes have a huge societal impact, and increasing exposure to tornado risk from increasing urbanization makes the assessment and prediction of changes in tornado frequencies and locations especially important. Although tornadoes have been documented on every continent except Antarctica, the United States, by far, has the most frequent and strongest tornadoes in the world, which is related to its unique geography. Diagnosing changes in tornado climatology is difficult because sampling biases, changes in reporting practices, and other nonmeteorological issues affect the US tornado database. However, several recent studies find convincing evidence that the variability of tornado frequencies is increasing, more tornadoes are occurring on fewer days, and the tornado season is shifting earlier.

14.1. INTRODUCTION

Tornadoes, broadly defined as a rotating column of air extending from the base of a thunderstorm to the Earth's surface, are considered the most intense and violent atmospheric circulations on the local scale. The wind speeds associated with tornadoes range from 30 to 135 m/s, and tornado width generally varies from just a few meters to 2 km. The highest measured wind speed in a tornado was 135 +/− 10 m/s on 3 May 1999 in Moore, Oklahoma, while the widest officially documented tornado was 4.2 km near El Reno, Oklahoma, on 31 May 2013. The high wind speeds in tornadoes can cause widespread destruction, especially when they occur in densely populated areas. Monetary loss from a single tornado can top $1.0 billion (USD). The most costly tornado in recent history was the Joplin, Missouri, tornado on 22 May 2011, which resulted in a loss of $2.8 billion. The deadliest tornado in recorded history occurred 26 April 1989 in Bangladesh where an estimated 1300 people were killed. In the United States, the year 2011 was a particularly damaging year for tornadoes with multiple outbreaks resulting in a total loss of $25 billion and 553 fatalities. The 2011 outbreaks included the "Super Outbreak" over the southeast United States on 27 April, one of the largest and deadliest outbreaks in the history of the United States [*NOAA*, 2012].

Most tornadoes, and especially the strongest tornadoes, are associated with a special class of thunderstorm known as supercells, which are broadly defined as storms with a deep, persistently rotating updraft known as a mesocyclone. Environmental conditions typically associated with supercells include strong instability and wind shear (change in wind speed and direction with height). Additional parameters that help identify favorable environments for significant tornadoes include strong low-level shear (i.e., lowest 1 km) and low lifting condensation level (LCL) heights [*Rasmussen and Blanchard,* 1998; *Thompson et al.,* 2003]. Although supercells contribute disproportionately to the numbers of tornadoes, tornadoes also frequently occur with other classes of

NOAA/OAR National Severe Storms Laboratory, Norman, Oklahoma, USA

Climate Extremes: Patterns and Mechanisms, Geophysical Monograph 226, First Edition.
Edited by S.-Y. Simon Wang, Jin-Ho Yoon, Christopher C. Funk, and Robert R. Gillies.

thunderstorms including quasilinear convective systems (QLCSs) and even clusters of nonrotating storms [*Smith et al.*, 2012; *Duda and Gallus*, 2010].

The intensity of tornadoes in the United States and Canada is assessed using the Enhanced Fujita Scale (EF-scale) [*WSEC*, 2006], which assigns a tornado to one of six categories ranging from zero to five that represent increasing levels of damage. The categories are assigned increasing levels of wind speed that, based on engineering studies and field experience, roughly match the damage that was produced. Thus, the EF-scale can be thought of as a proxy for actual wind speeds. The scale was designed this way because actual wind speeds in a tornado are so difficult to measure with instruments. Note, this design also means that it is possible (and likely happens quite often) for a very intense tornado that occurs over open fields and does not damage any structures to be assigned a relatively low value on the EF-scale. The Fujita Scale, which was introduced in 1971 by Tetsuya Theodore Fujita [*Fujita*, 1971] and adopted for rating tornadoes in their near-immediate aftermath in 1973, was replaced by the EF-scale in 2007.

Tornadoes can have a huge societal impact and increasing exposure to tornado risk from increasing urbanization makes the assessment and prediction of changes in tornado frequencies and locations especially important. Although tornadoes have been documented on every continent except Antarctica, the United States, by far, has the most frequent and strongest tornadoes in the world, which is related to its unique geography. Thus, this chapter will mostly emphasize tornado trends in the United States, where the impacts are disproportionately larger than other areas of the world. Examination of tornado trends in the United States is also aided by the fact that, although far from perfect quality, the United States has the most thorough and accessible database of observed tornadoes dating back to 1950 known as *Storm Data*, which is maintained by NOAA's National Centers for Environmental Information (NCEI) and the National Weather Service's (NWS) Storm Prediction Center (SPC) (http://www.ncdc.noaa.gov/stormevents/). Furthermore, a long-term study of US tornadoes by *Grazulis* [1993, 1997] includes all significant tornadoes, which include all tornadoes known to have produced a fatality from 1680–2005.

This chapter will cover several topics. First, before assessing tornado climatologies and trends, it is important to be aware of the limitations imposed by the nature of the tornado database so that one can properly frame the uncertainties and difficulties associated with assessing these changes. Thus, the limitations of the primary tornado database for the United States, known as *Storm Data*, will be discussed in the first section. Then, to provide a background against which changes are occurring, a general overview of the US tornado climatology will be given. Finally, recent research documenting changes in the tornado climatology will be presented.

14.2. *STORM DATA* TORNADO DATABASE

Before discussing long-term records of tornadoes, it is very important to be aware of the limitations and numerous nonmeteorological factors that have large influences on the tornado record. First, it was not until 1953 that a serious effort was made to formally document US tornado occurrences. This effort was motivated by the need to verify the tornado forecasts that began that year by the newly established Severe Local Storms (SELS) Unit. Before 1953, informal efforts by the Weather Bureau recorded about 200 tornadoes per year, but since then the average annual tornado count has been closer to 800 [e.g., *Grazulis,* 1993]. Second, population density and communication difficulties impact the number of reported tornadoes, especially in the earlier part of the record. For example, the first maps of tornado occurrences by *Finley* [1887] show relatively few tornadoes in areas of Texas and Oklahoma, an area that is now known to experience some of the most frequent occurrences of tornadoes in the United States. Because of increasing population and improved communication, as well as other factors like increasing public awareness, the proliferation of video cameras [e.g., *Brooks et al.*, 2003], and an explosion in the popularity of storm chasing, the average number of reported tornadoes has steadily increased since the 1950s [e.g., *Verbout et al.*, 2006]. The apparent increase is disproportionately contributed by weaker tornadoes and is very likely simply an artifact of more thorough reporting. Third, as discussed by *Brooks et al.* [2003] and references therein, there are basic issues with the accuracy and temporal consistency of tornado reports. These include errors in recording time and location information, differences in practices across NWS Weather Forecasting Offices (WFOs) for collecting reports for warning verification, and changes in the nature of detailed damage surveys [e.g., *Doswell and Burgess*, 1988; *Doswell et al.*, 2009]. The damage rating scale for tornadoes was not even introduced until 1973 by Tetsuya Fujita. Before this time, ratings were applied retroactively using resources like accounts and photographs from archived newspapers; thus, many of the pre-1973 ratings are highly subjective and uncertain. Furthermore, there are multiple lines of evidence indicating that the pre-1973 counts of significant tornadoes (i.e., EF2 and greater) are inflated, which has been hypothesized to occur because of newspapers overemphasizing damage [e.g., *Brooks and Craven*, 2002; *Verbout et al.*, 2006; and *Anderson et al.*, 2007].

Although the nonmeteorological influences on the raw numbers of tornadoes make long-term assessment of changes in the raw numbers extremely difficult, other aspects of the tornado record are less influenced by these factors and can thus be better leveraged for assessing changes. These aspects include number of tornadoes rated EF1 or higher, accumulated tornado path length, and number of "tornado days" (i.e., number of days with any number of tornadoes). The record of EF1 or greater tornadoes is less influenced by nonmeteorological factors because these tornadoes are much less likely to be missed in the earlier part of the record. According to the EF-scale, for a tornado to be rated EF1 or greater, it has to have caused at least moderate damage. Thus, by definition, lasting evidence of EF1 or greater tornadoes exists, which can be documented. On the other hand, EF0 tornadoes are defined as those that cause little or no damage. Thus, for an EF0 tornado to get in the record there has to be video or photographic evidence, and/or a credible eyewitness account. Because of communication difficulties in the earlier part of the record along with less public awareness of the need to report tornadoes, these tornadoes were documented much less frequently. Accumulated tornado path lengths are more stable because they are disproportionately contributed by long track length tornadoes, which tend to be well documented, and tornado days are more stable because, while the raw numbers of tornadoes can change depending on period of record, usually at least one tornado is detected during an event. Several recent studies, which will be discussed later in this chapter, have used these components of the tornado record to assess long-term changes.

14.3. US TORNADO CLIMATOLOGY

US tornado occurrence has coherent spatial, seasonal, and diurnal variability that is important to understand before assessing long-term changes. Spatially, and in terms of raw tornado frequencies, the areas of north Texas, Oklahoma, Kansas, eastern Colorado, and Nebraska, as well as Florida experience the most tornadoes on an annual basis. Tornadoes are also quite frequent farther north into the Dakotas, Minnesota, Iowa, and Illinois, as well as another area along the Gulf Coast in Alabama, Mississippi, Louisiana, and Arkansas (Fig. 14.1).

The areas over the High Plains are favored because of their location relative to the elevated terrain of the western United States and the Gulf of Mexico. The warm water of the Gulf of Mexico is a source region for very moist air masses that, because of the flat terrain of the Great Plains, can easily be transported by southerly low-level winds all the way to the northernmost US states like North Dakota and Minnesota. The mountains and higher terrain to the west of the Great Plains effectively

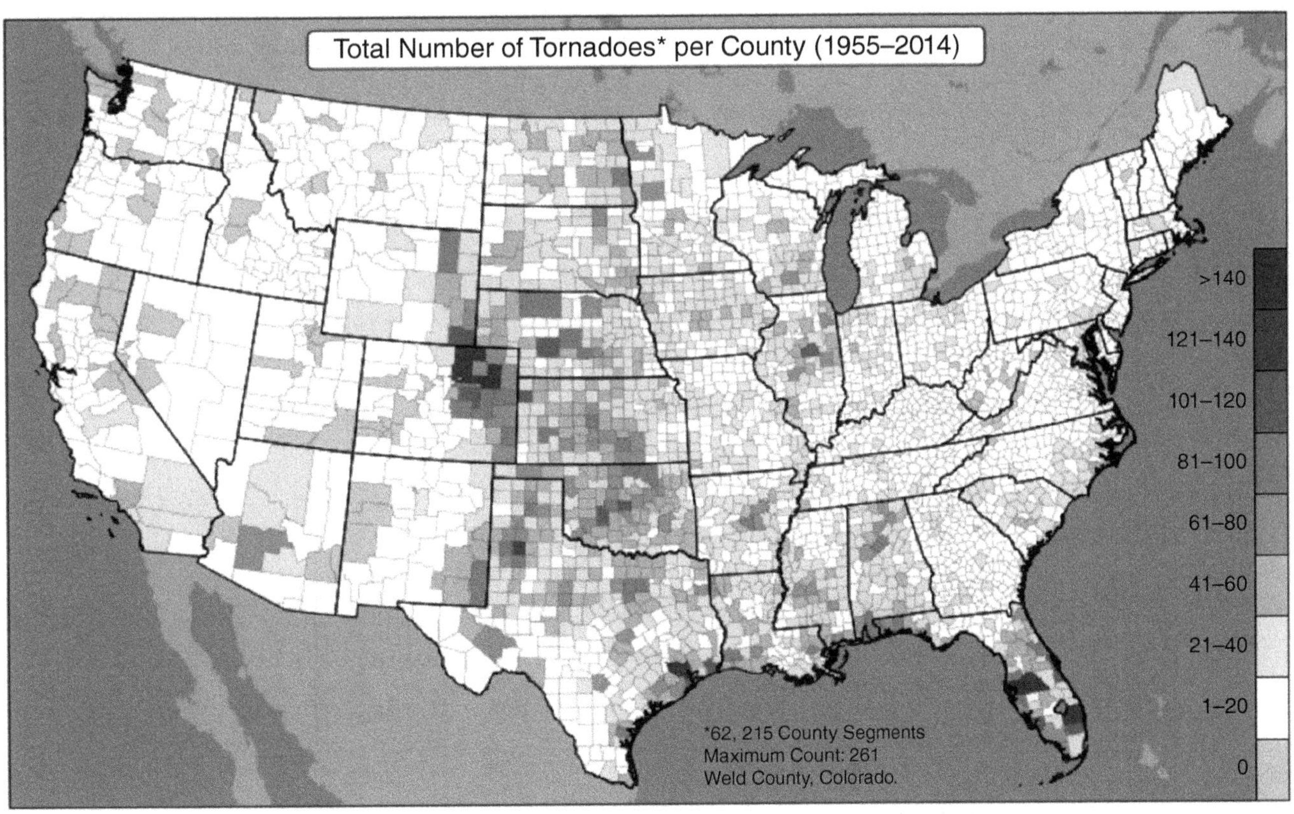

Figure 14.1 Total number of tornadoes per county in the United States. Courtesy of NOAA/SPC. *See electronic version for color representation.*

block Gulf of Mexico moisture from moving farther west. Furthermore, the higher terrain of northern Mexico and the southwest United States serves as an elevated heat source. When the relatively hot and dry air from this region is transported to the east over the Great Plains, a sharp moisture gradient known as a dry line forms where this air mass meets the relatively moist air originating from the Gulf of Mexico. Low-level convergence and subsequent upward motion along the dry line often helps to initiate storms that produce tornadoes when other favorable conditions are present. Also, the hot and dry air from the elevated terrain of the west is transported east of the dry line above the surface where it is known as an elevated mixed layer (EML). Along and to the east of the dry line, the relatively warm air of the EML can inhibit upward motion of air parcels to their level of free convection, thus, it is often referred to as a "cap." When this happens, large amounts of instability can build up, which is one of the main ingredients for tornadoes. If air parcels can overcome the cap, or if the cap erodes through various mechanisms, explosive storm development can occur. During the spring, when the average position of the upper tropospheric jet stream shifts from south to north over the Great Plains, another ingredient important for tornadoes, wind shear (i.e., change in speed and direction of wind with height), is often present and colocated with relatively high instability, making this time of year particularly favorable for tornadoes.

As opposed to the Great Plains, tornadoes occur in Florida simply because thunderstorms are so frequent, and localized sea breeze circulations can help spin up weak and short-lived tornadoes, as well as waterspouts that occur over the water just offshore from the coast. Although the raw frequencies of tornadoes are of similar magnitude to some areas of the Great Plains, intensities of tornadoes in Florida are on average much weaker, because sufficient vertical wind shear and instability for strong, long-track tornadoes is rarely present.

The Gulf Coast states are favored because they are so close to the moisture source of the Gulf of Mexico. Furthermore, the relatively warm temperatures over this region are supportive of thunderstorms for a large portion of the year. During the transition seasons of spring and fall when the upper tropospheric jet stream is positioned near and over this region, a combination of strong wind shear and instability is often associated with jet stream disturbances moving from west to east, which can lead to significant tornado outbreaks. In fact, two of the worst tornado outbreaks in US history, 3–4 April 1974 and 27–28 April 2011, occurred within this area.

Seasonally, and considering the entire CONUS, there is a strong annual cycle in tornado frequency with the peak occurring in May, with June a close second. The months October through February have the fewest relative frequencies of tornadoes. In terms of the number of days each month with tornadoes (i.e., tornado days), June experiences the peak. Thus, while June usually has more days with tornadoes, May usually has more tornadoes per day (Fig. 14.2). This reflects the tendency for more tornado outbreaks (i.e., days with high numbers of tornadoes over a specific region) during May.

US tornadoes are strongly tied to the diurnal cycle because solar heating generates instability, one of the necessary ingredients for tornadoes. However, tornadoes

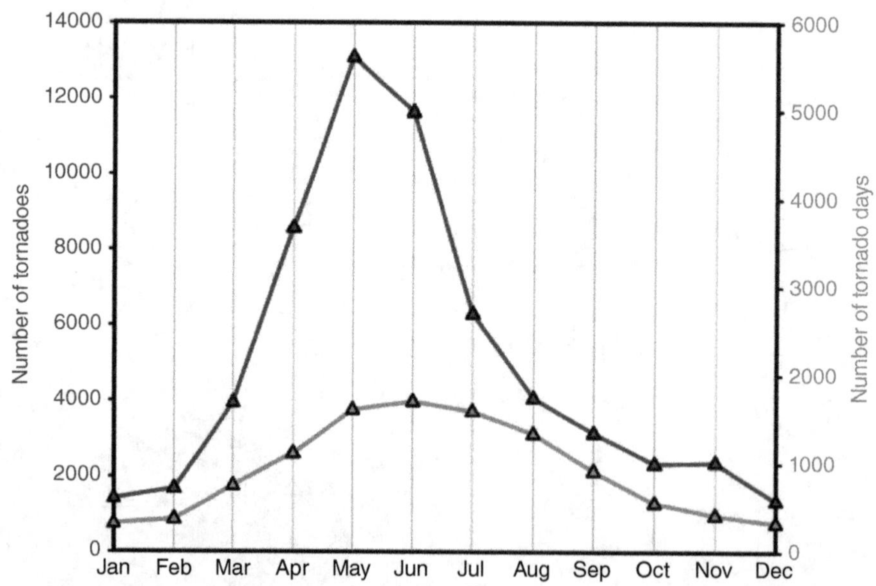

Figure 14.2 Total number of tornadoes 1950–2014 for each month (dark gray; axis on left), and total number of tornado days for each month (light gray; axis on right).

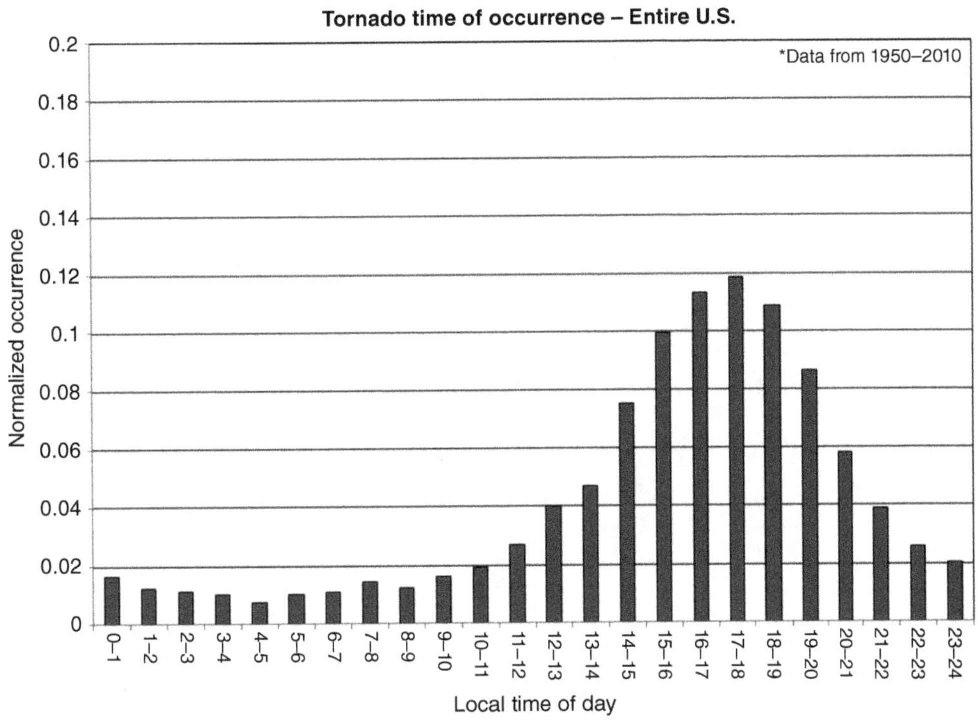

Figure 14.3 Normalized tornado occurrence at each hour of the day (local solar time) using data tornado reports from 1950–2010. Courtesy of NOAA's NCEI. *See electronic version for color representation.*

can occur at any time, and nighttime tornadoes can be some of the most dangerous because they are difficult to visually detect and occur while people are sleeping. The peak time of tornado occurrence is between 5 and 6 pm local time (Fig. 14.3), while there is a broad minimum from late at night through midmorning.

To quantify tornado risk in different regions during different times of the year, researchers [e.g., *Brooks et al.*, 2003] have estimated the probability of occurrence of "tornado days," where tornado days are defined as any tornado occurring within an 80 km × 80 km gridbox, which is consistent with how SPC probabilities of tornadoes are defined. Recall, tornado days is one of the statistics derived from the tornado record that likely has a relatively small influence from nonmeteorological factors. The probabilities are computed by mapping daily tornado reports onto an 80 km × 80 km grid covering the CONUS. For each day, a gridbox is assigned a value of "1" if one or more tornadoes occurs within the box and "0" if none occur. Then, raw frequencies for each day and grid box are found for the period being considered to obtain a raw annual cycle. The raw annual cycle is then smoothed in time and space to obtain the estimated probabilities.

The map for probability of tornado occurrence over an entire year (Fig. 14.4) resembles a C shape, with higher probabilities extending from the Gulf Coast states of Alabama and Mississippi, then through the central Great Plains including large portions of Texas, Oklahoma, and Kansas, and then extending eastward again including large portions of Nebraska, Iowa, and Illinois. According to this analysis, the area of the United States with the highest probability of experiencing a tornado within a 25 mile radius is central Oklahoma. Other local maxima occur in southwest Mississippi and Florida.

To illustrate how the areas of highest tornado risk vary spatially with season, Figure 14.5 displays these daily probabilities at the approximate midpoint of each month. From January through March the magnitude of highest risk gradually increases and is centered over the southeast United States. Then, during April and May, the highest risk magnitudes increase more sharply and the maximum shifts to the west so that it is centered over northern Texas, Oklahoma, and Kansas. During May and into June, the areas of highest risk begin to shift to the northern tier of US states. From July through September, the magnitude of highest risk gradually decreases and remains highest in the northern tier of states, while risk magnitude is also relatively high during this time in Florida. Finally, from October to December, the areas of highest risk shift back to the southeast United States.

It is apparent from Figure 14.5 that some areas of the United States like the Great Plains have a strong annual cycle in tornado risk, while others areas like the southeast

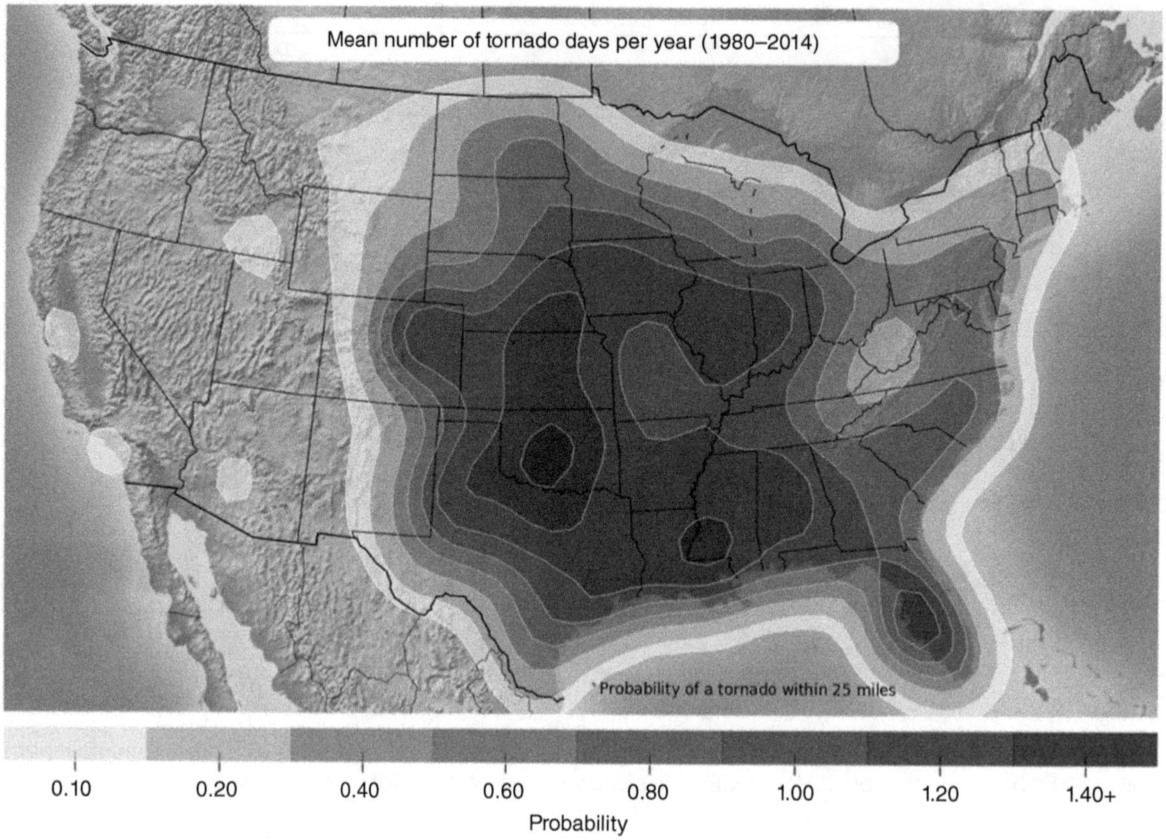

Figure 14.4 Mean annual probability of a tornado within 25 miles derived from 1980–2014 data. Courtesy Patrick Marsh NOAA/SPC. *See electronic version for color representation.*

United States have a weaker annual cycle. This means that while states like Oklahoma have the highest tornado risk in the United States for a relatively short time period during the spring, other states like Alabama have a lower peak risk, but one that is present for a longer duration. To quantify the variability in tornado season, *Brooks et al.* [2003] calculated a trimmed standard deviation of the date of maximum tornado risk for different regions (Fig. 14.6a). For a value of 10, which represents the minimum value that occurs over the Texas panhandle, this can be interpreted as there being an approximately 70% probability that that the maximum date of the tornado season occurs within a 20 day window. Using these trimmed standard deviations, *Brooks et al.* [2003] objectively defines "Tornado Alley" as the region with the most tornado days combined with the highest magnitude annual cycle, which occurs over the Great Plains (Fig. 14.6b). In other words, this is the region with the most tornado days that tend to occur within the shortest time interval. The region over the southeast United States with relatively high tornado risk that is spread out over a longer part of the year is commonly referred to as Dixie Alley. Other methods quantifying tornado risk that emphasize significant tornadoes and their path lengths since 1973, which are two other parts of the tornado base thought to have relatively small influence from nonmeteorological factors, indicate that the highest risk of tornadoes on an annual basis is centered over the southeast United States in Mississippi and Alabama [*Coleman and Dixon*, 2014]. Furthermore, the southeast United States is more prone to tornado fatalities (Fig. 14.7) because of a unique juxtaposition of physical and social vulnerabilities [*Ashley*, 2007], which include a high relative frequency of nocturnal tornadoes and large percentage of forest area making tornadoes difficult to visually detect, and relatively high population density that includes a large percentage of mobile homes, which are much more vulnerable to tornadoes than permanent homes.

14.4. CHANGES IN US TORNADO STATISTICS

The global average temperature has increased significantly during the last few decades, mostly as a result of anthropogenic greenhouse gas emissions [*Intergovernmental Panel on Climate Change,* 2013]. The US surface temperatures have also increased, but with greater year-to-year variability around the increasing trend than the planet as a whole, which is typical of regional trends. Given the large

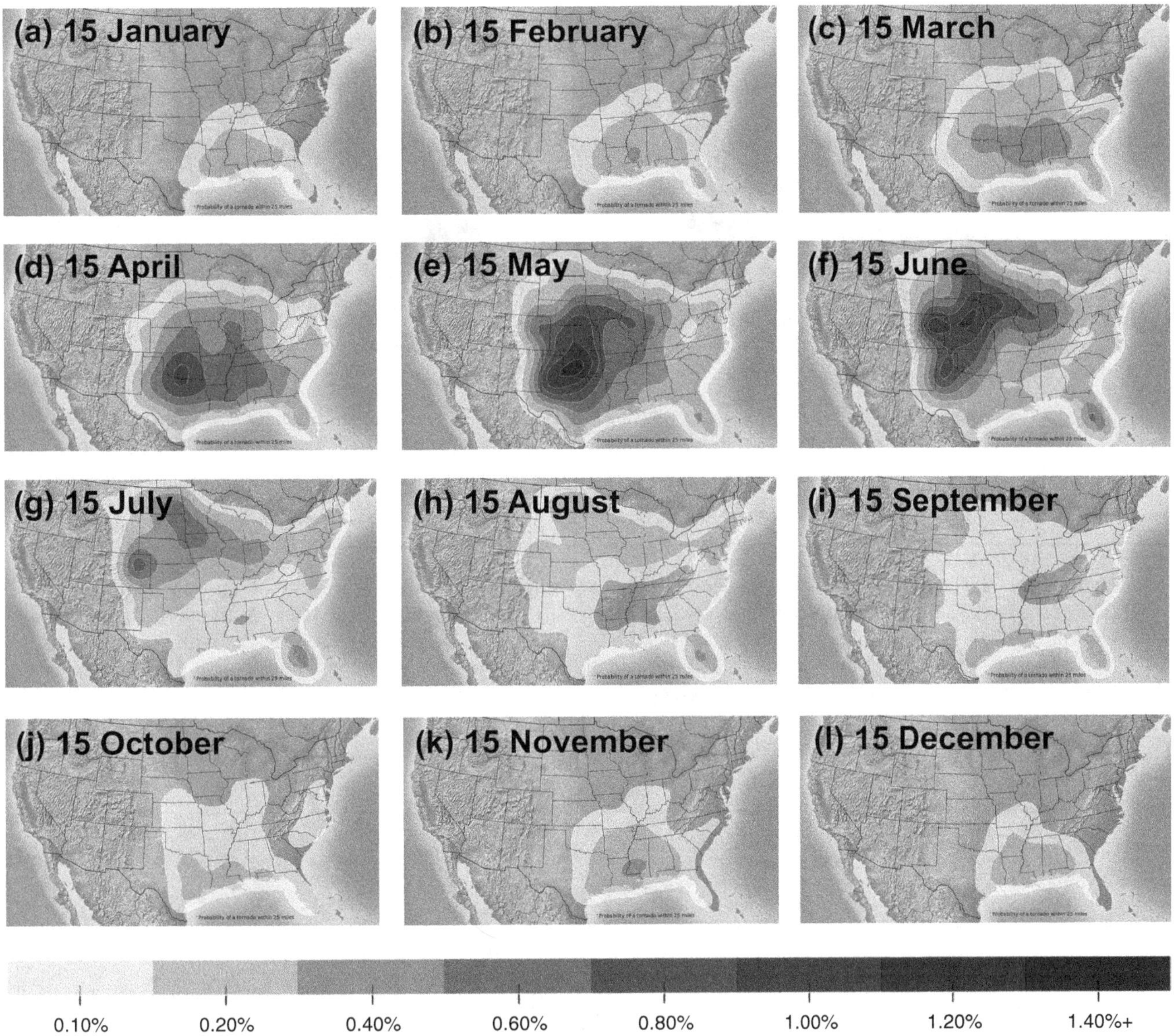

Figure 14.5 Daily probability of a tornado within 25 miles for the fifteenth day of each month derived using 1980–2014 data. Courtesy Patrick Marsh NOAA/SPC. (*See insert for color representation of the figure.*)

societal impact, it is important to investigate whether trends in tornadoes have changed and whether any changes can be attributed to anthropogenic climate change. As previously discussed, assessing changes in US tornado occurrence is especially challenging because of the uncertainties and many nonmeteorological impacts present in the tornado record. Furthermore, attribution of these changes is an additional challenge because it involves relating changes in tornadoes to long-term changes in the environments from which they form, which requires high quality, high-resolution reanalysis datasets.

Until the early 2010s, there were not any studies documenting changes in the US tornado climatology clearly not related to nonmeteorological issues with the tornado record. Part of the reason for this lack of research is that basic tornado statistics, such as annual frequencies of EF1 and greater tornadoes, exhibit large annual variability, but no significant trends (e.g., Fig. 14.8a). Furthermore, for many tornado variables that do exhibit trends (e.g., frequencies of EF0+ tornadoes), it is easy to attribute these trends to previously discussed reporting biases. However, since 2013 several studies have been conducted that indicate various aspects of the US tornado climatology are changing and that these changes are not artifacts of sampling issues in the tornado database.

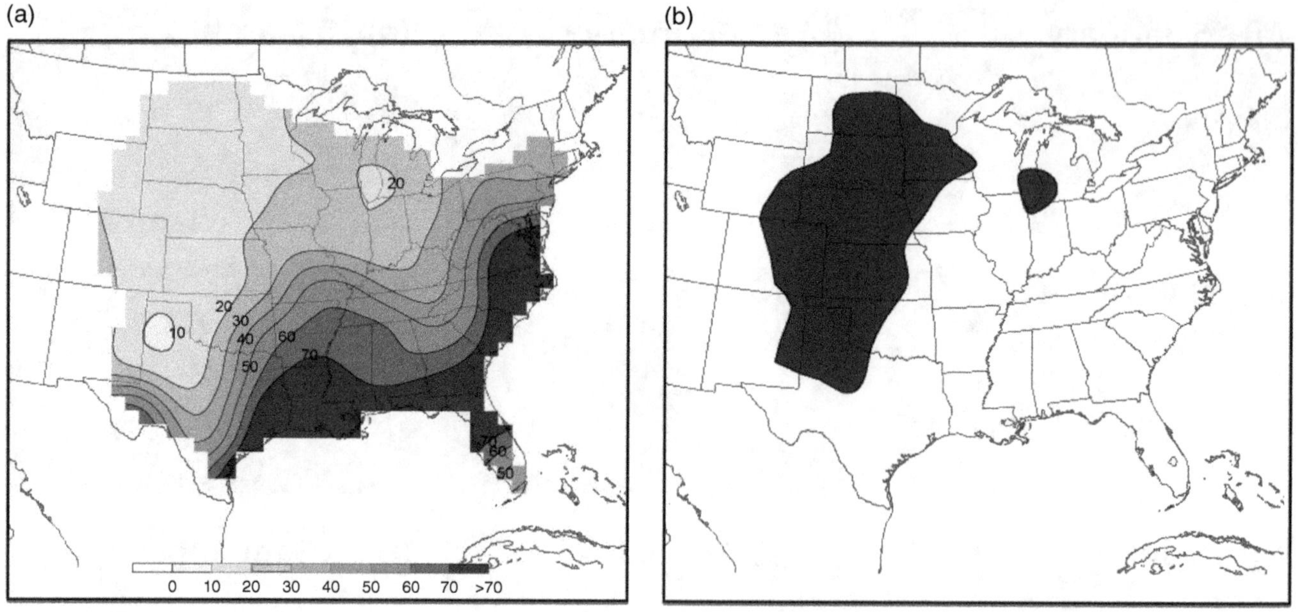

Figure 14.6 (a) The trimmed standard deviation of the date of maximum tornado threat for locations with at least 0.25 tornado days per year. The contour interval = 10 days, the minimum contour = 10 days, and the maximum contour = 70 days. The white areas indicate less than 0.25 tornado days per year. (b) The area with at least 0.5 tornado days per year, and the trimmed standard deviation for timing of annual maximum of less than 20 days shaded in gray. Figure provided by Harold Brooks.

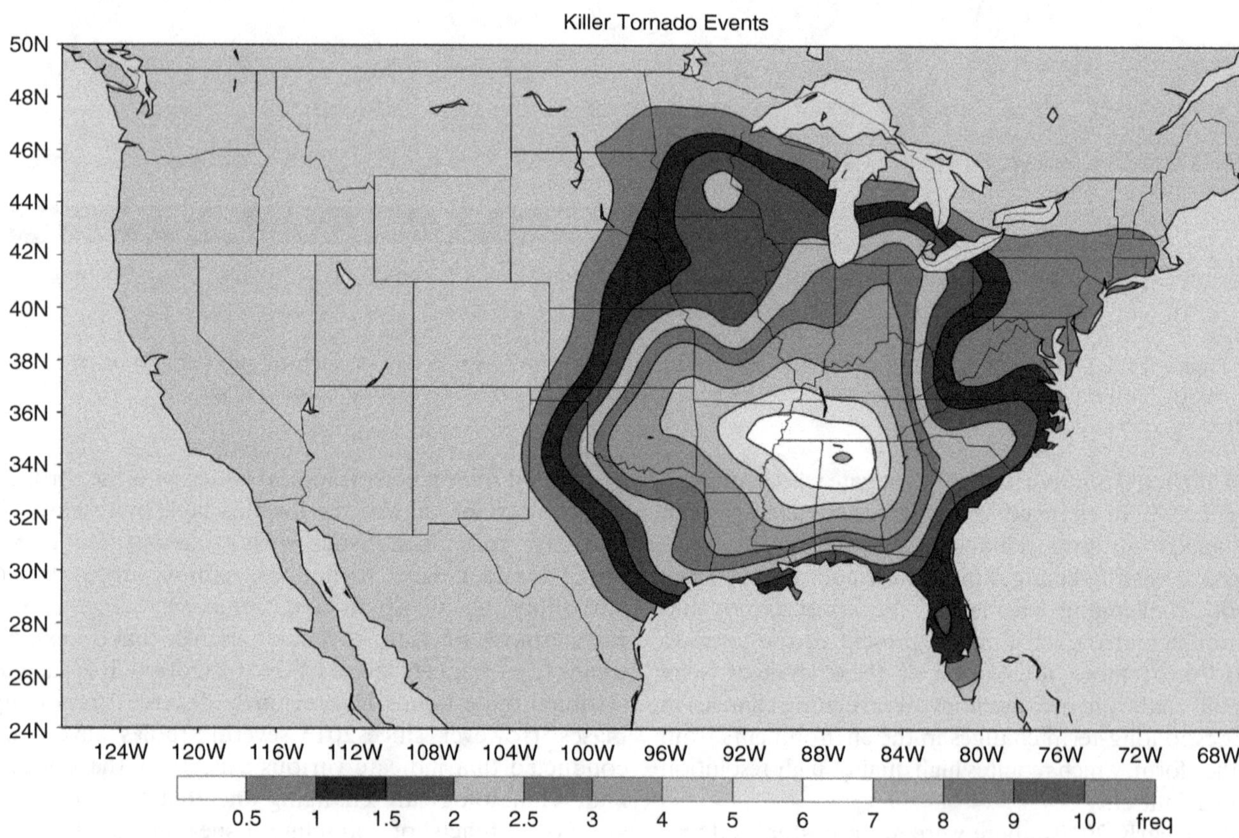

Figure 14.7 Smoothed frequency of the number of killer tornado events in a 80 × 80 km grid for 1950–2014. Data smoothed using a Gaussian kernel density weighting function with a standard deviation of 120 km.

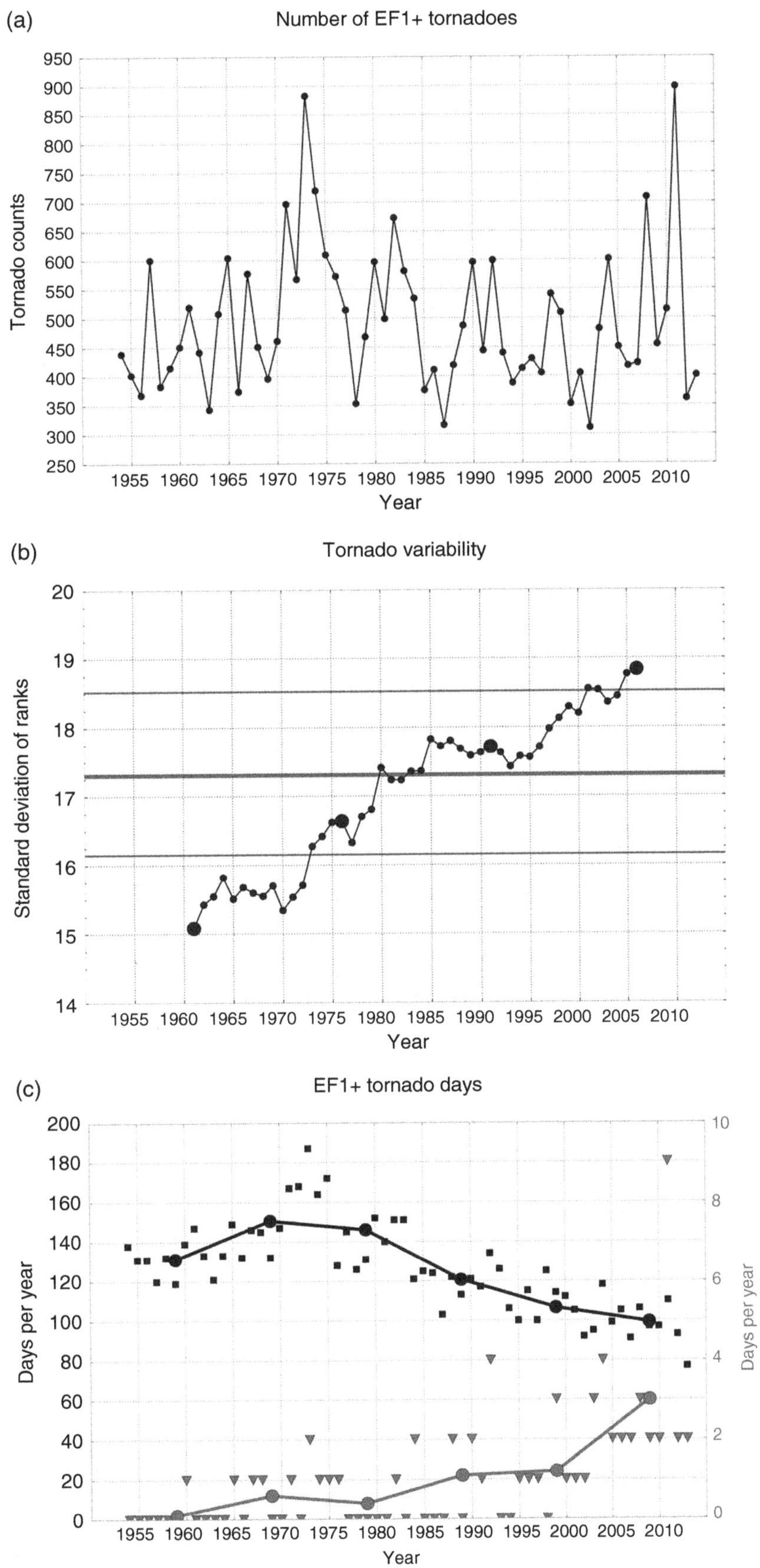

Figure 14.8 (a) Number of reported EF1+ tornadoes by year, 1953–2013. (b) Standard deviation of monthly tornado ranks for tornado counts over 1 year periods, where monthly tornado counts are ranked from 1 (fewest) to 60 (most) with ties assigned the mean value of ranks for the ties. Thick horizontal grey line is the mean value (17.3) and thinner grey lines are the 95% CI (16.1, 18.4). Large dots are associated with 15 year nonoverlapping periods. (c) Number of days per year with at least one (black squares) and more than 30 (grey triangles) EF1+ tornadoes. Large dots and lines indicate decadal means. (a), (b), and (c) adapted from Figures 1, 2, and 4, respectively, of *Brooks et al.* [2013].

First, by examining the variability of the standard deviation of monthly ranks for tornado counts, a clear increase in the variability of EF1+ tornado counts has been found (Fig. 14.8b) [*Brooks et al.*, 2014]. The interpretation of this finding is that extreme tornado months are becoming more common, where the extremes are equally balanced between months with abnormally low and high tornado counts. *Brooks et al.* [2014] also find increasing variability in the start date of tornado season (where start date is defined by the date at which 10% of the tornadoes for an average season have occurred), and find that while the number of days with at least one EF1 or greater tornado has decreased from 1954–2013, the number of days with at least 30 EF1 or greater tornadoes has increased (Fig. 14.8c). Thus, while the annual frequencies of EF1 or greater tornadoes have remained nearly constant, these tornadoes are increasingly occurring in bunches. *Brooks et al.* [2014] do not offer a physical hypothesis for the increased variability, but note that attribution would require examining the change in the areal distribution of favorable environmental conditions, as well as the probability of a tornado given favorable conditions.

Along similar lines as *Brooks et al.* [2014], *Tippett* [2014] derived a measure of year-to-year tornado volatility computed as the standard deviation of differenced annual tornado reports. *Tippett* [2014] found clear volatility change points in the early portion of the tornado record corresponding to known changes in reporting practices. For example, year-to-year tornado volatility decreased in 1975 when on-site tornado damage surveys began, which replaced the retrospective ratings. However, tornado volatility change points were also detected in the 2000s that were not associated with any known changes in tornado reporting practices. Although the exact year depends on the tornado intensity examined, each tornado intensity from EF0 to EF2 experienced a marked, statistically significant shift upward in volatility during the early to mid-2000s. To examine whether this upward shift was physical and not related to an unknown change in reporting, *Tippett* [2014] examined the correlation in annual tornado changes to that of a tornado environment index (TEI) [*Tippett et al.*, 2012, 2014]. The TEI is based on convective precipitation and storm relative helicity (i.e., environmental wind shear) as depicted by the North American Regional Reanalysis for the period 1979–2013, and calibrated against the annual cycle of EF0+ tornado reports. *Tippett* [2014] finds that the differenced TEI time series is highly correlated ($R = 0.68$) to the differenced annual tornado reports, giving confidence that the changes in tornado volatility are physical in nature.

Elsner et al. [2014] examined trends in tornado efficiency, which they define as the proportion of tornadoes occurring on "big" tornado days. Their results show a consistent decrease in the number of days with at least one tornado, but an increase in the number of days with many tornadoes (i.e., increasing tornado efficiency). *Elsner et al.* [2014] also find increases in tornado density, defined as the number of tornadoes over a localized region where tornadoes occurred.

Finally, using circular methods for analyzing temporal trends in peak tornado activity, *Long and Stoy* [2014] found that the tornado season over the central and southern US Great Plains is shifting earlier by an average of 1.55 days per decade (Fig. 14.9). Attribution of this trend was not analyzed, but the results are consistent with abundant evidence that spring in the United States is arriving earlier while autumn is lasting longer [e.g., *Wang et al.*, 2009].

Together, these studies present convincing evidence that aspects of the US tornado climatology are changing, and that these changes are not artifacts of nonmeteorological factors associated with the tornado database. Namely, annual tornado frequencies are becoming more variable, and although the relative number of tornadoes is remaining constant, more tornadoes are occurring on fewer days. Furthermore, the start date of the tornado season is becoming more variable, and is shifting earlier. These studies note that attribution of these changes is difficult, and that while it is certainly possible that anthropogenic global warming could be causing the changes, further analysis is needed for attribution.

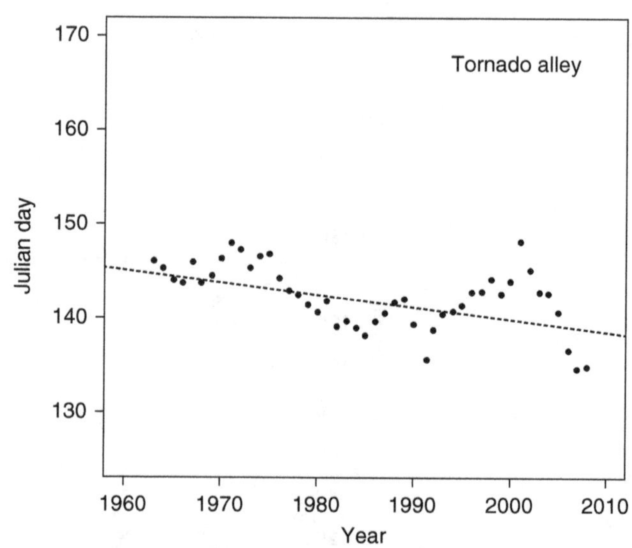

Figure 14.9 The Julian day of peak tornado activity for moving 10 year time windows over Tornado Alley (black dots) as defined by *Long and Stoy* [2014], with linear trend line (dashed line). The years on the *x*-axis represent the last year of the associated 10 year moving average (e.g., 1963 represents 1954–1963). From Figure 3 of *Long and Stoy* [2014].

14.5. CONCLUDING REMARKS

Despite problems with the US tornado record, including sampling biases, changes in reporting practices, and other nonmeteorological issues that affect the database, several recent studies find convincing evidence that the variability of tornado frequencies is increasing, more tornadoes are occurring on fewer days, and the tornado season is shifting earlier in the season. Given the large societal impact of tornadoes, monitoring these changes is particularly important. Furthermore, more work is needed to find attribution of these changes, and to find whether there is a link to anthropogenic factors.

ACKNOWLEDGMENTS

The author greatly appreciates feedback from Dr. Harold Brooks in preparing this chapter. Additionally, several helpful comments from an anonymous reviewer helped improve the chapter. Finally, Dr. Patrick Marsh's help was greatly appreciated in generating the images used to construct Figures 14.4 and 14.5 and updating the datasets upon which these figures were based.

REFERENCES

Anderson, C. J., C. K. Wikle, Z. Zhou, and J. A. Royle (2007), Population influences on tornado reports in the United States, *Weath. Forecast.*, *22*, 571–579.

Ashley, W. S. (2007), Spatial and temporal analysis of tornado fatalities in the United States: 1880–2005, *Weath. Forecast.*, *22*, 1214–1228.

Brooks, H. E., and J. P. Craven (2002), A database of proximity soundings for significant severe thunderstorms, 1957–1993, Prepreints, *21st Conference on Severe Local Storms*, San Antonio, TX, American Meteorological Society, 639–642.

Brooks, H. E., C. A. Doswell III, and M. P. Kay (2003), Climatological estimates of local daily tornado probability, *Weath. Forecast.*, *18*, 626–640.

Brooks, H., E., G. W. Carbin, P. T. Marsh (2014), Increased variability of tornado occurrence in the United States, *Science*, *346*, 349–352.

Coleman, T. A., and P. G. Dixon (2014), An objective analysis of tornado risk in the United States, *Weath. Forecast.*, *29*, 366–376.

Doswell, C. A., III, and D. W. Burgess (1988), On some issues of United States tornado climatology, *Monthly Weather Rev.*, *116*, 495–501.

Doswell, C. A., III, H. E. Brooks, and N. Dotzek (2009), On the implementation of the enhanced Fujita scale in the USA, *Atmos. Res.*, *93*, 554–563.

Duda, J. D., and W. A. Gallus Jr. (2010), Spring and summer midwestern severe weather reports in supercells compared to other morphologies, *Weath. Forecast.*, *25*, 190–206.

Elsner, J. B., S. C. Elsner, and T. H. Jagger (2014), The increasing efficiency of tornado days in the United States, *Climate Dyn.*; doi:10.1007/s00382-014-2277-3.

Finley, J. P. (1887), *Tornadoes: What They Are and How to Observe Them*, Insurance Monitor Press, New York.

Fujita, T. T. (1971), Proposed characterization of tornadoes and hurricanes by area and intensity, SMRP Research Paper 91, University of Chicago, Chicago, IL; available from Wind Engineering Research Center, Box 41023, Lubbock, TX 79409.

Grazulis, T. P. (1993), *Significant Tornadoes, 1680–1991*, Environmental Films.

Grazulis, T. P. (1997), *Significant Tornadoes: Update, 1992–1995*, Environmental Films.

IPCC (2013), *Climate Change 2013: The Physical Science Basis. Contribution of Working Group I to the Fifth Assessment Report of the Intergovernmental Panel on Climate Change*, edited by T. F. Stocker, D. Qin, G.-K. Plattner, M. Tignor, S. K. Allen, J. Boschung, A. Nauels, Y. Xia, V. Bex and P. M. Midgley, Cambridge University Press, Cambridge, UK, and New York, USA; doi:10.1017/CBO9781107415324.

Long, J. A., and P. C. Stoy (2014), Peak tornado activity in is occurring earlier in the heart of tornado alley, *Geophys. Res. Lett.*, *41*, 6259–6264.

NOAA (2012), Service assessment: The historic tornadoes of April 2011, National Weather Service; available online at www.nws.noaa.gov/os/assessments/pdfs/histororic_tornadoes.pdf.

Rasmussen, E. N., and, S. O. Blanchard (1998), A baseline climatology of sounding-derived supercell and tornado forecast parameters, *Weath. Forecast.*, *13*, 1148–1164.

Smith, B. T., R. L. Thompson, J. S. Grams, C. Broyles, and H. E. Brooks (2012), Convective modes for significant severe thunderstorms in the contiguous United States, Part I: Storm classification and climatology, *Weath. Forecast.*, *27*, 1114–1135.

Thompson, R. L., R. Edwards, and J. A. Hart (2003), Close proximity soundings within supercell environments obtained from the rapid update cycle, *Weath. Forecast.*, *18*, 1243–1261.

Tippett, M. K. (2014), Changing volatility of U.S. annual tornado reports, *Geophys. Res. Lett.*, *41*, 6956–6961.

Tippett, M. K., A. H. Sobel, and S. J. Camargo (2012), Association of U. S. tornado occurrence with monthly environmental parameter, *Geophys. Res. Lett.*, *39*, L02801; doi:10.1029/2011GL050368.

Tippett, M. K., A. H. Sobel, S. J. Camargo, and J. T. Allen (2014), An empirical relation between U.S. tornado activity and monthly environmental parameters, *J. Climate*, *27*, 2983–2999.

Verbout, S. M., H. E. Brooks, L. M. Leslie, and D. M. Schultz (2006), Evolution of the U.S. tornado database: 1954–2003, *Weath. Forecast.*, *21*, 86–93.

Wang, J., M. Hoerling, A. Kumar, and P. Pegion (2009), Attribution of the seasonality and regionality in climate trends over the United States during 1950–2000, *J. Climate Change*, *22*, 2571–2590; doi:10.1175/2008JCLI2359.1.

WSEC (2006), A recommendation for an enhanced Fujita scale (EF-scale), Wind Science and Engineering Center, Texas Tech University; available online at http://www.spc.noaa.gov/faq/tornado/ef-ttu.pdf.

15

Mechanisms Explaining Recent Changes in Australian Climate Extremes

Sophie C. Lewis[1], David J. Karoly[2], Andrew D. King[3], Sarah E. Perkins[4], and Markus G. Donat[4]

ABSTRACT

In 2010–2012, Australia had its record highest 2 year rainfall, as well as repeated shorter-term extreme rainfall events and damaging floods. Later in 2013, Australia experienced its record hottest year, together with repeated shorter-term temperature records and heat waves in different regions. In this chapter, we discuss the mechanisms explaining the extreme rainfall affecting Australia in 2010–2012 and the extreme temperatures affecting Australia in 2013 from a climate perspective. We focus on contributing factors such as El Niño–Southern Oscillation and human-caused climate change on the likelihood of such events. We find that the extreme 2010–2012 rainfall was primarily associated with the combination of the influence of two consecutive strong La Niña events and chaotic climate variability. Any influence of climate change is found to be weaker and harder to distinguish from natural climate variability. Conversely, human-caused climate change was an important contributor to the extreme temperatures across Australia in 2013, and the record temperature in 2013 was virtually impossible to occur due to natural variability alone. In the future, natural variability and ENSO will continue to greatly affect temperature and rainfall extremes. In addition, climate change is expected to continue to increase the frequency and intensity of extreme high temperatures across Australia.

15.1. INTRODUCTION

Australia lies between the Pacific Ocean, the Indian Ocean, and the Southern Ocean, spanning a wide range of climate zones from the tropics and subtropics to more temperate climates in the middle latitudes. These climate zones are influenced by a number of weather factors from monsoons and tropical cyclones in the north, persistent anticyclones over the arid interior, and the passage of cold fronts and mid-latitude storms driven by the strong westerly winds farther south over the Southern Ocean [e.g., *Risbey et al.*, 2009].

Australia has a highly variable climate, with the most variable annual rainfall of any inhabited continent [*Nicholls et al.*, 1997]. These extreme variations of rainfall are a key part of Australian life and have been described in Australian literature for more than a century. An often-quoted example is from the 1911 poem *My Country* by Dorothea Mackellar, which describes Australia as a land "of droughts and flooding rains" [*MacKellar*, 1911].This rainfall variability is primarily driven by the influence of El Niño–Southern Oscillation on eastern Australia [*McBride and Nicholls*, 1983; *Risbey et al.*, 2009].

[1]*ARC Centre of Excellence for Climate System Science, Fenner School of Environment and Society, The Australian National University, Acton, Australia*

[2]*ARC Centre of Excellence for Climate System Science, School of Earth Science, University of Melbourne, Melbourne, Australia*

[3]*ARC Centre of Excellence for Climate System Science, School of Earth Science, University of Melbourne, Melbourne, Australia*

[4]*ARC Centre of Excellence for Climate System Science, Climate Change Research Centre, University of New South Wales, Sydney, NSW, Australia*

Climate Extremes: Patterns and Mechanisms, Geophysical Monograph 226, First Edition.
Edited by S.-Y. Simon Wang, Jin-Ho Yoon, Christopher C. Funk, and Robert R. Gillies.

Most of the early research on mechanisms explaining variations of Australian climate focused on seasonal mean variations. However, there has been greater interest more recently on understanding variations in daily extremes because of the high impact of such extreme events and because of the greater availability of high quality observational and model data at higher space and temporal scales. The relationship between variations in annual mean temperature and rainfall across Australia and variations in some measures of daily extremes is quite strong. Cold years are associated with an increased number of cool days and a reduced number of warm days (e.g., 1956, 1974, 2011) and hot years are associated with fewer cool days and more warm days (e.g., 1959, 1980, 2013), as shown in Figure 15.1. These cold years are generally also very wet years across eastern Australia, with much greater than normal rainfall on very wet days (Fig. 15.1b,d), associated with strong La Niña events. On longer timescales, there has been a significant warming of about 0.8°C across Australia over the last 60 years, mainly due to anthropogenic climate change [*Karoly and Braganza*, 2005]. This mean warming is associated with more warm days and fewer cool days across Australia (Fig. 15.1c, e). There has also been a small increase in

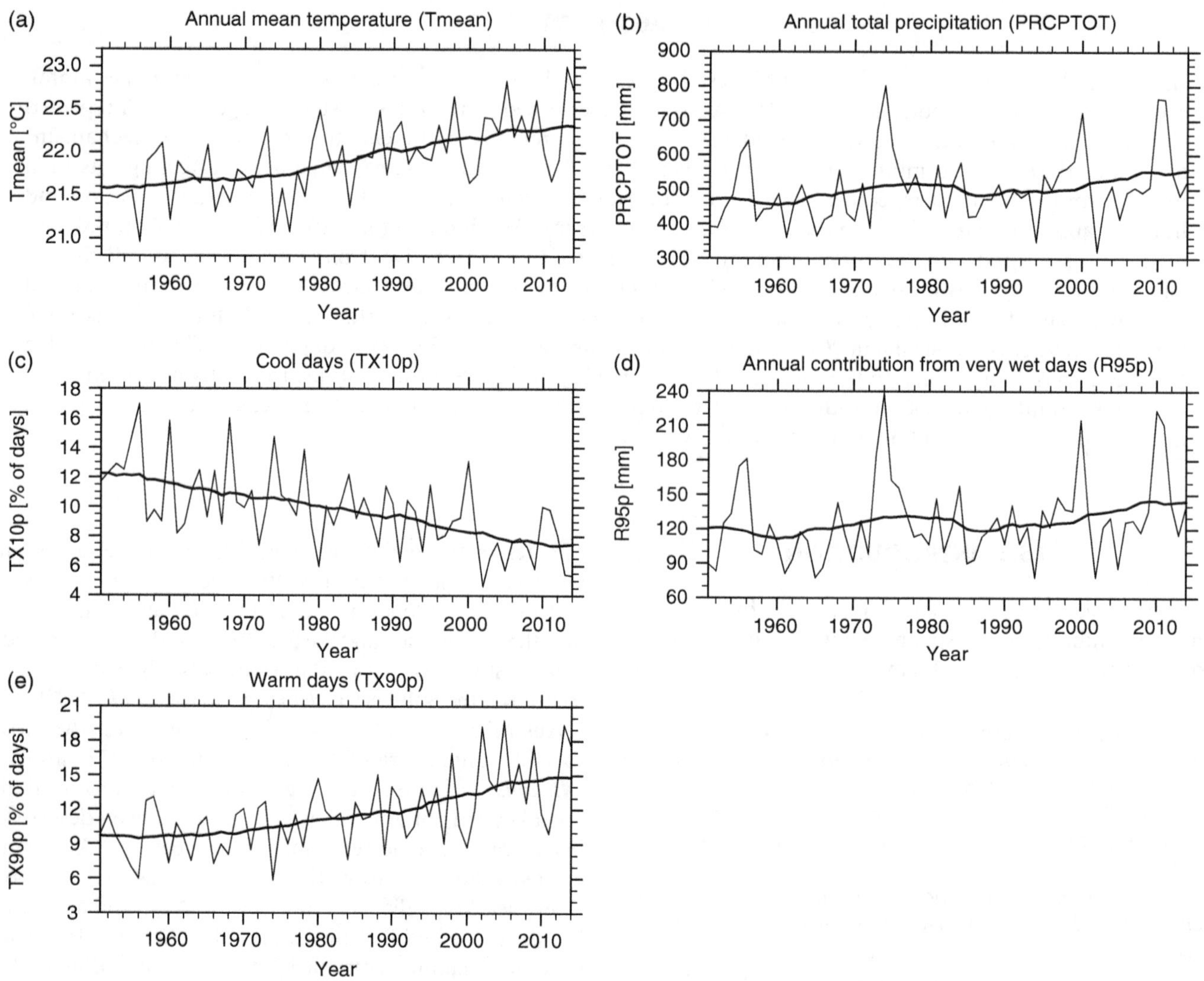

Figure 15.1 Time variations of some measures of mean climate and climate extremes averaged across Australia for the period 1950–2014. (a) Annual average mean temperature (Tmean), (b) total precipitation (PRCPTOT), (c) percentage of cool days (Tx10p), (d) precipitation from very wet days (R95p), and (e) percentage of warm days (Tx90p). These measures are based on indices defined by ETCCDI (Expert Team on Climate Change Detection and Indices, http://www.clivar.org/organization/etccdi), with data from the GHCNDEX global land gridded dataset of climate extremes [*Donat et al.*, 2013]. The thin line shows annual values and the thick line shows decadal variations.

total rainfall and rain on very wet days across Australia over the last 60 years but these changes are not significant. Rainfall changes are spatially inhomogeneous and dominated by the large interannual and decadal variations [*Nicholls et al.*, 1997].

The national weather service in Australia, the Bureau of Meteorology, has provided *Special Climate Statements* on significant climate events since 2007, typically releasing about four statements each year. A major new research initiative to improve understanding of the mechanisms explaining changes in Australian climate extremes at interannual and longer time scales was established in 2011 as part of the Centre of Excellence for Climate System Science funded by the Australian Research Council and bringing together researchers from five Universities across Australia.

In this chapter, the causes of recent temperature and rainfall extremes affecting Australia are discussed. While the immediate cause of any individual extreme weather event is the combination of multiple factors affecting the wind, temperature and moisture, Figure 15.1 shows that climatic factors such as El Niño and climate change can have major influences on extreme events. Often the magnitude of the event is primarily due to the random interference of different weather factors, with smaller changes in magnitude due to climate factors, while the frequency of occurrence of extreme events can be affected more by climate factors.

The next section describes studies of the causes of the extreme rainfall across Australia during 2010 and 2011, the highest 2 year rainfall total since records began in 1910. There were 12 *Special Climate Statements* on extreme rainfall events in 2011, more than in any other year. Section 15.3 describes studies of the extreme high temperatures and heat waves in Australia in 2013, the hottest year across Australia since high-quality temperature data were available in 1910. The final section draws together these results on recent temperature and rainfall extremes due to both natural variability and human-caused climate change and discusses possible future changes in climate extremes in Australia.

15.2. AUSTRALIAN RAINFALL EXTREMES OF 2010–2012

15.2.1. Climatological Context

Over the years of 2010 to 2012, Australia experienced a sustained period of extreme rainfall. Rainfall records were broken on daily through to seasonal and annual timescales for large regions [*Bureau of Meteorology*, 2012a; 2012b; *Ganter and Tobin*, 2013]. Averaged across Australia, a series of frequent heavy rain events from spring 2010 to autumn 2011, and again in late 2011, resulted in Australia's wettest 2 year period on record [*Bureau of Meteorology*, 2012a] (Fig. 15.2a, b). One feature of this extended 2 year rainfall record was an exceptionally heavy precipitation event recorded across southeastern (SE) Australia from 27 February to 5 March 2012 [*Bureau of Meteorology*, 2012b]. March 2012 was considered remarkable due to the record high daily rainfall accumulations (for any month) that were recorded over a broad swathe of southeastern Australia. During the February and March 2012 event, daily rainfall records were also broken at a number of stations, with some sites breaking all-time daily rainfall records for any month, in an observational record extending back to 1910.

The wet conditions in Australia during 2010–2012 were coincident with two consecutive strong La Niña events, which are typically associated with enhanced eastern Australian rainfall (Fig. 15.2d). In eastern Australia particularly, rainfall is highly variable through time [*Nicholls et al.*, 1997] and strongly linked to the El Niño–Southern Oscillation (ENSO) [*Risbey et al.*, 2009]. In general, the La Niña phase of ENSO is associated with increased rainfall over the east of Australia, and the El Niño phase with decreased rainfall in the region. Consequently, ENSO is associated with strong interannual variability in flood and drought risk in Australia. During the period of 2010 to early 2011, La Niña conditions were among the strongest on record, and while the La Niña event declined briefly through the austral winter of 2011, a second La Niña event emerged in spring and persisted through the 2011–2012 summer. These conditions were also associated with very warm sea surface temperatures around northern Australia and in the eastern Indian Ocean [*Bureau of Meteorology*, 2012a].

The persistent heavy rainfall of 2010–2012 ended the drought of the preceding decade, but resulted in severe flooding in northern and eastern Australia. This led to billions of dollars of flood damage, the evacuation of thousands and 35 deaths. The two consecutive years of flooding prompted questions about whether the risks of heavy rain in Eastern Australia were linked to either internal climate variability and/or externally forced climate change [*Christidis et al.*, 2013b]. Several studies addressed these questions and are discussed hereafter.

15.2.2. Assessing the Role of Pacific Variability in 2010–2012 Rainfall Extremes

Knowledge of the nature of variability in Australian extreme rainfall events and long-term changes in heavy rainfall events is limited, with a greater volume of research to date focused on variability of mean, rather than extreme, Australian rainfall [*Risbey et al.*, 2009]. However, a strong relationship between ENSO and extreme rainfall variability across the east and north of Australia is known

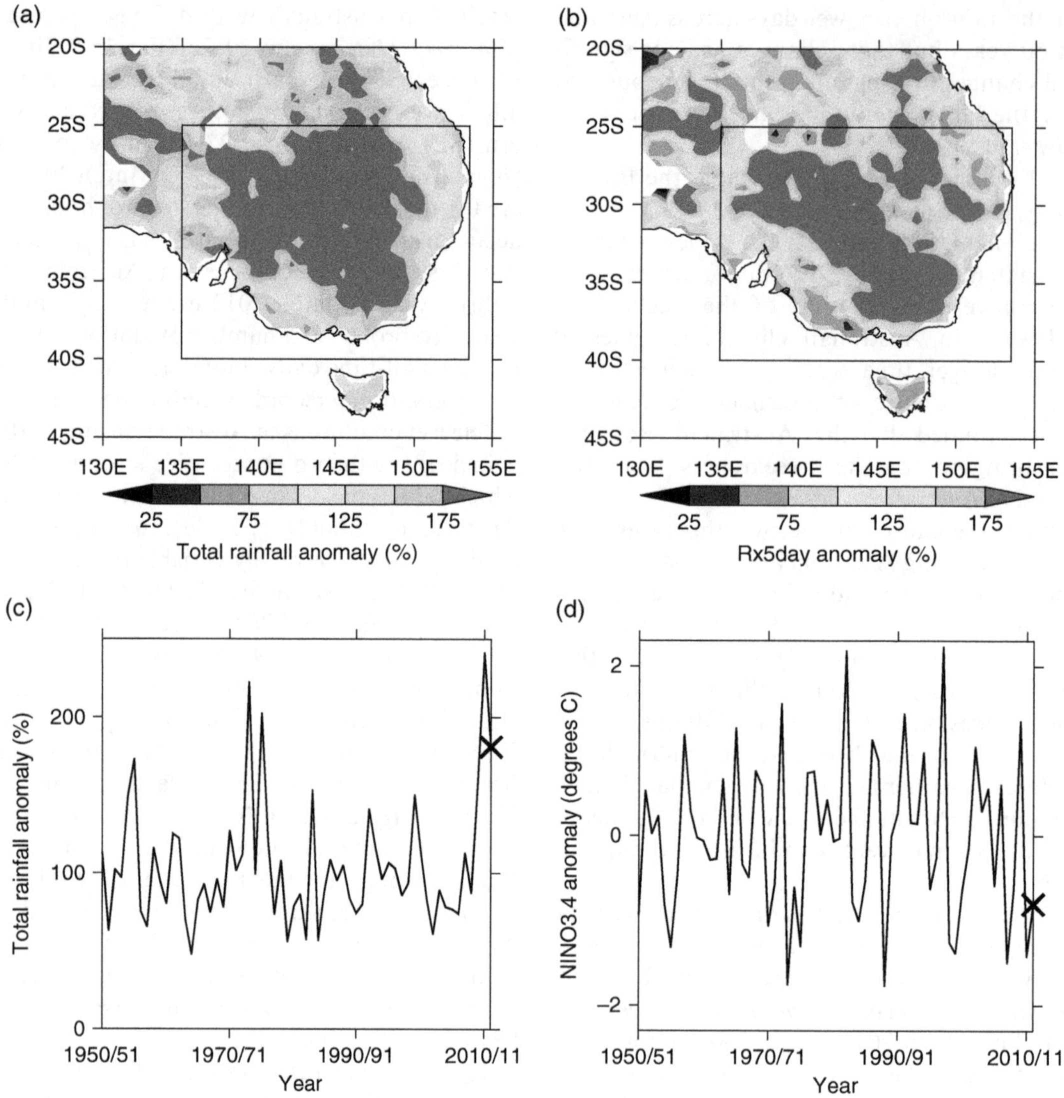

Figure 15.2 (a) Map of October 2011 to March 2012 total rainfall anomalies, relative to the period 1900–2012. The boxed region represents the area investigated by *King et al.* [2013b]. (b) As for panel (a) but showing highest 5 day total rainfall in the period (Rx5day) anomalies. Values in (a) and (b) are shown as a percentage of the October–March long-term mean values. Figures are modified from *King et al.* [2013b]. (c) Time series of area-average standardized total rainfall anomaly (%) for October–March mean seasonal rainfall for the box in (a). The observed 2011–2012 mean anomaly is shown by a cross. (d) As for panel (c) but showing Rx5day rainfall anomalies.

to exist [*Min et al.*, 2013; *King et al.*, 2014b], with heavier extreme rainfall in La Niña seasons than under El Niño or neutral conditions. Several studies focused on ENSO variability and the response of rainfall over Australia find a nonlinear effect, whereby the magnitude of cool (more La Niña-like) SST anomalies in the eastern equatorial Pacific Ocean has a greater influence on rainfall than the magnitude of warm (more El Niño-like) SST anomalies [*Power et al.*, 2006; *Cai et al.*, 2010]. *King et al.* [2013a] determined that the magnitude of a La Niña episode has a much greater influence on the intensity and duration of observed extreme eastern Australian rainfall than the magnitude of an El Niño episode.

Several recent studies have explicitly examined La Niña influences on extreme rainfall such as that occurring in eastern Australia in 2010–2012, including the March 2012 southeastern Australian extremes and the 2 year Australia-wide record rainfall. Using a suite of climate models contributing to the Coupled Model Intercomparison Project phase 5 (CMIP5) [*Taylor et al.*, 2012], *King et al.* [2013b] examined rainfall totals and maximum consecutive 5 day rainfall (Rx5day) across SE

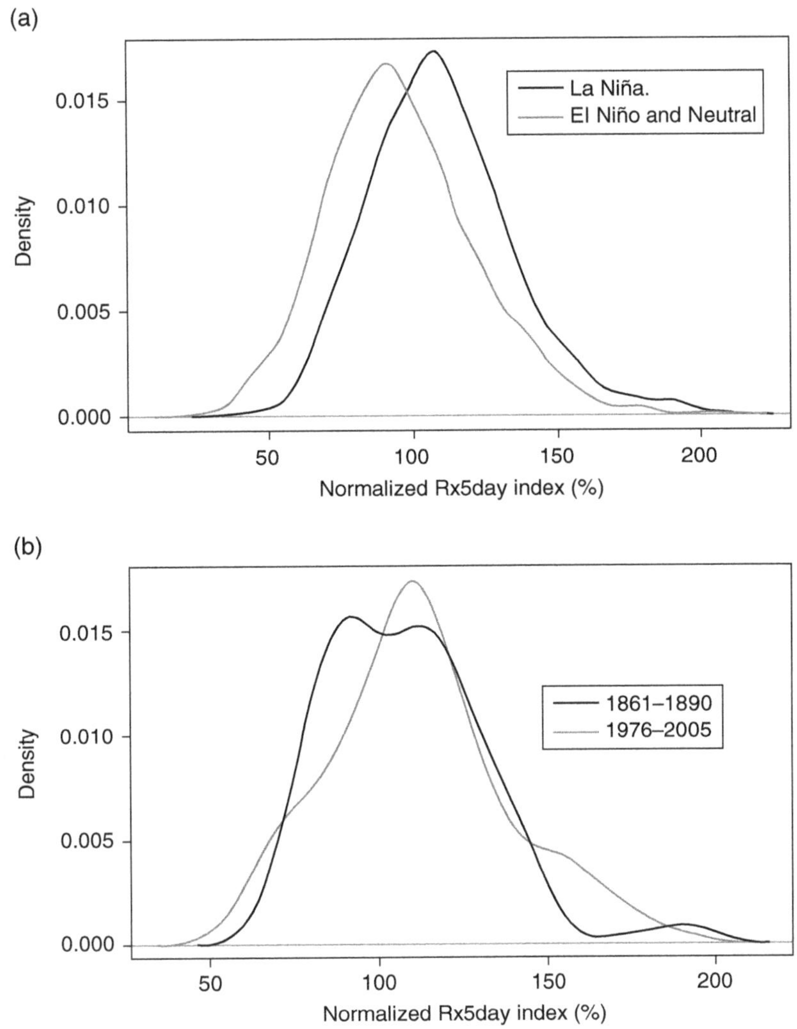

Figure 15.3 (a) Comparison of La Niña-only normalized Rx5day (black line) and El Niño and neutral phase (gray line) normalized Rx5day values for the entire simulated historical period from the five analyzed CMIP5 models. (b) Probability density (PDFs) of La Niña-only standardized Rx5day values in 1861–1890 (black line) and 1976–2005 (gray line) periods from five analyzed CMIP5 models. Adapted from *King et al.* [2013b]. *See electronic version for color representation.*

Australia during the October–March season. Historical period (1850–2005) simulations demonstrate that simulated interannual ENSO variability (NINO3.4 index) played a role in the magnitude of extreme Rx5day rainfall in southeastern Australia (Fig. 15.3a). Focusing on March 2012 rainfall totals (the third wettest March on record), *Christidis et al.* [2013b] concur that while the La Niña episode early in the 2012 increases the likelihood of a wet March, the extremely high rainfall amount is unlikely to be solely attributable to Pacific variability through ENSO. Seasonal, annual, and 2 year rainfall totals have also been investigated in CMIP5 simulations where model years were categorized as either La Niña years or El Niño/neutral [*Lewis and Karoly*, 2014a]. This study determined that there is a higher likelihood of heavy SE rainfall during simulated La Niña years (such as occurring in 2010–2011, and 2011–2012) than during El Niño/neutral years (Fig. 15.4a). For example, there is a greater than fivefold increase in the risk of heavy or extreme rainfall during April–March in this region attributable to La Niña SST anomalies.

This set of studies indicates that La Niña conditions *alone* do not wholly explain the extreme magnitude of Australian rainfall during this 2010–2012 period. The anomalously warm sea surface temperature (SST) conditions in northern Australia have been identified as a possible enhancing factor in the extreme rainfall during this period [*Evans and Boyer-Souchet*, 2012; *Christidis et al.*, 2013b; *Ummenhofer et al.*, 2015]. For example, *Ummenhofer et al.* [2015] use model experiments, with 2010-2011 ocean conditions conducted with and without long-term warming to identify enhanced onshore moisture transport and resultant precipitation in northeast Australia. Furthermore, *Hendon et al.* [2013] highlight a unique pattern and magnitude of tropical SST anomalies during the austral spring of 2010 that were associated with high rainfall across eastern Australia. Key aspects of this pattern were a strong cold anomaly in the central equatorial Pacific, and strong warm anomalies in the eastern Indian Ocean and the far western Pacific to the north of Australia. That is, the strong La Niña episode of 2010 coincided with an extreme positive excursion of the Southern Annular Mode (SAM). Hence, while Pacific SST variability contributed to the extreme Australian rainfall during this period, the extreme Australian rainfall events may arise from SST variability combined with unforced internal climatic variations [*Christidis et al.*, 2013b].

15.2.3. Evidence for an Anthropogenic Influence on 2010–2012 Rainfall Extremes

The extreme Australian rainfall of 2010–2012 has also been investigated in terms of the background warming trend due to anthropogenic forcings, such as long-lived greenhouse gases. A number of studies point to a general increasing trend in extreme rainfall, even in areas where mean rainfall changes are negligible

[*Karl and Knight*, 1998; *van den Besselaar et al.*, 2012]. It has been argued that this effect is associated with global warming and described by the Clausius-Clapeyron relation [*Held and Soden*, 2006], by which the saturation vapour pressure increases exponentially with temperature and the amount of water vapour held in the atmosphere increases by approximately 7% per 1°C rise in temperature. As a warmer atmosphere can hold more moisture, it has been argued that this effect can enhance extreme rainfall totals as the amount of water vapour available to be precipitated is increased. For example, averaged across many station observations globally, *Westra et al.* [2013] found extreme rainfall increases close to that expected by

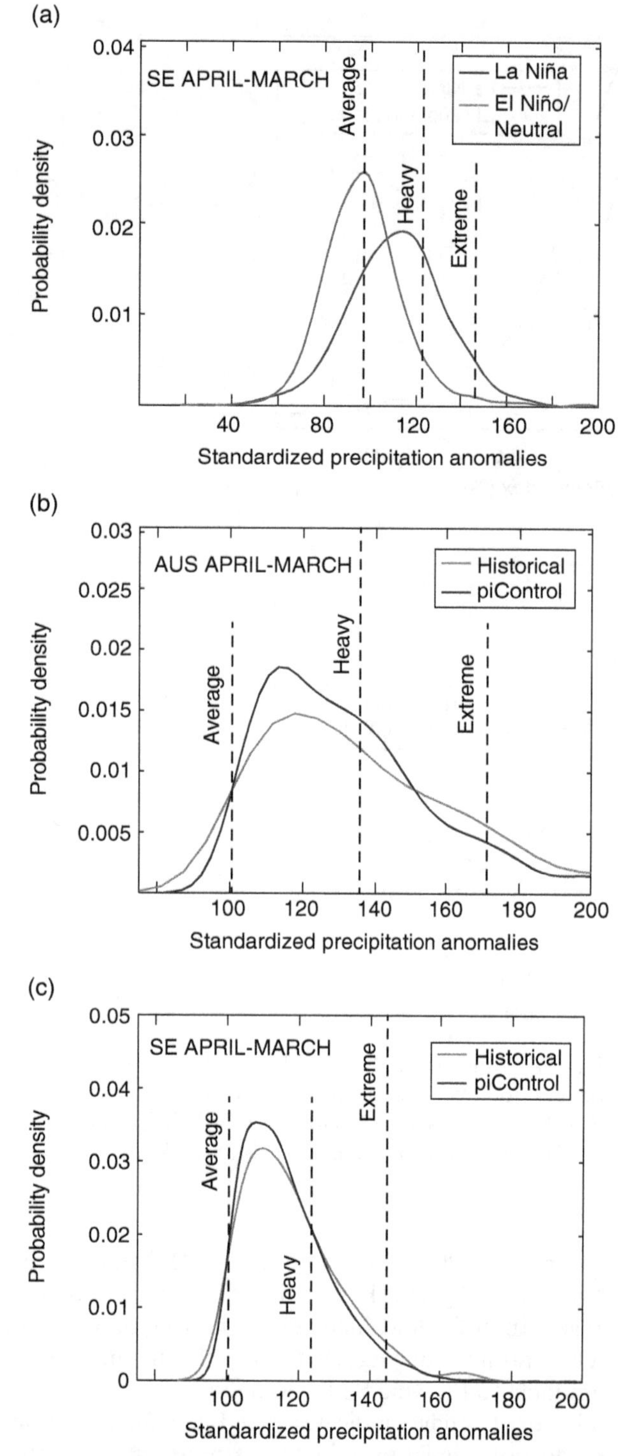

Figure 15.4 (a) Probability density functions (PDFs), comparing southeastern (SE) Australian standardized rainfall for April to March, during simulated La Niña (bold line) episodes with El Niño/neutral (gray line) episodes. ENSO conditions were defined as La Niña where sea surface temperature anomalies in the NINO3.4 region were < −0.5°C for six consecutive months in the year from April to March. Probabilities were derived from CMIP5 historical simulations, and rainfall anomalies were calculated as standardized precipitation anomalies (as a percentage of normal rainfall, as defined by the climatological mean). In all panels, vertical dashed lines show thresholds for mean standardized precipitation (average), one (heavy) and two (extreme) standard deviations above climatological mean determined from AWAP observational data for years 1910–2012. (b) PDFs for CMIP5 historical (red, 1976–2005 only) and piControl (bold line, all years shown relative to long-term mean) simulations for Australian April–March standardized precipitation for above average rainfall in simulated La Niña years. The thresholds are based on observed rainfall under all ENSO conditions, so they appear in the dry tail of the distribution of above average simulated rainfall associated with La Niña year. (c) As for panel (b) but showing southeastern (SE) Australian standardized precipitation. Adapted from *Lewis and Karoly* [2014a]. *See electronic version for color representation.*

the Clausius-Clapeyron relation. Given that Australian-average temperatures have increased by 0.9°C since 1910 [*Bureau of Meteorology*, 2014b], anthropogenic influences on March 2012 and 2010–2012 two-year rainfall extremes have also been explored.

In one example, SE Australian Rx5day and monthly total rainfall amounts were calculated in historical simulations from ten CMIP5 models [*King et al.*, 2013b] (Table 15.1). Ten models were selected based on both availability of data in the Australian node of the CMIP5 data repository and their ability to capture variability in ENSO and observed teleconnections to Australia. By comparing the probability of simulated extreme Rx5day rainfall events between 1861–1890 (during which anthropogenic warming was negligible), and 1976–2005, this study found limited evidence for of a change in extreme rainfall itself, or of a change in the relationship between ENSO and SE Australia extreme rainfall, that may be attributed to anthropogenic climate change (Fig. 15.3b). This result was largely confirmed by examining March 2012 monthly rainfall totals using a state-of-the-art system for Attribution of extreme weather and Climate Events (ACE) [*Christidis et al.*, 2013a]. This approach uses two large ensembles of simulations (600) over the period of late 2011 and early 2012. One ensemble represents the actual climate where the effect of all external forcings is accounted for and the other provides various possible representations of a hypothetical "natural" climate without the effect of human influences. In these experiments, prescribed anthropogenic forcings are long-lived greenhouse gases, aerosols, tropospheric and stratospheric ozone, and land use changes; and natural forcings are from volcanic aerosols and solar irradiance.

Comparing the occurrence of above average rainfall in these ensembles demonstrates that the overall effect of long-term human influences on the climate increases the chances of above average rainfall by 5–15% (best estimate), though the impact on extreme March precipitation is much more uncertain [*Christidis et al.*, 2013b]. The March 2012 finding was largely duplicated by another study analyzing the 2010–2012 seasonal to 2 year Australian rainfall extremes using a suite of CMIP5 and ACE-type ensembles (Table 15.1) [*Lewis and Karoly*, 2014a]. Furthermore, in this study, anthropogenic influences on heavy rainfall were equivocal and found to depend on the thresholds of rainfall, seasons, and Australian regions considered (e.g., comparison of Fig. 15.4b,c). This study suggests that multiple model datasets should be used to

Table 15.1 Details of Major Datasets, Experiments and Relevant Forcings Used in Analyzing Australian Rainfall and Temperature Extremes Observed over 2010–2013

Datasets	Experiment	Major forcings	Summary of analyses
CMIP5	historicalNat	Solar, volcanics	Australian rainfall: *King et al.* [2013], *Lewis and Karoly* [2014]
	historical	Anthropogenic (greenhouse gases, aerosols, ozone) and natural (solar, volcanics)	Australian temperatures: *Lewis and Karoly* [2013], *Lewis and Karoly* [2014b]
	RCP8.5	Anthropogenic (greenhouse gases, aerosols, ozone scenarios) and natural (solar)	
	piControl	Nonevolving preindustrial forcings	
CAM5.1	All-Hist	Evolving ocean surface temperatures, greenhouse gas concentrations, and aerosol concentrations over the last 50 years	Australian rainfall: *Lewis and Karoly* [2014]
	NonGHG-Hist	Greenhouse gas concentrations maintained at preindustrial levels, with estimates of anthropogenic SST warming removed	
HadGEM3-A	ALL	Evolving ocean surface temperatures, greenhouse gas concentrations, and aerosol concentrations over the last 50 years	Australian rainfall: *Christidis et al.* [2013], *Lewis and Karoly* [2014a]
	NAT	Greenhouse gas and aerosol concentrations maintained at pre-industrial levels, with estimates of anthropogenic SST warming removed	
CESM	historical	Anthropogenic (greenhouse gases, aerosols, ozone) and natural (solar, volcanics)	Australian temperatures: *Perkins et al.* [2014]
	RCP8.5	Anthropogenic (greenhouse gases, aerosols, ozone scenarios) and natural (solar)	
	piControl	Nonevolving preindustrial forcings	

investigate possible anthropogenically driven changes in rainfall extremes of various timescales.

These studies had distinct approaches and foci, and accordingly provide different quantifications of relative influences. However, together they do provide a coherent understanding of the various influences on Australia during the period of exceptional rainfall of 2010–2012. The heavy rainfall during this period was likely predominantly due to natural climate variability. The strong La Niña conditions that were coincident with the extreme rainfall were an important contributing factor, but do not wholly explain the record Australian rainfall. Hence, despite several potential contributing factors (anthropogenic warming, Pacific variability, and anomalous local sea surface temperatures), the extreme magnitude of the rainfall events arose partly from unforced internal climatic variations.

15.3. AUSTRALIA'S TEMPERATURE EXTREMES OF 2013

15.3.1. Climatological Context

Temperatures in Australia during 2013 were extreme. The year 2013 was the warmest year for Australia-wide temperatures since high-quality national records began in 1910 (Fig. 15.5a). Extreme heat was persistent throughout the year and record-breaking temperatures occurred on spatial scales from single sites through to national averages, and on daily through to annual timescales [*Bureau of Meteorology*, 2014a]. In 2013, new Australia-wide record temperatures were set for the hottest day, week, month, season (Fig. 15.5b), and year on record. The temperature anomaly calculated for the September 2013 record is the largest for any month since at least 1910 (Fig. 15.5c) [*Bureau of Meteorology*, 2013a]. The extreme temperatures occurring in Australia in 2013 were part of an extended period of well above average temperatures that began in the austral spring of 2012 and persisted through 2014.

Furthermore, in late December 2012 and early January 2013, an exceptionally extensive and long-lived heat wave affected large parts of Australia [*Bureau of Meteorology*, 2013b]. The heat wave was a major factor in January 2013 being the hottest month in the observational record Australia-wide. Heat waves are an important aspect of the climate system that warrants further discussion, because of their disastrous impacts on many systems, including human health [*Perkins and Alexander*, 2013]. In 2009, for example, a heat wave over SE Australia was associated with 374 additional deaths, which was double the number of people killed by the bushfire that followed [*Victorian Department of Health*, 2009]. In this section, however, we focus particularly on reviewing the extreme monthly, seasonal, and annual Australian temperatures of 2013, as many spatially coherent new temperature records were set that have been investigated quantitatively.

The mechanisms behind theses seasonal- to annual-scale extreme Australian temperatures of 2013 have been explored previously by investigating the contributing roles of anthropogenic climate change and natural variability to the observed 2013 record-breaking temperatures. First, the extreme 2013 heat is considered in the context that Australian area-average annual temperatures have increased by 0.9°C since 1910 [*Bureau of Meteorology*, 2014b], which may lead to very large percentage changes in the occurrence of extremes [*Trenberth*, 2012]. In addition, natural climatic variability plays an important role in extreme events. El Niño–Southern Oscillation is a key determinant of interannual variability in Australian temperatures (as it is for rainfall) that is explored here for the 2013 Australian case study.

15.3.2. Understanding Seasonal and Annual Extremes of 2013

The contributing factors to several Australia-wide monthly, seasonal, and annual temperature records set in 2013 have been examined using a uniform approach. In this approach, the probability of an event occurring is calculated in a large ensemble of climate models and then compared to the equivalent event probability in a parallel counterfactual model experiment, where only natural climate forcings (e.g., volcanic aerosols and solar irradiance) are imposed [*Allen*, 2003]. The fraction of attributable risk (FAR) value provides a quantitative estimate of the change in probability of the defined event occurring that can be attributed to a particular cause, specifically the difference between the model experiments (i.e., anthropogenic climate forcings). The fraction of attributable risk (FAR) was calculated as

$$FAR = 1 - \frac{P_{NAT}}{P_{ALL}}$$

where P_{NAT} denotes the probability of an event's occurring in a natural reference state and P_{ALL} under a parallel forced state [*Stone and Allen*, 2005].

The FAR approach has been used to quantify anthropogenic contributions to the record hot Australia-wide maximum (Tmax) and mean (Tmean) temperatures in the austral summer of 2012–2013 [*Lewis and Karoly*, 2013], as well as the record hot September, spring and annual 2013 temperature [*Lewis and Karoly*, 2014b]. These estimates of the attributable likelihood of temperature extremes used, for the first time, an ensemble composed of multiple global climate models contributing realizations of standardized experiments to CMIP5 [*Taylor et al.*, 2012]. Models were selected for inclusion in

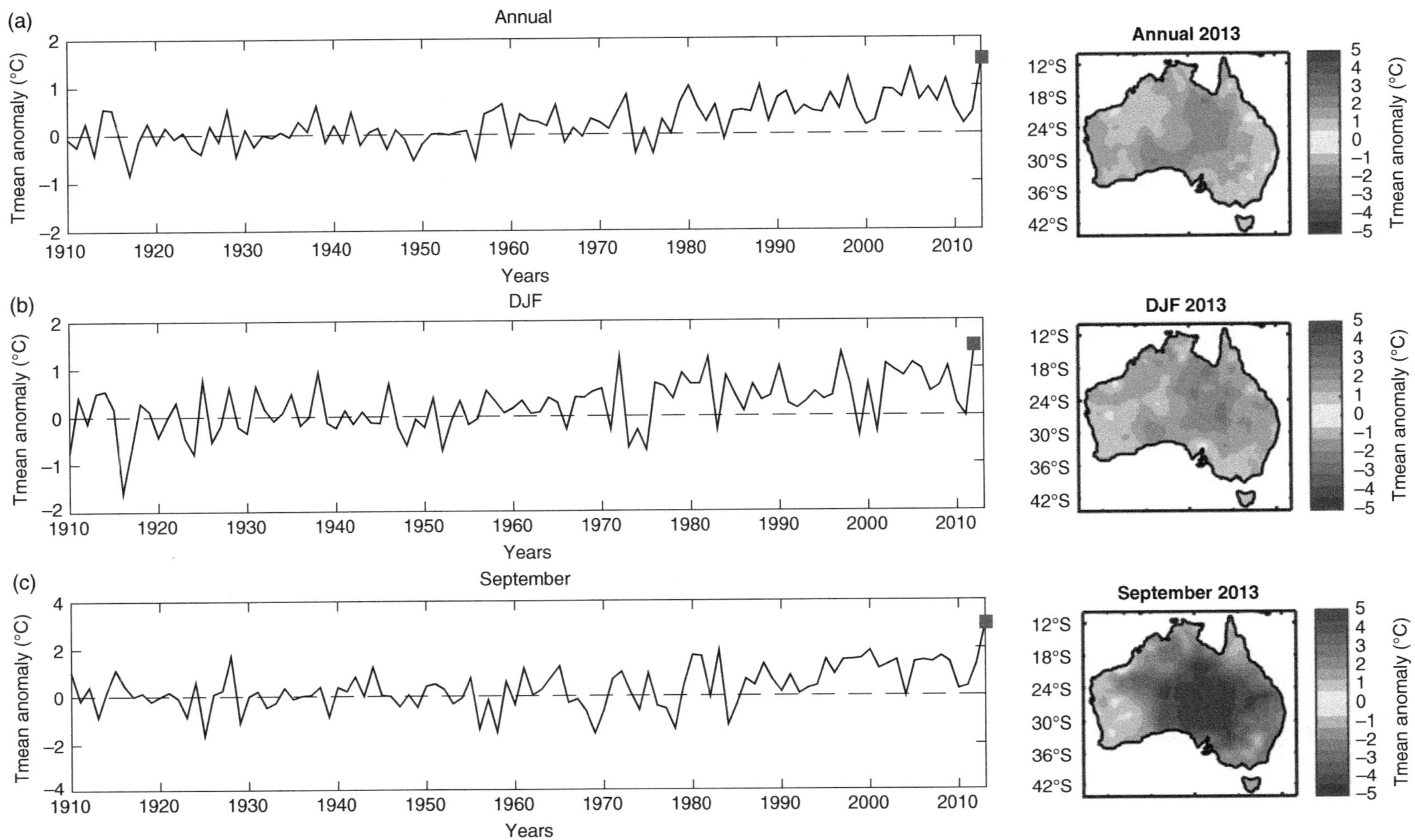

Figure 15.5 Observed Australian-average Tmean anomalies (°C, relative to 1911–1940) for (a) annual, (b) DJF, and (c) September averages from 1910–2013. The record anomalies of 2013 (square marker) are shown, together with spatial maps of average 2013 Australian conditions [*Jones et al.*, 2009]. Time series are derived from ACORN-SAT [*Trewin*, 2012]. *See electronic version for color representation.*

FAR analysis based on their skill in simulating observed Australian temperature variability [*Perkins et al.*, 2007]. The probability of Australian area-average temperatures exceeding an event magnitude threshold was calculated for simulations with natural forcings only (piControl and historicalNat) and compared to the probability in simulations with both anthropogenic (long-lived greenhouse gases, ozone, and aerosols) and natural (volcanic aerosols and solar) forcings imposed (historical and RCP8.5; Table 15.1).

Using a conservative temperature threshold (defined by the second hottest observed summer temperature anomaly), an estimate of the increase in likelihood of extreme Australian summer heat, attributable to anthropogenic influences, of at least fivefold (>90% confidence) was determined using model simulations for the most relevant years of 2006–2020 (Fig. 15.6a) [*Lewis and Karoly*, 2013]. Further analysis of the record hot spring (September–November) and September Australia-wide temperatures also indicated a substantial and statistically significant increase in likelihood of such extremes that is attributable to anthropogenic forcings [*Lewis and Karoly*, 2014b]. There is a 50-fold attributable increase in the likelihood of hot spring conditions and a 5-fold increase in the likelihood of hot September temperatures (Fig. 15.6b, c). The attributable change in likelihood of extreme annual average Australia-wide temperatures is even greater. The temperature anomaly observed in 2005 (the previous hottest Australian year on record) occurs only once in over 12 300 years of model simulations without anthropogenic forcings, but occurs approximately once in every 6 years in simulations with anthropogenic forcings imposed (Fig. 15.6d). The attributable risk is essentially 100%, and it is virtually impossible to reach such a temperature record due to naturally forced climate variability alone in these model simulations [*Knutson et al.*, 2014; *Lewis and Karoly*, 2014b].

Natural climate variability is also an important factor in extremes. ENSO is the main driver of Pacific climate variability [*Philander*, 1990], with El Niño conditions typically associated with warmer than average Australian temperature. For example, six of the eight hottest Australian summers on record, prior to 2013, were related to El Niño episodes. However, the record 2013 summer occurred following 2 years of exceptionally heavy rainfall associated with an extended La Niña event, after which sea surface temperatures in the NINO3.4 region returned to neutral conditions. Throughout 2013, ENSO conditions were predominantly neutral (observed NINO3.4 anomaly −0.16°C relative to 1981–2010) (http://www.ncdc.noaa. gov/sotc/global/2013/13), which is typically associated with near-normal Australian temperatures [*Power et al.*, 2006; *Arblaster and Alexander*, 2012]. Lewis *and Karoly* [2013] investigated changes in the likelihood of extreme Australian summer temperatures under varying ENSO conditions simulated in CMIP5 models. Modeled El Niño summers are characterized by generally warmer Australian average temperatures than neutral or La Niña summers. Using the conservative temperature anomaly threshold (the second hottest Australian summer observed), this study determined that extreme summer are at least three times more likely under El Niño conditions than under La Niña conditions. Hence, the ENSO neutral conditions during the 2013 are unlikely to explain the extreme seasonal and annual Australian heat in 2013.

Further analysis of the specific mechanisms behind the record Australian heat in September of 2013 using seasonal forecast models demonstrated that a combination of background warming, a strongly anomalous atmospheric circulation pattern, and dry and warm antecedent land-surface conditions contributed to this event [*Arblaster et al.*, 2014]. This alternative attribution approach supports the important role of anthropogenic climate change in the record Australian 2013 September heat.

15.3.3. Understanding Heat Waves of 2013

The exceptional heat waves of 2013 across Australia have also been investigated in terms of their relative contributing factors. Previous studies focused on observed temperature trends over the period of ~1950–2011 demonstrate a trend of increased frequency, intensity and duration of global heat waves and warm spells [*Perkins et al.*, 2012; *Perkins and Alexander*, 2013], including for Australia (Fig. 15.7a). The contributing influences on the Australian summer of 2012–2013 have specifically been investigated [*Perkins et al.*, 2014]. Heat waves were defined using the Excess Heat Factor (EHF) definition [*Nairn and Fawcett*, 2013; *Perkins and Alexander*, 2013] and two heat-wave characteristics, the total number of individual heat waves and the peak amplitude (hottest heat-wave day), were analyzed.

In this study, heat waves were calculated in a 21 member ensemble of simulations with the Community Earth System Model (CESM) [*Fischer and Knutti*, 2014] run under an experimental design including preindustrial control, historical and RCP8.5 simulations (Table 1) [*Taylor et al.*, 2012]. Using the fraction of attributable risk framework, *Perkins et al.* [2014] determined that the likelihood of both simulated heat-wave characteristics (heat-wave number and amplitude) has increased due to human activity (Fig. 15.7b,c). Specifically, the likelihood (> 90% confidence) of the total number of heat waves that occurred during the 2012–2013 Australian heat-wave summer has increased threefold due to anthropogenic forcings, while the likelihood of high heat-wave intensity

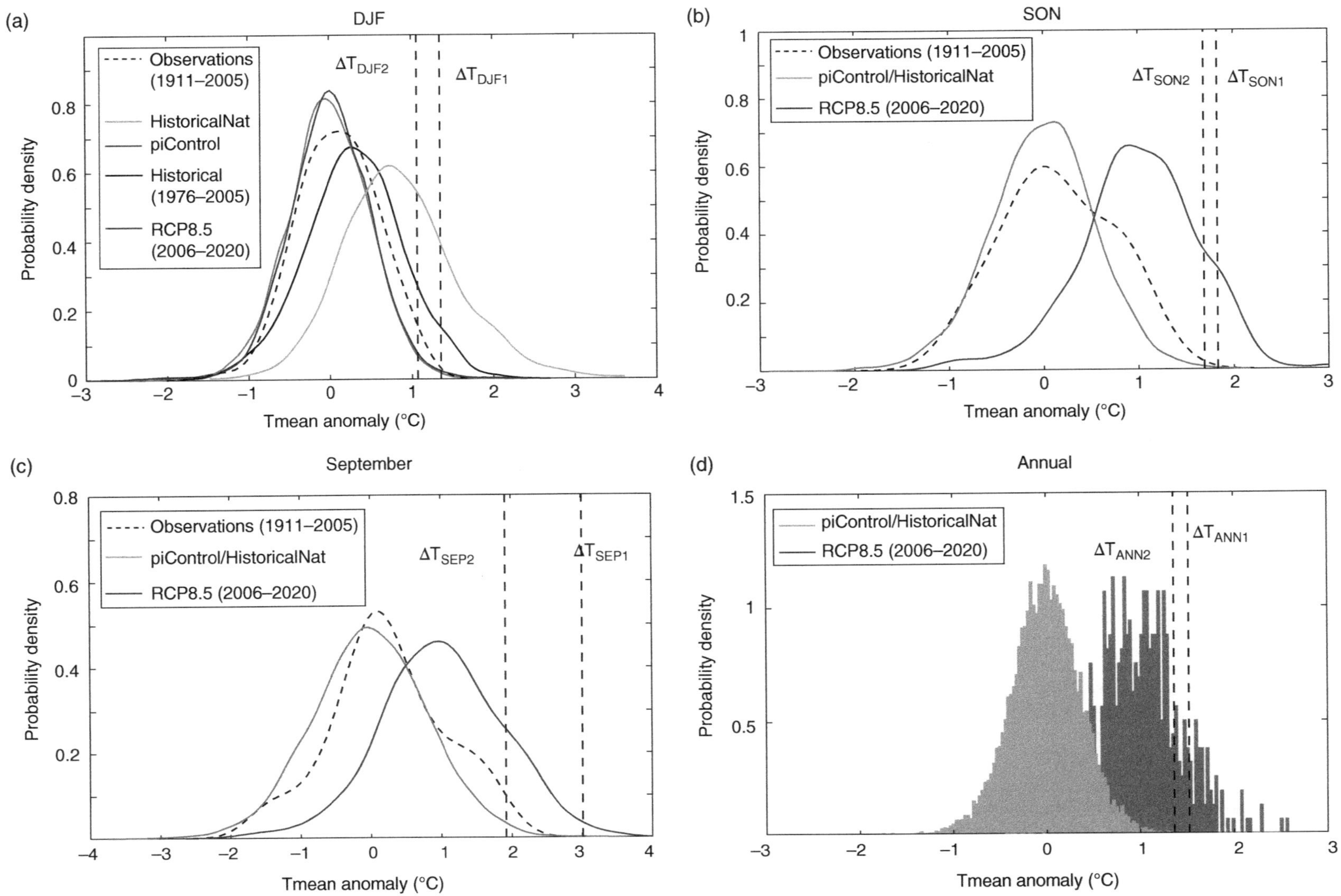

Figure 15.6 (a) Probability density functions for Australian summer (DJF) Tmean anomalies (°C, relative to 1911–1940) for observations (dashed), historicalNat (green, all years shown), piControl (dark blue, all years shown relative to long-term mean) and historical (black, 1976–2005 only) and RCP8.5 (black, 2006–2020 only) simulations. Vertical dashed lines show observed 2013 anomaly (ΔT_{DJF1}) and threshold of the second hottest summer on record (ΔT_{DJF2}). (b) As for panel (a) but showing Australian spring (SON) Tmean anomalies. (c) As for panel (a) but showing Australian September (SEP) Tmean anomalies. (d) Histograms for Australian annual average Tmean anomalies (°C, relative to 1911–1940) of RCP8.5 (red, years 2006–2020) and piControl and historicalNat (green, all years shown). Histograms are shown, rather than smoothed distributions, as there is only a single instance in the piControl and historicalNat simulations that exceeds ΔT_{ANN2}. Adapted from *Lewis and Karoly* [2013] and *Lewis and Karoly* [2014b]. (*See insert for color representation of the figure.*)

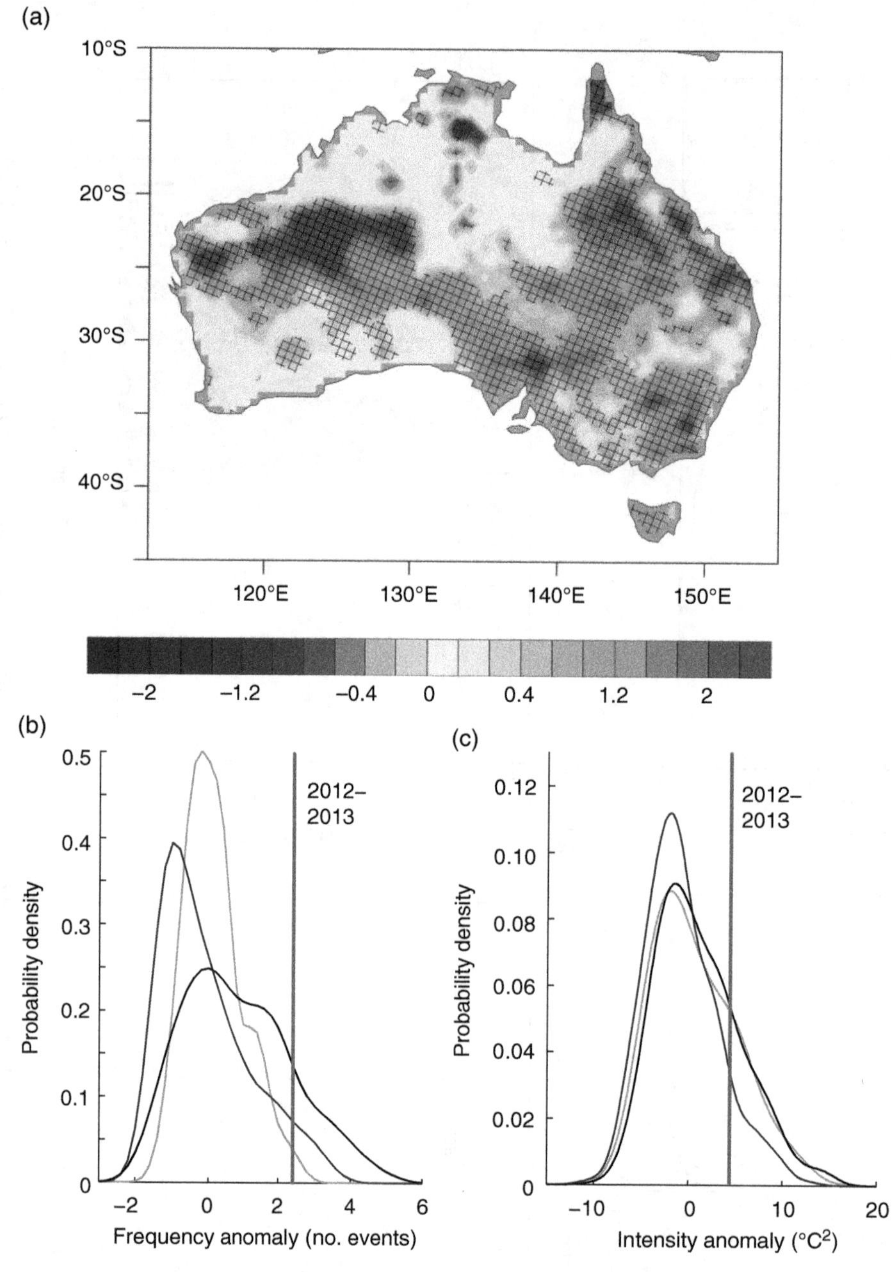

Figure 15.7 (a) Map of observed trends in the number of heat-wave days per season (November–March, days/decade) between 1950–2013. Hatching represents areas where trends are significant at the 5% level. (b) Probability density function for heat-wave frequency (number of heat waves), with light gray showing observations for 1955–2012, gray the CESM control ensemble, and solid black the historical/RCP8.5 simulations for 1984–2012. (c) As for panel (b) but showing heat-wave intensity (peak magnitude). Adapted from *Perkins et al.* [2012] and *Perkins et al.* [2014]. *See electronic version for color representation.*

increased twofold, compared to the simulated climate with no imposed anthropogenic forcings.

Heat waves are complex events that have multiple driving processes that should be considered in addition to anthropogenic contributions. These processes are generally not independent. Over southern Australia, where heat waves are the most intense [*Perkins and Alexander*, 2013] persistent high pressure systems adjacent to the area affected advect warm air from the interior of the continent [*Hudson et al.*, 2011; *Pezza et al.*, 2012; *Boschat et al.*, 2015]. In most cases, heat-wave-causing pressure systems over Australia are not considered classical blocking highs, but sit farther north from the blocking region and remain stationary until another, stronger system moves it [*Marshall et al.*, 2014]. Such a typically persistent high has been responsible for numerous extreme southeastern Australian heat waves, including the 2009 event over Victoria [*Hudson et al.*, 2011].

Heat waves over the southeast of Australia also have links to large-scale modes of climate variability. They are generally more common during particular phases (3–6) of the Madden-Julian Oscillation during summer, and La Niña phases of ENSO, when convection is enhanced over the north [*Parker et al.*, 2014]. Negative phases of the Southern Annular Mode (SAM) are also associated with southeastern heat waves during spring [*Marshall et al.*, 2014]. Over the north and east regions of Australia, heat waves are more frequent during El Niño [*Perkins et al.*, 2015]. Links between antecedent soil moisture and the intensity, frequency, and duration of heat waves have also been identified elsewhere [*Stott et al.*, 2004; *Dole et al.*, 2011].

15.4. SUMMARY AND CONCLUSIONS

We have discussed the mechanisms explaining the extreme rainfall affecting Australia in 2010 to 2012 and the extreme temperatures affecting Australia in 2013 from a climate perspective. The focus has been on considering factors such as El Niño–Southern Oscillation and human-caused climate change on the likelihood of such events. For the magnitude of extreme events, it is often the chaotic internal variability of weather and climate that is the major contributor.

Australia had its record highest 2 year rainfall in 2010–2011, as well as repeated shorter-term extreme rainfall events and damaging floods during the period 2010–2012. The extreme rainfall on these timescales was primarily associated with the combination of the influence of two consecutive strong La Niña events and chaotic climate variability. Any influence of climate change appears to be weaker and harder to distinguish from natural climate variability, as its estimated contribution varies between different climate models, different regions, and different periods. Other factors, such as the record high sea surface temperatures around northern Australia, may also have contributed but are hard to distinguish from the La Niña events and from human-caused climate change.

Australia experienced its record hottest year in 2013, together with repeated shorter-term temperature records and heat waves in different regions. Given that 2013 was not strongly influenced by El Niño or La Niña, ENSO was not a factor in these extreme temperatures, although El Niño is often associated with extreme temperatures and heat waves in northern and eastern Australia. Human-caused climate change was an important contributor to the monthly, seasonal, and annual extreme temperatures across Australia in 2013, together with chaotic natural climate variability. The relative importance of climate change or natural variability depends on the specific event being considered. Given that human-cased climate change has likely led to an increase of temperature across Australia by about 1°C, depending on the region, then its contribution can be estimated by considering the magnitude of the event and of natural variability. For the annual mean temperature averaged across Australia, the full range of natural temperature variations is less than about 2°C (Fig. 15.6d). Hence, the contribution to the record temperature in 2013 from climate change as well as natural variations in 2013 meant that it was virtually impossible to occur due to natural variability alone. However, for other temperature extremes at shorter time and space scales, such as monthly average temperatures (Fig. 15.6c) or heat waves (Fig.15.7c), the magnitude of natural variability is much larger and climate change makes a smaller contribution to the magnitude of the extreme events, although it makes a larger contribution to increasing their frequency. The drought across much of northeast Australia in 2013 and associated low soil moisture, parts of the natural climate variations, also played an important role in the Australian extreme temperatures in 2013 [*King et al.*, 2014a].

Given the understanding of roles of chaotic natural variability, El Niño and La Niña, and human-caused climate change on Australian climate extremes, we can consider likely future changes in climate extremes in Australia. Natural variability and ENSO will continue to greatly affect temperature and rainfall extremes. Climate change is expected to continue to increase the frequency and intensity of extreme high temperatures across Australia in most years [*CSIRO and Bureau of Meteorology*, 2015], although natural variability and strong La Niña events can overcome these trends and lead to cooler years. Climate change is also expected to lead to more intense heavy rainfall events at daily and subdaily timescales across Australia [*CSIRO and Bureau of Meteorology*, 2015], although in individual years the influence of ENSO is still likely to dominate.

With the major impacts of temperature and rainfall extremes on Australia, this knowledge of likely future changes in these extremes can be used to plan for and adapt to these anticipated changes in extremes.

ACKNOWLEDGMENTS

This research was supported by the ARC Centre of Excellence for Climate System Science (grant CE 110001028), an ARC DECRA to S.C.L (DE160100092). We acknowledge the NCI National Facility; the Bureau of Meteorology and CSIRO for providing AWAP and ACORN-SAT data; Dáithí Stone at the Lawrence Berkeley National Laboratory and Nikos Christidis and Peter Stott at the Met Office Hadley Centre for providing access to ACE-type data; the WCRP's Working Group on Coupled Modelling, which is responsible for CMIP. The US Department of Energy's PCMDI provides CMIP5 coordinating support.

REFERENCES

Allen, M. (2003), Liability for climate change, *Nature*, *421*(6926), 891–892; doi:10.1038/421891a.

Arblaster, J. M., and L. V. Alexander (2012), The impact of the El Niño–Southern Oscillation on maximum temperature extremes, *Geophys. Res. Lett.*, *39*(20), L20702; doi:10.1029/2012GL053409.

Arblaster, J. M., E.-P. Lim, H. H. Hendon, B. C. Trewin, M. C. Wheeler, G. Liu, and K. Braganza (2014), Understanding Australia's hottest September on record, in Explaining Extremes of 2013 from a Climate Perspective, *Bull. Amer. Meteor. Soc.*, *95*(9), S37–S41.

Boschat, G., A. Pezza, I. Simmonds, S. Perkins, T. Cowan, and A. Purich (2015), Large scale and subregional connections in the lead up to summer heat wave and extreme rainfall events in eastern Australia, *Climate Dyn.*, *44*(7–8), 1823–1840; doi:10.1007/s00382-014-2214-5.

Bureau of Meteorology (2012a), Australia's wettest two-year period on record; 2010–2011, *Special Climate Statement 38;* http://www.bom.gov.au/climate/current/statements/.

Bureau of Meteorology (2012b), Exceptional heavy rainfall across southeast Australia, *Special Climate Statement 39*; http://www.bom.gov.au/climate/current/statements/.

Bureau of Meteorology (2013a), Australia's warmest September on record, *Special Climate Statement 46;* http://www.bom.gov.au/climate/current/statements/.

Bureau of Meteorology (2013b), Extreme heat in January 2013, *Special Climate Statement 43;* http://www.bom.gov.au/climate/current/statements/.

Bureau of Meteorology (2014a), *Annual Climate Report 2013;* http://www.bom.gov.au/climate/annual_sum/2013/.

Bureau of Meteorology (2014b), *State of the Climate 2014*, edited by Bureau of Meteorology and CSIRO.

Cai, W., P. van Rensch, T. Cowan, and A. Sullivan (2010), Asymmetry in ENSO teleconnection with regional rainfall, its multidecadal variability, and impact, *J. Climate*, *23*(18), 4944–4955, doi:10.1175/2010JCLI3501.1.

Christidis, N., P. A. Stott, A. A. Scaife, A. Arribas, G. S. Jones, D. Copsey, J. R. Knight, and W. J. Tennant (2013a), A new HadGEM3-A-based system for attribution of weather- and climate-related extreme events, *J. Climate*, *26*(9), 2756–2783, doi:10.1175/JCLI-D-12-00169.1.

Christidis, N., P. A. Stott, D. J. Karoly, and A. Ciavarella (2013b), An Attribution Study of the Heavy Rainfall over Eastern Australia in March 2012, *Bull. Amer. Meteor. Soc*, *94*(9), S58–S61.

CSIRO and Bureau of Meteorology (2015), *Climate Change in Australia Information for Australia's Natural Resource Management Regions: Technical Report*, CSIRO and Bureau of Meteorology, Australia.

Dole, R., M. Hoerling, J. Perlwitz, and J. Eischeid (2011), Was there a basis for anticipating the 2010 Russian heat wave? *Geophys. Res. Lett.*; doi:10.1029/2010GL046582.

Donat, M. G., L. V. Alexander, H. Yang, I. Durre, R. Vose, and J. Caesar (2013), Global land-based datasets for monitoring climatic extremes, *Bull. Amer. Meteor. Soc.*, *94*(7), 997–1006, doi:10.1175/BAMS-D-12-00109.1.

Evans, J. P., and I. Boyer-Souchet (2012), Local sea surface temperatures add to extreme precipitation in northeast Australia during La Niña, *Geophys. Res. Lett.*, *39*(10), doi:10.1029/2012GL052014.

Fischer, E. M., and R. Knutti (2014), Detection of spatially aggregated changes in temperature and precipitation extremes, *Geophys. Res. Lett.*; doi:10.1002/2013GL058499.

Ganter, C., and S. Tobin (2013), Australia, in *State of the Climate in 2012*, edited by J. Blunden and D. S. Arndt, *Bull. Amer. Meteor. Soc.*, *94*(8), S196–S198.

Held, I. M., and B. J. Soden (2006), Robust responses of the hydrological cycle to global warming, *J. Climate*, *19*, 5686–5699; doi:10.1175/JCLI3990.1.

Hendon, H. H., E.-P. Lim, J. M. Arblaster, and D. L. T. Anderson (2013), Causes and predictability of the record wet east Australian spring 2010, *Climate Dyn.*; doi:10.1007/s00382-013-1700-5.

Hudson, D., A. G. Marshall, and O. Alves (2011), Intraseasonal forecasting of the 2009 summer and winter Australian heat waves using POAMA, *Weath. Forecast.*, *26*(3), 257–279; doi:10.1175/WAF-D-10-05041.1.

Jones, D. A., W. Wang, and R. Fawcett (2009), High-quality spatial climate data-sets for Australia, *Austral. Meteor. Oceanog. J.*, *58*(4), 233.

Karl, T. R., and R. W. Knight (1998), Secular trends of precipitation amount, frequency, and intensity in the United States, *Bull. Amer. Meteor. Soc.*, *79*(2), 231–241.

Karoly, D. J., and K. Braganza (2005), Attribution of recent temperature changes in the Australian region, *J. Climate*, *18*(3), 457–464, doi:10.1175/JCLI-3265.1.

King, A. D., D. J. Karoly, M. G. Donat, and L. V. Alexander (2014a), Climate change turns Australia's 2013 big dry into a year of record-breaking heat, in Explaining Extremes of 2013 from a Climate Perspective, *Bull. Amer. Meteor. Soc.*, *95*(9), S41–45.

King, A. D., L. V. Alexander, and M. G. Donat (2013a), Asymmetry in the response of eastern Australia extreme rainfall to low-frequency Pacific variability, *Geophys. Res. Lett.*, *40*(10), 1–7; doi:10.1002/grl.50427.

King, A. D., N. P. Klingaman, and L. V. Alexander (2014b), Extreme rainfall variability in Australia: Patterns, drivers, and predictability, *J. Climate*,, doi:10.1175/JCLI-D-13-00715.s1.

King, A. D., S. C. Lewis, S. E. Perkins, L. V. Alexander, M. G. Donat, D. J. Karoly, and M. T. Black (2013b), Limited evidence of anthropogenic influence on the 2011–2012 extreme rainfall over Southeast Australia, edited by T. C. Peterson, M. P. Hoerling, P. A. Stott, and S. C. Herring, *Bull. Amer. Meteor. Soc.*, *94*(9), S55–S58.

Knutson, T. R., F. Zeng, and A. T. Wittenberg (2014), Multimodel assessment of extreme annual-mean warm anomalies during 2013 over regions of Australia and the western tropical Pacific, in Explaining Extremes of 2013 from a Climate Perspective, *Bull. Amer. Meteor. Soc.*, *95*(9), S26–S30.

Lewis, S. C., and D. J. Karoly (2013), Anthropogenic contributions to Australia's record summer temperatures of 2013, *Geophys. Res. Lett.*, doi:10.1002/grl.50673.

Lewis, S. C., and D. J. Karoly (2014a), Are estimates of anthropogenic and natural influences on Australia's extreme 2010–2012 rainfall model-dependent?, *Climate Dyn.*, doi:10.1007/s00382-014-2283-5.

Lewis, S. C., and D. J. Karoly (2014b), The role of anthropogenic forcing in the record 2013 Australia-wide annual and spring temperatures, in Explaining Extremes of 2013 from a Climate Perspective, *Bull. Amer. Meteor. Soc.*, *95*(9), S31–S34.

MacKellar, D. (1911), *The Closed Door and Other Verses*, Australasian Authors' Agency.

Marshall, A. G., D. Hudson, M. C. Wheeler, O. Alves, H. H. Hendon, M. J. Pook, and J. S. Risbey (2014), Intra-seasonal drivers of extreme heat over Australia in observations and POAMA-2, *Climate Dyn.*, *43*(7-8), 1915–1937, doi:10.1007/s00382-013-2016-1.

McBride, J. L., and N. Nicholls (1983), Seasonal relationships between Australian rainfall and the southern oscillation, *Monthly Weather Rev.*, *111*, 1998–2004, doi:10.1175/1520-0493(1983)111<1998:SRBARA>2.0.CO;2.

Min, S. K., W. Cai, and P. Whetton (2013), Influence of climate variability on seasonal extremes over Australia, *J. Geophys. Res. Atmos.*, doi:10.1002/jgrd.50164.

Nairn, J. R., and R. Fawcett (2013), Defining heatwaves: Heatwave defined as a heat- impact event servicing all community and business sectors in Australia, CAWCR Technical Report No. 060.

Nicholls, N., W. Drosdowsky, and B. Lavery (1997), Australian rainfall variability and change, *Weather*, *52*(3), 66–72, doi:10.1002/j.1477-8696.1997.tb06274.x.

Parker, T. J., G. J. Berry, and M. J. Reeder (2014), The structure and evolution of heat waves in southeastern Australia, *J. Climate*, *27*(15), 5768–5785, doi:10.1175/JCLI-D-13-00740.1.

Perkins, S. E., and L. V. Alexander (2013), On the measurement of heat waves, *J. Climate*, doi:10.1175/JCLI-D-12-00383.1.

Perkins, S. E., A. J. Pitman, N. J. Holbrook, and J. McAneney (2007), Evaluation of the AR4 climate models' simulated daily maximum temperature, minimum temperature, and precipitation over Australia using probability density functions, *J. Climate*, *20*(17), 4356–4376, doi:10.1175/JCLI4253.1.

Perkins, S. E., D. Agueso, and C. J. White (2015), Relationships between climate variability, soil moisture and Australian heatwaves, *J. Geophys. Res. Atmos.*, *120* (16), 8144–8164.

Perkins, S. E., L. V. Alexander, and J. R. Nairn (2012), Increasing frequency, intensity and duration of observed global heatwaves and warm spells, *Geophys. Res. Lett.*, *39*(20), doi:10.1029/2012GL053361.

Perkins, S. E., S. C. Lewis, A. D. King, and L. V. Alexander (2014), Increased simulated risk of the hot Australian summer of 2012-2013 due to anthropogenic activity as measured by heatwave frequency and intensity, in Explaining Extremes of 2013 from a Climate Perspective, *Bull. Amer. Meteor. Soc.*, *95*(9), S34–S37.

Pezza, A. B., P. van Rensch, and W. Cai (2012), Severe heat waves in southern Australia: synoptic climatology and large scale connections, *Climate Dyn.*, *38*(1–2), 209–224, doi:10.1007/s00382-011-1016-2.

Philander, S. G. (1990), *El Niño, La Niña, and the Southern Oscillation*, Academic Press.

Power, S., M. Haylock, R. Colman, and X. Wang (2006), The predictability of interdecadal changes in ENSO activity and ENSO teleconnections, *J. Climate*, *19*(19), 4755–4771.

Risbey, J. S., M. J. Pook, P. C. McIntosh, M. C. Wheeler, and H. H. Hendon (2009), On the remote drivers of rainfall variability in Australia, *Monthly Weather Rev.*, *137*(10), 3233–3253, doi:10.1175/2009MWR2861.1.

Stone, D. A., and M. R. Allen (2005), The end-to-end attribution problem: From emissions to impacts, *Climatic Change*, *71*(3), 303–318, doi:10.1007/s10584-005-6778-2.

Stott, P. A., D. A. Stone, and M. R. Allen (2004), Human contribution to the European heatwave of 2003, *Nature*, *432*(7017), 610–614, doi:10.1038/nature03089.

Taylor, K. E., R. J. Stouffer, and G. A. Meehl (2012), An overview of CMIP5 and the experiment design, *Bull. Amer. Meteor. Soc.*, *93*(4), 485, doi:10.1175/BAMS-D-11-00094.1.

Trenberth, K. E. (2012), Framing the way to relate climate extremes to climate change, *Climatic Change*, *115*(2), 283–290, doi:10.1007/s10584-012-0441-5.

Trewin, B. (2012), A daily homogenized temperature data set for Australia, *Int. J. Climatol.*, doi:10.1002/joc.3530.

Ummenhofer, C. C., A. Sen Gupta, M. H. England, A. S. Taschetto, P. R. Briggs, and M. R. Raupach (2015). How did ocean warming affect Australian rainfall extremes during the 2010/2011 La Niña event?, *Geophys. Res. Lett.*, *42*, 9942–9951. http://doi.org/10.1002/(ISSN)1944-8007

van den Besselaar, E. J. M., A. M. G. Klein Tank, and T. A. Buishand (2012), Trends in European precipitation extremes over 1951–2010, *Int. J. Climatol.*; doi:10.1002/joc.3619.

Victorian Department of Health (2009), *January 2009 Heatwave in Victoria: An Assessment of Health Impacts*, Victorian Government Department of Human Services, Melbourne, Victoria.

Westra, S., L. V. Alexander, and F. W. Zwiers (2013), Global increasing trends in annual maximum daily precipitation, *J. Climate*, *26*(11), 3904–3918, doi:10.1175/JCLI-D-12-00502.1.

16

Unraveling East Africa's Climate Paradox

Bradfield Lyon[1] and Nicolas Vigaud[2]

ABSTRACT

An observed drying trend and projected increase in rainfall place East Africa in an apparent climate paradox. It is argued that the observational evidence, backed by climate modeling experiments, support the case that the recent regional scale drying trend in East Africa was dominated by an abrupt decline in March–May rainfall around 1998–1999 and associated with a shift of Pacific decadal variability (PDV) to its cold phase. While it is possible a warming trend in the western tropical Pacific contributed to the severity of recent droughts, the timing of the recent decline is closely tied to PDV. Thus, the paradox is seen to result from a mismatch of timescales on one hand (recent drying versus long-term wettening) and from taking climate change projections at face value on the other. The inability of coupled models to capture fundamental aspects of the current climate undermines confidence in the future climate projections in East Africa. Given substantial errors in coupled climate models' ability to simulate the current climate, it remains an open question as to whether or not the future climate in East Africa will become wetter as a result of anthropogenic climate change.

16.1. INTRODUCTION

Several East African countries are currently facing something of a climate paradox [*Lyon*, 2014]. For well over a decade, the region has been experiencing an increased frequency of drought (Fig. 16.1a), the most recent and severe of these events (in 2010–2011) triggering a humanitarian crisis and prompting a major international response [*FEWSNET*, 2011]. This rainfall decline has led to inevitable speculation that an anthropogenic climate change signal has begun to emerge in the region. Yet, paradoxically, climate model projections show a general consensus that increasing greenhouse gasses will in fact result in East Africa's climate becoming *wetter* during the current century (Fig. 16.1b) [*Christensen et al.*, 2007; *Shongwe et al.*, 2011]. Confounding this seeming paradox, in many parts of East Africa the rainfall seasonal cycle is bimodal, with the bulk of the recent rainfall decline occurring during the "long rains" season of March–May (MAM; other seasons denoted similarly) while the "short rains" of October–December (OND) show an upward trend over the past three decades [e.g., *Liebmann et al.*, 2014].

These divergent outcomes highlight the need to identify the physical mechanisms associated with both the observed and projected changes in East African climate. From a physical perspective, a true paradox is of course illusory and several hypotheses can be contemplated to account for the differing trends. For example, the coupled climate models (CMIP5) used to project the future climate may be missing, or misrepresenting, key physical processes important to East African rainfall. The future may indeed become drier and the model projections are simply wrong. Alternatively, the recent drying may be a

[1]School of Earth and Climate Sciences and Climate Change Institute, University of Maine, Orono, Maine, USA

[2]International Research Institute for Climate and Society, The Earth Institute, Columbia University, Palisades, New York, USA

Climate Extremes: Patterns and Mechanisms, Geophysical Monograph 226, First Edition.
Edited by S.-Y. Simon Wang, Jin-Ho Yoon, Christopher C. Funk, and Robert R. Gillies.

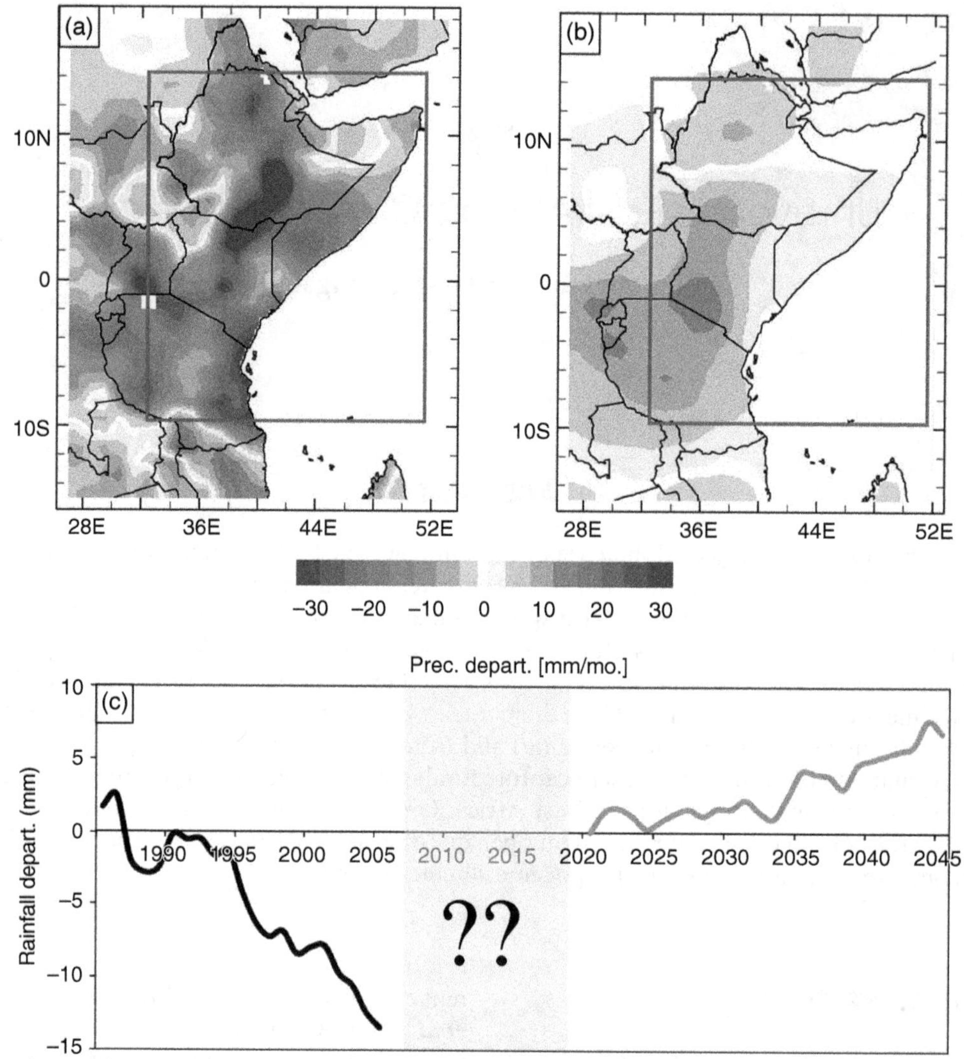

Figure 16.1 (a) Difference in MAM rainfall (mm/mo) between the periods 1999–2010 and 1977–1998. (b) Multimodel mean CMIP5 change in MAM rainfall for 2035–2050 minus 1970–2000. (c) 9 year moving averages of MAM rainfall anomalies for East Africa (boxed region) in observations (GPCC, black line) and in the multimodel mean of CMIP5 projections (grey line), the latter arbitrarily set to zero in 2020.

manifestation of lower-frequency natural climate variability, with the longer-term response to increasing greenhouse gases leading to more rainfall. A third hypothesis is that natural climate variations and anthropogenic climate change already underway may have constructively interfered, the latter increasing the severity of recent rainfall anomalies in East Africa while the future remains uncertain. Other hypotheses can of course be envisioned.

This chapter attempts to unravel some aspects of the apparent East African climate paradox by focusing primarily on the recent rainfall decline. The role of Pacific decadal variability will be discussed along with its possible interaction with long-term trends in sea surface temperatures (SSTs), particularly in the western equatorial Pacific. Some important errors in CMIP5 model simulations of the current East African climate will be detailed, which have substantial implications for interpreting future climate changes in the region. While these topics fall squarely within the realm of climate science, it is stressed that an increased understanding of East African climate variability and change has substantial ramifications outside of that sphere. East Africa is one of the most food-insecure areas of the world [*FAO et al.*, 2014]. The severe drought of 2010–2011 in Kenya, Somalia, and southeastern Ethiopia, for example, affected some 10 million people and was a contributing factor to more than 250,000 fatalities in Somalia alone [*UN OCHA*, 2011; *Checchi and Robinson*, 2013]. The hope is that an improved scientific understanding of East African climate

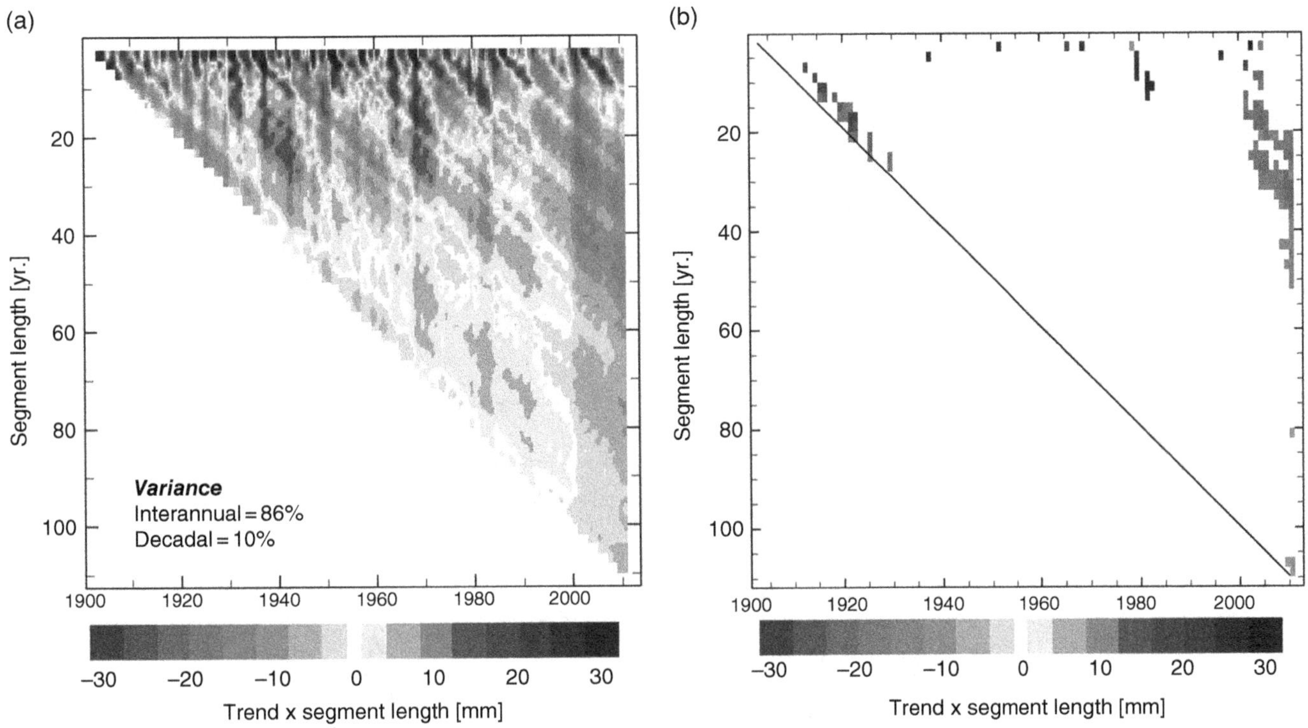

Figure 16.2 (a) Linear trend multiplied by segment length over which it was computed (shading, mm) for the East African MAM long rains in GPCC. Vertical axis is segment length, horizontal axis is the end date of fitted trend. The percent of variance for interannual and decadal scale fluctuations is shown at the lower left. (b) Statistically significant values ($p < 0.05$) based on a two-tailed t test. Adapted from *Lyon* [2014]. (*See insert for color representation of the figure.*)

will ultimately translate into better preparedness in order to mitigate impacts from the climate extreme of drought.

16.2. THE NATURE OF THE RECENT EAST AFRICAN LONG RAINS DECLINE

16.2.1. Evidence of an Abrupt Shift in Regional MAM Rainfall

Secular trends in climate data time series are ubiquitous as the climate system is continuously varying on multiple timescales. An illustration of this behavior for East African MAM seasonal rainfall is provided in Figure 16.2, which displays linear trends in precipitation averaged over the boxed region in Figure 16.1 (land areas only) as a function of (varying) segment length and end date. The underlying data are from the Global Precipitation Climatology Center (GPCC) version 6 [*Schneider et al.*, 2008] and the methodology follows that of *Liebmann et al.* [2010]. Large interannual variations in rainfall clearly affect short-term trends (e.g., segments < 10 years) in Figure 16.2, with decadal to multidecadal fluctuations also evident on longer timescales. In terms of rainfall variance (1901–2010), MAM rainfall is dominated by interannual fluctuations (86% of total variance) while decadal variations (based on a 9 year moving average) account for 10%. Of particular interest are the negative trends for segment lengths > ≈10 years starting around the year 2000. Statistically significant ($P < 0.05$) downward trends for time segments of around 15–30 years occur after that time (Fig. 16.2b). The abruptness of these changes suggests a rapid transition to very dry MAM seasons in recent years.

Evidence for an abrupt decline in MAM rainfall in East Africa is revealed in several different datasets. Since the number of stations in the GPCC dataset varies with time and may affect the analysis, outgoing longwave radiation (OLR) was also used here as a proxy for rainfall, the latter also having the desirable attribute of being a completely independent dataset. (Note, outgoing longwave radiation data are from the US Climate Prediction Center and downloaded via the International Research Institute for Climate and Society [IRI] Data Library at http://iridl.ldeo.columbia.edu/SOURCES/.NOAA/.NCEP/.CPC/.GLOBAL/.monthly/.olr.)

Results for both variables are shown in Figure 16.3, the top panel first showing the difference in monthly rainfall averaged across East Africa (boxed region in Fig. 16.1)

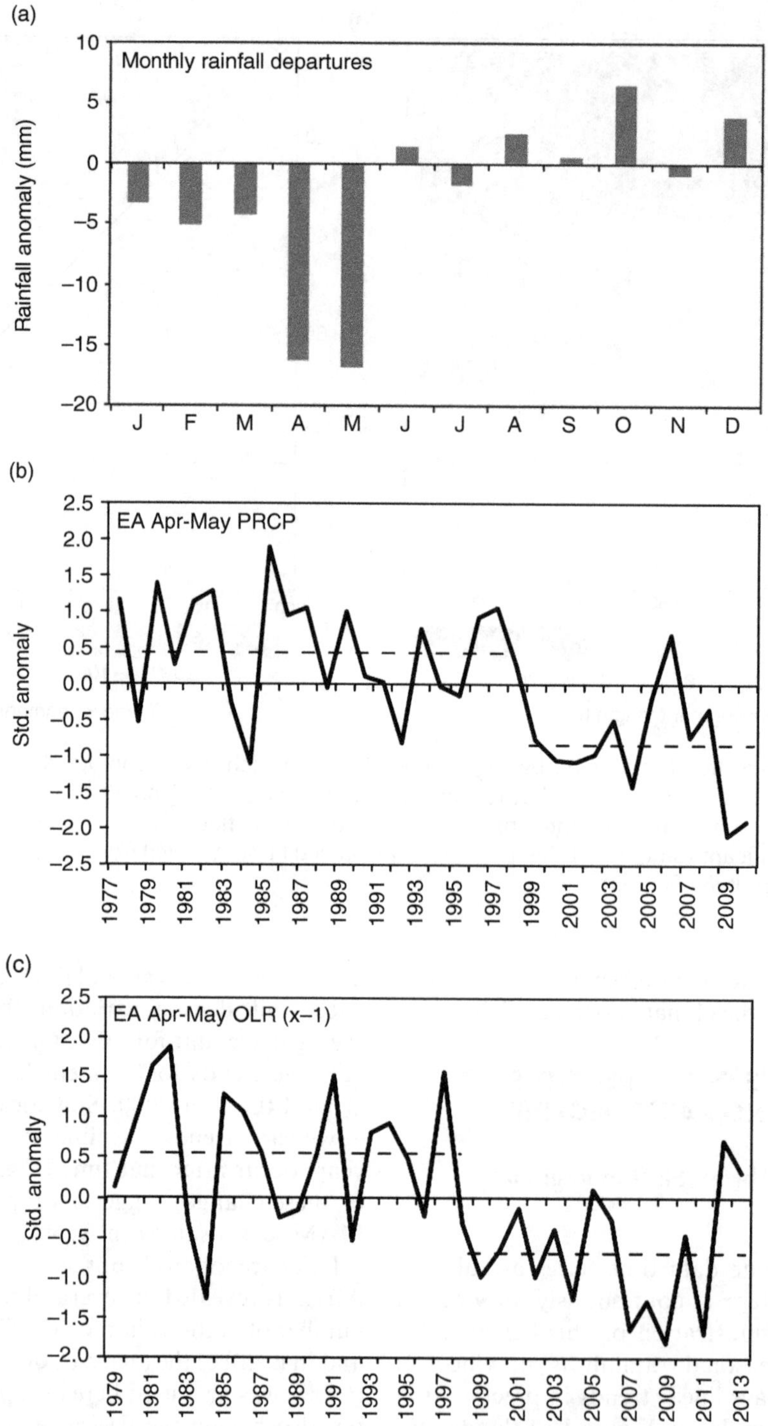

Figure 16.3 (a) Monthly rainfall differences for East Africa (mm) between the periods 1999–2013 and 1977–1998. (b) April–May season standardized rainfall departures from a 1977–2010 mean. Dashed horizontal lines indicate mean values for two epochs before and after a change point in 1999. c) As in (b), but for OLR anomalies (x -1). The most likely shift in (c) occurs in 1998.

between the periods 1999–2010 and 1977–1998 (using GPCC data). April and May are seen to have the greatest negative departures while the OND season, on average, exhibits relative *wettening* during the most recent period. Why were these two specific periods chosen? Figure 16.3b and c show time series of standardized MAM rainfall and OLR anomalies (multiplied by -1) averaged across East Africa (land areas only) for past three-plus decades. Apparent shifts in both these series occur around 1998–1999. As an objective means of identifying such potential change points, the RHtestV4 software package [*Wang*, 2008] was used. While this and related tools are typically used to identify nonclimate (and thus, spurious) break points in time series, here it is used to identify likely shifts that will be related to decadal variations in the climate system. Using this tool, a statistically significant change point ($p < 0.05$) was identified in GPCC rainfall, most likely occurring in 1999. The OLR data also exhibit a statistically significant shift ($p < 0.01$), most likely occurring in 1998. The linear correlation between the OLR and GPCC time series is $r = 0.8$ indicating the latter is able to capture a significant portion of the GPCC rainfall variance in addition to a near-synchronous shift in mean behavior. As mentioned, these shifts are argued to reflect physical changes in the climate system, with evidence for their association with Pacific decadal variability (PDV) summarized in section 16.3. The time series in Figure 16.3 provides a cautionary note when interpreting trends in temporally smoothed time series: the average across a "step" yields a "ramp," which can be misinterpreted as a longer-term trend rather than a more abrupt change.

16.2.2. Recent Changes in the Seasonality of the MAM Long Rains

Changes in the seasonal cycle of the long rains were examined using the Climate Hazards Group InfraRed Precipitation with Station data (CHIRPS) gridded daily rainfall dataset [*Funk et al.*, 2014]. The CHIRPS rainfall data are based on satellite estimates combined with in situ station data to provide some calibration. The data used were on a 0.25° lat. × lon. grid and covered the period 1981–2014. As a first step, a 1981–2010 base period monthly climatology was generated and used to identify grid points in the East African domain where the climatological mean MAM rainfall was less than 25% of the mean annual total (indicating a relative dry season). Such dry season locations (representing 23% of land area grid points) were dropped before undertaking subsequent computations. Figure 16.4a shows the accumulated, mean daily rainfall from 1 March to 31 May averaged over the periods 1981–1998 and 1999–2014 and for the seven driest MAM seasons during the post-1998 period identified in the CHIRPS data (2000, 2001, 2004, 2007, 2008, 2009, 2011). The figure indicates the expected lower rainfall accumulations for the post-1998 period, but differences in accumulations from the earlier period do not emerge until April, generally consistent with the results in Figure 16.3a, which was based on a different dataset (and where there was no masking for seasonality). For the seven driest MAM seasons in the post-1998 period, relative deficits emerge earlier, during the month of March.

Further changes in the seasonal cycle of the East African long rains are shown in Figure 16.4b, which displays average daily rainfall differences from a February to June mean value (using a 1981–1998 base period). A 15 day running average has been applied to smooth all three series shown in the figure. Using a zero threshold crossing as a simple benchmark for the regional onset of the long rains, this crossing is seen to occur in mid-March during both the 1981–1998 and 1999–2014 periods. Thus, by this definition, there is no systematic shift identified in the onset date of the long rains during the post-1998 period. What is seen, however, is an earlier cessation of the long rains during recent years. This is consistent with an earlier onset of the Indian summer monsoon [e.g., *Camberlin et al.*, 2010; *Kajikawa* et al,, 2012]. For the seven driest years in the post-1998 period, Figure 16.4b shows both a delayed onset and early demise of the long rains. *Camberlin and Okoola* [2003] found that a shorter rainy season is generally associated with reduced total seasonal rainfall, which was certainly the case in these recent drought years.

Overall, there is observational evidence from independent datasets that the recent decline in MAM rainfall in East Africa was associated with an abrupt downward shift around 1998–1999. Subseasonal changes in MAM are noted as well, most notably an earlier cessation of the rainy season (on the regional scale), which during the worst of the recent drought years was also accompanied by a delayed onset. The greatest contribution to the seasonal decline was a marked decrease in rainfall during the months of April and May. While abrupt and statistically significant shifts in East African rainfall and OLR are identified, this begs the question of an underlying cause. Their linkage to PDV is discussed in the next section.

16.3. LINKS TO PACIFIC DECADAL VARIABILITY

Supporting evidence that the abrupt shift in the East African long rains was associated with PDV is provided in Figure 16.5, which shows time series of detrended MAM rainfall anomalies (from GPCC) averaged over East African land areas (boxed region in Fig. 16.1) and the principal component time series of the leading mode of SST variations in the Pacific Ocean after removing the long-term trend and instantaneous influence of the El Niño–Southern Oscillation [from *Lyon*, 2014]. The latter SST index is significantly correlated ($r = 0.7$, $p < 0.05$) with the Pacific Decadal Oscillation (PDO) index of *Mantua et al.* [1997], with both indicating nearly synchronous phase shifts, including a shift to the cold phase in 1998–1999. Both time series in Figure 16.5 have been smoothed with a 9 year moving average and cover the period 1905–2005. Coherent temporal variations between these variables are seen across the full analysis period, the temporal correlation being $r = 0.72$, which is statistically significant at $p < 0.01$ based on a two-tailed t-test.

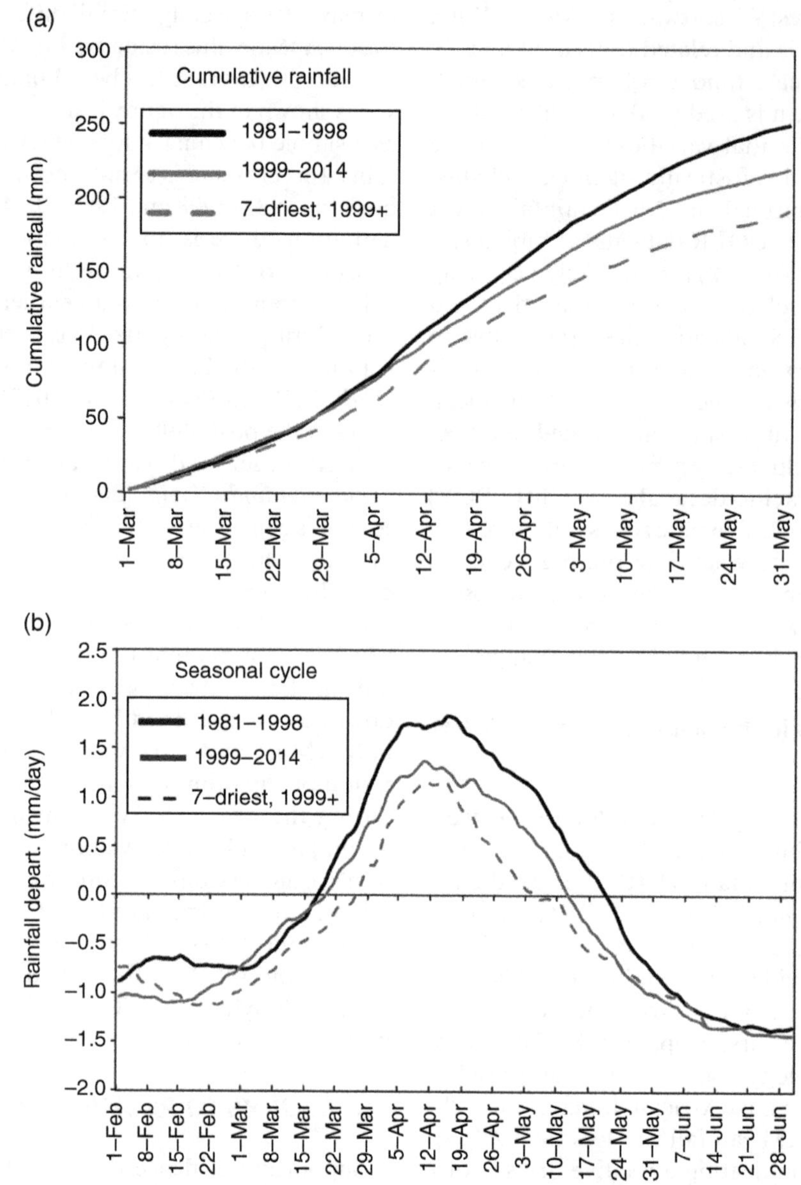

Figure 16.4 (a) Average, cumulative daily rainfall (mm) across East Africa land locations for 1 March–31 May based on CHIRPS data. Solid black (grey) line is the average for the period 1981–1998 (1999–2014) and dashed line for the seven driest MAM seasons during the post-1998 period. (b) Daily rainfall departures (mm/day) from a 1981–1998 average of 1 February–30 June area-average rainfall in CHIRPS for the periods 1981–1998 (solid black line), 1999–2014 (solid grey line) and the seven driest MAM seasons in the post-1998 period (dashed grey line).

Building on the above result, Figure 16.6 displays differences in average MAM precipitation (from GPCC) between the cold phases of PDV that have occurred over roughly the past 100 years and its most recent warm phase, during 1977–1998 [after *Lyon*, 2014]. The rainfall anomaly patterns are remarkably similar for all three periods, not just in Eastern Africa, but also across many other land areas of the globe. While there are certainly concerns regarding the quality of the GPCC data going back to the start of the twentieth century, it is exceedingly unlikely that the similar rainfall anomaly patterns in Figure 16.6 occurred by chance. The pattern correlation (55°S–55°N), for example, between rainfall anomalies for the periods 1999–2010 and 1914–1925 is $r=0.53$. Assuming (quite conservatively) 25 spatial degrees of freedom in MAM rainfall for this domain [e.g., *Van den Dool and Chervin*, 1986] the probability of a chance occurrence is $p<0.01$ based on a two-tailed t-test.

Additional evidence for abrupt changes in the climate system around 1998–1999 is provided in Figure 16.7,

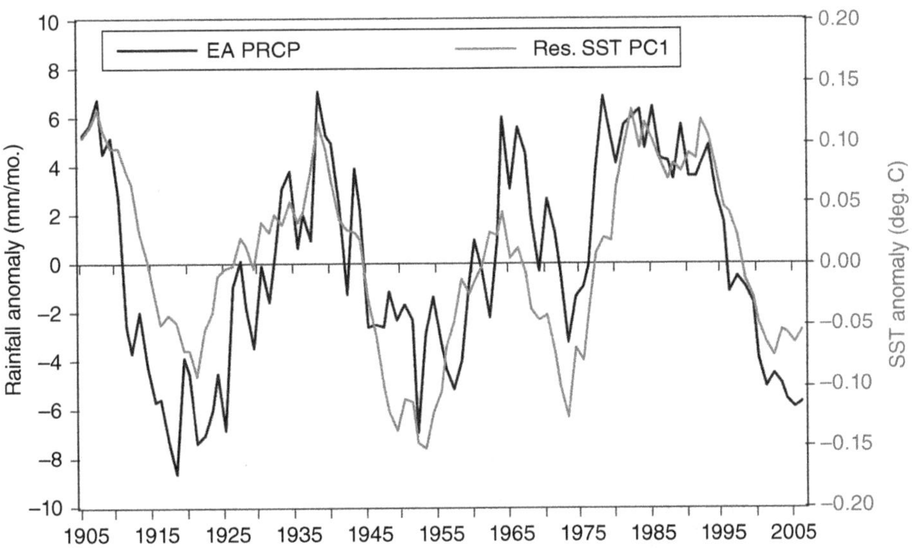

Figure 16.5 Time series of the leading PC (°C, right scale) from the residual SST EOF analysis from ERSST (multiplied by −1, gray line) and the detrended East African rainfall anomaly (mm/mo, left scale) for MAM. A 9 year moving average has been applied to both series. The temporal correlation between the series is r = 0.73 over the full 102 year period (1905–2006). Adapted from *Lyon* [2014]. *See electronic version for color representation.*

the top panel showing differences in SST between MAM seasons averaged over 1999–2014 and 1977–1998. These differences have been divided by the standard deviation of MAM SST anomalies (1981–2000) and are expressed in dimensionless, standardized units. East of the dateline, the tropical Pacific has cooled with anomalies exceeding minus one standard deviation in the eastern portion of the basin. Conversely, the western Pacific exhibits substantial positive SST anomalies, exceeding 1.5 standard deviations southeast of the Philippines and north of the equator (SSTs in this region also exhibit a substantial upward trend, which will be discussed further in section 16.4). Standardized SST differences are generally positive but of much smaller magnitude over the Indian Ocean.

Time series of anomalous (1979–1998 base period) MAM SST and rainfall averaged over the Pacific warm pool region (boxed area in Fig. 16.7a) are plotted in Figures 16.7b and c, respectively. The SST time series indicates an upward trend although there is considerably less variance in the series for the post-1998 period. MAM rainfall anomalies (from GPCPv2.2, *Huffman et al.* [2009], and CMAP, *Xie and Arkin* [1997]) over this region exhibit substantially different behavior than SST, with an abrupt upward shift in 1999, identified as a statistically significant ($p < 0.05$) change point in both datasets. Increased rainfall in the Pacific warm pool region has been associated with an intensification of the overturning Walker circulation across the Indian Ocean, with its descending branch favoring decreased rainfall over East Africa [*Funk et al.*, 2013; *Hoell et al.*, 2014; *Liebmann et al.*, 2014]. Chapter 4 of this book also discusses the western Pacific warming effect.

The above results suggest that a sustained intensification of the ascending branch of the Walker circulation occurred only after 1998, which coincides with the shift in PDV to its cold phase. By contrast, the warming of the Pacific warm pool has been fairly continuous over the past four decades and would thus not appear to be able to account for this abrupt behavior. The warming may have served to increase the *intensity* of the Walker circulation (and thus drying over East Africa) but it cannot by itself account for the *timing* of the transition. This will be discussed further in the next section.

16.4. PHYSICAL CONSIDERATIONS

As mentioned, a dynamical link between Pacific warm pool rainfall and drying in East Africa is an enhanced overturning Walker circulation [*Hoell and Funk*, 2013, 2014; *Hoell et al.*, 2014; *Liebmann et al.*, 2014], which is consistent with other studies [*Lyon and DeWitt*, 2012; *Lyon et al.*, 2014; *Yang et al.*, 2014]. Figure 16.8 shows the midtropospheric pressure vertical velocity for MAM along with the magnitude of the zonal wind component at 850 and 250 hPa as seen in the European Centre for Medium Range Weather Forecasts reanalysis (ERA-Interim, *Dee et al.* [2011]; negative values indicating *upward* motion). Values in the figure represent averages from 10°S to 10°N and are expressed as the difference between the periods 1999–2013 and 1979–1998. Enhanced upward motion is identified over the western tropical

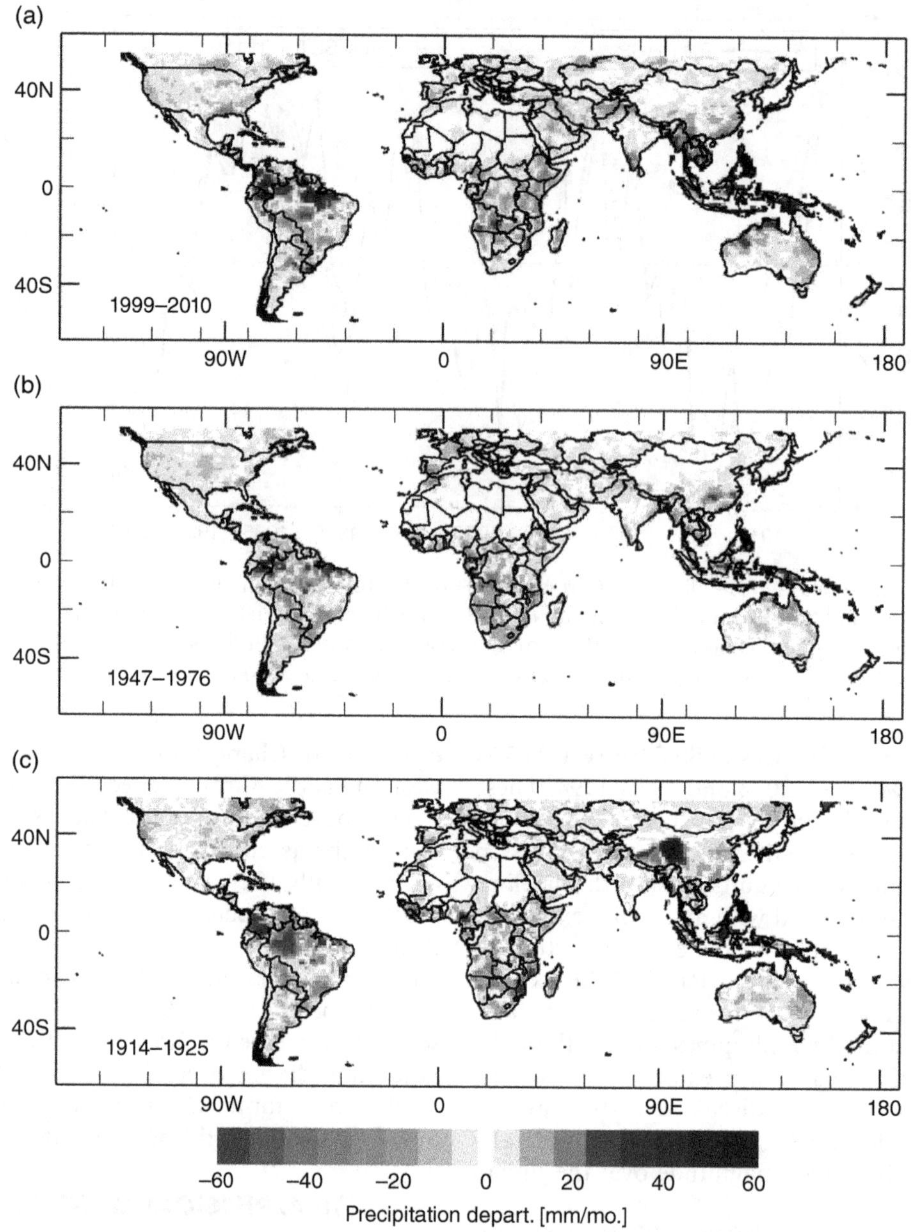

Figure 16.6 MAM precipitation differences (mm/mo.) between averages for (a) 1999–2010, (b) 1947–1976, and (c) 1914–1925 with the average for 1977–1998. Adapted from *Lyon* [2014].

Pacific centered near 135°E (the Pacific warm pool region), while enhanced descent relative to 1979–1998 is identified across most of the equatorial Indian Ocean to the west and the east-central Pacific to the east. The anomalous winds are also consistent with enhanced Walker cells. In the horizontal plane, the increase in rainfall in the Pacific warm pool region north of the equator (Fig. 16.7c) is consistent with previously identified anomalous 850 hPa westerlies across the northern Indian Ocean during the post-1998 period [*Lyon and DeWitt*, 2012; *Lyon et al.*, 2014]. Such an asymmetric (relative to the equator) wind pattern across the Indian Ocean is an expected Gill-type response to an off-equatorial heat source located farther east. This wind pattern is also consistent with an earlier cessation of the MAM long rains season since 1998, as increasingly divergent flow over East Africa becomes established earlier than occurs climatologically [*Yang et al.*, 2015].

While the recent and abrupt decline in East African rainfall exhibits linkages to PDV, this does not exempt a possible contribution from anthropogenic forcing. As mentioned, an upward trend in west Pacific warm pool SSTs has been observed over the past several decades. These higher SSTs have been associated with an increase

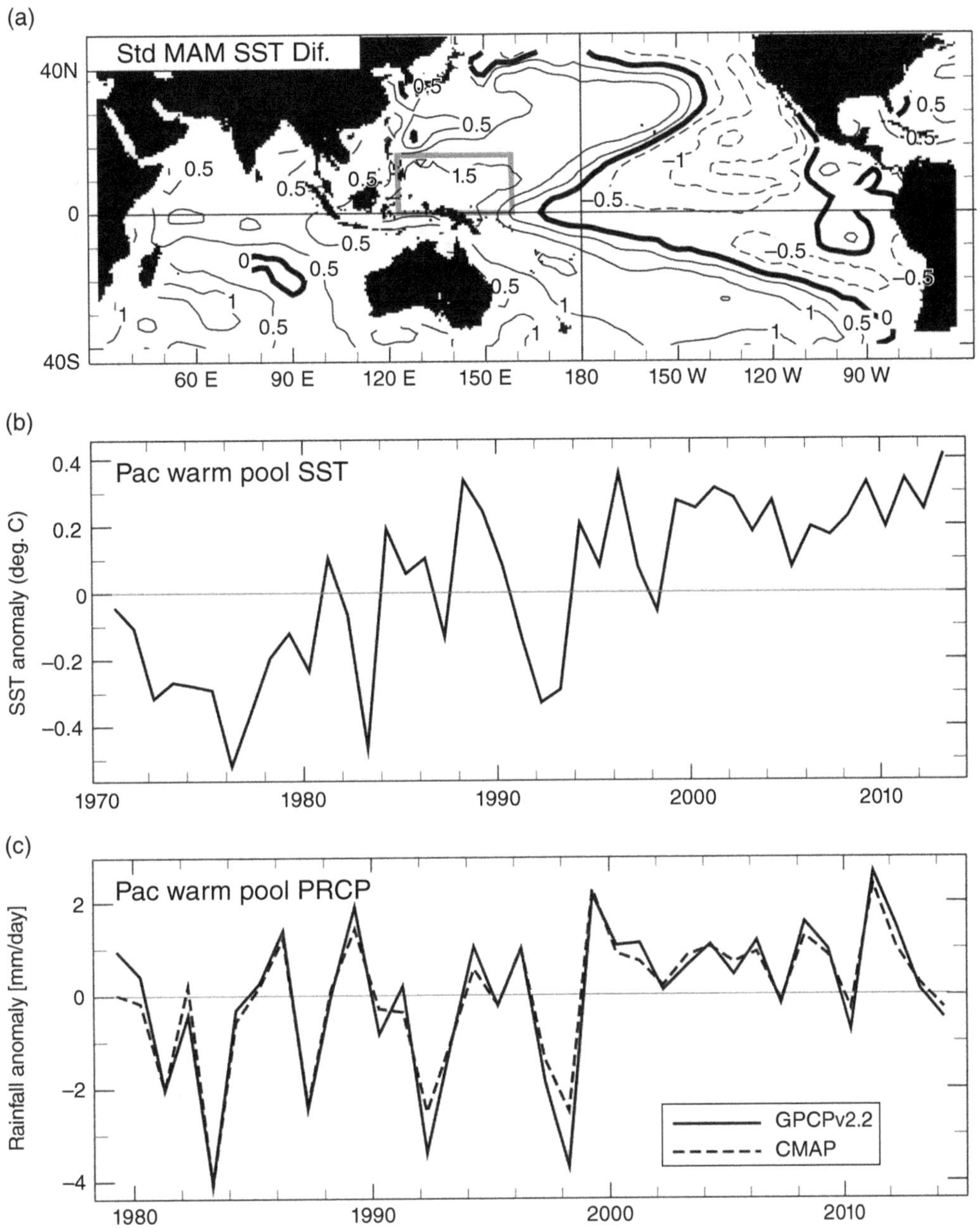

Figure 16.7 (a) Standardized difference in MAM SST between the periods 1999–2014 and 1977–1998 (dimensionless). (b) Time series (1970–2014) of MAM SST anomalies (deg. C) in the Pacific warm pool region (boxed region in [a]). (c) As in (b), but for rainfall (mm/day) in the GPCPv2.2 and CMAP datasets and for the period 1979–2014.

in the west equatorial Pacific zonal SST gradient and equatorial Walker cells [*Hoell and Funk*, 2013, 2014]. The combination of the warming trend of SSTs in the Pacific warm pool region and a recent shift of PDV to its cold phase might be expected to lead to a recent increase in the zonal SST gradient in the equatorial Pacific. Figure 16.9 displays anomalous values (1977–1998 base period) of the difference between MAM SSTs averaged over the Pacific warm pool and Niño4 regions in the Pacific based on the HADISST [*Rayner et al.*, 2003] dataset (the trend in the warm pool is lower in the ERSST data). (The warm pool region is defined as 10°S–10°N, 110°E–150°E, and Niño4 as 5°S–5°N, 160°E–150°W.)There is considerable interannual variability in this series, much associated with the El Niño Southern Oscillation [*Hoell and Funk*, 2013; *Hoell et al.*, 2014]. There is also an upward shift in 1999, which is identified as the most likely year of a change point ($p < 0.03$) with the timing again consistent with the shift of PDV to its cold phase. During earlier PDV cold phases (not shown), such jumps in the gradient are not as

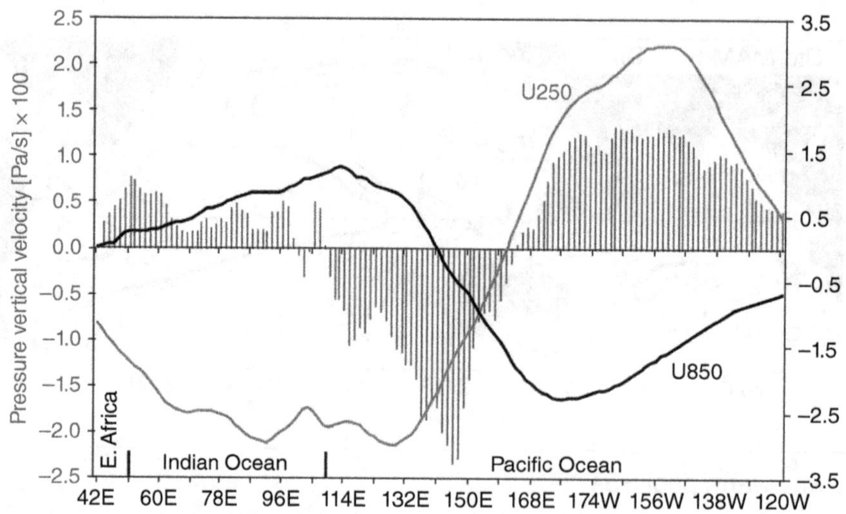

Figure 16.8 Vertical bars indicate the difference in ERA-interim pressure vertical velocity at 500 hPa (Pa/s x 100) between the periods 1999–2013 and 1979–1998 averaged from 10°S–10°N. Lines show similar differences in the zonal wind component (m/s) at 850 and 250 hPa.

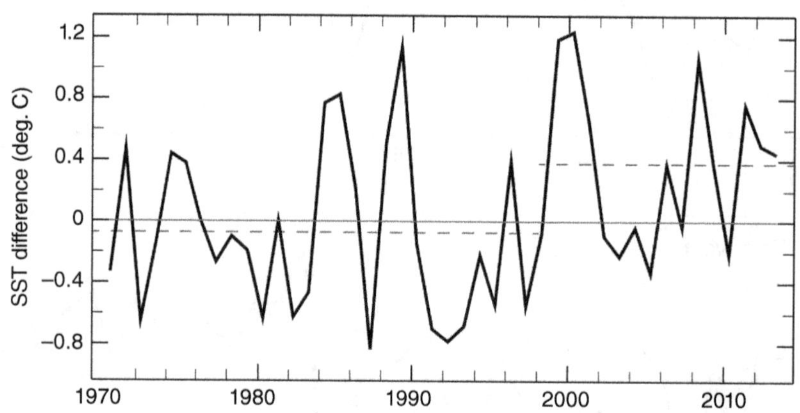

Figure 16.9 Anomalous values (deg. C) of the difference in area average SST between the Pacific warm pool and Nino4 regions during MAM 1970–2013. The dashed horizontal lines show the mean anomaly values for the periods prior to and after a change point in 1999.

readily observed. Issues of SST data quality going back 100 years aside, this suggests the greater influence of the west Pacific warming trend on the gradient in recent years. Again, however, while this could potentially increase the intensity of the rainfall response in the warm pool region in recent years, it does not account for the abrupt shift in behavior in 1998–1999.

In related work, a recent study by *Bayr et al.* [2014] has shown that an upward trend in the strength of Walker circulation (including over the Indian Ocean) since 1979 in the ERA-Interim Reanalysis is predominantly associated with a shift toward more La Niña-like conditions since 1998. This would also appear to be generally consistent with the study by *Jo et al.* [2014] who identify a statistically significant ($p<0.05$) upward shift in SST in 1998–1999 in the north equatorial current bifurcation region (8°N–18°N, 125°E–160°E) during DJF along with an abrupt upward shift in northward ocean heat transport (0–200 m) in the western Pacific at that time. They hypothesize that the more frequent occurrence of La Niña events in the post-1998 period played a role in maintaining the anomalously warm west Pacific SSTs. Other studies [e.g., *Hoell and Funk*, 2013; *Chen et al.*, 2008] have quantified the tendency for the western equatorial Pacific to warm during La Niña events. *Hoell and Funk* [2013] and *Hoell et al.* [2014] note that the recent warming in this region, while having a contribution from La Niña, also contains a contribution from a long-term upward trend (see Chapter 4).

Taken together, current evidence suggests that while long-term warming of SSTs in the tropical west Pacific, while possibly interacting with PDV to enhance the

Indian Ocean Walker cell in recent years, cannot account for abrupt changes in multiple fields that are observed in 1998–1999, including the regional rainfall decline in East Africa. That the near-global MAM rainfall anomaly *pattern* in the GPCC data for 1914–1925 is very similar ($p < 0.01$) to that in 1999–2010 strongly suggests PDV alone is capable of generating multiyear droughts, although the *intensity* of recent droughts appears to be greater than past decadal periods (GPCC data quality issues aside). The extent to which anthropogenic climate change may have contributed to the severity of the recent droughts is a fundamentally important question and is an area of active research.

As the attribution of natural versus anthropogenic forcing on East African droughts will necessitate the use of climate model experiments, the final section of this chapter briefly describes the ability of climate models to capture the recent drying in East Africa, linking it to SST anomalies primarily in the tropical Pacific. Some fundamental errors in CMIP5 model simulations of the current climate will also be briefly presented, along with their implications for climate projections.

16.5. CLIMATE MODEL SIMULATIONS OF EAST AFRICAN CLIMATE

16.5.1. AMIP Simulations and the Recent Drying

That the recent drying in East Africa is indeed related to SST variations has been confirmed in several modeling experiments. Some results summarizing these experiments are shown in Figure 16.10, which are all examples of running the atmospheric component of a climate model with forcing primarily from specified SST (so-called AMIP runs). Figure 16.10a shows the difference in MAM rainfall between ensemble mean (30 member) averages for the periods 1999–2013 and 1979–1998 in the ECHAM4.5 model forced with observed, global SSTs (GOGA). Drying is seen over East Africa, with a large-scale anomaly pattern generally quite similar to Figure 16.6a (over land) in observations. Figure 16.10b shows similar results for the same model but when observed SSTs are only specified in the tropical Pacific, with climatological values specified elsewhere. Drying in East Africa again appears, indicating the fundamental role of the tropical Pacific in its forcing. Rainfall anomalies in other regions are again similar to observations and the GOGA runs. Figure 16.10c is the same calculation, but based on an ensemble average of 16 runs from the ECHAM5 model where SSTs have been detrended and set to 1880 equivalent values, as have greenhouse gas concentrations. (These runs were obtained from the Earth Systems Research Laboratory [ESRL] website http://www.esrl.noaa.gov/psd/repository/alias/facts.)

Drying over East Africa is again reproduced, along with a similar large-scale anomaly pattern. This latter result suggests that long-term trends in SST are not a necessary condition to generate the recent drying in East Africa (at least in ECHAM5). This would seem consistent with the observational analysis showing earlier decades with drought in East Africa. However, the "residual" SST anomaly field identified after removing the long-term trend is sensitive to the specific methodology used (an example is given in section 16.5.2). As such, additional modeling studies are necessary to more rigorously examine the sensitivity of the model response to various SST anomaly patterns with and without the influence of anthropogenic forcing. This response also needs to be evaluated in more than one model. Figure 16.11, for example, shows the multimodel mean rainfall response (seven models) to observed global SSTs in the Pacific warm pool region. (These model runs are the same as those described in *Seager et al.* [2015].) Consistent with Figure 16.8 for observations, the multimodel mean also generates an abrupt increase in MAM rainfall over the warm pool in 1999, identified as the most likely year of a change point ($p < 0.05$).

16.5.2. CMIP Simulations

Here we very briefly describe a few important issues regarding the ability of CMIP5 models to simulate the current climate of East Africa, with implications for projections. Recent studies by *Yang et al.* [2014, 2015] have shown that the CMIP5 model simulations tend to reverse the relative magnitudes of the short (OND) and long (MAM) rains in East Africa, making the former too wet and the latter too dry climatologically. As shown in Figure 16.12, this appears related to a general bias in the CMIP5 model SST field, particularly over the Indian Ocean. The multimodel mean of simulations from 39 models (see *IPCC* [2013] for their description) generates a larger zonal gradient in equatorial *Indian Ocean* SSTs than is observed during OND (SSTs are too high in the west relative to the east), with the opposite tendency occurring in MAM. (These runs were obtained from the KNMI Climate Explorer at http://climexp.knmi.nl using all available ensemble members. Results were unaffected when using a single ensemble member per model [results not shown].)

There is also a 1 month delay in peak positive values of the gradient in CMIP5 relative to observations (April in observations, May in CMIP5). The models also lack virtually any cooling of the western Indian Ocean during the South Asian summer monsoon relative to observations, likely contributing to warmer conditions in OND there compared to observations. The model projections of SST for 2070–2099 (black bars in Fig. 16.12) tend to enhance

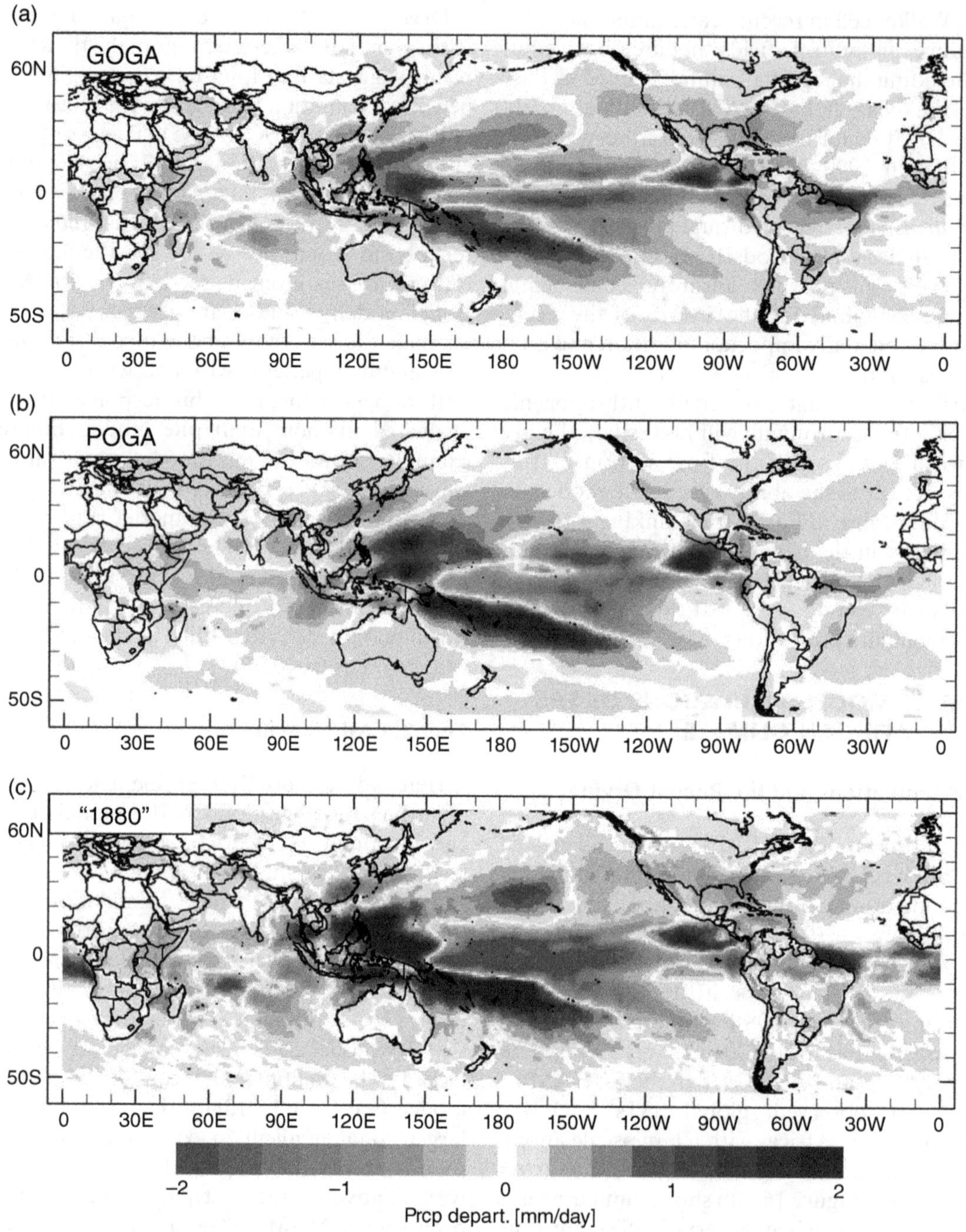

Figure 16.10 (a) Ensemble average MAM precipitation differences (mm/day) between 1999–2012 and 1979–1998 for runs of the ECHAM4.5 model forced with observed global SST. (b) As in (a) but with the observed SSTs limited to the tropical Pacific (120°E–80°W, 25°S–25°N). (c) As in (a) but for the ECHAM5 model forced with detrended SSTs and greenhouse gases fixed at 1880 levels.

these biases in the future and are consistent with wetter projected short and long rains [*Yang et al.*, 2015]. These results alone shed doubt on the reliability of climate projections for East Africa.

Of course, the tropical Pacific also has an important role to play in East African climate. Figure 16.13 provides some comparisons of CMIP5 simulations and observations over the period 1970–2014. The top panel first shows time series of the CMIP5 multimodel mean, global average (55°S–55°N) SST anomaly (1979–1998 base period) along with observations (from ERSST). Until the late 1990s, the warming trend in the models closely matched observations, followed by a recent warming hiatus in observations since around 2000 (it is not expected that the uninitialized

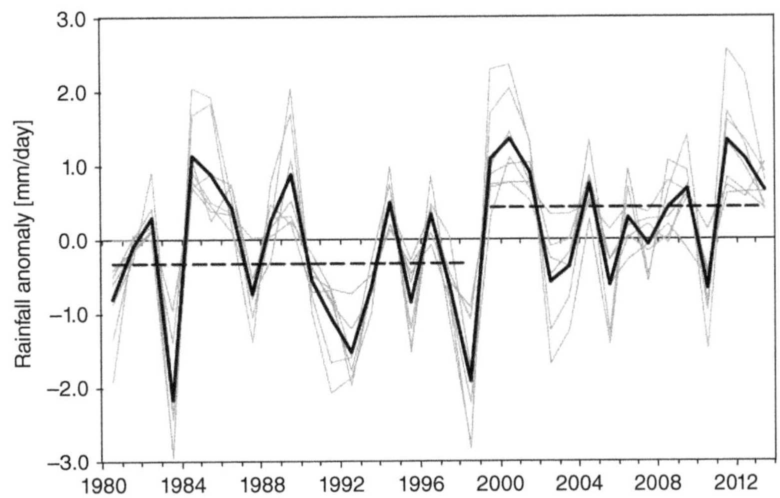

Figure 16.11 Ensemble mean values of anomalous Pacific warm pool precipitation (mm/day, grey lines) obtained from 7 AMIP style climate model simulations forced with observed global SST. Solid black line is the multimodel mean and the dashed horizontal lines indicate mean values for the pre- and post-1999 periods.

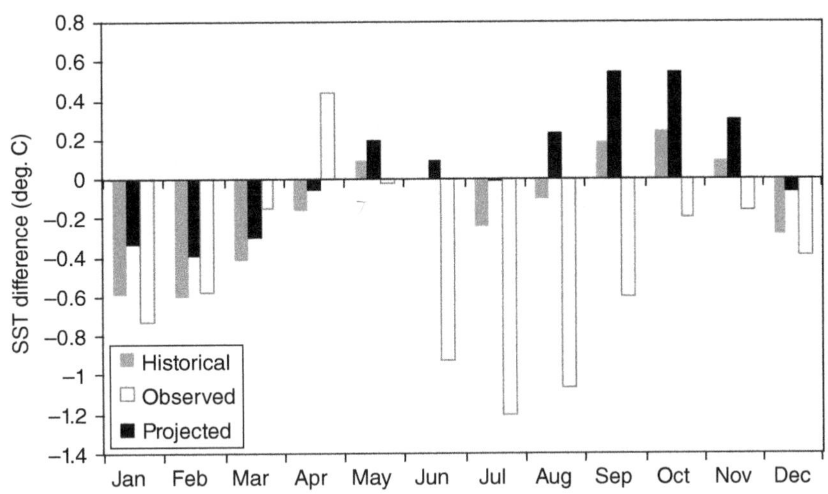

Figure 16.12 Monthly differences in Indian Ocean SST (deg. C) between averages over the western (50°E–70°E, 10°S–10°N) and eastern (90°E–110°E, 10°S–0°N) portions of the basin. White (grey) bars are climatological averages for the period 1971–2000 in observations (multimodel mean CMIP5 simulations). Black bars indicate the CMIP5 multimodel average values in projections (RCP8.5 scenario) for 2070–2099.

CMIP5 multimodel mean would capture PDV). Figure 16.13b shows a similar plot but for the Niño3.4 region in the east-central Pacific. While observations show no trend over the past four decades, the CMIP5 models show consistent warming over this period, a well-known tendency in the models even before the recent PDV shift [e.g., *Fyfe and Gillett*, 2014]. Whether the eastern Pacific will indeed begin to warm at a faster rate in the future has implications for climate changes in many regions of the world, including East Africa.

Does the observed lack of warming in the eastern Pacific relative to CMIP5 simulations indicate natural climate variations are acting to oppose anthropogenic warming or are the models simply generating too much warming? Figure 16.13c shows a simple but important example of attempting to quantify the contribution of anthropogenic climate change versus natural variability on SST changes in the west Pacific warm pool region. The figure shows a time series of observed SST anomalies in the region along with two estimates of the residual anomaly obtained after removing the trend. The residual values are computed in two ways: by subtracting the global average, observed SST anomaly (a proxy for the anthropogenic warming signal), and by removing the multimodel mean CMIP5 simulated warm pool anomaly. For the post-1998 period, the magnitude of the residuals

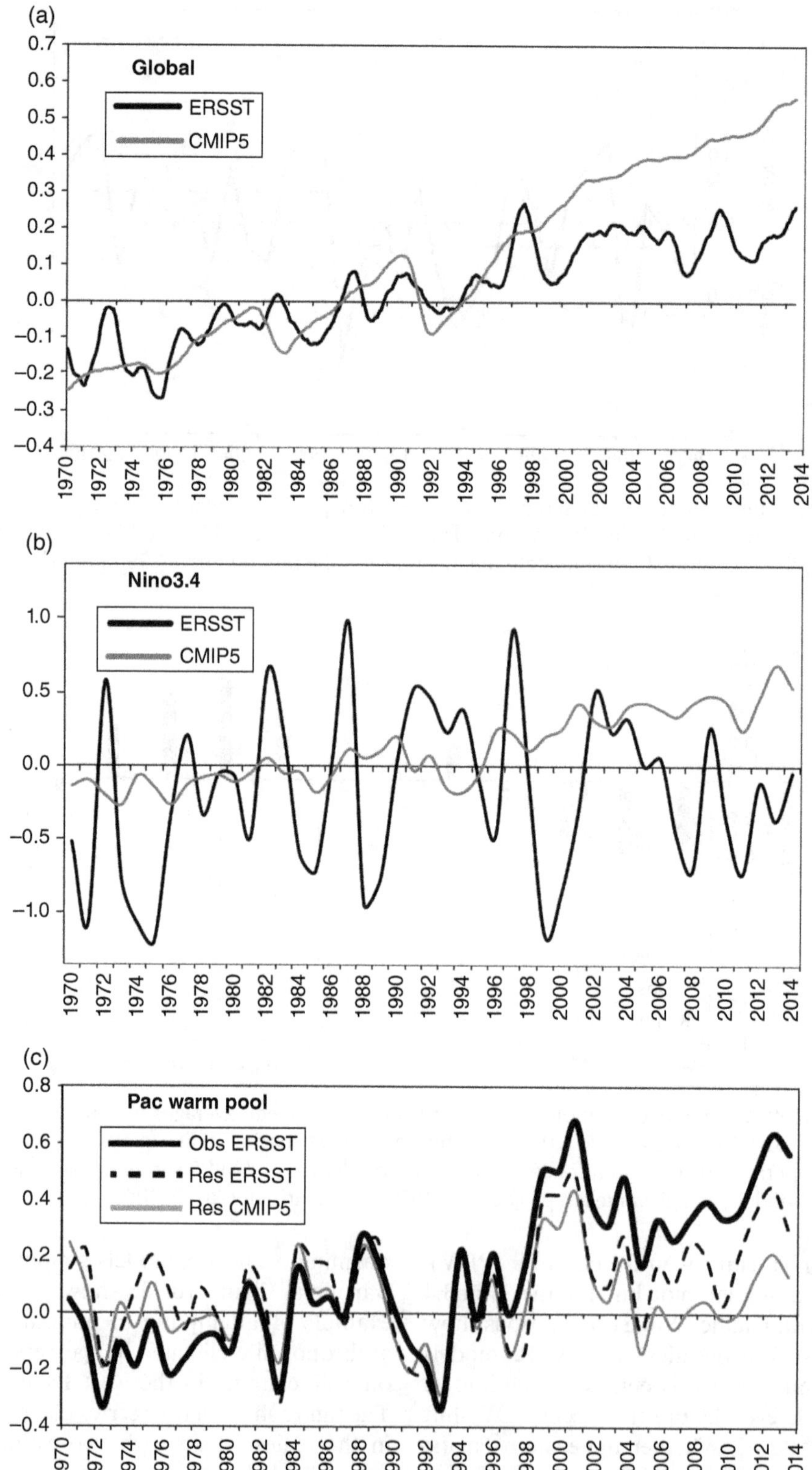

Figure 16.13 (a) Average MAM global (55°S–55°N) SST departure from a 1979–1998 base period mean (deg. C) for observations (black line) and the multimodel mean CMIP5 simulations (grey line). (b) As in (a) but for the Niño3.4 region in the Pacific (120°W–170°W, 5°S–5°N). (c) Observed MAM SST departure from a 1979–1998 base period mean in the Pacific warm pool region (solid black line) and the residual anomalies in this region obtained after removing the time-varying observed global average SST trend (black dashed line) and time-varying CMIP5 multimodel average warm pool trend (grey line).

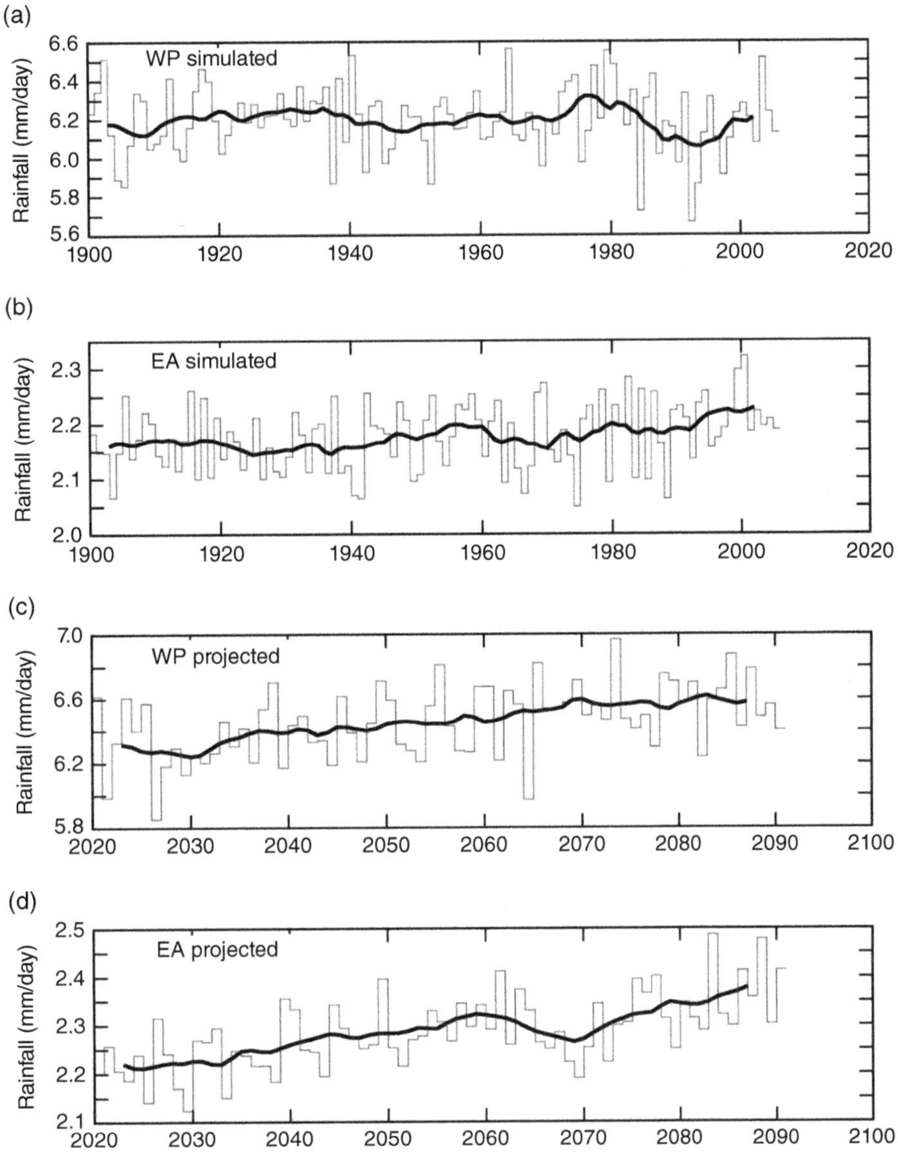

Figure 16.14 (a) Multimodel mean CMIP5 simulations of warm pool rainfall (mm/day) for MAM 1900–2005, with the solid black line representing the 10 year moving average. (b) As in (a) but for East African rainfall (land areas only). (c) and (d) as in (a) and (b) but for model projections (RCP8.5 scenario).

can differ substantially between methods, with the models showing a greater rate of SST increase in the warm pool than is observed in the observed global average value. Which residual estimate is a better representation of natural variations? Are the CMIP5 models realistically capturing the warming of the west Pacific while simultaneously overdoing the warming in the eastern part of the basin or is the former a fortuitous result? There is not a simple answer.

Finally, Figure 16.14 shows time series of the multimodel average CMIP5 simulated MAM rainfall for East Africa and the Pacific warm pool, 1901–2005. While the warm pool region has been warming at a rate generally consistent with observations (not shown), modeled precipitation for the region exhibits little trend over the past century (Fig. 16.14a) while simulated rainfall in East Africa (Fig. 16.14b) shows a slight increasing trend. The multimodel average projected rainfall for both regions (Fig. 16.14c, d) based on the RCP8.5 scenario shows increasing trends in both locations. In recent observations, the increase in rainfall in the Pacific warm pool region is associated with an enhanced Walker circulation over the Indian Ocean with associated subsidence over East Africa. This indicates that in the models, future changes in SST (and rainfall) in ocean locations outside the west Pacific are clearly important to the rainfall response in East

Africa. *Bayr et al.* [2014] show, for example, an eastward shift of the Pacific Walker cell in CMIP5 projections, generally favoring increased East African rainfall. Thus, while advances are being made in understanding the recent drying in East Africa, what future rainfall changes will be in response to increasing greenhouse gases remains uncertain.

16.6. CONCLUSIONS

What can be concluded about the East African climate paradox? It is argued that several pieces of observational evidence, with support from modeling studies, suggest the recent drying trend was dominated by an abrupt decline around 1998–1999 and associated with a shift of PDV to its cold phase. A long-term warming trend in the tropical Pacific warm pool region may possibly have acted to increase the severity of recent droughts but the timing of the rainfall decline is closely tied to PDV. Further modeling studies are required to investigate the influence of SST warming trends on the severity of recent East African droughts. From the perspective of recent climate behavior, the apparent paradox may be viewed as resulting from a mismatch of timescales, between decadal scale variability (possibly augmented by anthropogenic climate change) and distinct, longer-term climate changes in East Africa arising from increasing greenhouse gases, if the CMIP5 models are taken at face value. In the short term, the expectation is for PDV to retreat from its cold phase and thus a recovery of the long rains from their recent decline.

In the longer term, the impact of increasing greenhouse gases on the climate of East Africa is much less certain. Confidence that the future climate of East Africa will indeed become wetter is currently low. The CMIP5 coupled climate models exhibit significant errors in simulating key aspects of the current climate, not just in East Africa, but also in SST changes in the global oceans. The failure of the models to capture major features of the observed annual cycle of SSTs in the Indian Ocean is particularly problematic. Also critical is whether or not SSTs in the western equatorial Pacific will continue to increase faster than in the central Pacific in response to increasing anthropogenic forcing. If the observed asymmetric warming trends continue, the implication is they will enhance the future likelihood of drought in East Africa, particularly when the eastern portion of the basin cools in association with La Niña (and the cool phase of PDV). If the eastern Pacific in fact warms, the relative likelihood of drought will decrease. Overall, the current climate paradox in East Africa may also result from taking the model projections at face value. Whether or not the future climate in the region will indeed become wetter is considered very much an open question.

ACKNOWLEDGMENTS

This work was supported by grants from the National Science Foundation (award AGS 12-52301) and the National Aeronautical and Space Administration (award NNX12AR39G), which are gratefully acknowledged. Data used in the study were primarily accessed via the IRI Data Library (http://iridl.ldeo.columbia.edu/), with data for Figure 16.14 obtained from the Royal Netherlands Meteorological Institute (KNMI) Climate Explorer (https://climexp.knmi.nl/selectfield_cmip5.cgi).

REFERENCES

Bayr, T., D. Dommenget, T. Martin, and S. B. Power (2014), The eastward shift of the Walker Circulation in response to global warming and its relationship to ENSO variability, *Climate Dyn.*, *43*, 2747–2763; doi:10.1007/s00382-014-2091-y.

Camberlin, P., and R. E. Okoola (2003), The onset and cessation of the "long rains" in eastern Africa and their interannual variability, *Theor. App. Climatol.*, *75*, 43–54; doi:10.1007/s00704-002-0721-5.

Camberlin, P., B. Fontaine, S. Louvet, P. Oettli, and P. Valimba (2010), Climate adjustments over Africa accompanying the Indian monsoon onset, *J. Climate*, *23*, 2047–2064; doi:10.1175/2009JCLI3302.1.

Checchi, F., and W. C. Robinson (2013), Mortality among populations of southern and central Somalia affected by severe food insecurity and famine during 2010–2012, *FAO and FEWS NET*; http://www.fsnau.org/in-focus/study-report-mortality-among-populations-southern-and-central-somalia-affected-severe-food-.

Chen, J., A. D. Del Genio, B. E. Carlson, and M. G. Bosilovich (2008), The spatiotemporal structure of twentieth-century climate variations in observations and reanalyses, Part II: Pacific pan-decadal variability, *J. Climate*, *21*, 2634–2650.

Christensen, J. H., and coauthors (2007), Regional climate projections, in *Climate Change 2007: The Physical Science Basis*, edited by S. Solomon et al., Cambridge University Press, 866–872.

Dee, D. P., and coauthors (2011), The ERA-interim reanalysis: Configuration and performance of the data assimilation system, *Quart. J. Roy. Meteor. Soc.*, *137*, 553–597; doi: 10.1002/qj.828.

FAO, IFAD, and WFP (2014), The State of Food Insecurity in the World 2014, Strengthening the enabling environment for food security and nutrition, Rome, FAO.

FEWS NET (2011), Past year one of the driest on record in the eastern Horn, *Famine Early Warning System Network Report*, June 14, 2011, U.S. Agency for International Development, Washington, DC.

Funk, C., G. Husak, J. Michaelsen, S. Shukla, A. Hoell, B. Lyon, M. P. Hoerling, B. Liebmann, T. Zhang, J. Verdin, G. Galu, G. Eilerts, and J. Rowland (2013), Attribution of 2013 and 2003–2012 rainfall deficits in eastern Kenya and southern Somalia, in *Explaining Extreme Events of 2012 from a Climate Perspective*, *Bull. Amer. Meteor. Soc.*, *94*, S1–74.

Funk, C. C., P. J. Peterson, M. F. Landsfeld, D. H. Pedreros, J. P. Verdin, J. D. Rowland, B. E. Romero, G. J. Husak, J. C. Michaelsen, and A. P. Verdin, (2014), *A Quasi-Global Precipitation Time Series for Drought Monitoring*, U.S. Geological Survey Data Series 832; http://dx.doi.org/10.3133/ds832.

Fyfe, J. C., and N. P. Gillett (2014), Recent observed and simulated warming, *Nature Climate Change*, *4*, 150–151.

Hoell, A., and C. Funk (2013), The ENSO-related West Pacific sea surface temperature gradient, *J. Climate*, *26*, 9545–9562.

Hoell, A., and C. Funk (2014), Indo-Pacific sea surface temperature influences on failed consecutive rainy seasons over eastern Africa, *Climate Dyn.*, *43*, 1645–1660; doi:10.1007/s00382-013-1991-6.

Hoell, A., C. Funk, and M. Barlow (2014), La Niña diversity and northwest Indian Ocean rim teleconnections, *Climate Dyn.*, *43*, 2707–2724.

Huffman, G. J., R. F. Adler, D. T. Bolvin, and G. Gu (2009), Improving the global precipitation record: GPCP version 2.1, *Geophys. Res. Lett.*, *36*, L17808; doi:10.1029/2009GL040000.

IPCC (2013), Annex I: Atlas of Global and Regional Climate Projections, edited by G. J., van Oldenborgh, M. Collins, J. Arblaster, J. H. Christensen, J. Marotzke, S. B. Power, M. Rummukainen, and T. Zhou, in *Climate Change 2013: The Physical Science Basis, Contribution of Working Group I to the Fifth Assessment Report of the Intergovernmental Panel on Climate Change*, edited by T. F. Stocker, D. Qin, G.-K. Plattner, M. Tignor, S. K. Allen, J. Boschung, A. Nauels, Y. Xia, V. Bex, and P. M. Midgley, Cambridge University Press, Cambridge, UK and New York, USA.

Jo, H.-S., S.-W. Yeh, and B. P. Kirtman (2014), Role of the western tropical Pacific in the North Pacific regime shift in the winter of 1998–1999, *J. Geophys. Res. Oceans*, *119*, 6161–6170; doi:10.1002/2013JC009527.

Kajikawa, Y., T. Yasunari, S. Yoshida, and H. Fujinami (2012), Advanced Asian summer monsoon onset in recent decades, *Geophys. Res. Lett.*, *39*, L03803; doi:10.1029/2011GL050540.

Liebmann, B., M. P. Hoerling, C. Funk, I. Bladé, R.M. Dole, D. Allured, X. Quan, P. Pegion, J.K. Eischeid (2014), Understanding recent eastern Horn of Africa rainfall variability and change, *J. Climate*, *27*, 8630–8645.

Liebmann, B., R. M. Dole, C. Jones, I. Bladé, and D. Allured (2010), Influence of choice of time period on global surface temperature trend estimates, *Bull. Amer. Meteor. Soc.*, *91*, 1485–1491; doi:10.1175/2010BAMS3030.1.

Lyon, B (2014), Seasonal drought in the Greater Horn of Africa and its recent increase during the March-May long rains, *J. Climate*, *27*, 7953–7975.

Lyon, B., and D. G. DeWitt (2012), A recent and abrupt decline in the East African long rains, *Geophys. Res. Lett.*, *39*, L02702; doi:10.1029/2011GL050337.

Lyon, B., A. G. Barnston, and D. G. DeWitt (2014), Tropical Pacific forcing of a 1998–99 climate shift: Observational analysis and climate model results for the boreal spring season, *Climate Dyn.*, *43*, 893–909; doi:10.1007/s00382-013-1891-9.

Mantua, N. J., S. R. Hare, Y. Zhang, J. M. Wallace, and R. C. Francis (1997), A Pacific interdecadal climate oscillation with impacts on salmon production, *Bull. Amer. Meteor. Soc.*, *78*, 1069–1079.

Rayner, N. A., D. E. Parker, E. B. Horton, C. K. Folland, L. V. Alexander, D. P. Rowell, E. C. Kent, and A. Kaplan (2003), Global analyses of sea surface temperature, sea ice, and night marine air temperature since the late nineteenth century, *J. Geophys. Res.*, *108*, 4407; doi:10.1029/2002JD002670.

Schneider, U., T. Fuchs, A. Meyer-Christoffer, and B. Rudolf (2008), *Global Precipitation Analysis Products of the GPCC*, Global Precipitation Climatology Centre (GPCC), DWD, Internet Publication, 1–12; http://www.cec.org/Atlas/Files/Precipitation/GPCC_intro_products_2008.pdf.

Seager, R., M. Hoerling, S. Schubert, H. Wang, B. Lyon, A. Kumar, J. Nakamura, and N. Henderson (2015), Causes of the 2011 to 2014 California drought, *J. Climate*, *28*, 6997–7024.

Shongwe, M. E., G. J. van Oldenborgh, B. van den Hurk, and M. van Aalst (2011), Projected changes in mean and extreme precipitation in Africa under global warming, Part II: East Africa. *J. Climate*, *24*, 3718–3733; doi:10.1175/2010JCLI2883.1.

UN OCHA (2011), *Eastern Africa Drought, Humanitarian Rep. 4*, United Nations Organization for the Coordination of Humanitarian Affairs; http://reliefweb.int/sites/reliefweb.int/files/resources/Full_Report_1717.pdf.

Van den Dool, H. M., and R. M. Chervin (1986), A comparison of month-to-month persistence of anomalies in a general circulation model and in the earth's atmosphere, *J. Atmos. Sci.*, *43*, 1454–1466.

Wang, X. L. (2008), Accounting for autocorrelation in detecting mean-shifts in climate data series using the penalized maximal t or F test, *J. Appl. Meteor. Clim..*, *47*, 2423–2444.

Xie, P., and P. A. Arkin (1997), Global precipitation: A 17-year monthly analysis based on gauge observations, satellite estimates and numerical model outputs, *Bull. Amer. Meteor. Soc.*, *78*, 2539–2558.

Yang, W., R. Seager, M. A. Cane, and B. Lyon (2014), The East African long rains in observations and models, *J. Climate*, *27*, 7185–7202.

Yang, W., R. Seager, M. A. Cane, and B. Lyon (2015), The annual cycle of East African precipitation, *J. Climate*, *28*, 2385–2404.

17
A Physical Model for Extreme Drought over Southwest Asia

Andrew Hoell[1], Chris Funk[2], Mathew Barlow[3], and Forest Cannon[4]

ABSTRACT

The socioeconomic difficulties of southwest Asia, defined as the area bound by the domain 25°N–40°N and 40°E–70°E, are exacerbated by extreme precipitation deficits during the November–April rainy season. The precipitation deficits during many southwest Asia droughts have been examined in terms of the forcing by climate variability originating over the Pacific Ocean as a result of the El Niño–Southern Oscillation (ENSO), Pacific decadal variability (PDV), and the long-term warming of Pacific (LT) sea surface temperatures (SST). Here we examine how the most extreme November–April southwest Asia droughts relate to global SSTs and the associated large-scale atmospheric circulation anomalies and analyze the specific atmospheric forcing mechanisms responsible for changes in regional southwest Asian precipitation. The driest November–April seasons during 1948–2012 over southwest Asia are forced by subsidence and reductions of moisture fluxes as a result of the interaction of the mean flow with anomalous zonally symmetric high pressure throughout the Northern Hemisphere. The anomalous zonally symmetric high pressure throughout the Northern Hemisphere occurs simultaneously with cool central and eastern Pacific SST anomalies associated with La Niña and the negative phase of PDV and a warm west Pacific Ocean caused in part by the long-term warming of the west Pacific Ocean.

17.1. INTRODUCTION

The November–April wet season over southwest Asia (Fig. 17.1), defined as the region bound by the domain 25°N–40°N and 40°E–70°E, is prone to regular occurrences of significant precipitation deficits (Fig. 17.2) that result in hydrologic drought. These multiyear wet season droughts have substantial impacts upon the resilience of a region that is already socially, politically, and economically unstable and water and food insecure [*Gall*, 2008; *Eggerman and Panter-Brick*, 2010]. In this chapter, we examine the climatic forcing of the strongest southwest Asia droughts during the 1948–2012 period in terms of SST, global atmospheric circulations, and the dynamic atmospheric forcing of regional precipitation to develop a model for severe regional drought.

Extratropical cyclones guided by the jet stream centered between 25°N and 35°N latitude [*Krishnamurti*, 1961; *Schiemann et al.*, 2009] are primarily responsible for the November–April precipitation over southwest Asia. The spatial variations of precipitation throughout southwest Asia are largely governed by the complex topography of the region. The interaction of the large-scale west-to-east flow, which guides storm systems, with regional orography results in the largest precipitation amounts along the western slopes of the high mountains of southwest Asia. The highest regional precipitation amounts fall along the Zagros Mountains of Iran and Iraq [*Alijani*, 2008] and the western edge of the Tibetan

[1] *NOAA Earth System Research Laboratory, Physical Sciences Division, Boulder, Colorado, USA*

[2] *US Geological Survey and University of California Santa Barbara Geography, Santa Barbara, California, USA*

[3] *University of Massachusetts Lowell, Environmental Earth and Atmospheric Sciences, Lowell, Massachusetts, USA*

[4] *University of California Santa Barbara Geography, Santa Barbara, California, USA*

Climate Extremes: Patterns and Mechanisms, Geophysical Monograph 226, First Edition.
Edited by S.-Y. Simon Wang, Jin-Ho Yoon, Christopher C. Funk, and Robert R. Gillies.

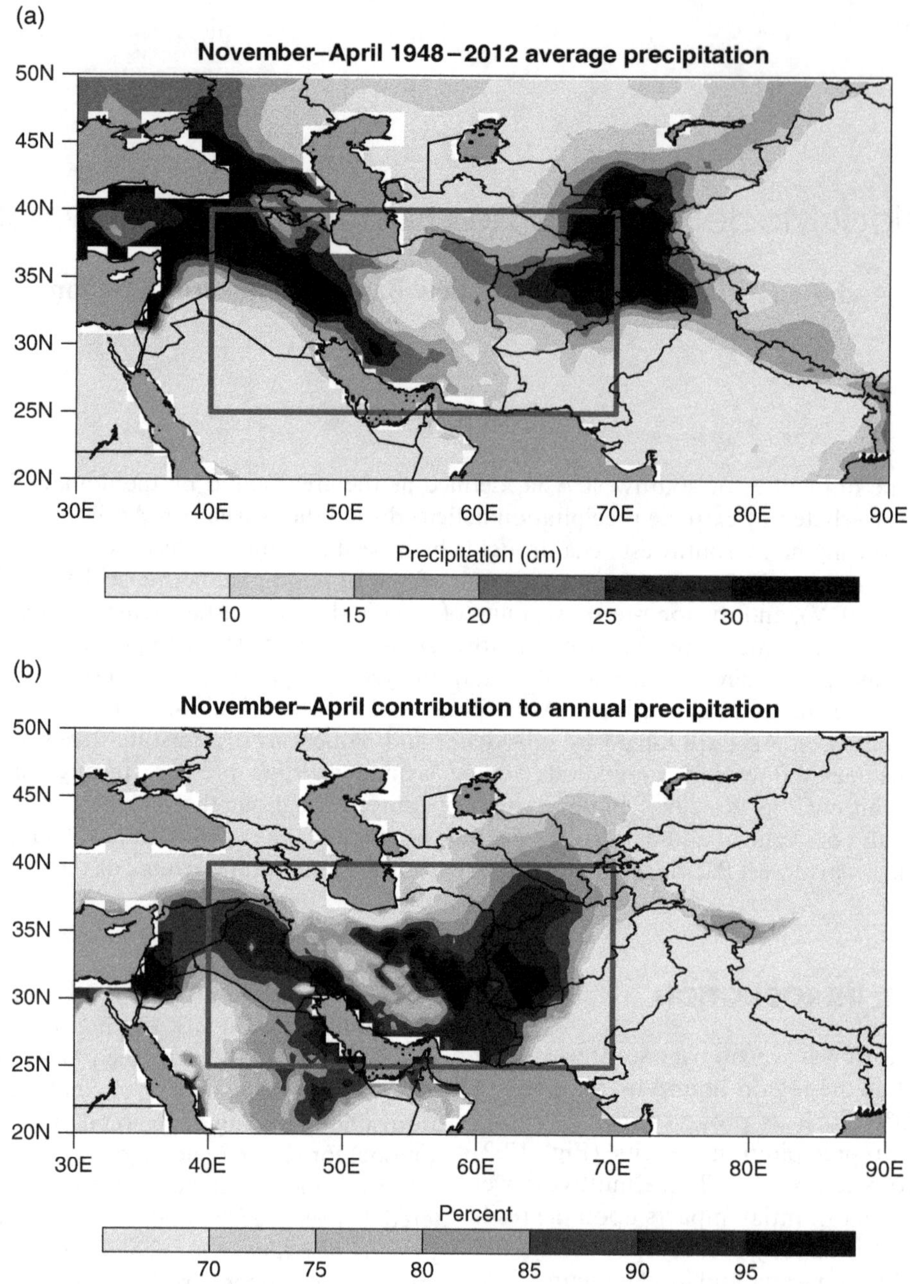

Figure 17.1 (a) November–April 1948–2012 average precipitation (cm) and (b) November–April contribution to the annual precipitation for 1948–2012 in percent.

Plateau (Fig. 17.1a). On the contrary, the lowest precipitation amounts during November–April are observed at low elevations, which include the Lut Desert of Iran, the Registan Desert of Afghanistan, and the Karakum Desert of Turkmenistan. On average, nearly 75% of the annual southwest Asia precipitation falls during November–April (Fig. 17.1b). Over eastern Iran and the Helmand and Farah provinces of Afghanistan, in excess of 90% of the annual precipitation falls during November–April (Fig. 17.1b).

Southwest Asia precipitation throughout the November–April season is strongly controlled by the atmospheric teleconnections driven by the four leading modes of global sea surface temperature (SST) variability. These four leading modes are the El Niño–Southern Oscillation (ENSO) [*Barlow et al.*, 2002; *Nazemosadat and Ghasemi*, 2004; *Mariotti et al.*, 2005; *Mariotti*, 2007; *Hoell and Funk*, 2013; *Hoell et al.*, 2013a; *Hoell et al.*, 2013b; *Hoell et al.*, 2014a], long-term warming of the global oceans (LT) [*Hoell et al.*, 2014b], Pacific decadal

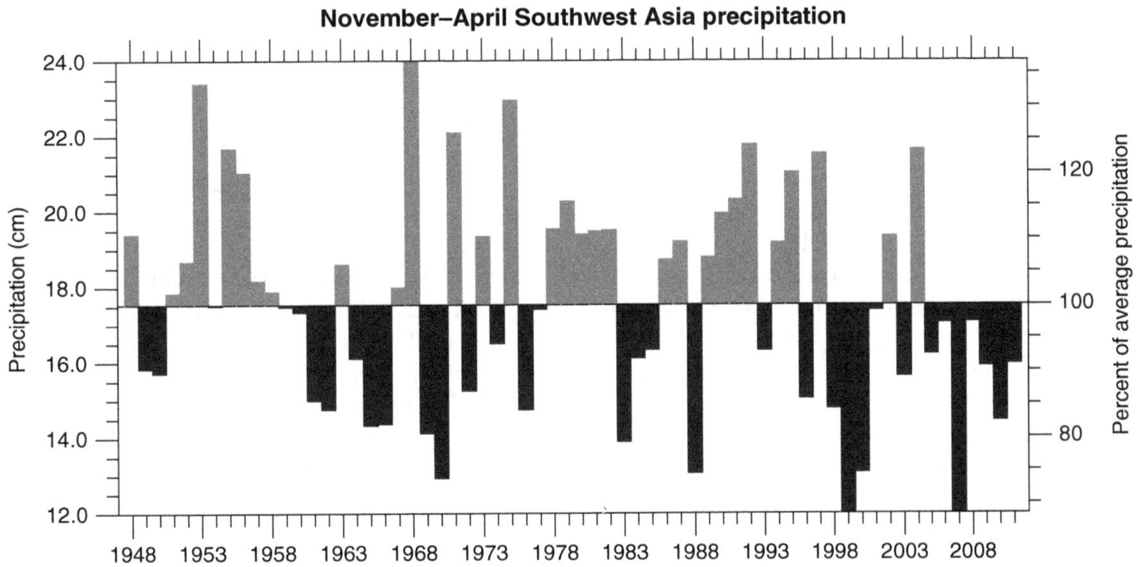

Figure 17.2 November–April precipitation averaged over southwest Asia for 1948–2012 expressed in cm (left ordinate) and percent of average (right ordinate).

variability (PDV) [*Meehl and Hu*, 2006; *Lyon et al.*, 2013; *Hoell et al.*, 2014b], and the Atlantic Multidecadal Oscillation (AMO) [*Sheffield and Wood*, 2008]. The AMO, which features multidecadal changes in North Atlantic SST, forces multidecadal changes in soil moisture over southwest Asia [*Sheffield and Wood*, 2008]. The warm (cold) phase of the AMO, characterized by warm (cold) north Atlantic SST, is related to decreases (increases) in southwest Asia soil moisture.

ENSO, LT, and PDV, which are SST modes that drive the tropical Indo-Pacific convection and diabatic heating, are especially impactful to the circulation and precipitation over southwest Asia during the entire November–April rainy season [*Barlow et al.*, 2002; *Hoell et al.*, 2012; *Hoell and Funk*, 2013; *Hoell et al.*, 2013a; *Hoell et al.*, 2013b; *Hoell et al.*, 2014a]. When Maritime Continent convection is increased and central tropical Pacific Ocean convection is suppressed associated with ENSO, LT, and PDV, the tropical diabatic heating changes drive a Rossby wave (high pressure) over southwest Asia. LT influences relate to a preferential warming of the western Pacific, which is associated with increased warm pool precipitation and diabatic forcing [*Funk and Hoell*, 2015], which influences Asia [*Hoell et al.*, 2014b]. The Rossby wave interacts with the mean climate to produce cold temperature advection that is balanced by precipitation-reducing subsidence [*Hoell et al.*, 2013b; *Hoell et al.*, 2014a]. The Rossby wave is also responsible for reducing the vertically integrated moisture flux into southwest Asia [*Mariotti*, 2007; *Hoell et al.*, 2013b; *Hoell et al.*, 2014a). The Madden-Julian Oscillation also has a similar effect on the atmospheric circulation and precipitation over southwest Asia, but operates on subseasonal timescales, generally for weeks to perhaps as long as a month [*Barlow et al.*, 2005; *Nazemosadat and Ghaedamini*, 2010; *Barlow*, 2011].

Drought over what is already a water scarce and semiarid region [*Oki and Kanae*, 2006] has significant deleterious effects upon the rain-fed subsistence farming throughout southwest Asia [e.g., *Ramankutty et al.*, 2008; *Ryan et al.*, 2012], which increase the potential for loss of life and property regionally [*Agrawala et al.*, 2001]. Since the majority of southwest Asia precipitation falls during November–April (Fig. 17.1), the winter season is the primary control on water resources over the region [*Barlow and Tippett*, 2008]. Over lower elevation areas where rainfall is more prevalent during November–April, below-average precipitation results in below-average soil moisture [*Hoell et al.*, 2015] that is harmful to wintertime cropping activities. Additionally, snowfall over higher elevation areas of southwest Asia is critically important to springtime and summertime runoff, which provides the necessary water to refill basins and sustain summertime cropping activities. Consequently, below average wintertime snowfall and a subsequent reduction in warm season runoff [*Hoell et al.*, 2015] place hundreds of millions of people at risk by diminishing annual water resources.

Southwest Asia precipitation demonstrated strong year-to-year variations during November–April 1948–2012. The strongest regional pluvials exceeded 135% of average seasonal rainfall while the strongest regional droughts fell below 65% of average seasonal rainfall. The region is prone to multiyear droughts and pluvials, as evidenced by the pluvial periods of the 1940s–1950s and early 1980s and the drought periods of the late 1990s to the present. The most extreme November–April drought seasons over southwest Asia during 1948–2012 were concentrated near the end of the period, 1999–2000, 2000–2001, and 2007–2008, and in the middle of the period, 1970–1971, 1983–1984, and 1988–1989. In this chapter, we examine the top six regional droughts over southwest Asia, which correspond to the driest 10% of precipitation seasons during 1948–2012, to better understand the blueprint for extreme regional drought in terms of SST, global atmospheric circulations, and the dynamic atmospheric forcing of regional precipitation.

In section 17.2, the spatial pattern of precipitation over southwest Asia and the surrounding regions is examined during the top November–April drought seasons to better understand the regional precipitation signatures of extreme droughts. In section 17.3, the SST during and preceding the extreme November–April southwest Asia drought seasons are examined to better understand the canonical SST expression related to precipitation deficits. In section 17.4, we examine how the atmospheric circulations during the driest southwest Asia precipitation seasons fit into a global framework. In section 17.4, we also examine the regional forcing of precipitation during the driest southwest Asia seasons in terms of atmospheric moisture availability and thermodynamically forced vertical motions. In section 17.5, we provide a summary.

17.2. PRECIPITATION PATTERNS

Precipitation anomalies during the six most extreme southwest Asia droughts of November–April 1948–2012 are displayed in Figure 17.3. During each of these drought seasons, widespread negative precipitation departures were observed throughout southwest Asia. The largest negative precipitation departures occurred over the climatologically rainiest areas, located in the region's high mountains. Positive precipitation anomalies were depicted during 1983–1984 and 2000–2001 over central Iran, a location in which there have been no precipitation observations historically [*Hoell et al.*, 2014b]. While these precipitation anomalies display positive values, it is uncertain whether these departures were in fact positive, or by how much, because of the lack of observations over these areas of Iran.

The most extreme southwest Asia droughts during November–April 1948–2012 are generally related to negative precipitation departures over the surrounding regions, which include the Middle East, Arabian Peninsula, and the Horn of Africa (Figs. 17.3 and 17.4). The 2000–2001 southwest Asia drought was the only true localized drought, with average to above average precipitation immediately to the west and southwest of southwest Asia (Fig. 17.3e). When all six of the droughts shown in Figure 17.3 are considered in aggregate (Fig. 17.4), it is clear that southwest Asia precipitation during the most extreme drought seasons behaves in synchrony with the larger region that includes the entire northwest Indian Ocean rim.

17.3. SST RELATIONSHIPS

Global SST anomalies during the six most extreme southwest Asia droughts of November–April 1948–2012 are displayed in Figure 17.5. The negative SST anomaly over the central tropical Pacific Ocean and the positive SST anomaly over the tropical and extratropical west Pacific Ocean are common to all November–April extreme drought seasons. The Indian Ocean SST anomaly patterns and Atlantic SST anomaly patterns for each November–April season demonstrate considerable interseason differences.

When the six November–April seasons in Figure 17.5 are considered in aggregate, as shown in Figure 17.6, it is clear that a La Niña combined with a warm west Pacific SST signature is the canonical SST pattern related to extreme southwest Asia drought. The negative phasing of the PDV is also prominent in the SST composite of extreme November–April southwest Asia droughts, having occurred during five of the six seasons. The combinations of PDV, La Niña, and a warm west Pacific shown in Figure 17.6 is very similar to the "Perfect Ocean for Drought," identified by *Hoerling and Kumar* [2003], that has been previously shown to strongly affect southwest Asia climate [*Barlow et al.*, 2002; *Hoell et al.*, 2013b; *Hoell et al.*, 2014a]. November–April southwest Asia droughts are related to warm SST over the South Atlantic Ocean and a dipole in SST between the North Atlantic and subtropical North Atlantic Ocean (Fig. 17.6). It is currently unknown how the entire Atlantic Ocean affects southwest Asia.

The aggregate SST anomaly of the months preceding extreme November–April droughts over southwest Asia are displayed in Figure 17.7 to identify the possible SST precursors of drought. Cool northeast Pacific SST preceded five of the six extreme southwest Asia droughts. A warm west Pacific Ocean also preceded five of the six droughts between May and August, and preceded all droughts after September. La Niña–related SST began to develop during the preceding July of all November–April extreme southwest Asia drought seasons. La Niña development prior to all drought seasons continued until November, where at that point the SST pattern and magnitude became phase locked through the following Northern Hemisphere spring.

17.4. ATMOSPHERIC TELECONNECTIONS

The atmospheric circulations during the extreme November–April drought seasons over southwest Asia for 1948–2012 are examined. First, we examine the global

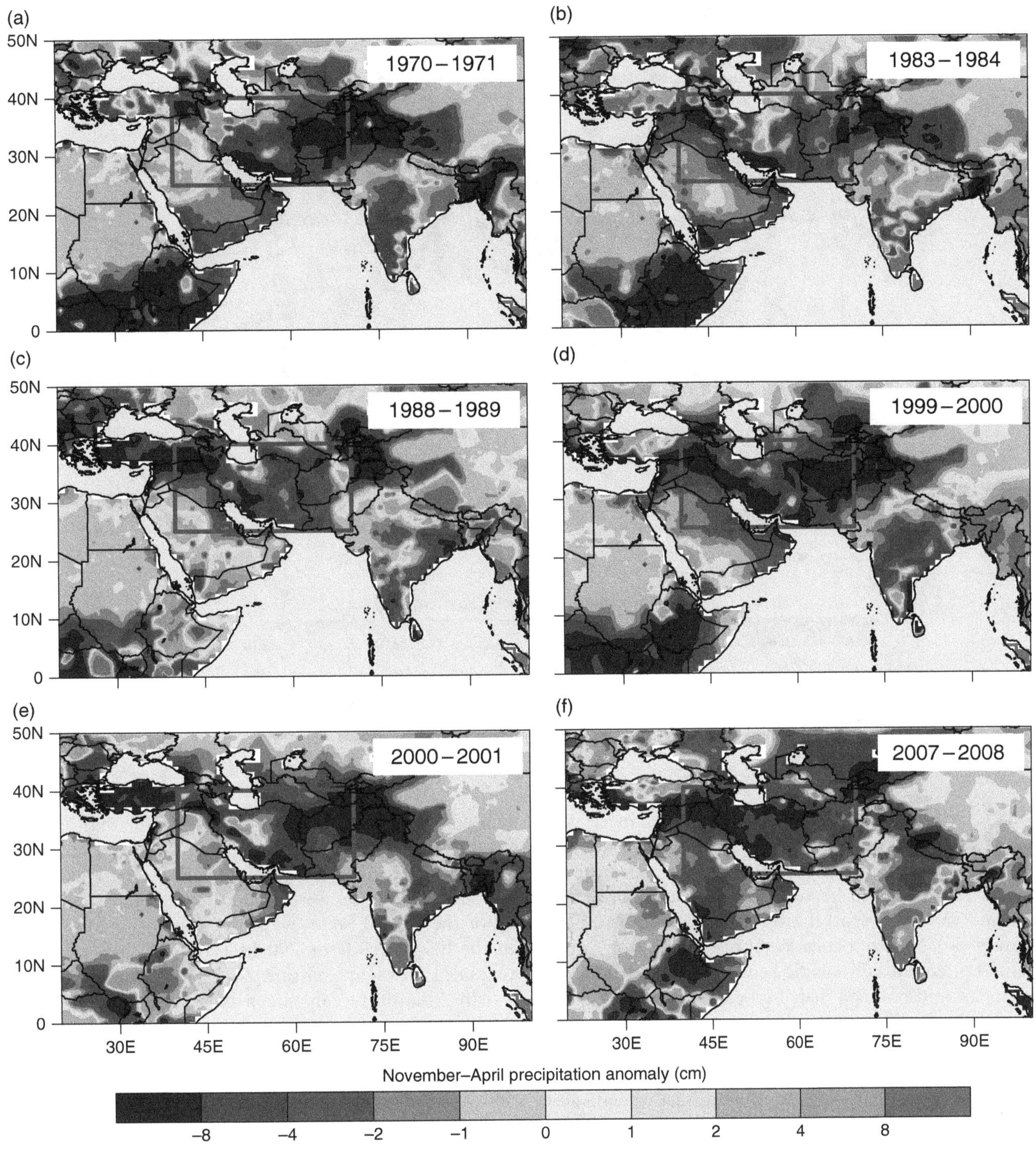

Figure 17.3 November–April precipitation anomaly (cm) for the six driest seasons areal averaged over southwest Asia. *(See insert for color representation of the figure.)*

atmospheric circulation to determine how extreme southwest Asia drought fits into a global framework. Second, we examine the atmospheric circulations in terms of atmospheric moisture availability and vertical motions to identify the mechanisms directly responsible for extreme southwest Asia drought.

17.4.1. Global Teleconnections

All six of the most extreme November–April southwest Asia droughts for 1948–2012 are related to a convection dipole between the Maritime Continent and tropical central Pacific Ocean (Fig. 17.8) consistent with the forcing

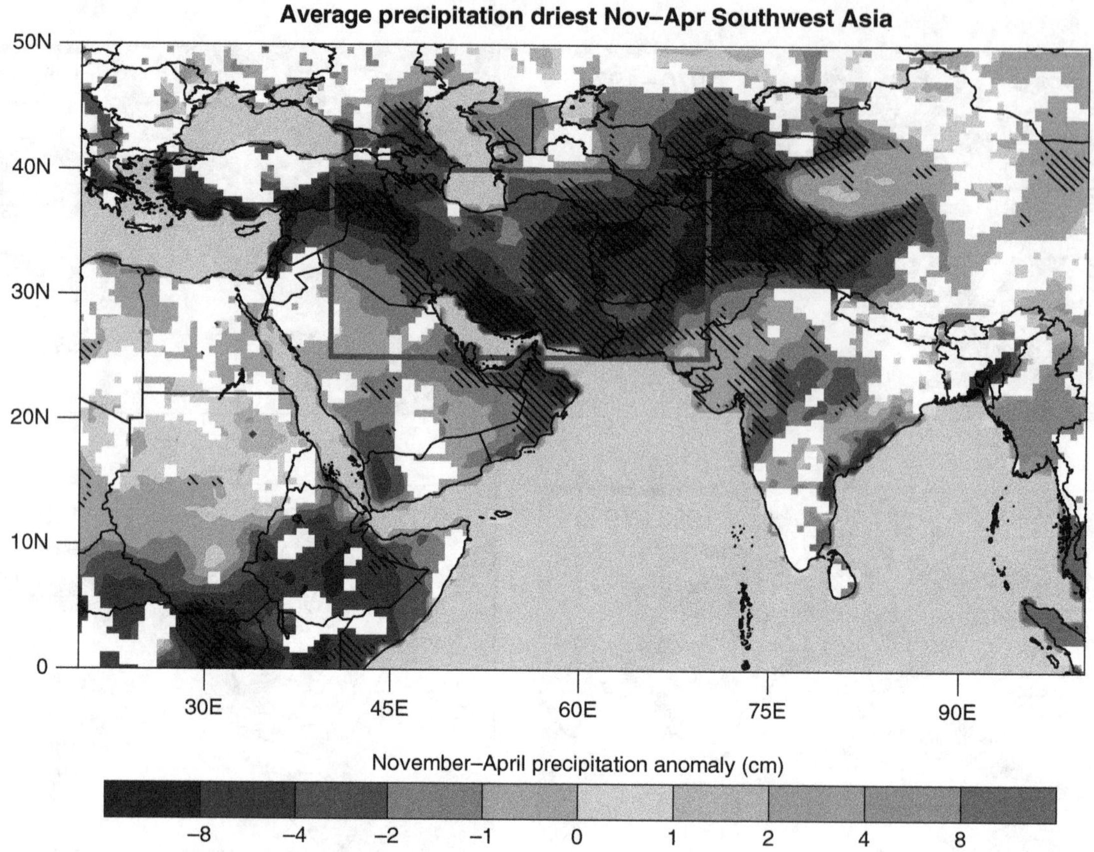

Figure 17.4 Composite precipitation anomaly (cm) for the six driest November–April seasons areal averaged over southwest Asia. Shading indicates that at least four November–April seasons had the same sign of precipitation anomaly while hatching indicates that all six November–April seasons had the same sign of precipitation anomaly.

by SST related to ENSO, LT, and PDV. Enhanced convection occurred over the Maritime Continent, extending to the northeast and southeast into the extratropical Pacific Ocean in both hemispheres. Reduced convection occurred over the central tropical Pacific Ocean. The changes in midtropospheric atmospheric convection resulted in enhanced diabatic heating over the Maritime Continent and reduced diabatic heating over the central tropical Pacific Ocean, which in turn forced the global atmosphere [*Barlow et al.*, 2002; *Barlow et al.*, 2007; *Hoell et al.*, 2013b; *Hoell et al.*, 2014a].

Extreme November–April southwest Asia droughts are related to zonally symmetric high pressure throughout the midlatitude Northern Hemisphere at 200 hPa (Fig. 17.9), which is a characteristic hemispheric circulation signal during La Niña events [*Hoell et al.*, 2013b]. The impacts of the teleconnections responsible for southwest Asia extreme droughts are indeed global.

Greater than four of the six extreme southwest Asia droughts during November–April 1948–2012 were related to equivalent barotropic high pressure across the North Pacific Ocean, western North America, and the southern United States, as demonstrated by greater than average heights at 200 hPa (Fig. 17.9) and 700 hPa (Fig. 17.10) across those regions. Only three of the six extreme southwest Asia droughts were related to high pressure across all of the United States, which is why no signal is shown in the height fields of 200 hPa and 700 hPa over that region. Equivalent barotropic high pressure over the North Atlantic Ocean, much of Europe, and northern Africa was also a common occurrence during extreme southwest Asia droughts.

The atmospheric circulation across Asia between 20°N and 40°N during the extreme November–April southwest Asia droughts of 1948–2012 is complex relative to the rest of the Northern Hemisphere. From east of the Mediterranean to the foothills of the Himalayas, the atmospheric circulation was equivalent barotropic during five of the six events, as high pressure was present in the upper (Fig. 17.9) and lower troposphere (Fig. 17.10). However, east of the Himalayas toward the Pacific Ocean, the atmospheric circulation was baroclinic during all six events, as high pressure was present in the upper troposphere (Fig. 17.9) and low pressure was present in

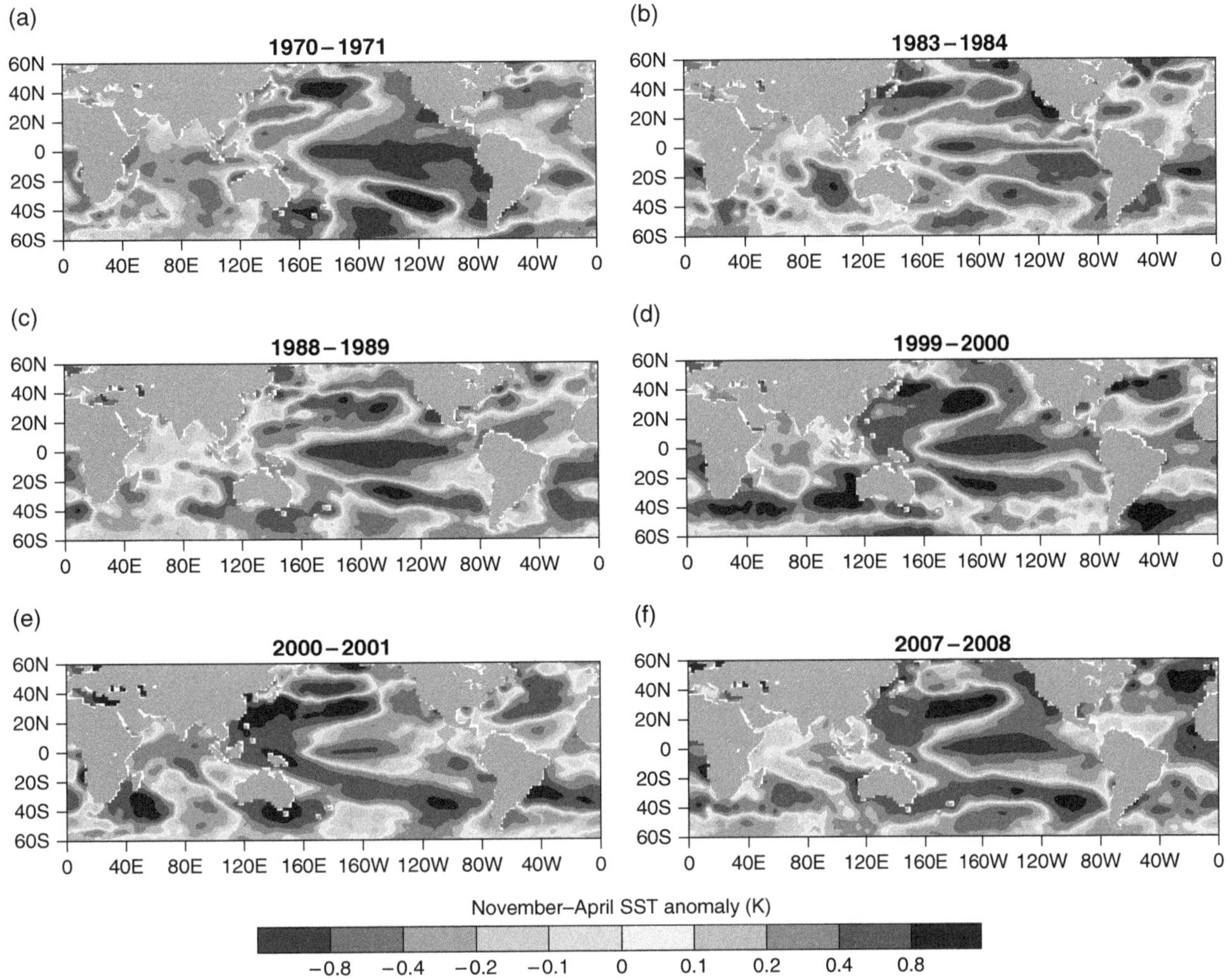

Figure 17.5 November–April SST anomaly (K) for the six driest seasons areal averaged over southwest Asia. *(See insert for color representation of the figure.)*

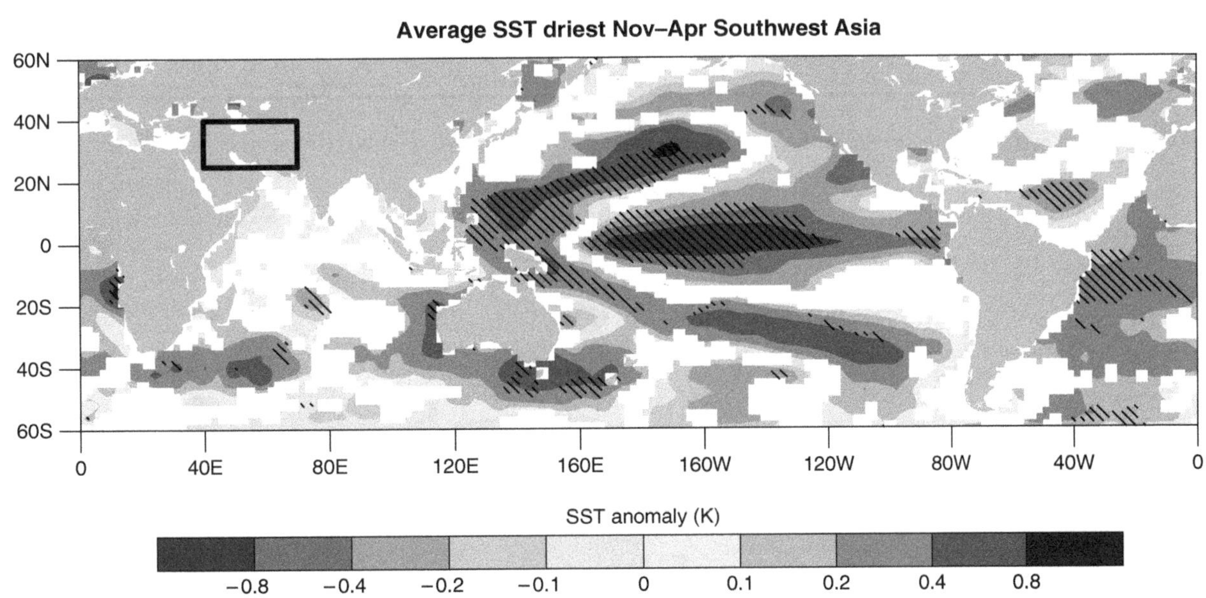

Figure 17.6 Composite SST anomaly (K) during the six driest November–April seasons areal averaged over southwest Asia. Shading indicates that at least four November–April seasons had the same sign of SST anomaly while hatching indicates that all six November–April seasons had the same sign of SST anomaly.

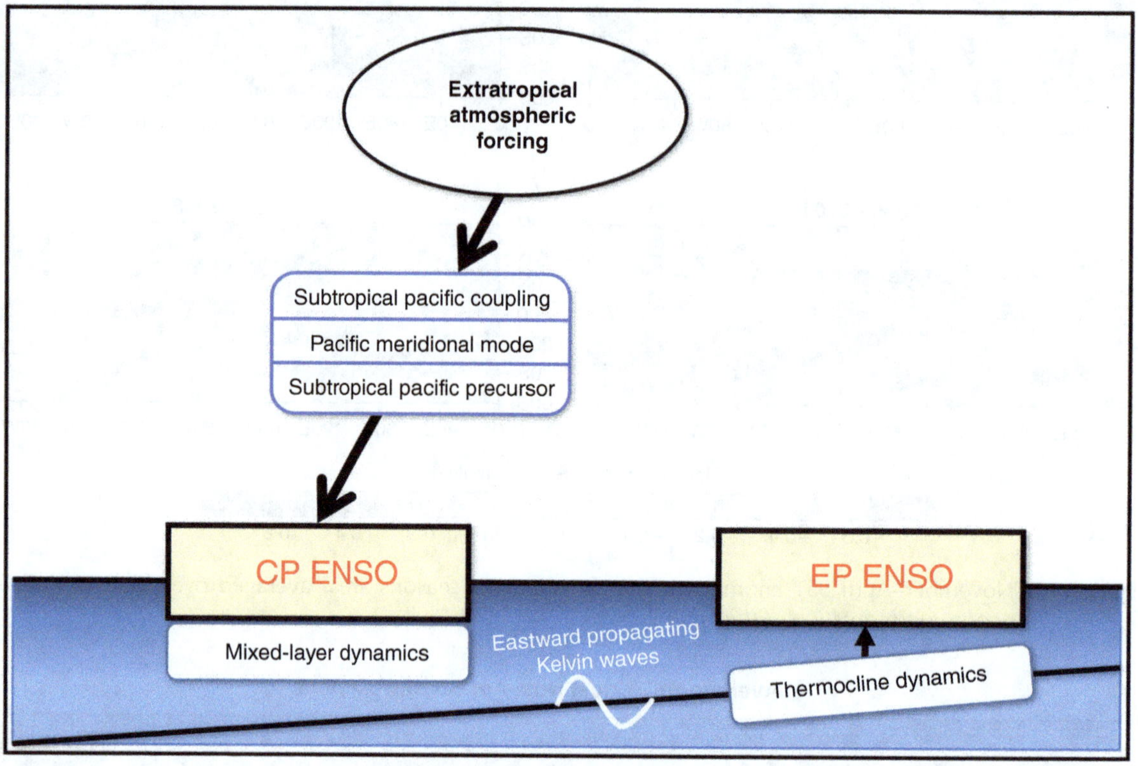

Figure 1.6 A schematic diagram illustrating the possible relationships between the subtropical Pacific precursors and the two types of ENSO (CP and EP).

Climate Extremes: Patterns and Mechanisms, Geophysical Monograph 226, First Edition.
Edited by S.-Y. Simon Wang, Jin-Ho Yoon, Christopher C. Funk, and Robert R. Gillies.

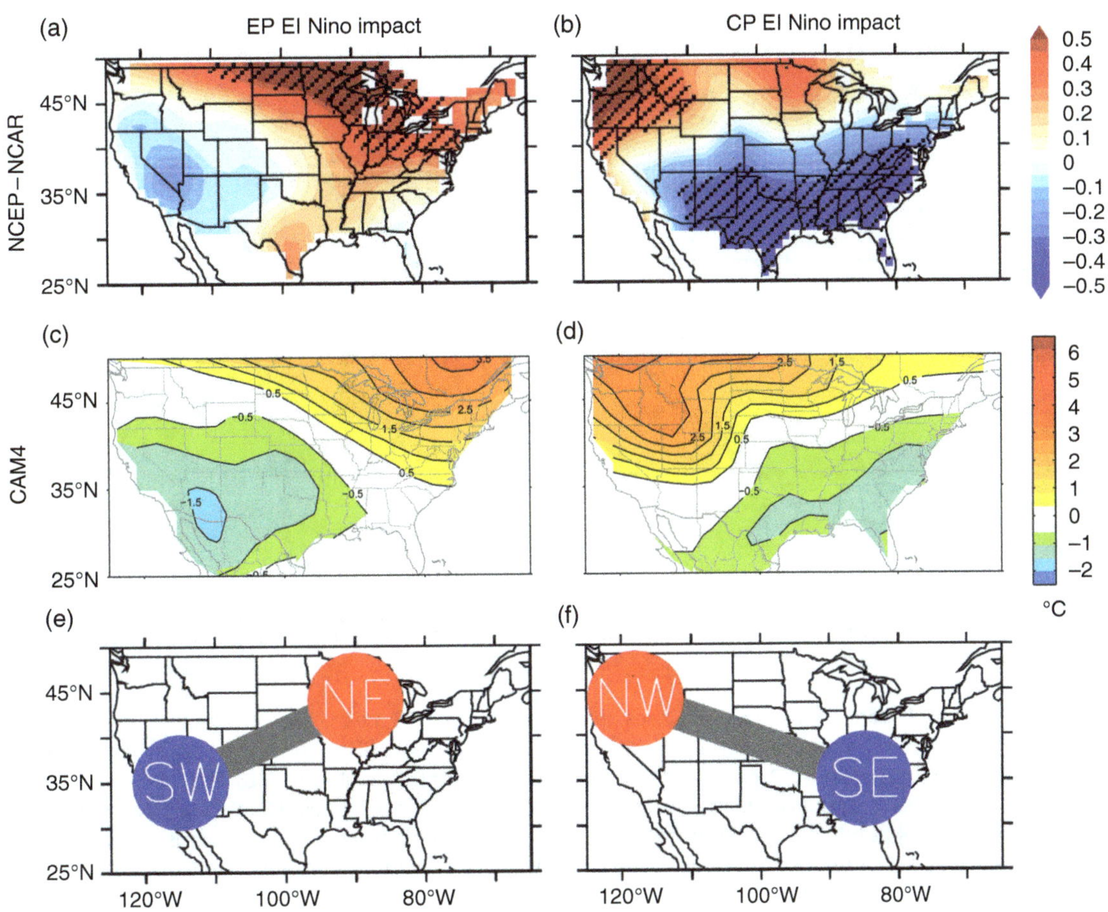

Figure 1.9 EP (left panels) and CP (right panels) El Niño impacts on US winter (January–March) surface air temperatures. Observed temperature anomalies regressed onto the (a) EP and (b) CP ENSO indices. Coefficients exceeding the 90% confidence level are dotted. The NCEP/NCAR reanalysis dataset for the period 1950–2010 is used. Temperature patterns reproduced by the Community Atmosphere Model version 4 (CAM4) in the (c) EP and (d) CP experiments. Only differences exceeding the 90% confidence level are shaded. Schematic diagrams of the (e) EP and (f) CP El Niño impacts on temperatures. Adapted from *Yu et al.* [2012b].

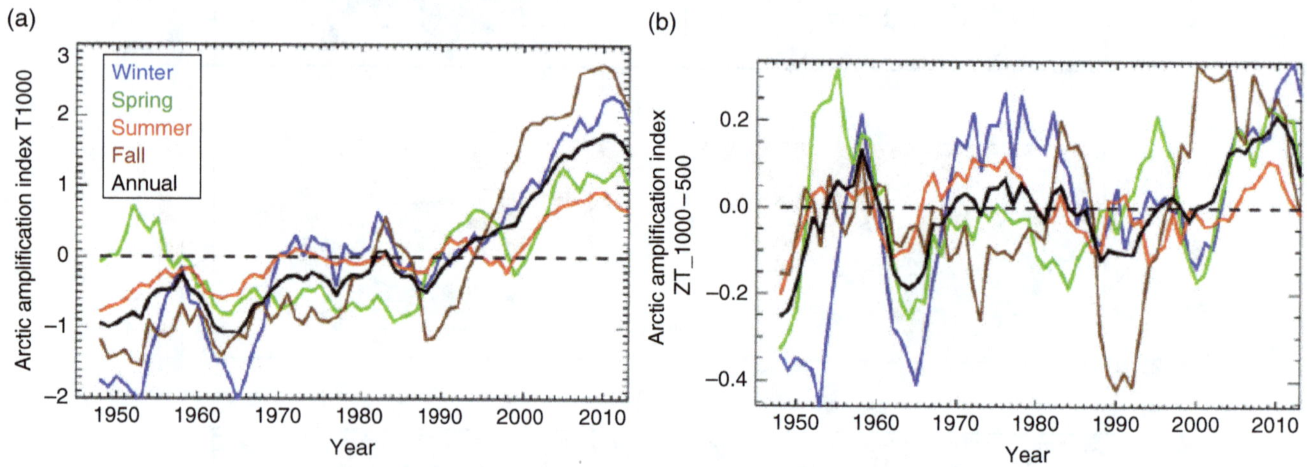

Figure 2.1 Arctic amplification seasonal time series (relative to 1948–2013 mean, °C) between the Arctic (70–90°N) and midlatitudes (30–60°N) based on two metrics: (a) differences in 1000 hPa temperature anomalies (relative to 1948–2013 mean, °C) between the Arctic (70–90°N) and midlatitudes (30–60°N), and (b) differences in 1000–500 hPa thickness anomalies (percent of mean for each zone) between the two zones. Data were obtained from the NCEP/NCAR Reanalysis at http://www.esrl.noaa.gov/psd/. Reproduced from *Francis and Vavrus* [2015], with permission via the Creative Commons Attribution 3.0 license http://creativecommons.org/licenses/by/3.0/legalcode.

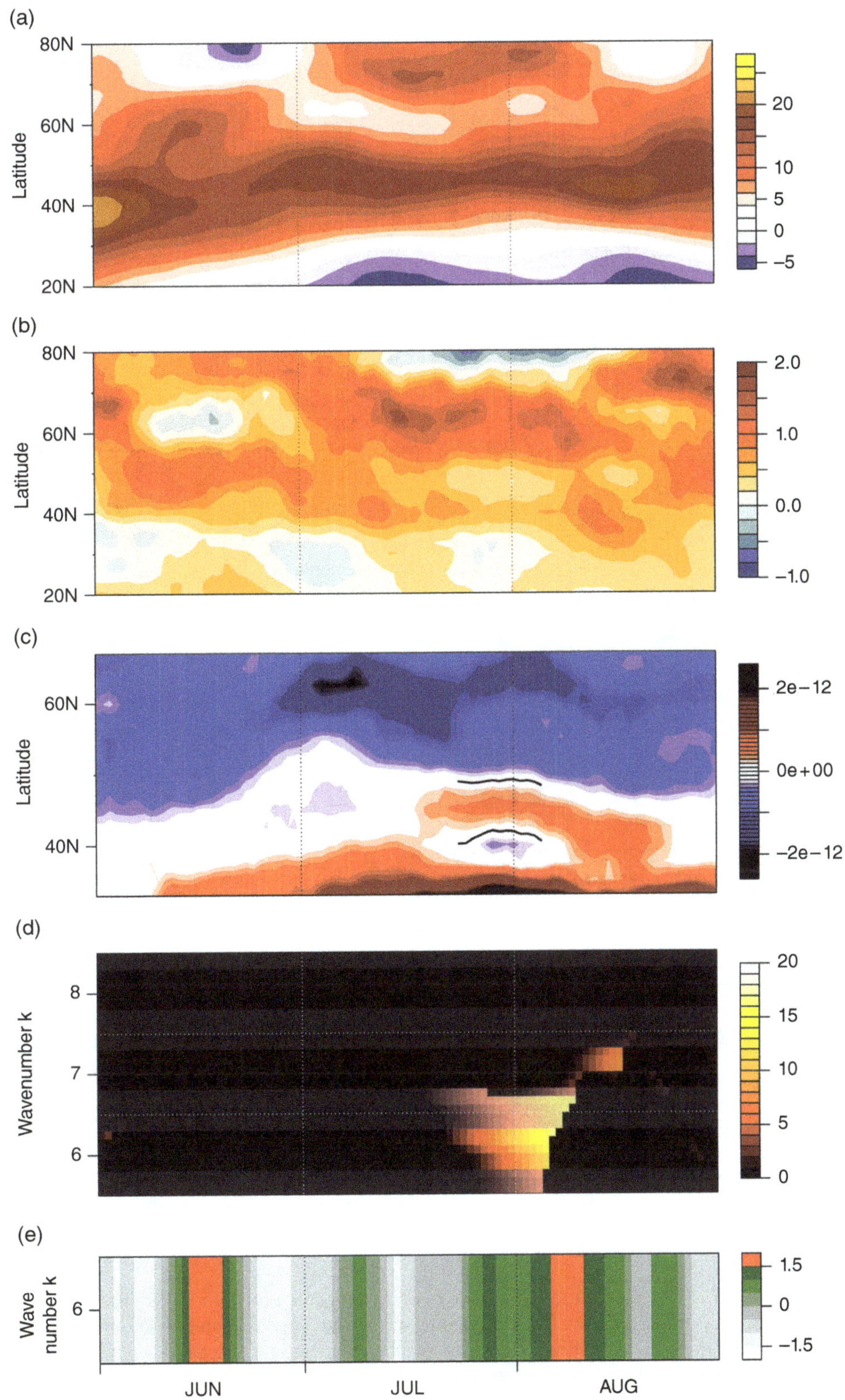

Figure 4.4 Double jet and resonant amplification during summer of 2010. Northern Hemisphere Hovmöller diagrams during summer 2010 for (a) zonal wind, (b) temperature anomaly, and (c) $l^2(k = 6)$. Variables are calculated from the ERA-interim reanalysis using a 15 day running mean low-frequency bandpass filter and averaged over all longitudes. Temperature anomalies from climatology (panel b) were calculated at each grid point before averaging over all longitudes. The black lines in panel (c) indicate the emergence of the two turning points of the waveguide for wave 6. Panel (d) plots the number of consecutive days for which resonance conditions were fulfilled for waves 6–8, and panel (e) plots the observed amplitude of wave 6.

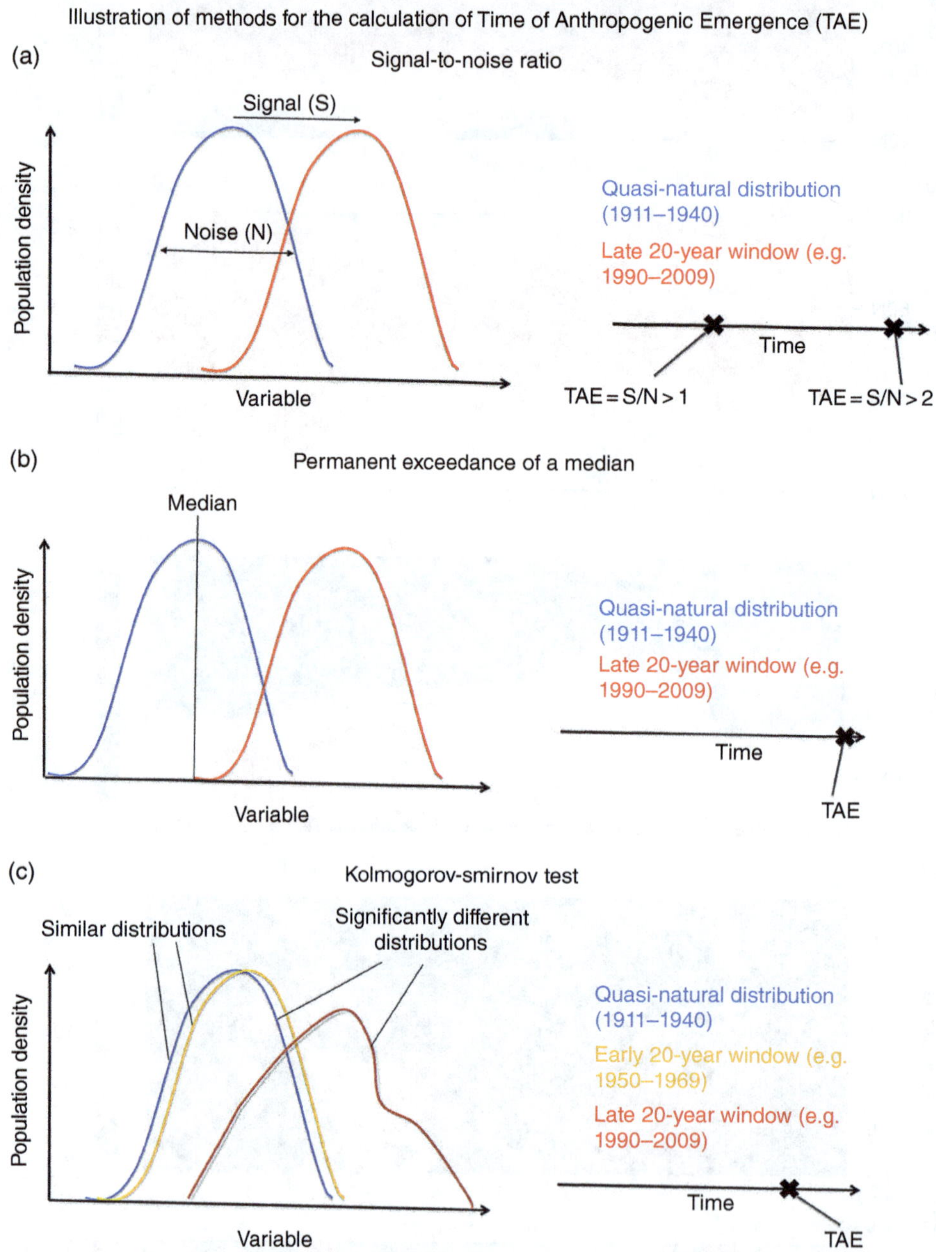

Figure 6.1 An illustration of the calculation of the time of anthropogenic emergence. As time passes, the climate warms further such that it is more different from its preindustrial state (approximated by the blue curve). The time of anthropogenic emergence can be defined as the point in time when (a) the signal-to-noise ratio exceeds a given threshold, (b) the statistical distribution exceeds the previous average (median), or (c) the statistical distributions for the baseline period and a later period are significantly different (see section 6.2, Defining Time of Emergence, for more details).

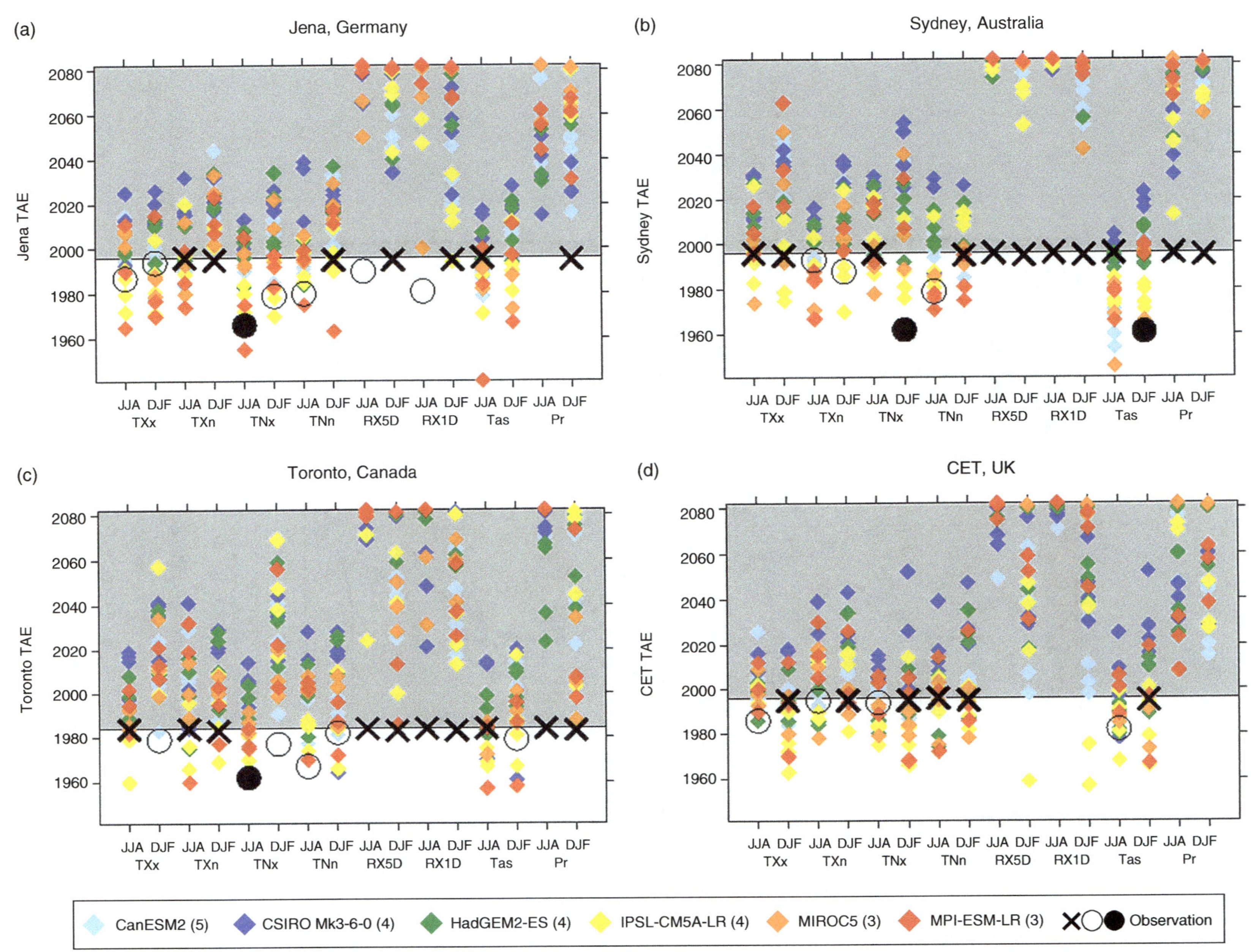

Figure 6.3 Time of anthropogenic emergence and model uncertainty. Graphs of TAE values for (a) Jena, (b) Sydney, (c) Toronto, and (d) Central England Temperature (CET) calculated for each station and corresponding model gridbox extreme and mean index in June–August and December–February shown for CanESM2 (light blue), CSIRO-Mk3-6-0 (dark blue), HadGEM2-ES (green), IPSL-CM5A-LR (yellow), MIROC5 (orange), and MPI-ESM-LR (red). The observed TAE is shown in black. Crosses indicate no emergence in the observations prior to the end of the time series, empty circles indicate emergence up to 20 yr prior to the end of the time series and filled black circles indicate emergence at least 20 yr prior to the end of the series. In observed time series with missing values, no symbol is shown.

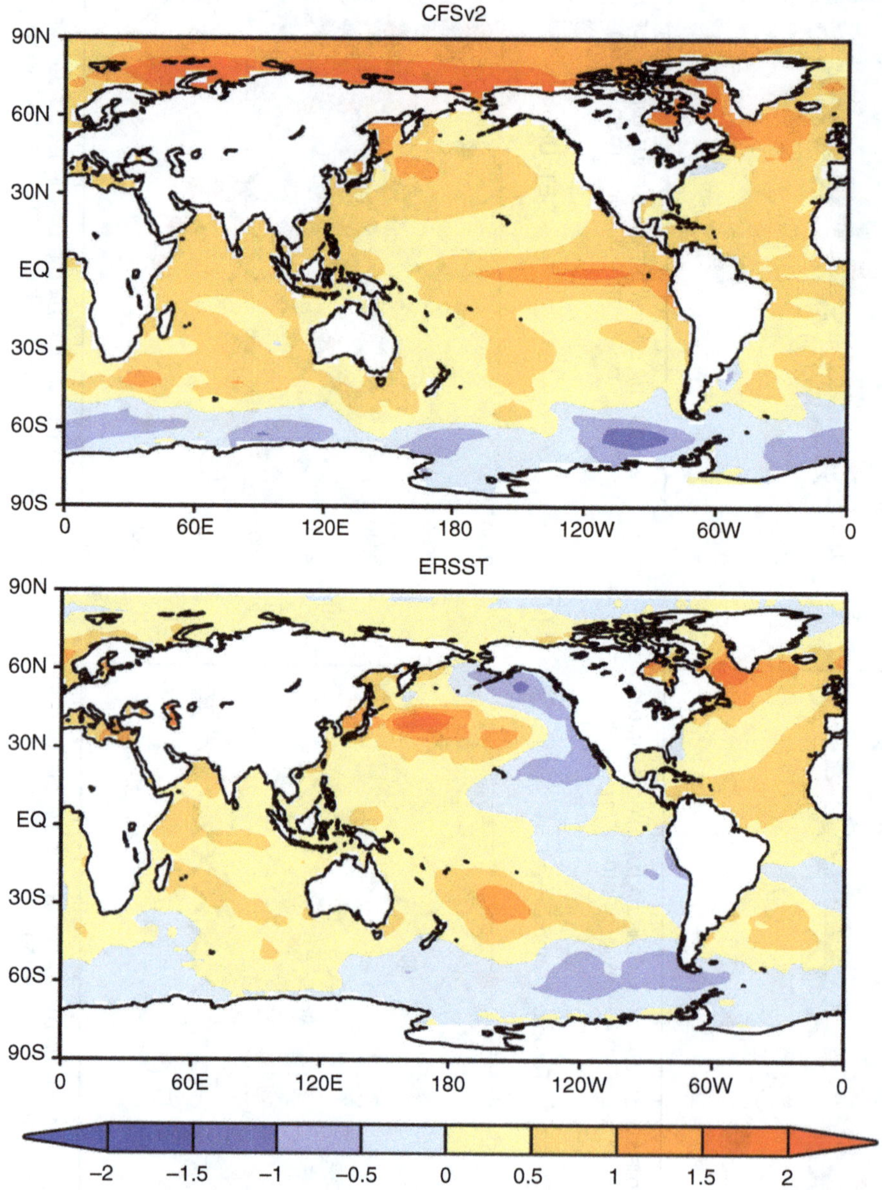

Figure 7.7 Global sea surface temperature anomaly differences for the periods of 2009–2012 minus 1983–1986 in the CFSv2 (upper) and the ERSSTv3 (lower).

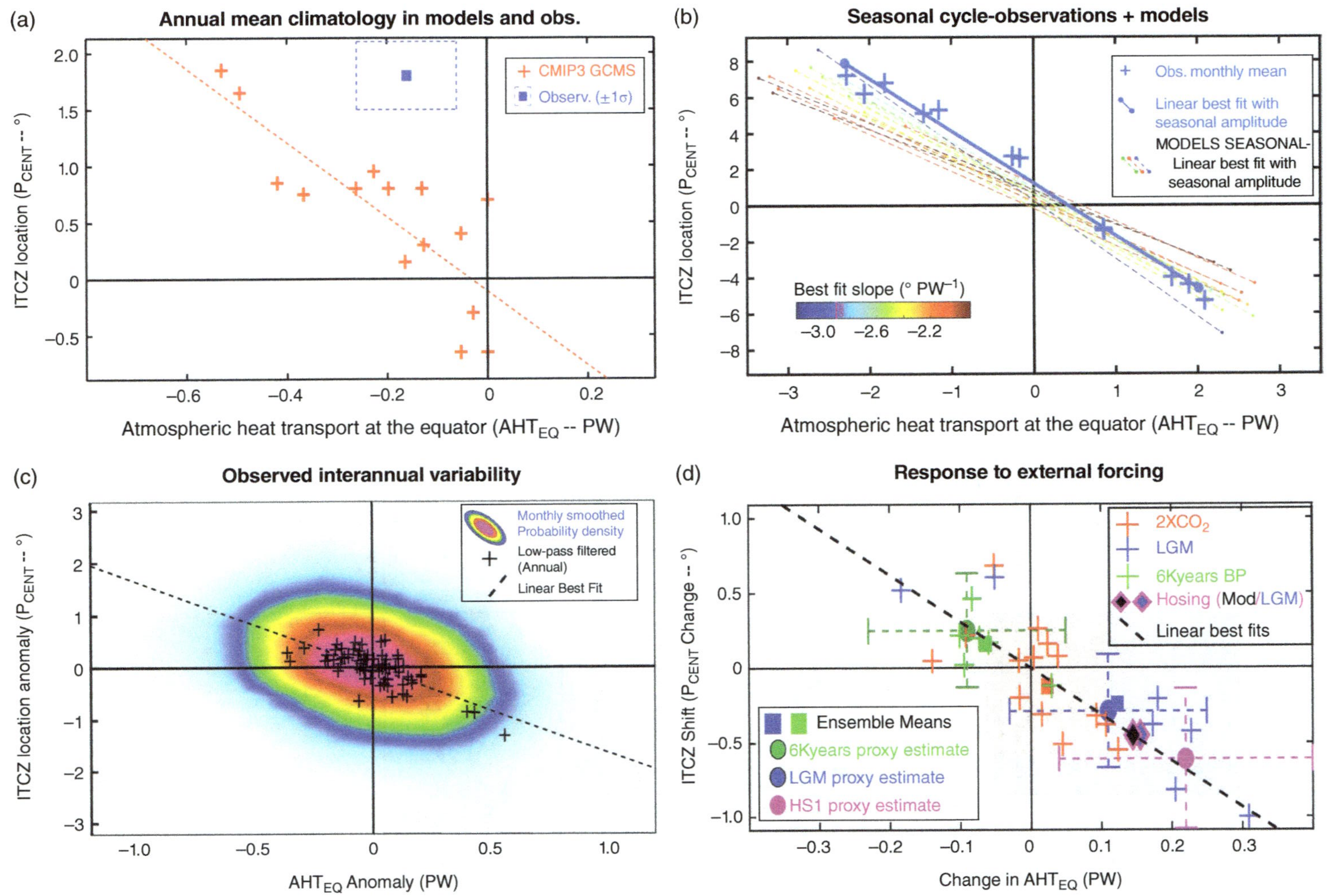

Figure 8.3 The relationship between ITCZ location and the atmospheric energy transport across the equator spanning different timescales in observation-based estimates, models, and paleoclimate reconstructions. In all cases, the centroid of tropical precipitation (P_{CENT}; between 20N and 20S) is used as a metric for the ITCZ location. Although the domain and range differ substantially between the panels, the relative scaling of the axes is constant throughout the panels. (a) Annual mean climatology in (red crosses) coupled CMIP3 preindustrial simulations and (blue box) observational estimates with uncertainty (1σ) indicated by the dashed blue rectangle, based on interannual variability and spread between the different sets of reanalyses [*Marshall et al.*, 2013]. (b) Observational estimate of seasonal cycle based on NCEP reanalysis energy fluxes and NOAA CPC merged analysis precipitation [*Donohoe et al.*, 2013b]. Monthly means are the blue crosses and the thick dashed blue line represents the linear best fit with length equal to the amplitude (first harmonic) of the seasonal cycle. Each of the dashed lines represents the slope and amplitude of a CMIP3 model with the color representing the slope value as given by the color bar inset in the figure. (c) Observational estimates of interannual variability [*Donohoe et al.*, 2014a] of the ITCZ (from NOAA CPC merged analysis) and AHT_{EQ} (from NCEP reanalysis). The colors are the smoothed probability distribution function of monthly anomalies, the crosses are the annual data (low pass filtered with cutoff period of 1 yr) and the dashed line is the linear best fit. (d) Response to anthropogenic and paleoclimate forcing [*Donohoe et al.*, 2013b] alongside proxy estimates based on sea surface temperature reconstructions [*Mcgee et al.*, 2014]. Each cross is a different simulation, color coded by experiment type (red=$2XCO_2$, blue=LGM, and green=mid-Holocene) and the solid squares and shaded areas represent the ensemble average and one standard deviation. Circles and dashed crosses are the paleo estimates and uncertainty.

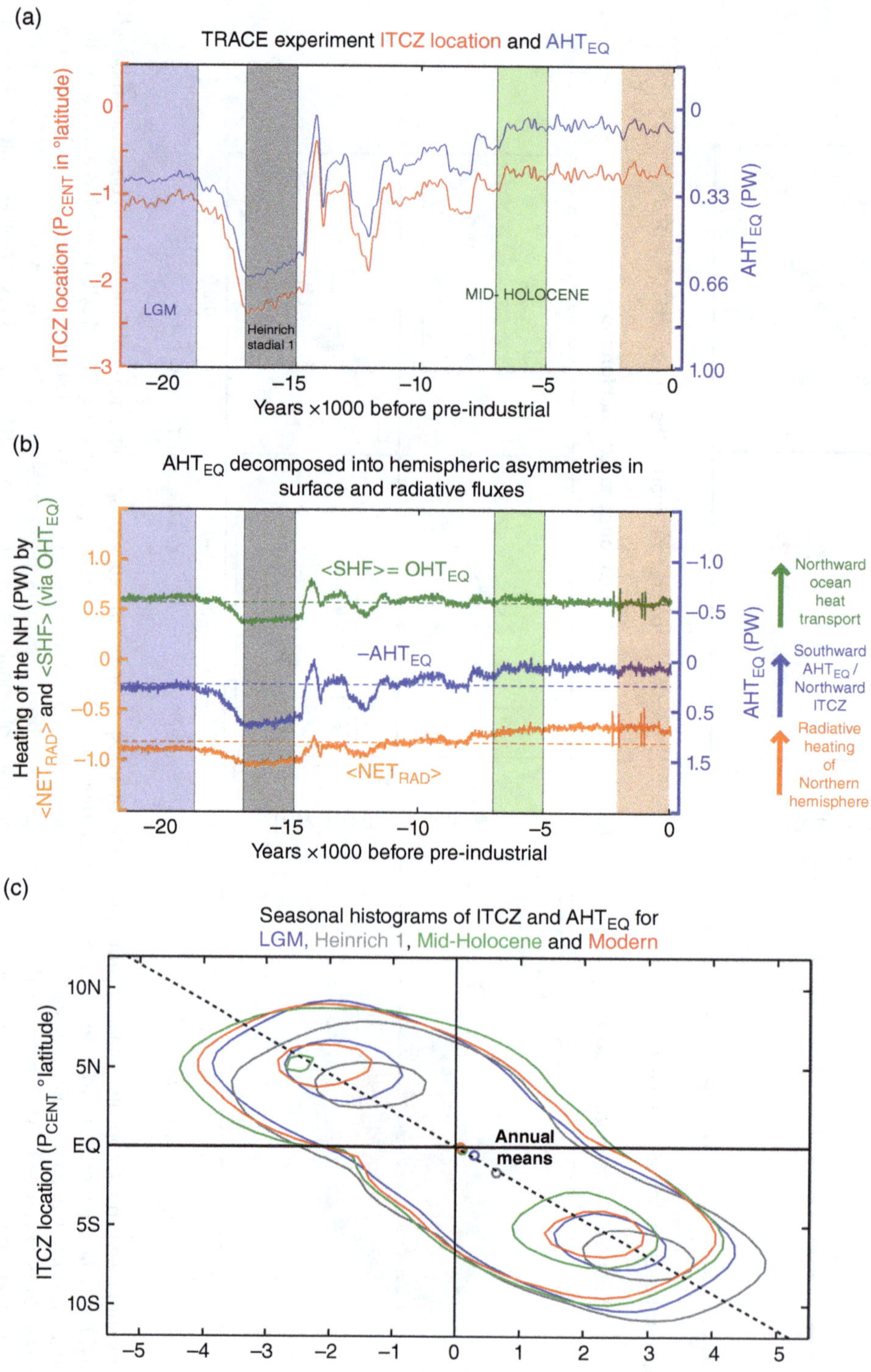

Figure 8.4 (a) Time series of 10 yr low pass filtered (red) ITCZ location (as measured by P_{CENT}) and (blue) AHT_{EQ} in the CCSM3 transient climate evolution (TraCE) experiment. The AHT_{EQ} has been negated and scaled by a factor of 3 as shown on the *y*-axis to the right. The vertical shading shows the time periods used to define the (blue) LGM, (gray) Heinrich Stadial 1, and (green) mid-Holocene. (b) The partitioning of AHT_{EQ} into (green) surface and radiative (orange) heating of the NH via equations (8.3) and (8.4). The AHT_{EQ} has been negated (right *y*-axis) so that the positive axis corresponds to radiative heating of the NH and a northward ITCZ shift. (c) The contours show smoothed probability distribution functions in the $AHT_{EQ}/ITCZ_{LOC}$ plane for the (blue) LGM, (gray) Heinrich Stadial 1, (green) mid-Holocene, and (red) modern. Annual mean values are shown with small circles. The dashed black line shows the linear best fit to all the data, which fall on the seasonal best fit.

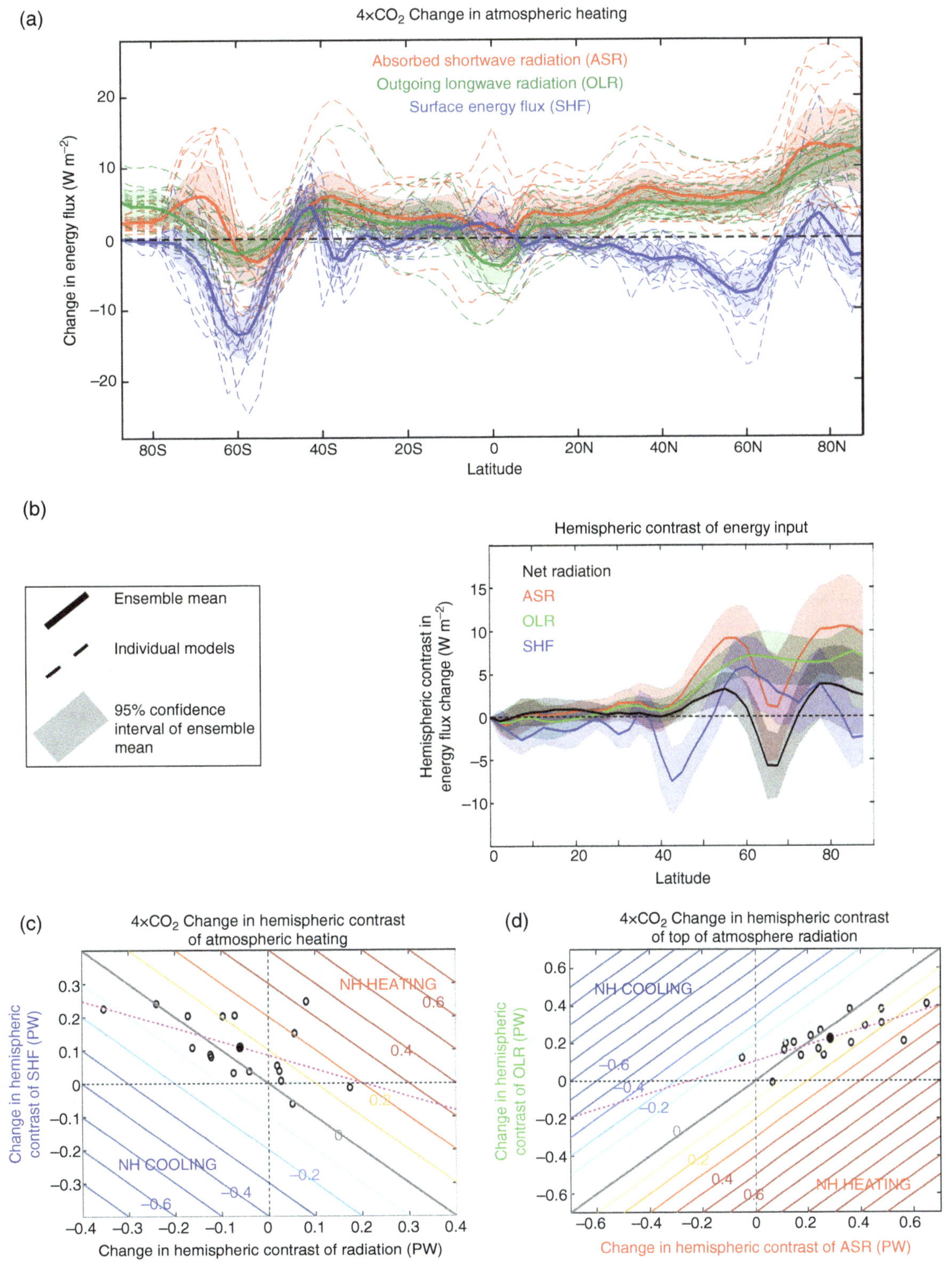

Figure 8.6 (a) The zonal mean change in energy input to the atmosphere in the 4XCO$_2$ CMIP5 simulations. ASR is shown in red, OLR in green, and the surface energy flux in blue. The dashed lines are the individual models, the solid line is the ensemble mean, and the shaded region is the 95% confidence interval of the ensemble mean. (b) As in (a) except the hemispheric contrast of the energy flux change (half the difference between the NH and SH) with total change in energy input to the atmosphere in black. (c) The spatially integrated hemispheric contrast of energy input to the NH atmosphere by SHF (abscissa) and TOA radiation (ordinate). The total change in energy input is the sum of the abscissa and ordinate and is indicated by the diagonal colored contours with slope negative unity. The zero contour is gray. The open circles are individual models and the solid circle is the ensemble mean. The dashed purple line is the linear best fit. (d) As in (c) but for the decomposition of the change in the hemispheric contrast of TOA radiation into ASR and OLR components. In this case, the total TOA change is the difference between ASR and OLR and is indicated by the slanted color lines with positive unity slope.

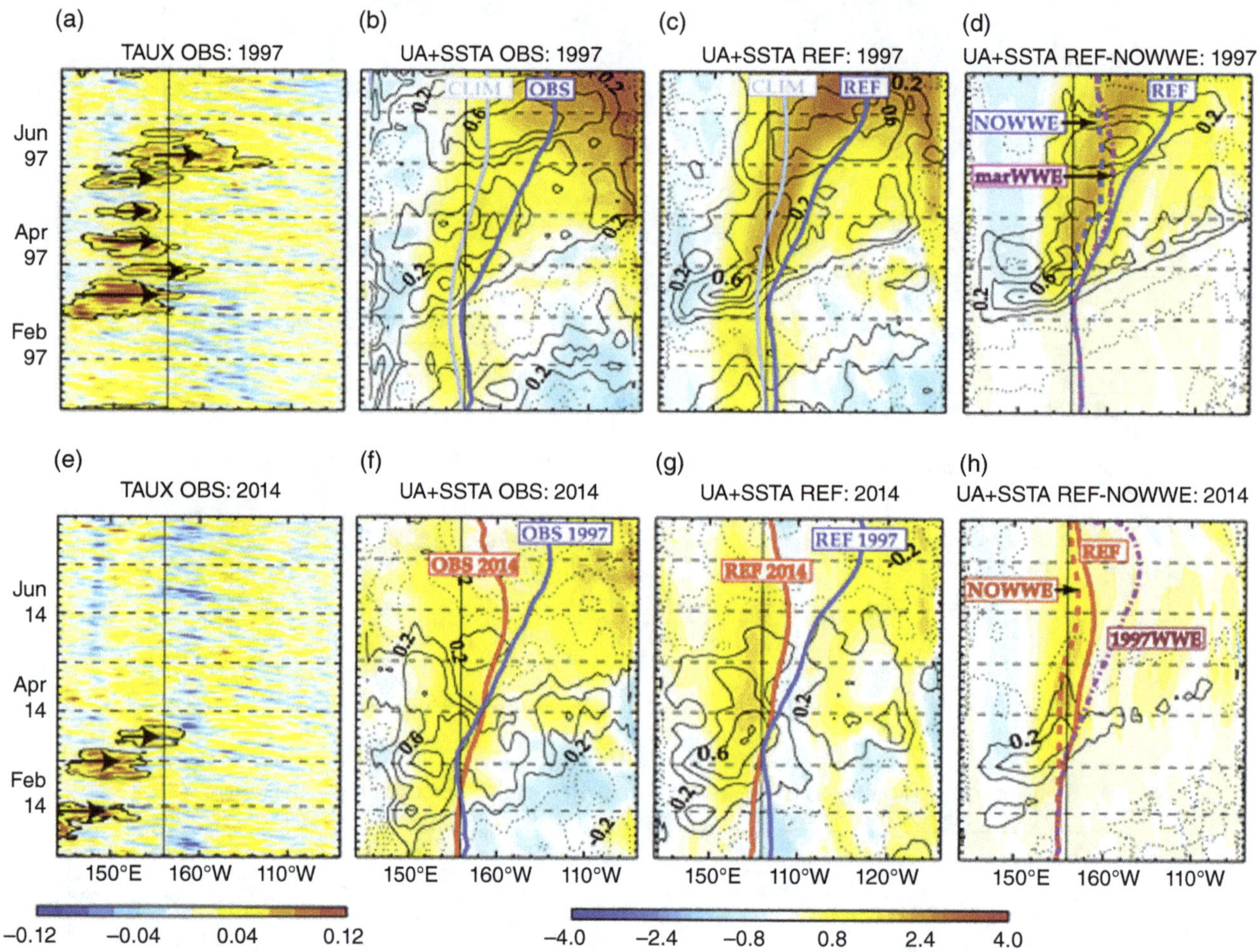

Figure 9.9 January–July 1997 3°N–3°S zonal time section of (a) zonal wind stress referred to as TAUX anomalies (colors with observations [OBS] referring to the NCEP2 model forcing) and WWEs (contours) and SST (color), surface currents (thin black contours) anomalies for (b) observations, (c) REF experiment, and (d) REF minus NO WWE experiments. (e–h) Same as (a–d) but for 2014. Plain blue lines indicate the EEWP position defined as the longitudinal position of the 28.5°C isotherm in the 3°S–3°N band. Plain grey lines indicate the climatological position of the EEWP. In (d), the dashed blue (respectively, dash-dotted purple) line indicates the EEWP position for the NOWWE (respectively, MarWWE) experiments. In (f) and (g), the EEWP is indicated in blue for 1997 and in red for 2014. In (h), the dashed red (respectively, dash-dotted purple) line indicates the EEWP position for the NO WWE (respectively, 1997WWE) experiments. Labels of the experiment EEWP are added for clarity on each panel. Reproduced from *Menkes et al.* [2014].

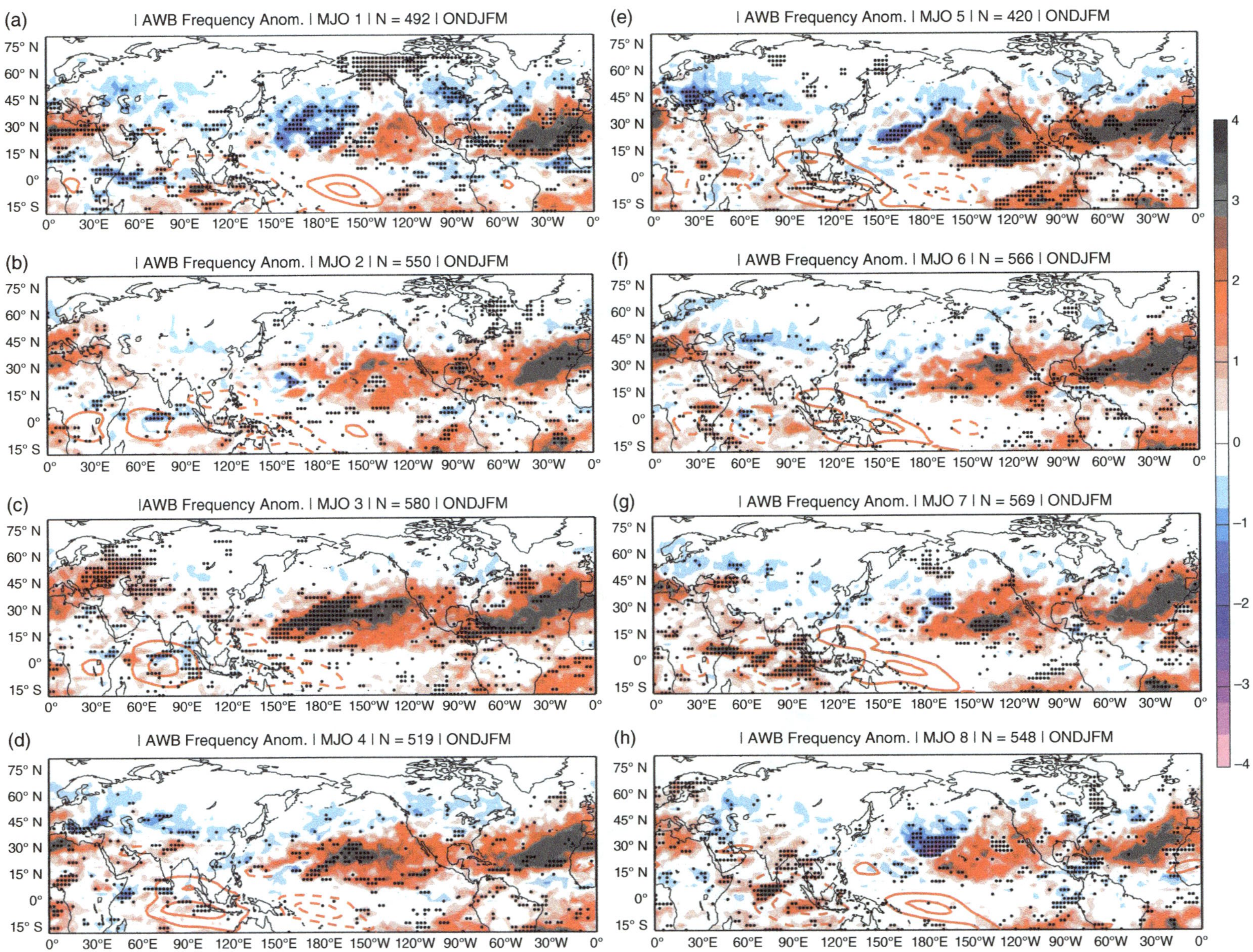

Figure 9.13 Composite during specified MJO phase and events that occur from October–March. Combined 330K and 350K AWB frequency departure from climatology (shaded) according to color bar in units of % day-1. Solid (dashed) red lines indicate areas of negative (positive) MJO filtered OLR starting at negative (positive) 6 W m-2 and contoured by 6 W m-2. Black dots indicate areas where the shaded values are statistically significantly different from climatology at the 90% level using a t-test.

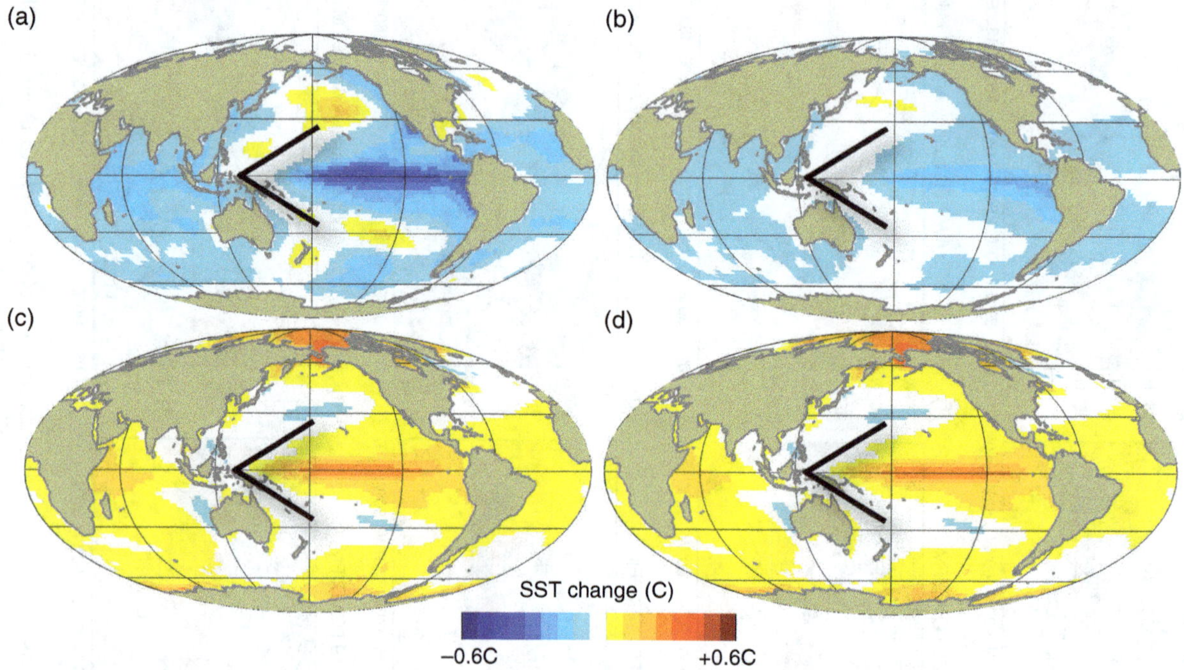

Figure 10.3 ENSO-related changes between 1999–2014 and 1979–1998 for (a) observed ERv4 DJFMAM SSTs and (b) JJASON SSTs and simulated (CESM1) SSTs for (c, d) DJFMAM and JJASON. Black lines denote schematic representation of the western V region.

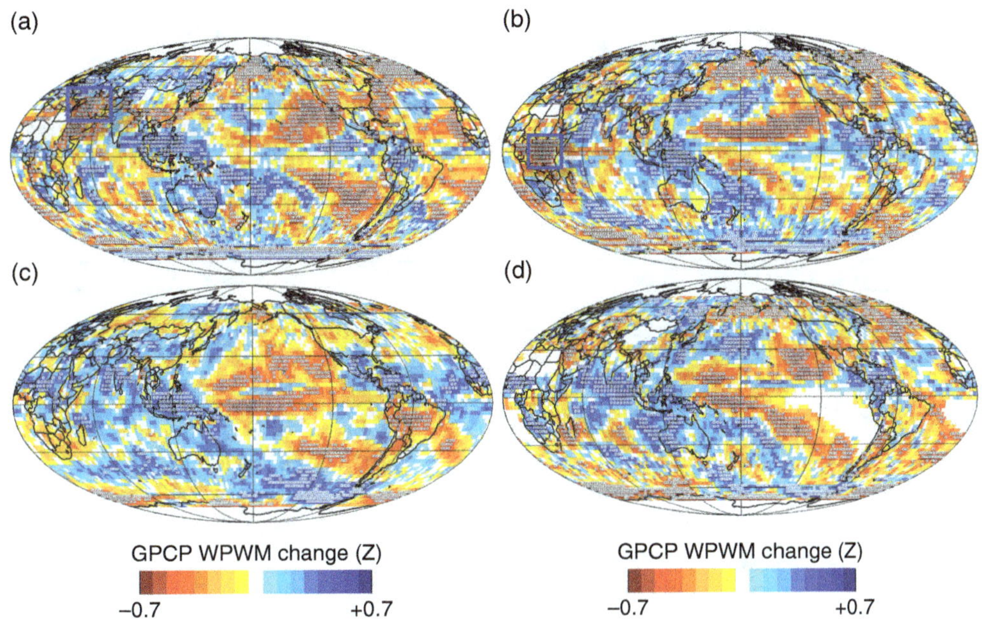

Figure 10.7 1999–2014 versus 1979–1998 GPCP changes associated with the WPWM EOF for (a) JFM, (b) AMJ, (c) JAS, and (d) OND.

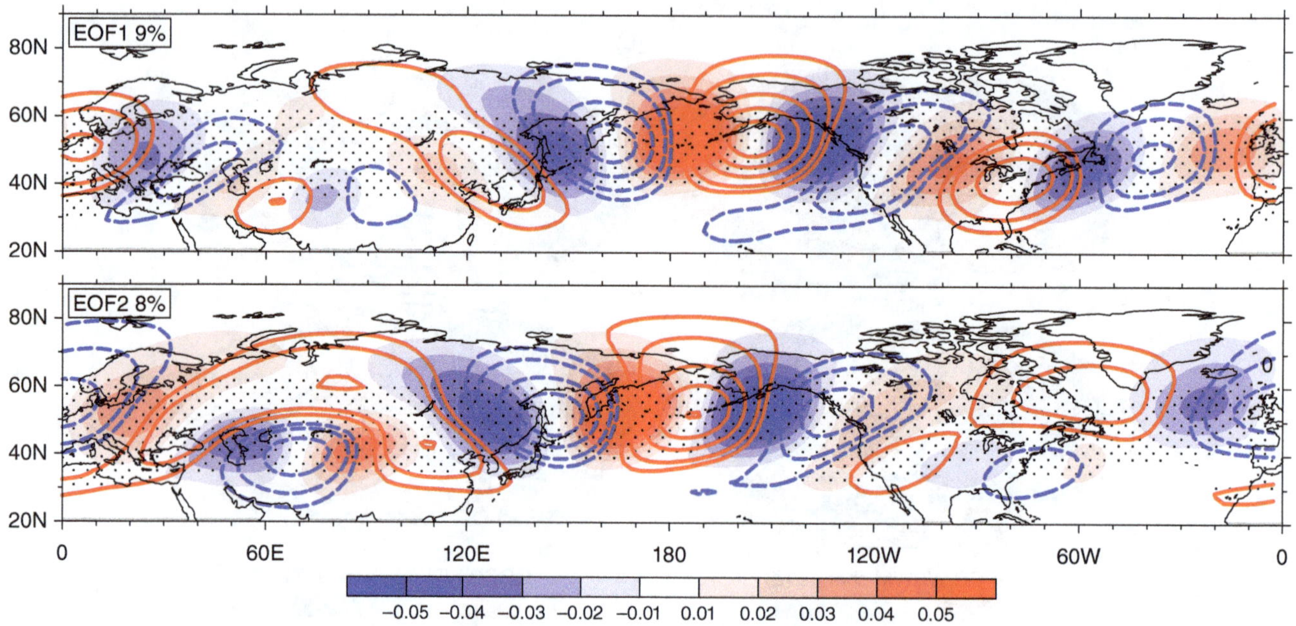

Figure 11.7 The leading two EOFs of 20–90 day band-pass filtered 300 hPa meridional wind (shading) and the corresponding regressed 300 hPa streamfunction anomalies (contour) during June to August from a 1000 year CAM3 integration. The contour intervals for the streamfunction anomalies are ±0.5, 1, 2, 3, 4 x 10^6 m^2s^{-1}. Stippling denotes regions with 300 hPa seasonal mean zonal wind exceeding 10 ms^{-1}.

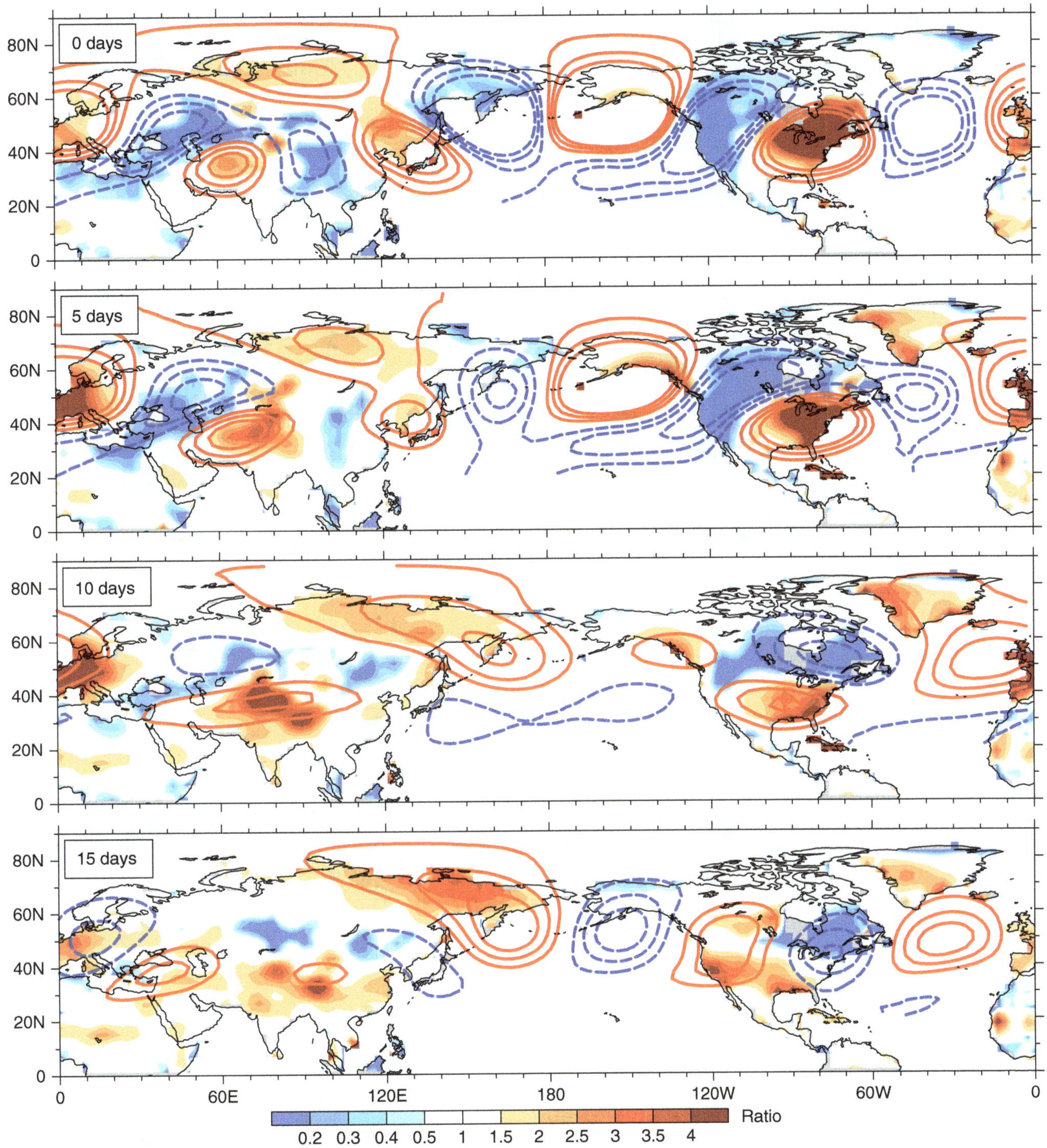

Figure 11.9 Shading represents ratio of probability of a heat wave occurring 0 days (top), 5 days (second row), 10 days (third row), and 15 days (bottom) after an extreme CGT1 event relative to probability of a heat wave without any precondition. Contour represents composite of 300 hPa streamfunction anomalies of those extreme CGT1 events on the corresponding days at ±1, 2, 3 x 10^6 m^2s^{-1} intervals.

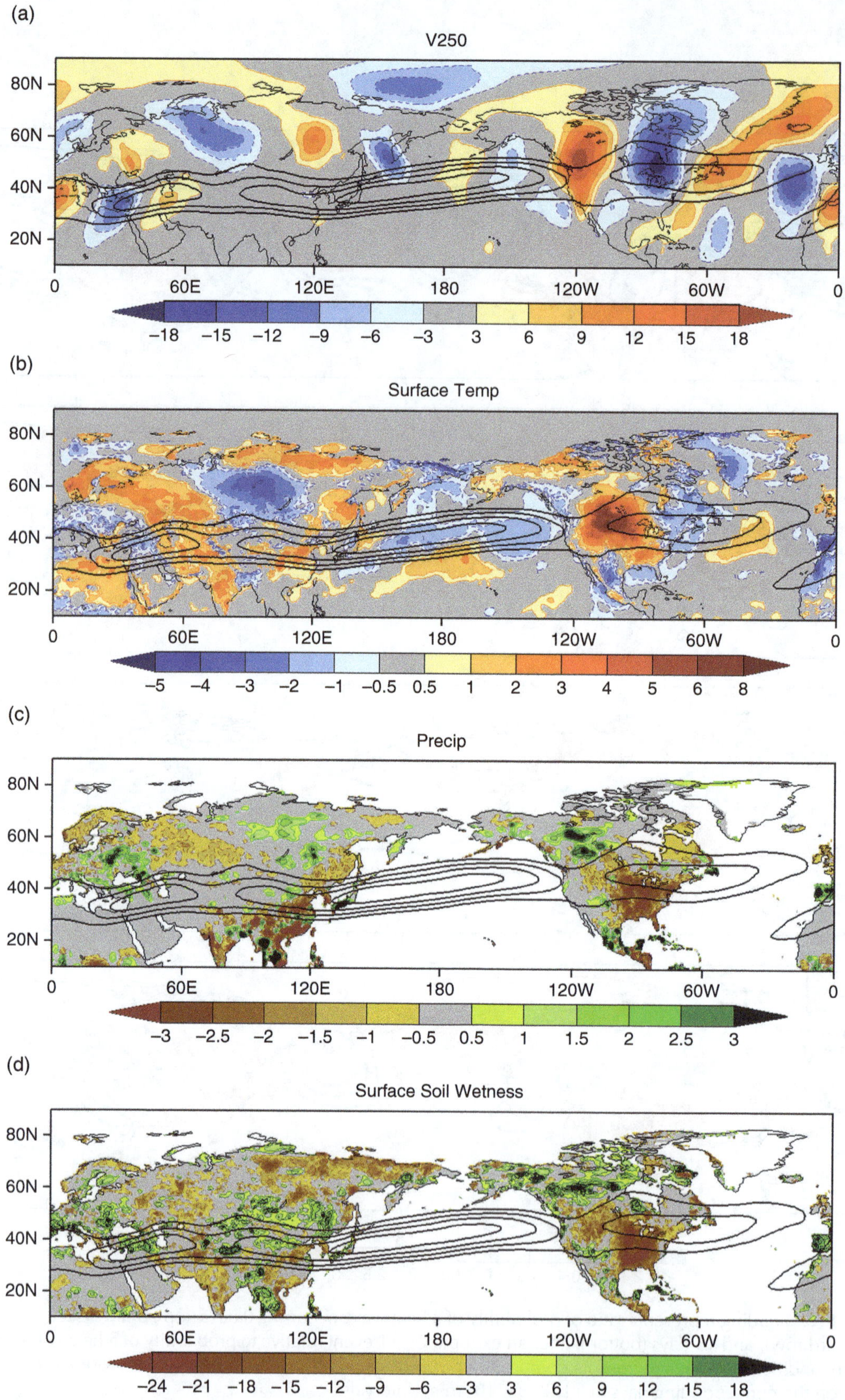

Figure 12.5 The June 1988 monthly anomalies for (a) MERRA 250 hPa meridional wind (m s^{-1}), (b) MERRA surface temperature (K), (c) MERRA-Land precipitation (mm day-1), and (d) MERRA-Land surface soil wetness (%). These anomalies are obtained as deviations from their 1980–2010 monthly means. The contours show the long-term mean (1980–2010) of June 250 hPa zonal wind, starting at 15 m s^{-1} with a 5 m s^{-1} interval.

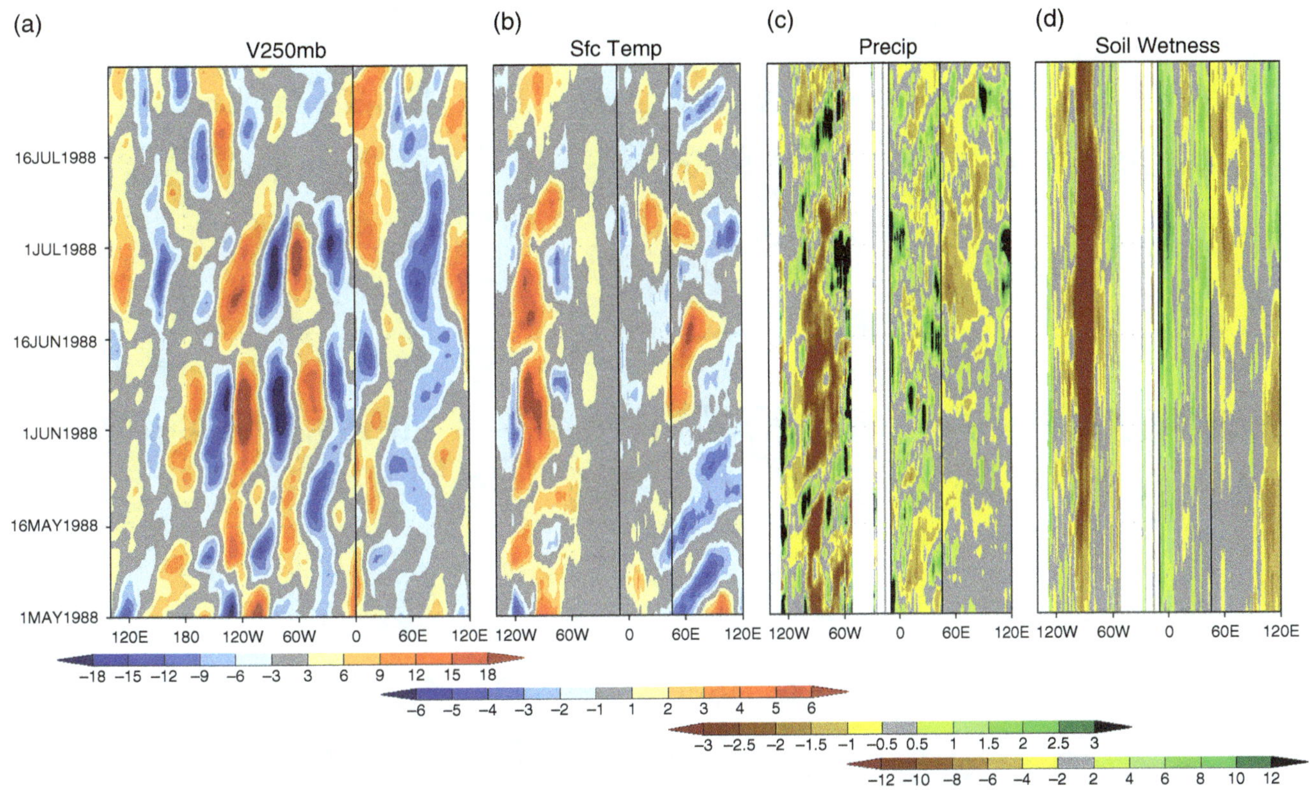

Figure 12.6 Evolution of (a) MERRA daily 250 hPa meridional wind (m s^{-1}) averaged between 40°N–60°N west of 0° and 50°N–70°N east of 0°, with a 10 day running mean smoother applied, (b) MERRA daily surface temperature anomalies (K), (c) MERRA-Land daily precipitation anomalies (mm day-1), and (d) MERRA-Land daily surface soil wetness anomalies (%), averaged between 30°N–50°N west of 10°W, between 35°N–55°N for 10°W–45°E, and between 50°N–70°N east of 45°E, with a 5 day running mean smoother applied, during the period 1 May–31 July 1998. The color shadings displayed from left to right are for panels (a) through (d), respectively.

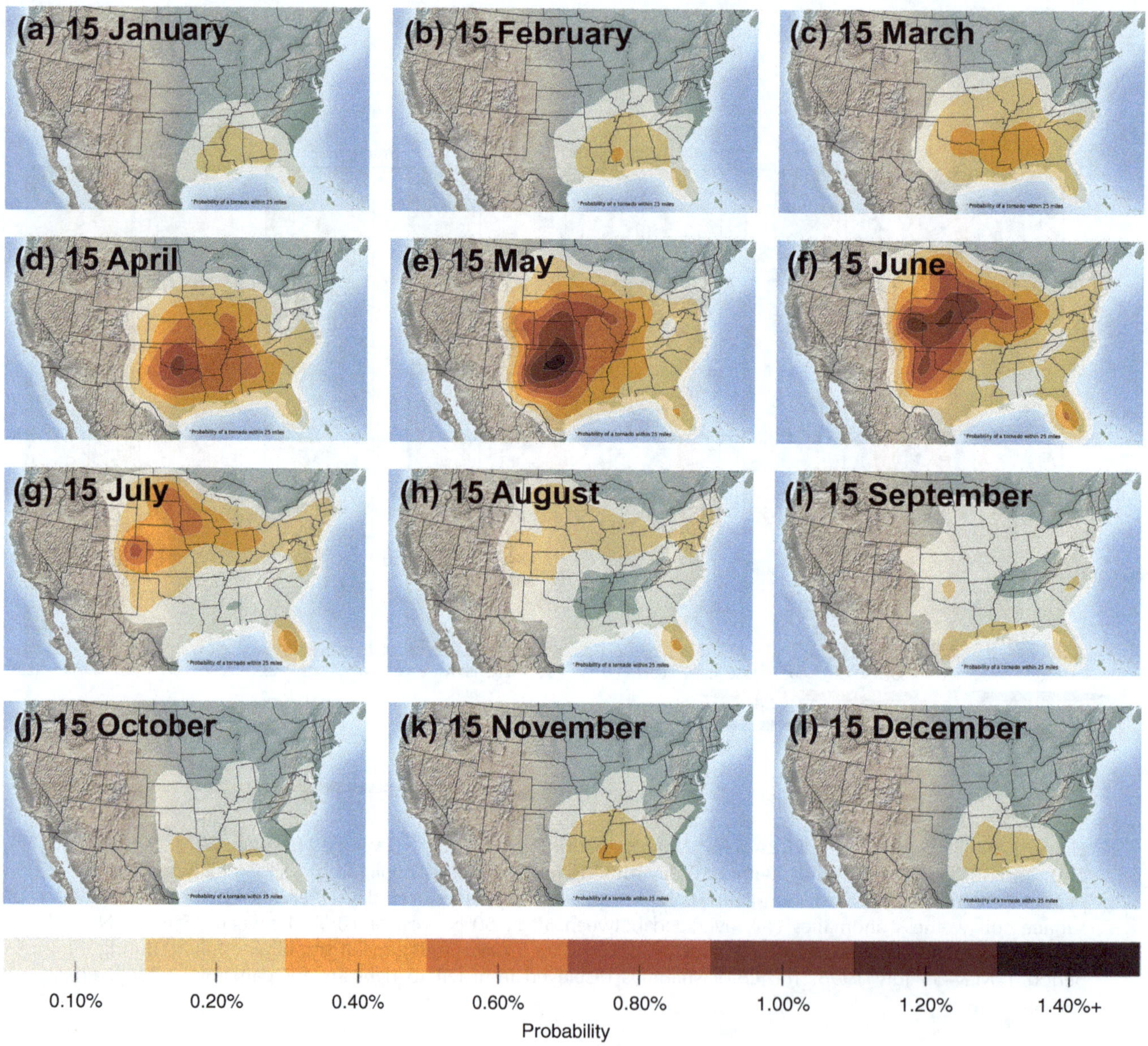

Figure 14.5 Daily probability of a tornado within 25 miles for the fifteenth day of each month derived using 1980–2014 data. Courtesy Patrick Marsh NOAA/SPC.

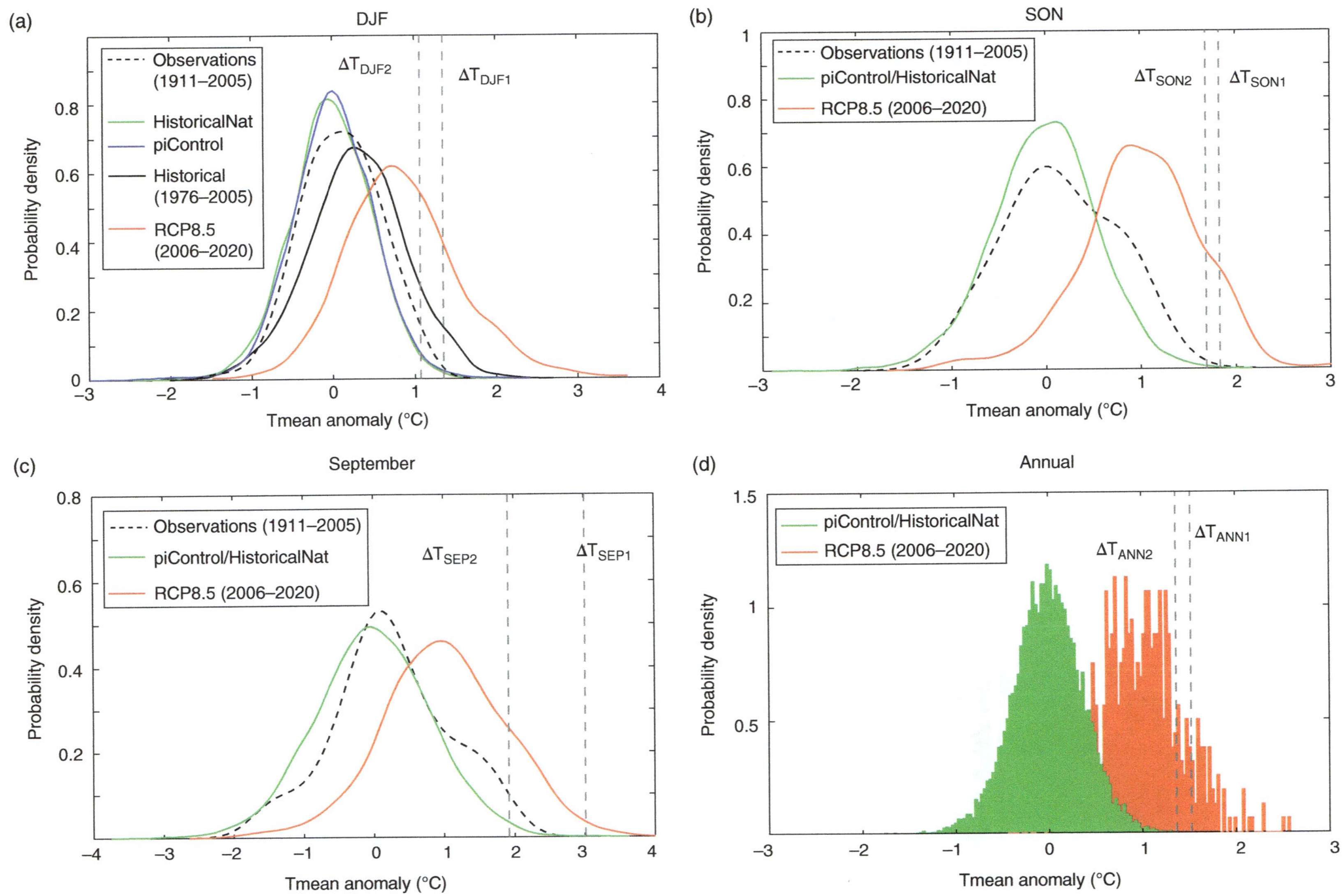

Figure 15.6 (a) Probability density functions for Australian summer (DJF) Tmean anomalies (°C, relative to 1911–1940) for observations (dashed black), historicalNat (green, all years shown), piControl (dark blue, all years shown relative to long-term mean) and historical (black, 1976–2005 only) and RCP8.5 (black, 2006–2020 only) simulations. Vertical dashed lines show observed 2013 anomaly (ΔT_{DJF1}) and threshold of the second hottest summer on record (ΔT_{DJF2}). (b) As for panel (a) but showing Australian spring (SON) Tmean anomalies. (c) As for panel (a) but showing Australian September (SEP) Tmean anomalies. (d) Histograms for Australian annual average Tmean anomalies (°C, relative to 1911–1940) of RCP8.5 (red, years 2006–2020) and piControl and historicalNat (green, all years shown). Histograms are shown, rather than smoothed distributions, as there is only a single instance in the piControl and historicalNat simulations that exceeds ΔT_{ANN2}. Adapted from *Lewis and Karoly* [2013] and *Lewis and Karoly* [2014b].

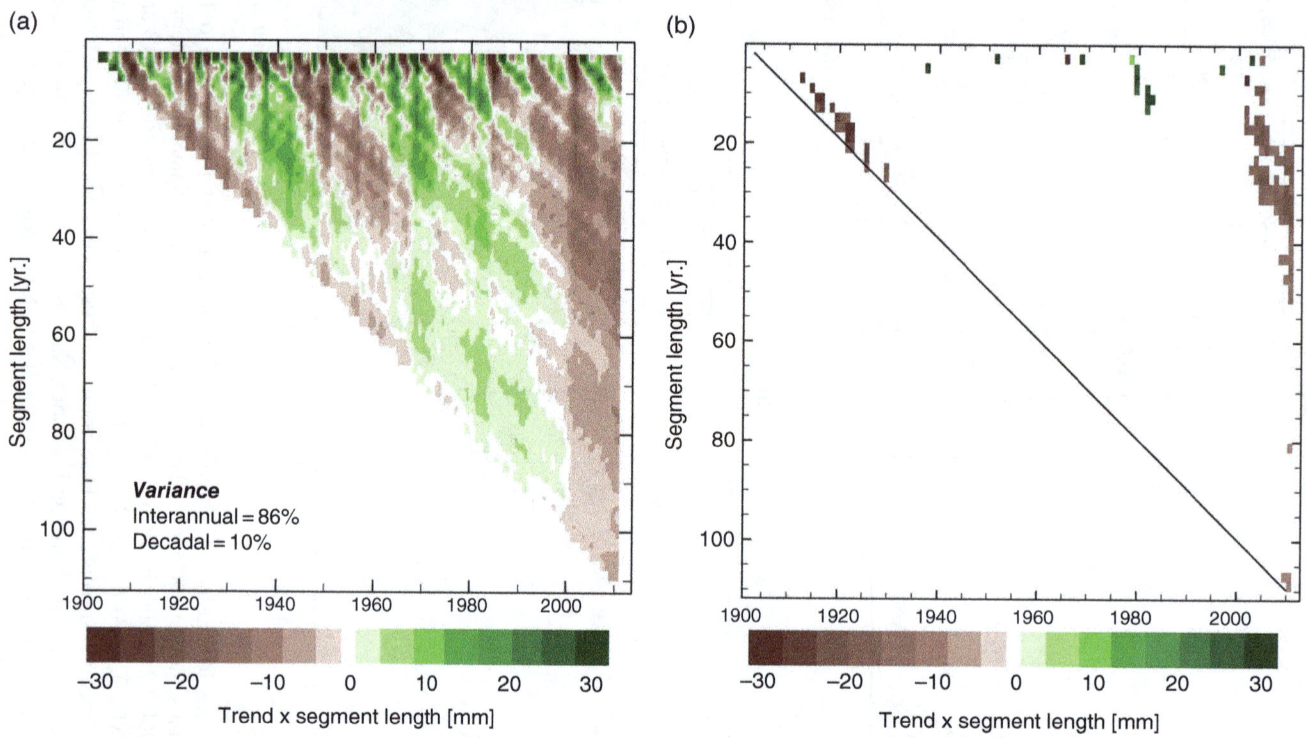

Figure 16.2 (a) Linear trend multiplied by segment length over which it was computed (shading, mm) for the East African MAM long rains in GPCC. Vertical axis is segment length, horizontal axis is the end date of fitted trend. The percent of variance for interannual and decadal scale fluctuations is shown at the lower left. (b) Statistically significant values ($p < 0.05$) based on a two-tailed t test. Adapted from *Lyon* [2014].

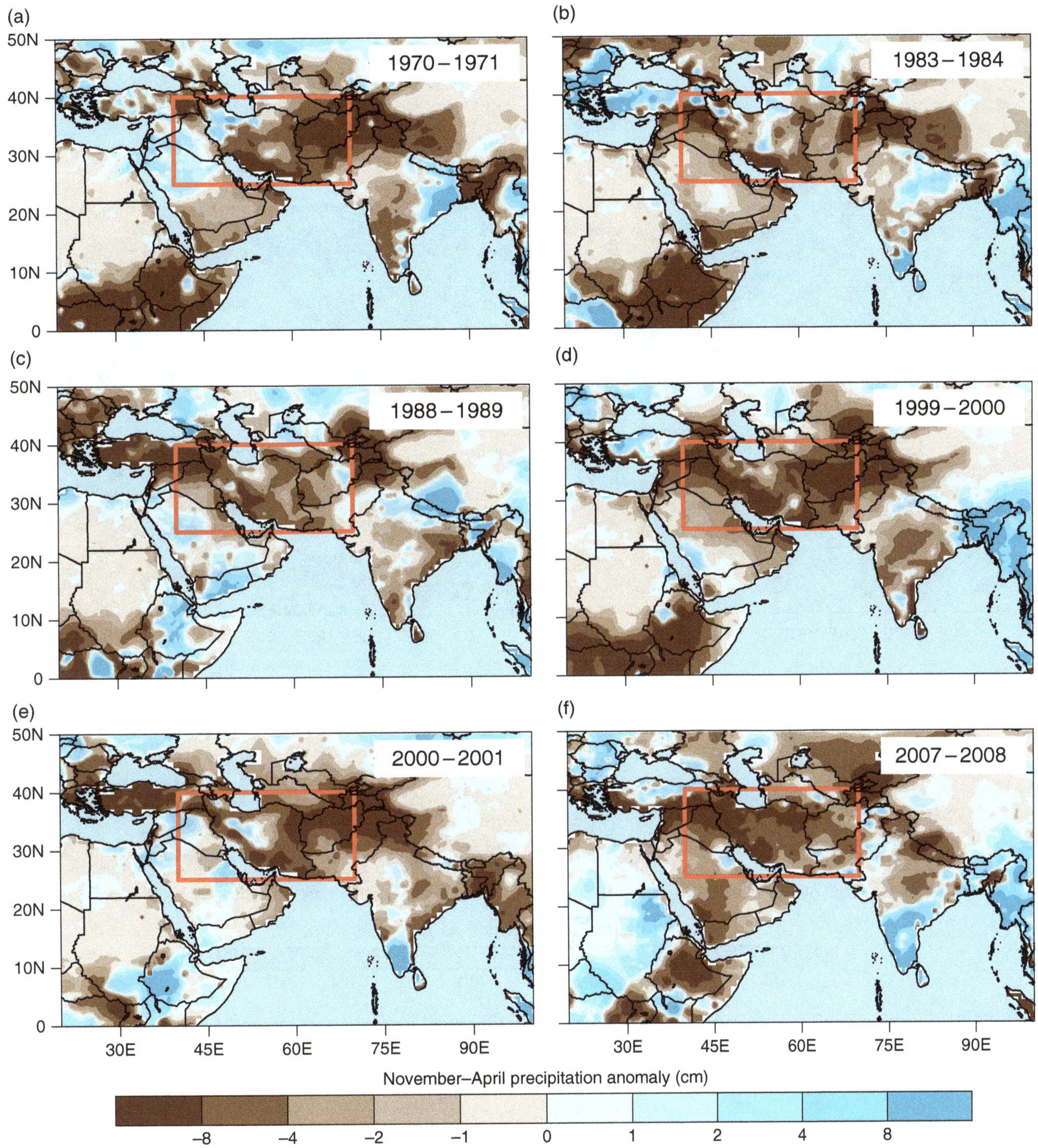

Figure 17.3 November–April precipitation anomaly (cm) for the six driest seasons areal averaged over southwest Asia.

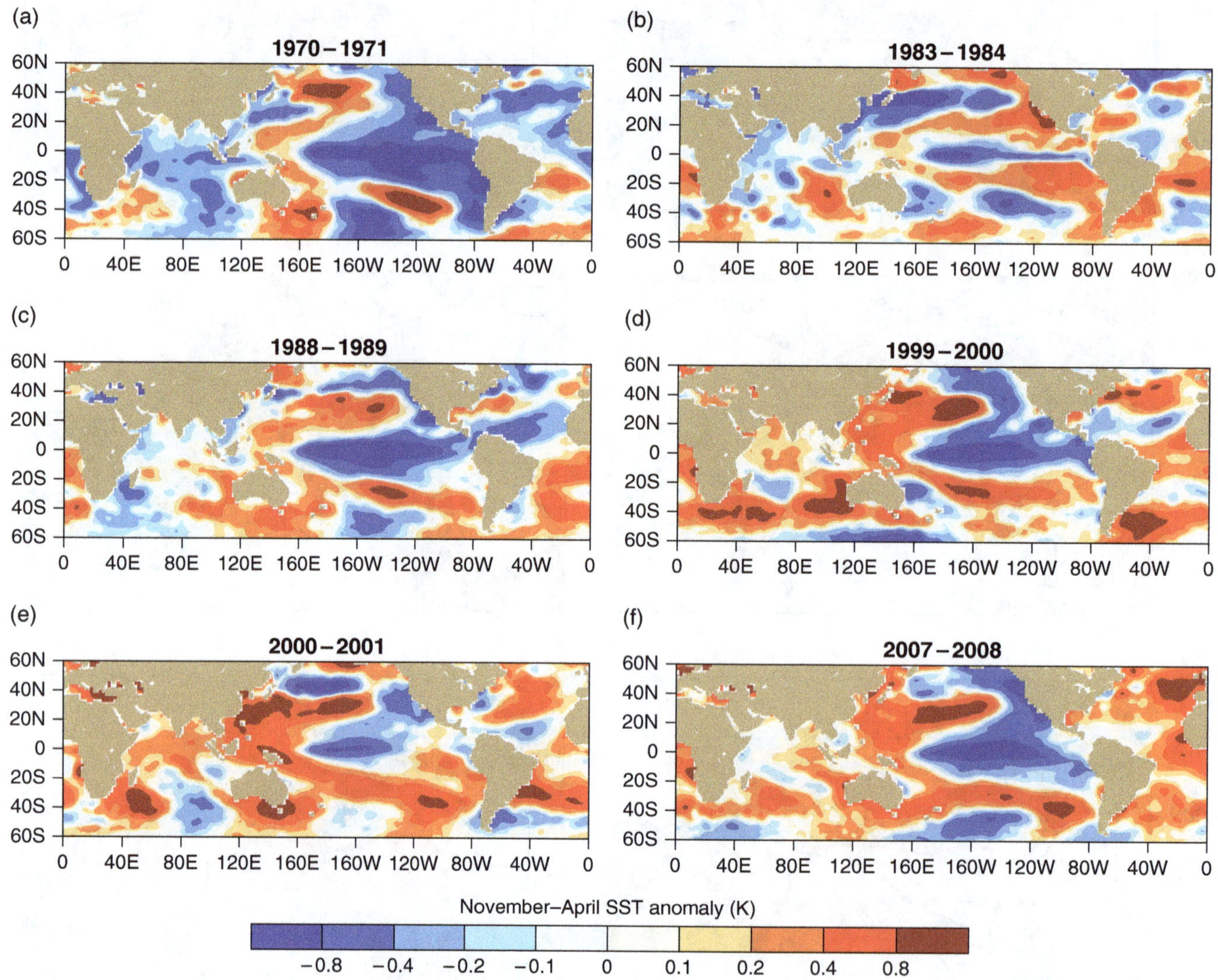

Figure 17.5 November–April SST anomaly (K) for the six driest seasons areal averaged over southwest Asia.

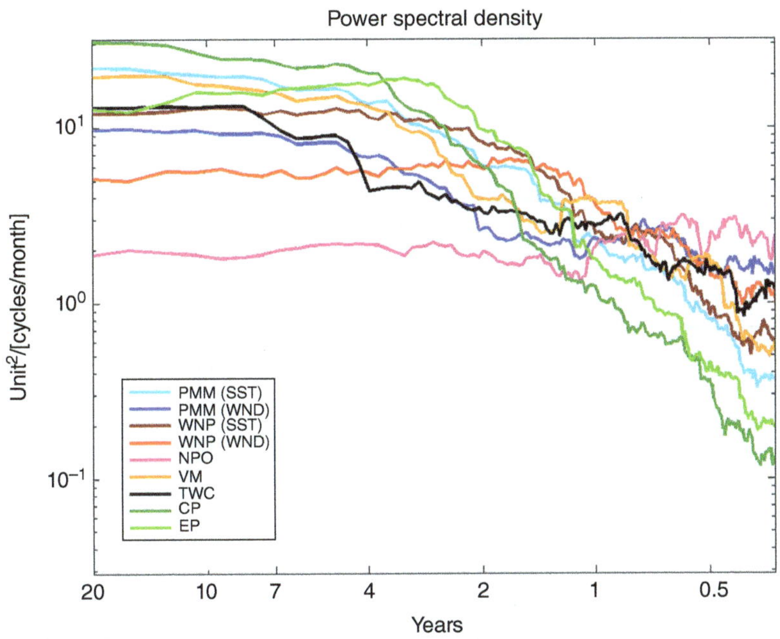

Figure 18.3 Power spectral density of ENSO precursor indices as a log-log plot. The *y*-axis represents the variance of the precursor in units^2 per cycle per month. The *x*-axis represents the period in years.

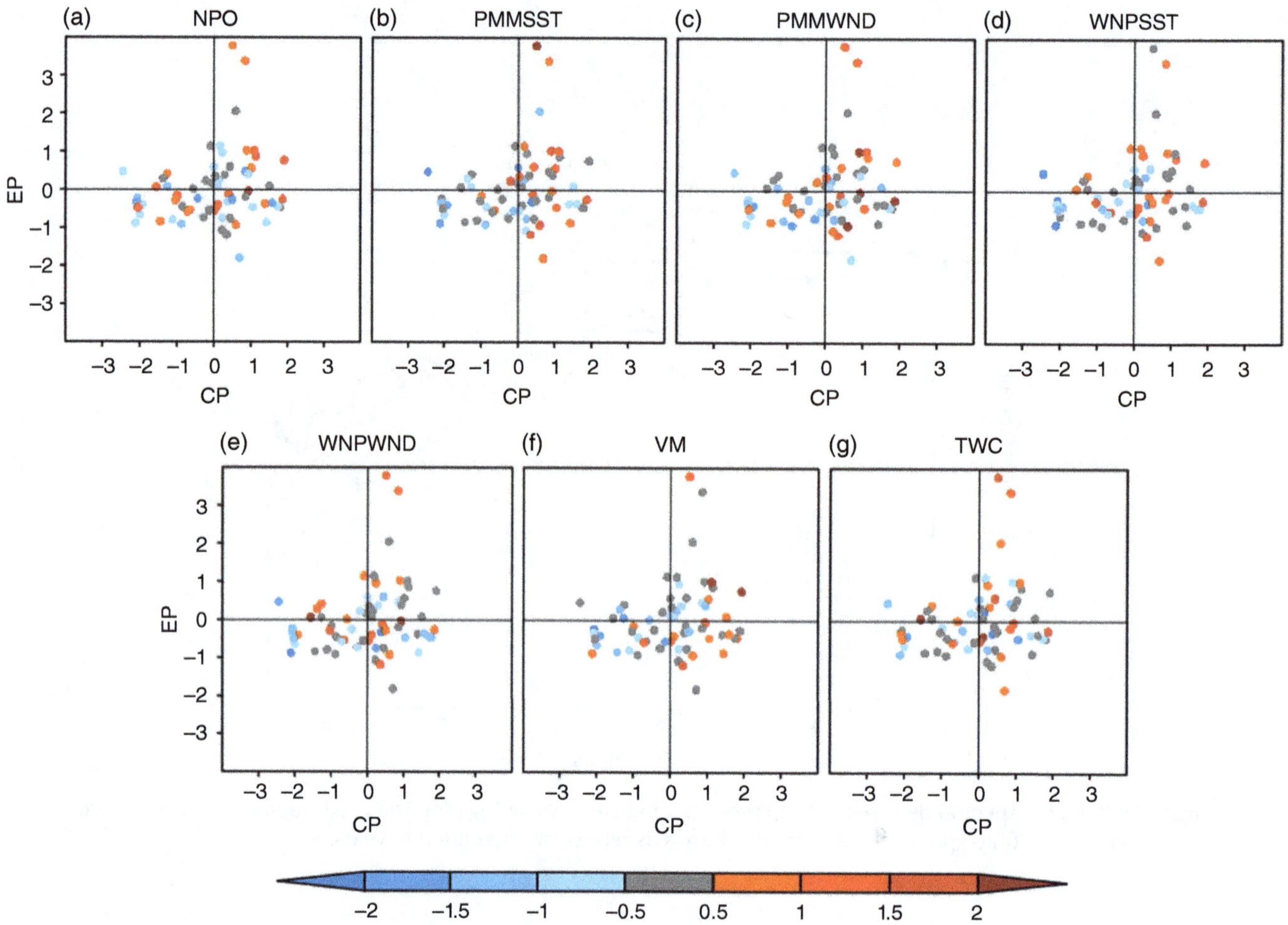

Figure 18.5 Scatterplot of CP vs. EP normalized values for Dec-Jan-Feb. Colors represent the normalized value of each precursor for the previous Dec-Jan-Feb.

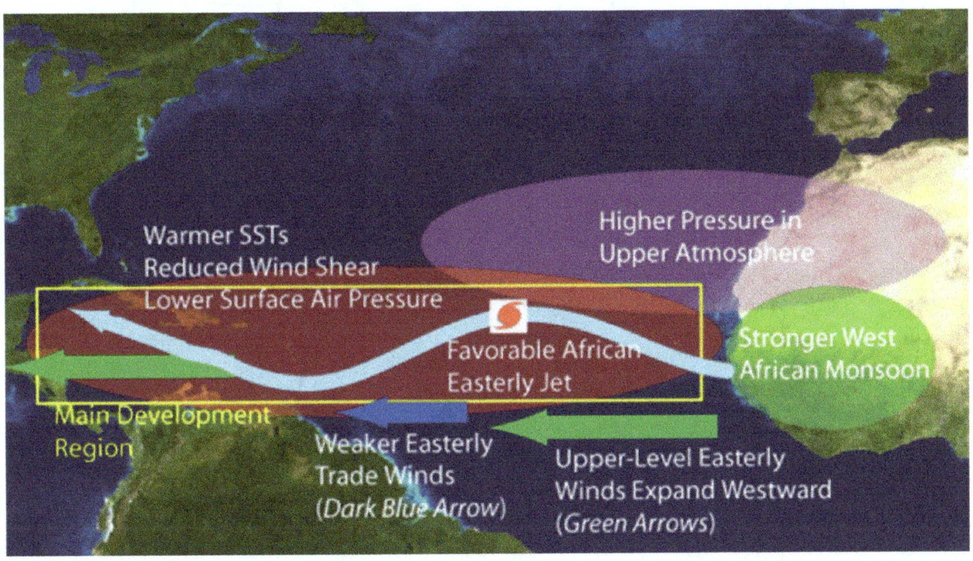

Figure 19.1 Schematic of atmospheric and oceanic anomalies during August–October associated with active Atlantic hurricane seasons and decades. Adapted from *Bell and Chelliah* [2006].

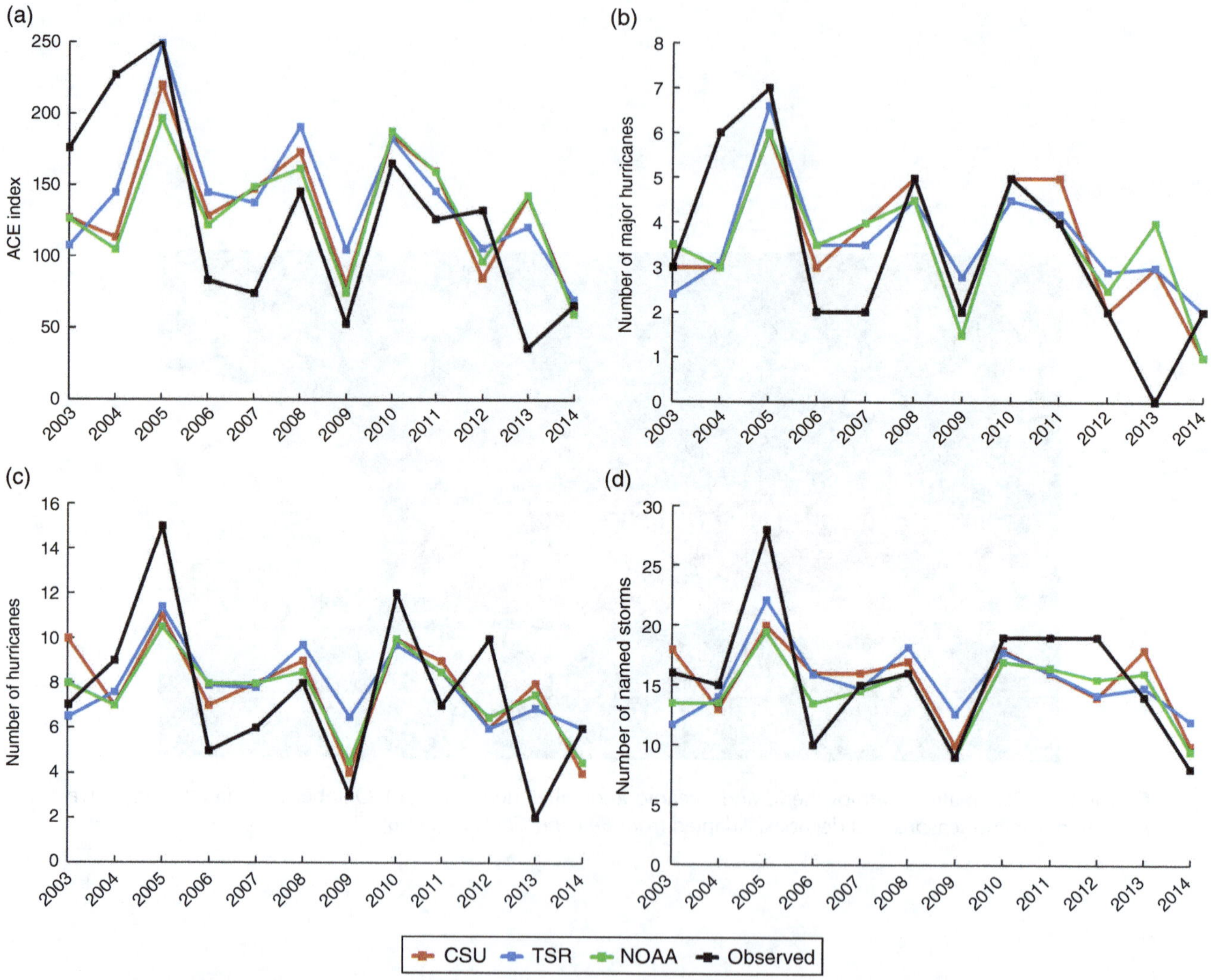

Figure 19.9 Time series comparing the seasonal outlook values issued in early August 2003–2014 by CSU, TSR, and NOAA with observed values. The comparison is made for (a) ACE, (b) major hurricane numbers, (c) hurricane numbers, and (d) named storm numbers.

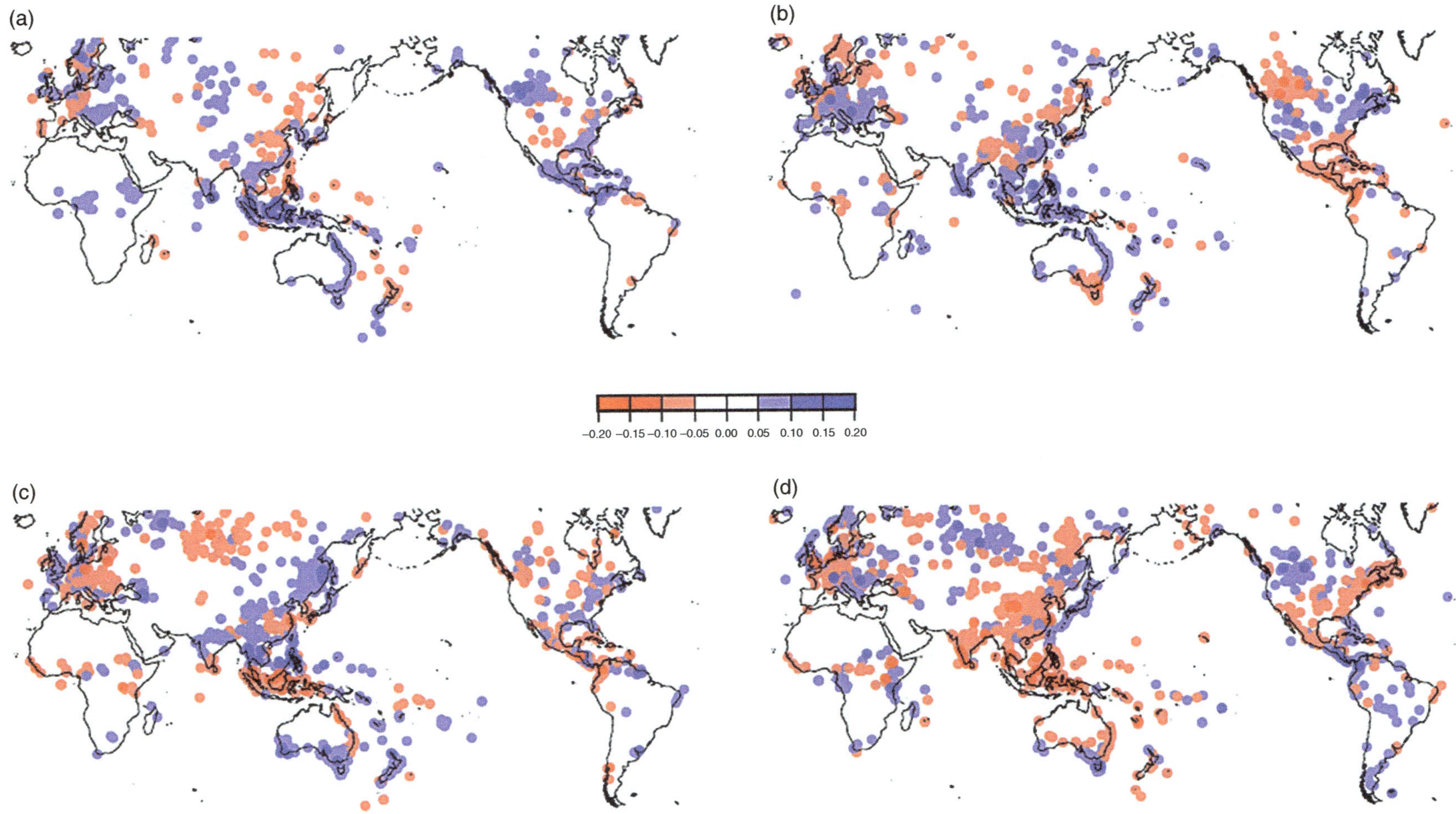

Figure 20.1 Worldwide distribution of daily rainfall anomalies during active MJO events when the center of enhanced convection is in phases (a) 2, (b) 4, (c) 6, and (d) 8 of the *Wheeler and Hendon* [2004] diagram. Colors represent differences between unconditional cumulative distribution function (CDF) from Monte-Carlo simulations and observed conditional CDF when the MJO was in the given phase. Enhanced (decreased) rainfall is expressed by blue (red) color dots over each station location used in the analysis. After *Donald et al.* [2006].

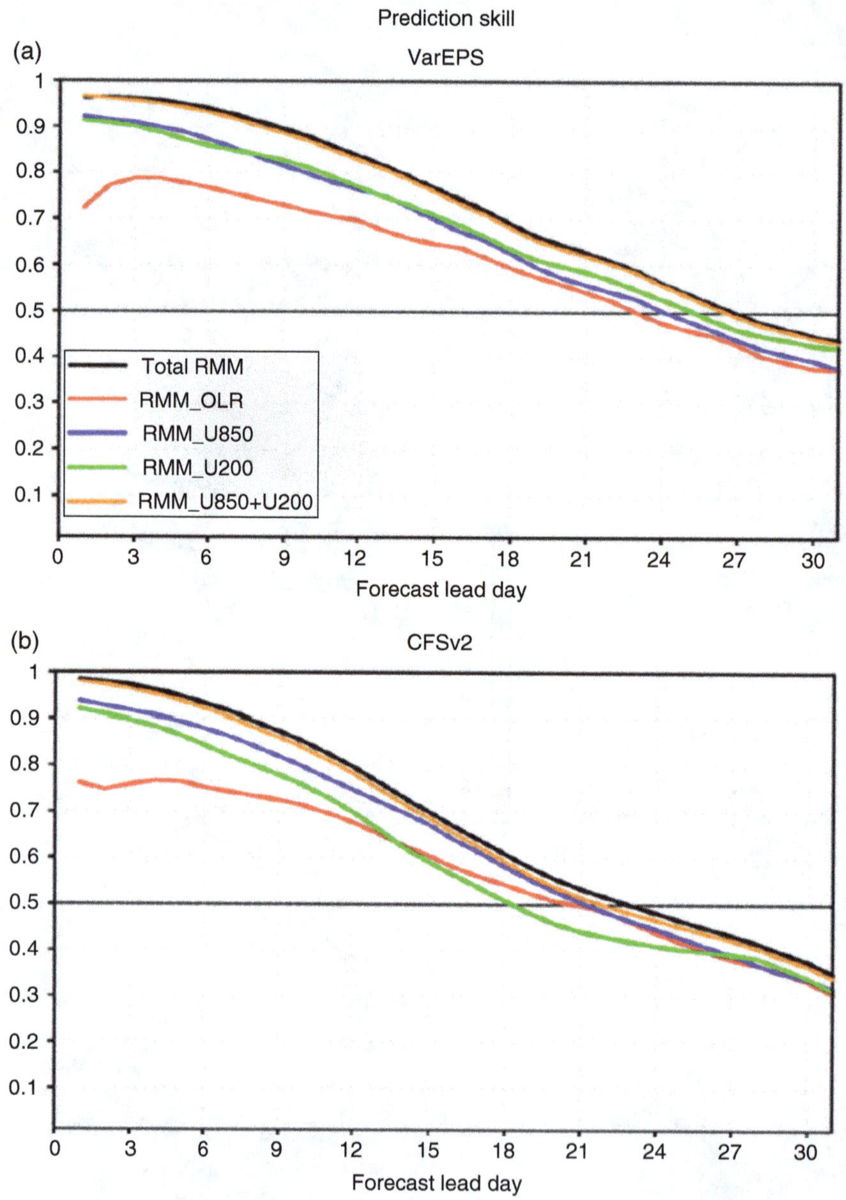

Figure 20.5 Prediction skill of the MJO according to different versions of the RMM index: OLR (red), U850 (blue), and U200 (green) separately, and U850 and U200 combined (brown). Verification is done for strong (high amplitude) MJO events in the forecasts from (a) ECMWF Variable Resolution Ensemble Prediction System (VarEPS), and (b) NCEP Climate Forecast System version 2 (CFSv2). After *Kim et al.* [2014].

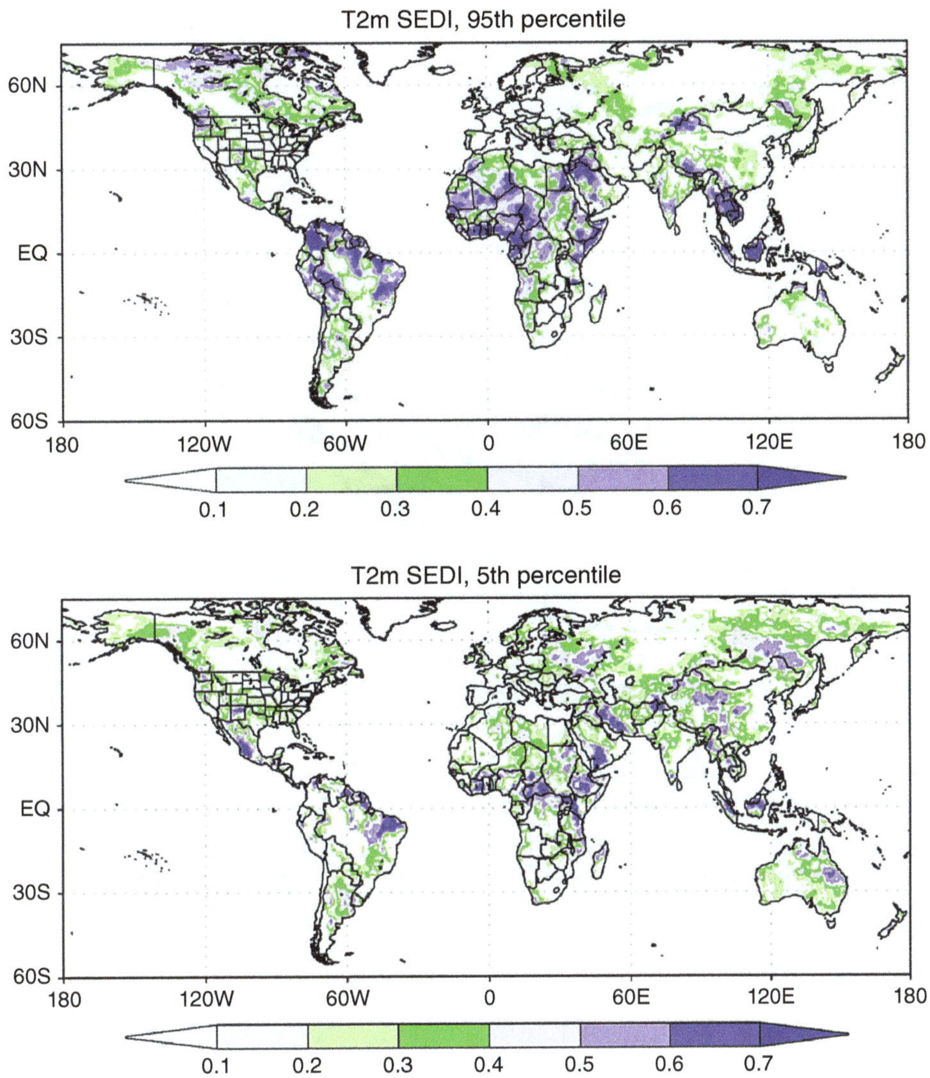

Figure 21.3 Symmetric External Dependence Index (SEDI) for seasonal forecasts of extreme 2 m temperature: SEDI for forecasts of positive extremes (95th percentile events) in the upper panel, negative extremes (5th percentile events) in the lower panel.

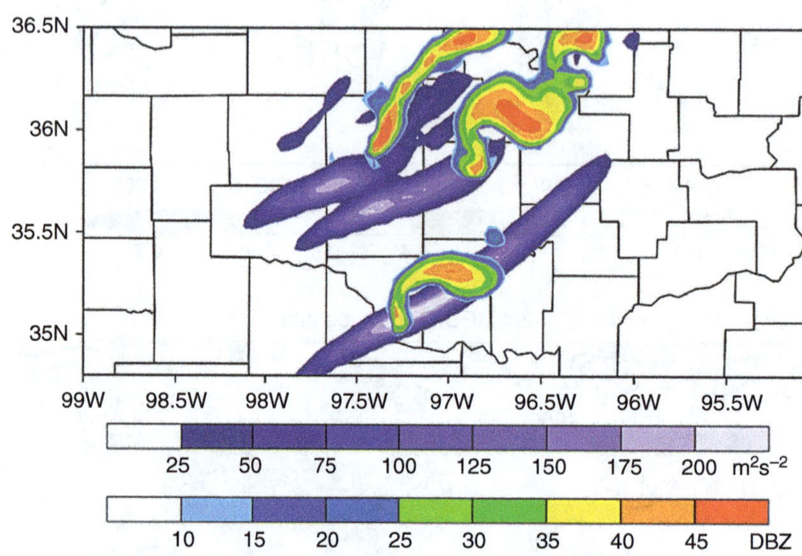

Figure 22.2 Hourly maximum updraft helicity valid 1800–0100 UTC 19–20 May 2013 as forecast by a CPM initialized at 0000 UTC 19 May at NOAA's National Severe Storms Laboratory (purple shading). The plot domain covers central Oklahoma. Simulated reflectivity valid at 2100 UTC is overlaid.

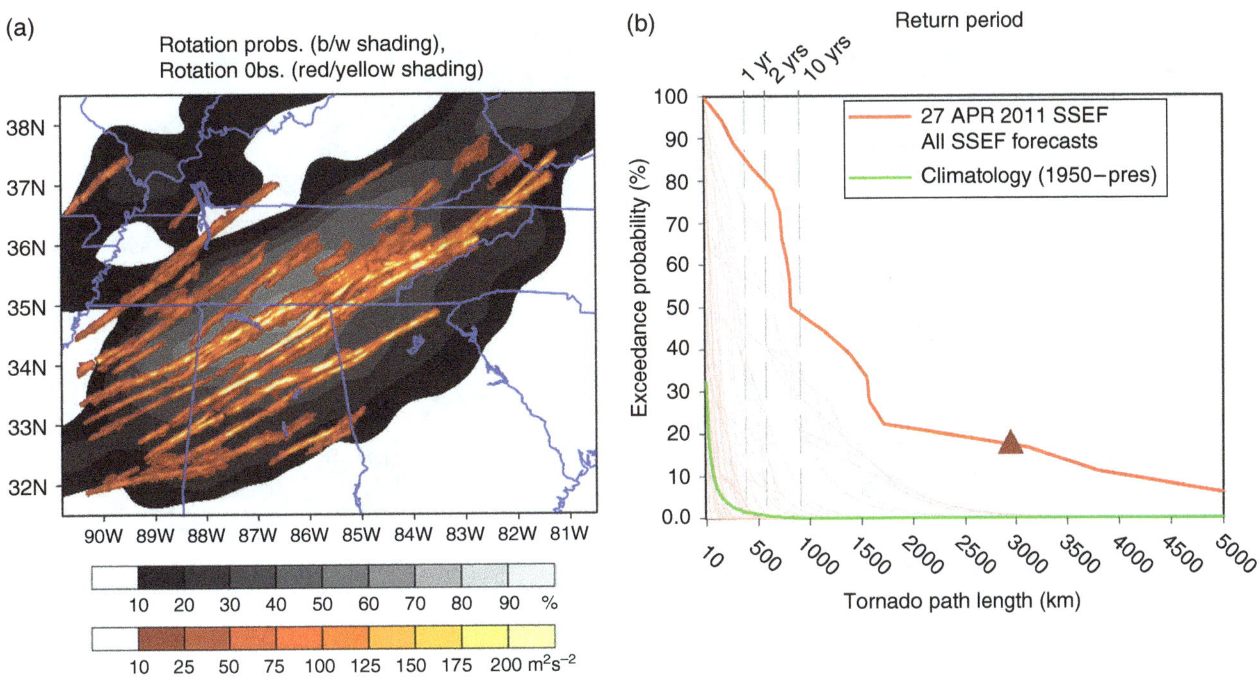

Figure 22.3 (a) Black and white shading shows probabilities of UH ≥ 100 m^2s^{-2} within 40 km of a point valid 1500–0300 UTC 27–28 April 2011 computed from a CPM-based ensemble initialized 0000 UTC 27 April 2011, which was used during the 2011 NOAA/Hazardous Weather Testbed Spring Forecasting Experiment. Probabilities were smoothed using a Gaussian kernel density weighting function with a standard deviation of 20 km. The red and yellow shading shows rotation tracks of observed rotating storms computing from 1.25 km grid-spacing, radar-data-assimilating, 3DVAR analyses constructed at 5 min intervals. (b) Exceedance probabilities as a function of cumulative tornado path length computed from the distribution of UH path-length forecasts from the CPM-based ensemble shown in (a). The dark red line is for 27 April, the light red lines are for 47 other cases in the dataset, and the green line is for the climatological exceedance probabilities computed from Storm Data for the period 1950–2011 (legend provided at top). The dark red triangle marks the actual cumulative tornado path length for 1200–0600 UTC 27–28 April. Adapted and modified from *Clark et al.* [2013].

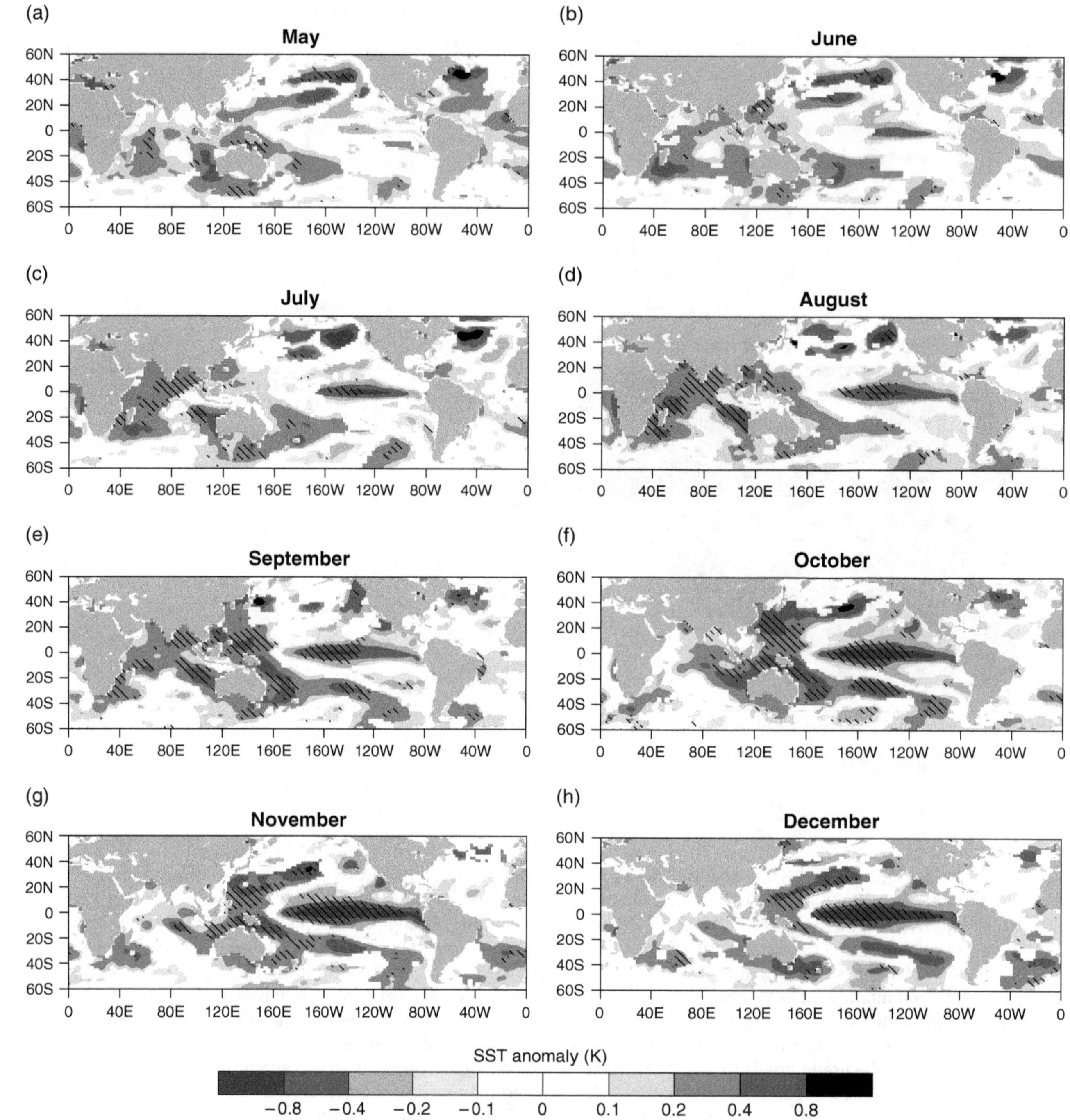

Figure 17.7 Composite SST anomaly (K) leading to the six driest November–April seasons areal averaged over southwest Asia. Shading indicates that at least 4 months had the same sign of SST anomaly while hatching indicates that all 6 months had the same sign of SST anomaly.

the lower troposphere (Fig. 17.10). This atmospheric circulation is a characteristic response to La Niña events with an SST expression that includes strong and opposing SST anomalies over the central and western tropical Pacific Ocean [*Hoell and Funk* 2013; *Hoell et al.*, 2013b; *Hoell et al.*, 2014a]. Currently, the mechanisms that force this particular atmospheric circulation pattern across Asia are unknown.

17.4.2. Regional Precipitation Forcing

The hemispheric atmospheric circulations during the strongest November–April southwest Asia droughts of 1948–2012 are very robust. Here we examine how the anomalous atmospheric circulations force precipitation over southwest Asia through changes to the available atmospheric moisture and forced vertical motions.

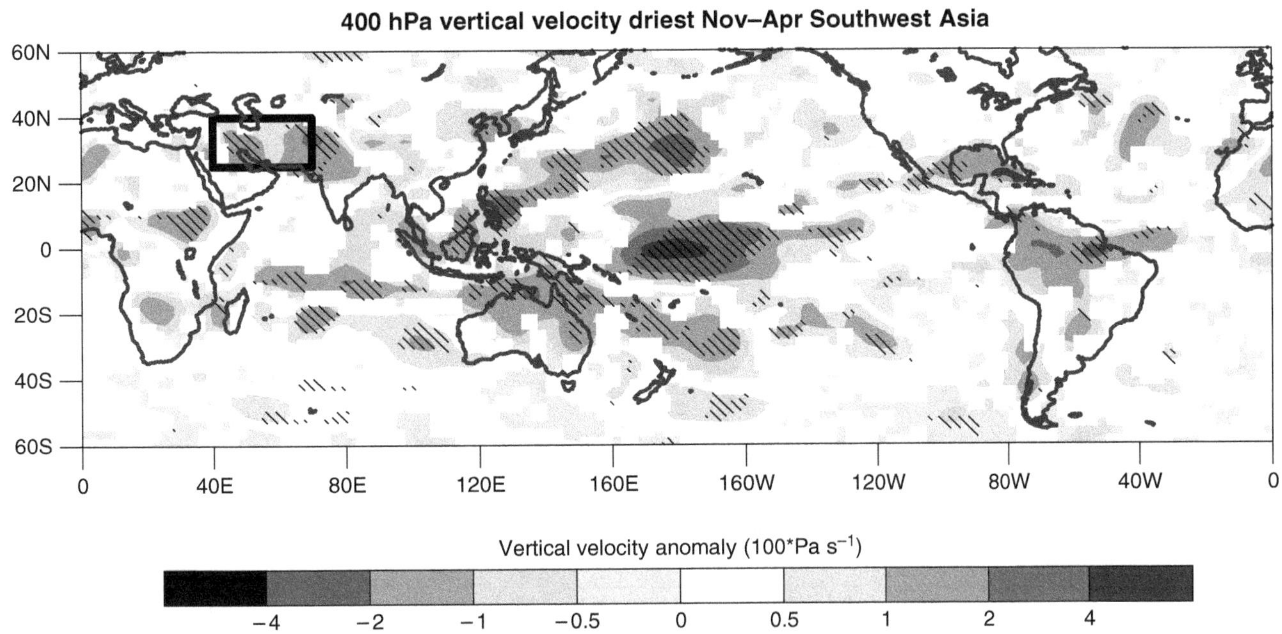

Figure 17.8 Composite vertical velocity anomaly (100 × Pa s^{-1}) for the six driest November–April seasons areal averaged over southwest Asia. Shading indicates that at least four November–April seasons had the same sign of vertical velocity anomaly while hatching indicates that all six November–April seasons had the same sign of vertical velocity anomaly.

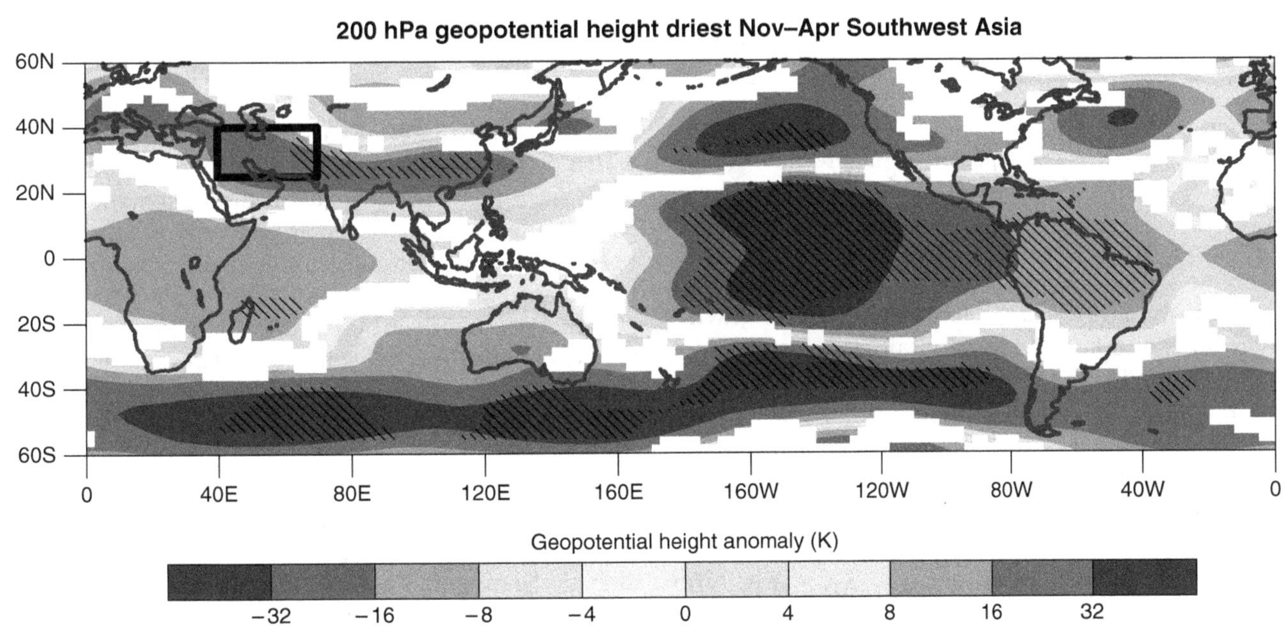

Figure 17.9 Composite 200 hPa geopotential height anomaly (m) for the six driest November–April seasons areal averaged over southwest Asia. Shading indicates that at least four November–April seasons had the same sign of geopotential height anomaly while hatching indicates that all six November–April seasons had the same sign of geopotential height anomaly.

We have used the vertically integrated moisture equation to examine the movement of moisture through the troposphere over southwest Asia to assess the possible impacts upon precipitation during La Niña events [*Hoell et al.*, 2013b; *Hoell et al.*, 2014a] and for individual months of the November–April season [*Hoell et al.*, 2015]. Here, we employ the same methodology to examine the flow of regional moisture during the most extreme southwest Asia droughts.

The vertically integrated moisture equation (equation [17.1]) states that evaporation minus precipitation in the atmosphere is balanced by the vertically integrated

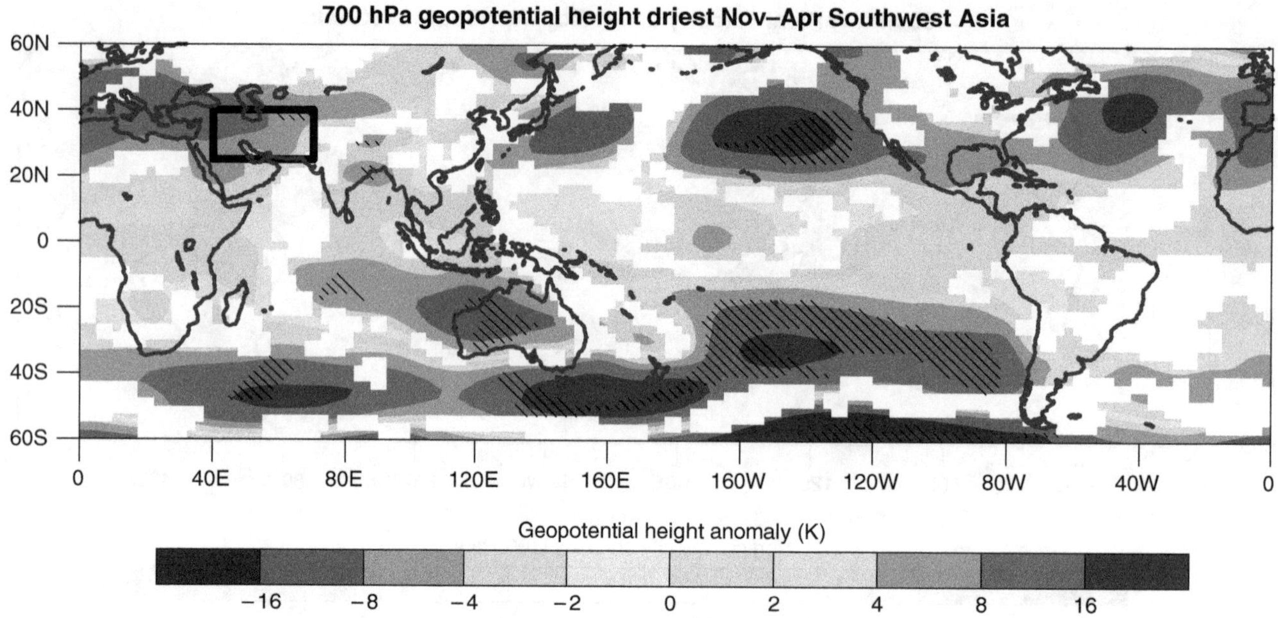

Figure 17.10 Composite 700 hPa geopotential height anomaly (m) for the six driest November–April seasons areal averaged over southwest Asia. Shading indicates that at least four November–April seasons had the same sign of geopotential height anomaly while hatching indicates that all six November–April seasons had the same sign of geopotential height anomaly.

precipitable water and the vertically integrated moisture flux divergence.

$$E - P = \frac{\partial}{\partial t}\frac{1}{g}\int_0^{p_s} q\,dp + \frac{\nabla .1}{g}\int_0^{p_s} q\vec{V}dp \qquad (17.1)$$

The terms of the vertically integrated moisture equation are as follows: E is evaporation, P is precipitation, q is specific humidity, g is the acceleration due to gravity, p is pressure, and $\vec{V}$ is the horizontal wind vector. Since we are interested in the movement of water in the atmosphere and its origins, we focus on the vertically integrated moisture flux, $\frac{1}{g}\int_0^{p_s} q\vec{V}dp$.

The region-wide changes in vertically integrated moisture flux are very similar across the six most extreme southwest Asia droughts of November–April 1948–2012 (Fig. 17.11). Anomalous northeasterly flow on the eastern side of the equivalent barotropic Rossby wave originated over the foothills of the Himalayas and advected air into southwest Asia (Fig. 17.11a). The anomalous northeasterly vertically integrated moisture flux advection throughout southwest Asia continued across the Arabian Peninsula, the Greater Horn of Africa, and finally into the northwest Indian Ocean. The average vertically integrated moisture flux for November–April (Fig. 17.11b) indicates that the moisture for the northwest Indian Ocean rim originates over the northwest Indian Ocean. Therefore, the anomalous northeasterly flow on the eastern side of the equivalent barotropic Rossby (Fig. 17.11a) reduced the amount of moisture available from the climatological source (Fig. 17.11a), which reduced the moisture available for precipitation.

We have used the hydrostatic thermodynamic energy equation to examine the forcing of vertical motions as a result of the interaction of anomalous circulations with the mean climate during La Niña events [*Hoell et al.*, 2013b; *Hoell et al.*, 2014a] and for individual months of the November–April season [*Hoell et al.*, 2015]. Here, we employ the same methodology to examine forced vertical motions during the most extreme southwest Asia droughts.

The thermodynamic energy equation (equation [17.2]) states that the sum of the local temperature change and the horizontal temperature advection is balanced by diabatic heating and a vertical term, which includes the effects of vertical temperature advection and adiabatic influences [*Holton*, 2004].

$$\frac{\partial T}{\partial t} + \vec{V}.\nabla T = \frac{J}{c_p} + S_p\varpi \qquad (17.2)$$

The terms of the thermodynamic energy equation are as follows: T is temperature, $\vec{V}$ is the horizontal wind vector, S_p is the static stability parameter, ϖ is the vertical velocity in pressure coordinates, c_p is the specific heat of dry air, and J is the diabatic heating.

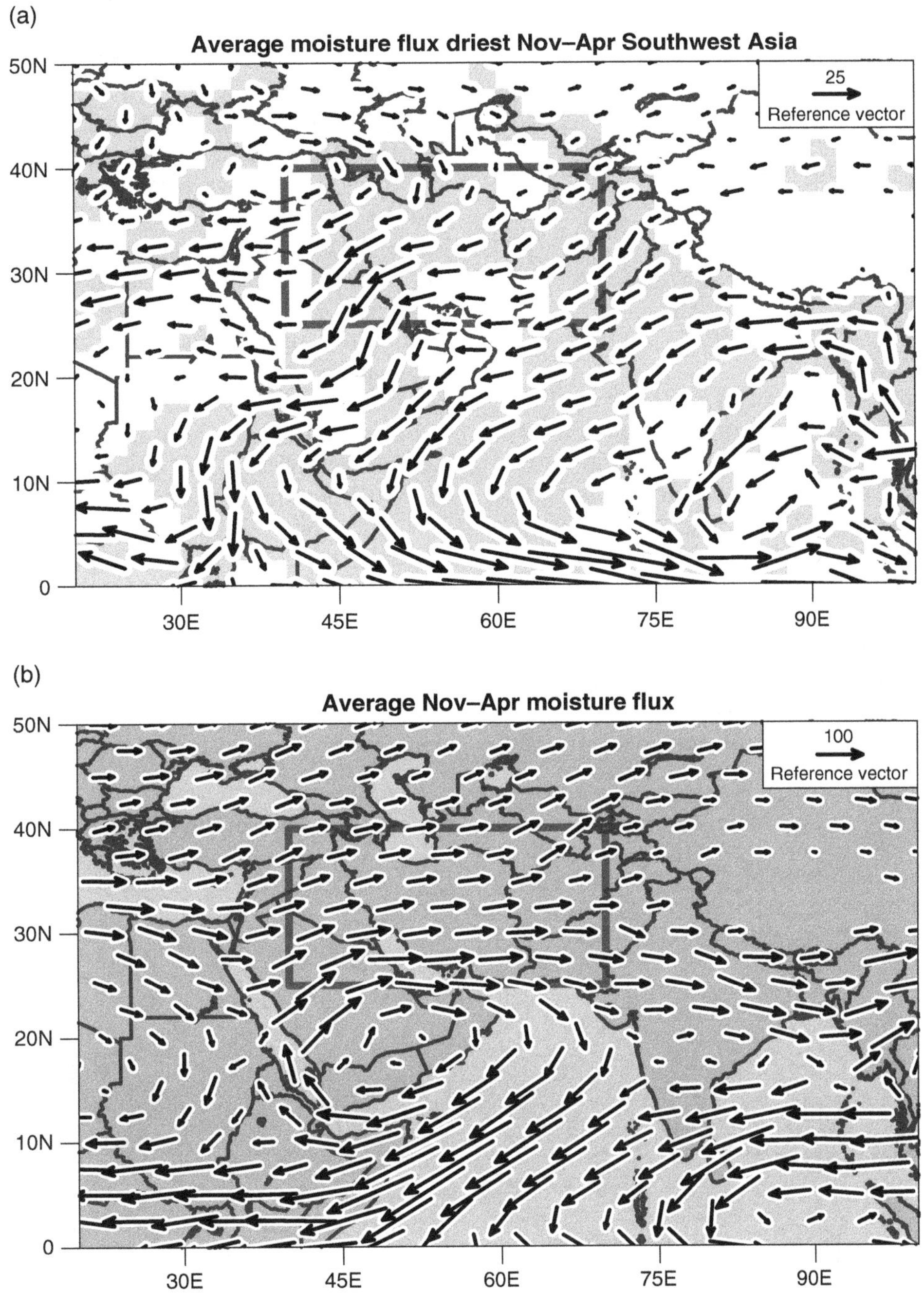

Figure 17.11 (a) Composite vertically integrated moisture flux anomaly (kg m^{-1} s^{-1}) for the six driest November–April seasons areal averaged over southwest Asia. Blue shading indicates that at least four November–April seasons had the same sign meridional component of vertically integrated moisture flux anomaly. (b) Average November–April vertically integrated moisture flux (kg m^{-1} s^{-1}). *See electronic version for color representation.*

Over subtropical and extratropical Asia, the static stability parameter is approximately constant in the horizontal, and the local temperature change is small relative to each of the other terms. The hydrostatic thermodynamic energy equation can be expressed as a balance between horizontal temperature advection and the vertical term, where the variability of the vertical term is due to only vertical motions. Equation (17.3) indicates that cold horizontal temperature advection is balanced by precipitation-reducing descending vertical motions.

$$\vec{V}.\nabla T = S_p\varpi \qquad (17.3)$$

In *Hoell et al.* [2014b], we separated the terms of equation (17.3) into means and anomalies to diagnose forced vertical motions as a result of the interaction of anomalous features with the mean climate. Using that methodology, the thermodynamic equation states that the changes in vertical motion are forced by the mean temperature advection by the anomalous wind and the anomalous temperature advection by the mean wind, which is shown in equation (17.4).

$$S_p \varpi_a = \vec{V}_a . \nabla T_m + \vec{V}_m . \nabla T_a \qquad (17.4)$$

We show how the interaction of anomalous circulations with the mean climate during extreme southwest Asia droughts generates horizontal temperature advection that is balanced by precipitation-modifying vertical motions at 400 hPa (Fig. 17.12). Since the atmospheric circulation over southwest Asia is equivalent barotropic and varies very little with height, 400 hPa is representative of the rest of the midtroposphere where the strongest precipitation-modifying vertical motions occur.

The horizontal distribution of mean temperature advection by the anomalous wind during extreme November–April 1948–2012 Southwest Asia droughts, $\vec{V}_a . \nabla T_m$ (contours in Fig. 17.12), is examined in terms of its contributing variables, the mean November–April temperature (shading in Fig. 17.12), and anomalous wind during extreme southwest Asia drought (vectors in Fig. 17.12). The anticyclonic circulation of the equivalent barotropic Rossby wave centered over the Persian Gulf and the northern Arabian Sea (~52°E, 35°N) is responsible for the mean temperature advection by the anomalous wind. Northwesterly flow across the mean meridional temperature gradient over the eastern half of southwest Asia, which includes Iran, Afghanistan, and Pakistan, results in cold horizontal temperature advection, which is associated with subsidence. Southerly flow across the mean meridional temperature gradient over the western half of southwest Asia, which includes Iraq, results in warm temperature advection.

The horizontal distribution of anomalous temperature advection by the mean wind during extreme November–April 1948–2012 southwest Asia droughts, $\vec{V}_m . \nabla T_a$ (contours in Fig. 17.13), is examined in terms of its contributing variables, the mean November–April wind (vectors in Fig. 17.13), and anomalous temperature during extreme southwest Asia drought (shading in Fig. 17.13). The anomalous warm temperatures associated with the equivalent

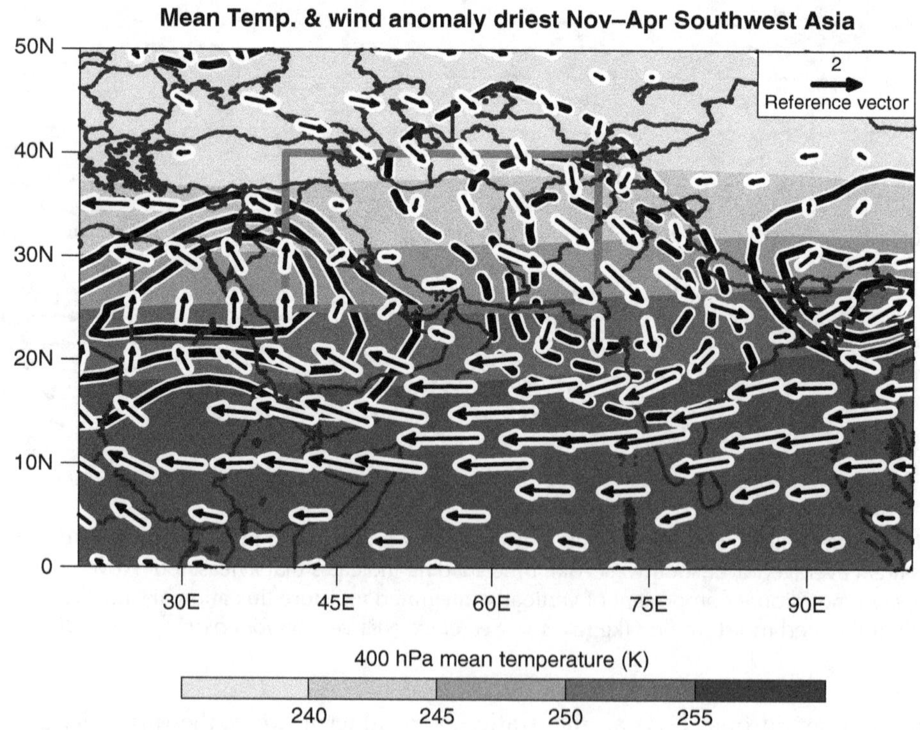

Figure 17.12 Composite 400 hPa November–April 1948–2012 mean temperature in K (shading) and 400 hPa wind anomaly (vector in units of m s^{-1}) and mean temperature advection by the anomalous wind (contoured every+-0.2, +-0.4, +-0.6 K d^{-1}) for the six driest November–April seasons areal averaged over southwest Asia. Vectors indicate that at least four November–April seasons had the same sign meridional component of 400 hPa wind anomaly.

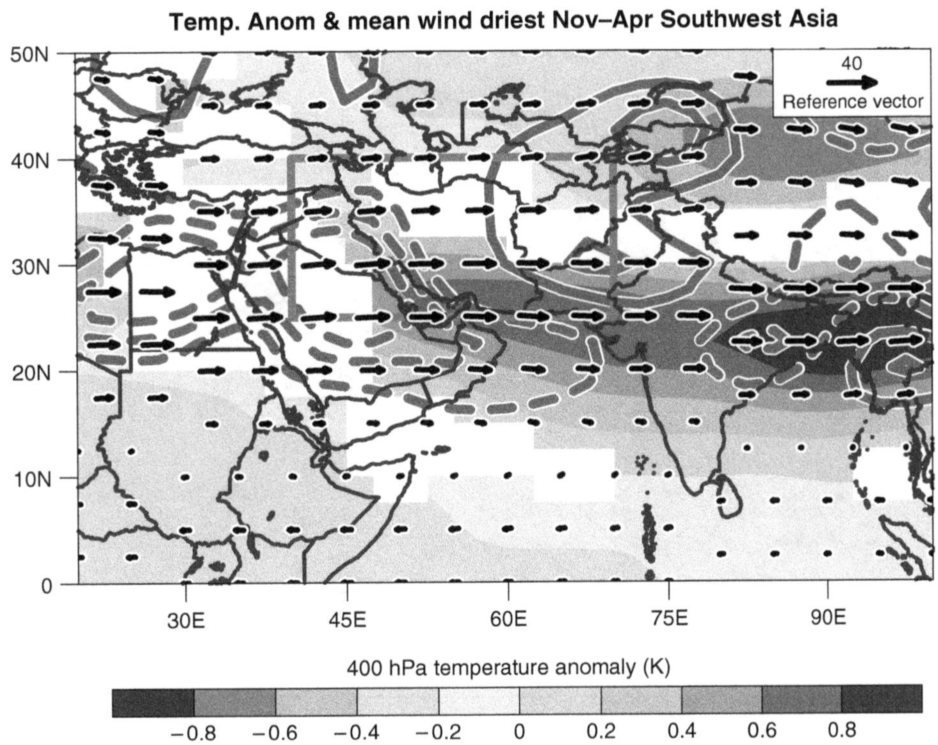

Figure 17.13 Composite 400 hPa air temperature anomaly (shaded in units of K) and 400 hPa November–April 1948–2012 mean wind anomaly in m s^{-1} (vector) and anomalous temperature advection by the mean wind (contoured every +-0.2, +-0.4, +-0.6 K d^{-1}) for the six driest November–April seasons areal averaged over southwest Asia. Shading indicates that at least four November–April seasons had the same sign air 400 hPa temperature anomaly.

barotropic Rossby wave is centered within the mean November–April jet stream, and is responsible for anomalous temperature advection by the mean wind. Zonal flow across the anomalous warm temperatures results in strong cold-temperature advection over the western half of southwest Asia and weak warm-temperature advection over the eastern half of southwest Asia.

The total anomalous temperature advection (Fig. 17.14) closely corresponds with the sum of the mean temperature advection by the anomalous wind (contours in Fig. 17.12) and the anomalous temperature advection by the mean wind (contours in Fig. 17.13). Cold anomalous temperature advection in the eastern portion of our study site was primarily due to the anomalous winds associated with the barotropic high pressure near the Persian Gulf (Fig. 17.12); this circulation feature is forced by Indo-Pacific diabatic forcing. Cold anomalous temperature advection in the western portion of our study site appears to be primarily due to the interaction of the anomalous temperature gradient and mean wind field (Fig. 17.13). An anomalous air gradient, below the subtropical westerly jet, results in subsidence-inducing cold advection. The total cold horizontal temperature advection (Fig. 17.14) was balanced by precipitation-reducing subsidence (Fig. 17.8) throughout southwest Asia during extreme drought. Subsidence throughout southwest Asia (Fig. 17.8) combined with the direction and divergence of anomalous vertically integrated moisture flux (Fig. 17.11) both contributed to extreme drought.

17.5. SUMMARY

In this chapter, we examined the atmospheric and oceanic conditions during the strongest six southwest Asia droughts during the November–April rainy season for 1948–2012. The atmospheric and oceanic features during all extreme southwest Asia droughts are similar, which has allowed us to develop a physical model for regional drought.

Southwest Asia receives in excess of 75% of its annual precipitation during November–April (Fig. 17.1b) primarily in the form of rainfall at low elevations and snowfall at high elevations. The rainfall at low elevations provides wintertime soil moisture while the snowfall at high elevations provide warm season runoff necessary to sustain water resources during the rest of the year [*Barlow and Tippett*, 2008; *Hoell et al.*, 2015]. Southwest Asia

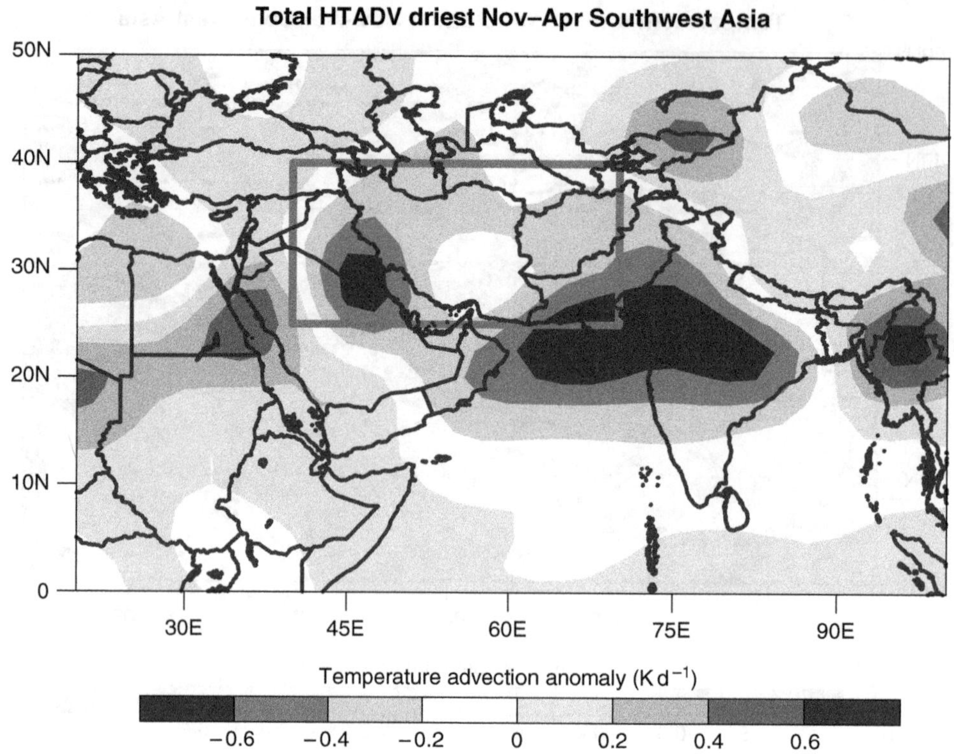

Figure 17.14 Composite 400 hPa total horizontal temperature advection (K d^{-1}) for the six driest November–April seasons areal averaged over southwest Asia.

precipitation exhibited strong year-to-year variability, including several multiyear drought and pluvial periods during November–April 1948–2012 (Fig. 17.2), which are linked to the ENSO, PDV, LT, and AMO SST modes. Here we examined six strongest regional droughts over southwest Asia in terms of SST, global atmospheric circulations, and the dynamic atmospheric forcing of regional precipitation to develop a physical model for regional extreme drought.

The canonical SST expression associated with extreme southwest Asia drought demonstrates a robust signal over the Pacific Ocean, with anomalously cold tropical and extratropical eastern Pacific SST as a result of La Niña and negative PDV, and a warm west Pacific Ocean primarily as a result of the positive phase of LT (Figs. 17.5 and 17.6). The characteristic "western V" LT warming pattern stretches from Indonesia into the northern and southern extratropical Pacific [*Solomon and Newman*, 2012; *Funk and Hoell*, 2015], and the strong agreement between observations and climate change simulations of warming in the West Pacific suggest a strong anthropogenic contribution to warming in this region [*Funk and Hoell*, 2015].

The combined effects of these modes (ENSO, PDV, and LT) produce an SST expression very similar to the "Perfect Ocean for Drought" pattern of *Hoerling and Kumar* [2003]. Extreme southwest Asia droughts occur simultaneously with a very strong global teleconnection, featuring a zonally symmetric high pressure throughout the midlatitude Northern Hemisphere at 200 hPa (Fig. 17.9), which is a characteristic hemispheric circulation signal during La Niña events [*Hoell et al.*, 2013b].

The anomalous atmospheric circulation, highlighted by an equivalent barotropic Rossby wave, during extreme southwest Asia droughts reduces the amount of available atmospheric moisture into the region (Fig. 17.11) and results in precipitation-reducing subsidence through interaction with the mean climate (Figs. 17.12–17.14). The anomalous northeasterly flow on the eastern side of the equivalent barotropic Rossby reduces the amount of moisture available from the climatological source, which is the northwest Indian Ocean (Fig. 17.11). The equivalent barotropic Rossby wave also interacts with the mean climate to produce midtropospheric cold-temperature advection that is balanced by precipitation subsidence (Figs. 17.12–17.14).

APPENDIX: DATA

Monthly 1948–2012 precipitation was drawn from the dataset initially developed by *Sheffield et al.* [2006] and updated to 2012 by *Nijssen et al.* [2014] on a 1.0 x 1.0

fixed latitude-longitude grid. The time period used to calculate the long-term monthly precipitation climatology is 1948–2012, and monthly precipitation anomalies were calculated by subtracting the long-term climatology from the total precipitation. The precipitation variability of the *Sheffield et al.* [2006] dataset is very similar to the precipitation variability of the Global Precipitation Climatology Centre [*Becker et al.*, 2013] for 1948–2010 over southwest Asia.

Atmospheric field variables were derived from the four times daily National Centers for Environmental Prediction-National Center for Atmospheric Research (NCEP-NCAR) reanalysis [*Kalnay et al.*, 1996] on a fixed 2.5 x 2.5 latitude-longitude pressure level grid for 1948–2012. The time period used to calculate the four times daily long-term climatology is 1948–2012, and four times daily anomalies were calculated by subtracting the long-term climatology from the total variable. The atmospheric field variables analyzed in this chapter are geopotential height, vertical velocity, terms of the hydrostatic thermodynamic energy equation, and terms of the moisture budget equation.

Monthly 1948–2012 SST was drawn from the Extended Reconstructed Sea Surface Temperature (ERSST) analysis version 3b [*Smith et al.*, 2008] on a fixed 2.0 x 2.0 latitude-longitude grid. The time period used to calculate the long-term monthly SST climatology is 1948–2012, and monthly SST anomalies were calculated by subtracting the long-term climatology from the total SST.

REFERENCES

Agrawala, S., M. Barlow, H. Cullen, and B. Lyon (2001), *The Drought and Humanitarian Crisis in Central and Southwest Asia: A Climate Perspective*, IRI Special Report, 01-11, International Research Institute for Climate and Society, Palisades, NY.

Alijani, B. (2008), Effect of the Zagros Mountains on the spatial distribution of precipitation, *J. Mt. Sci.*, *5* (3), 218–231; doi:10.1007/s11629-008-0126-8.

Barlow, M. (2011), The Madden-Julian Oscillation influence on Africa and West Asia, *Intraseasonal Variability in the Coupled Tropical Ocean-Atmosphere System*, edited by W. Lau and D. Waliser, Praxis, 477–493.

Barlow, M. A., and M. K. Tippett (2008), Variability and predictability of central Asia river flows: Antecedent winter precipitation and large-scale teleconnections, *J. Hydrometeor.*, *9*, 1334–1349; doi:10.1175/2008JHM976.1.

Barlow, M., A. Hoell, and F. Colby (2007), Examining the wintertime response to tropical convection over the Indian Ocean by modifying convective heating in a full atmospheric model, *Geophys. Res. Lett.*, *34*, n/a–n/a; doi:10.1029/2007GL030043.

Barlow, M., H. Cullen, and B. Lyon (2002), Drought in central and southwest Asia: La Niña, the warm pool, and Indian Ocean precipitation, *J. Climate*, *15*, 697–700; doi:10.1175/1520-0442(2002)015<0697:DICASA>2.0.CO;2.

Barlow, M., M. Wheeler, B. Lyon, and H. Cullen (2005), Modulation of daily precipitation over southwest Asia by the Madden–Julian Oscillation, *Monthly Weather Rev.*, *133*, 3579–3594; doi:10.1175/MWR3026.1.

Becker, A., P. Finger, A. Meyer-Christoffer, B. Rudolf, K. Schamm, U. Schneider, and M. Ziese (2013), A description of the global land-surface precipitation data products of the Global Precipitation Climatology Centre with sample applications including centennial (trend) analysis from 1901–present, *Earth Syst. Sci. Data*, *5*, 71–99; doi:10.5194/essd-5-71-2013.

Eggerman, M., and C. Panter-Brick (2010), Suffering, hope, and entrapment: Resilience and cultural values in Afghanistan, *Soc. Sci. Med.*, *71*, 71–83; doi:10.1016/j.socscimed.2010.03.023.

Funk, C., and A. Hoell (2015), The leading mode of observed and CMIP5 ENSO-residual sea surface temperatures and associated changes in Indo-Pacific climate, *J. Climate*; doi:10.1175/JCLI-D-00334.

Gall, C. (2008), War and Drought Threaten Afghan Food Supply, *New York Times*, September 19.

Hoell, A., and C. Funk (2013), The ENSO-Related West Pacific sea surface temperature gradient, *J. Climate*, *26*, 9545–9562; doi:10.1175/JCLI-D-12-00344.1.

Hoell, A., C. Funk, and M. Barlow (2013b), The regional forcing of northern hemisphere drought during recent warm tropical west Pacific Ocean La Niña events, *Climate Dyn.*, 1–23; doi:10.1007/s00382-013-1799-4.

Hoell, A., C. Funk, and M. Barlow (2014a), La Niña diversity and northwest Indian Ocean rim teleconnections, *Climate Dyn.*, 1–18; doi:10.1007/s00382-014-2083-y.

Hoell, A., C. Funk, and M. Barlow (2014b), The forcing of southwestern Asia teleconnections by low-frequency sea surface temperature variability during boreal winter, *J. Climate*, *28*, 1511–1526; doi:10.1175/JCLI-D-14-00344.1.

Hoell, A., M. Barlow, and R. Saini (2012), The leading pattern of intraseasonal and interannual indian ocean precipitation variability and its relationship with Asian circulation during the boreal cold season, *J. Climate*, *25*, 7509–7526; doi:10.1175/JCLI-D-11-00572.1.

Hoell, A., M. Barlow, and R. Saini (2013a), Intraseasonal and seasonal-to-interannual Indian Ocean convection and hemispheric teleconnections, *J. Climate*, *26*, 8850–8867; doi:10.1175/JCLI-D-12-00306.1.

Hoell, A., S. Shukla, M. Barlow, F. Cannon, C. Kelley, and C. Funk (2015), The forcing of monthy precipitation variability over southwest Asia during the boreal cold season, *J. Climate*; in revision.

Hoerling, M., and A. Kumar (2003), The perfect ocean for drought, *Science*, *299*, 691–694.

Holton, J. R. (2004), *An Introduction to Dynamic Meteorology*, Elsevier.

Kalnay, E., and coauthors (1996), The NCEP/NCAR 40-year reanalysis project, *Bull. Amer. Meteor. Soc.*, *77*, 437–471; doi:10.1175/1520-0477(1996)077<0437:TNYRP>2.0.CO;2.

Krishnamurti, T. N. (1961), The subtropical jet stream of winter, *J. Meteor.*, *18*, 172–191; doi:10.1175/1520-0469(1961)018<0172:TSJSOW>2.0.CO;2.

Lyon, B., A. Barnston, and D. DeWitt (2013), Tropical Pacific forcing of a 1998–1999 climate shift: Observational analysis and climate model results for the boreal spring season, *Climate Dyn.*, 1–17; doi:10.1007/s00382-013-1891-9.

Mariotti, A. (2007), How ENSO impacts precipitation in southwest central Asia, *Geophys. Res. Lett.*, *34*, L16706; doi:10.1029/2007GL030078.

Mariotti, A., J. Ballabrera-Poy, and N. Zeng (2005), Tropical influence on Euro-Asian autumn rainfall variability, *Climate Dyn.*, *24*, 511–521; doi:10.1007/s00382-004-0498-6.

Meehl, G. A., and A. Hu (2006), Megadroughts in the Indian monsoon region and southwest North America and a mechanism for associated multidecadal Pacific sea surface temperature anomalies, *J. Climate*, *19*, 1605–1623; doi:10.1175/JCLI3675.1.

Nazemosadat, M. J., and A. R. Ghasemi (2004), Quantifying the ENSO-related shifts in the intensity and probability of drought and wet periods in Iran, *J. Climate*, *17*, 4005–4018; doi:10.1175/1520-0442(2004)017<4005:QTESIT>2.0.CO;2.

Nazemosadat, M. J., and H. Ghaedamini (2010), On the relationships between the Madden-Julian Oscillation and precipitation variability in southern Iran and the Arabian Peninsula: Atmospheric circulation analysis, *J. Climate*, *23*, 887–904; doi:10.1175/2009JCLI2141.1.

Nijssen, B., and coauthors (2014), A prototype global drought information system based on multiple land surface models, *J. Hydrometeor.*, *15*, 1661–1676; doi:10.1175/JHM-D-13-090.1.

Oki, T., and S. Kanae (2006), Global hydrological cycles and world water resources, *Science*, *313*, 1068–1072.

Ramankutty, N., A. T. Evan, C. Monfreda, and J. A. Foley (2008), Farming the planet: 1. Geographic distribution of global agricultural lands in the year 2000, *Glob. Biogeochem. Cycles*, *22*, GB1003; doi:10.1029/2007GB002952.

Ryan, J., R. Sommer, and H. Ibrikci (2012), Fertilizer best management practices: A perspective from the dryland West Asia–North Africa region, *J. Agron. Crop Sci.*, *198*, 57–67; doi:10.1111/j.1439-037X.2011.00488.x.

Schiemann, R., D. Lüthi, and C. Schär (2009), Seasonality and interannual variability of the westerly jet in the Tibetan Plateau region, *J. Climate*, *22*, 2940–2957; doi:10.1175/2008JCLI2625.1.

Sheffield, J., and E. F. Wood (2008), Global trends and variability in soil moisture and drought characteristics, 1950–2000, from observation-driven simulations of the terrestrial hydrologic cycle, *J. Climate*, *21*, 432–458; doi:10.1175/2007JCLI1822.1.

Sheffield, J., G. Goteti, and E. F. Wood (2006), Development of a 50-year high-resolution global dataset of meteorological forcings for land surface modeling, *J. Climate*, *19*, 3088–3111; doi:10.1175/JCLI3790.1.

Smith, T. M., R. W. Reynolds, T. C. Peterson, and J. Lawrimore (2008), Improvements to NOAAs historical merged land–Ocean surface temperature analysis (1880–2006), *J. Climate*, *21*, 2283–2296; doi:10.1175/2007JCLI2100.1.

Solomon, A., and M. Newman (2012), Reconciling disparate twentieth-century Indo-Pacific ocean temperature trends in the instrumental record, *Nature Climate Change*, *2*, 691–699; doi:10.1038/nclimate1591.

Part IV
Prediction of Climate Extremes

18

Extratropical Precursors of the El Niño–Southern Oscillation

Kathy V. Pegion and Christopher Selman

ABSTRACT

Extratropical mechanisms thought to stochastically force ENSO are reviewed and investigated as potential ENSO predictors. Results indicate that these precursors have a stronger impact on the probability than amplitude of ENSO events and more strongly impact El Niño than La Niña. We identify the combination of precursors that best fit eastern and central Pacific ENSO indices using linear regression. The best models have poor fit to the EP index, representing only 4% of the variance of this index at 1 year lead time. However, the best model can represent 36% of the variance of the CP index at the same lead time. The extratropical precursors and their impact on the tropics can be seen a year in advance as opposed to events without an extratropical influence, which develop much more quickly beginning in the spring. As a result, extratropically forced ENSO events have a higher potential for long lead predictability. Although a subset of precursors is selected for the models, evidence of all of the precursors is present when fields are regressed with the modeled CP index. This demonstrates that the precursors are highly interrelated and should be evaluated in a holistic manner rather than as separate mechanisms.

18.1. INTRODUCTION

Current linear theories for the El Niño–Southern Oscillation (ENSO) rely on stochastic forcing to explain its existence or irregularity [e.g., *Suarez and Schopf*, 1988; *Zebiak and Cane*, 1987; *Penland and Sardeshmukh*, 1995; *Chang et al.*, 1996; *Moore and Kleeman*, 1999; *Thompson and Battisti*, 2000; *Thompson and Battisti*, 2001]. A number of different extratropical precursors have been proposed as mechanisms to stochastically force ENSO events up to 1 year in advance [e.g., *Chiang and Vimont*, 2004; *Vimont et al.*, 2001; *Vimont et al.*, 2003; *Alexander et al.*, 2010; *Anderson et al.*, 2013; *Wang et al.*, 2012; *Bond*, 2003, *Ding et al.*, 2015]. It seems reasonable that such precursors could provide useful information to long-lead ENSO prediction. It has also been shown that these precursors and their relationship with ENSO have changed on decadal and multidecadal timescales and may also be changing due to climate change [e.g., *Wang et al.*, 2013; *Yu et al.*, 2012; *Di Lorenzo et al.*, 2010].

While each of these precursors has been evaluated, mostly separately, in various papers, a comprehensive evaluation is needed to understand their potential to predict ENSO, especially given that the precursors are not independent. The purpose of this chapter is to provide an overview of the collection of extratropical ENSO precursor mechanisms in observations and evaluate the relationships between the precursors to determine which precursors are the best predictors of ENSO.

The precursors discussed here are primarily related to sea level pressure variability in the extratropical Pacific, associated with the North Pacific Oscillation (NPO), which can impact the tropics and stochastically force

Department of Atmospheric, Oceanic, and Earth Sciences and Center for Ocean-Land-Atmosphere Studies, George Mason University, Fairfax, Virginia, USA

Climate Extremes: Patterns and Mechanisms, Geophysical Monograph 226, First Edition.
Edited by S.-Y. Simon Wang, Jin-Ho Yoon, Christopher C. Funk, and Robert R. Gillies.

ENSO events. The North Pacific Oscillation (NPO) is a north-south sea level pressure (SLP) dipole in the north Pacific and is the second leading mode of SLP variability in the extratropical north Pacific [*Walker and Bliss*, 1932; *Rogers*, 1981; *Linkin and Nigam*, 2008]. The southern lobe of the NPO is centered at about 35°N and its circulation reaches into the tropics. Generally speaking, the processes associated with the precursors can be thought of as following the order NPO → precursor → ENSO. The seasonal footprinting mechanism (SFM; section 18.2b) and Pacific meridional mode (PMM; section 18.2a) link the NPO to the tropical Pacific through anomalous winds and/or SST on its subtropical southern lobe. The Victoria mode (VM; section 18.2e) and trade wind charging (TWC; section 18.2c) are thought to provide a more specific role for the ocean in bridging the NPO and ENSO through evolution of subsurface temperature anomalies and an additional "charging" of the warm water volume, respectively The final precursor discussed here may be an exception to the NPO → precursor → ENSO link. In this case, the east-west SST dipole in the western North Pacific (WNP; section 18.2d) is thought to directly induce Kelvin waves in the tropical western Pacific.

18.2. OVERVIEW OF PRECURSORS AND THEIR IMPACT ON ENSO

In this section, we provide a detailed description of each of the precursors and how they are thought to impact ENSO. For each of them, a schematic representation summarizing the main features and processes for impacting ENSO is shown in Figure 18.1.

18.2.1. The Pacific Meridional Mode (PMM)

The Pacific meridional mode (Fig. 18.1a) is the leading subtropical-tropical mode of coupled SST-wind variability in the Pacific once ENSO has been removed and is analogous to the meridional mode in the Atlantic [*Chiang and Vimont*, 2004]. It is characterized by an anomalous north-south SST gradient, displacement of the ITCZ toward the winter hemisphere, and anomalous trade wind variability in the northern subtropical Pacific. The associated wind anomalies peak in winter or early spring, followed by a corresponding peak of SST anomaly 1 month later.

The PMM influences the development of ENSO by establishing anomalous southwesterlies along the equatorial Pacific (Fig. 18.1a). In boreal winter, NPO variability induces a delayed response in SST variability, which produces a north-south dipole of SST anomalies in the subsequent spring (Step 1 in Fig. 18.1a). This established SST gradient then causes a northward displacement of the ITCZ [*Chang et al.*, 2007]. Anomalous southwesterlies accompany the displacement of the ITCZ (step 2 of Fig. 18.1a) and ultimately oppose and weaken the northeasterly trade winds, establishing conditions favorable for the development of an ENSO event.

The PMM has been shown to impact ENSO variability via wind-evaporation-SST (WES) feedback in both models and observations [*Zhang et al.*, 2009; *Vimont*, 2010; *Vimont et al.*, 2009; *Wu et al.*, 2009]. For example, *Chang et al.* [2007] found that more than 70% (12 of 17) of the ENSO events that occurred from 1958 to 2000 were preceded by PMM-related wind anomalies. *Zhang et al.* [2009] show in a simple coupled model that the PMM SST anomalies are associated with wind stress anomalies to the southwest, which reduce the heat flux and warm the SSTs, leading to southwestward propagation that eventually impacts the tropics. In a 400 year simulation of the CCM3-RGO model, *Zhang et al.* [2009] found that 66% (33%) of El Niño events are (not) preceded by like-signed PMM events. In a 40 year simulation from a high-resolution (i.e., eddy-resolving ocean) version of the NCAR/CCSM3.5 model, *Larson and Kirtman* [2013] found that the PMM is correlated with ENSO 7–9 months later at 0.7. There is also evidence for a corresponding PMM in the Southern Hemisphere (SPMM) that is able to impact the tropics through the same mechanisms as its Northern Hemisphere counterpart [*Zhang et al.*, 2014].

It is hypothesized that the PMM is related to other precursors. *Chiang and Vimont* [2004] and *Alexander et al.* [2010] suggested that the SFM (see section 18.2b), which is forced by the NPO, may excite the PMM and then act as a conduit through which the extratropical atmosphere can impact ENSO. This is consistent with the theoretical analysis of *Vimont* [2010] that identifies equatorially antisymmetric coupled modes, excited by variations in the trades that reach far into the tropics (e.g., the NPO).

18.2.2. Seasonal Footprinting Mechanism (SFM)

Similar to the PMM, Northern Hemisphere wintertime NPO-like patterns of SLP variability impart an anomalous SST variability that persists into the subsequent spring/summer seasons (Fig. 18.1b). SLP variability drives anomalous warming of SST by reducing wind speeds over the central equatorial Pacific. The resultant anomalous SST causes a northerly shift in the ITCZ accompanied by anomalous westerlies in the equatorial pacific. These anomalous westerlies are implicated in the generation of equatorial Kelvin waves that warm the eastern Pacific and ultimately establish an ENSO event [*Vimont et al.*, 2003].

The seasonal footprinting mechanism (SFM) provides a link between midlatitude atmospheric internal variability (i.e., the NPO) in the winter and tropical Pacific variability 1 year later [*Vimont et al.*, 2001; *Vimont et al.*,

2003; *Alexander et al.*, 2010]. There is evidence that this mechanism provides a stochastic forcing of the ENSO cycle and can at times initiate an El Niño or La Niña event [e.g., *Vimont et al.*, 2003; *Anderson*, 2003; *Anderson*, 2007; *Alexander et al.*, 2010; *Wang et al.*, 2013; *Pegion and Alexander*, 2013]. *Vimont et al.* [2009] demonstrate that the wind-evaporation-SST (WES) feedback and its seasonality are important to the development of SST anomalies associated with the NPO in simulations with an atmospheric model coupled to a slab ocean model. However, the WES feedback has not been identified in more sophisticated coupled models [e.g., *Deser et al.*, 2012].

Several aspects of the original SFM hypothesis have been difficult to identify in observational datasets. For example, in the original hypothesis, warm SST anomalies did not reach the equator until late spring/summer. However, warm SST anomalies are present in the tropics

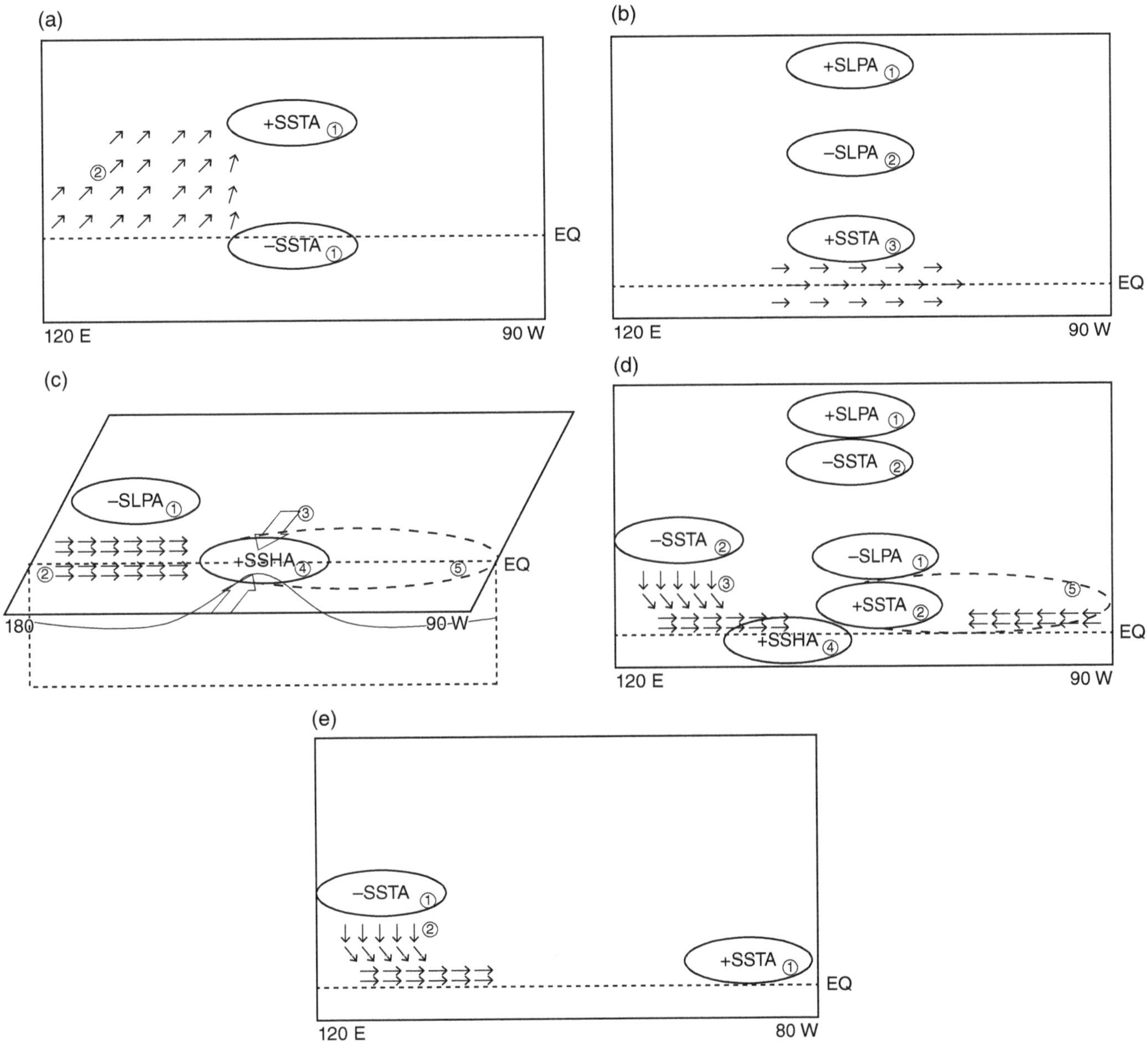

Figure 18.1 Schematic progression of events leading to an ENSO event for (a) the Pacific meridional mode, (b) seasonal footprinting mechanism, (c) trade wind charging, (d) Victoria mode, and (e) western North Pacific SST. Circled numbers within each diagram indicate the mechanistic order of events that occur within each paradigm. Vectors indicate direction of wind anomalies, while encircled anomalies indicate the location of driving maxima in anomalous conditions.

during winter prior to El Niño and even before in observations, the NCAR-CCSM4 model [*Deser et al.*, 2012] and the CFSv2 model [*Pegion and Alexander*, 2013] indicating that wind-evaporation-SST (WES) feedback through the SFM may not be the only mechanism providing a connection between the subtropics and tropics. As a result, additional mechanisms have been proposed to contribute to the subtropical-tropical connection that occurs in the SFM, including oceanic Rossby waves, which reflect off the western boundary [*Alexander et al.*, 2010] and a trade-wind charging mechanism [*Anderson et al.*, 2013], which will be discussed in the next section.

18.2.3. Trade Wind Charging (TWC)

Recent work by *Anderson et al.* [2013] indicates that the changes in winds associated with the NPO can directly force the subsurface temperature anomalies in the equatorial Pacific. They refer to this as "trade wind charging" (TWC) and show that subsurface temperature anomalies associated with the southern lobe of the NPO can be found as early as January in the central equatorial Pacific and along 5°N in ocean reanalyses. The TWC paradigm follows from the establishment of anomalously low sea level pressures in the west-central equatorial Pacific region (Step 1 of Fig. 18.1c). As in both PMM and TWC, NPO-like variability in boreal winter causes anomalously low SLP in the subtropics, producing anomalous westerlies (Step 2 of Fig. 18.1c). A key feature that distinguishes TWC from other precursors is the direct charging of subsurface temperature anomalies, which establish the warm water volume necessary for the development of ENSO [*Anderson et al.*, 2013]. Subsurface heat anomalies are first established in the wintertime (steps 3 and 4 of Fig. 18.2c), but slowly develop through time, expanding eastward and becoming more equatorially symmetric through the boreal spring (step 5 of Fig. 18.1c). Continued eastward propagation of sub- and sea-surface temperatures through boreal summer ultimately prime the conditions necessary for an ENSO event in the following winter.

Sensitivity experiments with an ocean model forced by wind stress anomalies from different regions indicate that the central Pacific zonal wind stress anomalies initiate both equatorial and off-equatorial subsurface temperature anomalies related to TWC and that surface heat flux anomalies (as proposed in the SFM and PMM) have little impact on the development of ENSO-related SST anomalies [*Anderson et al.*, 2013].

18.2.4. Western North Pacific SST (WNP)

The WNP pattern is associated with a dipole in SST anomalies between the western North Pacific and central to eastern equatorial Pacific in boreal winter (step 1 of

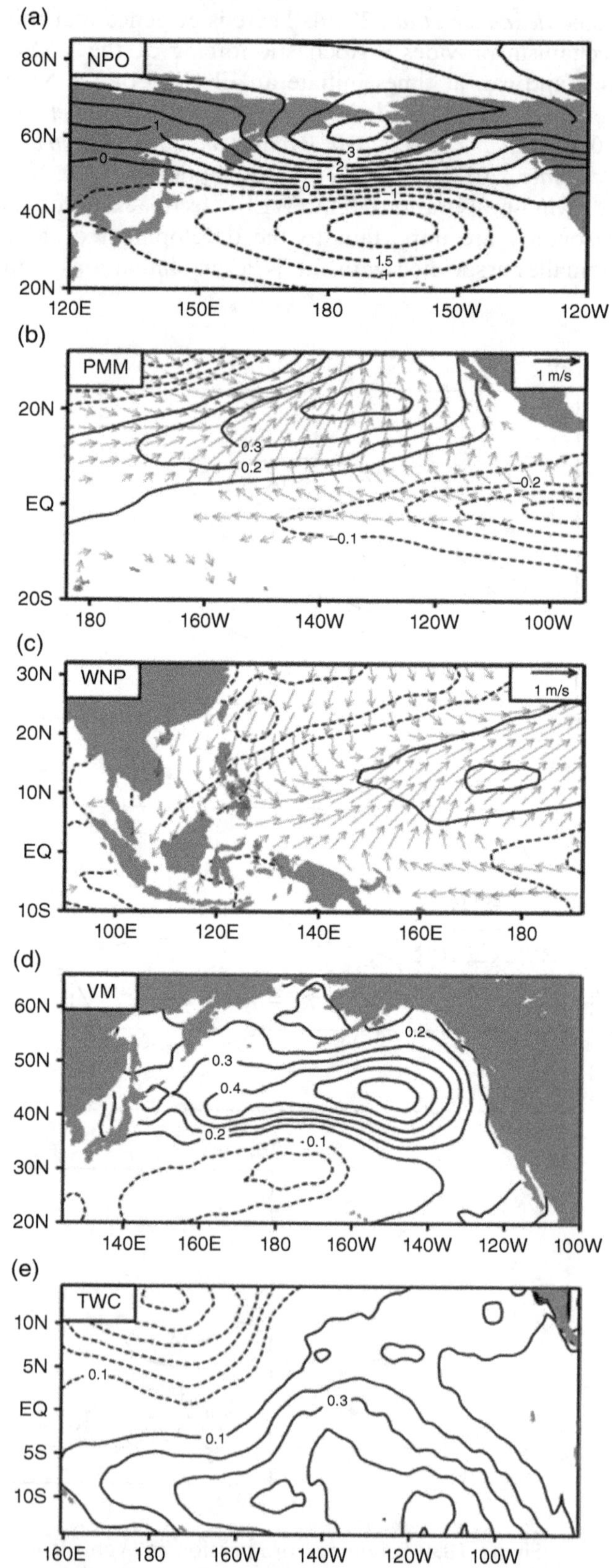

Figure 18.2 Spatial patterns for each precursor in units of (a) hPa, (b) °C for SST and m/s for winds, (c) °C for SST and ms^{-1} for winds, (d) °C, and (e) Nm^{-2}.

Fig. 18.1e). This dipole causes anomalous winds from the western North Pacific toward the central equatorial Pacific, into the eastern Pacific (step 2 of Fig. 18.1e). Imparting a stress on the ocean surface, these anomalous westerlies directly stimulate the development of oceanic Kelvin waves, which trigger the development of ENSO over the subsequent year. *Wang et al.* [2012] show this via two different methods. In the first, the WNP is calculated based on the average, detrended SSTA from 122°–132°E and 18°–28°N during Dec-Jan-Feb. Regressions with this index show a strong positive relationship with eastern Pacific SST anomalies and corresponding wind anomalies the following winter. The evolution of the SST and wind regressions with the WNP index is consistent with the evolution of SST and wind for the SFM and PMM, and leads to ENSO 1 year later. Similarly, the second pattern of maximum covariance analysis between boreal winter SST and 10 m wind anomalies in the western Pacific (90°E–170°W; 10°S–10°N) shows a similar pattern to the area averaged WNP index and is correlated with the PMM SST index at 0.74 during DJF. They argue that the WNP-related wind anomalies are able to directly induce Kelvin waves in the western Pacific, leading to ENSO events and that this direct forcing of Kelvin waves is more effective in initiating ENSO events than the PMM or SFM mechanisms. Modeling results also demonstrate that the relationship between the WNP and ENSO has become stronger in recent decades due to greenhouse gas forcing [*Wang et al.*, 2013].

18.2.5. Victoria Mode (VM)

The VM represents an integrated approach of other precursors, though with slightly different timings. As with all other precursors discussed so far, VM hinges on the establishment of an NPO-like circulation (step 1 of Fig. 18.1d), though there is also a simultaneous development of negative SST anomalies off the coast of the western North Pacific (step 2 of Fig. 18.1d). The simultaneous development of these two features implies that western North Pacific SST anomalies play a role in the development of the NPO [*Ding et al.*, 2015]. Continuing through the boreal winter into spring, both anomalous westerlies in the western equatorial Pacific and anomalous easterlies in the eastern equatorial Pacific promote the growth of positive SST anomalies in the central equatorial Pacific (step 3 of Fig. 18.1d). As in the TWC mechanism, there is also a strengthening of subsurface temperature anomalies, though they play less of a role in the development of ENSO (step 4 of Fig. 18.1d). The resultant dipole in SSTA between the western North Pacific and central equatorial Pacific serves to strengthen the anomalous westerlies throughout the summer season, ultimately producing the eastward expansion of positive SSTA along the equatorial Pacific, establishing the conditions necessary for ENSO in the winter.

The Victoria Mode (VM) is identified as the second leading mode of SST anomalies in the north Pacific and is considered a direct ocean response to atmospheric forcing by the NPO [*Bond*, 2003; *Ding et al.*, 2015]. The VM focuses on the basin-wide large-scale response of SST to the NPO forcing as opposed to the SFM, PMM, and WNP, which focus on more local responses. The VM peaks in spring and has its highest correlation with ENSO indices when it leads by ~11 months [*Ding et al.*, 2015]. Generally speaking, the local features and processes of the VM impact on ENSO are consistent with other precursors and occur via WES feedback in the tropical-subtropical Pacific, and evolution of subsurface ocean temperature along the equator [*Ding et al.*, 2015]. For example, the warm subtropical eastern Pacific SST anomalies are consistent with the PMM pattern, the air-sea coupling in the subtropical Pacific is consistent with the SFM, the cold western North Pacific SST pattern is consistent with the WNP precursor, and the evolution of subsurface temperature anomalies along the thermocline are consistent with the TWC mechanism. However, there appears to be some timing differences between the VM and other precursors in terms of its impact on ENSO. For example, *Ding et al.* [2015] argue that the impact of the VM on ENSO is stronger and more direct than the SFM, with equatorial wind anomalies that extend across the entire equatorial Pacific in late winter and early spring as opposed to the SFM-related winds, which are thought to impact the equatorial Pacific in the summer. There is also evidence that the VM and its relationship with ENSO has experienced decadal changes [e.g., *Bond*, 2003].

18.3. DATA AND DEFINITIONS

18.3.1. Datasets

The datasets used in this study are the sea level pressure (SLP), 10 m winds, and zonal momentum flux from the National Centers for Environmental Prediction-National Center for Atmospheric Research Reanalysis (NCEPR1) [*Kalnay et al.*, 1996] for the years 1948–2014. Sea surface temperatures (SST) are from the NOAA extended reconstructed SST, version 3b (ERSSTv3b) [*Reynolds et al.*, 2007] for 1948–2014. Anomalies are calculated by removing the average overall available years for each month. Data are detrended by removing a linear least squares fit in time.

18.3.2. ENSO Definitions

The indices used to represent ENSO follow the definitions by [*Takahashi et al.*, 2011] and are chosen because they represent ENSO events as a continuum. They are calculated by first performing EOF analysis of SST in the equatorial Pacific from 10°S–10°N. Following *Takahashi et al.* [2011], the ENSO indices are defined as follows:

$$CP = \frac{PC1 + PC2}{\sqrt{2}} \tag{18.1}$$

$$EP = \frac{PC1 - PC2}{\sqrt{2}} \tag{18.2}$$

where CP corresponds to central Pacific ENSO and EP corresponds to a so-called canonical ENSO [*Ashok et al.*, 2007; *Kug et al.*, 2009; *Larkin*, 2005; *Kao and Yu,* 2009]. PC1 and PC2 refer to the principal component time series associated with the first and second leading patterns of SST, respectively.

18.3.3. Precursor Definitions

A number of different definitions have been used to identify each of the precursors. For this analysis, we have chosen to use the definitions based on the second EOFs of SLP in the region 20°N–85°N; 120°E–120°W for the NPO, of SST in the region 20.5°N–65.5°N; 124.5°E–100.5°W for the VM, and zonal wind stress in the region 15°S–15°N; 160°E–80°W for the TWC. For the PMM, and WNP, we use MCA between SST and 10 m winds in the region 21°S–32°N; 175°E–95°W and 10°S–30°N; 90°E–170°W, respectively. A summary of the fields and methods used for each precursor is shown in Table 18.1 and are consistent with the studies indicated in the right-most column. Figure 18.2 shows the spatial patterns corresponding to each precursor. The signs of the patterns are shown such that they lead to El Niño events while their opposite sign leads to La Niña events. For the EOF-based precursors, a single time series is produced representing the temporal variability associated with each pattern. For the MCA-based precursors, two time series are produced and both time series will be considered when diagnosing the potential to predict ENSO (section 18.7).

18.4. EVALUATION OF PRECURSOR VARIABILITY AND COVARIABILITY

Given that each of the time series associated with the different precursors are determined from different fields and over different regions, it is not surprising that they contain variability across a range of timescales. Particularly, we would expect the precursors determined from winds, wind stress, or sea level pressure to have more weather "noise" and shorter timescale variability, while those determined from SST should contain more interannual to decadal variability. This is assessed in Figure 18.3 by evaluation of the power spectra. We compute power spectral density (PSD) from the various indices as one-sided periodograms with a half-width averaging window of 13 months. The results are displayed as a log-log plot such that the area under the curve represents the total variance (Fig. 18.3). Some indices demonstrate PSD similar to that of red noise, with most of the power (represented as squared units per cycle per month) being preferentially distributed to lower frequencies, but with no clear peaks over any return period (e.g., PMM SST). The NPO index, on the other hand, appears to mimic a white noise process, with power being more evenly distributed across all periods. The NPO does deviate slightly from this paradigm, however, in the 0.25 to 1 year range, indicating that there is higher-frequency variation in the NPO, likely associated with seasonal variability (i.e., stronger in winter versus summer). The VM index appears to mimic a red noise distribution up until the 0.75 to 1.5 year range, where a peak is found, indicating a component of interannual variability to the VM. There is also a small peak at the 0.5 year band, indicating some seasonal variability. There appear to be differences in the power distribution of WNP SST and wind indices, with WNP WND having a broad peak over the 1.5 to 2 year period. This is in stark contrast to the relatively red noise behavior of WNP SST. TWC deviates slightly from the typical red noise spectra, with small peaks around 5 and 7 year

Table 18.1 Definitions of Precursors

Name	Variables	Region	Method	Reference
PMM	SST, Wind	21°S–32°N; 175°E–95°W	2nd MCA	*Chiang and Vimont* [2004]
NPO	SLP	20°N–85°N; 120°E–120°W	2nd EOF/PC	*Vimont et al.* [2001] *Vimont et al.* [2003]
WNP	SST, Wind	10°S–30°N; 90°E–170°W	2nd MCA	*Wang et al.* [2012]
TWC	Taux	15°S–15°N; 160°E–80°W	2nd EOF/PC	*Anderson and Perez* [2015]
VM	SST	20.5°N–65.5°N; 124.5°E–100.5°W	2nd EOF/PC	*Bond* [2003] *Ding et al.* [2015]

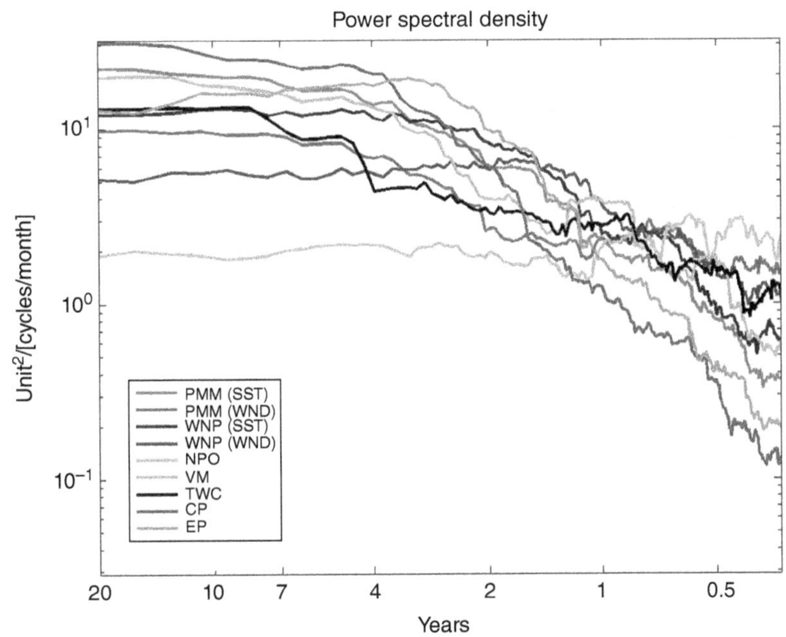

Figure 18.3 Power spectral density of ENSO precursor indices as a log-log plot. The *y*-axis represents the variance of the precursor in units² per cycle per month. The *x*-axis represents the period in years. (*See insert for color representation of the figure.*)

Table 18.2 Simultaneous Correlation Between Precursors and ENSO Indices

	PMM SST	PMM WND	WNP SST	WNP WND	NPO	VM	TWC	CP	EP
PMM SST		0.58	0.27	0.20	0.15	0.33	0.17	0.49	–0.74
PMM WND			0.26	0.36	0.52	0.27	0.27	–0.05	–0.33
WNP SST				0.62	0.17	0.26	0.27	–0.05	–0.33
WNP WND					0.32	0.13	0.39	0.0	–0.21
NPO						0.14	0.25	0.09	–0.10
VM							0.21	0.34	–0.14
TWC								0.11	–0.15
CP									0.0
EP									

periods, generally around the climatological ENSO return period. This is somewhat reinforced by the fact that the CP index has a small peak around the same time as TWC. The CP index also imitates red-noiselike behavior, partially implying that the CP index itself may be an expression of a natural red noise process [e.g., *Newman et al.*, 2011]. EP has a peak within the span of 3–4 years, consistent with the climatological ENSO return periods in observations.

Although we have presented each of the precursors here and described their processes separately, the precursors are not independent of each other. Table 18.2 shows the simultaneous correlation between each precursor with the other precursors. For example, PMM WND is correlated with all other precursors at least at 0.3. Additionally, WNP WND and TWC are correlated with each other ~0.4. In section 18.7, we will use a regression model to diagnose which combination of precursors best predict ENSO.

18.5. RELATIONSHIP BETWEEN PRECURSORS AND ENSO

In this section, we evaluate the relationship between each of the precursors and the development of ENSO. Figure 18.4 shows lag correlations between the Dec-Jan-Feb CP and EP ENSO indices with each of the precursors. The values on the *y*-axis indicates the correlation between the ENSO index and precursor, with negative signs on the *x*-axis indicating the precursors lead the ENSO index. For example, the CP ENSO index is highly correlated with its Dec-Jan-Feb value in the simultaneous Feb, with decreasing correlation to ~0.3 out to Dec the previous year. It is evident that some of the precursors have nonzero correlation at various times leading the CP ENSO index, but very small lagged correlations with EP ENSO. PMM SST and WND both have correlations with CP ENSO around 0.4 from the previous Dec through Sep, while WNP SST and WND have correlations between 0.4 and 0.6 with CP ENSO during the previous spring and summer. The VM shows the highest correlation of any precursor with CP ENSO at longer lead times and the most persistent, with correlations above 0.4 for 5–10 months.

Further evidence that extratropical precursors more strongly impact CP ENSO can be seen in Figure 18.5, which shows a scatterplot of CP vs. EP indices during Dec-Jan-Feb. The colors represent the normalized value of the respective precursor indices 1 year prior to the ENSO index. Shifts of colors from left to right along the *x*-axis indicate shifts in probabilities of CP ENSO events, while shifts in colors up or down the *y*-axis indicate changes in probabilities associated with EP ENSO. It is evident that there is a shift of red/orange colors to higher CP values for PMM SST, WNP SST, and VM. The top two points in each plot correspond to 1982-1983 and 1997-1998 El Niño events, which have high EP values, but also partially register on the CP spectrum as well. Depending on the precursor, these events were not preceded by the precursor (e.g., grey dots for WNP SST or VM) or were preceded by moderate (orange) or strong values (red/dark red) values of the precursors. It is also noted that although there is a shift in probabilities for CP El Niño events, it seems that the precursors have a much weaker impact on CP La Niña events with blue dots spanning the scatter plots across a range of CP values, perhaps with the exception of the VM. *Larson and Kirtman* [2014] have also shown no predictive relationship between La Niña events and the SSTs associated with the PMM.

18.6. DIAGNOSING PRECURSORS AS ENSO PREDICTORS

In order to address which precursors best predict ENSO, a linear regression model is constructed of the form:

$$y = \beta_1 x_1 + \beta_2 x_2 + \beta_3 x_3 + \cdots + \beta_n x_n + \varepsilon \quad (18.3)$$

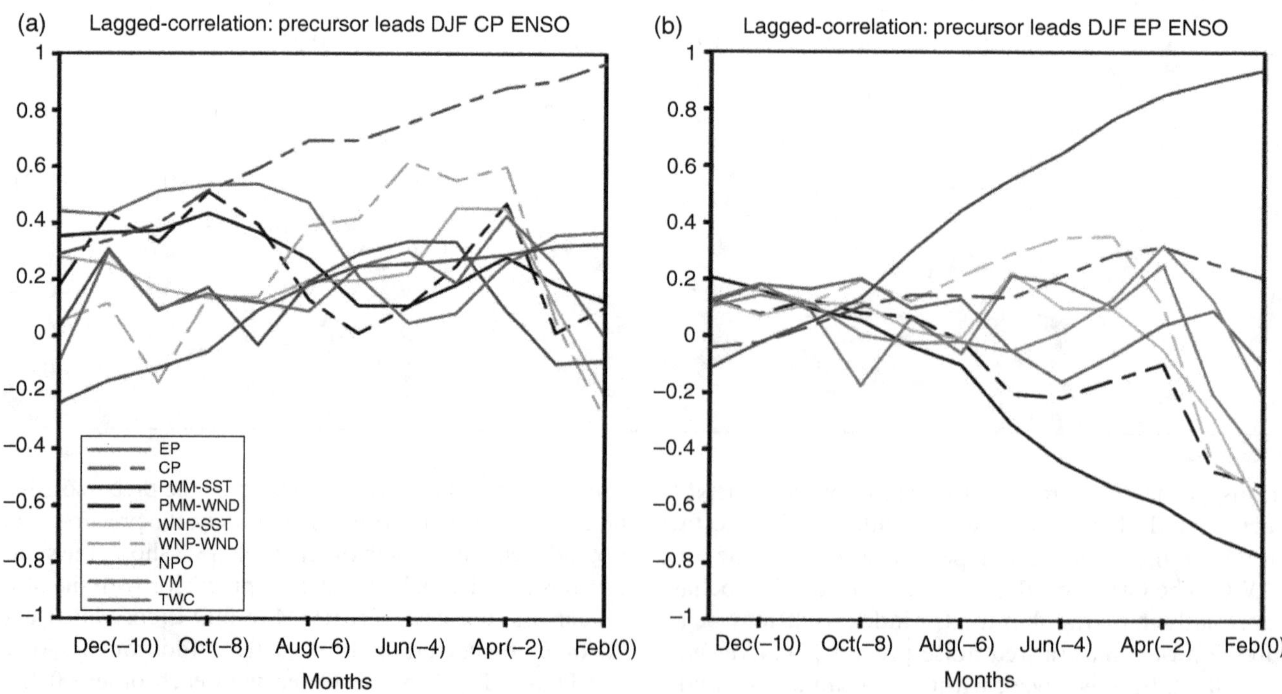

Figure 18.4 Lagged correlation in months between (a) DJF CP ENSO index and precursors and (b) DJF EP ENSO index and precursors. Negative lags indicate that the precursor leads the ENSO index.

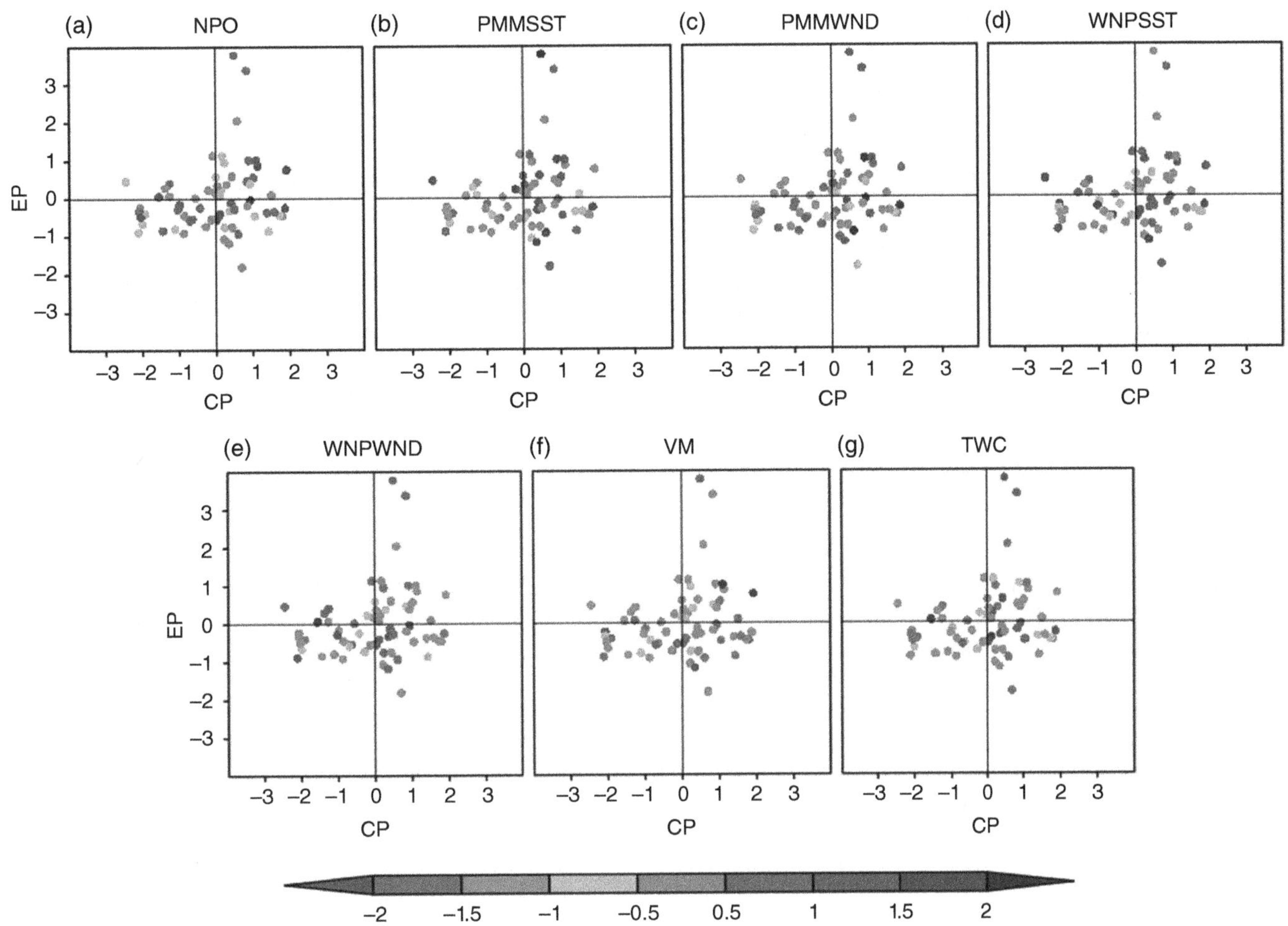

Figure 18.5 Scatterplot of CP vs. EP normalized values for Dec-Jan-Feb. Colors represent the normalized value of each precursor for the previous Dec-Jan-Feb. (*See insert for color representation of the figure.*)

Table 18.3 Best Regression Models for Each Index and Lead Time

Lead	Predictand	Model	r^2	RMSE
1 year	CP	0.34 PMMWND-0.5 NPO+0.27 WNPSST+0.61 VM	0.36	0.92
9 month	CP	0.42 PMMWND+0.53 VM	0.38	0.90
1 year	EP	0.25 TWC	0.04	0.91
9 month	EP	0.23 WNPWND	0.03	0.92

where the predictand, y, is an index representing ENSO (e.g., EP, CP) in the winter (Dec-Jan-Feb) season. The predictors (β_1, $\beta_{2,\ldots,}$ β_n) are the precursor indices (e.g., PMM SST, PMM WND, NPO, etc.) at 9 months (Mar-Apr-May) or 1 year (Dec-Jan-Feb) lead times and ε is the residual or error term. All possible models that consist of these predictors are considered. Selection of the best model is determined using the Akaike information criteria (AIC) [*Burnham and Anderson*, 2002], given by:

$$AIC = 2K - 2\ln L \tag{18.4}$$

where K represents the number of parameters and L is the maximum likelihood for the model given the observations. The two terms in the AIC equation consist of a measure of goodness of fit (-2lnL) and a penalty for increasing the number of parameters (2K). In this sense, the AIC can be used to identify the best model with the least number of parameters to minimize the problem of overfitting. This is particularly important here given that it is known that the predictors are not independent. The model with the lowest AIC score is selected as the best model (Table 18.3). Given K predictors, there are 2^K -1 candidate models. In our case, with K = 7, there are 127 candidate models. All coefficients for the models are tested to verify that the p-value < 0.05. Table 18.3 shows the best models for predicting the DJF CP and EP indices at 9 months and 1 year lead times. The r^2 values of 0.07 and 0.03 respectively indicate that the best models for EP explain very little of its variance and are not particularly skillful models. For CP ENSO, the most skillful model at 1 year lead time includes the PMM WND, NPO, WNP SST, and VM and explains 36% (i.e., correlation of 0.6). This tells us that the North Pacific sea level pressure in the form of the NPO, the subtropical winds associated with the PMM, western Pacific winds associated with the WNP region, and North Pacific SST all provide some unique, independent information to CP ENSO. The NPO curiously has a negative correlation coefficient in the regression model. This can be explained by the fact that the NPO is strongly present in other predictors and is in fact included in each of the other predictors (not shown). A negative coefficient acts as a correction term so that the NPO is not triple counted.

At 9 month lead times, it appears that the PMM WND and VM SST are sufficient to predict the CP index with the same level of skill. The fact that the NPO is not included as a predictor is not surprising since it is thought to lead the other predictors. As a basis for comparison, the skill of the operational model used by the National Centers for Environmental Prediction for seasonal predictions has a correlation of about 0.6–0.7 at 8 month lead for forecasts with target month in Jan and 0.5–0.6 for 8 month lead forecasts with target month in May [*Saha et al.*, 2014]. All models have high RMSE likely due to the fact that there is little relationship between the amplitude of the precursors and amplitude of the ENSO indices. This is evident in Figure 18.5. If there were a strong amplitude impact, then we would expect to see systematically higher (lower) values of the precursor indicated by darker red (blue) colors as the CP index increases (decreases). Essentially, the precursors shift the probabilities of ENSO events, but the strength of the precursor is not related to the strength of the ENSO event.

How does the evolution of ENSO events stochastically forced by these extratropical precursors compare to those that are not forced by these precursors? To evaluate this, we compare the regression of various fields with our predictand (y) to regressions of the same fields with the residual (ε). The residual explains 64% of the variance of the CP index at 1 year lead times and 62% at 9 month lead times. The results of comparing the predictand with the residual are similar for both lead times, therefore, we show the results of the regression model for 1 year lead time in the form of seasonal lagged regressions (Fig. 18.6). Comparison of the regression model (left panels) with the residual (right panels) shows the evolution of ENSO events forced by the extratropical precursors versus those events that are not forced by the precursors. It is clear that both the regression model and the residual evolve into El Niño over the course of the year. The evolution of events captured by the extratropical precursor regression model occurs more slowly than the events associated with the residual. This supports the idea that events forced by these precursors may be more predictable than events not forced from the extratropics. In the regression model, warm SST anomalies are present in the subtropical and equatorial western Pacific with opposite signed anomalies in the eastern Pacific. The warm subtropical SST anomalies occur on the southern lobe of the NPO and occur coincident with strong southwesterly winds in the western equatorial Pacific and southeasterly winds in the eastern equatorial Pacific that converge at about 160°W, consistent with the PMM. A SST maximum is also present in the north Pacific consistent with the VM. Cold SST anomalies are also present in the western Pacific with strong northerly winds that impact the tropical western Pacific, consistent with the WNP. Although not shown directly, charging of warm water onto the EQ can also be inferred by the wind stress curl, consistent with the TWC. The large-scale extratropical SST pattern remains throughout the year while the equatorial SST anomalies and winds continue to evolve. In the residual, the NPO dipole is not present and equatorial SST anomalies and westerly winds are weak and have small spatial scale in Dec-Jan-Feb. By spring, the westerly winds are strong in the western Pacific and SST anomalies have developed across the Pacific from about 180° to the South American coast. By fall and the following winter, both the model and residual have very similar patterns and amplitudes in terms of the average El Niño event that develops. Although only the PMM wind, NPO, and VM were selected as predictors in the best regression model, there is evidence of all of the predictors in Figure 18.6, further supporting the idea that they are interrelated.

18.7. RELATIONSHIP OF EXTRATROPICAL PRECURSORS TO 2014 AND 2015 EL NIÑO

As a final component of this analysis, we evaluate the role of the extratropical precursors in the development of two recent El Niño events, 2014–2015 and 2015–2016. This is done by projecting the 2013–2015 SST, winds, zonal wind stress, and sea level pressure anomalies onto the precursor spatial patterns to obtain the PC or MCA time series associated with each precursor during this time period. The PC and MCA spatial patterns are recomputed for 1948–2012 and we have verified that the regression model coefficients are consistent with those previously computed for the 1948–2014 period. We can then make legitimate predictions, which contain no future information, using the regression model for the 2014–2015 and 2015–2016 El Niño events. The values for each precursor included in the regression model at 1 year (Dec-Jan-Feb) prior to the 2014–2015 and 2015–2016 El

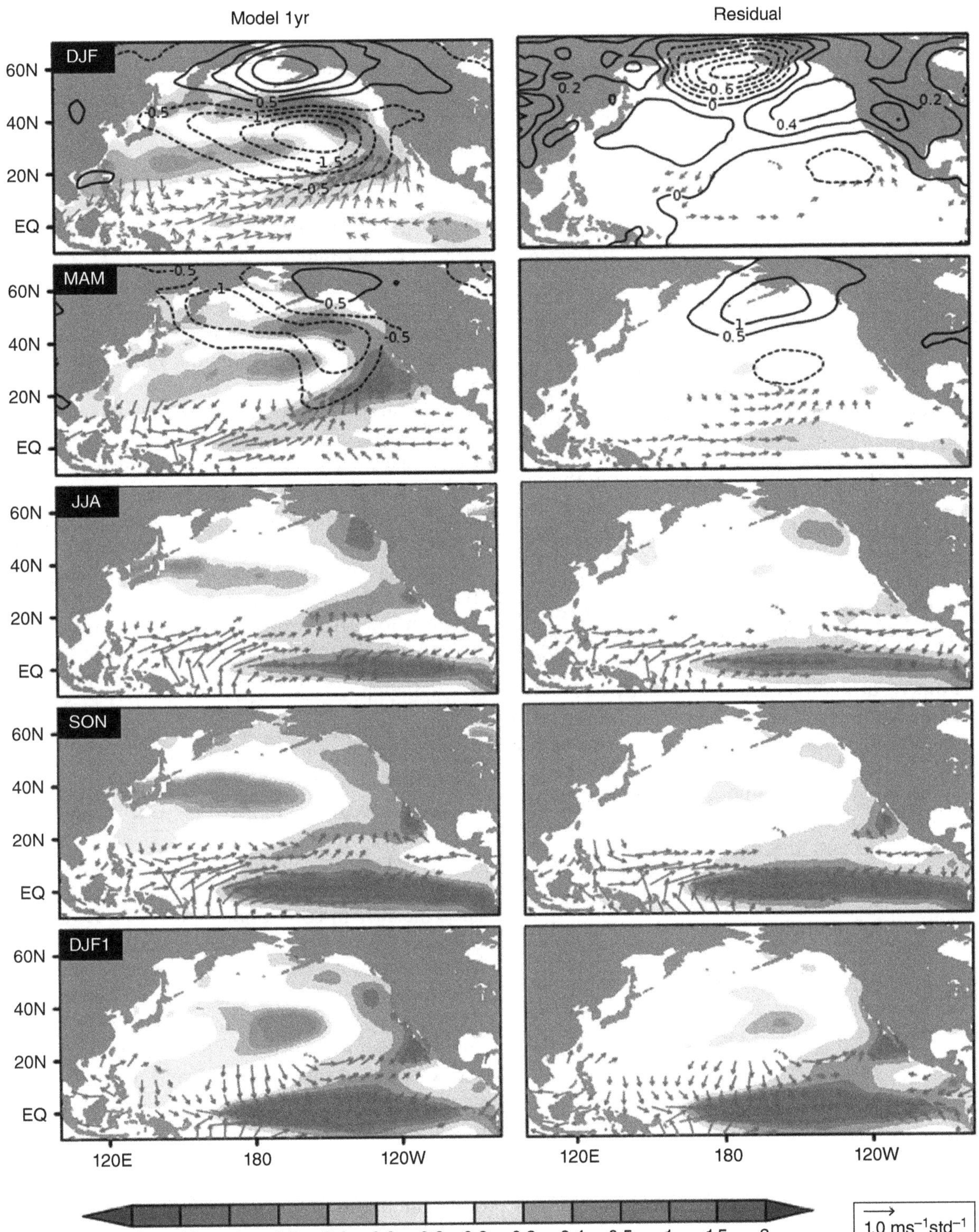

Figure 18.6 Seasonal lagged regressions of (shaded) SST (°C std^{-1}), (contours) SLP (hPa std^{-1}), (vectors) 10 m wind (ms^{-1}std^{-1}) anomalies with left modeled CP index from best regression model for 1 year lead, and right residual from best regression model for 1 year lead. SLP contour interval is 1 hPa per standard deviation and is shown only for Dec-Jan-Feb and Mar-Apr-May. Winds are shown only from 20°S–20°N. SST contour interval is 0.1°C std^{-1}. *See electronic version for color representation.*

Table 18.4 Normalized Values of Precursors Used as Predictors for Dec-Jan-Feb, Predicting Dec-Jan-Feb 2014–2015 and 2015–2016 CP Index One Year Later

	2014–2015	2015–2016
PMMWND	0.48	1.3
WNPSST	0.33	–0.5
NPO	–0.36	–1.6
VM	3.0	0.3
Predicted CP	0.5	1.2
Observed CP	0.43	2.8 (Dec only)

Note: Data are normalized by the standard deviation of the individual time series determined from 1948 to 2012. The predicted and observed normalized values of the CP index are shown in the bottom two rows. Note for 2015–2016, the observed CP index value is only reported for Dec 2015 as data for 2016 were not available at the time of analysis.

Niño events are shown in Table 18.4 along with the predicted and observed values of the CP index 1 year later. The regression model predicted a CP event of 0.5σ, while the observed value was 0.43σ. One year prior to this event, a strong VM (3.0σ) and moderate PMM WND (0.48σ) contributed to the positive value of the predicted CP index. For 2015–2016, the observed Dec CP value was 2.8σ (note that Dec 2015 is used here because Jan and Feb data were not yet available at the time of analysis) and the predicted Dec-Jan-Feb value was 1.2σ. The strong PMM WND appears important to predicting the strong CP event in 2015–2016. One explanation may be that the previous El Niño event had already primed the tropics and subtropics such that the subtropically focused precursor was already active and the higher latitude precursors (e.g., NPO and VM) were less important for the development of this El Niño. While these results show only two cases and much more analysis is needed to understand the details of how the precursors led to the development of these events, the precursors regression model is able to distinguish the stronger magnitude of the 2015–2016 El Niño as compared with the 2014–2015 event.

18.8. SUMMARY AND DISCUSSION

This chapter provides a review of extratropical ENSO precursors and identifies those precursors that are best predictors of ENSO at 9 month and 1 year lead times. It is shown that these extratropical precursors are more strongly related to CP ENSO events than EP events, consistent with *Yu and Kim* [2011] and *Yu and Kim* [2010], who indicate that CP ENSO events represent a mechanism that is excited from the subtropics, and *Vimont et al.* [2014], who demonstrate that the PMM preferentially excites CP ENSO events using a linear inverse modeling approach with norms that maximize CP vs. EP events. The evolution of CP ENSO events associated with the extratropical precursors shows that these events evolve more slowly than those not associated with the extratropical precursors and are therefore likely to have longer lead predictability. We have also provided evidence that the precursors are interrelated and each is evident in the seasonal evolution of extratropically forced events using a model with only a few precursors as predictors. This is not surprising given that each represents a regionally specific piece of Pacific variability.

In the future, increased understanding of how ENSO is forced by extratropical precursors and associated improvements in long lead prediction will require a holistic approach that addresses the full scope of processes associated with all of these precursor mechanisms rather than a regionally specific, individual precursor approach. Future work will also need to address questions associated with timescales of variability, namely how do climate change and decadal/multidecadal variability impact these extratropical precursors and their relationship with ENSO? This is important given that studies have shown changes in ENSO precursors due to global warming. *Wang et al.* [2013] demonstrate that the relationship between low-level winds associated with the WNP SST and ENSO the following year has become stronger since the midtwentieth century. This strengthening relationship is attributed to greenhouse gas (GHG) forcing by comparison of model simulations with and without GHGs. *Wang et al.* [2013] also show how some of the other precursors have changed in a sliding 25 year correlation between precursor indices in Dec-Jan-Feb and the Niño3.4 index the following year. Their results show that the PMM (both winds and SST) have become more highly correlated with ENSO the following year, with a large increase in correlation since 1995. By contrast, the NPO has become steadily less correlated with ENSO the following year over time. *Di Lorenzo et al.* [2015] offer an explanation for these changes based on fact that the PMM, WNP, and their impact on ENSO are highly dependent on the mean state SST. As the SSTs warm, it is not surprising that their impact on ENSO becomes stronger. Further, *DiLorenzo et al.* [2015] also discuss a feedback between the tropics and extratropics that links the interannual and decadal timescales. In both observations and a simple atmospheric model coupled to slab ocean model, they demonstrate that ENSO and PMM dynamics interact to feedback between tropical and extratropical mechanisms, which energize decadal variability of the tropical Pacific (e.g., the Pacific Decadal Oscillation and North Pacific Gyre Oscillation). Further studies are necessary to understand the feedback between interannual and decadal variability as well as the role of GHG forcing on the variability of the precursors and their impact on ENSO.

It is also important to note that the use of precursors as predictors in real-time forecast mode, even if they are shown to be decent predictors in nature, may still be constrained by the models' ability to properly represent observed precursor/ENSO predictor behavior. This is demonstrated by *Larson and Kirtman* [2014] who show that although the PMM SST is a precursor to ENSO events in a retrospective sense, the ability of models from the NMME to predict ENSO events based on Mar PMM SST is limited. Whether the forecast models and their respective reforecasts accurately represent the observed relationships may be just as important a problem to address as whether the precursor/ENSO relationships change with external forcing like greenhouse gases. Future work will need to address the ability of the models to represent the relationships between the extratropical precursors and the development of ENSO events in a consistent fashion similar to what we have done here with reanalysis data.

REFERENCES

Alexander, M. A., D. J. Vimont, P. Chang, and J. D. Scott (2010), The impact of extratropical atmospheric variability on ENSO: Testing the seasonal footprinting mechanism using coupled model experiments, *J. Climate*, *23*(11), 2885–2901; doi: 10.1175/2010JCLI3205.1.

Anderson, B. T. (2003), Tropical Pacific sea-surface temperatures and preceding sea level pressure anomalies in the subtropical North Pacific, *J. Geophys. Res.*, *108*(D23), 4732; doi: 10.1029/2003JD003805.

Anderson, B. T. (2007), On the joint role of subtropical atmospheric variability and equatorial subsurface heat content anomalies in initiating the onset of ENSO events, *J. Climate*, *20*, 1593– 1599; doi: 10.1175/JCLI4075.1.

Anderson, B. T., R. C. Perez, and A. Karspeck (2013), Triggering of El Niño onset through trade wind-induced charging of the equatorial Pacific, *Geophys. Res. Lett.*, *40*(6), 1212–1216; doi: 10.1002/grl.50200.

Ashok, K., S. K. Behera, S. A. Rao, H. Weng, and T. Yamagata (2007), El Niño Modoki and its possible teleconnection, *J. Geophys. Res.*, *112*(C11), C11007; doi: 10.1029/2006JC003798.

Bond, N. A. (2003), Recent shifts in the state of the North Pacific, *Geophys. Res. Lett.*, *30*(23), 2183; doi: 10.1029/2003GL018597.

Burnham, K. P., and D. R. Anderson (2002), *Model Selection and Multimodel Selection: A Practical Information Theoretic Approach*, 2d ed., Springer-Verlag.

Chang, P., L. Ji, H. Li, and M. Flügel (1996), Chaotic dynamics versus stochastic processes in El Niño– Southern Oscillation in coupled ocean-atmosphere models, *Physica D: Nonlinear Phenomena*, *98*(2–4), 301–320; doi: 10.1016/0167-2789(96)00116-9.

Chang, P., L. Zhang, R. Saravanan, D. J. Vimont, J. C. H. Chiang, L. Ji, H. Seidel, and M. K. Tippett (2007), Pacific meridional mode and El Niño–Southern Oscillation, *Geophys. Res. Lett.*, *34*(16), L16608; doi: 10.1029/2007GL030302.

Chiang, J. C. H., and D. J. Vimont (2004), Analogous Pacific and Atlantic meridional modes of tropical atmosphere–Ocean variability, *J. Climate*, *17*(21), 4143–4158; doi: 10.1175/JCLI4953.1.

Deser, C., A. S. Phillips, R. A. Tomas, Y. M. Okumura, M. A. Alexander, A. Capotondi, J. D. Scott, Y.-O. Kwon, and M. Ohba (2012), ENSO and Pacific Decadal Variability in the Community Climate System Model Version 4, *J. Climate*, *25*(8), 2622–2651; doi: 10.1175/JCLI-D-11-00301.1.

Di Lorenzo, E., G. Liguori, N. Schneider, J.C. Furtado, B. T. Anderson, and M. A. Alexander (2015), ENSO and meridional modes: A null hypothesis for Pacific climate variability, *Geophys. Res. Lett.*, *42*, 9440–9448; doi: 10.1002/2015GL066281.

Di Lorenzo, E., K. M. Cobb, J. C. Furtado, N. Schneider, B. T. Anderson, A. Bracco, M. A. Alexander, and D. J. Vimont (2010), Central Pacific El Niño and decadal climate change in the North Pacific Ocean, *Nat. Geosci.*, *3*(11), 762–765; doi: 10.1038/ngeo984.

Ding, R., J. Li, Y.-H. Tseng, C. Sun, and Y. Guo (2015), The Victoria mode in the North Pacific linking extratropical sea level pressure variations to ENSO, *J. Geophys. Res. Atmos.*, *120*, 27–45; doi: 10.1002/2014/JD022221.

Kalnay, E., M. Kanamitsu, and R. Kistler (1996), The NCEP/NCAR 40-year reanalysis project, *Bull. Amer. Meteor. Soc.*, *77*, 437–471; doi: 10.1175/1520 0477(1996)077<0437:TNYRP>2.0.CO;2.

Kao, H.-Y., and J.-Y. Yu (2009), Contrasting eastern-Pacific and central-Pacific types of ENSO, *J. Climate*, *22*(3), 615–632; doi: 10.1175/2008JCLI2309.1.

Kug, J.-S., F.-F. Jin, and S.-I. An (2009), Two types of El Niño events: Cold tongue El Niño and warm pool El Niño, *J. Climate*, *22*(6), 1499–1515; doi: 10.1175/2008JCLI2624.1.

Larkin, N. K. (2005), Global seasonal temperature and precipitation anomalies during El Niño autumn and winter, *Geophys. Res. Lett.*, *32*(16), L16705; doi: 10.1029/2005GL022860.

Larson, S., and B. P. Kirtman (2013), The Pacific meridional mode as a trigger for ENSO in a high-resolution coupled model, *Geophys. Res. Lett.*, *40*, 3189– 3194; doi: 10.1002/grl.50571.

Larson, S., and B. P. Kirtman (2014), The Pacific meridional mode as an ENSO precursor and predictor in the North American multimodel ensemble, *J. Climate*, *27*(18), 7018–7032; doi: 10.1175/JCLI-D-14-00055.1.

Linkin, M. E., and S. Nigam (2008), The North Pacific Oscillation–West Pacific teleconnection pattern: Mature-Phase Structure and Winter Impacts, *J. Climate*, *21*(9), 1979–1997; doi: 10.1175/2007JCLI2048.1.

Moore, A. M., and R. Kleeman (1999), Stochastic forcing of ENSO by the intraseasonal oscillation, *J. Climate*; doi: 10.1175/1520-0442(1999)012 < 1199:SFOEBT > 2.0.CO;2.

Newman, M., S. Shin, and M. A. Alexander (2011), Natural variation in ENSO flavors, *Geophys. Res. Lett.*, *38*, L14705; doi: 10.1029/2011GL047658.

Pegion, K., and M. Alexander (2013), The seasonal footprinting mechanism in CFSv2: simulation and impact on ENSO prediction, 41, 1671– 1683; doi 10.1007/s00382-013-1887-5

Penland, C., and P. D. Sardeshmukh (1995), The optimal growth of tropical sea surface temperature anomalies, *J. Climate*, *8*, 1999–2024.

Reynolds, R. W., T. M. Smith, C. Liu, D. B. Chelton, K. S. Casey, and M. G. Schlax (2007), Daily High-Resolution-Blended Analyses for Sea Surface Temperature, *J. Climate*, *20*, 5473–5496; doi: 10.1175/2007JCLI1824.1.

Rogers, J. C. (1981), The North Pacific Oscillation, J. Climatol., 1, 39–57; doi: 10.1002/joc.3370010106.

Saha, S., and Coauthors (2014), The NCEP Climate Forecast System Version 2, *J. Climate*, *2*(6), 2185–2208; doi: 10.1175/JCLI-D-12-00823.1.

Suarez, M. J., and P. S. Schopf (1988), A delayed action oscillator for ENSO, *J. Atmos. Sci.*, *45*(21), 3283–3287; doi: 10.1175/15200469(1988)045<3283:ADAOFE>2.0.CO;2.

Takahashi, K., and A. Montecinos, K. Goubanova, and B. Dewitte (2011), ENSO regimes: Reinterpreting the canonical and Modoki El Niño, *Geophys. Res. Lett.*, *38*, L10704, doi: 10.1029/2011GL047364.

Thompson, C. J., and D. S. Battisti (2000), A linear stochastic dynamical model of ENSO. Part I: Model development, *J. Climate*, *13*(15), 2818–2832; doi: 10.1175/1520-0442(2000)013<2818:ALSDMO>2.0.CO;2.

Thompson, C. J., and D. S. Battisti (2001), A linear stochastic dynamical model of ENSO. Part II: Analysis, *J. Climate*, *14*(4), 445–466; doi: 10.1175/1520-0442(2001)014<0445:ALSDMO>2.0.CO;2.

Vimont, D. J. (2010), Transient growth of thermodynamically coupled variations in the tropics under an equatorially symmetric mean state, *J. Climate*, *23*(21), 5771–5789; doi: 10.1175/2010JCLI3532.1.

Vimont, D. J., D. S. Battisti, and A. C. Hirst (2001), Footprinting: A seasonal connection between the tropics and mid-latitudes, *Geophys. Res. Lett.*, *28*(20), 3923; doi: 10.1029/2001GL013435.

Vimont, D. J., J. M. Wallace, and D. S. Battisti (2003), The seasonal footprinting mechanism in the Pacific: implications for ENSO, *J. Climate*, *16*(16), 2668–2675; doi: 10.1175/1520-0442(2003)016<2668:TSFMIT>2.0.CO;2.

Vimont, D. J., M. A. Alexander, and M. Newman (2014), Optimal growth of central and East Pacific ENSO events, *Geophys. Res. Lett.*, *41*(11), 4027–4034; doi: 10.1002/2014GL059997.

Vimont, D. J., M. Alexander, and A. Fontaine (2009), Midlatitude Excitation of Tropical Variability in the Pacific: The role of thermodynamic coupling and seasonality, *J. Climate*, *22*, 518–534; doi: 10.1175/2008JCLI2220.1.

Walker G.T., E. W. Bliss (1932): World weather V. Mem R Meteorol Soc 4, 36, 53–84.

Wang, S.-Y., M. L'Heureux, and H.-H. Chia (2012), ENSO prediction one year in advance using western North Pacific sea surface temperatures, *Geophys. Res. Lett.*, *39*(5), n/a–n/a; doi: 10.1029/2012GL050909.

Wang, S.-Y., M. L'Heureux, and J.-H. Yoon (2013), Are greenhouse gases changing ENSO precursors in the Western North Pacific? *J. Climate*, 130306100525002; doi: 10.1175/JCLI-D-12-00360.1.

Wu, S., L. Wu, Q. Liu, and S.-P. Xie (2009), Development processes of the Tropical Pacific Meridional Mode, *Adv. Atmos. Sci.*, *27*(1), 95–99; doi: 10.1007/s00376-009-8067-x.

Yu, J.-Y., and S. T. Kim (2010), Three evolution patterns of Central-Pacific El Niño, *Geophys. Res. Lett.*, *37*(8), L08706; doi: 10.1029/2010GL042810.

Yu, J. Y., and S. T. Kim (2011), Relationships between extratropical sea level pressure variations and the central Pacific and eastern Pacific types of ENSO, *J. Climate;* doi: 10.1175/2010JCLI3688.1.

Yu, J.-Y., M.-M. Lu, and S. T. Kim (2012), A change in the relationship between tropical central Pacific SST variability and the extratropical atmosphere around 1990, *Environ. Res. Lett.*, *7*(3), 034025; doi: 10.1088/1748-9326/7/3/034025.

Zebiak, S. E., and M. A. Cane (1987), A Model El Niño–Southern Oscillation, *Monthly Weather Rev.*, *115*(10), 2262–2278; doi: 10.1175/1520-0493(1987)115<2262:AMENO>2.0.CO;2.

Zhang, H., A. Clement, and P. Di Nezio (2014), The South Pacific Meridional Mode: A Mechanism for ENSO-like Variability, *J. Climate*, *27*(2), 769–783; doi: 10.1175/JCLI-D-13-00082.1.

Zhang, L., P. Chang, and L. Ji (2009), Linking the Pacific meridional mode to ENSO: Coupled model analysis, *J. Climate*, *22* (12); doi: 10.1175/2008JCLI2473.1.

19

North Atlantic Seasonal Hurricane Prediction: Underlying Science and an Evaluation of Statistical Models

Philip J. Klotzbach[1], Mark A. Saunders[2], Gerald D. Bell[3], and Eric S. Blake[4]

ABSTRACT

Statistically based seasonal hurricane outlooks for the North Atlantic were initiated by Colorado State University (CSU) in 1984, and have been issued every year since that time by CSU. The National Oceanic and Atmospheric Administration (NOAA) Climate Prediction Center and the UK-based Tropical Storm Risk (TSR) have the next longest records (1998-present) of continuous outlooks. This chapter describes how these three forecasts have evolved with time, and documents the approaches, the environmental fields, and the lead times which underpin the models' operation. Some of the environmental parameters used in early seasonal outlooks are no longer employed, but new predictive fields have been found that appear to be more important for seasonal hurricane prediction. An assessment is made of the deterministic skill of the seasonal hurricane outlooks issued in real time by CSU, NOAA, and TSR between 2003 and 2014. All methods show moderate-to-good skill for early August outlooks (prior to the most active portion of the hurricane season), low-to-moderate skill for outlooks issued in early June, and lesser skill for outlooks issued in early April. Overall, the TSR model has the most skillful predictions of Accumulated Cyclone Energy (ACE), while NOAA has the best named storm predictions issued in early August.

19.1. INTRODUCTION

Tropical cyclones (TC) are severe weather events that form in many parts of the tropics and impact continents including North America, Asia, Australia, and Africa. The damage caused by tropical cyclones can be catastrophic, and will only increase as coastal developments expand and populations grow [*Mendelsohn et al.*, 2012; *Peduzzi et al.*, 2012]. Improving our ability to predict seasonal tropical cyclone activity is one way to mitigate this increase in damage [*DeMaria et al.*, 2014].

The tropical climate system influences atmospheric dynamics and sea surface temperature (SST) anomaly patterns in all TC basins. Therefore, the climate system affects the strength of the hurricane seasons throughout the world. Because of this climate influence, some level of seasonal predictive skill is being achieved for most hurricane basins.

This chapter focuses on seasonal predictions of North Atlantic hurricane activity. The North Atlantic hurricane season lasts for 6 months from 1 June to 30 November. The season has a well-defined 3 month peak of August-September-October (ASO), during which 77% of all named storms, 84% of all hurricanes, and 93% of all major hurricanes have formed (1950–2014 data).

[1] *Department of Atmospheric Science, Colorado State University, Fort Collins, Colorado, USA*

[2] *Department of Space and Climate Physics, University College London, Holmbury St. Mary, Dorking, Surrey, UK*

[3] *Climate Prediction Center, NOAA/NWS/NCEP, College Park, Maryland, USA*

[4] *National Hurricane Center, National Oceanic and Atmospheric Administration, Miami, Florida, USA*

Climate Extremes: Patterns and Mechanisms, Geophysical Monograph 226, First Edition.
Edited by S.-Y. Simon Wang, Jin-Ho Yoon, Christopher C. Funk, and Robert R. Gillies.

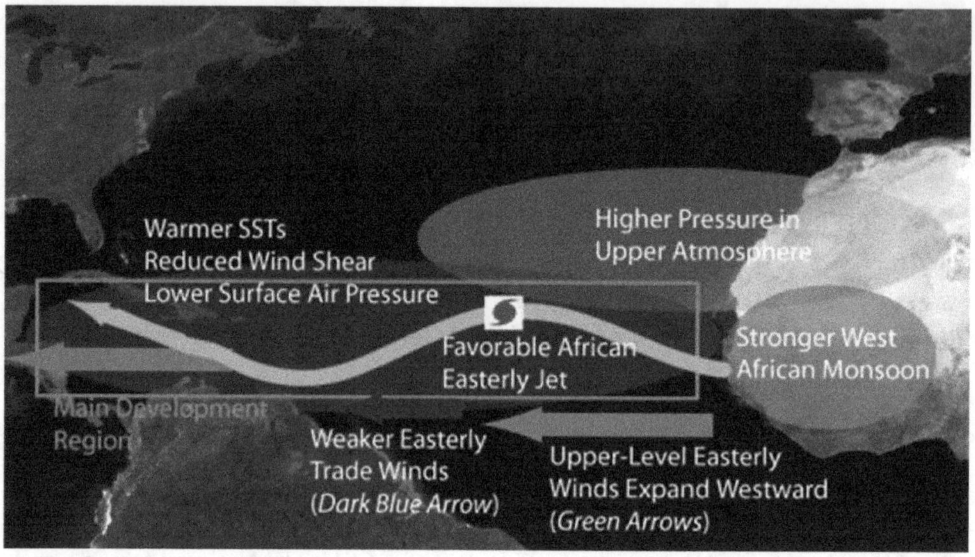

Figure 19.1 Schematic of atmospheric and oceanic anomalies during August–October associated with active Atlantic hurricane seasons and decades. Adapted from *Bell and Chelliah* [2006]. (*See insert for color representation of the figure.*)

Atlantic hurricane seasons feature large year-to-year and decade-to-decade fluctuations in strength, primarily in response to differing amounts of activity during ASO. Figure 19.1 shows the set of conducive conditions within the Atlantic hurricane Main Development Region (MDR) during ASO that produce a more active Atlantic hurricane season. Opposite conditions suppress hurricane formation and intensification in the MDR and produce a less active season. *Gray* [1984a, b], *Bell and Chelliah* [2006], and others have linked active and inactive hurricane seasons to seasonal fluctuations in oceanic and atmospheric conditions during ASO within the MDR (yellow box in Fig. 19.1), which spans the tropical North Atlantic Ocean and Caribbean Sea [*Goldenberg et al.*, 2001]. Such fluctuations often have strong climate links and involve a set of interrelated parameters, including SSTs, trade wind strength, vertical wind shear, atmospheric stability, and the strength of the west African monsoon. Therefore, above-normal and below-normal Atlantic hurricane seasons are typically not random occurrences. Instead, they often reflect a strong climate influence over a set of atmospheric and oceanic conditions within the MDR, which then collectively determine the overall strength of the hurricane season.

The seasonal hurricane outlooks are designed primarily to predict oceanic and atmospheric conditions within the MDR during ASO. Two large-scale climate phenomena, the El Niño–Southern Oscillation (ENSO) and the Atlantic Multidecadal Oscillation (AMO) account for much of the coherent variability observed across the MDR in the atmosphere and ocean on both interannual and multidecadal time scales [*Goldenberg et al.*, 2001; *Bell and Chelliah*, 2006]. This high degree of control exerted by the tropical climate system on Atlantic hurricane activity provides the underlying scientific basis for making seasonal Atlantic hurricane outlooks. Studies have established that by monitoring, understanding, and predicting these climate patterns and their associated regional circulation features, it is often possible to confidently predict the nature of the upcoming hurricane season.

One benefit of issuing seasonal outlooks is to anticipate the likelihood of extreme events. While weak tropical storms can form in marginally favorable environments, a set of very conducive conditions (Fig. 19.1) is required to produce powerful hurricanes and an exceptionally active season. Seasonal prediction models typically forecast an aggregate measure of overall seasonal activity such as the accumulated cyclone energy (ACE) index [*Bell et al.*, 2000]. The ACE index measures the combined intensity and duration of all named storms during the season, and it is therefore a measure of the overall strength of the hurricane season. ACE correlates strongly with major hurricanes (Category 3–5 on the Saffir-Simpson wind scale). For example, in seasons classified as below normal (<66 ACE units) by NOAA since 1966 (when daily geostationary satellite data became available), an average of 0.9 major hurricanes formed, compared with 3.9 major hurricanes in above-normal seasons (> 111 ACE units). This 4:1 ratio is especially important when one considers that major hurricanes cause approximately 80%–85% of TC-related damage on an annual basis [*Pielke et al.*, 2008].

This chapter evaluates the three longest-lived outlooks for North Atlantic hurricane activity. In order of longevity, these outlooks have been issued by Colorado State

University (CSU), the National Oceanic and Atmospheric Administration (NOAA), and Tropical Storm Risk (TSR). CSU started disseminating operational seasonal hurricane outlooks in 1984. NOAA's seasonal outlooks started in August 1998, and TSR began publishing seasonal outlooks in December 1998. Successful predictions of Atlantic basin seasonal hurricane activity are now also being made by dynamical models, such as those issued by the European Centre for Medium Range Weather Forecasts (ECMWF) [*Vitart and Stockdale*, 2001] and the UK Met Office [*Camp et al.*, 2015].

Building upon *Klotzbach* [2007], this chapter provides an updated review of statistically based seasonal hurricane outlooks for the North Atlantic basin, including an assessment of their skill. Section 19.2.1 summarizes the initial prediction scheme used by CSU in 1984. Section 19.2.2 discusses the development of CSU's seasonal hurricane outlooks since 1984. Section 19.2.3 describes the evolution of NOAA's outlooks since their original issuance in 1998. Section 19.2.4 provides a discussion of prediction development from TSR since 1999. In section 19.2.5, the real-time outlook skill of the three forecast models is evaluated and compared for the period 2003–2014. Potential future improvements to the statistical models are discussed in section 19.2.6. Section 19.3 concludes the chapter.

19.2. STATISTICALLY BASED SEASONAL HURRICANE OUTLOOK MODELS

The reason why the North Atlantic was chosen in 1984 for the first statistically based seasonal tropical cyclone outlook for the Northern Hemisphere was the greater year-to-year variability in TC activity present in this basin compared to the northeast Pacific or northwest Pacific basins [*Gray, personal communication*]. Based on 1986–2005 data, the coefficient of variation (the ratio of the standard deviation to the mean) is nearly twice as large for the Atlantic as for the northeast Pacific and about three times as large for the Atlantic as for the northwest Pacific [*Klotzbach*, 2007].

19.2.1. Early Research and Outlooks

Before 1984 there was little way of knowing how active an upcoming hurricane season would be. CSU issued the first statistically based seasonal hurricane outlooks for the North Atlantic basin in 1984 [*Gray*, 1984b]. Since then, the CSU outlooks have evolved and are currently available at http://tropical.colostate.edu. Early outlooks for the Atlantic basin were issued in June and updated in August. These outlooks utilized current and predicted strengths and phases of two large-scale climate phenomena: ENSO and the Quasi-Biennial Oscillation (QBO) [*Gray*, 1984a], along with forecasts of Caribbean basin

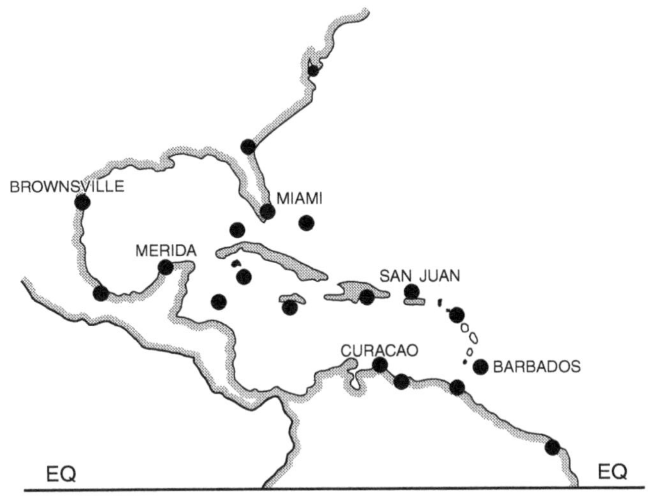

Figure 19.2 Locations of six stations used to estimate Caribbean basin sea level pressure anomalies in the original CSU outlook. From *Gray* [1984b].

sea level pressure (SLP). Figure 19.2 shows the six stations utilized to estimate Caribbean basin SLP anomalies. When an El Niño event was present, the predicted level of Atlantic hurricane activity was reduced, while both ENSO-neutral and La Niña events were treated equally. If the QBO was in its easterly phase at 30 hPa, or if the 30 hPa winds were increasing from the east, the predicted level of hurricane activity was reduced. If the QBO was in its westerly phase or the 30 hPa winds were increasing from the west, a stronger hurricane season was predicted. If SLP in the Caribbean basin was below average, a stronger hurricane season was predicted; and if SLP in this region was above average, a weaker hurricane season was predicted. This initial model showed considerable hindcast skill. The correlation between hindcast and observed named storms (tropical storms and hurricanes combined) was 0.82 for the period 1950–1982 (Fig. 19.3), and the correlation for hurricanes alone was 0.77.

19.2.2. CSU Model Development: 1984–Present

CSU's outlooks have undergone significant evolution since their original release. CSU began releasing early December predictions in 1990, while continuing to issue both June and August outlooks. *Gray et al.* [1992, 1993, 1994] detail the CSU prediction models used in the early 1990s. Figure 19.4 displays the predictors utilized in the early 1990s for their early August prediction scheme.

While all of CSU's seasonal outlooks still retain an ENSO component, other predictors have been added or removed over the years. For example, the models used in the early 1990s included new predictors that were closely related to West African rainfall. As discussed by *Landsea and Gray* [1992], when rainfall in the western Sahel is enhanced during June–July, Atlantic hurricane seasons tend to be more active. A stronger West African monsoon is associated with stronger and better defined easterly waves, weaker vertical wind shear, and warmer SSTs in the MDR, all of which favor more frequent and more intense tropical storms and hurricanes [*Bell and Chelliah*, 2006]. In addition, *Landsea and Gray* [1992] found a significant relationship between Gulf of Guinea rainfall during August–November and Atlantic hurricanes the following year. This relationship was a key predictor in CSU's original early December prediction model. Figure 19.5 displays the tracks of major hurricanes during the 10 wettest versus 10 driest years for the Gulf of Guinea region during the period from 1949 to 1989.

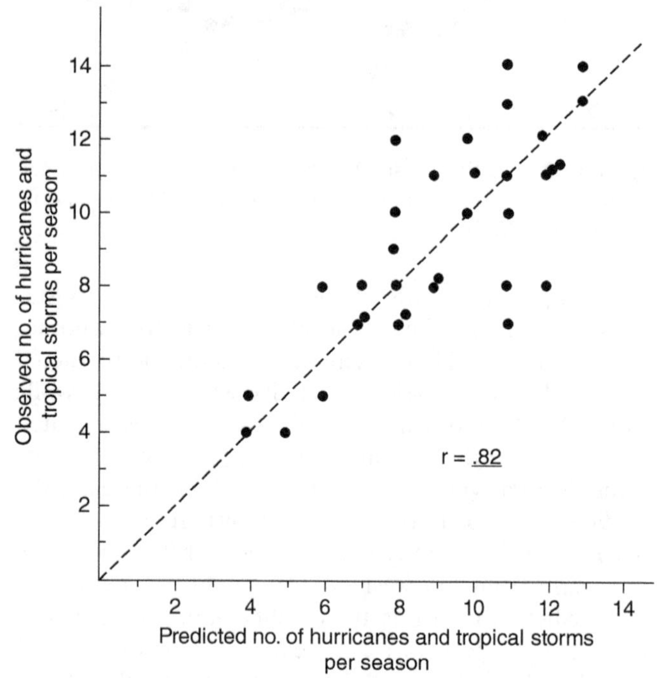

Figure 19.3 Hindcast skill based on the period 1950–1982 of the original early August outlook issued by CSU. The correlation (*r*) between the hindcast and observed number of hurricanes and tropical storms combined is 0.82. From *Gray* [1984b].

In the mid-1990s, the development of the NCEP/NCAR Reanalysis products [*Kistler et al.*, 2001] led to a transition from station-based predictors to grid-based predictors. In addition, the failure of previously used predictors, such as the QBO [*Camargo and Sobel*, 2010] and direct African rainfall measurements [*Klotzbach and Gray*, 2004] caused CSU to investigate other climate predictors. These failures also illustrated the challenges of making seasonal outlooks in an inherently nonstationary climate system.

While original forecast models were constructed using limited data (e.g., 1950–1980), longer periods of hindcast data are now available. In addition, with the development of the Twentieth Century Reanalysis [*Compo et al.*, 2011], a full three-dimensional realization of the atmosphere is now available back to 1851. Obviously, as one goes back in time, there is increased uncertainty both in atmospheric parameters and levels of hurricane activity. However, being able to evaluate predictor skill over 100+ years of prior data helps to avoid some of the pitfalls associated with predictor screening [*DelSole and Shukla,* 2009].

CSU discontinued its December outlooks following the 2011 hurricane season due to a lack of real-time predictive skill. The project currently issues outlooks in April, June, July, and August [*Klotzbach*, 2014]. Figure 19.6 displays the three predictors currently used in CSU's early August outlook.

As indicated, the CSU outlooks now utilize the low-level wind flow across the Caribbean Sea as an important predictor. Using the ERA-Interim Reanalysis [*Dee et al.*,

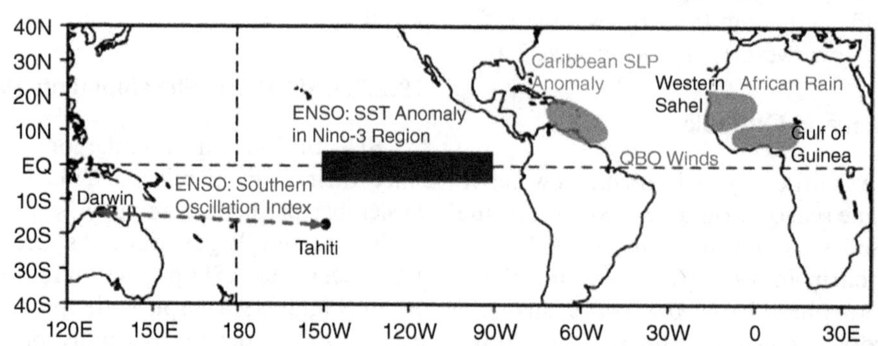

Figure 19.4 Predictors utilized by CSU in the early 1990s for their August seasonal outlook. Labels not mentioned in the text include (1) the Southern Oscillation Index, which is used to monitor ENSO and is a measure of the anomalous sea-level pressure difference between Darwin, Australia, and Tahiti, and (2) the Niño-3 region, which is an important area of the tropical Pacific used to monitor ENSO. Adapted from *Gray et al.* [1993].

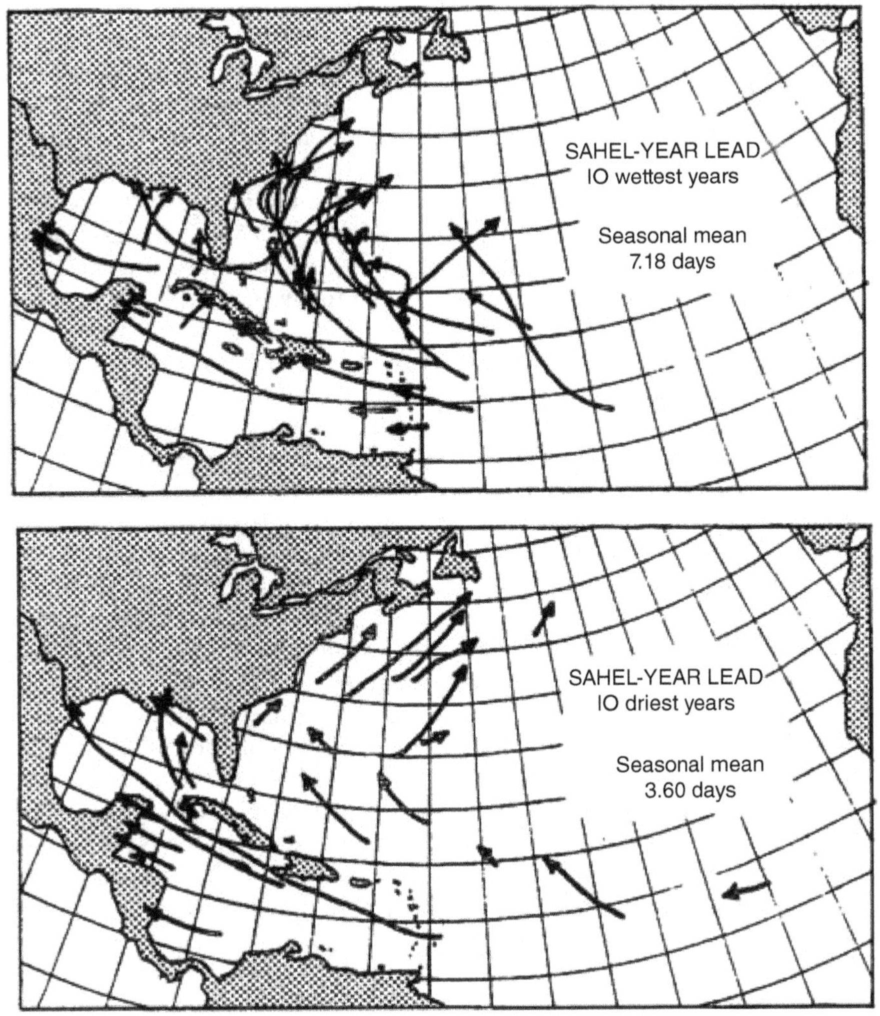

Figure 19.5 Tracks of major hurricanes in the year following the 10 wettest August–November periods in the Gulf of Guinea (top panel) and the 10 driest August–November periods from 1949–1989. This finding was why the previous August–November Gulf of Guinea rainfall was utilized in the initial early December outlook scheme issued by CSU. The seasonal mean is the average number of major hurricane days per year. From *Gray et al.* [1992].

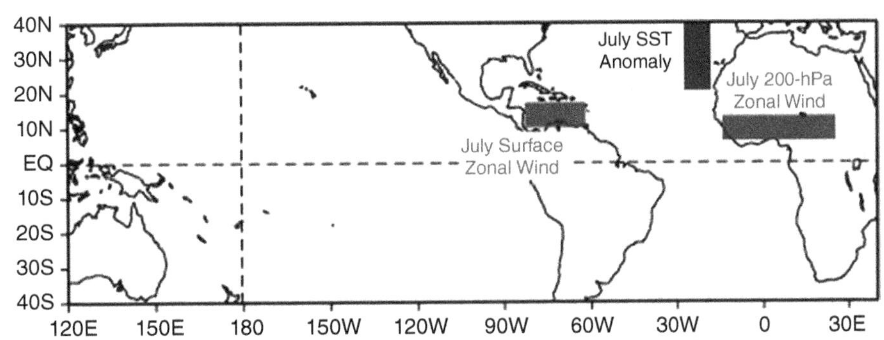

Figure 19.6 Location of predictors for CSU outlooks currently being issued in early August.

2011], July Caribbean trade winds correlate with post-1 August ACE at 0.78. This predictor had previously been noted by *Saunders and Lea* [2008] to explain a significant amount of variability of Atlantic TC activity. Indeed, TSR has used the predicted speed of the August-September Caribbean trade winds as one of its two main predictors for seasonal hurricane activity since 2001.

This trade-wind predictor is important because it is associated with a set of conditions which together influence Atlantic hurricane activity. For example, reduced trade wind strength over the Caribbean Sea implies higher than normal pressure in the eastern tropical Pacific, which is typically associated with La Niña conditions. Weaker trade winds are also typically associated with warmer than normal conditions in the tropical Atlantic and Caribbean Sea, along with an expanded Atlantic warm pool [*Wang and Lee*, 2007]. A larger warm pool generates a more conducive dynamic and thermodynamic environment for TC genesis and intensification.

Another predictor currently used by CSU is the SST anomaly in the northeastern subtropical Atlantic. The Atlantic tends to be more active when SSTs in this region are warmer than normal prior to the peak of the hurricane season [*Klotzbach*, 2011, 2014], likely because these warm anomalies tend to get advected into the Atlantic by the peak of the hurricane season [*Smirnov and Vimont,* 2012]. Additionally, warmer temperatures in this region are typically associated with weaker trade winds and a more conducive configuration of the African Easterly Jet (AEJ) during the peak months of the hurricane season.

19.2.3. NOAA Model Development: 1998–Present

NOAA's seasonal hurricane outlooks for the North Atlantic basin are an official product of the Climate Prediction Center and are made in collaboration with NOAA's National Hurricane Center and Hurricane Research Division. NOAA began issuing seasonal hurricane outlooks in August 1998. The outlooks, beginning with May 1999, are archived at www.cpc.ncep.noaa.gov/products/outlooks/hurricane-archive.shtml. These outlooks provide a general guide to the expected strength of the upcoming hurricane season. They are not a seasonal hurricane landfall outlook, and do not imply levels of activity for any particular location. NOAA's initial seasonal hurricane outlook is issued in late May and is then updated in early August.

For the outlooks issued from August 1998 to May 2000, NOAA indicated only the most likely season strength. Since August 2000, the outlooks have indicated the probabilities for the three season classifications: above, near, and below normal, as defined at www.cpc.nce.noaa.gov/products/outlooks/background_information.shtml#NOAADEF. Since August 2001, the outlooks have also include probabilistic statements for the likely ranges of named storms, hurricanes, major hurricanes, and ACE. However, there was flexibility during 2001–2002 in what was referred to as a "likely" range. Since May 2003, the likely ranges of activity have been specified with an estimated 70% probability of occurrence.

NOAA's seasonal hurricane outlooks reflect predictions of the combined impacts of three climate factors: ENSO [*Gray*, 1984a; *Goldenberg and Shapiro*, 1996], the AMO [*Gray et al.*, 1996; *Landsea et al.*, 1999], and the tropical multidecadal signal (TMS) [*Bell and Chelliah*, 2006]. The TMS is the leading multidecadal mode of tropical convective variability, and it captures the observed link between multidecadal fluctuations in Atlantic SSTs (i.e., the AMO), the West African monsoon system [*Hastenrath* 1990; *Gray*, 1990; *Landsea and Gray*, 1992; *Landsea et al.*, 1992; *Goldenberg and Shapiro*, 1996], and Amazon basin rainfall [*Chen et al.*, 2001; *Chu et al.*, 1994]. Together, these climate factors produce the interrelated set of atmospheric and oceanic conditions typically associated with both seasonal and multidecadal fluctuations in Atlantic hurricane activity (Fig. 19.1).

Three types of forecast tools provide guidance for the outlooks [*Bell and Blake*, 2015]. These include statistical tools, a hybrid statistical/dynamical ensemble forecast technique based on the NOAA Climate Forecast System (CFS) Version-2 (T-128), and purely dynamical model ensemble forecasts from the CFS high-resolution (T-382) model, the NOAA Geophysical Fluid Dynamics Laboratory (GFDL), the ECMWF, and the European Seasonal to Interannual Prediction (EUROSIP) model (Fig. 19.7). The updated outlook issued in August also incorporates predictive information such as anomalous early season activity and atmospheric and oceanic anomalies that may have developed which are not related to the dominant climate predictors.

One statistical prediction technique utilizes linear multiple regression equations to first establish the historical relationship between seasonal activity and the combined effects of the above climate factors. Forecasts of these climate factors are then input into the regression equations to predict the upcoming seasonal activity. In practice, the regression results for each prediction parameter are assembled into a look-up table [*Bell and Blake*, 2015], allowing forecasters to quickly assess a likely range of activity given uncertainties in the climate prediction itself. A second statistical technique uses climate-based analogues, which provide the forecaster with the observed ranges of activity in past seasons having similar climate conditions to those currently being predicted.

The hybrid statistical-dynamical technique [*Wang et al.*, 2009] uses regression equations to relate historical CFS-V2 model forecasts of anomalous seasonal Atlantic

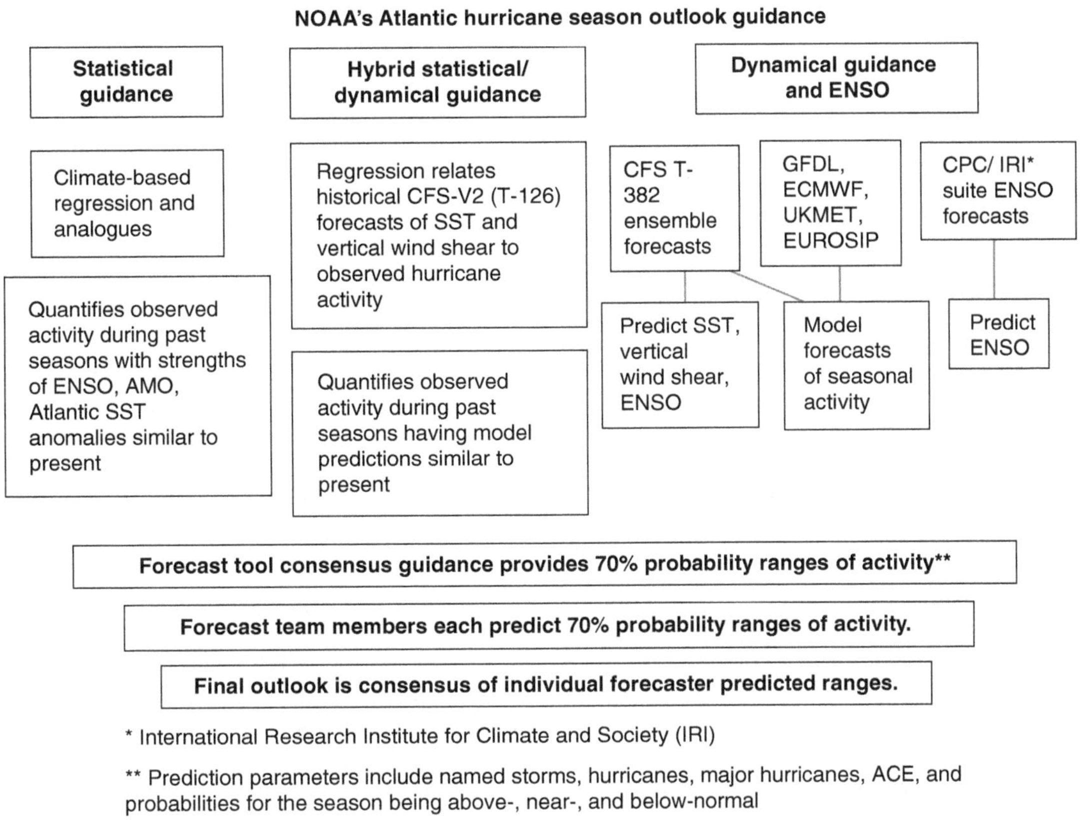

Figure 19.7 Schematic illustrating the tools that provide guidance for NOAA's Atlantic hurricane season outlooks.

SSTs and vertical wind shear to the observed seasonal hurricane activity in that year. The results are used to quantify the observed ranges of activity during past seasons having model predictions similar to the present.

One purely dynamical forecast tool in use since 2008 is the set of ensemble forecasts obtained from the CFS high-resolution model [*Schemm and Long*, 2009]. This tool provides guidance to the seasonal hurricane outlooks in three main ways. First, it aids in the prediction of the climate predictors themselves. Second, it aids in predicting the strength of the regional circulation anomalies associated with those climate predictors, which is especially important when there are competing climate factors or when there is an expectation for a significant evolution in those climate factors (such as ENSO) as the season progresses. Third, the model provides independent, bias-corrected predictions of seasonal activity based purely on model-generated hurricane tracks. Along similar lines, in 2010 the outlooks also began taking into account ensemble dynamical model predictions obtained from the GFDL, the ECMWF and the EUROSIP.

To arrive at the final seasonal hurricane outlook, all predicted ranges obtained from the various prediction tools are first assembled. Consensus guidance outlook ranges are then obtained by averaging separately, over all the prediction tools, the lower bounds and the upper bounds of the predicted ranges. The individual team forecasters then use this guidance to make predictions for the likely ranges of activity (~70% confidence) for each prediction parameter. The final Atlantic outlook reflects a consensus of these individual forecaster predictions.

19.2.4. TSR Model Development: 1999–Present

Tropical Storm Risk (TSR), based at University College London in the United Kingdom, has issued public outlooks for seasonal TC activity in the North Atlantic since December 1998. The TSR venture developed from a UK government-supported initiative called TSUNAMI, which ran from 1998 to 2000, and whose aim was to assist the competitiveness of the UK insurance industry.

TSR predicts basin-wide TC activity (namely numbers of storms of different strengths and the ACE index), U.S. landfalling TC activity, and Caribbean Lesser Antilles landfalling TC activity. Outlooks are issued in deterministic and tercile probabilistic form. The TSR prediction models are statistical in nature, but are underpinned by predictors that have sound physical links to contemporaneous TC activity. TSR issues seasonal outlooks in early December, April, June, July and August. All historical TSR seasonal TC outlooks are available online at www.tropicalstormrisk.com/forecasts.html, thereby allowing assessments to be made of the TSR real-time forecast skill. However, during the period from December 1998 through 2001 the TSR seasonal forecast models and their lead times of issuance were evolving. For a consistent assessment of TSR prediction skill at set lead times, it is recommended to use only outlooks starting with the 2002 hurricane season. TSR also provides within its seasonal outlooks the hindcast precision of each outlook parameter assessed over a prior 35 year period.

The TSR seasonal hurricane forecast model is sophisticated for a statistical model. The model divides the North Atlantic hurricane basin into three regions: (1) the tropical North Atlantic, (2) the Caribbean Sea and Gulf of Mexico, (3) the remainder of the North Atlantic outside regions (1) and (2). TSR employs separate outlook models for each of the three regions before summing the regional hurricane outlooks to obtain an overall North Atlantic hurricane outlook.

For regions (1) and (2), the model pools different environmental fields involving predictions of August-September SST anomalies and July–September trade wind speed to select the environmental field or combination of two fields, which gives the highest replicated real-time skill for individual predictands (number of tropical storms, number of hurricanes, number of major hurricanes, and ACE index) over the prior 10 year period. The nature of this process means that the details of the seasonal forecast model can vary subtly: (1) between individual predictands at the same lead time for a given year, (2) with lead time for the same predictand during the same year, and (3) from year-to-year for the same predictand at the same lead time. Separate forecast models are employed to predict: (1) July–September trade wind speed; (2) August–September SST anomalies for different regions in the tropical North Atlantic and Caribbean Sea; and (3) August–September SST anomalies for different Niño regions [*Lloyd-Hughes et al.*, 2004]. Finally, bias corrections are employed for each predictand based on the performance of that predictand over the prior 10 years.

Two environmental fields stand out among the fields that the TSR model pools in making its selection described above. These fields are (1) predicted speed of the trade winds for July-August-September for the region 7.5°– 17.5°N, 100°W– 30°W. The trade winds blow westward across the tropical Atlantic and Caribbean Sea and influence cyclonic vorticity and vertical wind shear over the MDR. (2) Predicted SST anomaly for August–September for the region 10°– 20°N, 60°W– 20°W between West Africa and the Caribbean, which includes the central and eastern MDR where many hurricanes develop during August and September. Waters here provide heat and moisture to help power the development of storms within the MDR. The nature of these two environmental fields and their anomalies, which are linked to active hurricane seasons, is shown in Figure 19.8. Further information on the TSR outlooks for North Atlantic TC activity and its underpinning methodology is described in *Lea and Saunders* [2004, 2006], *Saunders* [2006], and *Saunders and Lea* [2008].

TSR outlooks for US landfalling TC activity issued between December and July employ a historical thinning factor between tropical North Atlantic activity and US landfalling activity. The TSR outlook for US landfalling

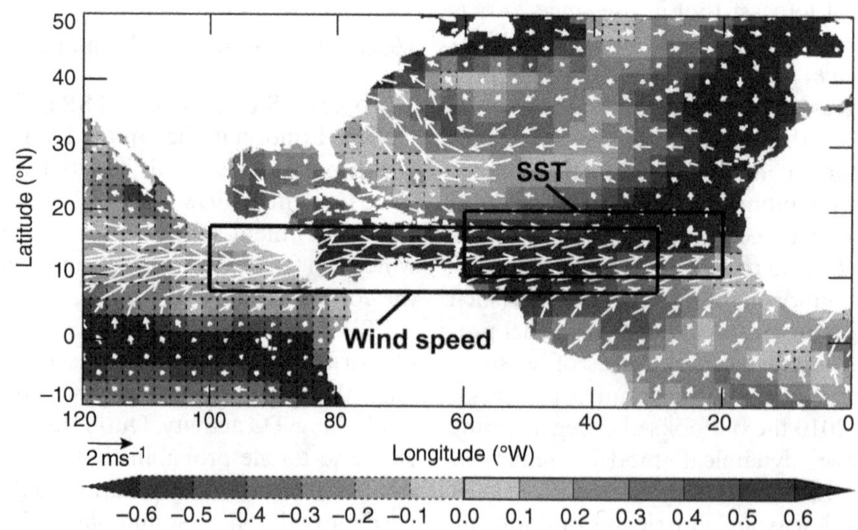

Figure 19.8 Nature of the TSR statistical model for replicating North Atlantic seasonal hurricane activity. The figure displays the two August–September environmental field areas that the TSR model employs most often in producing a seasonal hurricane outlook. The figure also displays the anomalies in August–September SST (shaded in °C) and 925 hPa wind (arrowed) linked to active Atlantic hurricane years. From *Saunders and Lea* [2008].

activity issued in early August employs the persistence of July steering winds [*Saunders and Lea*, 2005]. These winds either favor or hinder evolving hurricanes from reaching US shores during August and September. This model correctly anticipates whether US hurricane losses are above median or below median in ~75% of the years between 1950 and 2013. For the US ACE index, the TSR prediction skill increases from 3% (prior December) to 29% (early August) for the period 1980–2013.

19.2.5. Assessment of Seasonal Hurricane Outlook Skill: 2003–2014

An assessment and intercomparison of the real-time forecast skill of the CSU, NOAA, and TSR hurricane outlooks is performed for the 12 year period 2003–2014. This period is chosen because the forecast methodologies employed by each group have remained relatively stable over this period (see sections 19.2.2, 19.2.3, and 19.2.4), and because these outlooks are available and archived on their public websites.

The deterministic August outlooks for the four main measures of hurricane activity, ACE, major hurricane numbers, hurricane numbers, and named storm numbers, that were issued by each forecast group during 2003–2014 are shown in Figure 19.9. Since NOAA does not issue deterministic outlook values, but instead issues an outlook range having a 70% probability of occurrence, the midpoint of their outlook range is used as a proxy for their deterministic value.

Notable forecast successes are evident (e.g., 2005, 2010, and 2014) as well as forecast failures (e.g., 2007 and 2013).

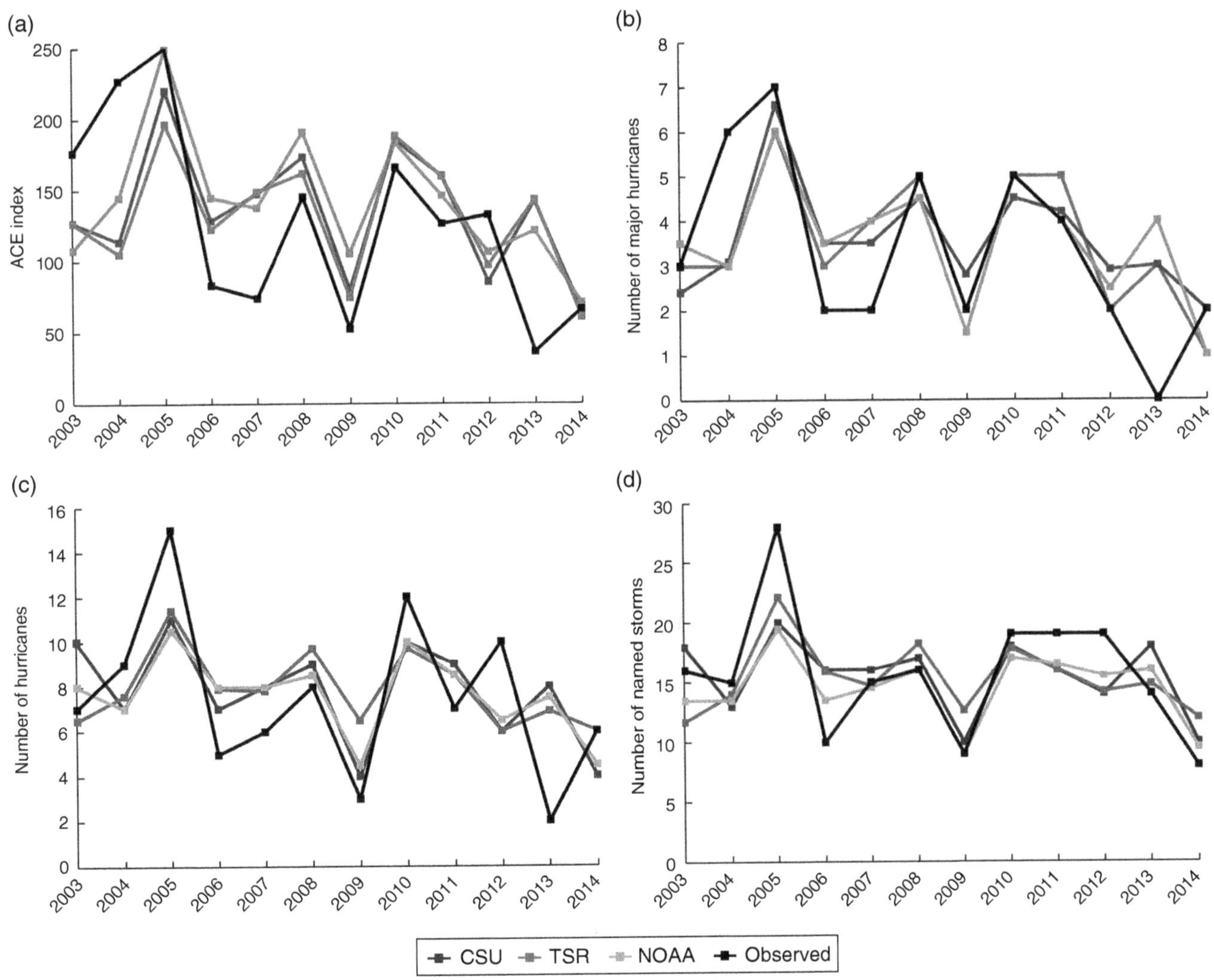

Figure 19.9 Time series comparing the seasonal outlook values issued in early August 2003–2014 by CSU, TSR, and NOAA with observed values. The comparison is made for (a) ACE, (b) major hurricane numbers, (c) hurricane numbers, and (d) named storm numbers. (*See insert for color representation of the figure.*)

The 12 year period includes the most active hurricane season on record (2005) as well as the quietest hurricane season since the early 1980s (2013). Thus, although the assessment period is relatively short, it provides a reasonable test of outlook performance.

The skill assessment and comparison are made separately for the four main measures of hurricane activity, and separately for the three outlook issue times of early April, early June (at the start of the official hurricane season), and early August (just prior to the main part of the hurricane season). It should be noted that only CSU and TSR issue outlooks in early April and that the NOAA outlook issued in late May is treated here as an early June outlook.

The assessment examines two measures of deterministic outlook skill. The first is the Spearman rank correlation (r_{rank}), which is a robust and resistant alternative to the Pearson product-moment correlation coefficient [*Wilks*, 2006]. The second skill measure is the mean square skill score (MSSS), defined as the percentage reduction in mean square error of the outlooks compared with outlooks made with a climatological mean. MSSS is the skill metric recommended by the World Meteorological Organization (WMO) for verification of deterministic seasonal outlooks [*WMO*, 2002; also see *Déqué*, 2003]. The MSSS is calculated here with respect to two different climatologies: a fixed 1951–2000 mean and a rolling prior 10 year mean. A prior 10 year mean is used, instead of the prior 5 year mean recommended by the WMO [*WMO*, 2008], because the 10 year mean is found to be a tougher benchmark to beat for all measures of hurricane activity.

Figures 19.10 and 19.11 display the real-time skill of the seasonal hurricane outlooks computed for the different lead times and activity measures. Figure 19.10 shows the skill using the Spearman rank correlation (r_{rank}), and Figure 19.11 shows the skill using the mean square skill score (MSSS). The findings from these two skill

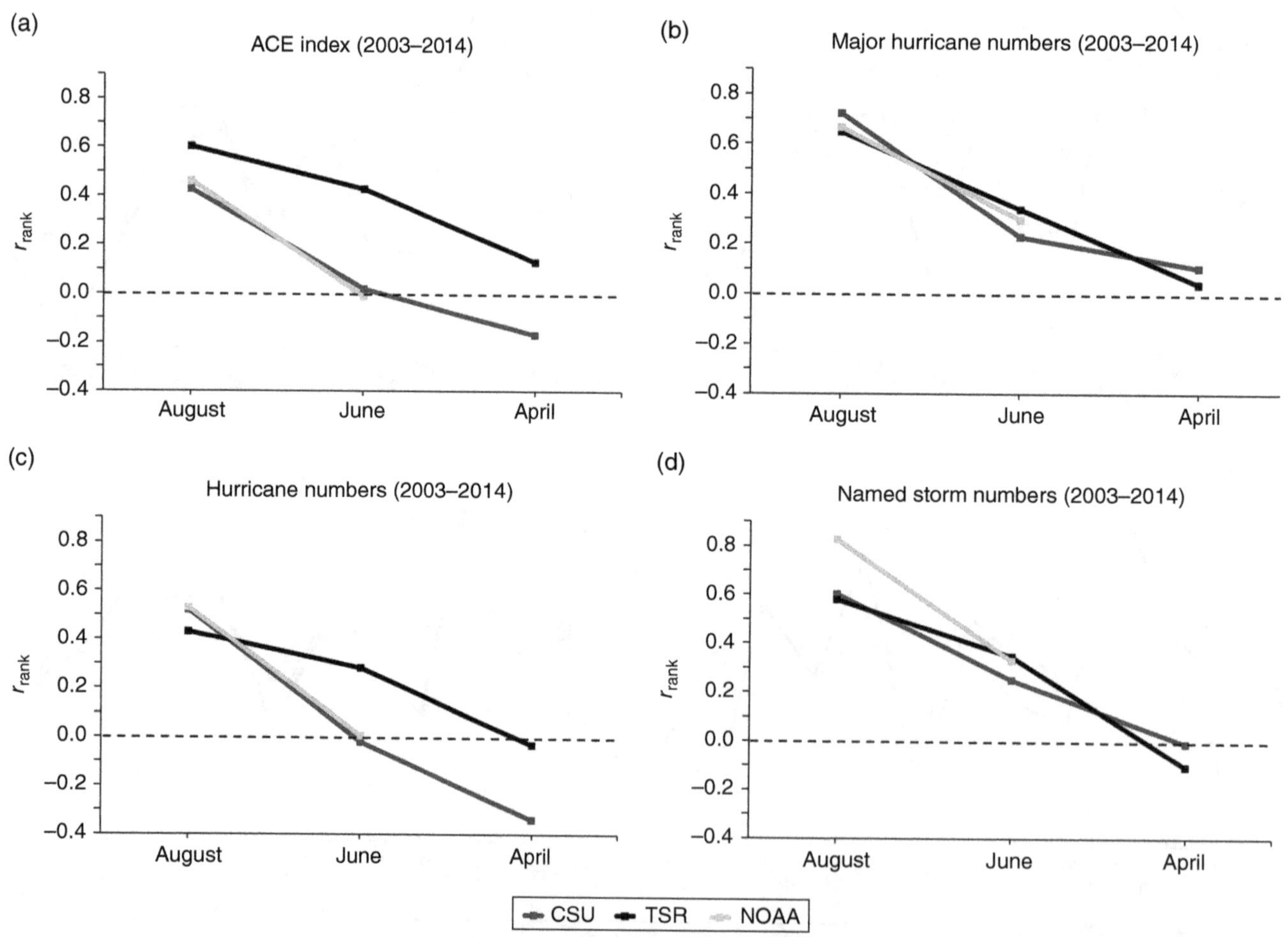

Figure 19.10 Skill of North Atlantic seasonal hurricane outlooks from 2003–2014 assessed using the Spearman rank correlation (r_{rank}) between the forecast and observed values. The assessment is made for (a) ACE, (b) major hurricane numbers, (c) hurricane numbers, and (d) named storm numbers. In each case the r_{rank} values are computed for CSU, TSR, and NOAA seasonal outlooks issued at lead times of early August, early June, and early April.

assessments are similar. For all models, the August outlooks are by far the most skillful, and the April outlooks are the least skillful. Overall, the TSR model is the most skillful predictor of the ACE index and is also the most skillful pre-season and early-season predictor for hurricane numbers. The NOAA model has the best August prediction for named storm numbers.

Benchmark skill values were then obtained by identifying the best performing statistical outlook model for each measure of hurricane activity based on the MSSS scores. The benchmark MSSS skill values for ACE, major hurricane numbers, and hurricane numbers are 10%–20% for early April outlooks, 20%–30% for early June outlooks, and 40%–60% for early August outlooks. For named storm numbers, the benchmark MSSS values are 0%–40% for early April outlooks, 20%–60% for early June outlooks, and then increase to 60%–80% for early August outlooks. These are the largest skill scores of all predicted parameters. The lower value in these ranges corresponds to the MSSS calculated with respect to the prior 10 year mean (dashed lines) and the larger value corresponds to the MSSS calculated with respect to the 1951–2000 mean.

The benchmark MSSS values show that the best performing statistical seasonal model offers skill for all measures of hurricane activity and that this skill extends out to early April. This skill may be described as moderate to good for early August outlooks, low to moderate for early June outlooks, and low for early April outlooks.

19.2.6. Future of Atlantic Basin Seasonal Hurricane Prediction

Although seasonal Atlantic hurricane outlooks are showing skill from early April, it is likely that there are untapped sources of seasonal predictability that can further enhance the predictive skill. These untapped sources of predictability may come, for example, from the identification of significant additional forcing factor(s) in years when ENSO is neutral and/or from further developments in dynamic modeling. We anticipate that as model resolution, data assimilation techniques, and model physics continue to improve, the utility of dynamic models for seasonal outlooks will continue to increase.

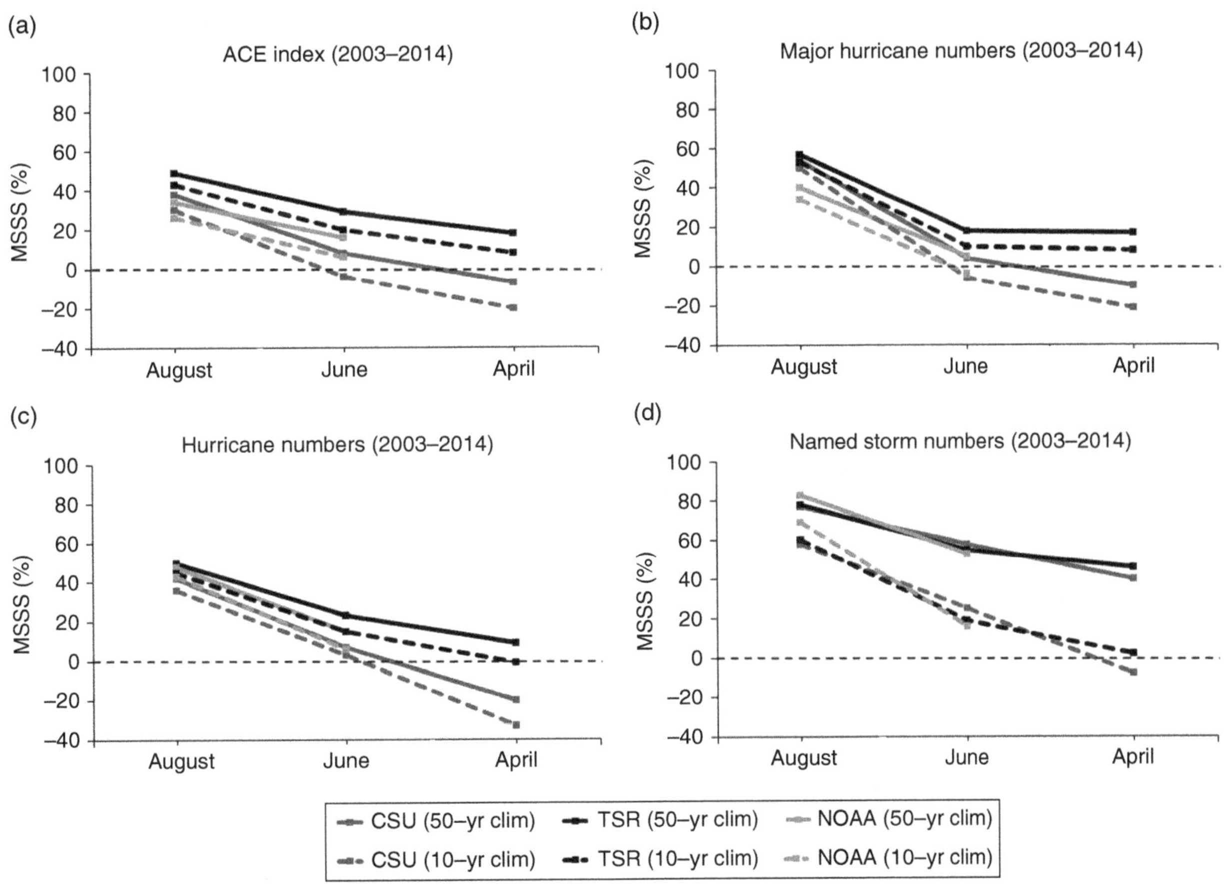

Figure 19.11 Skill of North Atlantic seasonal hurricane outlooks 2003–2014 assessed using the mean square skill score (MSSS) and displayed in the same format as Figure 19.10. The MSSS skill assessment is made with two different climatology forecasts: a fixed 1951–2000 mean and a rolling prior 10 year mean.

However, as this chapter focuses on statistical predictions, two areas of research related to future developments in statistical modeling are addressed. One promising area for further statistical model development is the ability to generate forecast models built over longer periods of data. In general, models built over long periods of data should prove to be more reliable in the future. The Twentieth Century Reanalysis developed by the Earth System Research Laboratory (ESRL) [*Compo et al.*, 2011] as well as the ERA-20C project from the ECMWF [*Stickler et al.*, 2014] provide gridded datasets since the start of the twentieth century. These datasets ingest surface data and then use an Ensemble Kalman filter (in the case of the Twentieth Century Reanalysis) and 4D variational data assimilation (in the case of the ERA-20C) to arrive at estimates of upper-air fields. The ECMWF is currently intensively involved in data rescue efforts from pibals and weather balloon data from the 1920s and 1930s in preparation for a fully coupled three-dimensional realization of the atmosphere dating back to 1900. There is obviously increased uncertainty as one heads back in time, but these datasets have proved and will likely continue to prove useful in better estimating the stability of relationships between predictors and Atlantic hurricane activity.

Another area that has helped with improving the accuracy of statistically based seasonal outlooks has been the reanalysis of the Atlantic basin hurricane database (HURDAT2) [*Landsea and Franklin*, 2013]. As is the case with large-scale fields, there is increased uncertainty in observed hurricane activity earlier in the record. This uncertainty becomes especially large prior to the mid-1960s when no geostationary satellite data were available. The reanalysis has attempted to reconstruct historical hurricane tracks back to 1851 using historical records from newspapers, ship logs, and other sources. This project is currently in the middle of the twentieth century and has likely provided more accurate estimates of historical ACE. *Vecchi and Knutson* [2008, 2011] have also provided an estimate of named storms and hurricanes, respectively, that were likely missed prior to 1965 through examination of ship traffic across the Atlantic basin. A similar adjusted ACE metric would be useful for continued improvement of statistically based models of seasonal hurricane activity.

19.3. CONCLUSIONS

This chapter has described how statistically based North Atlantic seasonal hurricane outlooks have developed since their inception in 1984. The first seasonal outlooks were issued by CSU, and were based on the phase of ENSO, the phase of the QBO, and Caribbean sea level pressure anomalies. The CSU model has evolved and now employs a variety of predictors derived from the latest global reanalysis products. In the late 1990s, statistically based seasonal hurricane outlooks were initiated by two other groups: NOAA and TSR. The NOAA model is statistical-dynamical in form and utilizes statistical techniques analyzing the state of the AMO and ENSO, combined hybrid statistical/dynamical techniques, and dynamical model output. The TSR model is sophisticated for a statistical model but primarily utilizes two predictors: (1) predicted tropical Atlantic sea surface temperatures and (2) predicted low-level trade wind flow across the tropical Atlantic and Caribbean Sea.

All three prediction models (CSU, NOAA, and TSR) show significant real-time skill for the 2003–2014 period, with the August outlooks being by far the most accurate. Overall, NOAA's August outlooks show the most skill in predicting named storm numbers. The TSR model shows the most skill in predicting ACE, and also has the highest preseason and early-season skill in predicting hurricane numbers.

The benchmark MSSS values show that the best performing statistical seasonal model offers skill for all measures of hurricane activity and that this skill extends out to early April. This skill may be described as moderate to good for early August outlooks, low to moderate for early June outlooks, and low for early April outlooks. It is likely that untapped sources of seasonal hurricane predictability remain to be discovered, and it is possible for statistical models to gain modest improvements upon the seasonal real-time outlook skills documented herein.

REFERENCES

Bell, G. D., and E. S. Blake (2015), NOAA's hurricane season outlooks, in *Global Guide to Tropical Cyclones,* edited by C. Guard; online at http://www.wmo.int/cycloneguide/.

Bell, G. D., and coauthors (2000), Climate assessment for 1999, *Bull. Amer. Meteor. Soc.*, *81*, S1–S50.

Bell, G. D., and M. Chelliah (2006), Leading tropical modes associated with interannual and multidecadal fluctuations in North Atlantic hurricane activity, *J. Climate*, *19*, 590–612.

Camargo, S. J., and A. H. Sobel (2010), Revisiting the influence of the Quasi-Biennial Oscillation on tropical cyclone activity, *J. Climate*, *23*, 5810–5825; doi:10.1175/2010JCLI3575.1.

Camp, J., and coauthors (2015), Seasonal forecasting of tropical storms using the Met Office high resolution seasonal forecasting system GloSea5, *Quart. J. Roy. Meteor. Soc.*; doi:10.1002/qj.2516.

Chen, T.-C, J.-H Yoon, K. J. St. Croix, and E. S. Takle (2001), Suppressing impacts of the Amazonian deforestation by the global circulation change, *Bull. Amer. Meteor. Soc.*, *82*, 2209–2216.

Chu, P.-S, Z.-P Yu, and S. Hastenrath, (1994), Detecting climate change concurrent with deforestation in the Amazon Basin: Which way has it gone? *Bull. Amer. Meteor. Soc.*, *75*, 579–583.

Compo, G. P., and coauthors (2011), The twentieth century reanalysis project, *Quart. J. Roy. Meteor. Soc.*, *137*, 1–28; doi:10.1002/qj.776.

Dee, D. P., and coauthors (2011), The ERA-Interim Reanalysis: Configuration and performance of the data assimilation system, *Quart. J. Roy. Meteor. Soc.*, *137*, 553–597; doi: 10.1002/qj.828.

DelSole, T., and J. Shukla (2009), Artificial skill due to predictor screening, *J. Climate*, *22*, 331–345.

DeMaria, M., C. R. Sampson, J. A. Knaff, and K. D. Musgrave (2014), Is tropical cyclone guidance improving? *Bull. Amer. Meteor. Soc.*, *95*, 387–398.

Déqué, M. (2003), Continuous variables, Chapter 5 in *Forecast Verification: A Practitioner's Guide in Atmospheric Science*, edited by I. T. Jolliffe and D. B. Stephenson, John Wiley & Sons, Chichester, England.

Goldenberg, S. B., and L. J. Shapiro (1996), Physical mechanisms for the association of El Niño and West African rainfall with Atlantic major hurricane activity, *J. Climate*, *9*, 1169–1187.

Goldenberg, S. B., C. W. Landsea, A. M. Mestas-Nunez, and W. M. Gray (2001), The recent increase in Atlantic hurricane activity: causes and implications, *Science*, *293*, 474–479.

Gray, W. M. (1984a), Atlantic seasonal hurricane frequency, Part I: El Niño and 30 mb quasi-biennial influence, *Monthly Weather Rev.*, *112*, 1649–1668.

Gray, W. M. (1984b), Atlantic seasonal hurricane frequency, Part II: Forecasting its variability, *Monthly Weather Rev.*, *112*, 1669–1683.

Gray, W. M. (1990), Strong association between West African rainfall and U.S. landfall of intense hurricanes, *Science*, *249*, 1251–1256.

Gray, W. M., C. W. Landsea, P. W. Mielke, and K. J. Berry (1992), Predicting Atlantic seasonal hurricane activity 6–11 months in advance, *Weath. Forecast.*, *7*, 440–455.

Gray, W. M., C. W. Landsea, P. W. Mielke, and K. J. Berry (1993), Predicting Atlantic basin seasonal tropical cyclone activity by 1 August, *Weath. Forecast.*, *8*, 73–86.

Gray, W. M., C. W. Landsea, P. W. Mielke, and K. J. Berry (1994), Predicting Atlantic basin seasonal tropical cyclone activity by 1 June, *Weath. Forecast.*, *9*, 103–115.

Gray, W. M., J. D. Sheaffer, and C. W. Landsea (1996), Climate trends associated with multi-decadal variability of Atlantic hurricane activity, in *Hurricanes: Climate and Socioeconomic Impacts*, edited by H. E. Diaz and R. S. Pulwarty, Springer-Verlag.

Hastenrath, S.(1990), Decadal-scale changes of the circulation in the tropical Atlantic sector associated with Sahel drought, *Int. J. Climatol.*, *10*, 459–472.

Kistler, R., and coauthors (2001), The NCEP-NCAR 50-year reanalysis: Monthly means CD-ROM and documentation, *Bull. Amer. Meteor. Soc.*, *82*, 247–267.

Klotzbach, P. J. (2007a), Recent developments in statistical prediction of Atlantic basin tropical cyclone activity, *Tellus A*, *59*, 511–518.

Klotzbach, P. J. (2007b), Revised prediction of seasonal Atlantic basin tropical cyclone activity from 1 August, *Weath. Forecast.*, *22*, 937–949.

Klotzbach, P. J. (2011), A simplified Atlantic basin seasonal hurricane prediction scheme from 1 August, *Geophys. Res. Lett.*, *38*, L16710; doi:10.1029/2011GL048603.

Klotzbach, P. J. (2014), Prediction of seasonal Atlantic basin accumulated cyclone energy from 1 July, *Weath. Forecast.*, *29*, 115–121; doi:10.1175/JCLI-D-13-00483.1.

Klotzbach, P. J., and W. M. Gray (2004), Updated 6-11-month prediction of Atlantic basin seasonal hurricane activity, *Weath. Forecast.*, *19*, 917–934.

Landsea, C. W., and J. L. Franklin (2013), Atlantic hurricane database uncertainty and presentation of a new database format, *Monthly Weather Rev.*, *141*, 3576–3592.

Landsea, C. W., and W. M. Gray (1992), The strong association between western Sahelian monsoon rainfall and intense Atlantic hurricanes, *J. Climate*, *5*, 435–453.

Landsea, C. W., R. A. Pielke, Jr., and A. M. Mestas-Nuñez (1999), Atlantic basin hurricanes: Indices of climate change, *Climate Change*, *42*, 89–129.

Landsea, C. W., W. M. Gray, P. W. Mielke, and K. J. Berry (1992), Long-term variations of western Sahelian monsoon rainfall and intense U.S. landfalling hurricanes, *J. Climate*, *5*, 1528–1534.

Lea, A. S., and M. A. Saunders (2004), Seasonal predictability of Accumulated Cyclone Energy in the North Atlantic, Proceedings of the 26th Conference on Hurricanes and Tropical Meteorology, Miami, USA, May 3–7, 2004, 419–420.

Lea, A. S., and M. A. Saunders (2006), How well forecast were the 2004 and 2005 Atlantic and US hurricane seasons? *Weather*, *61*, 245–249; doi:10.1256/wea.95.06.

Lloyd-Hughes, B., M. A. Saunders, and P. Rockett (2004), A consolidated CLIPER model for improved August-September ENSO prediction skill, *Weath. Forecast.*, *19*, 1089–1105; doi:http://dx.doi.org/10.1175/813.1.

Mendelsohn, R., K. Emanuel, S. Chonabayashi and L. Bakkensen (2012), The impact of climate change on global tropical cyclone damage, *Nature Climate Change*, *2*, 205–209; doi:10.1038/NCLIMATE1357.

Peduzzi, P., B. Chatenoux, H. Dao, A. De Bono, C. Herold, J. Kossin, F. Mouton and O. Nordbeck (2012), Global trends in tropical cyclone risk, *Nature Climate Change*, *2*, 289–294; doi:10.1038/NCLIMATE1410.

Pielke, R. A., Jr., and coauthors (2008), Normalized hurricane damage in the United States: 1900–2005, *Nat. Haz. Rev.*, *9*, 29–42.

Saunders, M. A. (2006), Winds of change, Post Magazine Risk Report, 28–29, 9 November 2006; online at http://tropicalstormrisk.com/docs/Hurricanes-Post09112006.pdf.

Saunders, M. A., and A. S. Lea (2005), Seasonal prediction of hurricane activity reaching the coast of the United States, *Nature*, *434*, 1005–1008; doi:10.1038/nature03454.

Saunders, M. A., and A. S. Lea (2008), Large contribution of sea surface warming to recent increase in Atlantic hurricane activity, *Nature*, *451*, 557–560; doi:10.1038/nature06422.

Schemm, J.-K. E., and L. Long (2009), Dynamic hurricane season prediction experiment with the NCEP CFS CGCM, Workshop on High Resolution Climate Modeling, ICTP, Trieste, Italy, August 9–13, 2009.

Smirnov, D., and D. J. Vimont (2012), Extratropical forcing of tropical Atlantic variability during boreal summer and fall, *J. Climate*, *25*, 2056–2076.

Stickler, A., and coauthors (2014), ERA-Clim: Historical surface and upper-air data for future reanalyses, *Bull. Amer. Meteor. Soc.*, *95*, 1419–1430; doi:http://dx.doi.org/10.1175/BAMS-D-13-00147.1.

Vecchi, G. A., and T. R. Knutson (2008), On estimates of historical North Atlantic tropical cyclone activity, *J. Climate*, *21*, 3580–3600.

Vecchi, G. A., and T. R. Knutson (2011), Estimating annual numbers of Atlantic hurricanes missing from the HURDAT database (1878–1965) using ship track density, *J. Climate*, *24*, 1736–1746.

Vitart, F., and T. N. Stockdale (2001), Seasonal forecasting of tropical storms using coupled GCM integrations, *Weath. Forecast.*, *129*, 2521–237.

Wang, C., and S-K. Lee (2007), Atlantic warm pool, Caribbean low-level jet, and their potential impact on Atlantic hurricanes, *Geophys. Res. Lett.*, *34*, L02703; doi:10.1029/2006GL028579.

Wang, H., and coauthors (2009), A statistical forecast model for Atlantic seasonal hurricane activity based on the NCEP dynamical seasonal forecast, *J. Climate*, *22*, 4481–4500.

Wilks, D. S. (2006), *Statistical Methods in the Atmospheric Sciences*, Academic Press, San Diego.

WMO (2002), Standardized verification system for long-range forecasts (LRF), *New Attachment II-9 to the Manual on the GDPS (WMO-No. 485)*, Vol. *1*, Geneva, Switzerland.

WMO (2008), Report from expert meeting to evaluate skill of tropical cyclone seasonal forecasts, World Meteorological Organization Technical Document No. 1455, Geneva, Switzerland.

20

Predicting Subseasonal Precipitation Variations Based on the Madden-Julian Oscillation

Charles Jones

ABSTRACT

The Madden-Julian Oscillation (MJO) is the most important source of intraseasonal variability in the coupled ocean-atmosphere system and has a significant influence on weather and climate variability on global scales. This chapter reviews the significance of the MJO in the variability of precipitation, including occurrences of heavy precipitation and the evolution of the predictive skill of the MJO in operational models. Additionally, the chapter focuses on the importance of the MJO in modulating the forecast skill and predictability of precipitation.

20.1. INTRODUCTION

Initially named the "40–50 day oscillation," the Madden-Julian Oscillation (MJO) [*Madden and Julian*, 1971, 1972, 1994] was discovered at a time when many fewer meteorological observations were available compared to today's global observing network. Using rigorous and clever spectral and cross-spectral analyses of surface and radiosonde observations, *Madden and Julian* [1971, 1972] were able not only to detect a significant intraseasonal variability in the atmosphere but also to infer that the oscillation has a large-scale structure and eastward propagating characteristics. Subsequent studies since the 1980s allowed an improved understanding of the nature of the MJO [*Lau and Waliser*, 2012; *Madden and Julian*, 1994; *Zhang*, 2005].

The oscillation, which is confined to the global tropics, has a large-scale structure (zonal wavenumbers 1–4), timescales of about 20–100 days, propagates eastward with phase speeds of 5–10 m s^{-1}, and involves significant variations in convective clouds, precipitation, tropospheric winds, temperature and moisture, surface pressure, and heat and radiative fluxes [*Adames and Wallace*, 2014; *Flatau et al.*, 1997; *Hendon and Salby*, 1994; *Jones and Weare*, 1996; *Jones et al.*, 1998; *Jones et al.*, 2004c; *Kiladis et al.*, 2005; *Matthews*, 2000; *Shinoda et al.*, 1998; *Weickmann and Sardeshmukh*, 1994; *Weickmann et al.*, 1985; *Wheeler and Kiladis*, 1999].

Since the MJO involves substantial latent heat release due to deep convective activity and changes in the large-scale tropospheric circulation, the MJO influences the variability on monsoon systems [*Carvalho et al.*, 2002b; *Chi et al.*, 2015; *Goswami and Mohan*, 2001; *Hendon and Liebmann*, 1990; *Higgins and Shi*, 2001; *Jones and Carvalho*, 2002; *Muza et al.*, 2009; *Pohl and Camberlin*, 2006], upper-ocean variability [*Isoguchi and Kawamura*, 2006; *Jin et al.*, 2013; *Matthews and Meredith*, 2004; *Matthews et al.*, 2007; *Waliser et al.*, 2005], tropical cyclones and hurricane activity [*Bessafi and Wheeler*, 2006; *Diamond and Renwick*, 2015; *Duvel*, 2015; *Klotzbach and Oliver*, 2015; *Liebmann et al.*, 1994; *Maloney and Hartmann*, 2000; *Zhao et al.*, 2015], US tornado and hail

Department of Geography and Earth Research Institute, University of California Santa Barbara, Santa Barbara, California, USA

Climate Extremes: Patterns and Mechanisms, Geophysical Monograph 226, First Edition.
Edited by S.-Y. Simon Wang, Jin-Ho Yoon, Christopher C. Funk, and Robert R. Gillies.

variability [*Barrett and Gensini*, 2013; *Barrett and Henley*, 2015; *Thompson and Roundy*, 2012], Northern Hemisphere circulation and snow depth, Arctic temperature and sea-ice changes [*Barrett et al.*, 2015; *Flatau and Kim*, 2012; *Henderson et al.*, 2014; *Oliver*, 2015; *Vecchi and Bond*, 2004], Antarctic Oscillation [*Carvalho et al.*, 2005], El Niño–Southern Oscillation (ENSO) [*Lau*, 2012; *McPhaden*, 1999; *McPhaden et al.*, 2006; *Weickmann*, 1991], and North Atlantic Oscillation (NAO) [*Cassou*, 2008; *Lin et al.*, 2009]. The *Zhang* [2013] recent review paper highlights the fundamental role of the MJO linking weather and climate variability on global scales.

While important advances have been achieved in the theoretical knowledge of the MJO (see *Wang* [2005] for a review), there are still many aspects that need to be resolved. Similar to many other oscillations found in the climate system, the MJO is far from behaving like a perfect oscillator. The MJO shows significant variability in seasonality, amplitude (or intensity), duration, and phase evolution [*Lau and Waliser*, 2012; *Madden and Julian*, 1994; *Zhang*, 2005]. The MJO occurs irregularly in time as single isolated episodes or in sequences of events in which a primary event is followed by successive events [*Jones*, 2009; *Jones and Carvalho*, 2011a; *Matthews*, 2000; 2008]. Recently, the Dynamics of the MJO (DYNAMO) field campaign in the Indian Ocean was implemented to improve the understanding of the initiation mechanisms of the MJO [*Chen et al.*, 2015; *Li et al.*, 2015; *Nasuno et al.*, 2015; *Powell and Houze*, 2015; *Xu et al.*, 2015]. Furthermore, the MJO shows significant variations on interannual and decadal timescales [*Jones and Carvalho*, 2006, 2011a; *Kessler*, 2001; *Pohl and Matthews*, 2007; *Weickmann*, 1991] and possibly long-term changes associated with global warming [*Chang et al.*, 2015; *Jones and Carvalho*, 2011b; *Liu*, 2013; *Liu et al.*, 2013; *Maloney and Xie*, 2013; *Takahashi et al.*, 2011].

Since the MJO has a slow evolution (20–100 days) relative to synoptic (1–10 days) timescales and influences the variability of the atmosphere not only in the tropical regions but also globally, the MJO is regarded as an important source of potential predictability of weather, particularly in lead times beyond the deterministic limits of predictability (i.e., beyond about 10–14 days) [*Gottschalck et al.*, 2010; *Vitart and Molteni*, 2010; *Waliser et al.*, 2003]. The main goal of this chapter is to review the influence of the MJO in the variability of precipitation, including occurrences of heavy (or extreme) events and the evolution of the predictive skill of the MJO in operational models. Additionally, this chapter focuses on the importance of the MJO in modulating the forecast skill and predictability of precipitation and discusses ongoing research to explore subseasonal forecasting.

20.2. THE MJO INFLUENCE ON THE VARIABILITY OF PRECIPITATION

The canonical life cycle of the MJO involves enhancement of deep convective clouds over eastern Africa that progressively intensifies and propagates through the Indian Ocean, Indonesia, and western Pacific. Coupled to the region of large-scale ascending motion and deep convective activity, an extensive region of descending motion and suppressed convection also propagates eastward. Consequently, this propagating dipole structure regulates the variability of precipitation in the tropics, and many observational studies have found clear and robust signals linked to the MJO in Africa [*Berhane and Zaitchik*, 2014; *Pohl and Camberlin*, 2006; *Pohl et al.*, 2007], Maritime Continent [*Hidayat and Kizu*, 2010; *Marzuki et al.*, 2016], Australia [*Hendon and Liebmann*, 1990; *Wheeler et al.*, 2009], and western Pacific [*Carbone and Li*, 2015]. Indeed, as the region of deep convective anomalies propagates from the Indian Ocean to the western Pacific, the MJO triggers Rossby wave trains toward the extratropics and midlatitudes of both hemispheres that also modulate the variability of precipitation over Asia [*Barlow et al.*, 2005; *Cannon et al.*, 2015; *Jeong et al.*, 2005; *Jeong et al.*, 2008; *Zhang et al.*, 2009], Central America [*Barlow and Salstein*, 2006], South America [*Carvalho et al.*, 2002a; *Carvalho et al.*, 2004; *De Souza and Ambrizzi*, 2006; *Muza et al.*, 2009; *Nogues-Paegle and Mo*, 1997; *Nogues-Paegle and Mo*, 2002], and North America [*Becker et al.*, 2011; *Jones*, 2000; *Jones and Carvalho*, 2012; *Mo*, 1999; *Mo and Higgins*, 1998a, b, c; *Zhou et al.*, 2012].

The global influence of the MJO on precipitation is seen in Figure 20.1 [*Donald et al.*, 2006]. The anomalies were calculated with station data from many locations around the world and composited on four phases of the MJO life cycle as defined by the real-time monitoring index of the MJO (RMM) described by *Wheeler and Hendon* [2004]. The phases correspond to enhanced convective anomalies over: tropical Indian Ocean phase 2 (Fig. 20.1a), Maritime Continent phase 4 (Fig. 20.1b), western Pacific phase 6 (Fig. 20.1c), and central Pacific phase 8 (Fig. 20.1d).

The significance of the MJO in controlling the distribution of precipitation is particularly important in the occurrence of heavy precipitation [*Jones*, 2000; *Jones and Carvalho*, 2012; *Zhang*, 2013]. In a recent study, *Jones and Carvalho* [2014] analyzed the boreal winter season of 1 Nov 2004–31 Mar 2005 to investigate the sensitivity of heavy precipitation over the contiguous United States to the amplitude of the MJO. The 2004–2005 winter was characterized by a weak warm ENSO event, but extremely wet conditions over most of the western and midwestern United States. Figure 20.2 (top)

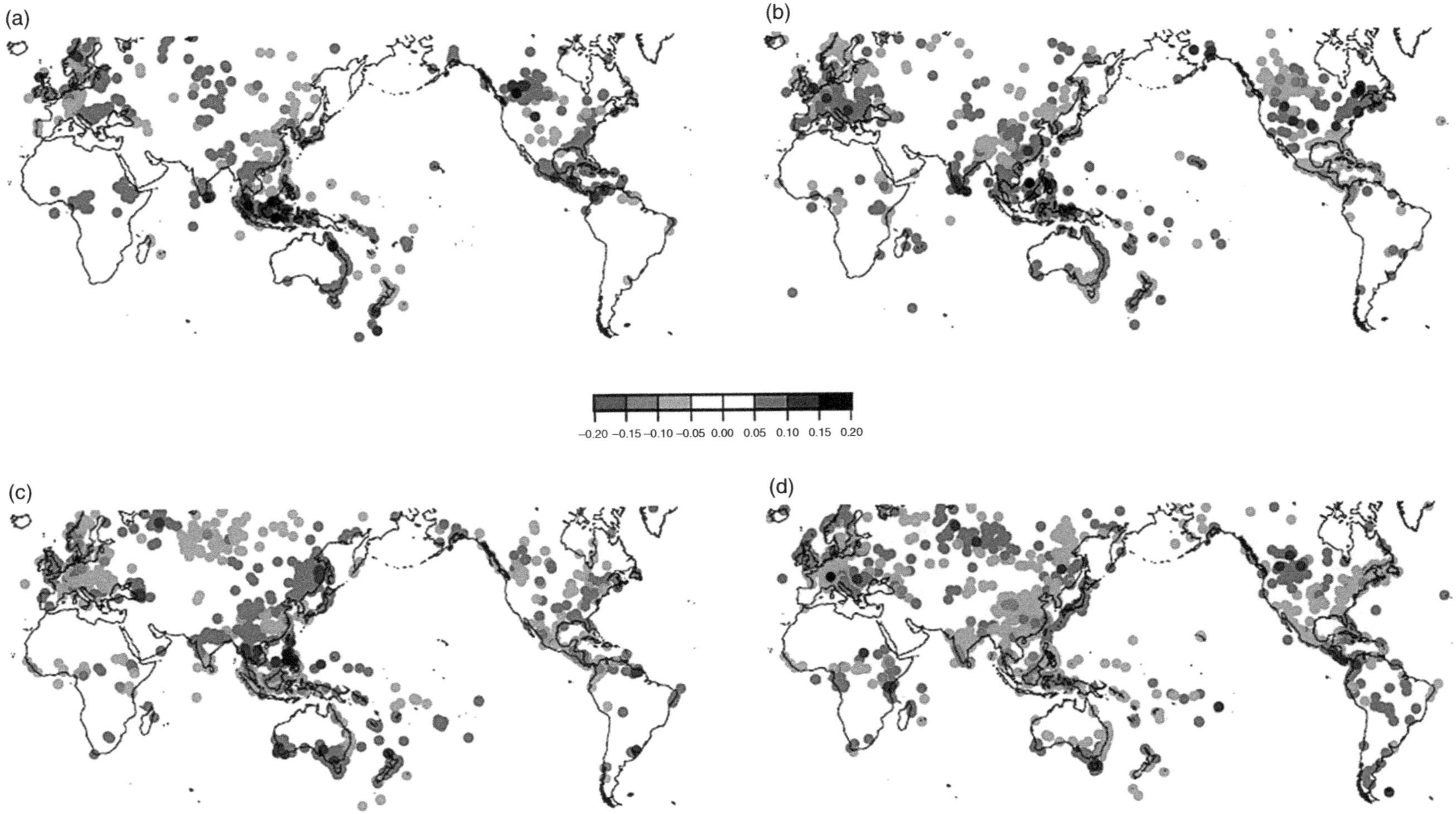

Figure 20.1 Worldwide distribution of daily rainfall anomalies during active MJO events when the center of enhanced convection is in phases (a) 2, (b) 4, (c) 6, and (d) 8 of the *Wheeler and Hendon* [2004] diagram. Colors represent differences between unconditional cumulative distribution function (CDF) from Monte-Carlo simulations and observed conditional CDF when the MJO was in the given phase. Enhanced (decreased) rainfall is expressed by blue (red) color dots over each station location used in the analysis. After *Donald et al.* [2006]. (*See insert for color representation of the figure.*)

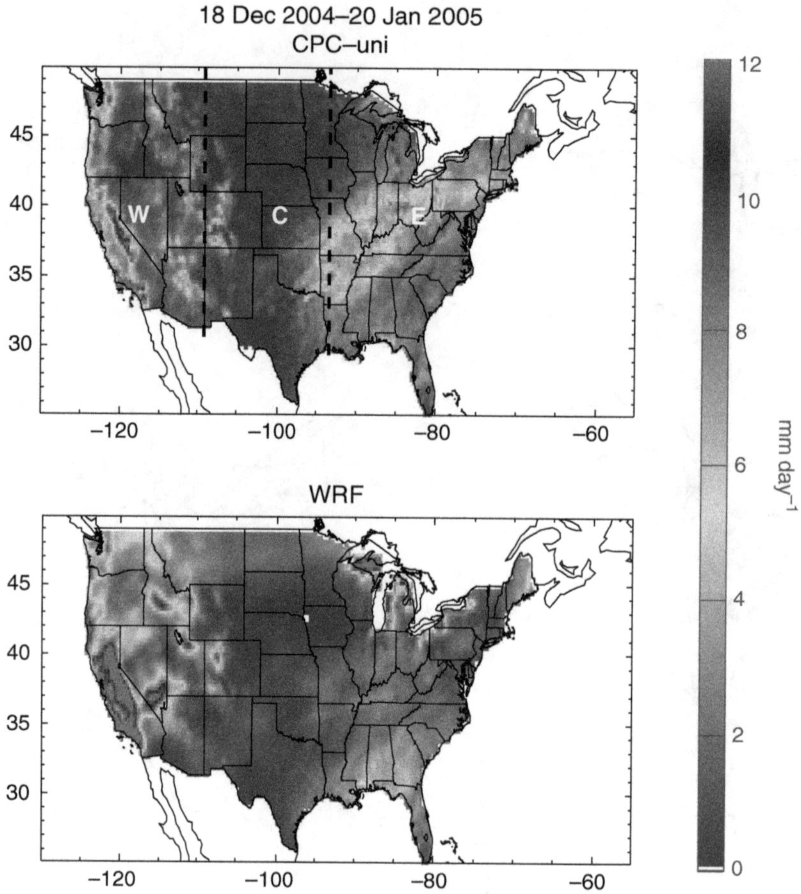

Figure 20.2 (top) Observed mean daily precipitation and (bottom) mean daily precipitation simulated with the WRF model (bias removed) during 18 Nov 2004–20 Jan 2005. Thick dashed longitudinal lines in the top figure indicate western, central, and eastern sectors over the contiguous United States used to analyze statistics of heavy precipitation. After *Jones and Carvalho* [2014].

shows the total precipitation during 1 Nov 2004–31 Mar 2005 expressed as departure from the 1 Nov–31 Mar, 1948–2011 climatology. The southwest and midwest United States received 150%–250% above normal and several places in California, Nevada, Arizona, New Mexico, and Texas registered 300%–400% above normal. January 2005 was a specially wet month and contributed significantly to the winter season total (Fig. 20.2, bottom). The wet conditions were associated with a blocking ridge in the Gulf of Alaska, an amplified trough over the southwestern CONUS, and southward displaced jet stream and storm track.

Several major storms contributed to heavy precipitation. The first storm produced 25–50 cm of snow in the Midwest (Ohio valley) on 23 December, whereas heavy precipitation over California happened during 28–29 December. Atmospheric rivers were very important during the 29–31 December precipitation events [*Jankov et al.*, 2011; *Smith et al.*, 2010]. California was hit by heavy rainfall and snow in high elevations during 7–11 January, while the last major storm of the month contributed with heavy precipitation over the Midwest during 11–13 January.

Jones and Carvalho [2014] used the Weather Research and Forecasting (WRF) model forced with National Centers for Environmental Prediction (NCEP) Climate Forecast System Reanalysis (CFSR) to simulate the precipitation distribution over the contiguous U.S. during the 2004–2005 winter season. Two types of experiments were performed: a control case and perturbation experiments in which the amplitude of the MJO was increased or decreased. The intraseasonal signal associated with the MJO was defined as 20–100 day band-pass filtered anomalies and zonal wavenumbers 1–5. In the perturbation experiments, the observed MJO signal was modified to different amplitudes: reduced to 25% and 50% and increased by 25%, 50%, and 75%. In addition, another experiment was performed to completely remove the effect of the MJO (NoMJO experiment) on the precipitation distribution over the United States (see *Jones and Carvalho* [2014] for additional details).

For the western, central, and eastern sectors of the contiguous United States (defined in Fig. 20.2), Figure 20.3 shows the importance of the MJO in controlling the distribution of precipitation. The plot is constructed by taking precipitation time series ($P > 0.01\,\text{mm day}^{-1}$) from all grid points in each sector and calculating statistics (interquartile range, median, 90th, and 95th percentiles). The horizontal axis indicates WRF experiments where the amplitude increases from the NoMJO case to the MJO amplitude being increased by 75%. In the western sector (Fig. 20.3 top), the control (CTRL) experiment shows median precipitation of about $3.5\,\text{mm day}^{-1}$, interquartile range of $0.5\text{–}12\,\text{mm day}^{-1}$, and 90th and 95th percentiles of $24\,\text{mm day}^{-1}$ and $34\,\text{mm day}^{-1}$, respectively. As the MJO amplitude increases from NoMJO to 75% increase, the median and lower quartiles show moderate increases. In contrast, there is an almost linear and significant increase in the upper quartile and extreme values (i.e., 90th and 95th percentiles) as the MJO amplitude changes from NoMJO to CTRL experiments. As the MJO signal is further increased in the +25%, +50%, and +75% experiments, the upper quartile and extreme values show further nonlinear increases. Similar behavior is found for the statistics in the central sector (Fig. 20.3, middle) indicating nearly linear increases with MJO amplitude. Over the eastern sector (Fig. 20.3, bottom), the interquartile range responds differently with increases in MJO amplitudes compared with other sectors; only the extremes (90th and 95th percentiles) in precipitation appear to have a linear relationship with the MJO signal.

The overview discussed above clearly shows the influence of the MJO in the global distribution of precipitation, especially heavy precipitation. In this context, some questions are particularly important and are discussed in the next two sections. Have operational numerical weather

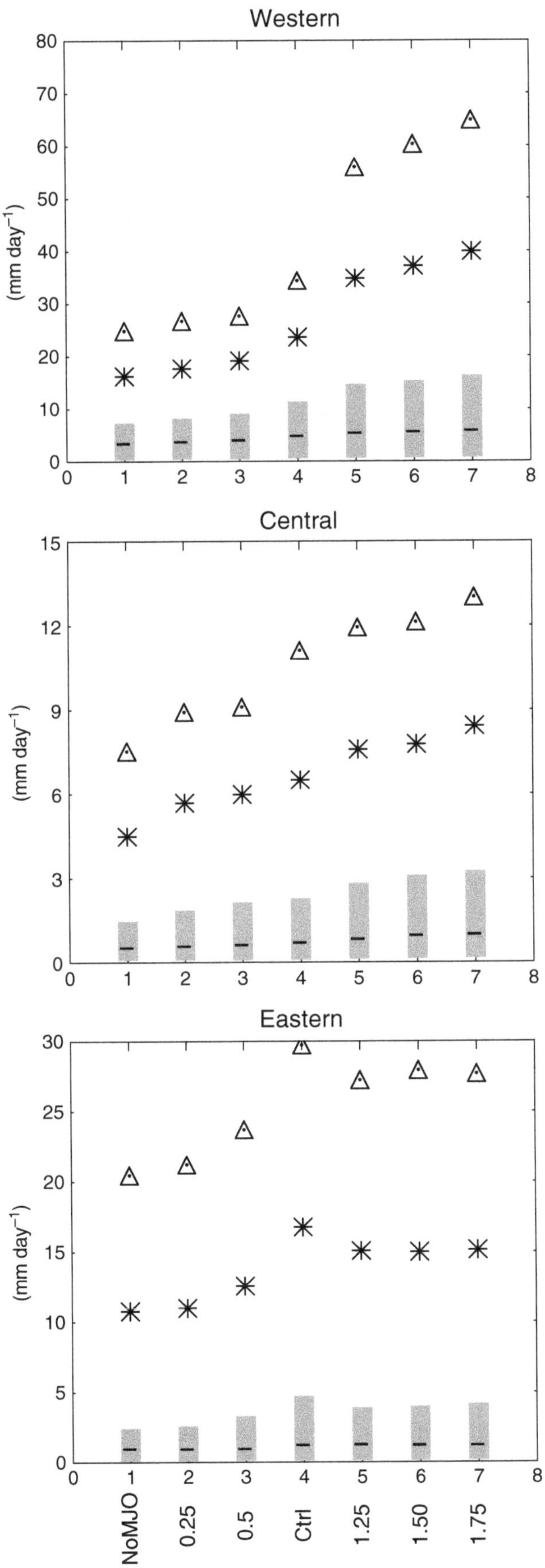

Figure 20.3 Sensitivity of daily precipitation (mm day^{-1}) during 18 Dec 2004–20 Jan 2005 simulated with the WRF model. Horizontal axis shows sensitivity experiments with different MJO amplitude signals: NoMJO, reduced to 0.25, 0.5, control (CTRL), increased by 0.25, 0.5, and 0.75. Statistics are computed over the western (top), central (middle), and eastern (bottom) sectors of the contiguous United States (defined in Fig. 20.2). Shaded boxes indicate interquartile range, median value (tick), 90th (star), and 95th (triangle) percentiles. Statistics are calculated from all grid points in each sector. After *Jones and Carvalho* [2014].

prediction models improved the forecast skill of the MJO over the years? How does the MJO influence the predictability of precipitation, especially heavy precipitation? How does the variability of the MJO impact the forecast skill of precipitation?

20.3. FORECASTING THE MJO

Since the MJO influences global weather variability, interest in forecasting the MJO and its influence in tropical and extratropical regions dates back to the late 1980s and early 1990s. It was quickly realized that operational models at that time had significant systematic errors with weak MJO amplitudes and unrealistic eastward propagation of 30–60 day variability. *Chen and Alpert* [1990] analyzed the former National Meteorological Center's (NMC) medium-range operational forecasts and noticed that forecast skill was particularly good when the amplitude of the MJO was large. *Ferranti et al.* [1990] showed that the skill of the European Centre for Medium-Range Weather Forecast model in the extratropics was significantly improved when systematic errors associated with the MJO were minimized. *Lau and Chang* [1992] analyzed one season of 30 day global forecasts of a Dynamical Extended Range Forecasts (DERF) experiment and concluded that the amplitude of the MJO was important in the forecast skill of tropical and extratropical atmospheric circulation patterns.

Other studies based on subsequent DERF data showed moderate progress in the forecast skill of the MJO. *Hendon et al.* [2000] concluded that forecasts in the tropics and midlatitudes of the Northern Hemisphere during boreal winter had less skill when the model was initialized during or prior to periods of active MJO as opposed to quiescent episodes of the oscillation. Using the same DERF data, *Jones et al.* [2000] analyzed the skill of the medium-range forecast model in forecasting the MJO during January 1985–December 1989. Figure 20.4 shows the zonal average along the equator of anomaly correlations of intraseasonal variations in the zonal component of the wind at 200 hPa. The forecasts were verified for periods of active and inactive periods of the MJO. Phases were defined when the MJO enhanced convection was situated over the Indian Ocean (PC1-), Maritime Continent (PC2-), western Pacific (PC1+), and central Pacific (PC2+). The results show that forecast skill in the tropical region was higher during active MJO (correlations of 0.4 at 7 day lead) than in inactive periods (correlation of 0.3 at 7 day lead). However, it is clear that useful forecast skill of the MJO in early models extended only to about 5 to 7 days lead time.

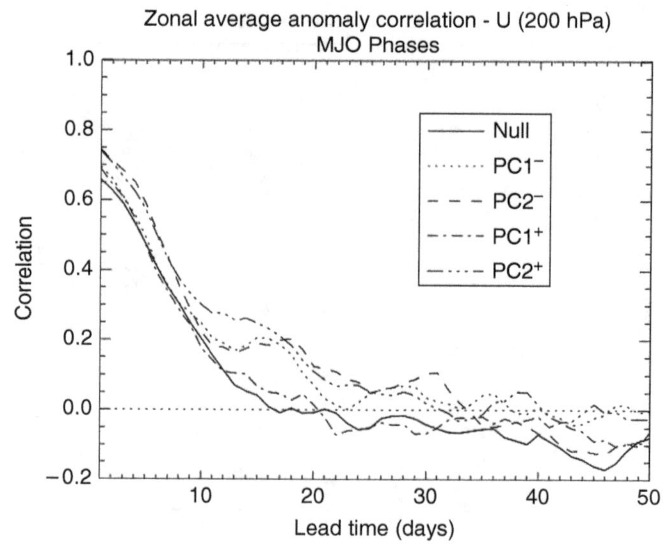

Figure 20.4 Zonal average along the equator of anomaly correlations in U (200 hPa) during different phases of the MJO: PC1 (dotted), PC2 (dashed), PC1+ (dash-dotted), PC2+ (long dash-short dash), and Null cases (solid). After *Jones et al.* [2000].

Because of the limited forecast skill of operational numerical models, several studies developed statistical forecast models of the MJO using a variety of approaches. *Waliser et al.* [1999] developed a statistical model based on singular value decomposition of 30–70 days outgoing longwave radiation (OLR) anomalies. Winter and summer season validations against band-passed OLR anomalies indicated useful skills out to 5–20 days over most of the Eastern Hemisphere. Likewise, *Lo and Hendon* [2000] used an autoregressive model of filtered OLR and upper level streamfunction, which provided skillful forecasts out to 15 days of lead time. *Mo* [2001] used a combination of singular spectral analysis and maximum entropy method that provided forecast skills out to 20 day lead time. *Wheeler and Weickmann* [2001] developed a tropical wave theory filtering technique to forecast tropical convectively coupled modes with forecast skill in the range of 15–20 days. *Jones et al.* [2004d] used 20–90 day anomalies of OLR and zonal winds at 200 hPa (U200) and 850 hPa (U850) to develop a forecast model of the MJO based on combined empirical orthogonal functions, which provided useful skill out to 50 days. However, the skill of statistical models in forecasting the MJO is much lower when verified in real time, and are typically in the range of 10–15 day lead time [*Jiang et al.*, 2008; *Love and Matthews*, 2009]. Whether or not statistical models have useful skill in forecasting the initiation of MJO events needs yet to be determined.

Given the challenges related to the MJO, substantial efforts have been dedicated to improve the representation of the MJO in global models as well as increase the forecast skill of the MJO in operational models. One such effort was to develop specific metrics to evaluate the characteristics of the MJO in observations and models [*Gottschalck et al.*, 2010]. In addition, since the MJO involves substantial interactions among convective processes, large-scale atmospheric circulation, and air-sea interaction, substantial efforts were also dedicated to improve the representation of the MJO in global models [*Hung et al.*, 2013; *Zhang*, 2005; *Zhang et al.*, 2006].

To illustrate the skill of current operational models, Figure 20.5 shows correlations between forecasts and observations of the RMM index for two forecast systems: the ECMWP Variable Resolution Ensemble Prediction System (VarEPS) and the NCEP Climate Forecast System version 2 (CFSv2) [*Kim et al.*, 2014]. Different measures

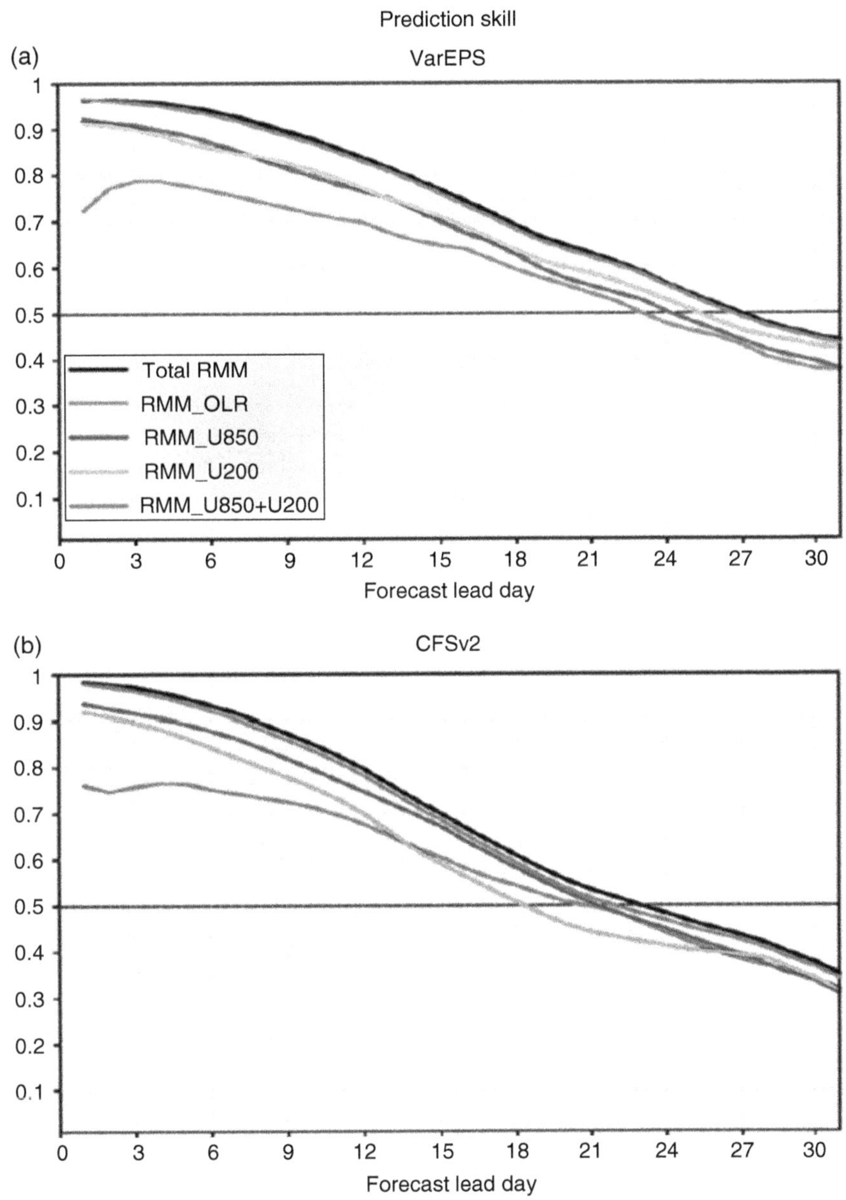

Figure 20.5 Prediction skill of the MJO according to different versions of the RMM index: OLR (red), U850 (blue), and U200 (green) separately, and U850 and U200 combined (brown). Verification is done for strong (high amplitude) MJO events in the forecasts from (a) ECMWF Variable Resolution Ensemble Prediction System (VarEPS), and (b) NCEP Climate Forecast System version 2 (CFSv2). After *Kim et al.* [2014]. (*See insert for color representation of the figure.*)

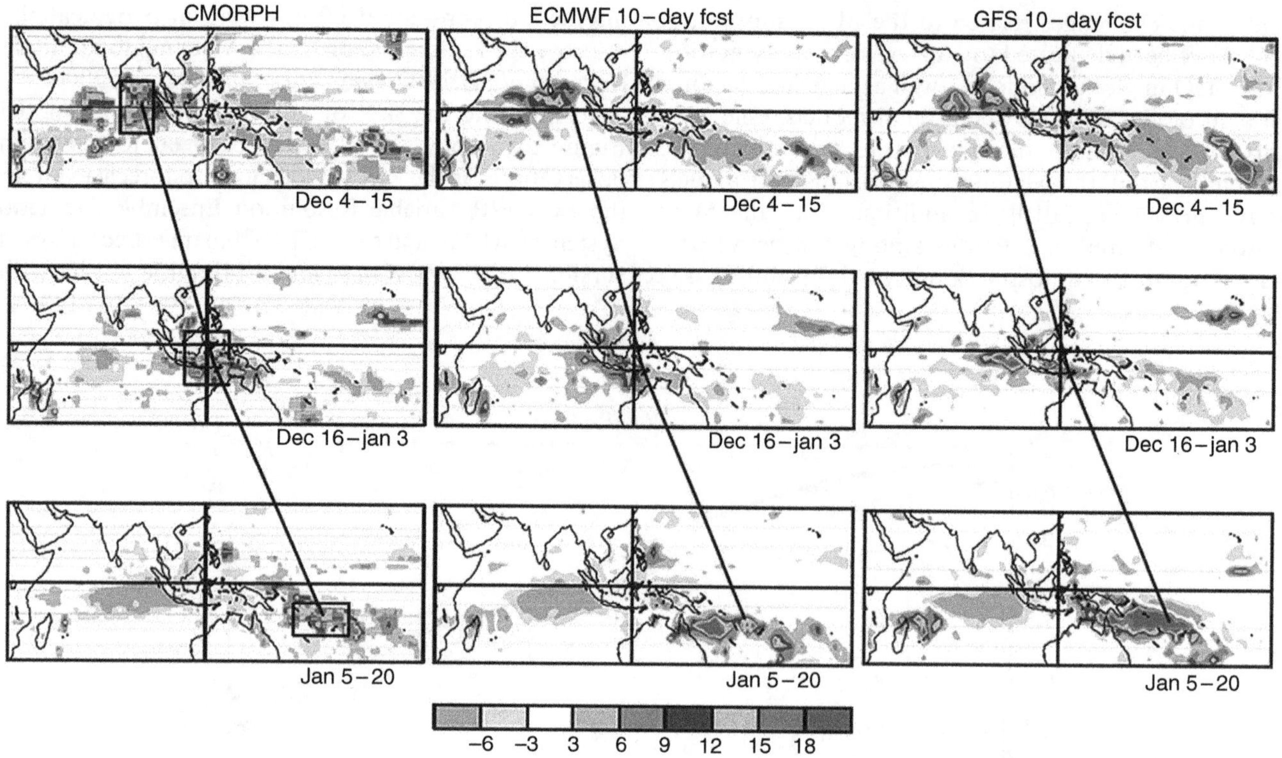

Figure 20.6 Precipitation anomalies (mm day^{-1}) observed in CMORPH satellite data (left) and 10 day forecasts from ECMWF (middle) and GFS models (right). Anomalies are averaged on three different periods 4–15 Dec 2007 (top row), 16 Dec 2007–3 Jan 2008 (middle row), and 5–20 Jan 2008 (bottom row). Solid lines connect the eastward progression of precipitation anomalies calculated as departures from the 4 month mean November 2007–February 2008. After *Janowiak et al.* [2010].

are included depending on the calculation of the RMM index: combination of OLR, U200, U850 (total), only OLR, only U850, only U200, and U850 combined with U200 (see *Kim et al.* [2014] for details). Despite some variations among the different RMM indexes, the predictability of the MJO extends beyond 32 days lead time in both hindcasts systems, which shows a major improvement in forecast skill compared with 1990s and early 2000s (e.g., Fig. 20.4).

Figure 20.6 gives another example of significant advances in forecasting the MJO [*Janowiak et al.*, 2010]. The left column shows the evolution of time average anomalies of tropical precipitation from Climate Prediction Center (CPC) morphing procedure (CMORPH) satellite data during a period of active MJO. The middle and right columns show 10 day forecasts from the ECMWF and NCEP Global Forecast System (GFS) models, respectively, and averaged on the same verification dates. Despite some shortcomings in capturing details in the spatial structure, both models show remarkable skill in forecasting the eastward propagation of the precipitation anomalies.

Additional studies that discuss progress in improving the forecast skill of the MJO in operational forecast models include *Hirons et al.* [2013b], *Hirons et al.* [2013a], *Lin et al.* [2008], *Rashid et al.* [2010], *Vitart* [2014], *Vitart and Jung* [2010], *Vitart and Molteni* [2010], *Vitart et al.* [2007], and *Matsueda and Endo* [2011]. Similarly, coordinated efforts to estimate the predictability of the MJO in current global climate models have been developed. *Neena et al.* [2014], for instance, analyzed extended-range hindcasts from eight different coupled models of the Intraseasonal Variability Hindcast Experiment (ISVHE) and determined the predictability of the MJO to be 20–30 days for single-member forecasts and 35–45 days for ensemble mean forecasts.

20.4. THE MJO AND PREDICTABILITY OF PRECIPITATION

Several studies have demonstrated important linkages between the MJO and variations in the extratropical atmospheric circulation during boreal winter [*Gloeckler and Roundy*, 2013; *Lee and Lim*, 2012; *Moore et al.*, 2010; *Riddle et al.*, 2013]. Therefore, this section discusses the significance of the MJO in controlling the forecast skill of the large-scale atmospheric circulation in the extratropics and influences in the predictive skill of precipitation.

Jones et al. [2004a] used the NASA Goddard Laboratory for the Atmosphere (GLA) global climate model to conduct potential predictability experiments in a 10 year period forced with fixed climatological sea surface temperatures (SST) to generate a control dataset. Periods of active and inactive MJO events were identified, perturbations added to initial conditions, and the model was integrated for 90 days. Figure 20.7 shows forecast verifications expressed as mean anomaly correlations in geopotential height (500 hPa), streamfunction (200 hPa), and zonal winds (850 hPa) over the Northern Hemisphere midlatitudes (20°–60°N). Phases were defined when the MJO enhanced convection was situated over the Indian Ocean (PC1-), Maritime Continent (PC2-), western Pacific (PC1+), and central Pacific (PC2+). The results show that the mean anomaly correlation is systematically higher during active MJO periods. For instance, the predictability of geopotential height (500 hPa) extends to about 16–18 days during active MJO but only to 15 days during Null cases. In particular, the MJO phase denoted by INDIAN shows the largest mean anomaly correlation for all three variables.

Recently, *Jones et al.* [2015] used 45 day forecasts of geopotential height (500 hPa, Z500) from NCEP Climate Forecast System version 2 (CFSv2) reforecasts (1 November–31 March, 1999–2010) to investigate the role of the MJO on the forecast skill in the Northern Hemisphere extratropics during boreal winter. Figure 20.8a shows the mean anomaly correlation of Z500 during active MJO events four phases, PI, PII, PIII, and PIV. These phases were grouped according to the phases of *Wheeler and Hendon* [2004] (PI = 1, 2, PII = 3, 4, PIII = 5, 6, and PIV = 7, 8). In additional, the mean anomaly correlation during null events is also indicated. Figure 20.8b shows the test statistic used to detect differences in mean anomaly correlation between phases PI, PII, PIII, and PIV, and inactive.

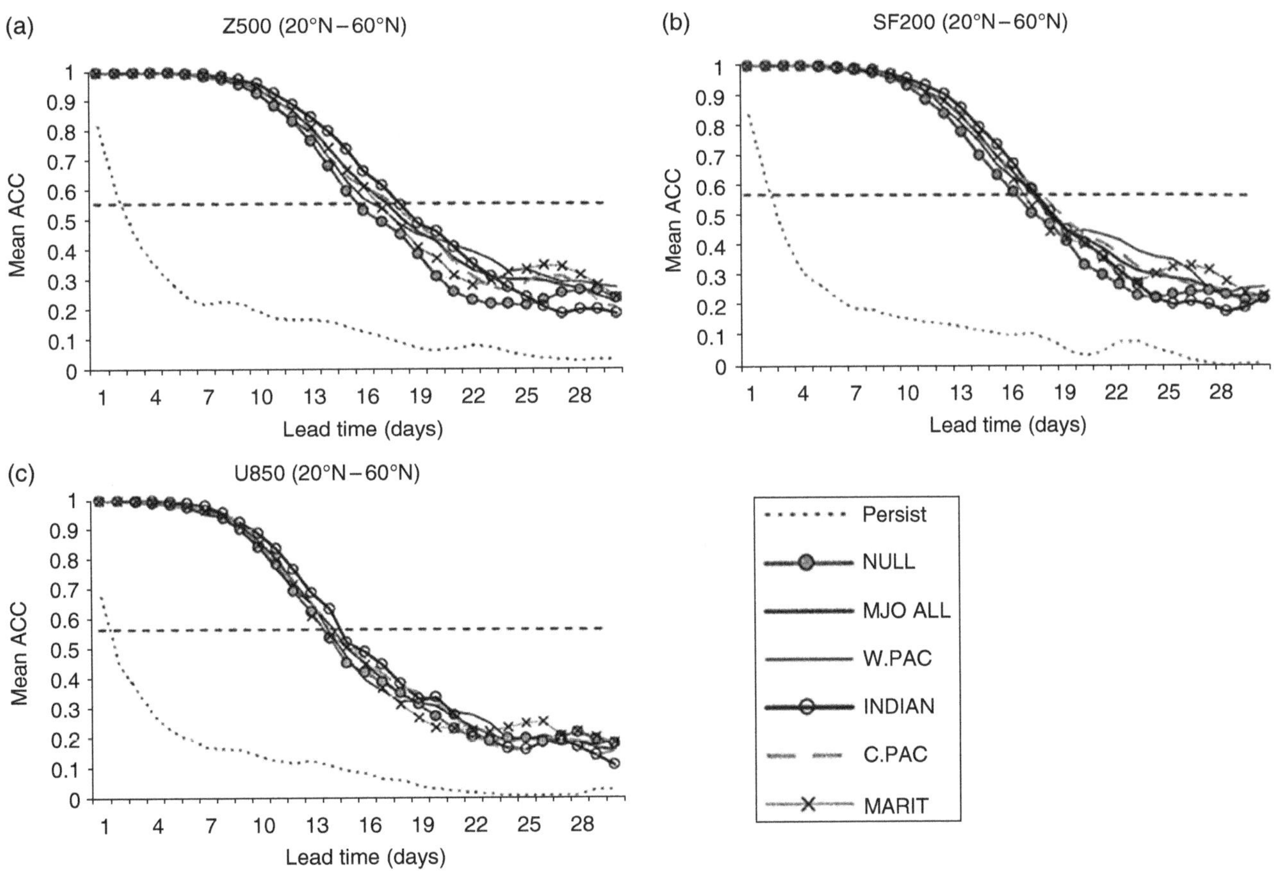

Figure 20.7 Mean anomaly (pattern) correlation over the Northern Hemisphere (20°–60°N) during boreal winter: (a) geopotential height (500 hPa), (b) streamfunction (200 hPa), and (c) zonal component of the wind (850 hPa). Anomaly correlations are averaged over all model run cases in each phase of the MJO and null (inactive) cases (see text for details). After *Jones et al.* [2004].

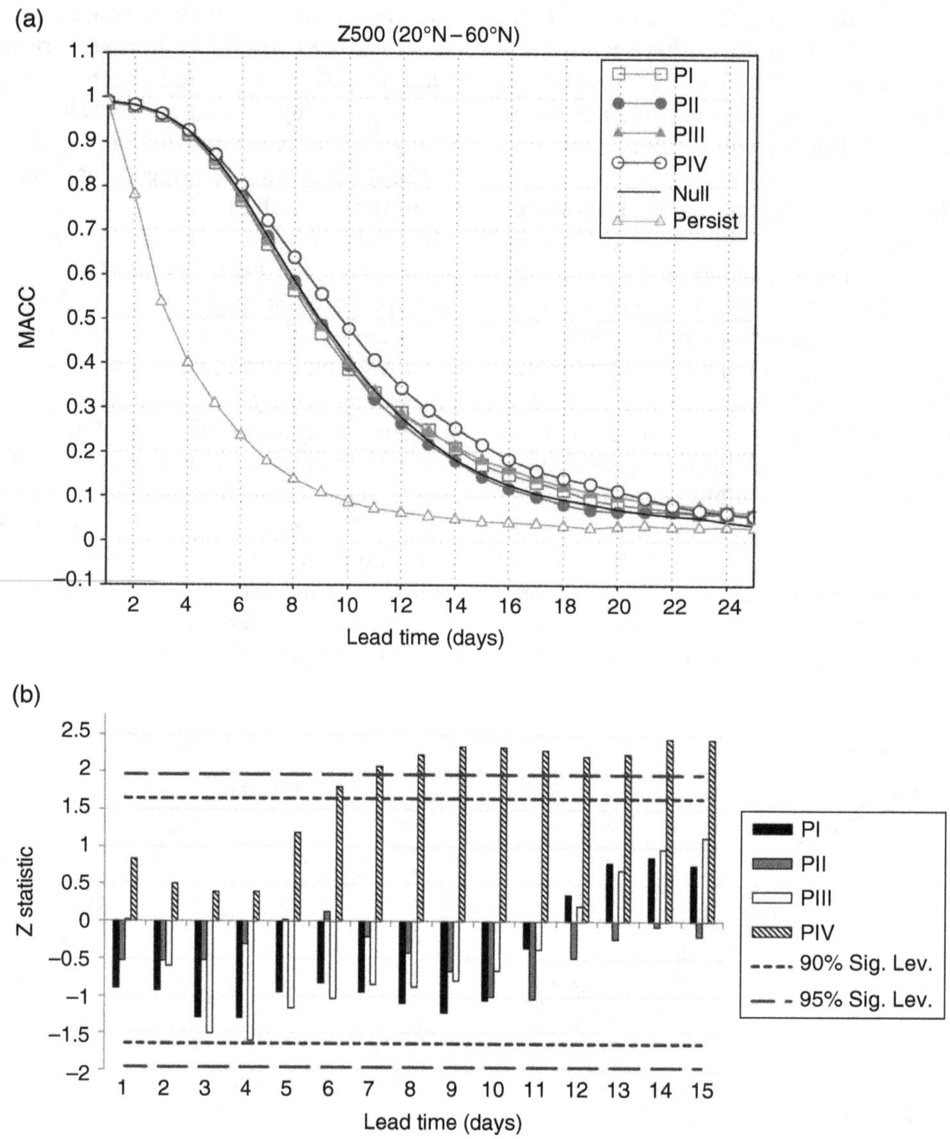

Figure 20.8 (a) Mean anomaly correlation in geopotential height (500 hPa) during each group of MJO phases and null case (indicated in the inset; see text for phase definitions). Persistence forecast is indicated with triangles. (b) Z statistic of differences in forecasts between active MJO and null cases. Short (long) dashed horizontal lines indicate statistical differences at 90% (95%) significance level. Mean anomaly correlations were computed over 20°–60°N. After *Jones et al.* [2015].

First, a Fisher transformation is first applied to the mean anomaly correlation at each lead time to ensure Gaussian distribution:

$$\Gamma(\tau) = 0.5 \ln \frac{(1+\rho)}{(1-\rho)} \tag{20.1}$$

Where ρ is the mean anomaly correlation at the lead time τ. Next, the test statistic Z is computed for differences between Γ in MJO phases and inactive cases at each lead time:

$$Z = \frac{(\Gamma_{MJO} - \Gamma_{null})}{\sqrt{\frac{\sigma^2_{MJO}}{N_{MJO}} + \frac{\sigma^2_{null}}{N_{null}}}} \tag{20.2}$$

Where σ is the variance of Γ, and N is the number of cases. Values of Z greater than 1.64 (1.96) indicate that differences in Γ are statistically significant at 90% (95%) level (see *Jones et al.* [2015] for details).

The CFSv2 forecast skills of Z500 extend to about 8 days lead time (correlations = 0.6; Fig. 20.8a) for both active MJO in the PI, PII, and PIII phases and in null days. In contrast, the forecast skill during MJO in the PIV phase (enhanced convection over the western Pacific and Western Hemisphere) is higher than in the other cases and extends to 7–15 day lead times. Also, the differences in skill between MJO in PIV phase and null cases grow rapidly from 4 to 7 days lead time (Fig. 20.8b) and are statistically significant (95% level) from 7 to 15 days lead.

These results (Fig. 20.8) differ from *Jones et al.* [2004a] (Fig. 20.7). While it is difficult to perform a direct comparison between the two studies, some important points are noteworthy. *Jones et al.* [2004a] used twin predictability experiments with prescribed climatological SST, a low resolution, old generation, global model, and a small sample of MJO events. *Jones et al.* [2015] used a fully coupled climate model (CFSv2), real forecasts, and a much larger MJO sample size. This brings the issue that the ability to maintain the amplitude and eastward propagation of the MJO during the forecast period is highly model dependent [*Matsueda and Endo*, 2011], which, consequently, impacts the forecast skill in the extratropics.

As discussed before, the MJO is regarded as a major source of potential predictability on subseasonal timescales and significant interest has been devoted to monitor its variability and influence on the forecast skill in the extratropics. Interestingly, relatively few studies have specifically investigated the role of the MJO on the predictive skill and potential predictability of precipitation in the extratropics. Using the same NASA GLA global model experiments described in *Jones et al.* [2004a], *Jones et al.* [2004b] investigated the potential predictability of extreme precipitation associated with the MJO. They concluded that the prediction of extreme precipitation during an active MJO is higher than in quiescent periods of the oscillation. Overall, the predictability experiments indicated the mean number of correct forecasts of extremes during active MJO periods to be nearly twice the correct number of extremes during quiescent phases of the oscillation in locations of statistically significant signals.

As an example of latitudinal differences in the predictive skill of precipitation, Figure 20.9 shows the temporal correlation between precipitation from CMORPH data and forecasts [*Janowiak et al.*, 2010]. Forecasts from four different models were used for a period of active MJO (November 2007–February 2008) and correlations were averaged over three latitudinal bands.

For the tropical and Southern Hemisphere zonal regions, the ECMWF forecasts have higher average spatial correlation at each forecast lead, with GFS performing next best, followed by ERA-Interim, and CFS. Over the 30°–60°N zonal band, the averaged correlation for ECMWF and GFS is practically identical at all forecast leads. During the Northern Hemisphere winter season, variations in forecast skill between the ECMWF and GFS models are similar when precipitation is largely stratiform.

In another study, *Jones et al.* [2011a] used the first version of the NCEP CFS model (CFS.v1) to evaluate the skill of nonprobabilistic and probabilistic forecasts of extreme precipitation over the contiguous United States during boreal winter. However, due to the design of the CFSv1 reforecasts, *Jones et al.* [2011a] were only able to analyze deterministic forecasts to investigate the

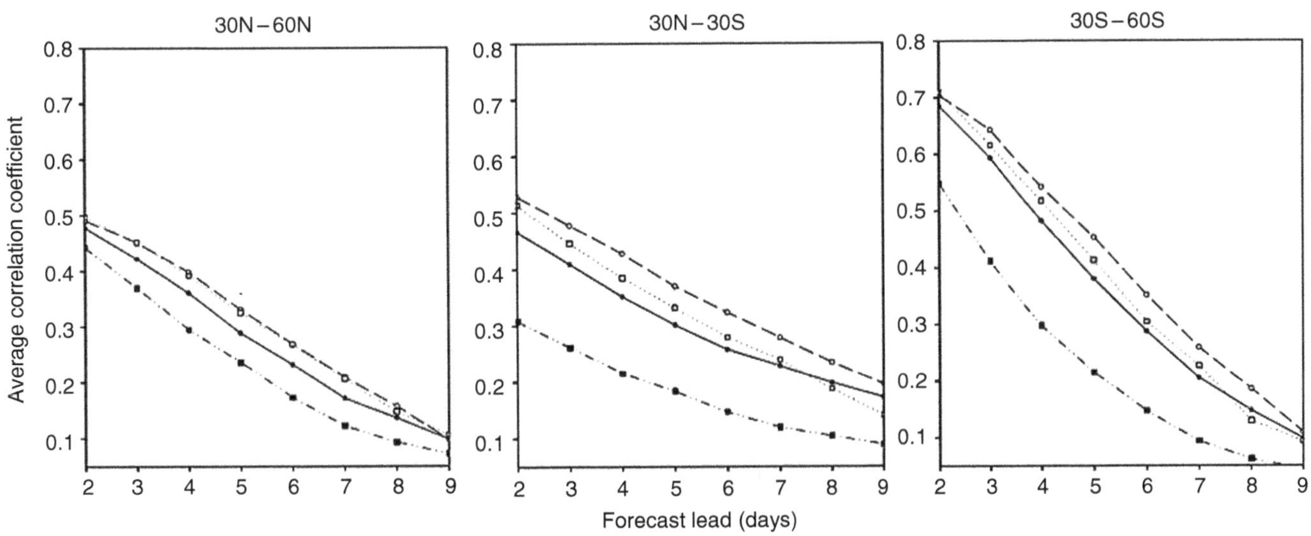

Figure 20.9 Temporal correlations between model forecasts and CMORPH daily precipitation. Correlations are averaged over three latitudinal bands: 30 N–60 N (left), 30 N–30S (middle), and 30S–60S (right). Each curve is for a given model: ECMWF (dashed), ERA Interim (solid), GFS (dotted), and CFS (dashed-dotted). Period of verification: November 2007–February 2008. After *Janowiak et al.* [2010].

influence of the MJO on the predictive skill of extreme precipitation. Nevertheless, they concluded that the deterministic forecast skill of extreme precipitation is usually higher when the MJO is active and has enhanced convection occurring over the Western Hemisphere, Africa, and/or the western Indian Ocean than in quiescent periods of the oscillation.

Their analyses show that when the MJO is active, Heidke skill scores (HSS) greater than 0.1 extend to 13–14 day lead times, almost a week longer than the overall skill assessment of the CFSv1. Figure 20.10 summarizes the spatial influence of the MJO on the forecast skill of extreme precipitation over the contiguous United States. The calculation was done for two types of extreme precipitation (e.g., daily precipitation exceeding the 75th and 90th percentiles). The solid dots represent the percentage of the contiguous United States with HSS≥0.1 averaged over (top) days 1–7 and (bottom) days 8–14. The statistic was computed as the number of grid points with HSS≥0.1 and significance level relative to inactive MJO conditions≥5% divided by the total number of points in the domain. The upper and lower tips of the bars show the spread (i.e., maximum and minimum percentages of the CONUS with HSS≥0.1). The results clearly show that when the MJO is active, the largest influence on forecast skill of extreme precipitation corresponds to phases 1–2 and 7–8. For 75th percentile extremes, about 15% of the CONUS has HSS≥0.1 for week 1 lead time (Fig. 20.10, top left) and up to about 20% if one considers 1–7 day lead times individually. For week 2 lead time, phase 8 stands out with about 15% average and up to 29% of the CONUS with HSS≥0.1 for 8–14 days analyzed individually (Fig. 20.10, bottom left). For 90th percentile extremes and week 1 lead time (Fig. 20.10, top right), the separation between phases 1 and 2, 7, and 8, and the remaining ones is greater, such that phases 1 and 8 have approximately the same influence on forecast skill. While the percentage of the contiguous United States with HSS≥0.1 tends to decrease for week 2

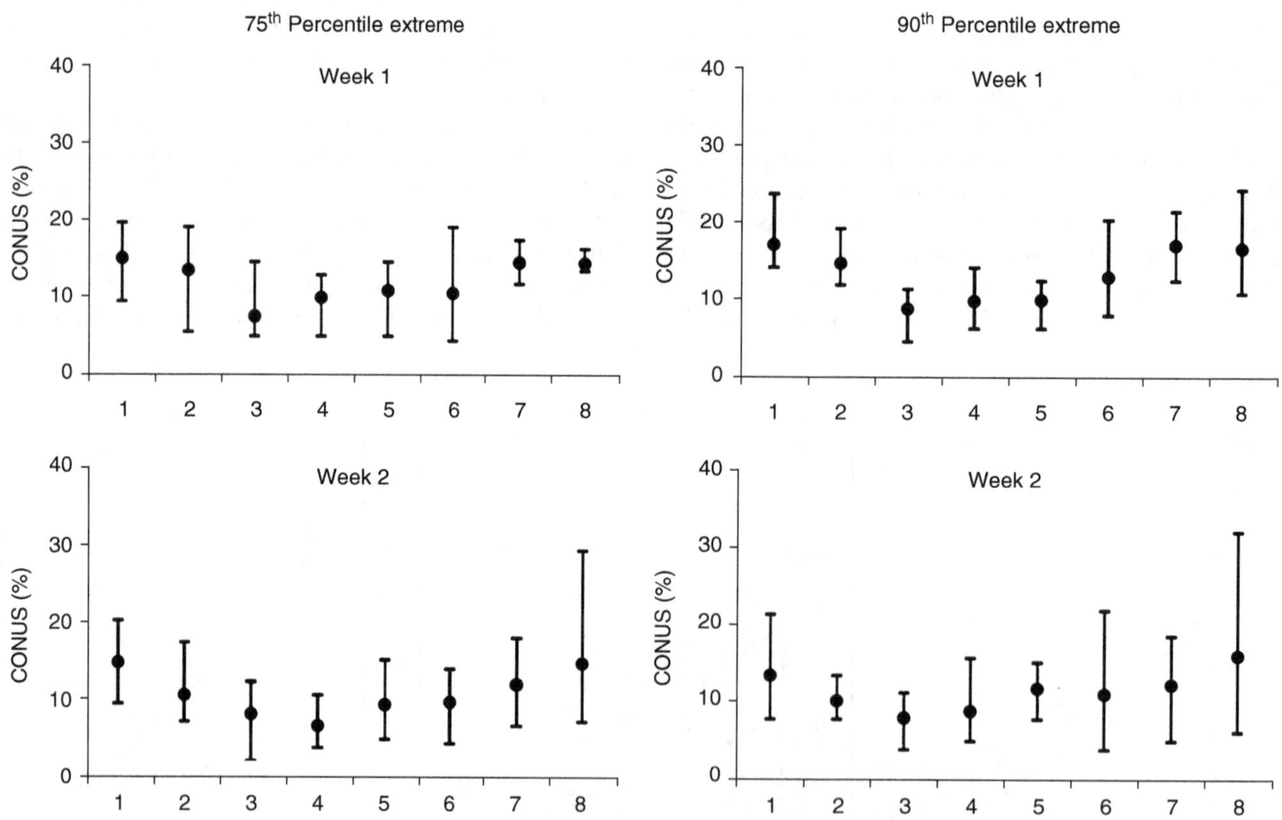

Figure 20.10 Statistics of Heidke skill scores (HSS) during active MJO and obtained with NCEP Climate Forecast System version 1 (CFSv1) model. Horizontal axis shows MJO phases. Left (right) column shows statistics for 75th (90th) percentiles of heavy precipitation. Rows indicate statistics for 1 week (top) and 2 week (bottom) lead times. Dots represent percentage of CONUS covered by HSS>0.1 and averaged during (top row) 1–7 days and (bottom row) 8–14 days. Upper and lower tips of the bars show max and min percentages of CONUS with HSS>0.1 within the 1 and 2 week lead times. Statistics were computed only over gridpoints where HSS during active MJO was statistically different than HSS in inactive MJO (5% significance level). After *Jones et al.* [2011].

lead time (Fig. 20.10, bottom right), phase 8 has a substantial influence (average of 16% and maximum of 32%) (see *Jones et al.* [2011b] for additional details).

The significant influence of the MJO on forecasts of extreme precipitation was further investigated by *Jones et al.*[2011b] using a simple decision model to estimate the value of deterministic forecasts of extreme precipitation over the contiguous United States. The model was applied to CFSv1 reforecasts and the results show that when enhanced convection associated with the MJO is in phases 1–2 and 7–8, the relative values in the forecasts are larger and extend to longer leads than when the MJO is inactive.

20.5. CONCLUSIONS

Since the MJO was discovered more than 35 years ago, significant advances have been accomplished to understand the mechanisms of the MJO and its role in weather and climate variability. The main goal of this chapter is to review the role of the MJO in controlling the variability of precipitation on global scales. In addition, the chapter reviews the literature that investigated the relationships between the activity of the MJO and the predictive skill and predictability of precipitation including extremes or heavy precipitation.

Recently, the World Weather Research Program, the Observing System Research and Predictability Experiment (THORPEX), and World Climate Research Program (WWRP/THORPEX/WCRP) initiated a joint research effort to improve and understand the physical processes controlling forecast skills on subseasonal-to-seasonal timescales (http://www.wmo.int). While the MJO is recognized to be a significant source of predictability on subseasonal scales, the MJO also shows a large degree of variability from event to event, seasonal and interannual timescales. In addition, the MJO is known to interact with other modes of low-frequency variability in the climate system (e.g., NAO, ENSO). Major advances to understand and improve the predictability on subseasonal-to-seasonal scales will likely come in the next few years.

ACKNOWLEDGMENTS

The author is grateful to many helpful discussions and encouragement from Dr. Leila M. V. Carvalho (UCSB). The research support from the Climate and Large-Scale Scale Dynamics Program from the National Science Foundation (AGS-1053294) is gratefully acknowledged. Any opinions, findings, and conclusions or recommendations expressed in this material are those of the author and do not necessarily reflect the views of the National Science Foundation.

REFERENCES

Adames, A. F., and J. M. Wallace (2014), Three-dimensional structure and evolution of the vertical velocity and divergence fields in the MJO, *J. Atmos. Sci.*, *71*(12), 267–287.

Barlow, M., and D. Salstein (2006), Summertime influence of the Madden-Julian Oscillation on daily rainfall over Mexico and Central America, *Geophys. Res. Lett.*, *33*(21); doi: 10.1029/2006GL027738.

Barlow, M., M. Wheeler, B. Lyon, and H. Cullen (2005), Modulation of daily precipitation over southwest Asia by the Madden-Julian Oscillation, *Monthly Weather Rev.*, *133*(12), 3579–3594.

Barrett, B. S., and V. A. Gensini (2013), Variability of central United States April-May tornado day likelihood by phase of the Madden-Julian Oscillation, *Geophys. Res. Lett.*, *40*(11), 2790–2795.

Barrett, B. S., and B. N. Henley (2015), Intraseasonal variability of hail in the contiguous United States: Relationship to the Madden-Julian Oscillation, *Monthly Weather Rev.*, *143*(4), 1086–1103.

Barrett, B. S., G. R. Henderson, and J. S. Werling (2015), The influence of the MJO on the intraseasonal variability of Northern Hemisphere spring snow depth, *J. Climate*, *28*(18), 7250–7262.

Becker, E. J., E. H. Berbery, and R. W. Higgins (2011), Modulation of cold-season U.S. daily precipitation by the Madden-Julian Oscillation, *J. Climate*, *24*(19), 5157–5166.

Berhane, F., and B. Zaitchik (2014), Modulation of daily precipitation over East Africa by the Madden-Julian Oscillation, *J. Climate*, *27*(15), 6016–6034.

Bessafi, M., and M. C. Wheeler (2006), Modulation of south Indian ocean tropical cyclones by the Madden-Julian oscillation and convectively coupled equatorial waves, *Monthly Weather Rev.*, *134*(2), 638–656.

Cannon, F., L. M. V. Carvalho, C. Jones, A. Hoell, J. Norris, G. N. Kiladis, and A. A. Tahir (2015), The Influence of tropical forcing on extreme winter precipitation in the western Himalaya, *Climate Dyn.*; Submitted.

Carbone, R. E., and Y. Li (2015), Tropical oceanic rainfall and sea surface temperature structure: Parsing causation from correlation in the MJO, *J. Atmos. Sci.*, *72*(7), 2703–2718.

Carvalho, L. M. V., C. Jones, and B. Liebmann (2002a), Extreme Precipitation Events in Southeastern South America and Large-Scale Convective Patterns in the South Atlantic Convergence Zone, *J. Climate*, *15*(17), 2377–2394.

Carvalho, L. M. V., C. Jones, and B. Liebmann (2004), The South Atlantic convergence zone: Intensity, form, persistence, and relationships with intraseasonal to interannual activity and extreme rainfall, *J. Climate*, *17*(1), 88–108.

Carvalho, L. M. V., C. Jones, and M. A. F. Silva Dias (2002b), Intraseasonal large-scale circulations and mesoscale convective activity in tropical South America during the TRMM-LBA campaign, *J. Geophys. Res.Atmos.*, *107*(D20), 10.1029/2001JD000745.

Carvalho, L. M. V., C. Jones, and T. Ambrizzi (2005), Opposite phases of the antarctic oscillation and relationships with intraseasonal to interannual activity in the tropics during the austral summer, *J. Climate*, *18*(5), 702–718.

Cassou, C. (2008), Intraseasonal interaction between the Madden-Julian Oscillation and the North Atlantic Oscillation, *Nature*, *455*(7212), 523–527.

Chang, C. W. J. N., W. L. Tseng, H. H. Hsu, N. Keenlyside, and B. J. Tsuang (2015), The Madden-Julian Oscillation in a warmer world, *Geophys. Res. Lett.*, *42*(14), 6034–6042.

Chen, S., et al. (2015), A study of CINDY/DYNAMO MJO suppressed phase, *J. Atmos. Sci.*, *72*(10), 3755–3779.

Chen, T. C., and J. C. Alpert (1990), Systematic-errors in the annual and intraseasonal variations of the planetary-scale divergent circulation in Nmc medium-range forecasts, *Monthly Weather Rev.*, *118*(12), 2607–2623.

Chi, Y. Z., F. Q. Zhang, W. Li, J. H. He, and Z. Y. Guan (2015), Correlation between the Onset of the East Asian Subtropical Summer Monsoon and the Eastward Propagation of the Madden-Julian Oscillation, *J. Atmos. Sci.*, *72*(3), 1200–1214.

De Souza, E. B., and T. Ambrizzi (2006), Modulation of the intraseasonal rainfall over tropical brazil by the Madden-Julian oscillation, *Int. J. Climatol.*, *26*(13), 1759–1776.

Diamond, H. J., and J. A. Renwick (2015), The climatological relationship between tropical cyclones in the southwest pacific and the Madden-Julian Oscillation, *Int. J. Climatol.*, *35*(5), 676–686.

Donald, A., H. Meinke, B. Power, A. D. N. Maia, M. C. Wheeler, N. White, R. C. Stone, and J. Ribbe (2006), Near-global impact of the Madden-Julian Oscillation on rainfall, *Geophys. Res. Lett.*, *33*(9).

Duvel, J. P. (2015), Initiation and intensification of tropical depressions over the southern Indian Ocean: Influence of the MJO, *Monthly Weather Rev.*, *143*(6), 2170–2191.

Ferranti, L., T. N. Palmer, F. Molteni, and E. Klinker (1990), Tropical extratropical interaction associated with the 30–60 Day Oscillation and its impact on medium and extended range prediction, *J. Atmos. Sci.*, *47*(18), 2177–2199.

Flatau, M., and Y.-J. Kim (2012), Interaction between the MJO and polar circulations, *J. Climate*, *26*(11), 3562–3574.

Flatau, M., P. J. Flatau, P. Phoebus, and P. P. Niller (1997), The feedback between equatorial convection and local radiative and evaporative processes: The implications for intraseasonal oscillations, *J. Atmos. Sci.*, *54*(19), 2373–2386.

Gloeckler, L. C., and P. E. Roundy (2013), Modulation of the Extratropical Circulation by Combined Activity of the Madden-Julian Oscillation and equatorial Rossby waves during boreal winter, *Monthly Weather Rev.*, *141*(4), 1347–1357.

Goswami, B. N., and R. S. A. Mohan (2001), Estimate of predictability of monthly means in tropics from observations, *Current Sci.*, *80*(1), 56–63.

Gottschalck, J., et al. (2010), A framework for assessing operational Madden-Julian Oscillation forecasts a clivar MJO working group project, *Bull. Amer. Meteor. Soc.*, *91*(9), 1247–1258.

Henderson, G. R., B. S. Barrett, and D. M. Lafleur (2014), Arctic sea ice and the Madden-Julian Oscillation (MJO), *Climate Dyn.*, *43*(7–8), 2185–2196.

Hendon, H. H., and B. Liebmann (1990), The intraseasonal (30–50 Day) oscillation of the Australian summer monsoon, *J. Atmos. Sci.*, *47*(24), 2909–2923.

Hendon, H. H., and M. L. Salby (1994), The life-cycle of the Madden-Julian Oscillation, *J. Atmos. Sci.*, *51*(15), 2225–2237.

Hendon, H. H., B. Liebmann, M. Newman, J. D. Glick, and J. E. Schemm (2000), Medium-range forecast errors associated with active episodes of the Madden-Julian oscillation, *Monthly Weather Rev.*, *128*(1), 69–86.

Hidayat, R., and S. Kizu (2010), Influence of the Madden-Julian Oscillation on Indonesian rainfall variability in austral summer, *Int. J. Climatol.*, *30*(12), 1816–1825.

Higgins, R. W., and W. Shi (2001), Intercomparison of the principal modes of interannual and intraseasonal variability of the North American monsoon system, *J. Climate*, *14*(3), 403–417.

Hirons, L. C., P. Inness, F. Vitart, and P. Bechtold (2013a), Understanding advances in the simulation of intraseasonal variability in the ECMWF model. Part II: The application of process-based diagnostics, *Quart. J. Roy. Meteor. Soc.*, *139*(675), 1427–1444.

Hirons, L. C., P. Inness, F. Vitart, and P. Bechtold (2013b), Understanding advances in the simulation of intraseasonal variability in the ECMWF model. Part I: The representation of the MJO, *Quart. J. Roy. Meteor. Soc.*, *139*(675), 1417–1426.

Hung, M. P., J. L. Lin, W. Q. Wang, D. Kim, T. Shinoda, and S. J. Weaver (2013), MJO and Convectively coupled equatorial waves simulated by CMIP5 climate models, *J. Climate*, *26*(17), 6185–6214.

Isoguchi, O., and H. Kawamura (2006), MJO-related summer cooling and phytoplankton blooms in the South China Sea in recent years, *Geophys. Res. Lett.*, *33*(16).

Jankov, I., et al. (2011), An Evaluation of Five ARW-WRF Microphysics Schemes Using Synthetic GOES Imagery for an Atmospheric River Event Affecting the California Coast, *J. Hydrometeor.*, *12*(4), 618–633.

Janowiak, J. E., P. Bauer, W. Wang, P. A. Arkin, and J. Gottschalck (2010), An evaluation of precipitation forecasts from operational models and reanalyses including precipitation variations associated with MJO activity, *Monthly Weather Rev.*, *138*(12), 4542–4560.

Jeong, J. H., B. M. Kim, C. H. Ho, and Y. H. Noh (2008), Systematic variation in wintertime precipitation in East Asia by MJO-induced extratropical vertical motion, *J. Climate*, *21*(4), 788–801.

Jeong, J. H., C. H. Ho, B. M. Kim, and W. T. Kwon (2005), Influence of the Madden-Julian Oscillation on wintertime surface air temperature and cold surges in east Asia, *J. Geophys. Res. Atmos.*, *110*; doi:10.1029/2004JD005408.

Jiang, X., D. E. Waliser, M. C. Wheeler, C. Jones, M.-I. Lee, and S. D. Schubert (2008), Assessing the skill of an all-season statistical forecast model for the Madden-Julian Oscillation, *Monthly Weather Rev.*, *136*, 1940–1956.

Jin, D., D. E. Waliser, C. Jones, and R. Murtugudde (2013), Modulation of tropical ocean surface chlorophyll by the Madden-Julian Oscillation, *Climate Dyn.*, *40*(1–2), 39–58.

Jones, C. (2000), Occurrence of extreme precipitation events in California and relationships with the Madden-Julian oscillation, *J. Climate*, *13*(20), 3576–3587.

Jones, C. (2009), A homogeneous stochastic model of the Madden-Julian Oscillation, *J. Climate*, *22*(12), 3270–3288.

Jones, C., and B. C. Weare (1996), The role of low-level moisture convergence and ocean latent heat fluxes in the Madden and Julian oscillation: An observational analysis using ISCCB data and ECMWF analyses, *J. Climate*, *9*(12), 3086–3104.

Jones, C., and L. M. V. Carvalho (2002), Active and break phases in the South American Monsoon system, *J. Climate*, *15*(8), 905–914.

Jones, C., and L. M. V. Carvalho (2006), Changes in the activity of the Madden-Julian oscillation during 1958–2004, *J. Climate*, *19*, 6353–6370.

Jones, C., and L. M. V. Carvalho (2011a), Stochastic simulations of the Madden-Julian Oscillation activity, *Climate Dyn.*, *36*, 229–246; doi:210.1007/s00382-00009-00660-00382.

Jones, C., and L. M. V. Carvalho (2011b), Will global warming modify the activity of the Madden-Julian Oscillation?, *Quart. J. Roy. Meteor. Soc.*, *137*, 544–552; doi:510.1002/qj.1765.

Jones, C., and L. M. V. Carvalho (2012), Spatial-intensity variations in extreme precipitation in the contiguous United States and the Madden-Julian Oscillation, *J. Climate*, *25*(14), 4898–4913.

Jones, C., and L. M. V. Carvalho (2014), Sensitivity to Madden-Julian Oscillation variations on heavy precipitation over the contiguous United States, *Atmos. Res.*, *147–148*(0), 10–26.

Jones, C., A. Hazra, and L. M. V. Carvalho (2015), The Madden-Julian Oscillation and boreal winter forecast skill: an analysis of NCEP CFSv2 reforecasts, *J. Climate*, *28*, 6297–6307.

Jones, C., D. E. Waliser, and C. Gautier (1998), The influence of the Madden-Julian oscillation on ocean surface heat fluxes and sea surface temperature, *J. Climate*, *11*(5), 1057–1072.

Jones, C., D. E. Waliser, J. K. E. Schemm, and W. K. M. Lau (2000), Prediction skill of the Madden and Julian Oscillation in dynamical extended range forecasts, *Climate Dyn.*, *16*(4), 273–289.

Jones, C., D. E. Waliser, K. M. Lau, and W. Stern (2004a), The Madden-Julian oscillation and its impact on Northern Hemisphere weather predictability, *Monthly Weather Rev.*, *132*(6), 1462–1471.

Jones, C., D. E. Waliser, K. M. Lau, and W. Stern (2004b), Global occurrences of extreme precipitation and the Madden-Julian oscillation: Observations and predictability, *J. Climate*, *17*(23), 4575–4589.

Jones, C., J. Gottschalck, L. M. V. Carvalho, and W. R. Higgins (2011a), Influence of the Madden-Julian Oscillation on forecasts of extreme precipitation in the contiguous United States, *Monthly Weather Rev.*, *139*, 332–350.

Jones, C., L. M. V. Carvalho, J. Gottschalck, and W. R. Higgins (2011b), The Madden-Julian Oscillation and the relative value of deterministic forecasts of extreme precipitation in the contiguous United States, *J. Climate*, *24*, 2421–2428.

Jones, C., L. M. V. Carvalho, R. W. Higgins, D. E. Waliser, and J. K. E. Schemm (2004c), Climatology of tropical intraseasonal convective anomalies: 1979–2002, *J. Climate*, *17*(3), 523–539.

Jones, C., L. M. V. Carvalho, R. W. Higgins, D. E. Waliser, and J. K. E. Schemm (2004d), A statistical forecast model of tropical intraseasonal convective anomalies, *J. Climate*, *17*(11), 2078–2095.

Kessler, W. S. (2001), EOF representations of the Madden-Julian oscillation and its connection with ENSO, *J. Climate*, *14*(13), 3055–3061.

Kiladis, G. N., K. H. Straub, and P. T. Haertel (2005), Zonal and vertical structure of the Madden-Julian oscillation, *J. Atmos. Sci.*, *62*(8), 2790–2809.

Kim, H. M., P. J. Webster, V. E. Toma, and D. Kim (2014), Predictability and prediction skill of the MJO in two operational forecasting systems, *J. Climate*, *27*(14), 5364–5378.

Klotzbach, P. J., and E. C. J. Oliver (2015), Modulation of Atlantic basin tropical cyclone activity by the Madden-Julian Oscillation (MJO) from 1905 to 2011, *J. Climate*, *28*(1), 204–217.

Lau, K. M., and F. C. Chang (1992), Tropical Intraseasonal Oscillation and its prediction by the NMC operational model, *J. Climate*, *5*(12), 1365–1378.

Lau, W. K. M. (2012), *El Nino Southern Oscillation connection*, in *Intraseasonal Variability in the Atmosphere-Ocean Climate System*, edited by W. K. M. Lau and D. E. Waliser, 2nd ed., Springer, Chichester, UK.

Lau, W. K. M., and D. E. Waliser (2012), *Intraseasonal Variability in the Atmosphere-Ocean Climate System*, 2nd ed., Springer, Chichester, UK.

Lee, Y. Y., and G. H. Lim (2012), Dependency of the North Pacific winter storm tracks on the zonal distribution of MJO convection, *J. Geophys. Res. Atmos.*, *117*.

Li, T., C. B. Zhao, P. C. Hsu, and T. Nasuno (2015), MJO initiation processes over the tropical Indian Ocean during DYNAMO/CINDY2011, *J. Climate*, *28*(6), 2121–2135.

Liebmann, B., H. H. Hendon, and J. D. Glick (1994), The relationship between tropical cyclones of the western Pacific and Indian Oceans and the Madden-Julian Oscillation, *J. Meteor. Soc. Japan*, *72*(3), 401–412.

Lin, H., G. Brunet, and J. Derome (2008), Forecast skill of the Madden-Julian Oscillation in two Canadian atmospheric models, *Monthly Weather Rev.*, *136*(11), 4130–4149.

Lin, H., G. Brunet, and J. Derome (2009), An observed connection between the North Atlantic Oscillation and the Madden-Julian Oscillation, *J. Climate*, *22*(2), 364–380.

Liu, P. (2013), Changes in a modeled MJO with idealized global warming, *Climate Dyn.*, *40*(3–4), 761–773.

Liu, P., T. Li, B. Wang, M. Zhang, J.-j. Luo, Y. Masumoto, X. Wang, and E. Roeckner (2013), MJO change with A1B global warming estimated by the 40-km ECHAM5, *Climate Dyn.*, *41*(3–4), 1009–1023.

Lo, F., and H. H. Hendon (2000), Empirical extended-range prediction of the Madden-Julian oscillation, *Monthly Weather Rev.*, *128*(7), 2528–2543.

Love, B. S., and A. J. Matthews (2009), Real-time localised forecasting of the Madden-Julian Oscillation using neural network models, *Quart. J. Roy. Meteor. Soc.*, *135*(643), 1471–1483.

Madden, R. A., and P. R. Julian (1971), Detection of a 40–50 day Oscillation in the zonal wind in the tropical Pacific, *J. Atmos. Sci.*, *28*(5), 702–708.

Madden, R. A., and P. R. Julian (1972), Description of global-scale circulation cells in the tropics with a 40–50 day period, *J. Atmos. Sci.*, *29*(6), 1109–1123.

Madden, R. A., and P. R. Julian (1994), Observations of the 40–50-Day Tropical Oscillation-A Rev., *Monthly Weather Rev.*, *122*(5), 814–837.

Maloney, E. D., and D. L. Hartmann (2000), Modulation of eastern North Pacific hurricanes by the Madden-Julian oscillation, *J. Climate*, *13*(9), 1451–1460.

Maloney, E. D., and S.-P. Xie (2013), Sensitivity of tropical intraseasonal variability to the pattern of climate warming, *J. Adv. Model. Earth Syst.*, *5*(1), 32–47.

Marzuki, H. Hashiguchi, T. Kozu, T. Shimomai, Y. Shibagaki, and Y. Takahashi (2016), Precipitation microstructure in different Madden-Julian Oscillation phases over Sumatra, *Atmos. Res.*, *168*, 121–138.

Matsueda, M., and H. Endo (2011), Verification of medium-range MJO forecasts with TIGGE, *Geophys. Res. Lett.*, *38*.

Matthews, A. J. (2000), Propagation mechanisms for the Madden-Julian Oscillation, *Quart. J. Roy. Meteor. Soc.*, *126*(569), 2637–2651.

Matthews, A. J. (2008), Primary and successive events in the Madden-Julian Oscillation, *Quart. J. Roy. Meteor. Soc.*; doi:10.1002/qj.1224.

Matthews, A. J., and M. P. Meredith (2004), Variability of Antarctic circumpolar transport and the southern annular mode associated with the Madden-Julian Oscillation, *Geophys. Res. Lett.*, *31*(24); doi:10.1029/2004GL021666.

Matthews, A. J., P. Singhruck, and K. J. Heywood (2007), Deep ocean impact of a Madden-Julian Oscillation observed by Argo floats, *Science*, *318*(5857), 1765–1769.

McPhaden, M. J. (1999), Genesis and evolution of the 1997–98 El Nino, *Science*, *283*(5404), 950–954.

McPhaden, M. J., X. B. Zhang, H. H. Hendon, and M. C. Wheeler (2006), Large scale dynamics and MJO forcing of ENSO variability, *Geophys. Res. Lett.*, *33*(16).

Mo, K. C. (1999), Alternating wet and dry episodes over California and intraseasonal oscillations, *Monthly Weather Rev.*, *127*(12), 2759–2776.

Mo, K. C. (2001), Adaptive filtering and prediction of intraseasonal oscillations, *Monthly Weather Rev.*, *129*(4), 802–817.

Mo, K. C., and R. W. Higgins (1998a), The Pacific-South American modes and tropical convection during the Southern Hemisphere winter, *Monthly Weather Rev.*, *126*(6), 1581–1596.

Mo, K. C., and R. W. Higgins (1998b), Tropical influences on California precipitation, *J. Climate*, *11*(3), 412–430.

Mo, K. C., and R. W. Higgins (1998c), Tropical convection and precipitation regimes in the western United States, *J. Climate*, *11*(9), 2404–2423.

Moore, R. W., O. Martius, and T. Spengler (2010), The modulation of the subtropical and extratropical atmosphere in the Pacific basin in response to the Madden-Julian Oscillation, *Monthly Weather Rev.*, *138*(7), 2761–2779.

Muza, M. N., L. M. V. Carvalho, C. Jones, and B. Liebmann (2009), Intraseasonal and interannual variability of extreme dry and wet Events over southeastern South America and subtropical Atlantic during the austral summer, *J. Climate*, *22*, 1682–1699.

Nasuno, T., T. Li, and K. Kikuchi (2015), Moistening processes before the convective initiation of Madden-Julian Oscillation events during the CINDY2011/DYNAMO period, *Monthly Weather Rev.*, *143*(2), 622–643.

Neena, J. M., J. Y. Lee, D. Waliser, B. Wang, and X. N. Jiang (2014), Predictability of the Madden-Julian oscillation in the intraseasonal variability hindcast experiment (ISVHE), *J. Climate*, *27*(12), 4531–4543.

Nogues-Paegle, J., and K. C. Mo (1997), Alternating wet and dry conditions over South America during summer, *Monthly Weather Rev.*, *125*(2), 279–291.

Nogues-Paegle, J., and K. C. Mo (2002), Linkages between summer rainfall variability over south America and sea surface temperature anomalies, *J. Climate*, *15*(12), 1389–1407.

Oliver, E. C. J. (2015), Multidecadal variations in the modulation of Alaska wintertime air temperature by the Madden-Julian Oscillation, *Theor. App. Climatol.*, *121*(1–2), 1–11.

Pohl, B., and P. Camberlin (2006), Influence of the Madden-Julian Oscillation on East African rainfall. I: Intraseasonal variability and regional dependency, *Quart. J. Roy. Meteor. Soc.*, *132*(621), 2521–2539.

Pohl, B., and A. J. Matthews (2007), Observed changes in the lifetime and amplitude of the Madden-Julian oscillation associated with interannual ENSO sea surface temperature anomalies, *J. Climate*, *20*(11), 2659–2674.

Pohl, B., Y. Richard, and N. Fauchereau (2007), Influence of the Madden-Julian oscillation on Southern African summer rainfall, *J. Climate*, *20*(16), 4227–4242.

Powell, S. W., and R. A. Houze (2015), Effect of dry large-scale vertical motions on initial MJO convective onset, *J. Geophys. Res.-Atmospheres*, *120*(10), 4783–4805.

Rashid, H. A., H. H. Hendon, M. C. Wheeler, and O. Alves (2010), Prediction of the Madden-Julian oscillation with the POAMA dynamical prediction system, *Climate Dyn.*; doi:10.1007/s00382-00010-00754-x.

Riddle, E. E., M. B. Stoner, N. C. Johnson, M. L. L'Heureux, D. C. Collins, and S. B. Feldstein (2013), The impact of the MJO on clusters of wintertime circulation anomalies over the North American region, *Climate Dyn.*, *40*(7–8), 1749–1766.

Shinoda, T., H. H. Hendon, and J. Glick (1998), Intraseasonal variability of surface fluxes and sea surface temperature in the tropical western Pacific and Indian Oceans, *J. Climate*, *11*(7), 1685–1702.

Smith, B. L., S. E. Yuter, P. J. Neiman, and D. E. Kingsmill (2010), Water Vapor Fluxes and Orographic Precipitation over Northern California Associated with a Landfalling Atmospheric River, *Mon Weather Rev*, *138*(1), 74–100.

Takahashi, C., N. Sato, A. Seiki, K. Yoneyama, and R. Shirooka (2011), Projected Future Change of MJO and its Extratropical Teleconnection in East Asia during the Northern Winter Simulated in IPCC AR4 Models, *Sola*, *7*, 201–204.

Thompson, D. B., and P. E. Roundy (2012), The Relationship between the Madden-Julian Oscillation and U.S. Violent Tornado Outbreaks in the Spring, *Monthly Weather Rev.*, *141*(6), 2087–2095.

Vecchi, G. A., and N. A. Bond (2004), The Madden-Julian Oscillation (MJO) and northern high latitude wintertime surface air temperatures, *Geophys. Res. Lett.*, *31*(4).

Vitart, F. (2014), Evolution of ECMWF sub-seasonal forecast skill scores, *Quart. J. Roy. Meteor. Soc.*, *140*(683), 1889–1899.

Vitart, F., and F. Molteni (2010), Simulation of the Madden-Julian Oscillation and its teleconnections in the ECMWF forecast system, *Quart. J. Roy. Meteor. Soc.*, *136*(649), 842–855.

Vitart, F., and T. Jung (2010), Impact of the Northern Hemisphere extratropics on the skill in predicting the Madden Julian Oscillation, *Geophys. Res. Lett.*, *37*.

Vitart, F., S. Woolnough, M. A. Balmaseda, and A. M. Tompkins (2007), Monthly forecast of the Madden-Julian oscillation using a coupled GCM, *Monthly Weather Rev.*, *135*(7), 2700–2715.

Waliser, D. E., C. Jones, J. K. E. Schemm, and N. E. Graham (1999), A statistical extended-range tropical forecast model based on the slow evolution of the Madden-Julian oscillation, *J. Climate*, *12*(7), 1918–1939.

Waliser, D. E., K. M. Lau, W. Stern, and C. Jones (2003), Potential predictability of the Madden-Julian oscillation, *Bull. Amer. Meteor. Soc.*, *84*(1), 33–50.

Waliser, D. E., R. Murtugudde, P. Strutton, and J. L. Li (2005), Subseasonal organization of ocean chlorophyll: Prospects for prediction based on the Madden-Julian Oscillation, *Geophys. Res. Lett.*, *32*(23), L23602, 23610.21029/22005GL024300.

Wang, B. (2005), Theory, in *Intraseasonal Variability in the Atmosphere-Ocean Climate System*, edited by W. K. M. Lau and D. E. Waliser, Springer, Chichester, UK.

Weickmann, K. M. (1991), El-Nino Southern Oscillation and Madden-Julian (30–60 Day) Oscillations during 1981–1982, *J. Geophys. Res.Oceans*, *96*, 3187–3195.

Weickmann, K. M., and P. D. Sardeshmukh (1994), The atmospheric angular-momentum cycle-associated with a Madden-Julian Oscillation, *J. Atmos. Sci.*, *51*(21), 3194–3208.

Weickmann, K. M., G. R. Lussky, and J. E. Kutzbach (1985), Intraseasonal (30–60 Day) fluctuations of outgoing longwave radiation and 250 mb streamfunction during northern winter, *Monthly Weather Rev.*, *113*(6), 941–961.

Wheeler, M., and G. N. Kiladis (1999), Convectively coupled equatorial waves: Analysis of clouds and temperature in the wavenumber-frequency domain, *J. Atmos. Sci.*, *56*(3), 374–399.

Wheeler, M., and K. M. Weickmann (2001), Real-time monitoring and prediction of modes of coherent synoptic to intraseasonal tropical variability, *Monthly Weather Rev.*, *129*(11), 2677–2694.

Wheeler, M., and H. H. Hendon (2004), An all-season real-time multivariate MJO index: Development of an index for monitoring and prediction, *Monthly Weather Rev.*, *132*(8), 1917–1932.

Wheeler, M., H. H. Hendon, S. Cleland, H. Meinke, and A. Donald (2009), Impacts of the Madden-Julian Oscillation on Australian rainfall and circulation, *J. Climate*, *22*(6), 1482–1498.

Xu, W., S. A. Rutledge, C. Schumacher, and M. Katsumata (2015), Evolution, properties, and spatial variability of MJO convection near and off the equator during DYNAMO, *J. Atmos. Sci.*, *72*(11), 4126–4147.

Zhang, C. D. (2005), Madden-Julian oscillation, *Rev. Geophys.*, *43*(2), 1–36.

Zhang, C. D. (2013), Madden-Julian Oscillation bridging weather and climate, *Bull. Amer. Meteor. Soc.*, *94*(12), 1849–1870.

Zhang, C. D., M. Dong, S. Gualdi, H. H. Hendon, E. D. Maloney, A. Marshall, K. R. Sperber, and W. Q. Wang (2006), Simulations of the Madden-Julian Oscillation in four pairs of coupled and uncoupled global models, *Climate Dyn.*, *27*(6), 573–592.

Zhang, L., B. Wang, and Q. Zeng (2009), Impact of the Madden-Julian Oscillation on summer rainfall in southeast China, *J. Climate*, *22*(2), 201–216.

Zhao, H. K., R. Yoshida, and G. B. Raga (2015), Impact of the Madden-Julian Oscillation on western North Pacific tropical cyclogenesis associated with large-scale patterns, *J Appl Meteorol Clim*, *54*(7), 1413–1429.

Zhou, S. T., M. L'Heureux, S. Weaver, and A. Kumar (2012), A composite study of the MJO influence on the surface air temperature and precipitation over the continental United States, *Climate Dyn.*, *38*(7–8), 1459–1471.

21

Prediction of Short-Term Climate Extremes with a Multimodel Ensemble

Emily J. Becker

ABSTRACT

This chapter focuses on the prediction of monthly and seasonal climate extremes at leads of one to several months. Earlier studies that assess the forecast skill of short-term climate extremes (STCE), primarily using a single ensemble global climate model, are reviewed. This chapter details an assessment of the skill of STCE prediction by a state-of-the-art multimodel ensemble, the North American Multi-Model Ensemble (NMME), including forecasts for 5th and 95th percentile seasonal events in maximum, minimum, and mean 2 m temperature. Area-aggregate forecast-observation gridpoint anomaly correlations for forecasts of extremes are found to be higher than area-aggregate anomaly correlations for all forecasts, confirming the results of earlier studies. Assessment of the forecast skill for extremes using the Symmetric Extremal Dependence Index reveals several geographic areas where model forecasts are substantially more skillful than a climatological forecast. This chapter also includes a discussion of potential predictability of STCE, examined using perfect-model studies, and sources of predictability, including major climate modes, that may present windows of forecast opportunity.

21.1. INTRODUCTION

Climate has its greatest effects on humans, plants, and animals through departures from the normal. As most living organisms are acclimated to the mean climate, substantial deviations from this normal can result in stresses. Short-term climate extremes (STCE), such as a seasonal mean maximum temperature much above normal, or an extreme 1 month average precipitation rate, can have large consequences; these consequences could potentially be mitigated by preparation, if the events can be skillfully forecast. The frequency of extremes may be increasing, as several other chapters in this book have detailed, and this subject has been of substantial interest for some time [e.g., *Easterling et al.*, 2000a,b].

Climate Prediction Center/NOAA and INNOVIM LLC, College Park, Maryland, USA

While seasonal climate prediction has become a standard practice, and many national meteorological services around the world have robust operational climate forecasting centers, there are currently relative few outlooks for specific extreme events on the monthly and seasonal timescale. In most operational contexts, seasonal climate outlooks are concerned with shifts in the distribution relative to the long-term climatological mean. A common format for seasonal climate outlooks is a probabilistic forecast for where the seasonal mean will fall relative to the long-term average, often for three categories: above normal, below normal, and near normal. These categories (or terciles) are equally likely for a climatological forecast. A major difficulty in developing skillful forecasts for STCE, and furthermore in assessing prediction skill, is that the distribution is by definition poorly sampled when examining rare events. Accurately assessing the underlying distribution of a monthly or seasonal

Climate Extremes: Patterns and Mechanisms, Geophysical Monograph 226, First Edition.
Edited by S.-Y. Simon Wang, Jin-Ho Yoon, Christopher C. Funk, and Robert R. Gillies.

variable from 30 or so years of data is challenging, but there is potential for developing useful forecasts.

This chapter will focus on the prediction of extremes on monthly and seasonal timescales, at leads of one to several months. Relatively few studies have examined the prediction of extremes on these timescales, particularly using current operational forecast structures. One recent assessment of real-time outlooks for seasonal extremes, primarily based on dynamical atmospheric prediction models, found some skill was present [*Barnston and Mason*, 2011]. These forecasts, from the International Research Institute (IRI), present the likelihood of precipitation and temperature falling into the lower or upper 15th percentiles of the climatological distribution for the 3 month period at 0.5 month lead, that is, a half-month following the forecast issuance.

On longer timescales, several recent studies have examined the potential for prediction of the number or intensity of extreme events on decadal timescales, including heat waves and extreme precipitation, with some encouraging conclusions [*Eade et al.*, 2012; *Hanlon et al.*, 2013a, b; *Hanlon et al.*, 2014]. And on the other end of the timescale, the operational model used at Australia's Bureau of Meteorology has been found to simulate subseasonal (2–3 weeks) heat extremes with some success [*White et al.*, 2014; *Hudson et al.*, 2015].

When focusing on extremes, some subtleties arise that are not relevant in studies of overall forecast skill. Several different definitions of "extreme" are available, including but not limited to a value that falls above or below an absolute threshold (e.g., a mean temperature below 0 °C), the tails of a climatological distribution, or a percentile threshold based on ranked historical events. This variety has complicated studies of trends in extremes [*Nicholls*, 1995; *Easterling et al.*, 2000b], and imposes difficulties on assessments of forecasts for extremes. A workshop on extreme events [*Garrett and Müller*, 2008] concluded that, faced with a limited dataset and asked to predict future extremes, the wisest approach is to determine which of the canonical distributions best fits the extremes from the existing data, and then to use the chosen distribution for predictions. They go on to say that this is a highly uncertain process, however, and that other constraints are desirable.

Another consideration resulting from the analysis sample being a subset of the entire record is that the forecast skill for extremes can be evaluated from two different angles. First, the sample of forecast/observation gridpoint pairs to be analyzed can be conditioned based on the forecasts; that is, the question being asked is "an extreme was forecast, did it occur?" Alternatively, the sample can be conditioned based on the occurrence of an extreme in the observed record ("an extreme occurred, was it forecast?") The first scenario is the common approach of forecast verification. The results of a comparative analysis of these two approaches found little difference in respective skill scores [*Becker et al.*, 2013].

21.2. PREDICTION SKILL

Becker et al. [2013] examined skill for forecasts of short-term climate extremes (STCE) in 2 m temperature (T2m), precipitation (Prate), and sea surface temperature (SST) using 29 years of retrospective forecasts (1982-2010) from NCEP's Climate Forecast System version 2 (CFSv2) [*Saha et al.*, 2014]. CFSv2 is a state-of-the-art coupled global climate model, and the current operational climate model at NOAA. Monthly mean forecasts out to 9 month leads were available, initialized each month. Anomaly correlations (AC) and root-mean-square errors (RMSE) for forecasts of T2m and Prate extremes were aggregated over two land-only regions: North and South America. SST extremes were evaluated in the Niño3.4 region and in the area of the Atlantic hurricane main development region.

In all regions and fields, AC for forecasts of extremes was found to be higher than those of all forecasts. When extremes were defined using a sliding scale, that is, the threshold defining an extreme was raised and the sample was restricted to progressively rarer events, both the AC and the RMSE increased, but the anomaly correlation increased faster than the RMSE. As the AC is a measure of the signal, and RMSE can be considered a signal-to-noise ratio, this indicates the signal grows despite increased noise. For forecasts of the highest skill (SST in the Niño3.4 region), the RMSE does not increase at all as the sample is restricted to more and more extreme events, while RMSE of lowest-skill forecasts (Prate in North America) increases rapidly.

The current study builds on the earlier work of *Becker et al.* [2013] and uses a suite of state-of-the-art coupled models to predict temperature extremes over land. To this end, the focus herein is the prediction of extremes in mean temperature (T2m), maximum temperature (Tmax), and minimum temperature (Tmin). The mechanisms that govern maximum and minimum temperature are different from each other, and it is not expected that the prediction skill will be identical for the three temperature datasets. Consistent with earlier studies, "lead 1" is defined as the first full month or season after the initial month, for example, the lead 1 seasonal forecast from January initial conditions is February-March-April. This study focuses on temperature, given the very low skill for dynamical model forecasts of precipitation over extratropical land found in *Becker et al.* [2013, 2014].

21.2.1. Methods and Data

The North American Multi-Model Ensemble (NMME) Phase I was developed and implemented in 2011, with real-time monthly mean forecasts starting August 2011

[*Kirtman et al.*, 2014]. When an ensemble is created from multiple models, the result is higher-quality forecasts than any single ensemble model [*Palmer et al.*, 2004; *Hagedorn et al.*, 2005; *Doblas-Reyes et al.*, 2005; *Smith et al.*, 2013]. The real-time Multi-Model Ensemble has incorporated between six and eight state-of-the-art coupled GCMs from operational and research centers in the United States and Canada; to date, 11 models from six modeling centers have participated. See www.cpc.ncep.noaa.gov/products/NMME/Phase1models.png for details on all the models.

All models are required to adhere to the NMME Phase I protocol, which specifies that all forecasts must be available on the same 1.0° longitude × 1.0° latitude horizontal grid, and include leads up to 9 months. Initialization, data assimilation, and model physics are not specified, and differ between modeling centers. Retrospective forecasts from each model are required for at least the period of 1982–2010, each month for a total of 12 initial conditions. These retrospective forecasts (or hindcasts) are used for model calibration, bias correction, skill assessment, and diagnostics. Several dozen studies have been completed using the NMME database. Go to http://www.nws.noaa.gov/ost/CTB/nmme_pub.htm for a list of publications.

Real-time and archived forecast graphics from Aug. 2011 to the present are available at www.cpc.ncep.noaa.gov/products/NMME. Hindcast and forecast data are archived at the International Research Institute for Climate and Society (IRI), accessible at http://iridl.ldeo.columbia.edu/SOURCES/.Models/.NMME/. Monthly mean and seasonal (three-month-mean) real-time forecasts have been delivered since August 2011 for 2 m surface temperature (T2m), precipitation rate (precip), and sea surface temperature (SST). Skill for NMME seasonal forecasts of these fields was assessed in *Becker et al.* [2014]. The spread of ensemble member forecasts around the ensemble mean is fairly realistic in the models of the NMME, in that the spread is comparable to the interannual variability of the observed fields. An additional "reality check" on the NMME models examined the persistence of anomalies in the models, which was found to be similar to that of the observations.

Retrospective forecast data were obtained for six NMME models, all of which actively contributed to the real-time system between 2014 and early 2016. These are the NCEP-CFSv2 [*Saha et al.*, 2014], CMC-CanCM3 and CanCM4 [*Merryfield et al.*, 2013], GFDL-CM2.1 [*Zhang et al.*, 2007], GFDL-FLOR [*Vecchi et al.*, 2014], and NASA-GEOS5 [*Vernieres et al.*, 2012]. All models have retrospective forecasts initialized each month over the period 1982–2010. See Table 21.1 for more detail about the six models used in this study. The hindcasts from several older and newer models, not used for this study, are still available for research.

Mean 2 m temperature forecasts were verified using the monthly-mean, station-observation based Global Historical Climatology Network + Climate Anomaly Monitoring System (GHCN + CAMS) [*Fan and van den Dool*, 2008]. GHCN + CAMS is a combination of two station observation datasets, and has a native resolution of 0.5° latitude × 0.5° longitude, regridded to 1° × 1° for the purposes of this study. Maximum and minimum temperature verification uses a recently developed daily, 0.5° latitude × 0.5° longitude gridded dataset derived from CAMS, developed at the Climate Prediction Center. For use in this study, the daily Tmax and Tmin were averaged into monthly means and regridded to 1° × 1°.

A Gaussian distribution was fit to all individual model hindcast ensemble means, and the resulting standard deviation used to determine the threshold of extremes: STCE were defined as +/- 1.645 * local gridpoint standard deviation. These thresholds are approximately equivalent to the 95th and 5th percentile; in a 29-year sample, this results in two to three extreme events per gridpoint per target month/season. The Gaussian distribution may not be the best fit for every field in every geographic location, and the resulting STCE are unlikely to all be events of great natural or social impact. However, this definition will capture most of the extremes in the record. An earlier study experimented with the area-aggregate anomaly correlation using several different methods of defining approximately 5th/95th percentile STCE, and found only slight variation in the results [*Becker et al.*, 2013].

This study primarily uses the ensemble mean of each model and the NMME multimodel ensemble mean. Observed fields used can be represented by $O(s, j, m)$, where s represents the local gridpoint index, j is the year, and m is the target month. A single model ensemble member forecast can be represented by $F(s, j, m, n, \tau)$, where n is the ensemble member number, and τ is forecast lead in months. The first step is calculating the ensemble mean from the N ensemble members for each model:

$$F_{ens} = \frac{1}{N}\sum_{n=1}^{N} F\left(s, j, m, n, \tau\right) \quad (21.1)$$

An independent dataset is constructed from the 1982–2010 hindcasts with cross-validation. Anomalies, defined by the prime (′) for each year j, are formed by

$$F_{ens}'\left(s, j, m\right) = F_{ens}\left(s, j, m\right) - \left\{F_{ens}\left(s, m\right)\right\} \quad (21.2)$$

where {F} denotes the mean overall years not including j. This method, wherein anomalies are constructed by removing the individual model's climatology (as opposed to subtracting an observed climatology), performs a systematic error correction for bias in the model's mean.

Table 21.1 North American Multi-Model Ensemble (NMME) Models Used in This Study

Center/Model	Hindcast period	Number of members	Arrangement of members	Lead (mon)	Model resolution atmos.	Model resolution ocean	Reference
NCEP/CFSv2	1982–2010	24 (28 Nov)	4 members (0, 6, 12, 18z)	0-9	T126L64	MOM4L40 0.25° Eq	*Saha et al.* [2006]
GFDL/CM2.1	1982–2010	10	All first of the month 0Z	0-11	2 x 2.5deg L24	MOM4L50 0.3° Eq	*Delworth* [2006]
GFDL/CM2.5 (FLOR)	1982–2015	24	All first of the month 0Z	0-11	C18L32 (50 km)	MOM5 L50 0.30° Eq 1°Polar1.5	*Vecchi et al.* [2014]
Environment Canada/ CMC1-CanCM3	1981–2010	10	All first of the month 0Z	0-11	CanAM3 T63L31	CanOM4 L40 0.94° Eq	*Merryfield et al.* [2013]
Environment Canada/ CMC1-CanCM4	1981–2010	10	All first of the month 0Z	0-11	CanAM4 T63L35	CanOM4 L40 0.94° Eq	*Merryfield et al.* [2013]
NASA/GEOS5	1981–2010	11	4 members every fifth day; 7 members on the last day of last month	0-9	1 x 1.25 deg L72	MOM4L40 0.25° Eq	*Vernieres et al.* [2012]

Debate about if this 1-year-out cross-validation method adequately resolves the uncertainty in the models' climatologies is left for another time.

The NMME multimodel ensemble mean anomaly is the average of the models' ensemble mean anomalies:

$$F'_{NMME} = \frac{1}{K}\sum_{k} F'_{ens,k}(s,j,m) \tag{21.3}$$

where K is the number of models (seven in the case of this study). The NMME multimodel ensemble mean is constructed with equal weighting, meaning each model's average contributes equally to the multimodel mean. This is the practice in CPC's real-time forecasts, and several studies have concluded that it is very difficult to determine successful weighting schemes, especially based on a relatively short set of retrospective forecasts [*Hagedorn et al.*,, 2005; *Tebaldi and Knutti*, 2007] and the high colinearity of the forecasts [*Peña and van den Dool*, 2008].

The area-aggregate anomaly correlation, a measure of the association between gridpoint forecast and observation pairs [*Wilks*, 1995; *van den Dool*, 2007], is

$$AC(m,\tau) = \frac{\sum_s \sum_j F'(s,j,m,\tau)O'(s,j,m)}{\left(\sum_s \sum_j \left[F'(s,j,m,\tau)\right]^2 \sum_s \sum_j \left[O'(s,j,m)\right]^2\right)^{\frac{1}{2}}} \tag{21.4}$$

Latitude weighting is applied to the area aggregate. For small correlation values, the sampling error in a correlation is $\frac{1}{\sqrt{N_{eff}-2}}$, where N_{eff} is the effective sample size. The use of large area aggregations here achieves statistical significance for all but the very lowest correlations, even when the sample is restricted to, for example, the 5th percentile, which occurs at a rate of approximate 1.5 events per 29 year sample. The disadvantage of area aggregations is obvious, however, in that a metric assessed for the Northern Hemisphere extratropics will obscure substantial regional variations.

Most skill metrics tend toward degenerate when assessing rare events. *Ferro and Stephenson* [2011] proposed the Symmetric Extremal Dependence Index (SEDI), which uses the hit rate *H* and false alarm rate *F* from a 2 x 2 contingency table [*Wilks*, 1995; *White et al.*, 2014]. A hit is counted if, for a forecast of an extreme event, the corresponding observation is also an extreme, and a false alarm is recorded if the corresponding observation is not an extreme. The SEDI normalizes the hit rate with the false alarm rate:

$$SEDI = \frac{lnF - lnH - \ln(1-F) + \ln(1-H)}{lnF + lnH + \ln(l-F) + \ln(1-H)} \tag{21.5}$$

and is nondegenerate for rare events. A positive SEDI can be interpreted as an improvement over a climatological forecast.

21.2.2. Results

The lead-1 seasonal (3-month-mean) forecast was first evaluated for all 12 initial conditions separately from the six models individually and the NMME multimodel mean. Confirming the earlier conclusion of *Becker et al.* [2013], anomaly correlations are substantially higher when the sample is restricted to extreme events than when all forecasts are sampled (Figs. 21.1, 21.2). Focusing on Tmin, T2m, and Tmax in the Northern Hemisphere (Fig. 21.1), the average anomaly correlation for all forecasts (*x*-axis) is in the approximate range of 0.15–0.4 for all models and seasons, with a few points falling outside that range; AC for Tmin and T2m forecasts is generally slightly higher than for Tmax. Forecasts from the six-model NMME (black-outlined squares in Fig. 21.1) generally earn the highest anomaly correlations for each season.

AC when only extremes are considered (*y*-axis) is higher than for the entire sample, with Tmin and T2m STCE forecasts generally in the range of 0.3–0.7. Tmax STCE forecasts have somewhat lower anomaly correlations, with some STCE forecasts earning lower AC than their all-cases counterpart. The increase is greater for STCE forecasts that already had higher AC, and there is a greater range of AC for STCE forecasts. For example, the AC for all Tmin forecasts ranges from 0.1 to slightly more than 0.4, but AC for forecasts of positive extremes ranges from about 0.1 to nearly 0.7.

A seasonal cycle in the skill of temperature forecasts in the Northern Hemisphere is noticeable in T2m, Tmax, and Tmin, for both positive and negative extremes, with highest anomaly correlations occurring during the late spring and summer. This is likely in part a result of soil moisture memory and initialization of soil moisture in the models [*Huang and van den Dool*, 1993; *Hamilton et al.*, 2012]. Positive (95th percentile) and negative (5th percentile) STCE generally have similar patterns, but ACs of forecasts for winter negative extremes for Tmin, T2m, and Tmax are mostly lower than forecasts for positive extremes. Another noticeable difference is the AC of forecasts for summer negative extremes, which are somewhat higher than of forecasts for summer positive extremes. In other words, warm winter lows are somewhat better predicted than cold winter lows, and cool summer highs are somewhat better predicted than hot summer highs.

Results for analysis of Southern Hemisphere extratropics (Fig. 21.2) are noisier than for Northern Hemisphere extratropics, as the sample size is substantially smaller. Colors in these figures reflect the season relevant to the

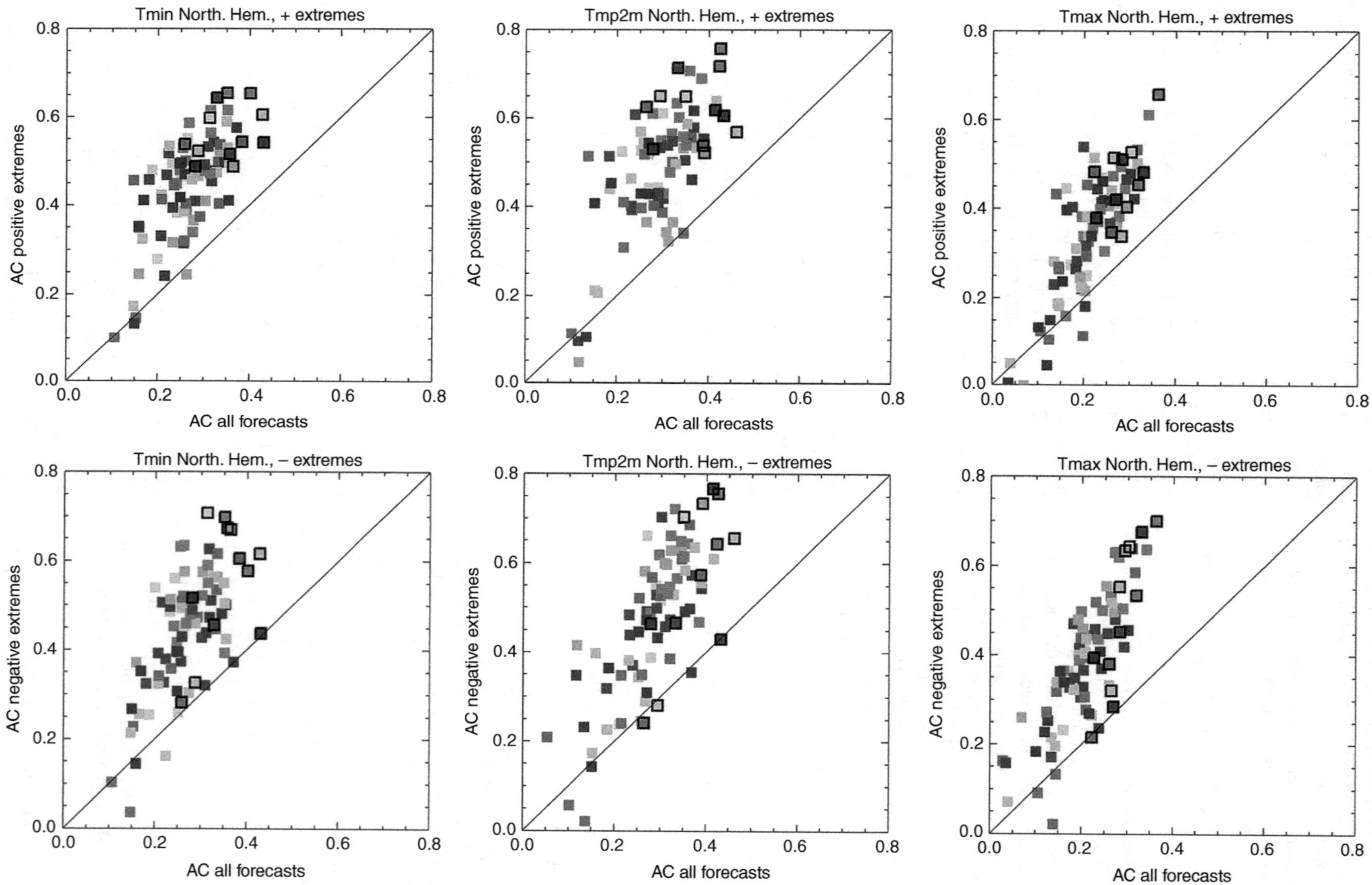

Figure 21.1 Anomaly correlation (AC) for lead-1 seasonal forecasts of minimum temperature (Tmin, left), mean 2 m temperature (Tmp2m, center), and maximum temperature (Tmax, right), area-aggregated over Northern Hemisphere extratropics (23°N–75°N) land from six individual NMME models and NMME multimodel ensemble mean (denoted with black outlines). AC for all forecasts is on the horizontal axis; AC for forecasts of extremes is on the vertical axis. Results are shown for positive extremes (95th percentile events) in the top row, negative extremes (5th percentile events) in the lower row. Colors indicate season: Green = spring seasons (FMA, MAM, AMJ), red = summer, orange = autumn, and blue = winter. *See electronic version for color representation*.

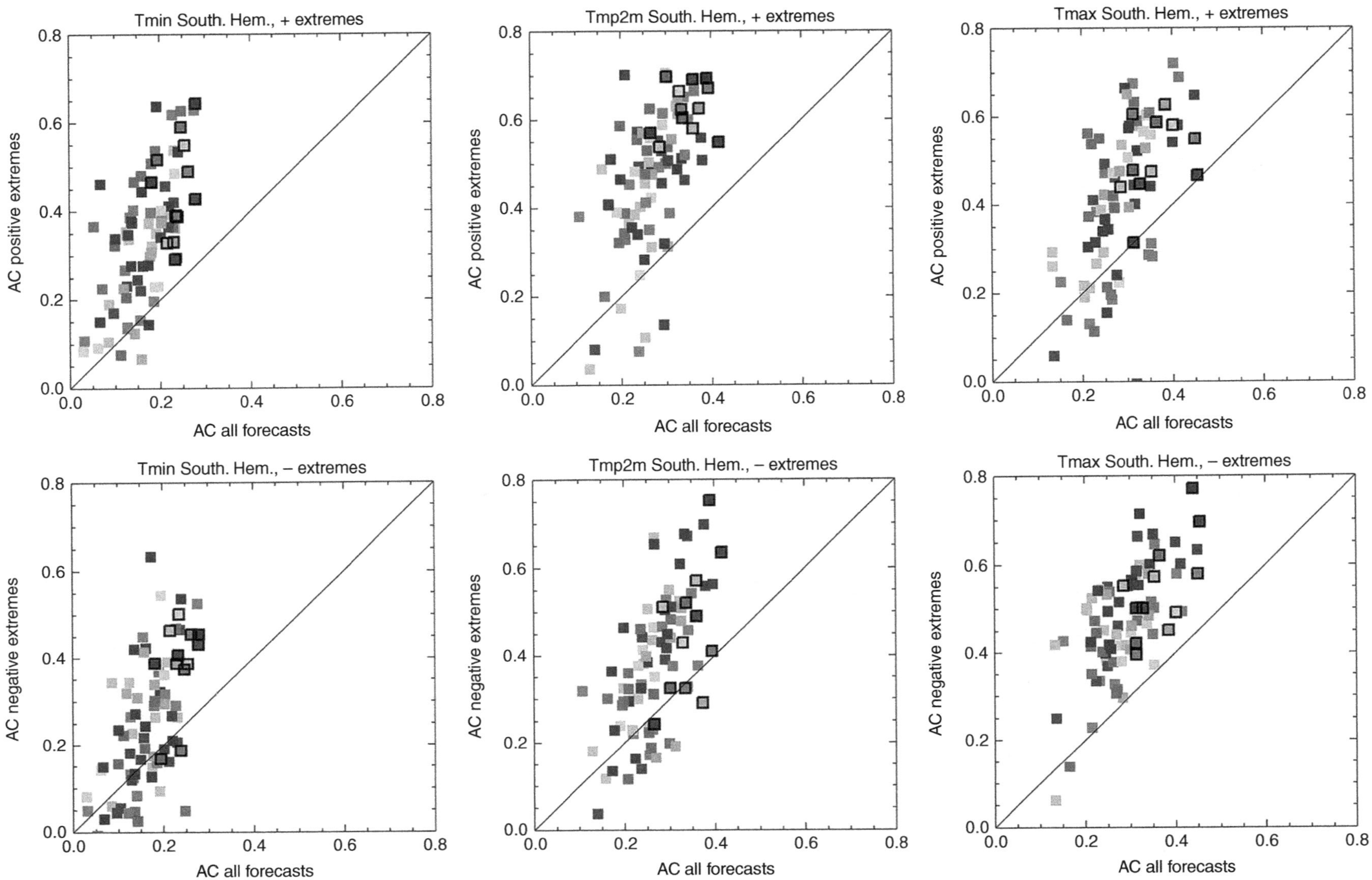

Figure 21.2 as in Figure 21.1, for Southern Hemisphere extratropical land. Colors are for Southern Hemisphere seasons: Orange = autumn (FMA, MAM, AMJ), blue = winter, green = spring, red = summer. *See electronic version for color representation.*

hemisphere (i.e., blue colors are for winter seasons) centered on June, July and August. In this hemisphere, overall forecasts for Tmin have lower AC (generally < 0.3) than forecasts for T2m and Tmax (approximately 0.1–0.5). Seasonality is not particularly clear, except for maximum temperatures, when both extremes and overall forecasts are better predicted in the spring, with lowest ACs generally occurring during the fall.

Higher anomaly correlations for overall forecasts and for extremes bear some relationship to stronger trends. For example, trends in minimum temperature have been somewhat stronger than trends in maximum temperature in the Northern Hemisphere [*Vose et al.*, 2005], and the anomaly correlations for Tmin forecast are higher than for Tmax in this hemisphere (Fig. 21.1). Similarly, Tmax exhibits a stronger overall trend than Tmin for most seasons in the Southern Hemisphere [*Vose et al.*, 2005], where AC of Tmax forecasts is generally higher than for Tmin forecasts. However, this relationship is merely anecdotal, and needs more exploration before conclusions can be drawn.

The Symmetric Extremal Dependence Index (SEDI) indicates that regional patterns for the 5th and 95th percentile events are not identical for T2m (Fig. 21.3). This index uses the hit rate and false alarm; a forecast for a 5th percentile event is considered a hit if a 5th percentile event occurs. A SEDI of 0.3, for example, can be considered a 30% improvement over the climatological forecast

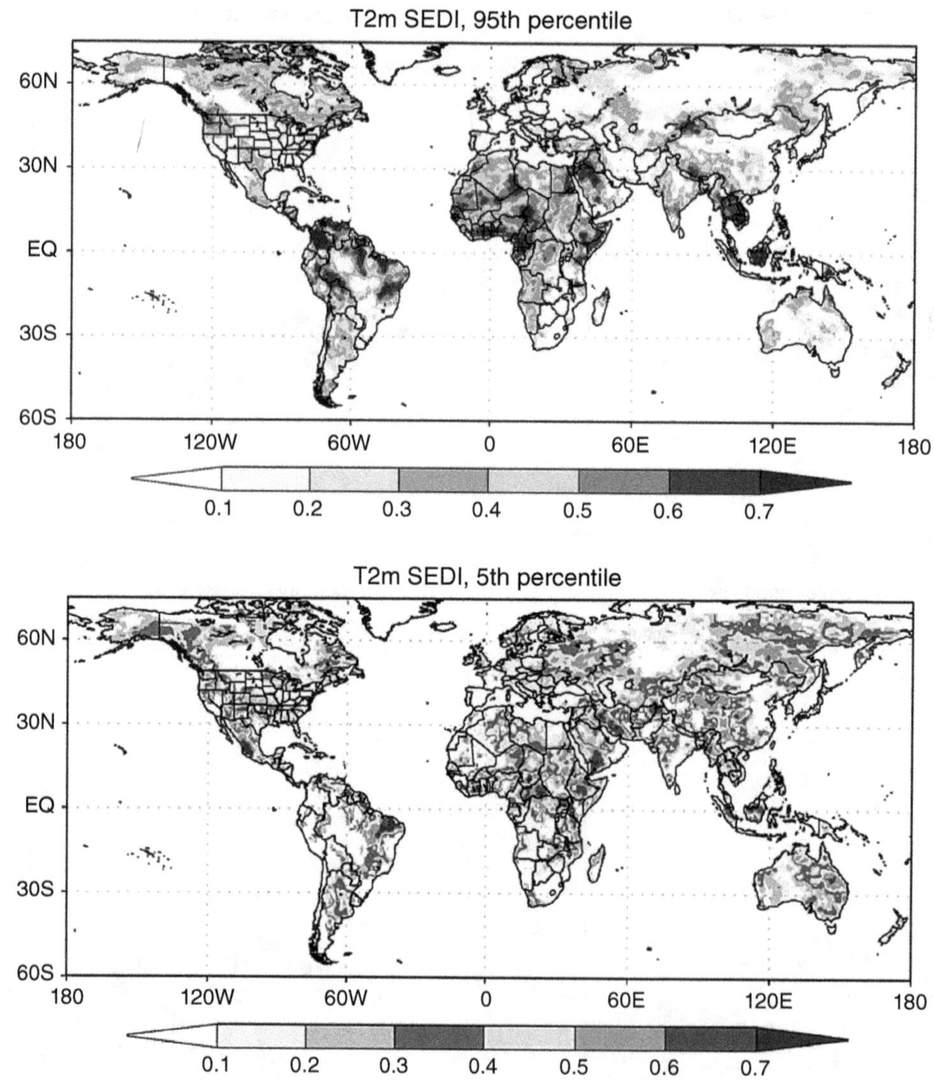

Figure 21.3 Symmetric External Dependence Index (SEDI) for seasonal forecasts of extreme 2 m temperature: SEDI for forecasts of positive extremes (95th percentile events) in the upper panel, negative extremes (5th percentile events) in the lower panel. (*See insert for color representation of the figure.*)

(a 5% chance of a 5th percentile event). Overall, forecasts for T2m 95th percentile events (positive extremes) show higher SEDI, but with less coverage over Asia and Australia. The highest scores for forecasts of T2m 95th percentile events are in the tropics, a finding that bears further exploration before a satisfactory explanation can be obtained.

For Tmax positive extremes (95th percentile), that is, forecasts that potentially include heat extremes, SEDI is generally positive over most of the same areas as T2m, including much of North America, some areas of Asia, and Australia (Fig. 21.4). Forecasts for 5th percentile Tmax have less coverage in the far northern latitudes, but somewhat more in the Northern Hemisphere midlatitudes. Forecasts for Tmin extremes (Fig. 21.5) have generally very low SEDI scores over the Southern Hemisphere extratropics, not a surprise given the lower area-aggregated anomaly correlations for this field seen in Fig. 21.2. In the Northern Hemisphere, forecasts for positive Tmin extremes (95th percentile, that is, high monthly-mean minimum temperatures) are better than climatology over much of the northern tier of North America, a region where both the trend has been strong and ENSO teleconnections tend to warm anomalies.

21.3. PREDICTABILITY

The precise meaning of "predictability" can be elusive, as this term is variously applied to mean realized prediction skill, potential prediction skill (i.e., the physical limit to which the system may ever be predicted), or factors that could contribute to prediction skill. This discussion focuses on the theoretical limit of prediction, that is, potential prediction skill, primarily from studies

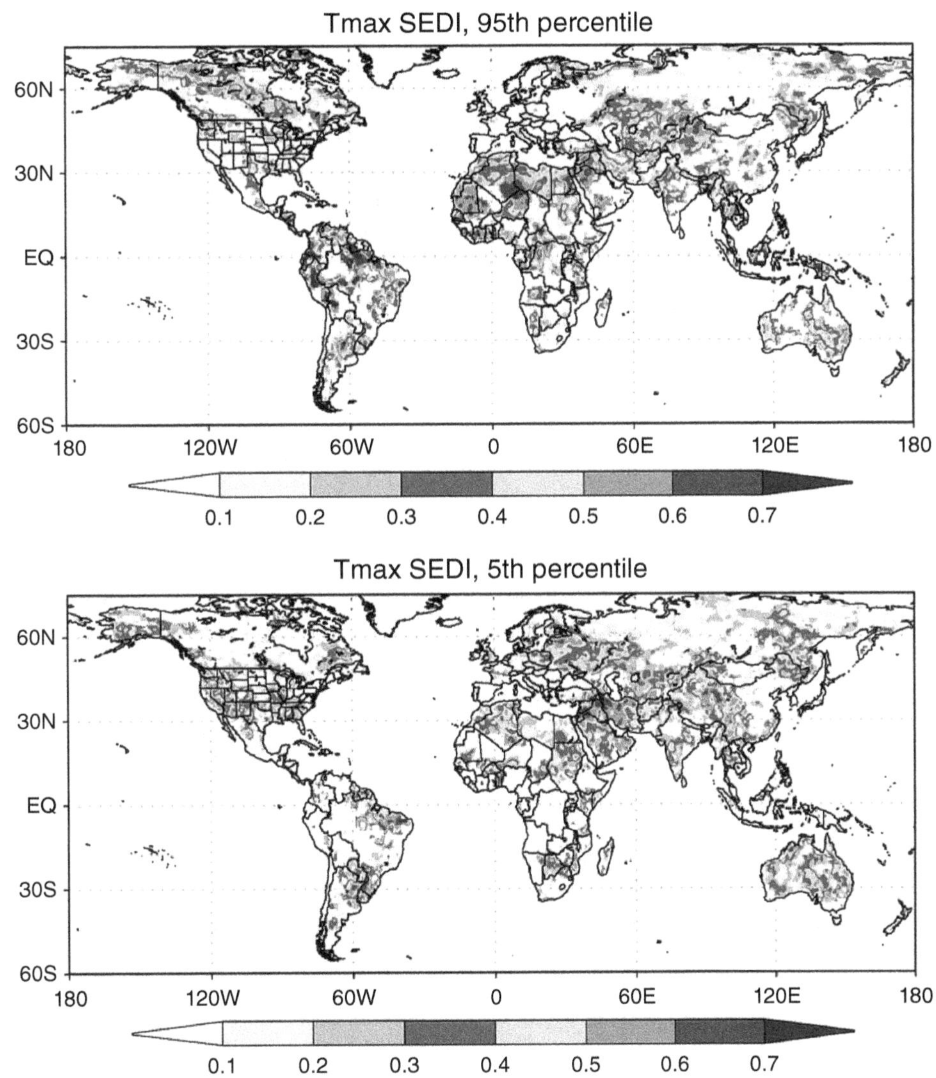

Figure 21.4 As in Figure 21.3, for maximum temperature.

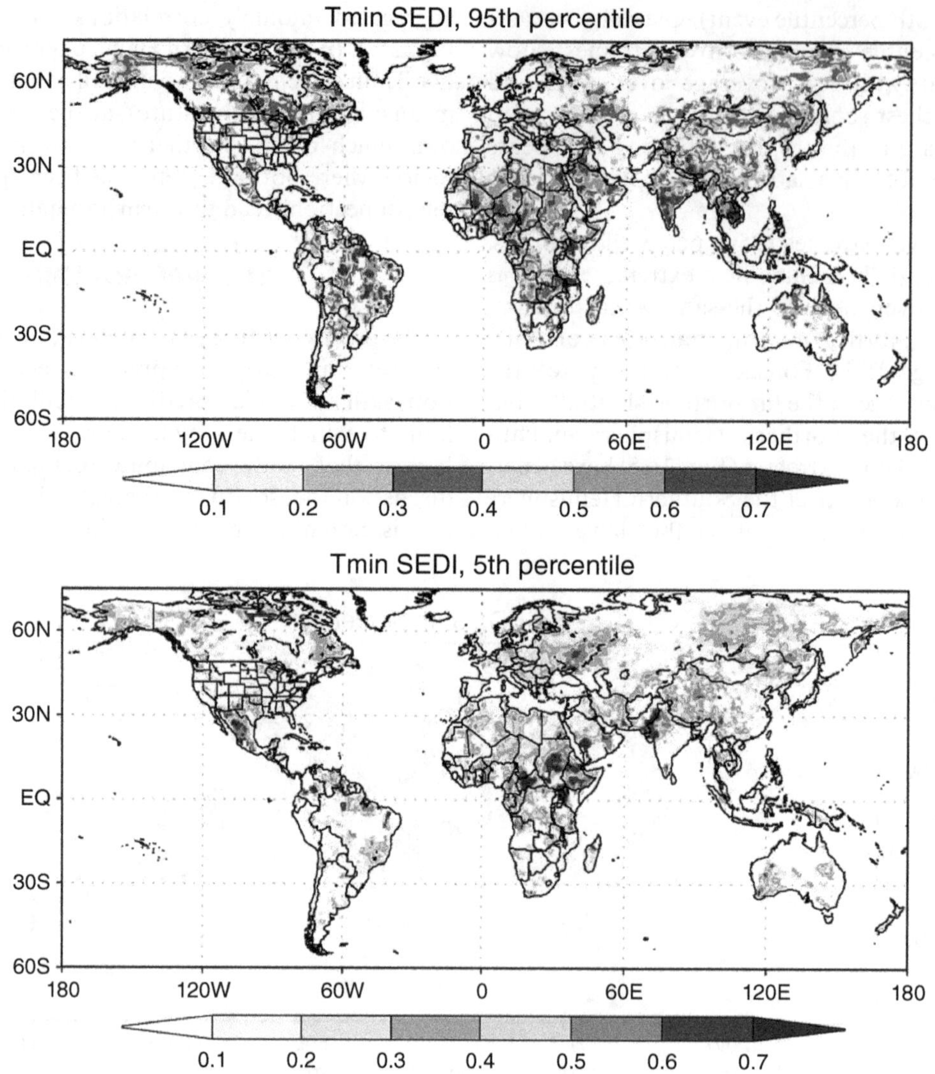

Figure 21.5 As in Figure 21.3, for minimum temperature.

using perfect model assumptions, and (somewhat) predictable climate modes that have been found to affect extremes, therefore presenting the potential for a shift in the probabilities: windows of forecast opportunity [*White et al.*, 2014].

One method of assessing predictability takes advantage of an ensemble model with N members to use one member as a substitute observation, with the ensemble mean of the remaining N-1 as the forecast [*Lorenz*, 1982; *Boer*, 2004]. This method assumes that the forecast and the observation are from the same world, and there are no systematic biases to be corrected, the "perfect model assumption."

Predictability in seasonal T2m, Prate, and SST from the NMME vary somewhat depending on the model used, but not greatly, and suggests that there is some unrealized skill to be found in forecasts for all three variables [*Becker et al.*, 2014]. Both prediction skill and predictability exhibit seasonality, with skill and predictability dependent on the target season [*Becker et al.*, 2014]. Perfect model predictability can be applied to extremes in the same manner as to prediction of the seasonal mean, and has similar results: the potential predictability of STCE is higher than current prediction skill of STCE, with similar regional and seasonal patterns [*Becker et al.*, 2013].

Often, short-term climate extremes can be attributed to external forcings, such as major climate variability modes [*Kenyon and Hegerl*, 2008, 2010, and references therein]. These include the El Niño–Southern Oscillation (ENSO) [*Rasmusson and Carpenter*, 1982; *Wolter et al.*, 1999], the North Atlantic Oscillation (NAO) [*Hurrell*, 1995; *Scaife et al.*, 2008], the Pacific–North Atlantic Oscillation (PNA) [*Wallace and Gutzler*, 1981], or the Madden-Julian

Oscillation (MJO) [*Madden and Julian,* 1972; *Jones,* 2000; *Jones et al.,* 2011], or climate change [e.g. *Easterling et al.,* 2000b]. By shifting the mean of the distribution, these modes shift the probability of extremes (if extremes are defined by climatology). If they also change the shape of the distribution, they will further affect the probability. If these changes to the mean and/or shape can be quantified, predictability can be characterized (a little.) These external modes, many of which are at least somewhat predictable on varying timescales, present the potential for "forecasts of opportunity." There is some suggested linearity in the influence of climate variability modes on extremes [*Kenyon and Hegerl,* 2008].

Use of these opportunities relies on prediction of the climate modes themselves, and many assessments are hampered by short observational records that may see only a few active phases, especially for interannual modes such as ENSO. Beyond the issue of potential predictability of the modes of variability, however, it is important to note that extreme events do not always arise from extreme forcing such as a strong climate mode, with some extreme climate events arising from noise internal to the system [*Kumar et al.,* 2013]. The probability of an extreme event is affected by many different factors, some more predictable than others [e.g., *Hoerling et al.,* 2013; *Trenberth et al.,* 2015]. It is likely that any potentially successful early warning system for increased probability of extreme climate events will incorporate several sources of information.

21.4. SUMMARY AND DISCUSSION

While extreme climate events have been a topic of great interest for decades, the science of successfully predicting extreme monthly and seasonal climate events is still in its infancy, and there is no overwhelming evidence of the existence or potential for very skillful forecasts or for short-term climate extreme temperature events. However, some skill has been found in dynamical model forecasts, and predictability studies give hope that more skill could potentially be achieved, although the relationship between potential and actual skill is far from clear [*Kumar et al.,* 2014].

Most of the assessments of dynamical model forecasts for STCE presented in this chapter are of deterministic forecasts, that is, forecasts for a specific anomaly, whether from a single model or a multimodel ensemble. Probabilistic forecasts allow a greater understanding of the uncertainty in a prediction, but when used in the case of extremes, present another layer of variation, as a threshold of probability must be chosen to define an elevated chance of extreme. An assessment would need to be done to discover how the choice of threshold affects the skill of forecasts.

Forecasts for increased risk of extreme weather events have been in use for years, and the US National Weather Service has recently deployed sophisticated severe weather warning systems with several risk levels associated with elevated probabilities of rare events. It is hard to imagine a similar system could ever be developed for short-term climate extreme events. If reliable forecast information for STCE were available, collaboration with social scientists and users would be necessary to identify the best way to deploy this information. For example, is a 10% probability of a 5th percentile minimum temperature event in North America in January a threshold that would trigger changes to energy budgeting? What about a 20% chance of a 10th percentile event? The value of an early warning system is widely recognized, especially in light of the likelihood that extreme events will increase with climate change. Many questions remain about what form such a system would take.

REFERENCES

Barnston, A. G., and S. J. Mason (2011), Evaluation of IRI's seasonal climate forecasts for the extreme 15% tails, *Weath. Forecast.*, *26*(4), 545–554.

Becker, E. J., H. van den Dool, and M. Peña (2013), Short-term climate extremes: Prediction skill and predictability, *J. Climate*, *26*, 512–531.

Becker, E. J., H. van den Dool, and Q. Zhang (2014), Predictability and forecast skill in NMME, *J. Climate*, *27*, 5891–5906; doi: http://dx.doi.org/10.1175/JCLI-D-13-00597.1.

Boer, G. (2004), Long timescale potential predictability in an ensemble of coupled climate models, *Cimate Dyn.*, *23*, 29–44.

Delworth, T. L., A. J. Broccoli, A. Rosati, R. J. Stouffer, V. Balaji, J. A. Beesley, and J. W. Durachta (2006), GFDL's CM2 global coupled climate models. Part I: Formulation and simulation characteristics, *J. Climate*, *19*(5), 643–674.

Doblas-Reyes, F. J., R. Hagedorn, and T. N. Palmer (2005), The rationale behind the success of multimodel ensembles in seasonal forecasting, II, Calibration and combination, *Tellus A*, *57*, 234–252; doi:10.1111/j.1600-0870.2005.00104.x.

Eade, R., E. Hamilton, D. M. Smith, R. J. Graham, and A. A. Scaife (2012), Forecasting the number of extreme daily events out to a decade ahead, *J. Geophys. Res.*, *117*, D21110; doi:10.1029/2012JD018015.

Easterling, D. R., G. A. Meehl, C. Parmesan, S. A. Changnon, T. R. Karl, and L. O. Mearns (2000b), Climate extremes: Observations, modeling, and impacts, *Science*, *289*(5487), 2068–2074.

Easterling, D. R., J. L. Evans, P. Y. Groisman, T. R. Karl, K. E. Kunkel, and P. Ambenje (2000a), Observed variability and trends in extreme climate events: A brief review, *Bull. Amer. Meteor. Soc.*, *81*, 417–425.

Fan, Y., and H. van den Dool (2008), A global monthly land surface air temperature analysis for 1948-present, *J. Geophys. Res.*, *113*, D01103; doi:10.1029/2007JD008470.

Ferro, C. A. T., and D. B. Stephenson (2011), Extremal dependence indices: Improved verification measures for deterministic forecasts of rare binary events, *Weather Forecast.*, *26*, 699–713.

Garrett, C., and P. Müller (2008), Extreme events, *Bull. Amer. Meteor. Soc.*, *89*, 1733–1733.

Hagedorn, R, F. J. Doblas-Reyes, and T. N. Palmer (2005), The rationale behind the success of multimodel ensembles in seasonal forecasting, I, Basic concept, *Tellus Series A-Dyn. Meteor. Oceanog.*, *57*, 219–233; doi: 10.1111/j.1600-0870.2005.00103.x.

Hamilton, E., R. Eade, R. J. Graham, A. A. Scaife, D. M. Smith, A. Maidens, and C. MacLachlan (2012), Forecasting the number of extreme daily events on seasonal timescales, *J. Geophys. Res.*, *117*, D03114; doi:10.1029/2011JD016541.

Hanlon, H. M., G. C. Hegerl,, S. F. B. Tett, and D. M. Smith (2014), Near-term prediction of impact-relevant extreme temperature indices, *Climatic Change*, 1–16.

Hanlon, H. M., G. C. Hegerl, S. F. Tett, and D. M. Smith (2013a), Can a decadal forecasting system predict temperature extreme indices?, *J. Climate*, *26*(11), 3728–3744.

Hanlon, H. M., S. Morak, and G. C. Hegerl (2013b), Detection and prediction of mean and extreme European summer temperatures with a multimodel ensemble, *J. Geophys. Res. Atmos.*, *118*(17), 9631–9641.

Hoerling, M., A. Kumar, R. Dole, J. W. Nielsen-Gammon, J. Eischeid, J. Perlwitz, X.-W. Quan, T. Zhang, P. Pegion, and M. Chen (2013), Anatomy of an extreme event, *J. Climate*, *26*, 2811–2832;doi:http://dx.doi.org/10.1175/JCLI-D-12-00270.1.

Huang, J., and H. M. van den Dool (1993), Monthly precipitation-temperature relations and temperature prediction over the United States, *J. Climate*, *6*, 1111–1132.

Hudson, D., A. G. Marshall, O. Alves, G. Young, D. Jones, and A. Watkins (2015), Forewarned is forearmed: Extended range forecast guidance of recent extreme heat events in Australia, *Weath. Forecast*.

Hurrell, J. W. (1995), Decadal trends in the North Atlantic oscillation: Regional temperatures and precipitation, *Science*, *269*, 676–679.

Jones, C. (2000), Occurrence of extreme precipitation events in California and relationships with the Madden-Julian Oscillation, *J. Climate*, *13*, 3576–3587.

Jones, C., L. M. V. Carvalho, J. Gottschalck, and W. Higgins (2011), The Madden-Julian Oscillation and the relative value of deterministic forecasts of extreme precipitation in the contiguous United States, *J. Climate*, *24*, 2421–2428.

Kenyon, J., and G. C. Hegerl (2008), Influence of modes of climate variability on global temperature extremes, *J. Climate*, *21*(15), 3872–3889; doi:10.1175/2008JCLI2125.1.

Kenyon, J., and G. C. Hegerl (2010), Influence of modes of climate variability on global precipitation extremes, *J. Climate*, *23*, 6248–6262; doi: 10.1175/2010JCLI3617.1.

Kirtman, B. P., D. Min, J. M. Infanti, J. L. Kinter, D. A. Paolino, Q. Zhang, H. van den Dool, S. Saha, M. P. Mendez, E. Becker, P. Peng, P. Tripp, J. Huang, D. G. DeWitt, M. K. Tippett, A. G. Barnston, S. Li, A. Rosati, S. D. Schubert, Y.-K. Lim, Z. E. Li, J. Tribbia, K. Pegion, W. Merryfield, B. Denis, and E. Wood (2014), Phase-1 seasonal to interannual prediction, phase-2 toward developing intra-seasonal prediction, *Bull. Amer. Meteor. Soc.*, *95*, 585–601; doi:10.1175/BAMS-D-12-00050.1.

Kumar, A., M. Chen, M. Hoerling, and J. Eischeid (2013), Do extreme climate events require extreme forcings?, *Geophys. Res. Lett.*, *40*(13), 3440–3445.

Kumar, A., P. Peng, and M. Chen (2014), Is there a relationship between potential and actual skill?, *Monthly Weather Rev.*, *142*, 2220–2227;doi:http://dx.doi.org/10.1175/MWR-D-13-00287.1.

Lorenz, E. N. (1982), Atmospheric predictability experiments with a large numerical model, *Tellus*, *23*, 243–254.

Madden, R. A., and P. R. Julian (1972), Description of global-scale circulation cells in the tropics with a 40–50 Day Period, *J. Atmos. Sci.*, *29*, 1109–1123.

Merryfield, W. J., and coauthors (2013), The Canadian seasonal to interannual prediction system, part I: Models and initialization, *Monthly Weather Rev.*, *141*, 2910–2945; doi:10.1175/MWR-D-12-00216.1.

Nicholls, N. (1995), Long-term climate monitoring and extreme events, *Climatic Change*, *31*, 231–245.

Palmer, T. N., and coauthors (2004), Development of a European multimodel ensemble system for seasonal-to-interannual prediction (DEMETER), *Bull. Amer. Meteor. Soc.*, *85*, 853–872.

Peña, M., and H. van den Dool (2008), Consolidation of multimodel forecasts by ridge regression: Application to Pacific sea surface temperature, *J. Climate*, *21*, 6521–6538.

Rasmusson, E. M., and T. H. Carpenter (1982), Variations in tropical sea surface temperature and surface wind fields associated with the Southern Oscillation–El Niño, *Monthly Weather Rev.*, *110*, 354–384.

Saha, S., S. Nadiga, C. Thiaw, J. Wang, W. Wang, Q. Zhang, H. M. Van den Dool et al. (2006), "The NCEP climate forecast system", *J. Climate*, *19*(15), 3483–3517.

Saha, S., S. Moorthi, X. Wu, J. Wang, S. Nadiga, P. Tripp, D. Behringer, Y.-T. Hou, H.-Y. Chuang, M. Iredell, M. Ek, J. Meng, R. Yang, M. Peña Mendez, H. van den Dool, Q. Zhang, W. Wang, M. Chen, and E. Becker (2014), The NCEP climate forecast system version 2, *J. Climate*, *27*, 2185–2208.

Scaife, A. A., C. Folland, L. V. Alexander, A. Moberg, and J. R. Knight (2008), European climate extremes and the North Atlantic Oscillation, *J. Climate*, *21*, 72–83.

Smith, D. M., and coauthors (2013), Real-time multi-model decadal climate predictions, *Climate Dyn.*, *41*(11–12), 2875–2888.

Tebaldi, C., and R. Knutti (2007), The use of the multi-model ensemble in probabilistic climate projections, *Phil. Trans. R. Meteor. Soc.*, *365*, 2053–2075.

Trenberth, K. E., J. T. Fasullo, and T. G. Shepard (2015), Attribution of climate extreme events, *Nature Climate Change*, *5*, 725–730.

Van den Dool, H. (2007), *Empirical Methods in Short-Term Climate Prediction*, Oxford University Press.

Vecchi, G. A., T. Delworth, R. Gudgel, S. Kapnick, A. Rosati, A. T. Wittenberg, F. Zeng, W. Anderson, V. Balaji, K. Dixon, L. Jia, H.-S. Kim, L. Krishnamurthy, R. Msadek, W. F. Stern, S. D. Underwood, G. Villarini, X. Yang, and S. Zhang (2014),

On the seasonal forecasting of regional tropical cyclone activity, *J. Climate*, *27*, 7994–8016.

Vernieres, G., M. M. Rienecker, R. Kovach, and C. L. Keppenne (2012), The GEOS-iODAS: Description and evaluation, NASA Tech. Rep. NASA/TM-2012-104606, vol. 30; http://gmao.gsfc.nasa.gov/pubs/docs/Vernieres589.pdf.

Vose, R. S., D. R. Easterling, and B. Gleason (2005), Maximum and minimum temperature trends for the globe: An update through 2004, *Geophys. Res. Lett.*, *32*, L23822; doi:10.1029/2005GL024379.

Wallace, J. M., and D. S. Gutzler (1981), Teleconnections in the geopotential height field during the Northern Hemisphere winter, *Monthly Weather Rev.*, *109*, 784–812.

White, C. J., D. Hudson, and O. Alves (2014), ENSO, the IOD and the intraseasonal prediction of heat extremes across Australia using POAMA-2, *Climate Dyn.*, *7*, 1791–1810.

Wilks, D. (1995), *Statistical Methods in the Atmospheric Sciences*, Academic Press.

Wolter, K., R. M. Dole, and C. A. Smith (1999), Short-term climate extremes over the continental United States and ENSO, Part 1, Seasonal temperatures, *J. Climate*, *12* 3255–3272.

Zhang, S., M. J. Harrison, A. Rosati, and A. Wittenberg (2007), System design and evaluation of coupled ensemble data assimilation for global oceanic climate studies, *Monthly Weather Rev.*, *135*, 3541–3564.

22

Toward Predicting US Tornadoes in the Late 21st Century

Adam J. Clark

ABSTRACT

Projecting changes in US tornado frequencies is difficult because, in addition to large uncertainty associated with projections of regional climate change, current global and regional climate models use grid spacing too coarse to resolve tornadoes or even individual tornado-producing storms. Thus, environmental proxies are used for predicting changes, which comprise a set of resolvable large-scale weather parameters that correlate well with tornadoes. Recent studies using this approach generally find an increase in the frequency of days with favorable tornado environments through the end of the 21st century, especially during spring and autumn. However, it is unknown whether these favorable environments would be associated with storm development, which depends on the strength of synoptic scale forcing and the presence of localized lifting. To reduce uncertainty with environmental proxies, downscaling experiments are just beginning to use convection-permitting resolution (i.e., fine enough grid spacing to resolve individual storms), so that storm-based proxies can be used for predicting tornado trends. The first of these studies finds evidence for an increase in severe weather frequency (not just tornadoes) over the United States toward the end of the 21st century adding to growing evidence that late 21st century severe weather will be more frequent, driven by increased low-level moisture.

22.1. PROJECTING CHANGES IN US TORNADO ACTIVITY USING ENVIRONMENTAL PROXIES

Global climate model ensembles like those used in the 2013 Intergovernmental Panel on Climate Change (IPCC) report as part of the Coupled Model Intercomparison Project, Phase 5 (CMIP5), are commonly used to project future climate change under various greenhouse gas emissions scenarios. However, because of computational limitations, the horizontal and vertical grid spacing of these models is much too coarse to resolve tornadoes or even individual tornado-producing storms. Even the highest resolution regional climate model downscaling experiments use grid spacing much too coarse to depict tornadoes. Thus, for predicting changes in tornado climatology, as well as severe weather in general, it is necessary to use environmental proxies, that is, a set of resolvable large-scale weather parameters that correlates well with tornado occurrence. In the United States, there is a long history of using these environmental proxies, or "ingredients-based" methods, for short-term severe weather forecasting at agencies like NOAA's Storm Prediction Center [e.g., *Beebe*, 1955; *Brooks et al.*, 1994; *Rasmussen and Blanchard*, 1998; *Thompson et al.*, 2003]. For short-term severe weather forecasting, instability and deep-layer (0–6 km) vertical wind shear are important for discriminating environments supportive of any type of severe weather (i.e., wind, hail, and tornadoes) from nonsevere environments. However, for discriminating significant tornado (i.e., EF2+) environments, additional variables need to be considered, specifically, large values of vertical wind shear and moisture in the lowest 1 km of the troposphere [*Thompson et al.*, 2003].

Brooks et al. [2003b] first used relationships between observed tornadoes and their associated environments as

NOAA/OAR National Severe Storms Laboratory, Norman, Oklahoma, USA

Climate Extremes: Patterns and Mechanisms, Geophysical Monograph 226, First Edition.
Edited by S.-Y. Simon Wang, Jin-Ho Yoon, Christopher C. Funk, and Robert R. Gillies.

depicted by global reanalysis data over the United States to estimate annual frequencies of tornado environments. Finding these relationships involves using discriminant analysis [*Wilks*, 1995] to calculate empirically derived relationships that indicate whether a combination of environmental parameters is likely to be associated with observed tornadoes. Using these relationships, *Brooks et al.* [2003b] estimated the frequency of tornado-supporting environments in Europe finding, not surprisingly, that the number of days per year with favorable tornado parameters was much smaller compared to the United States. However, extreme caution must be used when applying these empirically derived relationships to other parts of the world because it involves making a very large assumption, namely, that the probability of a favorable environment being associated with an observed event does not depend on the region or time of year. This can be described as the "efficiency" of the atmosphere in producing a tornado given favorable conditions. Indeed, further research by *Brooks* [2009] using observed severe weather databases over the United States and Europe finds that the probability of significant severe storms over Europe are higher than the United States given the same set of favorable environmental conditions (favorable severe weather conditions occur much more frequently in the United States so that the overall number of events is still higher there). Furthermore, *Brooks et al.* [2003b] find that probabilities of severe storms given favorable environments are higher in the southeastern United States during the cool season, speculating that this results from differences in the frequency of convective initiation (i.e., storm formation). Analysis of environmental conditions cannot identify convective initiation processes, since these depend on the strength of synoptic scale forcing along with more localized lifting mechanisms such as low-level convergence along boundaries. Thus, *Brooks et al.* [2003b] speculate that higher severe weather probabilities given similar environments result from stronger synoptic forcing and boundaries being present more frequently.

These studies suggest that, while it would be straightforward to apply their results to estimate the change in the frequency of environments supportive of tornadoes and severe weather from climate model simulations, it is very uncertain what the actual frequency of severe weather would be given these environments. Furthermore, the ability of the climate models to represent the future distribution of tornado environments is unknown and requires thorough verification using hindcasts. With these caveats in mind, *Diffenbaugh et al.* [2013] examined the frequency of US severe storm environments in CMIP5 global climate model simulations, finding robust increases in severe environments for the eastern United States during all seasons, but especially spring and autumn. Their examination included an analysis of days with parameters especially favorable for tornadoes, namely, high convective instability and strong low-level vertical wind shear. The number of days with the favorable tornado parameters also exhibited a robust increase through the end of the 21st century.

Another approach for estimating US tornado activity using environmental proxies involves computing monthly average parameters that are combined into a calibrated index known as the Tornado Environmental Index (TEI) [*Tippett et al.*, 2014]. The use of monthly averages is a significant distinction from the previous works described above that use the frequency of conditions at much shorter timescales. Furthermore, the success of the monthly index suggests a connection between quantities varying on climate timescales and tornado activity. Put another way, changes in the frequencies in extreme environments are likely accompanied by changes in the monthly average of those environments. *Tippett et al.* [2014] calculated the TEI using Poisson regression of numbers of monthly US tornadoes against monthly averages of co-located atmospheric parameters from the North American Regional Reanalysis (NARR) [*Mesinger et al.*, 2006], finding that the ideal combination of parameters in the regression was convective precipitation and storm relative helicity.

Recent studies applying these kinds of environmental proxies to output from regional climate models (RCMs) and global climate models (GCMs) over the United States generally find an increase in the number of days with environments supporting severe weather [e.g., *Trapp et al.*, 2007]. However, the environmental proxies do not consider whether storms actually occur, thus, current research is beginning to emphasize the use of model grid spacing in downscaling experiments fine enough to depict individual storms. These works are discussed in the next section.

22.2. SHORT-TERM TORNADO PREDICTION USING HIGH RESOLUTION MODELS AND APPLICATIONS TO DYNAMICAL DOWNSCALING

One common method to obtain more detailed predictions of climate change at the regional level is through downscaling, which involves using output from a GCM to drive a regional numerical model at higher spatial resolution. For predicting changes in severe weather and tornado climatology, higher resolution is very important since convective storm processes occur at much smaller scales than that resolved by GCMs. However, only until very recently have computational capabilities made it possible to run downscaling experiments at grid scales small enough to reasonably depict individual storms that produce tornadoes and severe weather. Furthermore, research is evolving rapidly and much is still being learned on using high-resolution models for short-term weather predictions, let alone longer-term climate predictions. Similar to how ingredients-based approaches based on forecast environments have been extended from short-term tornado forecasting to tornado predictions at longer timescales, short-term weather forecasting approaches using models that explicitly depict storms will soon be applied in regional downscaling experiments to make longer timescale predictions. This downscaling approach has the potential to greatly improve regional scale predictions, but the approaches are still in their infancy. To provide some background, the rest of this section describes how models with high enough resolution to explicitly depict storms have been used in the weather forecasting community and how they are initially being applied in climate downscaling experiments.

A major breakthrough in the weather forecasting community occurred with the advent of convection-permitting models (CPMs), that is, models that use high enough resolution so that convective overturning can be represented on the model grid without the need for convective parameterization. CPMs use horizontal grid spacing ≤ 4 km, which is about the coarsest grid spacing at which mesoscale convective systems can be adequately depicted. To illustrate the difference between CPMs and models that parameterize convection, Figure 22.1 shows a conceptual model of a supercell overlaid with a 32 km grid-mesh and with a 4 km grid-mesh. Only one gridpoint covers the supercell with the 32 km mesh while several gridpoints, enough for an adequate representation, overlay the storm with the 4 km grid-mesh.

Much of the initial research on using CPMs for severe weather forecasting occurred as part of activities within annual NOAA/Hazardous Weather Testbed Spring Forecasting Experiments in the mid-2000s [e.g., *Kain et al.*, 2006; *Clark et al.*, 2012a], and these annual experiments continue to test and refine their use. Through real-time experimental forecasting activities completed in collaboration with SPC forecasters, it was found that CPMs could depict realistic convective-scale storm structures, accurately distinguish dominant convective modes, and at times provide extraordinarily accurate forecasts of convective system location and timing up to 36 hours in advance [e.g., *Done et al.*, 2004]. For example, *Weisman et al.* [2013] presents an example of an intense, high-impact convective system simulated with 27 hour lead time on 8 May 2009. This was an important breakthrough because, prior to the mid-2000s, there was widespread skepticism about the potential utility of CPMs for operational prediction. In fact, there was no mention of CPMs in NOAA's Environmental Modeling Center's 5–10 year production-suite plans. However, as of this writing, several operational CPMs such as the High Resolution Rapid Refresh (HRRR) model are run in the United States, and long-term plans for EMC's future suite of modeling systems heavily emphasize development of operational CPM-based ensemble systems.

An important realization from early work on CPMs was that fully exploiting them requires new and innovative model diagnostics, verification, and visualization strategies. This is because potential hazards from coarser models that parameterize convection are inferred through aspects of forecast synoptic and mesoscale environments (i.e., ingredients-based forecasting methods), whereas the added value from CPMs come from their ability to explicitly simulate convection and provide much more direct information on potential hazards. Thus, a new paradigm of forecast diagnostics would be needed that is more focused on attributes of forecast storms, rather than the environment of the storms. Of course, accurate forecasts of environments remain important and challenging, but given the wide range of storm modes observed in similar regions of severe weather parameter space [e.g., *Thompson et al.*, 2003; *Grams et al.*, 2012], direct information on forecast storms adds significant forecast value.

For tornado forecasting, a storm-based model diagnostic known as updraft-helicity (UH) has been developed, which identifies simulated mesocyclones from CPM forecasts. Because CPMs are still too coarse to resolve tornadoes, but most tornadoes are produced by strong mesocyclones that are resolved by CPMs, UH serves as a storm-based tornado proxy. However, extreme caution is needed when applying UH as a tornado predictor since, just like in nature, not all mesocyclones produce tornadoes. As presented in *Kain et al.* [2008], helicity is defined by

$$H = V \cdot \nabla \times V \tag{22.1}$$

The *horizontal* component of environmental helicity is often used by forecasters to assess the potential for rotating storms from models too coarse to resolve convection. However, in CPMs, rotation in simulated storms can be measured directly using the vertical component of helicity, H_v, given by

$$H_v = \left(\frac{\partial v}{\partial x} - \frac{\partial u}{\partial y} \right) \tag{22.2}$$

Then, to measure rotation over the depth of a storm, H_v is integrated over a layer giving UH, which is expressed as

$$UH = \int_{z_1}^{z_2} w \zeta \, dz \tag{22.3}$$

where ζ is the vertical component of the relative vorticity. Although ζ and w can be positive or negative, severe

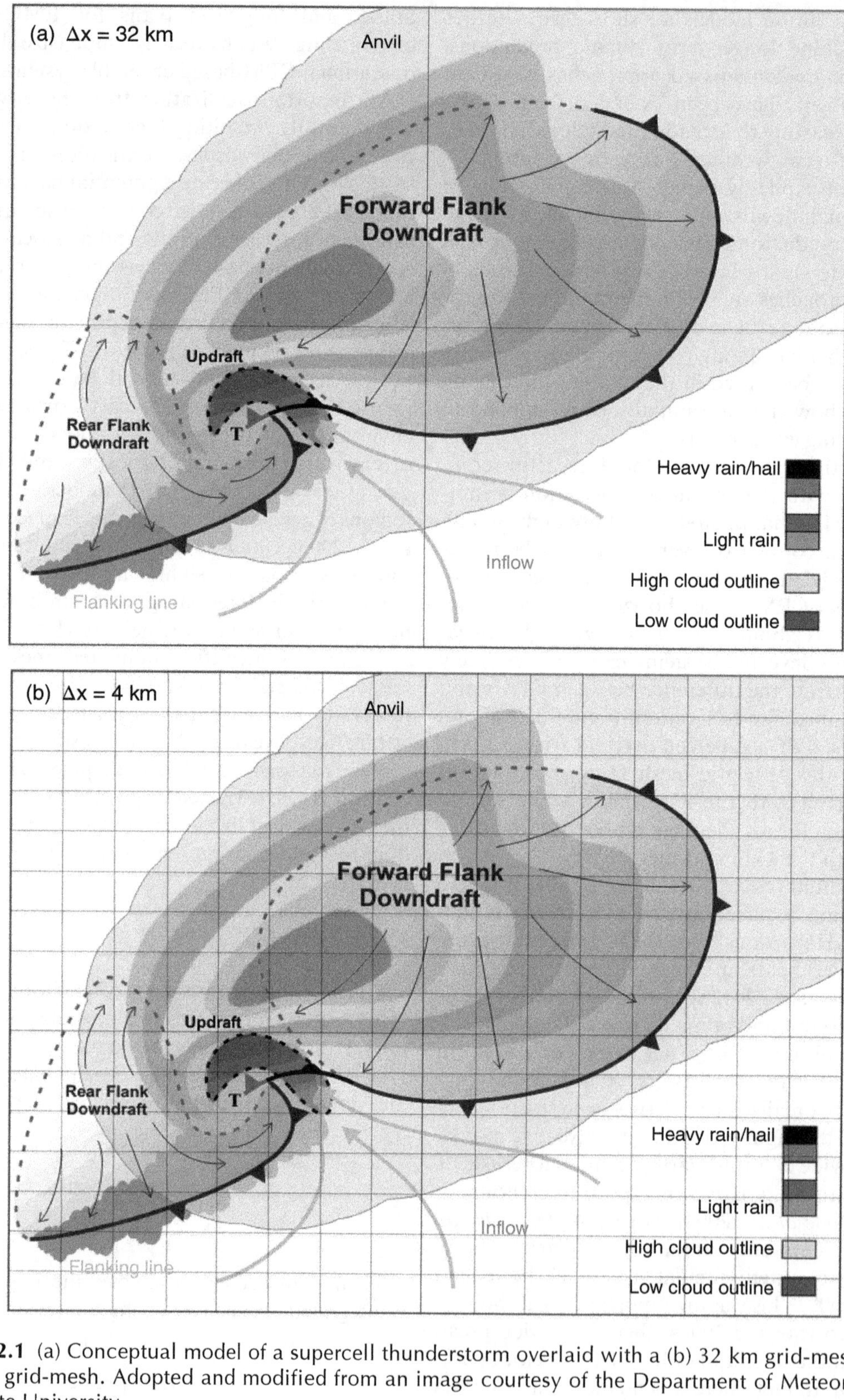

Figure 22.1 (a) Conceptual model of a supercell thunderstorm overlaid with a (b) 32 km grid-mesh, and a (b) 4 km grid-mesh. Adopted and modified from an image courtesy of the Department of Meteorology at Penn State University.

weather applications of UH only consider positive values of these variables, which represent cyclonic rotation within an updraft. Using model data at 1 km increments in the vertical direction and only considering the 2 to 5 km layer (i.e., lower to midtroposphere) where mesocyclones typically occur, UH is typically calculated using the expression

$$UH \approx w \sum_{z=2000m}^{z=5000m} \overline{w\zeta\Delta z} = \left(\overline{w\zeta_{2,3}} + \overline{w\zeta_{3,4}} + \overline{w\zeta_{4,5}}\right) \times 1000 \quad (22.4)$$

where the overbar indicates a layer average and the subscripts indicate the bottom and top of the layer in kilometers.

One of the challenges in providing useful information on forecast storm attributes such as UH is that convective storms evolve on timescales of order minutes, which are much shorter than typical model output frequencies of 1–3 hours. To monitor fields between model output times, a strategy was developed in which maximum values at a gridpoint from any model time step within a specified time period (typically the time interval between model output times) are stored as a two-dimensional array and then output at regular model output times [*Kain et al.*, 2010]. These fields are called hourly maximum fields (HMFs), and provide useful information on peak intensity and track of simulated storm features without storing models fields at every time step. An example of hourly maximum UH from a CPM forecast is shown in Figure 22.2.

Recent research has begun to explore methods for relating cumulative properties of hourly maximum UH to that of tornadoes. For example, *Clark et al.* [2012b, 2013] used an algorithm to measure the path length of coherent hourly maximum UH "swaths" that occurred in forecast models integrated over the United States. The cumulative UH path lengths (i.e., the sum of all UH path lengths within the model domain) over 18 and 24 hour periods were then compared to cumulative tornado path lengths over the same periods, with the hope that total UH path length could be used as a reliable predictor of total tornado path length. Path length is a preferred metric because it can be precisely measured during damage surveys [e.g., *Edwards et al.*, 2004], and cumulative path length is a very good indicator for the overall severity of a tornado outbreak [e.g., *Shafer and Doswell*, 2010]. The forecast UH path lengths work better for predicting tornado path lengths during spring months (March–May) than summer (June–August), which is because UH path lengths tend to greatly overpredict tornado path lengths during summer. It is likely that UH produced by frequently occurring organized convective systems in the CPMs during the summer largely explains the overprediction. UH associated with these systems is oftentimes not indicative of strong rotation, rather, it can reflect strongly sheared updrafts, which are common in strong MCSs. Thus, in these cases, it can actually serve as a useful proxy for severe wind [e.g., *Sobash et al.*, 2011], but if not distinguished from true mesocyclones, it will result in a false alarm for tornado occurrence.

Other findings from comparing UH to tornado path length include that utilizing information from an ensemble of CPMs can improve the tornado path length predictions, and that using information from the environment in which the simulated UH occurs can help filter out UH not likely to be associated with observed tornadoes. For

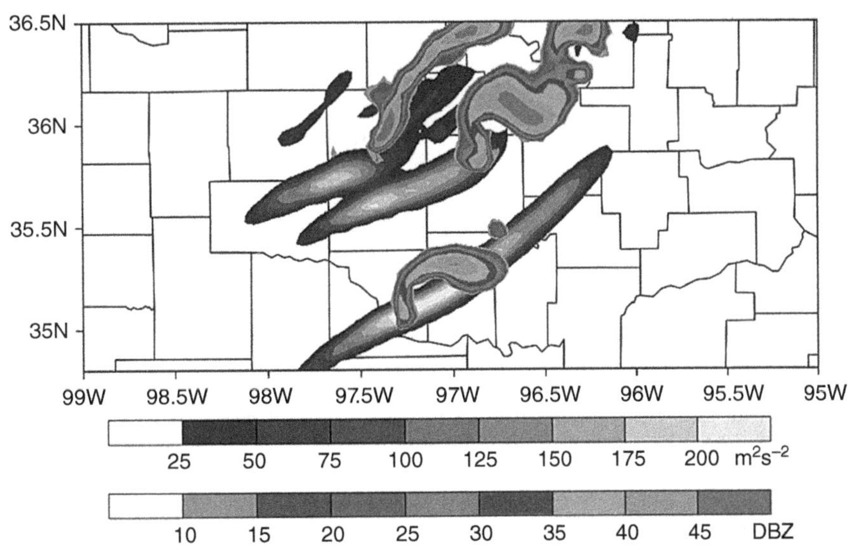

Figure 22.2 Hourly maximum updraft helicity valid 1800–0100 UTC 19–20 May 2013 as forecast by a CPM initialized at 0000 UTC 19 May at NOAA's National Severe Storms Laboratory (purple shading). The plot domain covers central Oklahoma. Simulated reflectivity valid at 2100 UTC is overlaid. (*See insert for color representation of the figure.*)

example, tornadoes are much more likely in environments with low lifting condensation level heights and in environments that support surface-based convection (i.e., storm inflow is draw from near surface layer). An example of UH forecasts being used to derive probabilistic tornado forecasts is shown in Figure 22.3. These probabilities were derived from forecasts produced by the University of Oklahoma's Center for Analysis and Prediction of Storms for the 27 April 2011 super-outbreak of tornadoes.

This case was notable because it was one of the worst tornado outbreaks in the history of the United States with a cumulative tornado path length for the period 1200 UTC 27 April–0600 UTC 28 April of 2949 km. The only other comparable outbreak since 1950 was on 3–4 April 1974, which had a cumulative path length of 3964 km for the corresponding period. The CPM-based ensemble illustrated in Figure 22.3 was able to recognize high probabilities of an extreme event. In fact, out of 69 cases examined, the UH path length forecasts of the 27–28 April case were by far the most extreme. Using the relationship derived between UH and tornado path lengths, the forecasts were indicating about a 50% probability of an event with cumulative tornado path lengths at or exceeding a 10 year return period (Fig. 22.3b).

It is intuitive to extend relationships found between storm-based diagnostics in weather prediction models and observed convective hazards such as tornadoes to longer, climate timescales via dynamical downscaling. The obvious advantage to this approach is that it isn't necessary to make assumptions on whether favorable severe environments will actually be associated with severe weather. CPMs allow for direct diagnosis of simulated storms. Thus, as long as reliable simulated-storm-based proxies for hazardous weather can be developed, the environments are not as important as the characteristics of the storms themselves.

The first study to attempt dynamical downscaling of GCM simulations using a CPM for generating long-term climate predictions was *Gensini and Mote* [2015], in which 2080–2090 March–May hazardous convective weather was simulated for the eastern United States. The study builds on previous works finding that modeled severe weather proxies from CPMs forced by coarse resolution GCM-scale conditions can accurately capture the

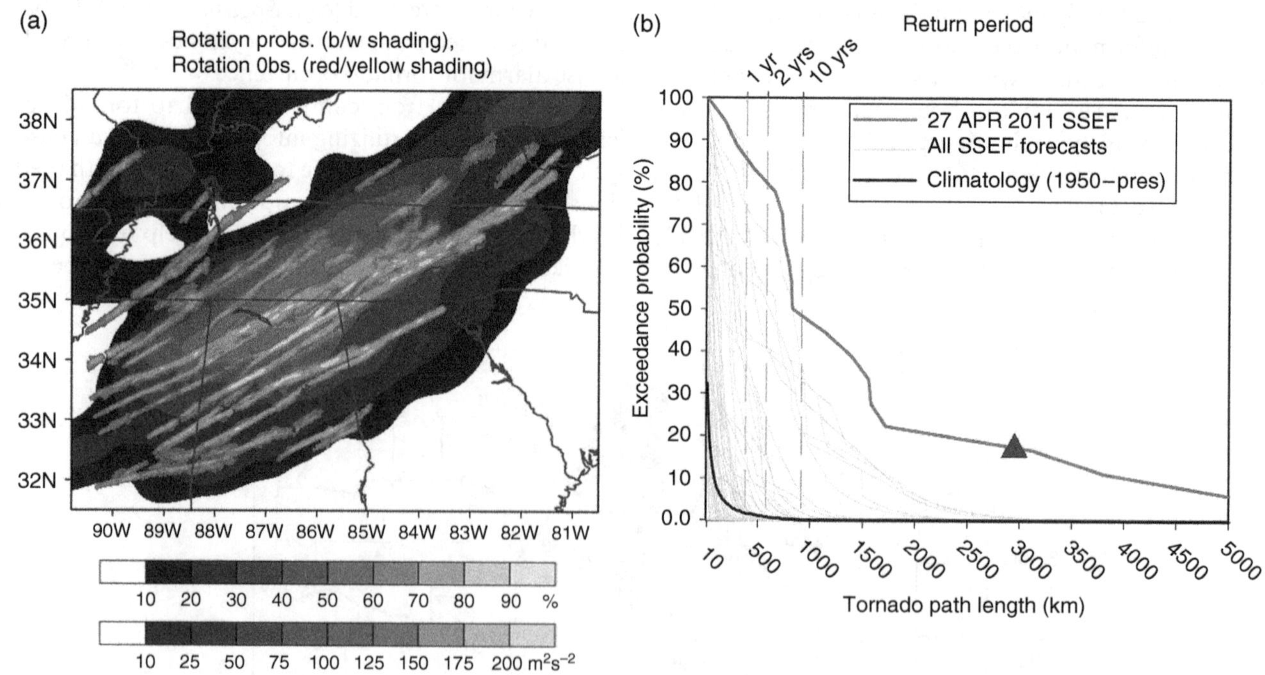

Figure 22.3 (a) Black and white shading shows probabilities of UH ≥ 100 m²s⁻² within 40 km of a point valid 1500–0300 UTC 27–28 April 2011 computed from a CPM-based ensemble initialized 0000 UTC 27 April 2011, which was used during the 2011 NOAA/Hazardous Weather Testbed Spring Forecasting Experiment. Probabilities were smoothed using a Gaussian kernel density weighting function with a standard deviation of 20 km. The red and yellow shading shows rotation tracks of observed rotating storms computing from 1.25 km grid-spacing, radar-data-assimilating, 3DVAR analyses constructed at 5 min intervals. (b) Exceedance probabilities as a function of cumulative tornado path length computed from the distribution of UH path-length forecasts from the CPM-based ensemble shown in (a). The dark red line is for 27 April, the light red lines are for 47 other cases in the dataset, and the green line is for the climatological exceedance probabilities computed from Storm Data for the period 1950–2011 (legend provided at top). The dark red triangle marks the actual cumulative tornado path length for 1200–0600 UTC 27–28 April. Adapted and modified from *Clark et al.* [2013]. (*See insert for color representation of the figure.*)

spatiotemporal variability of observed reports [e.g., *Trapp et al.*, 2011; *Robinson et al.*, 2013; *Gensini and Mote*, 2014]. Note that because these studies considered hazardous weather in general, which includes tornadoes, wind, and hail, it is uncertain whether the trends for tornadoes would necessarily follow the trends for general severe weather. It is possible that an increase in overall severe weather could result from a decrease in tornadoes that is more than offset by an increase in other hazards. Thus, caution is urged extrapolating results for all severe to that for individual types of severe weather.

To identify proxy severe weather reports (not just tornadoes) over the United States, *Gensini and Mote* [2015] use hourly thresholds of $UH \geq 60 m^2 s^{-2}$ and simulated radar reflectivity ≥ 40 dBZ, which was found to best approximate severe weather in historical simulations from the 1980–1990 period. In terms of raw counts of proxy reports, a 27% increase is predicted for the 2080–2090 period relative to the period 1980–1990, with most of the increase occurring in March (Fig. 22.4). Additionally, an increase in the annual variability of proxy severe events is predicted. Interestingly, this increase in variability is consistent with changes in the tornado record that have already been observed and were discussed in the previous section. *Gensini and Mote* [2015] find that the increase in proxy reports is related to an increase in the frequency of favorable environments for severe weather in future climate, namely, an increase in the juxtaposition of large instability and deep-layer shear (Fig. 22.5). The results are consistent with several earlier studies [e.g., *Trapp et al.*, 2007; *Diffenbaugh et al.*, 2013; *Gensini et al.*, 2014] and add to growing evidence suggesting that late 21st century severe weather over the United States will be more frequent and that the increase will be driven by increases in low-level moisture, which increases instability. While previous studies have also suggested that the decreasing north-south temperature gradient because of polar amplification will on average reduce deep-layer wind shear, the current body of work suggests that the increase in instability will more than make up for the decrease in shear to result in increased severe weather frequency. Furthermore, *Diffenbaugh et al.* [2013] find that decreases in shear are concentrated on days with low instability and therefore do not contribute to a decrease in severe weather environments.

22.3. CONCLUDING REMARKS

Because tornadoes can have such a large societal impact and increasing urbanization will result in increasing exposure to tornado risk, accurately projecting long-term changes in the frequency and locations of tornadoes is especially important. Although the current body of research suggests severe weather frequencies (including tornadoes) will likely increase through the end of the 21st century, this type of research is still in its infancy and the results are accompanied by large uncertainties. These uncertainties can be better quantified and perhaps reduced through thorough verification of downscaled regional climate simulations using hindcasts, improvement in model physics, and further development of storm-based proxies for tornadoes and severe weather.

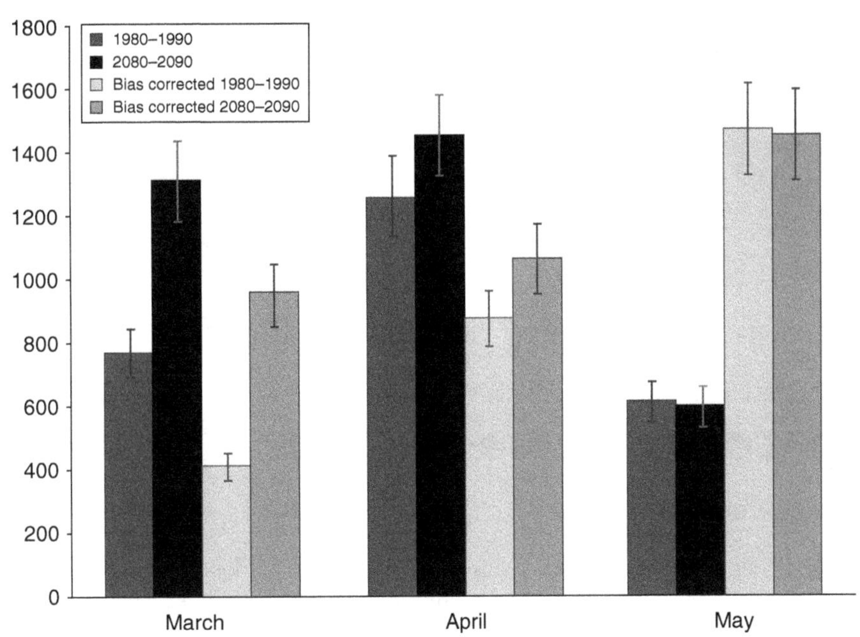

Figure 22.4 Average monthly comparisons of raw and bias corrected synthetic severe weather reports. Error bars indicate the standard error. From Figure 2 of *Gensini and Mote* [2015].

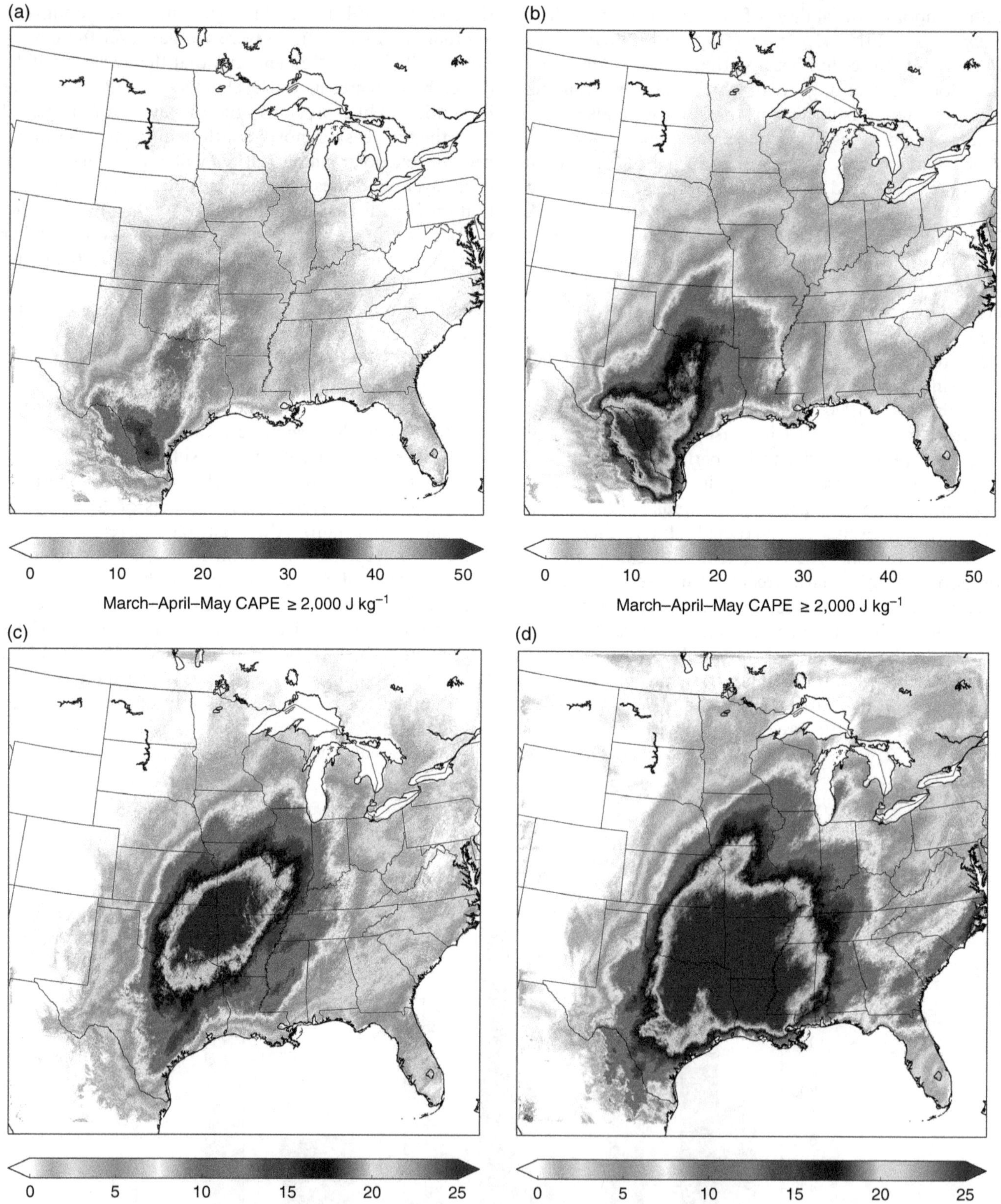

Figure 22.5 Average frequency of March–May historical (a) CAPE ≥ 2000 J kg^{-1} and (c) the product of CAPE and 0–6-km BWD exceeding 20,000. Panels (b) and (d) represent the future period, respectively. In the case of CAPE × 0–6 km BWD, CAPE is constrained to only evaluate events with ≥ 100 J kg^{-1} . From Figure 10 of *Gensini and Mote* [2015].

ACKNOWLEDGMENTS

The author greatly appreciates feedback from Dr. Harold Brooks in preparing this chapter. Additionally, several helpful comments from an anonymous reviewer helped improve the chapter.

REFERENCES

Beebe. R. G. (1955), Types of air masses in which tornadoes occur, *Bull. Amer. Meteor. Soc.*, *36*, 349–350.

Brooks, H. E., C. A. Doswell III, and J. Cooper (1994), On the environments of tornadic and non-tornadic mesocyclones, *Weath. Forecast.*, *10*, 606–618.

Brooks, H. E., J. W. Lee, and J. P. Craven (2003b), The spatial distribution of severe thunderstorm and tornado environments from global reanalysis data, *Atmos. Res.*, *67–68*, 73–94.

Brooks, H. E. (2009), Proximity soundings for Europe and the United States from reanalysis data, *Atmos. Res.*, *93*, 546–553; doi:10.1016/j.atmosres.2008.10.005.

Clark, A., J., and coauthors (2012a), An overview of the 2010 hazardous weather testbed experimental forecast program spring experiment, *Bull. Amer. Meteor. Soc.*, *93*, 55–74.

Clark, A. J., J. S. Kain, P. T. Marsh, J. Correia, M. Xue, and F. Kong (2012b), Forecasting tornado path lengths using a 3-dimensional object identification algorithm applied to convection-allowing forecasts, *Weath. Forecast.*, *27*, 1090–1113.

Clark, A. J., J. Gao, P. T. Marsh, T. Smith, J. S. Kain, J. Correia, M. Xue, and F. Kong (2013), Tornado path length forecasts from 2010 to 2011 using ensemble updraft helicity, *Weath. Forecast.*, *28*, 387–407.

Diffenbaugh, N. S., M. Scherer, and R. J. Trapp (2013), Robust increases in severe thunderstorm environments in response to greenhouse forcing, *Proc. Nat. Acad. Sci. USA*, *110*, 16 361–16 366; doi:10.1073/pnas.1307758110.

Done, J., C. A. David, and M. L. Weisman (2004), The next generation of NWP: Explicit forecasts of convection using the Weather Research and Forecast (WRF) model, *Atmos. Sci. Lett.*, *5*, 110–117.

Edwards, R., R. L. Thompson, K. C. Brosbie, J. A. Hart, and C. A. Doswell III (2004), Proposals for modernizing the definitions of tornado and severe thunderstorm outbreaks, Preprints, *22nd on Conf. Severe Local Storms*, Hyannis, MA, American Meteorological Society, 7B.2; http://ams.confex.com/ams/pdfpapers/81342.pdf.

Gensini, V. A., and T. L. Mote (2014), Estimations of hazardous convective weather in the United States using dynamical downscaling, *J. Climate*, *27*, 6581–6589.

Gensini, V. A., C. Ramseyer, and T. L. Mote (2014), Future convective environments using NARCCAP, *Int. J. Climatol.*, *34*, 1699–1705.

Gensini, V. A., and T. L. Mote (2015), Downscaled estimates of late 21st century severe weather from CCSM3, *Climate Change*, *129*, 307–321.

Grams, J. S., R. L. Thompson, D. V. Snively, J. A. Prentice, G. M. Hodges, and L. J. Reames (2012), A climatology and comparison of parameters for significant tornado events in the United States, *Weath. Forecast.*, *27*, 106–123.

Kain, J. S., S. J. Weiss, J. J. Levit, M. E. Baldwin, and D. R. Bright (2006), Examination of convection-allowing configurations of the WRF Model for the prediction of severe convective weather: The SPC/NSSL spring program 2004, *Weath. Forecast.*, *21*, 167–181.

Kain, J. S., and coauthors (2008), Some practical considerations regarding horizontal resolution in the first generation of operational convection-allowing NWP, *Weath. Forecast.*, *5*, 931–952.

Kain, J. S., S. R. Dembek, S. J. Weiss, J. L. Case, J. J. Levit, and R. A. Sobash (2010), Extracting unique information from high resolution forecast models: Monitoring selected fields and phenomena every time step, *Weath. Forecast.*, *25*, 1536–1542.

Mesinger, F., and coauthors (2006), North American regional reanalysis, *Bull. Amer. Meteor. Soc.*, *87*, 343–360.

Rasmussen, E. N., and S. O. Blanchard (1998), A baseline climatology of sounding-derived supercell and tornado forecast parameters, *Weath. Forecast.*, *13*, 1148–1164.

Robinson, E. D., R. J. Trapp, and M. E. Baldwin (2013), The geospatial and temporal distributions of severe thunderstorms from high resolution dynamical downscaling, *J. Appl. Meteor.*, *52*, 2147–2161.

Shafer, C. M., and C. A. Doswell III (2010), A multivariate index for ranking and classifying severe weather outbreaks, *E-J. Sev. Storms Meteor.*, *5* (1), 1–39.

Sobash, R. A., J. S. Kain, D. R. Bright, A. R. Dean, M. C. Coniglio, and S. J. Weiss (2011), Probabilistic forecast guidance for severe thunderstorms based on the identification of extreme phenomena in convection-allowing model forecasts, *Weath. Forecast.*, *26*, 714–728.

Thompson, R. L., R. Edwards, and J. A. Hart (2003), Close proximity soundings within supercell environments obtained from the Rapid Update Cycle, *Weath. Forecast.*, *18*, 1243–1261.

Tippett, M. K., A. H. Sobel, S. J. Camargo, and J. T. Allen (2014), An empirical relation between U.S. tornado activity and monthly environmental parameters, *J. Climate*, *27*, 2983–2999.

Trapp, R. J., N. S. Diffenbaugh, H. E. Brooks, M. E. Baldwin, E. D. Robinson, and J. S. Pal (2007), Changes in severe thunderstorm environment frequency during the 21st century caused by anthropogenically enhanced global radiative forcing, *Proc. Nat. Acad. Sci. USA*, *104*, 19719–19723.

Trapp, R. J., E. Robinson, M. Baldwin, N. Diffenbaugh, and B. Schwedler (2011), Regional climate of hazardous convective weather through high-resolution dynamical downscaling, *Climate Dyn.*, *37*, 677–688.

Weisman, M. L., W. C. Skamarock, and J. B. Klemp (1997), The resolution dependence of explicitly modeled convective storms. *Mon. Wea. Rev.*, *125*, 527–548.

Weisman, M. L., C. Evans, and L. Bosart (2013), The 8 May 2009 Superderecho: Analysis of a real-time explicit convective forecast, *Weath. Forecast.*, *28*, 863–892.

Wilks, D. S. (1995), *Statistical Methods in the Atmospheric Sciences*, Academic Press, New York.

INDEX

Climate Extremes: Patterns and Mechanisms, Geophysical Monograph 226, First Edition.
Edited by S.-Y. Simon Wang, Jin-Ho Yoon, Christopher C. Funk, and Robert R. Gillies.

www.ingramcontent.com/pod-product-compliance
Lightning Source LLC
LaVergne TN
LVHW082109200926
848484LV00008B/1933